Transplantation

A subject collection from *Cold Spring Harbor Perspectives in Medicine*

OTHER SUBJECT COLLECTIONS FROM *COLD SPRING HARBOR PERSPECTIVES IN MEDICINE*

Cystic Fibrosis: A Trilogy of Biochemistry, Physiology, and Therapy
Hemoglobin and its Diseases
Addiction
Parkinson's Disease
Type 1 Diabetes
Angiogenesis: Biology and Pathology
HIV: From Biology to Prevention and Treatment
The Biology of Alzheimer Disease

SUBJECT COLLECTIONS FROM *COLD SPRING HARBOR PERSPECTIVES IN BIOLOGY*

Signaling by Receptor Tyrosine Kinases
Mitochondria
DNA Repair, Mutagenesis, and Other Responses to DNA Damage
Cell Survival and Cell Death
Immune Tolerance
DNA Replication
Endoplasmic Reticulum
Wnt Signaling
Protein Synthesis and Translational Control
The Synapse
Extracellular Matrix Biology
Protein Homeostasis
Calcium Signaling
The Golgi
Germ Cells
The Mammary Gland as an Experimental Model
The Biology of Lipids: Trafficking, Regulation, and Function
Auxin Signaling: From Synthesis to Systems Biology
The Nucleus
Neuronal Guidance: The Biology of Brain Wiring
Cell Biology of Bacteria
Cell–Cell Junctions
Generation and Interpretation of Morphogen Gradients
Immunoreceptor Signaling
NF-κB: A Network Hub Controlling Immunity, Inflammation, and Cancer
Symmetry Breaking in Biology
The Origins of Life
The p53 Family

Transplantation

A subject collection from *Cold Spring Harbor Perspectives in Medicine*

EDITED BY

Laurence A. Turka

Harvard Medical School
Massachusetts General Hospital

Kathryn J. Wood

University of Oxford

COLD SPRING HARBOR LABORATORY PRESS
Cold Spring Harbor, New York • www.cshlpress.org

Transplantation

A Subject Collection from *Cold Spring Harbor Perspectives in Medicine*
Articles online at www.perspectivesinmedicine.org

Printed in the United States of America

Executive Editor	Richard Sever
Managing Editor	Maria Smit
Project Manager	Barbara Acosta
Permissions Administrator	Carol Brown
Production Editor	Diane Schubach
Production Manager/Cover Designer	Denise Weiss
Publisher	John Inglis

Front cover artwork: The Healing of Justinian by Saint Cosmas and Saint Damian as depicted by the artist Fra Giovanni Angelico (1395–1455). According to Christian beliefs, Saints Cosmas and Damian were twin brothers, born in Arabia, who were well known for their skill as physicians. Their most famous miraculous deed, depicted in this painting, was the grafting of a leg from a recently deceased Ethiopian to replace Deacon Justinian's gangrenous leg. (This work is in the public domain and its use is free of known restrictions under copyright law, including all related and neighboring rights.)

Library of Congress Cataloging-in-Publication Data

Transplantation : a subject collection from Cold Spring Harbor perspectives in medicine / edited by Laurence A. Turka, Harvard Medical School, Massachusetts General Hospital, and Kathryn J. Wood, University of Oxford.
pages cm -- (Cold Spring Harbor perspectives in medicine)
Includes index.
ISBN 978-1-936113-88-0 (hardcover : alk. paper)
1. Transplantation of organs, tissues, etc. 2. Transplantation immunology. 3. Graft rejection–Prevention. I. Turka, Laurence A., editor of compilation. II. Wood, Kathryn J., editor of compilation.

RD120.7.T668 2013
617.9′54–dc23

2013020596

10 9 8 7 6 5 4 3 2 1

All World Wide Web addresses are accurate to the best of our knowledge at the time of printing.

For a complete catalog of all Cold Spring Harbor Laboratory Press publications, visit our website at www.cshlpress.org.

Contents

Preface

THE TITLE OF THIS VOLUME, *Transplantation*, reflects a long-held dream of physicians and their patients. Mythological references to xenogeneic transplantation have occurred in Hindu legends that are likely at least 5 millennia old, although it is not likely that these were performed for health-related reasons. Chinese references to therapeutic heart transplantation date back almost 2500 years. These were clearly therapeutic in nature (being performed reciprocally between two individuals to balance their spirits and energies) and are documented both in a collection of oral legends and as a work of art.

Perhaps the most famous representation of transplantation in art is the iconic painting reproduced on the cover of this volume: Fra Angelico's depiction of the legendary transplantation, by Saints Cosmas and Damian, of a leg from a "Moor" to replace the gangrenous limb of the church deacon Justinian. One of the most notable aspects of the painting is the pair of shoes in the lower left corner. Clearly, the surgeons were planning for success!

We will not recapitulate the history of transplantation here—it is a fascinating one that deserves more lengthy treatment, and indeed the chapter in this volume by Barker and Markmann is an outstanding review. Suffice it to say that conception to reality required roughly 5000 years, but since then, progress has been brisk. In our own lifetimes we have witnessed the birth of clinical transplantation across histocompatibility barriers, and its development from an experimental procedure to a routine treatment performed worldwide.

In accepting the invitation to edit this volume, our goal was to provide a comprehensive view of the field as it exists in 2013. We selected a series of highly relevant basic and clinical topics—at least highly relevant today. We recognize that many will feel that an important area was omitted, and for this we plead a combination of ignorance and slavish attention to the space dictates of the publisher. We also fully appreciate that many new, unimagined areas will emerge in the coming years, and some, which seem timely today, will become anachronisms. *C'est la vie.*

Many leading authorities in the field have contributed to this volume. We are well aware of the numerous demands on their time and are grateful that they managed to set some of it aside for this effort. In providing instructions to authors, we deliberately provided little more than chapter titles and a rough guideline of how to treat the topic. What follows then is what we desired—that is, a highly individual treatment of each area. Some authors are more clinically oriented, and others more research focused; some provide very practical information, and others chose to give a "treetops" view. We feel this is a strength of the volume, given the interrelatedness of the topics, and trust that the interested reader will peruse many of the chapters, which taken together, provide a rather complete view of the field.

In addition to thanking the authors, we would like to acknowledge many helpful staff at Cold Spring Harbor Laboratory Press—in particular, Barbara Acosta, who managed the project from inception (only slightly easier than herding cats). She did a superb job, and our own jobs would have been impossible without her. Special thanks as well go to Richard Sever, who initiated the project by inviting us as editors. Last, we would like to thank the many individuals (including our mentors, colleagues, and trainees) who, over the years, have helped teach us and shape our view of the field. In many ways, our ability to edit this volume is a tribute to all of them.

LAURENCE A. TURKA
KATHRYN J. WOOD

Historical Overview of Transplantation

Clyde F. Barker[1,2] and James F. Markmann[2]

[1]Department of Surgery, Hospital of the University of Pennsylvania, Philadelphia, Pennsylvania 19104

[2]Department of Surgery, Harvard Medical School, and Division of Transplantation, Massachusetts General Hospital, Boston, Massachusetts 02114

Correspondence: clyde.barker@uphs.upenn.edu

Except for legends and claims of miracles, most histories of transplantation cover only the last 60 years because there were no earlier successes. However, the story of even this era has been documented in such rich detail that a full account would fill several volumes. Thus, this brief summary must be limited to highly selected "landmarks." Some landmarks had an immediate impact, but the importance of others went unrecognized for decades. Some findings that deserved landmark status were overlooked or forgotten, whereas others of no biological significance had major impact. Placing these events in perspective is challenging. Several of transplantation's pioneers are still alive, and most of the others are within living memory. Virtually all of them have produced their own accounts. For the most part, they agree on what the "landmarks" are, but their differences in emphasis and perspective make an interesting story.

PRE-HISTORY: THE ERA OF MYTHS AND MIRACLES

The idea of replacing diseased or damaged body parts has been around for millennia. Envisioned were complex transplants such as the "successful" transplantation of an entire leg by the 3rd century sainted physicians Cosmos and Damian, which is depicted in several famous paintings (Zimmerman 1998). As early as 600 B.C., the use of autogenous skin flaps to replace missing noses was conceived, and by the sixteenth century, Gaspare Tagliacozzi (Tagliacozzi 1597) and other pioneering plastic surgeons were successful with such procedures. The obvious extension of these methods was to use detached or "free" grafts of the patient's own tissue or that of other donors. But not until the twentieth century was it ever mentioned that grafts might fail. Even the great eighteenth century experimentalist John Hunter, who transplanted human teeth and autotransplanted cocks' spurs into their combs, seemed unaware that homografts would fail (Martin 1970). Only in the last half of the 20th century has there been a consensus that the outcome of homografts differs from that of autografts.

For a long time, proponents of skin homografts refused to admit that they would not work. Success was even claimed for grafts of whole ears and noses. After centuries of sloppy observation and self-deception, the realization crept in that detached (free) skin grafts were useless. In retrospect, the technical failure of even the early autografts was not surprising because at first

Cite this article as *Cold Spring Harb Perspect Med* doi: 10.1101/cshperspect.a014977

full-thickness skin grafts were used. These thick grafts never became established because their underlying layer of fat and other tissue prevented revascularization. The first major technical advance (almost a "landmark") came only in 1869, when Jacques-Louis Reverdin discovered that small, thin (split thickness) grafts would heal (Reverdin 1869). His autogenous "pinch grafts" successfully covered burns, ulcers, or open wounds. Others caught on, and soon an extensive experience with both autografts and homografts was accumulated. Surprisingly, neither Reverdin nor other enthusiasts (including well-respected surgeons such as Joseph Lister) noticed that homografts were inferior to autografts (Goldman 1987). Only one report of the time suggested otherwise. In 1871, the British surgeon George Pollock described a set of successful autogenous grafts, while on the same patient's wound, homografts both from himself and another donor soon "disappeared" (Pollock 1871). This report was ignored while for additional decades claims of successful homografts continued. In one such case during the Boer War, Winston Churchill donated skin to help heal the open wound of a fellow officer. Years later, Churchill (1944) reported that this graft was still successful.

TRANSPLANTATION RESEARCH OF A LOST ERA

Although consensus on the fate of homografts would not be reached for another 50 years, during the first decades of the twentieth century several well-known investigators established not only the inevitability of homograft failure but most of the other basic principles of transplantation immunology. In 1903, Paul Ehrlich studied transplantation of tumors in mice without bothering to determine the response to transplants of normal tissue (Medawar 1958). The use of tumor homografts confused the issue because they would sometimes overwhelm their recipients before being rejected. In the same year, the Danish biologist Carl Jensen (1903) perceived that the failure of tumor homografts was an immune reaction, but this explanation was discounted by Ehrlich because no antibody (the accepted hallmark of immunity) could be detected.

Georg Schöne may deserve recognition as the first transplantation immunologist. Working in Ehrlich's laboratory in 1912, he studied grafts of skin rather than tumors. He determined that homografts always failed and that subsequent grafts from the same donor failed more rapidly than the first (Schöne 1912). Three decades later, several surgeons would be squabbling for the credit of discovering this "second set" response, suggesting that they deserved a share of Medawar's Nobel Prize, which in part was based on this landmark observation.

By the end of the 1920s, scientists at the Rockefeller Institute firmly established other tenets of transplantation immunology including the central role of the lymphocyte. James B. Murphy's work was especially prescient (Murphy 1914a; Silverstein 2001). He showed that resistance to tumor homografts was dependent on the lymphoid system. He sought to extend graft survival by getting rid of lymphocytes with irradiation, splenectomy, or benzol, the first chemical immunosuppressive agent, noting that these methods decreased the lymphocytic infiltration he observed in failing homografts (Murphy 1914b). Murphy was convinced that these cells were responsible for homograft destruction. He could not explain how, because of the firmly prevailing notion of the time that lymphocytes were fixed cells that lacked mobility. All of these findings were published in widely read scientific journals, but they were ignored by most and eventually were largely forgotten. David Hamilton in his excellent recent history dubs this the "lost era" of transplantation (Hamilton 2012, pp. 105–125).

PIONEERS OF ORGAN GRAFTING

Half a century before proponents of skin homografts finally conceded their futility, surgeons using the more complex model of kidney transplantation recognized that homograft failure was inevitable. Alexis Carrel is commonly credited with originating both vascular suturing and its use in organ transplantation (Fig. 1). Although the award of the 1912 Nobel Prize for

Figure 1. Alexis Carrel, whose pioneering work on blood vessel suturing and organ transplantation was recognized by the 1912 Nobel Prize. (Photograph ca. 1907 from the collection of the American Surgical Association.)

his development of these techniques was well deserved, he was actually not the first in either endeavor. Mathieu Jaboulay, the Chief of Surgery in Lyon, where Carrel trained, and the German surgeon Julius Dörfler introduced the full-thickness blood vessel suturing technique (Dörfler 1895; Jaboulay and Brain 1896). Carrel only adopted this method a decade later on the advice of Charles Guthrie after he initially advocated partial thickness suturing.

Technically successful kidney transplants were accomplished first not by Carrel but by Emerich Ullmann, who in 1902 performed a dog autotransplant and a dog-to-goat xenograft (Ullmann 1914). In 1906, the first two renal transplants in humans were performed by Jaboulay using a pig donor for one and a goat donor for the other (Jaboulay 1906). Ernst Unger, after first performing more than 100 kidney transplants in animals, performed the third and fourth human transplants in 1909 using monkey donors (Unger 1910). None of these early human kidney xenografts functioned for more than a few days, and all of the patients soon died.

In 1904, Carrel left France after failing in several examinations to qualify for a faculty position there. After a brief sojourn in Montreal, he moved to Chicago, where he partnered with the physiologist Charles Guthrie. They collaborated for barely 12 months, but during this time, they successfully transplanted the kidney, thyroid, ovary, heart, lung, and small bowel, averaging a publication on this work every 14 days (Malanin 1979). Carrel's success with organ grafts was not dependent on a new method of suturing but on his use of fine needles and suture material, his exceptional technical skill, and his obsession with strict asepsis.

Carrel's relationship with Guthrie soon cooled because Guthrie objected to Carrel's seven single-author papers about their joint work and to Carrel's habit of advancing his fame by reports in the newspapers. After Carrel left Chicago in 1906 for the Rockefeller Institute in New York, Guthrie published in *Science* a criticism of Carrel and the contention that he, rather than Carrel, deserved most of the credit for their joint accomplishments. He wrote: "It is a singular fact that up until the time Carrel and I engaged in the work together, his experiments did not yield good results, and that our results almost from the beginning of our work together were excellent!" (Guthrie 1909). This was only one of several instances in which the Nobel Prize Committee saw it differently than some of the candidates, and Guthrie was not given a share of the Nobel Prize.

Carrel's extensive experience with organ transplants in animals left no doubt that, although autografts could be consistently successful, homografts never were. In view of the stubborn ongoing claims of successful skin homografts, this was one of Carrel's most important findings, in itself a landmark. Carrel did not know why homografts failed, but he began to explore methods to avoid this such as matching of donor and recipient. Under the influence of his colleague James B. Murphy, he irradiated recipients in unpublished and now forgotten experiments, finding that this improved results (Flexner 1914).

World War I interrupted the productive transplantation research at the Rockefeller Institute by Carrel, Murphy, and their colleagues. Carrel, after spending the war in France treating wounded soldiers, returned to the Rockefeller

Institute but not to research in transplantation. Instead, he formed an unlikely partnership with the aviator Charles Lindbergh, who approached him to discuss the possibility of a heart operation on a relative. Carrel responded that open heart surgery would require a pump oxygenator. Lindbergh offered to build such a device, and Carrel provided laboratory space for the project (Lindbergh 1978). Lindbergh's pump was not used for heart surgery but for perfusion of organs and tissues. It allowed preservation of organs for as long as 3 weeks (Anonymous 1931). The publicity-seeking Carrel made sure that the pump was prominently exhibited at the 1939 New York World's Fair.

Carrel was the originator of tissue culture, another technique that subsequently played an important role in transplantation. He incubated small fragments of embryonic chicken heart in dilute plasma. Carrel claimed that by 1919, this "immortal" tissue had been cultured for 1939 passages and was still normal and pulsating. But it was subsequently determined that embryonic cells with a normal diploid set of chromosomes cannot be maintained in culture for more than 50 doublings unless they undergo malignant transformation. Carrel's laboratory technician eventually admitted the fraud, saying that because Dr. Carrel would be upset if the strain was lost, she added new embryo cells when they were needed (Witkowski 1980).

Despite the continued claims of success for skin homografts, the well-documented failure of experimental organ homografts by Carrel discouraged further research in this field, which during the 1920s and 1930s was continued by only a few. Frank Mann at the Mayo Clinic conducted extensive studies of canine renal and heart homografts but failed to extend Carrel's earlier findings or explore Carrel's suggestions for preventing rejection (Mann 1932).

LEO LOEB, A FORGOTTEN HERO

In the 1930s, Leo Loeb, an émigré from Nazi Germany working at Washington University in St. Louis, was one of only a few transplantation researchers. He determined that the strength and timing of rejection of skin homografts in rats was governed by the extent of genetic disparity of donor and recipient. He also showed that the lymphocyte was involved (Loeb 1945). As Chief of Pathology, he influenced his plastic surgeon colleagues James B. Brown and Earl Padgett in studies determining that identical twins would accept exchanged skin grafts (Brown 1937). Loeb reported his finding that grafts exchanged between (inadequately) inbred mice of the same strain would fail, mistakenly claiming that this was due to some mystical "finer mechanism." For this error (later retracted after further inbreeding of his mice), he was viciously ridiculed by his famous faculty colleague, the geneticist C.C. Little, and by others including Peter Medawar, whose unfair antipathy for Loeb influenced him to dismiss the importance of the lymphocyte and for years espouse the humoral theory of rejection (Hamilton 2012, p. 160). This dispute forever tarnished the reputation of Leo Loeb, a forgotten hero of transplantation research.

UNMODIFIED ANIMAL AND HUMAN KIDNEY TRANSPLANTS

In 1933, the Soviet surgeon Yu Yu Voronoy performed the first human-to-human kidney transplant. That the kidney was not procured until 6 hours after the donor's death and that it was transplanted across a major blood group mismatch probably accounted for its prompt failure (Voronoy 1937). Four other human homografts that Voronoy performed between 1933 and 1949 also failed rapidly. Published in Russian, this experience remained unknown in the West until the 1950s.

In the 1940s and early 1950s, experimental dog kidney transplantation was actively conducted by surgeons in Paris and Boston and also by Morton Simonsen in Denmark and William Dempster in London. The "second set" phenomenon was again observed, but no new insights emerged. Simonsen searched in vain for an antibody response, being unaware of Loeb's work on the lymphocyte (Simonsen 1953). Dempster, who joined Medawar in denouncing Loeb and his cellular theory of immunity, may have been the first to use radiation in organ transplant recipients (Dempster 1953a). He also treated dog homograft recipients with cortisone (Dempster 1953b), which Rupert Billingham had found to prolong survival of skin homografts in rodents (Billingham and Krohn 1951). In dogs, neither treatment had much effect on rejection. Dempster like others despaired of further progress in organ transplants and advised against any attempt in humans.

The faint hope that homografts might fare better in humans than animals was spectacularly rekindled in 1950 when Chicago urologist Richard Lawler performed a human kidney transplant (Lawler et al. 1950). In this patient whose renal function was impaired but not terminal, the homograft was claimed to be successful. It may have produced urine briefly, but when it was removed after several months, it was found to be shrunken. This procedure was of no scientific significance and of no benefit to the patient, who survived for several years with declining renal function of her own kidney. Within the medical profession, this "maverick attempt" was widely criticized, and Lawler's urology colleagues treated him with distain. However, because enthusiastic reports in the lay media generated public support, the impact of this transplant was substantially positive. The French surgeon René Küss said that it provided him the excuse to start a program in human kidney transplantation (Küss 1991).

Because no method was available to prevent rejection, it was recognized that the chances of success were remote, but the lack of dialysis or any other treatment for renal failure was used to justify trials of transplantation in otherwise doomed patients. In 1951, two teams working separately in Paris performed nine kidney transplants (Küss and Bomget 1992). Most of the donors were guillotined criminals. The kidneys were placed retroperitoneally in the pelvis revascularized by iliac vessels with the ureter anastamosed to the bladder, a method devised by Küss that is still the standard operation. None of these transplants showed meaningful function, and all of the patients died within days or weeks (Küss et al. 1951). The ninth transplant in this series was the first to use a living relative as the donor, the patient's mother. Unlike the others, this kidney promptly functioned, but it was rejected after 3 weeks.

In the concurrent Boston program at the Peter Bent Brigham Hospital, David Hume performed nine kidney transplants between 1951 and 1953 (Hume et al. 1955). The donors were patients who had died during surgery or hydrocephalus patients who had a normal kidney removed so that its ureter could be used as a conduit to drain their excess cerebrospinal fluid to the bladder. Except for one orthotopic transplant, they were placed in the anterior thigh with the ureter brought out to the skin. Some of these recipients were treated with ACTH, cortisone, and testosterone. Only four kidneys showed any function, and in three of these, function was brief. But, remarkably, one transplant functioned for 5.5 months before it was rejected. This helped sustain the hope that in humans the outcome of homografts might be better than predicted from animal transplants, thus possibly justifying continued attempts made in as many as six more patients that were never reported. But in the opinion of most physicians of the time, the Paris and Boston experience confirmed the futility of kidney homografts, proving that such human experimentation was unwarranted and unethical. However, two landmark events would soon conspire to brush aside this pessimism and launch the modern era of transplantation.

PRELUDE TO THE MODERN ERA

The prelude to the modern era of transplantation began by chance. During World War II, Peter Medawar, a young Oxford zoologist who had no previous interest in transplantation, was assigned to join plastic surgeon Thomas Gibson in exploring the use of skin homografts for treatment of burned aviators (Gibson and Medawar 1943). Working in the Burn Unit of Glasgow's Royal Infirmary, they soon reconfirmed that homografts always failed. Medawar attributed to Gibson the credit for their rediscovery of the "second set phenomenon" previously described in animals by several others including Schöne in 1912 and in a human patient by Emile Holman in 1921 (Schöne 1912; Holman 1924). Like the earlier researchers, they realized that this identified rejection as an immunological event. This important observation and its interpretation are often credited to Medawar as an original discovery and a novel insight. They were not, but Medawar's many subsequent contributions were so important, his experiments so precise, and his speaking and writing so masterful that he rightfully deserved to be considered the central figure in the emerging field of transplantation.

Returning to Oxford after the war, Medawar conducted extensive studies of skin homografts in rabbits, more firmly characterizing the timing, histological morphology, and immunological nature of rejection (Medawar 1944). Partly because Medawar was unaware of James B. Murphy's work on the lymphoid system 20 years earlier and refused to believe Leo Loeb's similar findings of the 1930s, he remained convinced for another decade that grafts failed because of humoral rather than cellular immunity (Brent 1997). Frustrated that no antibody could be detected, Medawar then turned his attention away from transplant rejection. Instead, he and his first graduate student, Rupert Billingham, began to study the esoteric phenomenon that, in spotted guinea pigs, the pigmented areas of skin autografts gradually encroach on the surrounding white skin (Billingham and Medawar 1948).

Soon after this, Medawar accepted the Chair of Zoology at the University of Birmingham, recruiting Billingham to join his faculty. Together they continued to study pigment spread.

In 1949, serendipity assumed importance in the story. At a cocktail party Medawar talked with his faculty colleague Hugh Donald, who was studying twin cattle. Donald asked whether identical twins could be distinguished from fraternal twins. Medawar responded that skin grafts exchanged between twins would be accepted only by the identical ones.

When Donald requested that he perform such skin grafting experiments, Medawar was reluctant because he was uncomfortable with the prospect of handling large animals. Therefore, he enlisted the aid of his junior colleague, Billingham, who as the grandson of a dairy farmer was not afraid of cows. Billingham and Medawar had little scientific interest in the outcome of the skin grafts, which they felt was entirely predictable. However, the results were unexpectedly interesting (Billingham 1991) (Fig. 2).

They found that most cows accepted their twin's graft, a surprising result because they knew that most cattle twins are fraternal and some of the twin pairs they grafted were of different genders (Anderson et al. 1951). Totally puzzled by this finding, they discussed it with Hugh Donald, who suggested that they read a paper published 4 years earlier in *Science*. When they did so, the significance of their results suddenly became clear.

To place the cattle twin chapter of the story of tolerance in proper context, it is necessary to go back 200 years. In 1779, the English surgeon John Hunter provided the first anatomical description of the freemartin, a term used for the generally sterile female of a pair of cattle twins of unlike sex. Hunter dissected freemartins, finding that they had masculinized sex organs (Palmer 1835). That Hunter was unable to explain this curious phenomenon is ironic because he was an authority on the circulation of the placenta.

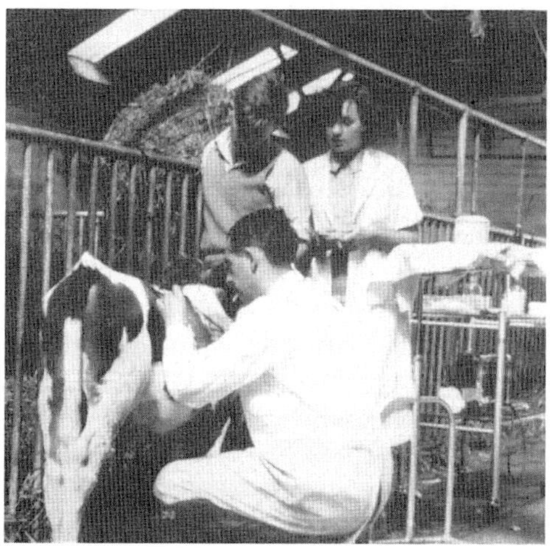

Figure 2. Peter Medawar skin-grafting a cow. The unexpected acceptance of grafts exchanged between chimeric bovine fraternal twins was the key to understanding tolerance. (The photograph is a gift from the private collection of Rupert Billingham, who was the photographer.)

The next important link in the story was provided in 1916 when Frank Lillie, an embryologist at the University of Chicago, dissected a pair of unborn cattle twins (Lillie 1916). He found that the chorions of the twins' placentas were fused, causing a common intrauterine circulation that would allow blood to be exchanged freely between the twins. Like John Hunter, Lillie also found that when cattle twins were of unlike sex, the gonads of the female were usually rudimentary. He reasoned that male hormones circulating through the female embryo inhibited the development of its reproductive organs.

Three decades later, in 1945, Ray Owen at the University of Wisconsin wrote the next chapter. In studying erythrocytes in cattle, Owen found that in this species fraternal twins frequently had a mixture of two red blood cell types (Owen 1945). Recalling Lillie's finding of placental fusion of bovine twin embryos, Owen concluded that not only hormones but also cellular elements of the blood must be exchanged in utero by twin cattle. He realized that persistence of red blood cell chimerism in adulthood must depend on intrauterine transfer not only of short-lived red blood cells but also of stem cells that would perpetuate them.

Six years after the publication of Owen's largely forgotten paper, Billingham and Medawar read it with fascination and suddenly understood why cattle accepted their fraternal twin's graft. They realized that, like the freemartins studied by Hunter, Lillie, and Owen before them, their twins must have exchanged blood in utero and that the donor cell chimerism persisted in adulthood. They reasoned that the stem cells exchanged in utero by these twins would be not only those for red blood cells but also for leukocytes, and that these were probably responsible for skin graft acceptance. They also realized at once with considerable excitement that it might be easy for them to repeat *Nature*'s bovine twin experiment in other species. In 1951, Billingham and Medawar moved to University College, London. Medawar said, "Thank God we've left those cows behind."

In London, Billingham, Medawar, and graduate student Leslie Brent set out to induce chimerism and homograft acceptance in mice by inoculating intrauterine fetuses with donor strain spleen cells (Fig. 3). In retrospect, they were quite lucky to have achieved any successes. By chance, the inbred strains they chose for the experiments were CBA and A, virtually the only H-2-incompatible combination available to them in which neither graft-versus-host disease nor incompatibility of skin-specific antigens would cause death or rejection of the graft.

In adulthood, survivors of their intrauterine inocula accepted skin grafts but only from the

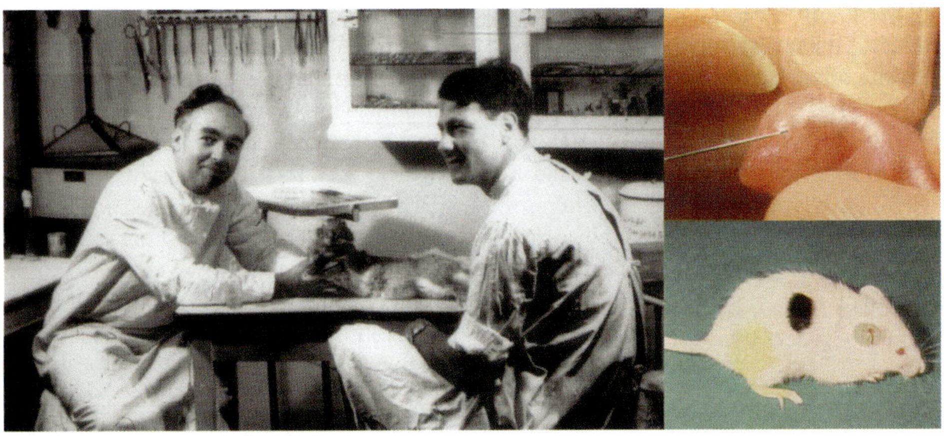

Figure 3. Rupert Billingham and Leslie Brent in their laboratory, where they inoculated neonatal mice with spleen cells (*upper insert*). In adulthood, the mice accepted skin homografts from the donor strain (*lower insert*). (The photographs are a gift from Rupert Billingham's private collection.)

spleen cell donor strain (Billingham et al. 1953). Like the cattle work, it showed that allograft rejection was not inevitable. Surprisingly, neither the publication of the work in October 1953 nor its presentation at a New York Transplantation meeting the next year had sudden dramatic impact, in part because of Medawar's statement that it had no clinical implication (Hamilton 2012, pp. 225–226). Over time, however, it is clear that no other experiment in the field has approached the importance of their demonstration that induction of chimerism can prevent graft rejection. It resulted in Medawar's 1966 Nobel Prize, and it reverberates in present-day clinical trials to induce tolerance.

During their experiments to induce tolerance, Billingham and Brent made another quite unexpected observation. Many of their chimeric mice were sickly "runts." They soon determined that this was because immunocompetent cells in the neonatal inocula migrated to and attacked the lymphoid tissues of their new hosts—that is, graft-versus-host disease (GVHD). Although they briefly reported this in 1957, publication of their conclusive proof was delayed until 1959 by Brent's extended sabbatical with Ray Owen (Brent 1991). Meanwhile, in 1957, Morton Simonsen had independently discovered and published his evidence for GVHD in chickens that he had injected as embryos with allogeneic lymphoid cells (Simonsen 1957, 1985). That to cause GVHD, lymphocytes must be mobile was among the findings belatedly inducing Medawar to accept the importance of cellular immunity, of which he became the foremost proponent. Even better evidence of cellular immunity was provided by Avrion Mitchison's demonstration that immunity to tumor grafts could be transferred by cells but not antibody (Mitchison 1954). Final proof of lymphocyte mobility came in 1959 when James Gowans (Gowans 1957) showed that lymphocytes recirculate from blood to lymph and back again.

THE FIRST SUCCESSES

In Boston, barely 14 months after the initial report of tolerance in chimeric mice by Billingham, Brent, and Medawar, came the next landmark. On December 23, 1954, Joseph Murray bypassed the barrier of rejection by using the patient's identical twin as the donor of a human kidney transplant (Fig. 4) (Murray et al. 1955; Merrill et al. 1956). In retrospect, the success of this transplant had no real scientific significance because technical accomplishment of human kidney transplants was nothing new and it had also been known for decades that skin grafts exchanged between identical twins were not rejected (Brown 1937). But, the impact of this first successful human transplant was immediate and profound. Widespread enthusiastic reports were an important stimulus for surgeons to pursue further efforts in transplantation. But because induction of chimerism in human recipients by neonatal treatment was clearly impossible, another approach would be necessary. The next year, one was found by Joan Main and Richmond Prehn, who showed that weakening the immune system of adult mice by radiation allowed them to induce chimerism by inoculating bone marrow cells. Skin grafts were then accepted if they came from the bone marrow donor strain (Main and Prehn 1955). This and the similar success of the method in one dog kidney homograft (Mannick et al. 1959) encouraged transplantation teams in Paris and Boston to pursue this approach for preventing rejection of human kidney homografts.

In 1958, Murray's team used the Main–Prehn strategy in two human kidney recipients that they conditioned with lethal total body irradiation (TBI) and donor bone marrow. Ten other patients were treated with sublethal TBI without bone marrow. Disappointingly, 11 of the 12 irradiated patients died within a month (Murray et al. 1962). The survivor (who was not given bone marrow) maintained adequate function of his fraternal twin brother's kidney for 20 years. Scientifically, this was a more important accomplishment than the identical twin case because it was the first time the genetic barrier to human kidney transplantation had been breached (Merrill et al. 1960). Five months later in Paris, Jean Hamburger and colleagues (Hamburger et al. 1959) using the same irradiation treatment were successful with another fraternal twin transplant, which functioned until the

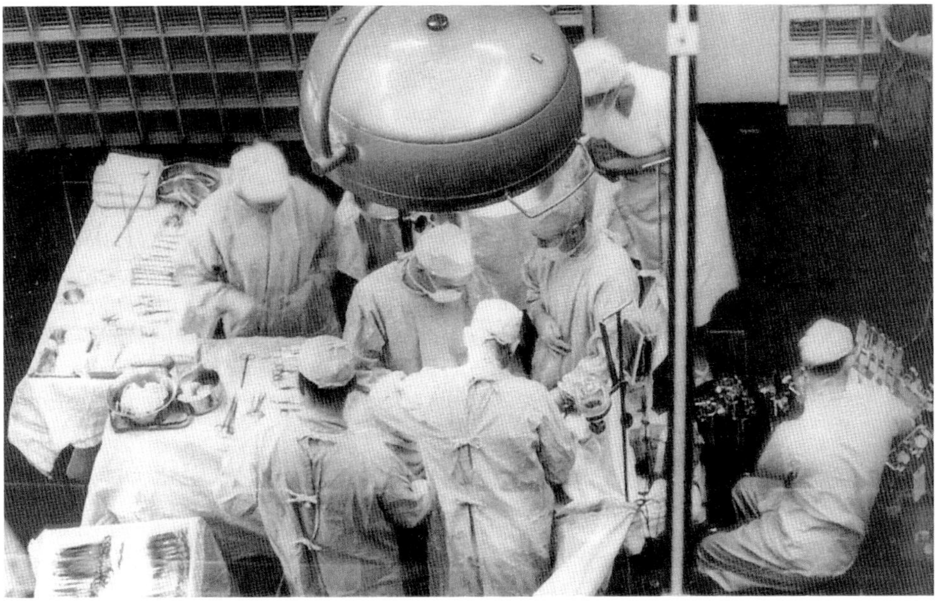

Figure 4. Joseph Murray and his team performing the first successful kidney transplant in 1954 using as a donor the recipient's identical twin.

patient's death from unrelated causes 26 years later.

In these two dizygotic twin cases, there was speculation that the donor and recipient, like twin cattle, had become chimeric by exchanging tolerogenic blood cells during gestation. However, between 1960 and 1962 in Paris, Jean Hamburger and René Küss showed that this was unnecessary by performing four successful transplants in nontwin recipients conditioned by total body irradiation (Küss 1962). This French experience was the principal (and perhaps the only) justification for continuing human kidney transplantation. Because bone marrow inocula were not used in these patients, it was assumed that chimerism was not necessary for success.

CHEMICAL IMMUNOSUPPRESSION

Although Hamburger and Küss both used adrenal cortical steroids as an adjunct to TBI and Küss secondarily administered 6-mercaptopurine (6-MP) to one of his irradiated patients, there was as yet no systematic investigation of drug-based immunosuppression as a substitute for radiation. In the 1950s, oncologists were evaluating drugs including nitrogen mustard and 6-MP for treatment of malignancies. In 1959, interest was drawn to the use of such drugs for transplantation by the report of an experiment by Robert Schwartz and William Dameshek at Tufts University. They found in rabbits that 6-MP reduced the antibody response to bovine albumen (Schwartz and Dameshek 1959). In 1960, they reported that the drug modestly extended the survival of skin homografts (Schwartz and Dameshek 1960). When Roy Calne, a surgical trainee in London, learned of these experiments, he tested the effect of 6-MP on rejection of dog kidney homografts and found that it significantly prolonged their survival. He promptly reported this in *Lancet* (Calne 1960). Simultaneously, Charles Zukoski working in Richmond with David Hume independently made the same observation, but his report in the *Surgical Forum* did not appear until the following year (Zukoski et al. 1960). Calne also treated three human kidney recipients with 6-MP, but they all died without showing any function of the transplant (Hopewell et al. 1964).

In 1960, with Medawar's help, Calne obtained a research fellowship with Joseph Murray in Boston. Although Murray advised him to pursue the Brigham's ongoing experiments with whole body irradiation, Calne began working instead with 6-MP and its derivative azathioprine, which he obtained from Gertrude Elion and George Hitchings (Hitchings and Elion 1954), who were subsequently awarded the Nobel Prize for development of these immunosuppressive agents. Calne's demonstration that in dogs kidney transplant rejection could sometimes be delayed substantially with these drugs stimulated the Brigham team to begin using them in human kidney recipients.

NATIONAL RESEARCH COUNCIL CONFERENCE

In 1963, one of the most important landmarks in the history of transplantation was revealed during a small conference organized by the National Research Council (NRC). About 25 individuals including most of the world's active transplant clinicians and scientists assembled in Washington to review the status of human kidney transplantation. The results presented by these acknowledged experts were extremely discouraging. Less than 10% of their several hundred allograft recipients had survived as long as 3 months (Goodwin and Martin 1963). Of patients treated with total body irradiation, only six had approached or achieved 1 year survival (Starzl 2000). Hope was expressed that immunosuppressive drugs might be more effective. Murray reported his first 10 patients treated with 6-MP and azathioprine instead of irradiation (Murray et al. 1963). One had survived for a year, although at the time of the conference it was failing. The others died within 6 months. Thus, at this point, drugs seemed no more effective than radiation. The mood at the conference was so gloomy that some participants questioned whether continued activity in human transplantation could be justified (Küss 1992).

The gloom was dispelled by only one presentation given by Tom Starzl, a virtually unknown newcomer to the field, who was invited to the conference as an afterthought. He described a new immunosuppressive protocol that had allowed >70% 1-year renal graft survival. He had more surviving patients than the rest of the world's better known participants combined. At first, his audience was incredulous. Tapes that recorded the subsequent sometimes acrimonious discussions were lost, but eventually Starzl's unprecedented results had to be believed because he had brought with him charts detailing the daily progress of each patient—including laboratory tests, urine output, and immunosuppressive drug doses (Hamilton 2012, pp. 279–280, 487). Starzl's innovation based on his consistent success in reversing homograft rejection in dogs was to add prednisone to azathioprine. Although rejection usually occurred in patients on azathioprine alone, it was usually reversible with large doses of prednisone. In most patients, drug doses could then be diminished without provoking rejection. This presentation caused a sensation (Küss 1992). The formal report of the conference consolidated the results of all participants. This failed to stress and seemed almost to obfuscate the dramatic reaction to Starzl's presentation. In fact, the impact on those present was extraordinary. They would have been still more impressed if they could have known that half a century later, some of the patients Starzl described would be off immunosuppression with the same functioning allografts and that they would have been found microchimeric with their donors' lymphoid cells.

The outlook for renal transplantation was completely changed by Starzl's report. Transplant historian Nick Tilney described it as "letting the genie out of the bottle" (Tilney 2003). Many of the conference attendees promptly visited Starzl in Denver to learn how to adopt his immunosuppressive protocol (Starzl 1990). The news of the breakthrough spread quickly by its publication 5 weeks later (Starzl et al. 1963). Before the NRC conference, there had been only three active renal transplant centers in North America (Boston, Denver, and Richmond). As the effectiveness of Starzl's innovative immunosuppression became known, within a year 50 new transplant programs began in the United

States alone. All of them and others that began subsequently adopted the Starzl "cocktail immunosuppression." In fact, this protocol remained the virtual world standard for almost the next two decades (Brent 1997).

A PERIOD OF CONSOLIDATION

The next quarter century, 1964–1980, is often classified as a period of consolidation or plateau. During this period, except for the development of antilymphocytic serum (in large part responsible for the first albeit marginal success of extrarenal transplants), there were no landmarks. Importantly, however, there was steady progress. Practical innovations and accomplishments took place that were necessary for maturation of kidney transplantation into a useful clinical service. These included availability of dialysis; Medicare funding of end-stage renal disease; antibody screening to avoid hyperacute rejection; the importance of tissue typing for related donor transplants; acceptance of brain death; ex vivo preservation, allowing donor organs to be transported and shared;, and, perhaps most importantly, the accumulation of experience in patient management that led to the avoidance of over-immunosuppression, thereby decreasing resultant infections and deaths.

Hemolysis

Hemodialysis for renal failure was pioneered in Holland by Willem Kolff during World War II, but chronic renal failure could be treated only after Belding Scribner in 1960 devised Teflon arteriovenous conduits for long-term vascular access (Kapoian 1997). Before that, each dialysis treatment used up an accessible artery and vein until they ran out. In 1966, James Cimino and Michael Brescia introduced subcutaneous anastamoses of an artery and vein at the wrist to arterialize superficial arm veins that could then be accessed by simple needle puncture. But even in the late 1960s, chronic dialysis was available in only a few centers, and in these expense limited its use to a small number of patients. Proliferation of centers able to offer chronic dialysis and transplantation took place only after Congress in 1972 approved Medicare funding for patients of any age with end-stage renal disease.

Brain Death

Before the mid 1960s, utilization of kidneys from deceased donors was limited by the ischemic damage that set in as soon as the heart stopped beating, this being the time-honored definition of death. During the next few years, there was gradual although controversial acceptance that irreversible loss of brain function was also a form of death and one that would allow removal of organs from a "heart-beating cadaver." In 1968, the "Harvard ad hoc Committee on Brain Death" published its recommendation that irreversible loss of brain function be accepted as death (Harvard Medical School 1968). Uneasiness that this might be influenced by the desire of transplanters to recover organs from possibly salvageable patients and criticism over the poor outcomes of the first heart transplants that were being performed during this same period fueled controversy over this concept. But eventually, brain death was widely accepted, a crucial factor in increasing numbers of transplants, especially of extra renal organs.

Organ Preservation

As early as 1905, Carrel's colleague Charles Guthrie had advocated cooling to protect donor organs before transplantation (Hamilton 2012, p. 101). That it was not used in early human transplants probably in part accounted for their poor results. Initially, Starzl used total body hypothermia to protect donor organs, but by 1960 switched to infusing cold solution into the portal vein to protect donor livers (Marchioro et al. 1963). By 1963, pretransplant infusion of cold solution into the renal artery of donor kidneys became standard (Collins 1969). When multiple organs were procured from a donor, in situ cooling by infusion of cold solution to the aorta was used. Ex vivo perfusion of isolated kidneys by a pump (reminiscent of Lindberg's machine in the 1930s) was shown to extend preservation of kidneys for

2–3 days (Belzer et al. 1967), but the popularity of this method waned after 1987 when Folkert Belzer introduced University of Wisconsin solution, which when simply infused into blood vessels of donor organs allowed preservation almost as long (Belzer et al. 1992).

Donor Organ Sharing

After donor kidney preservation of up to 6 hours was accomplished in the mid-1960s, sharing of organs between centers became practical. At first, sharing was local and informal. In 1967, Paul Terasaki started the first sharing organization in Los Angeles (Terasaki 1990). The Boston Interhospital Organ Bank followed in 1968. Subsequently, perceptions arose that local control allowed inequity of donor organ allocation. There was concern that U.S. organs were sent abroad or transplanted at U.S. centers into foreign nationals. Recognition of a need to formalize distribution of donor organs at a national level led Congress to pass the National Transplant Act in 1984. The Southeastern Organ Procurement Foundation (SEOPF), founded in 1969 and eventually composed of 12 hospitals in several cities, served as the template for the resultant national entity that now controls organ allocation and placement, monitors performance of transplant centers and organ procurement organizations, collects data, and controls quality—the United Network for Organ Sharing (UNOS) (McDonald 1988). UNOS has been a force for order and good, but to some its effectiveness has seemed compromised by the excessive scale of its tasks and the number and size of its committees and inevitable bureaucracy. Adjustments in appropriate organ allocation in response to changes in the relative importance of histocompatibility and issues of equity have been difficult, prolonged, and politicized.

Histocompatibility Typing

Although tissue matching was suggested by Alexis Carrel and studied in animals by George Snell and by Peter Gorer, it could not begin to emerge as a reality for human transplants until 1958, when Jean Dausset discovered the first human leukocyte antigen (HLA) (Snell 1948; Dausset 1958). Antibodies against this antigen were identified in transfused patients and multiparous women by Rose Payne (1957) and Jon van Rood (van Rood et al. 1958) soon after. Testing for antibodies (by agglutination techniques) was cumbersome and inconsistent until Paul Terasaki in 1964 developed a microcytotoxicity assay (Terasaki 1964). His test, which mixed recipient serum and donor lymphocytes in tiny wells, quickly became the standard. For several years, Terasaki did the typing for most U.S. transplant centers. Several of his early findings were of lasting importance (Terasaki 1990): (1) a positive cross-match test identifying donor-specific antibodies in the serum of a prospective kidney recipient predicts hyperacute rejection; and (2) matching can reliably identify the optimal donor within a family. It was next assumed by histocompatibility experts that matching would be just as important in selection of unrelated donors. However, in 1970, when Terasaki examined his large database (1216 kidney transplant patients from 52 centers) to relate typing with outcome of cadaver renal allografts, he found no correlation. Terasaki's announcement of this at the Transplantation Congress in the Hague elicited consternation in members of the tissue typing community, who contended that his methods must be faulty (Terasaki 1990). His paper was the only one not accepted for publication in the conference proceedings. NIH made an emergency site visit to Terasaki's laboratory and abruptly withdrew his grant, closing down most of his research. His funding was subsequently restored when others confirmed that his findings were correct. Since that time, typing has improved, with identification of many additional histocompatibility antigens including those of the important Class II locus (D or DR) (Ting and Morris 1978). Histocompatibility matching remains crucial in bone marrow transplantation and important in selection of family donors. But even now for unrelated donor organ transplants, the benefit remains much less unless there is a perfect match.

Additional important findings that took place during the consolidation period are listed because space precludes their full discussion:

1. Blood transfusions, rather than decreasing the chance of kidney allograft survival, were found during the 1970s to improve it. This still-not-understood observation by Gerhard Opelz and others led to protocols of deliberate pretransplant transfusions (Opelz et al. 1973). However, this strategy was abandoned when the even greater improvement due to cyclosporine obscured the benefits of transfusions.
2. In some pigs and rats, liver homografts survived without immunosuppression and also induced acceptance of other types of graft from the liver donor (Calne 1969; Kamada 1980).
3. William Summerlin's report that pretransplant culture of skin allografts allowed their acceptance without immunosuppression was greeted with excitement but was subsequently proven to be fraudulent. Summerlin was painting the grafts to make them look viable (Medawar 1996).
4. The origin of multiple transplantation societies, journals, national and international meetings, and registries of clinical results ensured prompt dissemination of scientific and clinical findings to a degree not seen in other fields.

EXTRARENAL ORGAN TRANSPLANTS

During the period of consolidation, transplantation of nonrenal organs (liver, heart, and pancreas) began. These landmarks are subjects of other articles in this collection. Their initial success (albeit marginal) was promoted by improvement in immunosuppression by antilymphocyte serum (ALS), which thus in itself deserves landmark status. Further improvement awaited cyclosporine.

ANTILYMPHOCYTE SERUM

Antilymphocyte serum (ALS) had a prolonged evolution. At the end of the nineteenth century, Elie Metchnikoff proposed the use of antiserum to mitigate cellular immunity (Metchnikoff 1899), a concept reexplored six decades later by Byron Waksman (Waksman et al. 1961). Depletion of lymphocytes with antiserum was a logical extension of Gowans's finding (Gowans et al. 1963) that thoracic duct drainage delayed skin graft rejection in rats. In 1963, Michael Woodruff reported that ALS was remarkably effective in extending skin allograft survival in rodent models (Woodruff and Anderson 1963). Monaco et al. (1966) and Levey and Medawar (1966) reported similar success. In 1966, Starzl was the first to use ALS clinically. After finding it effective for kidney allografts, he also credited it with allowing his first successful human liver transplants in 1967 (Starzl et al. 1967a,b). Many others, including Anthony Monaco, John Najarian, and A.G. Sheil, treated patients with homemade ALS from rabbits or horses immunized with lymph node cells or cultured lymphoblasts before pharmaceutical companies began to produce it. Variability in effectiveness of different batches of this polyclonal antiserum was troublesome before the advent of monoclonal derivatives of ALS that were directed first at all T lymphocytes and subsequently at their subsets or their interleukin 2 receptors (Cosimi et al. 1981). These agents have become a mainstay of the modern immunosuppressive armamentarium, especially for induction therapy.

Cyclosporine and Tacrolimus

The next landmark was the "wonder drug" cyclosporine. After a slow start, it revolutionized transplantation by substantially improving kidney transplant results and greatly facilitating successful extrarenal transplants. Cyclosporine is a fungal derivative first reported in 1976 by Jean-François Borel to have immunosuppressive qualities (Borel et al. 1976). Following several years of encouraging animal experiments, Calne began in 1979 to use it as a single agent in treatment of human kidney recipients, finding it to be more potent than azathioprine but also toxic in higher doses, leading to infections, lymphomas, and renal failure (Calne et al. 1979). Results of initial trials in Boston and Canada rated it as unimpressive to poor, causing some to believe the drug should be abandoned. Once again as he had before with azathioprine, Starzl by adding prednisone to the new drug developed a protocol that strikingly improved outcomes of kidney transplants (Starzl et al. 1980). In addition, this transformed transplantation of extrarenal organs into a practical clinical service (Starzl et al. 1981).

Cyclosporine soon became the standard baseline immunosuppressant and remained so until 1989, when it was shown by Starzl that rejection of liver and other organ allografts resistant to treatment by cyclosporine, steroids, and antibodies could often by reversed by an even more potent drug, tacrolimus (Starzl et al. 1989). Tacrolimus has now in large part replaced cyclosporine as the usual baseline agent.

Chimerism

Modern immunosuppression with agents such as cyclosporine/tacrolimus and T-cell antibodies now allows excellent short- and midterm survival of allografts, for 1 year exceeding 90% for kidney and approaching this for other organs. Nevertheless, because of ongoing morbidity from drug toxicity and graft loss from chronic rejection, achievement of drug-free immunosuppression remains the ultimate goal. Plans for inducing tolerance invariably start with review of the 1953 demonstration by Billingham, Brent, and Medawar that chimerism induced in neonatal mice by lymphoid cell inocula allows acceptance of donor strain skin grafts. Main and Prehn's subsequent success by infusion of donor cells to irradiated adult mice inspired pioneer transplanters in Boston and Paris to test this method in a small number of human kidney allograft recipients, but with no success. When in several irradiated human recipients, graft survival was achieved without donor cell inocula and in others achieved with immunosuppressive drugs alone, it appeared that donor cell chimerism was irrelevant to long-term allograft success.

In animal models, there has been continued exploration of donor cell inocula (combined with immunosuppression) for inducing chimerism and tolerance of allografts (Monaco et al. 1966; Lance and Medawar 1969; Thomas et al. 1987). But in humans between 1959 and 1990, there were only a few trials of this strategy, the first by Monaco et al. (1976) and the largest by W.H. Barber and A.G. Diethelm, who treated 57 kidney recipients with an induction course of ALG and cryopreserved donor bone marrow given 17 days after transplantation (Barber et al. 1991). Their results were equivocal, although early graft rejection seemed less than in the control group.

In 1992, attention was dramatically refocused on the role of chimerism in human allograft survival. Tom Starzl discovered that donor leukocyte chimerism was present in patients who had maintained successful kidney or liver grafts for up to three decades (Starzl et al. 1993). Sensitive immunochemical and molecular assays were necessary to detect donor cells, which in some patients were not found in blood but only in biopsies of skin, lymph nodes, and other tissues. This extensive search determined that a microchemic state was present in all 30 patients studied.

Because these recipients had not been given donor cells, the chimeric cells could only have reached them as passengers migrating from the donor organ. Many of the patients appeared to be tolerant, because they were off all immunosuppression. This finding was the basis of Starzl's belief that chimerism is an important cause (not the consequence) of successful transplantation, culminating in his hypothesis of a two-way paradigm for tolerance—that is, successful engraftment is the result of the responses of coexisting donor and recipient cells each to the other causing reciprocal clonal exhaustion followed by peripheral clonal deletion (Starzl et al. 1992, 1996). This interpretation has been rebutted by Kathryn Wood, David Sachs, and others who argue that persistent microchimerism induced by a donor organ is as likely to be the effect as the cause of tolerance (Wood and Sachs 1996). Whatever the resolution of this debate, it is clear that Starzl's demonstration of microchimerism in his patients has been an important stimulus for reexploration of this approach to allograft tolerance by trials described in Markmann and Kawai 2012).

The evolution of organ transplantation in the last half century is one of Medicine's great

Figure 5. Participants in the March 1999 meeting of the third Annual Longmire Surgical Society Scientific Day Symposium at UCLA: *Milestones in Transplantation*. This group arrived at a consensus on the Historical Landmarks in Clinical Transplantation. (*Front row, left to right*) Robert Good, Norman Shumway, E. Donnell Thomas, Joseph Murray, Jean Dausset, and Jon Van Rood. (*Back row, left to right*) Leslie Brent, Roy Calne, Paul Terasaki, Robert Schwartz, Carl Groth, and Thomas Starzl.

stories. In support of this contention are the five Nobel Prizes given for transplantation (19, if related immunology is included). In addition, this year's Lasker Award to Tom Starzl and Roy Calne brings to six the number of these prestigious awards for transplantation or related work. The distribution of these Prizes has not been without controversy. Largely forgotten rivalries that enrich the story are those between Alexis Carrel and his colleague Charles Guthrie; between the proponents of cellular immunity (Elie Metchnikoff, James B. Murphy, Leo Loeb, and Avrion Mitchison) and the champions of humoral immunity (Paul Ehrlich, Peter Gorer, and others, even including Peter Medawar during his early career); between transplant pioneers in Boston, Denver, and Paris, even between two separate teams of transplanters. However, it is notable and reassuring that in 1999, 12 of the field's surviving pioneers met to select and agree on the "historical landmarks" briefly reviewed in this work (Groth et al. 2000) (Fig. 5). The promise of further progress toward the ultimate achievement of tolerance is described in subsequent work.

REFERENCES

Anderson D, Billingham RE, Lampkin GH, Medawar PB. 1951. The use of skin grafting to distinguish between monozygotic and dizygotic twins in cattle. *Heredity* **5:** 379–397.

Anonymous. 1931. Apparatus to circulate liquid under constant pressure in a closed system. *Science* **73:** 566.

Barber WH, Mankin JA, Laskow DA, Deierhoi MH, Julian BA, Curtis JJ, Diethelm AG. 1991. Long term results of a controlled prospective study with transfusion of donor specific bone marrow in 57 cadaveric renal allograft recipients. *Transplantation* **51:** 70–75.

Belzer FO, Ashby BS, Dunphy JE. 1967. 24-hour and 72-hour preservation of canine kidneys. *Lancet* **2:** 536–538.

Belzer FO, Alessandro AM, Hoffman RM, Knechtle SJ, Reed A, Pirsch JD, Kalayoglum M, Sollinger HW. 1992. The use of UW solution in clinical transplantation—A four year experience. *Ann Surg* **215:** 579–585.

Billingham RE. 1991. Reminiscenses of a transplanter. In *History of transplantation: Thirty-five recollections* (ed. Terasaki P), pp. 73–91. UCLA Tissue Typing Laboratory, Los Angeles.

Billingham RE, Medawar PB. 1948. Pigment spread and cell heredity in guinea pigs. *Heredity* **2:** 29–47.

Billingham RE, Krohn PL, Medawar PB. 1951. Effect of cortisone on survival of skin homografts in rabbits. *BMJ* **1:** 1157–1158.

Billingham RE, Brent L, Medawar PB. 1953. "Actively acquired tolerance" of foreign cells. *Nature* **172:** 603–606.

Borel JF, Feurer C, Gubler HU, Stahelin H. 1976. Biological effects of cyclosporin A: A new antilymphocytic agent. *Agents Actions* **6:** 468–475.

Brent LA. 1991. Tolerance and GVHD: An exciting decade. In *History of transplantation, thirty-five recollections* (ed. Terasaki P), pp. 93–109. UCLATissue Typing Laboratory, Los Angeles.

Brent L. 1997. *A history of transplantation immunology*, pp. 70–71, 313. Academic Press, Los Angeles.

Brown JB. 1937. Homografting of skin: With report of success in identical twins. *Surgery* 558–563.

Calne RY. 1960. The rejection of renal homografts: Inhibition in dogs by 6-mercaptopurine. *Lancet* **1:** 417–418.

Calne RY, Sells RA, Penna JR, Davis DR, Millard PR, Herbertson BM, Binns RM, Davies DA. 1969. Induction of immunological tolerance by porcine liver allografts. *Nature* **223:** 472–476.

Calne RY, Rolles K, White DJG, Thiru S, Evans DB, McMaster P, Dunn DC, Craddock GN, Henderson RG, et al. 1979. Cyclosporin A initially as the only immunosuppressant in 34 recipients of cadaveric organs; 32 kidneys, 2 pancreases, and 2 livers. *Lancet* **2:** 1033–1036.

Churchill W. 1944. *My early life*, p. 211. MacMillan, London.

Collins GM, Bravo-Shugarman M, Terasaki PI. 1969. Kidney preservation for transportation: Initial perfusion and 30 hours' ice storage. *Lancet* **2:** 1219–1222.

Cosimi AB, Colvin RB, Burton RC, Rubin RH, Goldstein G, Kung PC, Hansen WP, Delmonico FL, Russell PS. 1981. Use of monoclonal antibodies to T-cell subsets for immunological monitoring and treatment in recipients of renal allografts. *N Engl J Med* **305:** 308–314.

Dausset J. 1958. Isoleuco-anticorps. *Acta Haematol* **20:** 156–166.

Dempster WJ. 1953a. Kidney homotransplantation. *Br J Surg* **40:** 447–465.

Dempster WJ. 1953b. The effects of cortisone on the homotransplanted kidney. *Arch Int Pharmacodyn Ther* **95:** 253–282.

Dörfler J. 1895. Uber die Naht von Arterienwunden 261. *Chir* **22:** 1113.

Flexner S. 1914. *Report on Cancer. Report of the Director of Laboratories to the Scientific Directors. RURG 439. Rockefeller Archives*, p. 322. Rockefeller Archives Center, Sleepy Hollow, NY.

Gibson T, Medawar PB. 1943. The fate of skin homografts in man. *J Anat* **77:** 299–310.

Goldman M. 1987. *Lister Ward*, pp. 118–119. Hilger, Bristol, UK.

Goodwin WE, Martin EC. 1963. Transplantation of the kidney. *Urol Surv* **13:** 229–248.

Gowans JL. 1957. The effect of the continuous reinfusion of lymph and lymphocytes on the output of lymphocytes from the thoracic duct of unanesthetized rats. *Br J Exp Pathol* **38:** 67–81.

Gowans JL, McGreggor DD, Cowan DM. 1963. The role of small lymphocytes in the rejection of homografts of skin. In *The immunologically competent cell* (ed. Ciba Foundation Study Group), Vol. 16, p. 20. Churchill, London.

Groth CG, Brent LB, Calne RY, Dausset JB, Good RA, Murray JE, Shumway NE, Schwartz RS, Starzl TE, Terasaki PI. 2000. Historical landmarks in clinical transplantation: Conclusions from the consensus conference at the University of California, Los Angeles. *World J Surg* **24:** 755–843.

Guthrie CC. 1909. On misleading statements. *Science* **29:** 29–31.

Hamburger J, Vaysse J, Crosnier J, Tubiana M, Lalanne CM, Antoine B, Auvert J, Soulier JP, Dormont J, Salmon C, et al. 1959. Transplantation of a kidney between non-monozygotic twins after irradiation of the receiver: Good function at the fourth month. *Presse Med* **67:** 1771–1775.

Hamburger J, Vaysse J, Crosnier J, Auvert J, Lalanne CL, Hopper J Jr. 1962. Renal homotransplantation in man after radiation of the recipient. *Am J Med* **32:** 854–871.

Hamilton D. 2012. *A history of organ transplantation*. University of Pittsburgh Press, Pittsburgh.

Harvard Medical School. 1984. A definition of irreversible coma, 1968. Report of the Ad Hoc Committee of the Harvard Medical School to examine the definition of brain death. *JAMA* **252:** 677–679.

Hitchings GH, Elion GB. 1954. The chemistry and biochemistry of purine analogs. *Ann NY Acad Sci* **60:** 195–199.

Holman E. 1924. Protein sensitization in iso-skin grafting. Is the latter of practical value? *Surg Gyne Obs* **38:** 100.

Hopewell J, Calne RY, Beswick I. 1964. Three clinical cases of renal transplantation. *BMJ* **1:** 411–413.

Hume DM, Merrill JP, Miller BF, Thorn GW. 1955. Experiences with renal homotransplantation in the human: Report of nine cases. *J Clin Invest* **34:** 327–382.

Jaboulay M. 1906. Kidney grafts in the antecubital fossa by arterial and venous anastomosis [in French]. *Bull Lyon Med* **107:** 575–577.

Jaboulay M, Brian E. 1896. Recherches expérimentales sur la suture at la greffe artérielles. *Lyon Medicale* **81:** 97–99.

Jensen CO. 1903. Experimentelle untersuchungen über Krebs bei Mäusen. *Zentralblatt Bacteriol Parasitenkinde* **34:** 28–34.

Kamada N, Brons G, Davies HFFS. 1980. Fully allogeneic liver grafting in rats induces a state of systemic nonreactivity to donor transplantation antigens. *Transplantation* **29:** 429–431.

Kaporian T, Sherman RA. 1997. A brief history of vascular access for hemodialysis: An unfinished story. *Semin Nephrol* **17:** 239–245.

Kostakis AJ, White DJG, Calne RY. 1977. Prolongation of rat heart allograft survival by cyclosporin A. *Int Res Commun Syst Med Sci* **5:** 280.

Küss R. 1991. Human renal transplantation memories, 1951 to 1981. In *History of transplantation: Thirty-five recollections* (ed. Terasaki P), pp. 37–59. UCLA Tissue Typing Laboratory, Los Angeles.

Küss R, Bourget P. 1992. *An illustrated history of organ transplantation: The great adventure of the century*, p. 53. Sandos, Rueil-maloncison, France.

Küss R, Teinturier J, Milliez P. 1951. Quelques essais de greffe rein chez l'homme. *Mem Acad Chir* **77:** 755–764.

Küss R, Legrain M, Mathe G, Nedey R, Camey M. 1962. Homologous human kidney transplantation: Experience with six patients. *Postgrad Med J* **38:** 528–531.

Lance EM, Medawar PB. 1969. Quantitative studies on tissue transplantation immunity. IX. Induction of tolerance with antilymphocytic serum. *Proc R Soc B* **173:** 447–473.

Lawler RH, West JW, McNulty PH, Clancy EJ, Murphy RP. 1950. Homotransplantation of the kidney in the human. *JAMA* **144:** 844–845.

Levey RH, Medawar PB. 1966. Nature and mode of action of antilymphocytic antiserum. *Proc Natl Acad Sci* **56:** 1130–1137.

Lillie FR. 1916. The theory of the Freemartin. *Science* **43:** 611–613.

Lindbergh CA. 1978. *Autobiography of values*, p. 32. Harcourt Brace Jovanovich, New York.

Loeb L. 1945. *The biological basis of individuality*, pp. 98–106. Thomas, Springfield, IL.

Main JM, Prehn RT. 1955. Successful skin homografts after the administration of high dosage x radiation and homologous bone marrow. *J Natl Cancer Inst* **15:** 1023–1029.

Malinin TI. 1979. *Surgery and life, the extraordinary career of Alexis Carrel*, pp. 22–29. Harcourt Brace Jovanovich, New York.

Mann FC. 1932. Transplantation of organs. In *Contributions to medical science: In honor of Emanuel Liebman*. International Press, New York.

Mannick JA, Lochte HL, Ashley CA, Thomas ED, Ferrebee JW. 1959. A functioning kidney homotransplant in the dog. *Surgery* **46:** 821–828.

Marchioro TL, Huntley RT, Waddell WR, Starzl TE. 1963. Extra-corporeal perfusion for obtaining postmortem homografts. *Surgery* **54:** 900–911.

Markmann JF, Kawai T. 2012. The quest for transplantation tolerance: How we finally sipped from the cup. *Sci Transl Med* **4:** 2–3.

Martin CE. 1970. John Hunter and tissue transplantation. *Surg Gyn Obstet* **131:** 306–310.

McDonald JC. 1988. The national organ procurement and transplantation network. *JAMA* **259:** 725–726.

Medawar PB. 1944. The behavior and fate of skin autografts and skin homografts in rabbits. *J Anat* **78:** 176–199.

Medawar PB. 1958. The immunology of transplantation. In *Harvey Lecture Series, 1956–1957*, Vol. 52, pp. 144–176. Academic Press, New York.

Medawar PB. 1996. *The strange case of the spotted mice and other classic essays on science*. Oxford University Press, New York.

Merrill JP, Murray JE, Harrison JH, Guild WR. 1956. Successful homotransplantation of the human kidney between identical twins. *JAMA* **160:** 277–282.

Merrill JP, Murray JE, Harrison JH, Friedman EA, Dealy JB Jr, Dammin GJ. 1960. Successful homotransplantation of the kidney between nonidentical twins. *N Engl J Med* **262:** 1251–1260.

Metchnikoff I. 1899. Etude sur la resorption des cellules. *Ann Inst Pasteur* **13:** 737.

Mitchison NA. 1954. Passive transfer of transplantation immunity. *Proc R Soc Series B* **142:** 72–87.

Monaco AP, Wood ML, Russell PS. 1966. Studies on heterologous antilymphocyte serum in mice. III. Immunologic tolerance and chimerism produced across the H-2 locus with adult thymectomy and antilymphocyte serum. *Ann NY Acad Sci* **129:** 190–209.

Monaco AP, Clark AW, Wood ML, Sahyoun AJ, Codish SD, Brown RW. 1976. Possible active enhancement of a human cadaver renal allograft with antilymphocyte serum (ALS) and donor bone marrow. *Surgery* **79:** 384–392.

Murphy JB. 1914a. Factors of resistance to heteroplastic tissue-graftings. III. Studies in tissue specificity. *J Exp Med* **19:** 513–522.

Murphy JB. 1914b. Heteroplastic tissue grafting effected through Roentgen-ray lymphoid destruction. *JAMA* **62:** 1459.

Murray JE, Merrill JP, Harrison JH. 1955. Renal homotransplantation in identical twins. *Surg Forum* **6:** 432–436.

Murray JE, Merrill JP, Dammin GJ, Dealy JB, Alexandre GPJ, Harrison JH. 1962. Kidney transplantation in modified recipients. *Ann Surg* **156:** 337–355.

Murray JE, Merrill JP, Harrison JH, Wilson RE, Dammin GJ. 1963. Prolonged survival of human kidney homografts by immunosuppressive drug therapy. *N Eng J Med* **268:** 1315–1323.

National Academy of Sciences. 1964. National Research Council. *Transplantation* **2:** 147–165.

Opelz G, Dharmendra PS, Senger DPS, Mickey MR, Terasaki PI. 1973. Effect of blood transfusions on subsequent kidney transplantation. *Trans Proc* **4:** 253–259.

Owen RD. 1945. Immunogenetic consequences of vascular anastomoses between bovine twins. *Science* **102:** 400–407.

Palmer JF, ed. 1835. *The complete works of John Hunter*, Vol. 2, pp. 56, 97–104. J. & H.G. Langley, London.

Payne R. 1957. Leukocyte agglutination in human sera: Correlation between blood transfusions and their development. *JAMA* **99:** 587–606.

Pollock G. 1871. Cases of skin grafting and skin transplantation. *Trans Clinic Soc* **4:** 37–47.

Reverdin JL. 1869. Greffe épidermique. *Bull De La Société Impériale Chirurgicale de Paris* **1869:** 511–515.

Schöne G. 1912. *Die heteroplastische und homooplastische*. Springer, Berlin.

Schwartz R, Dameshek W. 1959. Drug-induced immunological tolerance. *Nature* **183:** 1682–1683.

Schwartz R, Dameshek W. 1960. The effects of 6-mercaptopurine on homograft reactions. *J Clin Invest* **39:** 952–958.

Silverstein AM. 2001. The lymphocyte in immunology: From James B. Murphy to James L. Gowans. *Nat Immunol* **2:** 569–571.

Simonsen M. 1953. Biological incompatibility in kidney transplantation in dogs. II. Serological investigations. *Acta Pathol Microbiol Scand* **32:** 35–84.

Simonsen M. 1957. The impact on the developing embryo and newborn animal of adult homologous cells. *Acta Pathol Microbiol Scand* **40:** 480–500.

Simonsen M. 1985. Graft vs. host reactions: The history that never was and the way things happen to happen. *Immunol Rev* **88:** 5–23.

Snell GD. 1948. Methods for the study of histocompatibility genes. *J Genet* **49:** 87–108.

Starzl TE. 1990. My thirty-five year view of organ transplantation. In *History of clinical transplantation. Thirty-five recollections* (ed. Terasaki PI), pp. 145–172. UCLA Tissue Typing Laboratory, Los Angeles.

Starzl TE. 2000. History of clinical transplantation. In *Historical landmarks in clinical transplantation* (ed. Groth CG, Longmire WP Jr), pp. 759–782. Springer, New York.

Starzl TE, Marchioro TL, Waddell WR. 1963. The reversal of rejection in human renal homografts with subsequent development of homograft tolerance. *Surg Gynecol Obstet* **117:** 385–395.

Starzl TE, Marchioro TL, Porter KA, Iwasaki Y, Cerilli GJ. 1967a. The use of heterologous antilymphoid agents in canine renal and liver homotransplantation and in human renal homotransplantation. *Surg Gynecol Obstet* **124:** 301–308.

Starzl TE, Groth CG, Brettschneider L, Penn I, Fulginiti VA, Moon JB, Blanchard H, Martin AJ Jr, Porter KA. 1967b. Orthotopic transplantation of the human liver. *Ann Surg* **168:** 392–415.

Starzl TE, Weil R III, Iwatsuki S, Klintmalm G, Schroter GPJ, Koep LJ, Iwaki Y, Terasaki PI, Porter KA. 1980. The use of cyclosporin A and prednisone in cadaver kidney transplantation. *Surg Gynecol Obstet* **151:** 17–26.

Starzl TE, Klintmalm GBG, Porter KA, Iwatsuki S, Schroter GPJ. 1981. Liver transplantation with use of cyclosporin A and prednisone. *N Engl J Med* **305:** 266–269.

Starzl TE, Todo S, Fung J, Demetris AJ, Venkataramanan R, Jain A. 1989. FK 506 for human liver, kidney and pancreas transplantation. *Lancet* **2:** 1000–1004.

Starzl TE, Demetris AJ, Murase N, Ildstad S, Ricordi C, Trucco M. 1992. Cell migration, chimerism, and graft acceptance. *Lancet* **339:** 1579–1582.

Starzl TE, Demetris AJ, Trucco M, Zeevi A, Ramos H, Terasaki P, Rudert WA, Kocova M, Ricordi C, Ildstad S, et al. 1993. Chimerism and donor-specific nonreactivity 27 to 29 years after kidney allotransplantation. *Transplantation* **55:** 1272–1277.

Starzl TE, Demetris AJ, Murase N, Trucco M, Thomson AW, Rao AS. 1996. The lost chord: Microchimerism. *Immunol Today* **17:** 577–588.

Tagliacozzi G. 1597. *Decurtorum cirugia per insitionum.* Bindonum, Venice, Italy.

Terasaki PI. 1990. Histocompatibility. History of transplantation. In *History of transplantation: Thirty-five recollections* (ed. Terasaki P), pp. 513–538. UCLA Tissue Typing Laboratory, Los Angeles.

Terasaki PI, McClelland JD. 1964. Microdroplet assay of human serum cytoxins. *Nature* **204:** 998–1000.

Thomas J, Carver M, Cunningham P, Park K, Gonder J, Thomas F. 1987. Promotion of incompatible allograft acceptance in rhesus monkeys given posttransplant antithymocyte globulin and donor bone marrow. I. In vivo parameters and immunohistologic evidence suggesting microchimerism. *Transplantation* **43:** 332–338.

Tilney NL. 2003. *Transplant: From myth to reality.* Yale University Press, New Haven.

Ting A, Morris PJ. 1978. Matching for B cell antigens of the HLA-DR series in cadaveric renal transplantation. *Lancet* **311:** 575–577.

Ullmann E. 1914. Tissue and organ transplantation. *Ann Surg* **60:** 195–219.

Unger E. 1910. Kidney transplantation [in German]. *Wien Klin Wochenschr* **47:** 573.

van Rood JJ, Eernisses JG, van Leeuwen A. 1958. Leucocyte antibodies in sera from pregnant women. *Nature* **181:** 1735–1736.

Voronoy U. 1937. Blocking the reticuloendothelial system in man in some forms of mercuric chloride intoxication and the transplantation of the cadaver kidney as a method of treatment for the anuria resulting from the intoxication [in Spanish]. *Siglo Med* **97:** 296.

Waksman BY, Arbouys S, Arnason BG. 1961. The use of specific "lymphocyte" antisera to inhibit hypersensitive reactions of the "delayed" type. *J Exp Med* **114:** 997–1022.

Witkowski JA. 1980. Carrel's immortal cells. *Med Hist* **24:** 129–142.

Wood K, Sachs DH. 1996. Chimerism and transplantation tolerance: Cause and effect. *Immunol Today* **17:** 541–590.

Woodruff MFA, Anderson NF. 1963. Effect of lymphocyte depletion by thoracic duct fistula and administration of anti-lymphocytic serum on the survival of skin homografts in rats. *Nature* **200:** 702.

Zimmerman KW. 1998. *One leg in the grave: The miracle of transplantation of the black leg by Saint Cosmos and Damian.* Elsevier, Maarssen, The Netherlands.

Zukowski CF, Lee HM, Hume DM. 1960. The prolongation of functional survival of canine renal homograftts by 6-mercaptopurine. *Surg Forum* **11:** 470–472.

Why Is Organ Transplantation Clinically Important?

Josep M. Grinyó

Department of Nephrology, Hospital Universitari de Bellvitge, IDIBELL, University of Barcelona, Barcelona 08907, Spain

Correspondence: jgrinyo@bellvitgehospital.cat

Solid organ transplantations save lives in patients affected by terminal organ failures and improve quality of life. Organ transplantations have gradually ameliorated in the last two decades, usually provide excellent results in children and young adults, and are increasingly challenged by the growing proportion of elderly transplant patients with comorbidities. Renal transplantation increases patient survival over dialysis, and lifesaving transplants are indispensible to treat patients with liver, heart, or lung irreversible diseases. Solid organ transplant program activity has been steadily growing but is still far from global needs, with great differences among countries. Solid organ transplantations are essential for developed and mature health care systems.

The introduction of antibiotics, massive vaccinations of the general population to prevent infectious diseases, and organ transplantation may be considered the miracles of twentieth century medicine. Organ transplantation is the best therapy for terminal and irreversible organ failure. Kidney transplantation introduced in the 1950s was the pioneer solid organ transplant to treat patients with end-stage renal disease (ESRD) in an era when renal replacement therapies were in their first steps. The gradual introduction of dialysis increased the recipients' pool of renal transplant candidates, who may abandon dialysis after a successful transplantation. In the so-called era of conventional immunosuppression based on the use of azathioprine and steroids, the rejection rates and early graft failure were high, and nonrenal solid organ transplant programs without supportive replacement therapies had poor results with very few of them having regular activity until the early 1980s. The introduction of cyclosporine 30 years ago improved transplant outcomes, and solid organ transplant programs were generalized all over the developed world.

According to activity data reported to the Global Observatory on Donation and Transplantation (GODT) (World Health Organization 2012a), analysis from 2010 transplant activity for 95 countries, representing nearly 90% of the worldwide population, shows that ~106,879 solid organ transplants were performed worldwide: 73,179 kidney transplants (46% from living donors), 21,602 liver transplants (15% from living donors), 5582 heart transplants, 3927 lung transplants, 2362 pancreas transplants, and 227 small bowel. This activity increased 2.12% during 2009, but it is estimated

Cite this article as *Cold Spring Harb Perspect Med* doi: 10.1101/cshperspect.a014985

that it is far from global needs. On the other hand, there are huge geographical differences on transplant activity ranging from >70 per million population (p.m.p.) in developed countries to 0–2.4 p.m.p. in developing countries. The implementation of transplant programs requires social altruism for organ donation, optimization of transplant clinical practices, and an economic effort by the distinct health care systems. Such commitments around organ transplantation aim to improve the life expectancy, clinical condition, and quality of life of solid organ transplant recipients.

The clinical importance of solid organ transplantations may be evaluated according to their impact on patient survival, the reduction of comorbidity, improvement of labor life, and global quality of life of transplant population.

ESRD patients may survive with supportive dialysis, and diabetic patients with insulin therapy. In contrast, in the absence of long-lasting artificial organ support or inefficient pharmacological treatment, solid organ transplantation becomes a lifesaving therapy in patients with irreversible liver, heart, and respiratory failures.

As the distinct solid organ transplants will be addressed in specific articles in this collection, this introductory article will be mainly focused on transplant outcomes and benefits derived from organ transplantations, which obviously give sense to these therapies.

KIDNEY TRANSPLANTATION

Despite that renal transplantation is considered the best therapeutic option for the treatment of ESRD, access to transplantation varies among centers (Ravanan et al. 2010) and countries (Stel et al. 2005). The mean percentage of ESRD-prevalent patients included in the waiting list is ~20%–25%, but some centers and countries may have lower percentages of ESRD patients listed. Comorbidity and demographic factors may account for differences in the access to transplantation. Activation on the waiting list in the first 2 years of the start of renal replacement therapy (RRT) significantly decreases with older age, nonwhite ethnicity, and diabetes mellitus, and the probability of receiving a kidney transplant follows a similar trend. According to the United Kingdom Transplant Registry, the probability of receiving a kidney allograft is reduced by >40% in older recipients in comparison with young patients, and by >30% in nonwhites in comparison with Caucasians. Differences among centers on the time taken to activate inclusion on the waiting list and the proportion of listed patients with ESRD may widely vary, and a similar observation also applies regarding the access to transplantation (Ravanan et al. 2010). These center disparities may not be strictly explained by patient-related factors and might reflect distinct center criteria on a candidate's evaluation for renal transplantation. A patient starting dialysis in a nontransplanting renal center was less likely to be registered for transplantation (odds ratio 0.85, 0.77 to 0.94) or receive a transplant from a donor after cardiac death or a living kidney donor (0.69, 0.59 to 0.80) compared with patients cared for in transplanting renal centers but with similar chances of receiving a transplant from a donor after brain stem death. In another study conducted in England and Wales, specific variables significantly associated with listing were age, primary renal disease, graft number, social deprivation, and ethnicity but not gender. In contrast with the previous publication, other investigators have reported that whether the renal unit was also a transplant unit was not significant (Dudley et al. 2009). A multinational European study based on the ERA-EDTA Registry (2010) analyzing the access to transplantation by country and comorbid status showed that the relative risk of receiving a first transplant within 4 years after the start of renal replacement in four European countries varies from 0.26 to 4.10 in patients without comorbidity and from 0.12 to 3.19 in patients with comorbidity. These international differences in access to transplantation were attributed to differences in the availability of donor kidneys in relation to the number of dialysis patients, and were only partly owing to comorbid variability (Stel et al. 2005). This and other previous studies (Jager et al. 2001; Miskulin et al. 2002) show that sicker patients remain on dialysis programs and the healthier patients are selected for renal transplantation to have acceptable graft and patient survivals with this therapeutic modality, taking into account the limited number of organs available nowadays for renal transplantation. Clinical guidelines for the evaluation, selection, and preparation of the potential kidney transplant recipient may help to standardize access to renal transplantation (Dudley and Harden 2011). This approach is especially relevant with the current high incidence and prevalence of patients with ESRD. According to the recently published annual 2009 renal replacement report of the ERA-EDTA Registry (van de Luijtgaarden et al. 2012), in Europe the overall incidence rate of ESRD patients was 125 p.m.p., and the overall prevalence was 730 p.m.p., with wide variations among countries. From 2005 to 2009, the relative change of the hemodialysis (HD), peritoneal dialysis (PD), and transplantation distribution (at day 91 after the start of RRT) was an overall 0.5% decrease in HD, 1.4% decrease in PD utilization, and, interestingly, a 1.8% increase of patients living on a functioning graft. The assessment of waiting list dynamics may help to provide a better understanding of the limitations and contributions of renal transplantation to the treatment of ESRD. The Renal Patients Registry of Catalonia (2010) shows that for new cases in 1990–2010 that were not excluded from the waiting list at the start of RRT, at 10 years, 54% had a functioning transplant, 5.2% were on the waiting list, 2.3% were pending study, 8.1% were excluded, and the remaining 30.2% had died. According to the same registry, despite the older age and the increase of incident and prevalent ESRD patients in the last two decades, the waiting list size has not increased, the proportion of patients excluded because of clinical reasons remains quite constant, the exclusion for age has decreased by 50%, and the proportion of patients with a functioning transplant has increased from 30% in 1990 to 54% in 2010. These data suggest new clinical attitudes to offer renal transplantation to a larger number of ESRD patients despite their older age and comorbidity. In some European countries with very active transplant programs such as Norway, 70% of the patients on RRT were living with a functioning graft (591 p.m.p.). Also, the number of kidney transplants performed p.m.p. in 2009 widely varied among countries, the highest being in some regions of Spain (Cantabria, Catalonia) with ≥ 70 transplants p.m.p. These wide variations on the incidence and prevalence of ESRD and transplant activities across Europe strongly suggest that there is room for improving an integrated approach in the diagnosis and treatment of ESRD patients, including active policies to promote organ donation and transplantation.

The selection of transplant candidates with better clinical conditions may contribute to the better patient survival in comparison to the distinct dialysis modalities. In the cohort of patients between 2000 and 2004, 5-yr patient survival on dialysis adjusted for age, gender, and primary renal disease was 48%. Adjusted patient survivals for first deceased donor transplantation were 91%, and for living donor 94%. Adjusted 5-year graft survival during this same time period were 79% and 85% for deceased and living donor kidney transplants, respectively (Fig. 1) (van de Luijtgaarden et al. 2012). Children have excellent transplant outcomes. In 2011, according to the Scientific Registry for Transplant Recipients, for deceased and living donor transplants, the estimated 1-year conditional half-lives were 11.9 years and 15.3 years in pediatric populations.

According to the 2012 USRD Annual Data Report, ~117,000 new patients began treatment for ESRD in the United States in 2010 (incidence rate 348 p.m.p.), 10 times more patients than 30 years ago. The majority of ESRD patients started on HD (91%), 7% on PD, and 2% received a kidney allograft. The number of prevalent patients was almost 600,000. Of the ESRD population, 30% had a kidney transplant, 65% of them are on HD, and 7% are treated with PD. Importantly, the mortality of patients on dialysis has decreased 26% since 1985. However, mortality for dialysis patients is still far higher than in the general population and transplant patients (USRDS Annual Data Report 2012). In the 2005 patients' cohort, 5-year adjusted survival probabilities were 35% for HD, 41% for PD, and 73% for kidney transplant recipients. The adjusted 10-year graft survival

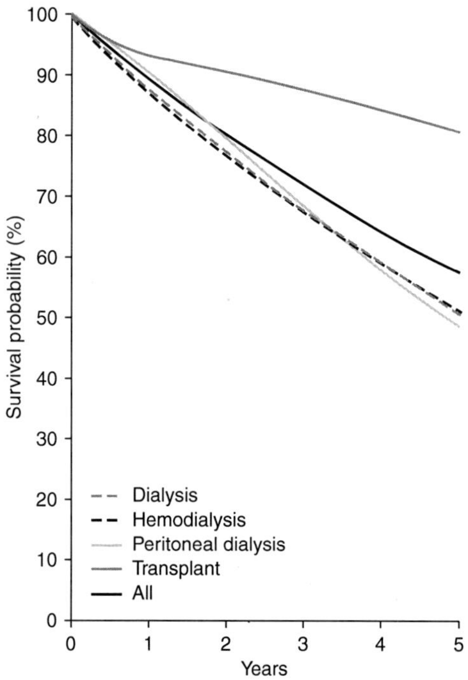

Figure 1. Survival of incident dialysis patients and first transplant recipients between 2000 and 2004, by treatment modality, adjusted for age, gender, and primary renal disease. (From van de Luijtgardeen et al. 2012; reprinted, with permission.)

for deceased donor transplants has increased from 18.8% in 1980 to 42.7% in 2000, and from 46.2% in 1980 to 58.6% in 2000 for living donor transplants. This gradual amelioration of transplant results is especially relevant taking into account the permanent increase of age of incident and prevalent patients with RRT, which has augmented by ~10 years from the mid-1980s to 2010 up to 64.8 years and 60.8 years for incident and prevalent patients, respectively (Renal Patients Registry of Catalonia 2010).

Another benefit of renal transplantation may be the reduction of hospital readmissions. Rates of hospitalization for ESRD patients are double those of the general population. Renal transplantation has importantly reduced hospitalization indexes in comparison with HD, for which the rate is 36% for all-cause rehospitalization. The total admission rates for dialysis patients have barely changed in the last decade as per 1000 patient years; period-prevalent patients by age, gender, ethnicity, and primary diagnosis were 1889 in 1993 and 1856 in 2010. For transplant patients, the rates were 1020 in 1993 and 841 in 2010.

Preemptive Renal Transplantation

Among the distinct alternatives for the treatment of ESRD, renal transplantation may also be the first therapeutic option of RRT to avoid dialysis. Prolonged dialysis time is considered a detrimental factor for transplant survival (Meier-Kriesche et al. 2000; Meier-Kriesche and Schold 2005; Kasiske et al. 2010) and increases the risk of hospital admission after transplantation (Gram et al. 2012). Preemptive renal transplantation (PRT) may avoid dialysis-related comorbidity, improve quality of life, and provide better transplant outcomes. Data reported a decade ago showed that longer waiting times on dialysis negatively impact on posttransplant graft and patient survival. In a paired kidney study on the impact of pretransplant dialysis on graft survival, dialysis time before transplantation longer than 2 years was associated with 29% 10-year graft survival rates in comparison with 63% in patients with 0–6 months on dialysis (Meier-Kriesche and Kaplan 2002). As PRT is associated with better patient and graft survival than transplantation after dialysis, this therapeutic option might be favored in all ESRD patients at the time of initiation of RRT and has been more common in the last 15 years. It has also been performed at higher levels of estimated glomerular filtration rate (eGFR), particularly among recipients of live-donor transplants (Friedewald and Reese 2012). In the United States, preemptive listing for kidney transplantation has increased in the last decade (Davis 2010) and occurs in 16.6%–21.3% of listed patients. Patients with less comorbid clinical entities, such as patients with polycystic kidney disease (41.9%) and glomerulonephritis (24.6%), were more often listed before dialysis than those with hypertension (12.2%) and diabetes (14.4%) (Keith et al. 2008).

Because of the better outcomes of pancreas kidney transplantation (PKT), it has been claimed as the optimal therapy for patients with ESRD, especially if a living donor is available. PKT might be favored with an improved patient access to nephrology and transplant care, and better education of the community and health-care personnel (Davis 2010). Nevertheless, recent studies have raised new concerns about the consequences of early versus late PKT (Friedewald and Reese 2012) and how to find an adequate "balance of benefit (maximization of utility) and justice (fairness in organ allocation)" (Petrini 2009).

Owing to the benefits of PKT, we might see a wider adoption of this therapeutic alternative in the coming years (Huang and Samaniego 2012).

Renal Transplantation in Older Recipients

The age of patients with ESRD initiating RRT has gradually increased in the last decade. According to the ERA-EDTA Registry, in 2009, the mean age of incident patients was 64.1 years. This implies that the proportion of patients older than 60 years on RRT in some areas may be higher than 60% and patients older than 70 years account for more than 50% (Renal Patients Registry of Catalonia 2010). This results in an increasing proportion of older patients on waiting lists and, in the last years of the past decade, >30% of kidney transplant recipients were 60 years or older. This translates to an increment of prevalent older patients with a functioning allograft that was very low in the mid-1980s, whereas in 2010 it accounted for 34% of patients with a kidney transplant. The growing proportion of elderly patients has also been observed in other solid organ transplant programs, especially in lung and heart transplants (Abecassis et al. 2012). Although it is well known that graft and patient survival are poorer in older suitable patients than in younger recipients, renal transplantation from deceased or living donor decreases mortality in older patients by >30% over those patients remaining on dialysis (Table 1) (ERA-EDTA Registry 2010). Older transplant recipients frequently suffer from comorbid conditions such as cardiovascular disease, infections, malignancies, physical limitations, cognition alterations, and overall reduced quality of life. Older age is associated with an increased frequency of early readmissions after transplantation (McAdams-Demarco et al 2012). The risk of death is also exacerbated in older recipients. The mortality rate in patients older than 60 years in the first year after transplantation is more than double that in younger recipients (10.5% vs. 4.4%), which is aggravated with comorbidities and whether the patient received an ECD kidney. Nevertheless, older recipients have shorter life expectancies than younger patients and might benefit from receiving kidneys from ECD donors to increase access to transplantation. The potential utility of allocating kidneys from donors ≥65 years to recipients of the same age (old to old), regardless of human leukocyte antigen (HLA) matching and with short cold ischemia time, was addressed in the Eurotransplant organization 10 years ago in the so-called Eurotransplant Senior Program (ESP) (Frei et al. 2008). In this program, graft and patient survival were not negatively affected by the ESP allocation when compared with standard allocation, suggesting that donor/recipient age matching could increase the number of elderly recipients transplanted, which might result in a longer life expectancy compared with dialysis.

Table 1. Five-year survival probabilities (cohort 2001–2005) in incident dialysis patients and transplant recipients (survival rates in %)

Age (years)	Dialysis	1st DCD	1st LDT
0–19		95.3	95.9
20–44	77.2	93.9	96.9
45–64	54.5	84.2	91.2
65–74	35.8	69.1	79.5
75+	19.9		

1st DCD, first deceased donor transplant; 1st LDT, first living donor transplant (ERA-EDTA Registry 2010).

SOLID ORGAN TRANSPLANTATION IN DIABETIC PATIENTS

The World Health Organization (WHO) estimates that 347 million people worldwide have diabetes, >80% of people with diabetes live in low- or middle-income countries, and projects that diabetes deaths will double between 2005 and 2030 (World Health Organization 2012b).

The most severe and well-known complications of diabetes such as retinopathy, cardiovascular disease, neuropathy, and nephropathy greatly deteriorate the diabetic patient clinical condition and reduce life expectancy. According to the WHO, a total of 57 million deaths occurred in the world during 2008; 36 million (63%) were attributable to noncommunicable diseases, principally cardiovascular diseases (17 million), cancer (7.6 million), chronic respiratory diseases (4.2 million), and diabetes (1.3 million). Diabetes entails a high risk of ESRD and premature mortality, even higher than in cancer patients. The mortality risk of patients with diabetes and nephropathy is higher than the average mortality rate of all types of cancer (Lambers Heerspink and de Zeeuw 2011). Diabetic nephropathy is now not only the leading cause of ESRD in much of Australasia, Europe, and North America, but also in less wealthy countries, including India, several Latin American countries, Malaysia, and Turkey (White et al. 2012). Diabetic nephropathy accounts for ~35% of ESRD in Australasia, ~38% in the United States, and ~28% in Europe.

RRT therapy in diabetic patients with ESRD consists of supportive dialysis, and, in suitable candidates, renal transplantation, and also simultaneous kidney/pancreas transplantation or pancreas alone, reverse diabetes and attenuate or prevent diabetic complications. In Europe, in the cohort of incident patients in 2001–2005, diabetics on dialysis programs had the worst 5-year survival, either on PD or on HD, with an average rate of 39%, below the 50.9% observed in all patients (ERA-EDTA Registry 2010). In this registry, in eligible patients, renal transplantation increased 5-year survivals up to 85.1% in deceased donor transplants and up to 85.6% in living donor transplants, although below the rates reported for all patients (91% for deceased donor and 94.2% for living donor transplants).

Besides kidney transplantation alone in ESRD diabetic patients, the beneficial effects of pancreas transplantation have been a matter of discussion. Pancreas transplant programs were initiated more than 20 years ago in several countries and have progressively ameliorated their results. According to a recent summary of the International Pancreas Registry (Gruessner 2011), "at the end of 2010 more than 35,000 pancreas transplantations have been reported; 24,000 transplantations in the United States and more than 12,000 outside the United States. The types of pancreas transplantations performed have been simultaneously with a kidney (SPK) (75%), those given after a previous kidney transplantation (PAK) (18%), and pancreas transplantation alone (PTA) (7%)." In these last three decades, optimization of surgical techniques and the introduction of low-toxicity immunosuppression have gradually improved the outcomes of pancreas transplantation, regarding patient survival and graft function. "Patient survival now reaches over 95% at 1 year post-transplant and over 83% after 5 years. The best graft survival was found in SPK with 86% pancreas and 93% kidney graft function at 1 year. PAK pancreas graft function was 80%, and PTA pancreas graft function reached 78% at 1 year."

The beneficial effects of euglycemia on diabetic complications may not be clearly evident in the short term after transplantation but may be noticed in the long term. In nonuremic patients with long-term type 1 diabetes, with mild to advanced diabetic nephropathy lesions at baseline, no significant changes were observed after 5 years of pancreas transplantation. But in most patients, glomerular, tubular, and interstitial lesions markedly improved at 10 years, indicating that a prolonged time of normoglycemia may heal diabetic nephropathy (Fioretto and Mauer 2011).

Diabetic retinopathy is a leading cause of blindness. Despite initial controversies on the beneficial effects of pancreas transplantation on this severe complication, PTA seems also to improve or stabilize diabetic retinopathy in >50% of patients 2 years after transplantation (Giannarelli et al. 2006), suggesting that a functioning pancreas graft may prevent visual loss in diabetic patients. Successful pancreas transplantation may also improve sensory or motor neuropathy according to electrophysiology studies (Allen et al. 1997, Martinenghi et al. 1997; Navarro et al. 1997), findings recently supported by experimental data (Spadella et al. 2012).

 Cite this article as *Cold Spring Harb Perspect Med* doi: 10.1101/cshperspect.a014985

Islet transplantation may be a therapeutic alternative to pancreas transplantation to restore normal glucose metabolism in type 1 diabetes, providing a 59% insulin independence rate in comparison with 75% with PTA, which might suggest that islets transplantation should be offered as a first alternative only in patients with high surgical risk, although the debate is still open on the role of these two types of transplantation in the treatment of diabetes (Maffi et al. 2011).

LIFESAVING SOLID ORGAN TRANSPLANTS

Liver Transplantation

Liver transplantation (LT) is the second most frequently performed transplant after kidney transplantation. According to the 2010 report of the European Liver Transplant Registry, the main indications for LT are cirrhosis (58%), cancers (14%), cholestatic diesases (10%), acute hepatic failure (8%), metabolic diseases (6%), and other diseases (4%). Data from the same registry show that 1-year and 5-year patient survivals according to the number of liver transplantations were 87% and 62% in patients with first LT, 75% and 47% for second LT, and 75% and 44% for third LT, respectively. 5-Year graft survivals according to the number of transplants were 55%, 29%, and 36%, for first, second, and third LT, respectively (European Liver Transplant Registry 2012). In the increasing proportion of elderly recipients (>60 years), patient survivals are slightly lower, but acceptable, than in all patients with same indications for transplantation. Liver transplant activity has increased in Europe since the 1980s, peaking in 2007 with >6000 LT, probably owing to wider acceptance donor criteria, including older donors. Advanced donor age significantly reduces 10-year-graft survivals from 55% for donors <55 years, 47% for donors 55–65 years, to 45% for donors ≥65 years, although these rates are considered clinically acceptable. Pediatric LT recipients have excellent outcomes with significantly higher 10-year survival rates (76%) over adults (59%). Among the adult LT recipients population, increasing age has a gradual negative impact on survivals from 65% 10-year patient survival in recipients aged 15–45 years, 59% in 45–60 years, and 49% in patients ≥60 years. Despite the negative impact of donor and recipient age on LT outcomes, survival rates reported are considered clinically relevant in patients with life-threatening irreversible liver failure. Other transplant registries show similar data on LT (Scientific Registry of Transplant Recipients and OPTN Annual Data Report 2011).

Heart and Lung Transplantation

Heart transplantation is indicated in patients suffering from refractory cardiac failure owing to cardiomyopathy (53.5%), coronary artery disease (30.8%), congenital heart diseases (9%), retransplant because of previous graft failure (2.6%), valvular heart disease (1.4%), and for other entities (2.7%) (International Society for Heart and Lung Transplantation Registry 2011). Heart transplant activity and outcomes have greatly improved. This accounts for the growing number of patients living with a functioning heart transplant, which in the United States is more than 20,000 people. Short-term graft failure (6 months and 1 year, adjusted for age, sex, and race) has been declining over the past decade, and in general is low, 0.07 at 6 months and 0.09 at 1 year for patients who underwent transplant in 2010, and graft failure at 3, 5, and 10 years posttransplant has steadily declined. Half-life increased from 11.5 years in 1991 to 14 years in 2009, and overall 5-year graft survival was 74.9%, and similar among all disease groups (Scientific Registry of Transplant Recipients and OPTN Annual Data Report 2011). The highest survival rates are observed among young and pediatric recipients. In the United States, graft survival for heart transplants performed in children in 2005 was 87.5% at 6 months, 84.6% at 1 year, and 72.1% at 5 years. In these transplant patients, for heart transplants performed in 2009–2010, the 1-year conditional graft half-life was 17.4 years in comparison with 9.8 years for those performed in 1991–1992. As in other solid organ transplants, older donor and recipient age negatively influence patient survival.

The main indication for lung transplantation is respiratory insufficiency secondary to idiopathic pulmonary fibrosis (30%), emphysema/chronic obstructive pulmonary disease (26%), and cystic fibrosis (14%) (International Society for Heart and Lung Transplantation Registry 2011). Lung transplant activity has gradually increased. The number of adults living with a transplanted lung was 9000 in the United States last year, more than twofold than in 2000, and the number of pediatric recipients has remained stable for the last decade. In lung transplantation, a new organ allocation scheme was introduced 7 years ago, the so-called Lung Allocation Score (LAS), which takes into account the risk of death without transplant and posttransplant, regardless of the waiting time, disincentivizing early listing. This implies that "patients listed are those with more immediate need of transplant" (Scientific Registry of Transplant Recipients and OPTN Annual Data Report 2011). After an initial decline following the implementation of LAS, the waiting list has steadily increased. The LAS has increased as well, from 36.6 in 2006 to 40.8 in 2011, reflecting the increased illness severity of candidates, which may account for the increasing mortality rates among wait-listed candidates. LAS correlates with patient survival, being around 50% at 5 years in recipients with LAS <50 and reduced by $\sim 10\%$ in higher scores. Despite this more complex scenario in lung transplantation and the growing number of older recipients, early and late graft failure have diminished. Early graft failure dropped to 5.3% in 2011, and long-term outcomes seem to have also improved, as indicated by the increment in conditional survival of at least 1-year adult half-life from 5.2 years in 1991 to 7.7 years in 2009. This positive evolution of lung transplantation has occurred despite the well-identified comorbidity accompanying lung transplant patients. Concerning pediatric lung transplantation, the number of listed and transplanted patients has declined the last few years, and the highest 5-year patient survival, around 60%, is observed in children >1 year old, which is reduced by 20% in patients <1 year.

Heart–lung transplantation has remained almost stable in the last three decades and the major indications are congenital heart disease (35.7%), pulmonary arterial hypertension (27.6%), and cystic fibrosis (14%) (Christie et al. 2012). According to the corresponding registries, during the last part of 2011, 106 heart–lung transplants were performed in the United States and 50 in Europe. The results of heart–lung transplantation have progressively ameliorated in the last 30 years, mainly owing to the reduction of early mortality. Recipients who survive the first year have generally good outcomes, with a half-life of 10 years conditional on surviving at least 1 year.

QUALITY OF LIFE AND COMORBIDITY IN SOLID ORGAN TRANSPLANTATION

All solid organ transplant recipients suffer from transplant-inherent-related comorbidities, such as hypertension, new-onset diabetes mellitus, cardiovascular events, infections, and cancer, and in nonrenal transplants a high percentage of patients develop chronic kidney disease (CKD). Despite these limitations, solid organ transplants increase life expectancy and generally improve quality of life.

In a systematic review of clinically relevant outcomes in adult kidney transplantation compared with dialysis, in addition to the reduction of mortality, renal transplantation compared with dialysis reduced cardiac events, heart failure, ischemic heart disease, hospitalization, and hospitalization for infection (Tonelli et al. 2011). A quality of life analysis using the SF-36 also significantly favored transplantation. In another study (Wyld et al. 2012) on patients with advanced CKD on utility-based (0 represents death and 1 full health) quality of life, transplant patients had a mean utility of 0.82, which was 0.11 lower in dialysis patients and 0.20 lower in patients under conservative therapy. Interestingly, mean utility in transplant patients increased from 0.66 in the 1980s to 0.85 in the 2000s. The growing number of elderly transplant candidates with poorer clinical conditions than young patients, challenges solid organ transplant programs, and data on quality of life in older patients are scarce (Kniepeiss et al. 2012).

Despite that the quality of life in pediatric renal transplant patients is slightly lower than in the general population, and school attendance and performance is also lower, employment rates in adulthood are similar to the general population (Dehennault 2011). However, in LT, young adults still have physical and psychological sequelae that affect health status two decades after transplantation (Mohammad et al. 2012). Employment rate may also reflect quality of life in transplant recipients. Unfortunately, labor insertion in solid organ transplanted patients is usually lower than in the general population, and some studies have reported employment rates <50% (Kristen et al. 2010; Huda et al. 2012). Social coordinated efforts should be made to reintroduce transplant patients to labor life.

CONCLUSIONS

Solid organ transplantations save lives in patients affected by terminal organ failures and improve quality of life. Solid organ transplant programs provide excellent results in children and young adults and are increasingly challenged by the expanding proportion of elderly transplant patients. Solid organ transplant program activity has been growing in the last two decades, and is essential for developed and mature health care systems.

REFERENCES

Abecassis M, Bridges ND, Clancy CJ, Dew MA, Eldadah B, Englesbe MJ, Flessner MF, Frank JC, Friedewald J, Gill J, et al. 2012. Solid-organ transplantation in older adults: Current status and future research. *Am J Transplant* **12:** 2608–2622.

Allen RD, Al-Harbi IS, Morris JG. 1997. Diabetic neuropathy after pancreas transplantation: Determinants of recovery. *Transplantation* **63:** 830–838.

Christie JD, Edwards LB, Kucheryavaya AY, Benden C, Dipchand AI, Dobbels F, Kirk R, Rahmel AO, Stehlik J, Hertz MI, International Society of Heart, et al. 2012. The Registry of the International Society for Heart and Lung Transplantation: 29th adult lung and heart–lung transplant report—2012. *J Heart Lung Transplant* **31:** 1073–1086.

Davis CL. 2010. Preemptive transplantation and the transplant first initiative. *Curr Opin Nephrol Hypertens* **19:** 592–597.

Dehennault M. 2011. Paediatric features in childhood renal transplantation: Quality of life, school, adherence, transfer to adult nephrologists. *Nephrol Ther* **7:** 599–603.

Dudley CR, Harden P. 2011. Assessment of the potential kidney transplant recipient, http://www.renal.org/clinical/guidelinessection.

Dudley CR, Johnson RJ, Thomas HL, Ravanan R, Ansell D. 2009. Factors that influence access to the national renal transplant waiting list. *Transplantation* **88:** 96–102.

ERA-EDTA Registry. 2010. Annual Report, http://www.era-edta-reg.org.

European Liver Transplant Registry. 2012. http://www.eltr.org.

Fioretto P, Mauer M. 2011. Effects of pancreas transplantation on the prevention and reversal of diabetic nephropathy. *Contrib Nephrol* **170:** 237–246.

Frei U, Noeldeke J, Machold-Fabrizii V, Arbogast H, Margreiter R, Fricke L, Voiculescu A, Kliem V, Ebel H, Albert U, et al. 2008. Prospective age-matching in elderly kidney transplant recipients—A 5-year analysis of the Eurotransplant Senior Program. *Am J Transplant* **8:** 50–57.

Friedewald JJ, Reese PP. 2012. The kidney-first initiative: What is the current status of preemptive transplantation? *Adv Chronic Kidney Dis* **19:** 252–256.

Giannarelli R, Coppelli A, Sartini MS, Del Chiaro M, Vistoli F, Rizzo G, Barsotti M, Del Prato S, Mosca F, Boggi U, et al. 2006. Pancreas transplant alone has beneficial effects on retinopathy in type 1 diabetic patients. *Diabetologia* **49:** 2977–2982.

Grams ME, McAdams Demarco MA, Kucirka LM, Segev DL. 2012. Recipient age and time spent hospitalized in the year before and after kidney transplantation. *Transplantation* **15:** 750–756.

Gruessner AC. 2011. 2011 Update on pancreas transplantation: Comprehensive trend analysis of 25,000 cases followed up over the course of twenty-four years at the international pancreas transplant registry (IPTR). *Rev Diabet Stud* **8:** 6–16.

Huang Y, Samaniego M. 2012. Preemptive kidney transplantation: Has it come of age? *Nephrol Ther* **8:** 428–432.

Huda A, Newcomer R, Harrington C, Blegen MG, Keeffe EB. 2012. High rate of unemployment after liver transplantation: Analysis of the United Network for Organ Sharing database. *Liver Transpl* **18:** 89–99.

International Society for Heart and Lung Transplantation Registry. 2011. http://ishlt.org.

Jager KJ, Merkus MP, Boeschoten EW, Dekker FW, Tijssen JGP, Krediet RT, Necosad Study Group. 2001. What happens to patients starting dialysis in The Netherlands? *Netherlands J Med* **58:** 163–173.

Kasiske BL, Israni AK, Snyder JJ, Skeans MA, Peng Y, Weinhandl ED. 2010. A simple tool to predict outcomes after kidney transplant. *Am J Kidney Dis* **56:** 947–960.

Keith D, Ashby VB, Port FK, Leichtman AB. 2008. Insurance type and minority status associated with large disparities in prelisting dialysis among candidates for kidney transplantation. *Clin J Am Soc Nephrol* **3:** 463–470.

Kniepeiss D, Wagner D, Pienaar S, Thaler HW, Porubsky C, Tscheliessnigg KH, Roller RE. 2012. Solid organ transplantation: Technical progress meets human dignity: A review of the literature considering elderly patients' health related quality of life following transplantation. *Ageing Res Rev* **11:** 181–187.

Kristen AV, Katus HA, Dengler TJ. 2010. Return to work after heart transplantation. *Versicherungsmedizin* **62:** 67–72.

Lambers Heerspink HJ, de Zeeuw D. 2011. The kidney in type 2 diabetes therapy. *Rev Diabet Stud Fall* **8:** 392–402.

Maffi P, Scavini M, Socci C, Piemonti L, Caldara R, Gremizzi C, Melzi R, Nano R, Orsenigo E, Venturini M, et al. 2011. Risks and benefits of transplantation in the cure of type 1 diabetes: Whole pancreas versus islet transplantation. A single center study. *Rev Diabet Stud Spring* **8:** 44–50.

Martinenghi S, Comi G, Galardi G. 1997. Amelioration of nerve conduction velocity following simultaneous kidney/pancreas transplantation is due to the glycemic control provided by the pancreas. *Diabetologia* **40:** 1110–1112.

McAdams-Demarco MA, Grams ME, Hall EC, Coresh J, Segev DL. 2012. Early hospital readmission after kidney transplantation: Patient and center-level associations. *Am J Transplant* **12:** 3283–3288.

Meier-Kriesche HU, Kaplan B. 2002. Waiting time on dialysis as the strongest modifiable risk factor for renal transplant outcomes: A paired donor kidney analysis. *Transplantation* **74:** 1377–1381.

Meier-Kriesche HU, Schold JD. 2005. The impact of pretransplant dialysis on outcomes in renal transplantation. *Semin Dial* **18:** 499–504.

Meier-Kriesche HU, Port FK, Ojo AO, Rudich SM, Hanson JA, Cibrik DM, Leichtman AB, Kaplan B. 2000. Effect of waiting time on renal transplant outcome. *Kidney Int* **58:** 1311–1317.

Miskulin DC, Meyer KB, Athienites NV, Martin AA, Terrin N, Marsh JV, Fink NE, Coresh J, Powe NR, Klag MJ, et al. 2002. Comorbidity and other factors associated with modality selection in incident dialysis patients: The CHOICE Study. *Am J Kidney Dis* **39:** 324–336.

Mohammad S, Hormaza L, Neighbors K, Boone P, Tierney M, Azzam RK, Butt Z, Alonso EM. 2012. Health status in young adults two decades after pediatric liver transplantation. *Am J Transplant* **12:** 1486–1495.

Navarro X, Sutherland DE, Kennedy WR. 1997. Long-term effects of pancreatic transplantation on diabetic neuropathy. *Ann Neurol* **42:** 727–736.

Petrini C. 2009. Preemptive kidney transplantation: Ethical issues. *Ann Ist Super Sanita* **45:** 173–177.

Ravanan R, Udayaraj U, Ansell D, Collett D, Johnson R, O'Neill J, Tomson CR, Dudley CR. 2010. Variation between centres in access to renal transplantation in UK: Longitudinal cohort study. *BMJ* **341:** c3451.

Renal Patients Registry of Catalonia. 2010. Statistical Report, http://www20.gencat.cat/docs/canalsalut.

Scientific Registry of Transplant Recipients and OPTN Annual Data Report. 2011. http://srtr.transplant.hrsa.gov.

Spadella CT, Lucchesi AN, Alberti S, Resende LA. 2012. Early pancreas transplant improves motor nerve conduction in alloxan-induced diabetic rats. *Exp Clin Endocrinol Diabetes* **120:** 567–572.

Stel VS, van Dijk PC, van Manen JG, Dekker FW, Ansell D, Conte F, Kramar R, Leivestad T, Vela E, Briggs JD, et al. 2005. Prevalence of co-morbidity in different European RRT populations and its effect on access to renal transplantation. *Nephrol Dial Transplant* **20:** 2803–2811.

Tonelli M, Wiebe N, Knoll G, Bello A, Browne S, Jadhav D, Klarenbach S, Gill J. 2011. Systematic review: Kidney transplantation compared with dialysis in clinically relevant outcomes. *Am J Transplant* **11:** 2093–2109.

USRDS. 2012. Annual Data Report. http://www.usrds.org.

van de Luijtgaarden M, Nordijiz M, Wanner C, Jager K, on behalf of the European Renal Registry investigators. 2012. Renal replacement therapy in Europe—A summary of the 2009 ERA–EDTA Registry Annual Report. *Clin Kidney J* **5:** 209–219.

White SL, Chadban SJ, Jan S , Chapman JR, Cass A. 2012. How can we achieve global equity in provision of renal replacement therapy? Bulletin of the World Health Organization, http://www.who.int/bulletin.

World Health Organization. 2012a. Global observatory on donation and transplantation, http://www.transplant-observatory.org.

World Health Organization. 2012b. Diabetes programme, http://www.who.int/diabetes.

Wyld M, Morton RL, Hayen A, Howard K, Webster AC. 2012. A systematic review and meta-analysis of utility-based quality of life in chronic kidney disease treatments. *PLoS Med* **9:** e1001307.

Origin and Biology of the Allogeneic Response

Fadi G. Lakkis[1] and Robert I. Lechler[2]

[1]Thomas E. Starzl Transplantation Institute, University of Pittsburgh School of Medicine, Pittsburgh, Pennsylvania 15238

[2]King's College London, Counting House, Guy's Hospital, Great Maze Pond SE1 9RT, United Kingdom

Correspondence: lakkisf@upmc.edu

The recognition by the immune system of nonself determinants on cells, tissues, or organs transplanted between genetically disparate members of the same species can lead to a potent allogeneic response that is responsible for rejection. We review here fundamental concepts that underlie the origins and biology of allorecognition in the mammalian immune system. We examine why and how T cells are alloreactive and discuss emerging evidence of allorecognition by innate immune cells. The nature of T cells (naïve vs. memory) and the alloantigen presentation pathways (direct, indirect, and semidirect) that initiate the allogeneic response are outlined.

The defense systems of all studied metazoans, ranging from sponges to humans, possess the inherent ability to recognize and reject cells or tissues of genetically disparate members of the same species (Burnet 1971; Hildemann 1974; Rosengarten and Nicotra 2011). This phenomenon, referred to as the allogeneic response, constitutes the principal barrier to the spontaneous acceptance of organs transplanted from one human to another. In this article, we address two fundamental problems in transplantation immunology. First is the origin of allorecognition (or *why* we are alloreactive), and second is the mechanism of allorecognition (or *how* the immune system recognizes a transplanted organ). Satisfactory resolution of these problems is key to uncovering the primary steps required for initiating allograft rejection and to advancing therapy in clinical transplantation.

ORIGINS OF ALLORECOGNITION: *WHY* ARE WE ALLOREACTIVE?

A defining feature of the mammalian allogeneic response is the central role that T cells play in both the afferent and efferent phases of the response. Studies in neonatally thymectomized and subsequently irradiated adult mice established that T cells are the *sine qua non* of allograft rejection (Miller 1962; Hall et al. 1978a,b). The subsequent and well-justified T-cell focus of transplantation researchers led to most of the advances in our understanding of the allogeneic response and to effective antirejection therapies in the clinic, but at the same time overshadowed potentially important allorecognition phenomena that predate T cells. In this section, we discuss the basis and origins of allorecognition by both T cells and innate immune cells. We define what is being recognized in the allogeneic

Cite this article as *Cold Spring Harb Perspect Med* doi: 10.1101/cshperspect.a014993

response, why T cells are alloreactive, and venture into the less charted territory of allorecognition by innate immune cells.

What Is Allogeneic Nonself?

In the broadest sense, allogeneic nonself is any antigen or group of antigens expressed on donor but not recipient tissues that is responsible for graft rejection. Major histocompatibility complex (MHC) proteins occupy a central position among alloantigens because they are highly polymorphic, ubiquitously expressed, and capable of eliciting unusually large polyclonal T-cell responses (Sherman and Chattopadhyay 1993) (see "Why Are T Cells Alloreactive?"). Careful matching at MHC loci between donors and recipients prolongs allograft survival in immunosuppressed recipients but does not lead to spontaneous graft acceptance (Takemoto et al. 2000). Other polymorphic molecules known as minor histocompatibility antigens (mHAgs) that are unrelated to the MHC also trigger graft rejection (Dierselhuis and Goulmy 2009). mHAgs range from proteins encoded by mitochondrial DNA to those encoded virtually anywhere in the nuclear genome (e.g., the H-Y antigen encoded on the Y chromosome). Although a single nonself-mHAg induces a much more limited T-cell response than that caused by a single nonself-MHC antigen, the presence of numerous mHAg mismatches in otherwise MHC-matched donor–recipient pairs almost invariably leads to rapid graft rejection (Peugh et al. 1986). These fundamental observations underscore the fact that allogeneic nonself is not restricted to foreign MHC molecules but that it encompasses non-MHC molecules expressed in the donor but not shared by the recipient.

Why Are T Cells Alloreactive?

The mammalian immune system has presumably evolved to recognize and eliminate microbial pathogens (Janeway 2001). It is therefore not surprising that the T-cell repertoires of all studied mammals have the capacity to respond to countless microbes, but why is it that these same repertoires also react vigorously to allogeneic grafts? The answer lies in our understanding of the molecular mechanisms by which T cells recognize foreign antigens. T cells detect foreign antigens via T-cell receptors (TCRs) for antigens generated by random somatic gene arrangement during T-cell development. TCRs do not engage whole antigens but instead bind to antigen-derived peptides complexed to self-MHC molecules, a phenomenon known as self-restricted antigen recognition (Yin et al. 2012). The TCR:MHC + peptide interaction is dependent on recognition by the TCRs of amino acid residues present in both the bound peptide and the regions of the MHC molecule surrounding the peptide-binding groove. The immense variety of randomly generated TCRs and the fact that multiple MHC genes and both alleles of each MHC gene are coexpressed in any given individual enable the T-cell repertoire to react to a vast array of self-MHC + nonself–peptide combinations. This arrangement confers the organism with protection against microbial pathogens but at the same time enables it to respond to nonmicrobial foreign peptides complexed to self-MHC. The latter include allopeptides derived from nonself-mHAg or nonself-MHC molecules that elicit graft rejection. Therefore, alloreactivity is an unavoidable consequence of the random process by which TCRs arise.

Another and perhaps more striking feature of the T-cell repertoire is its ability to react to cells bearing intact foreign MHC molecules via recognition of nonself-MHC + peptide complexes (Sherman and Chattopadhyay 1993). This property, referred to as "direct" alloreactivity, has been the subject of intense study because it defies the accepted paradigm that during ontogeny only T cells that recognize self-MHC are selected in the thymus to populate the mature repertoire (so-called positive selection). Although multiple hypotheses were originally proposed to account for the presence of these unusual T cells (Sherman and Chattopadhyay 1993), it is now established that they are part of the usual self-restricted T cell repertoire. Direct alloreactivity, it turns out, is the result of cross-reactivity (or degeneracy) of TCRs that bind self-MHC +

nonself-peptide with nonself-MHC + peptide complexes (Yin and Mariuzza 2009). The majority of peptides in the latter are derived from endogenous self-proteins, implying that foreignness can also be dictated by surrounding amino acid residues in the MHC molecule. Polymorphic residues located within the peptide-binding groove of the MHC molecule also profoundly affect selection of bound peptides, further amplifying differences between histoincompatible individuals. Molecular modeling and more recent X-ray crystallographic analyses have confirmed that the portion of the TCR involved in recognition of nonself-MHC + peptide is the same as that involved in the recognition of self-MHC + nonself-peptide (Macdonald et al. 2009). At the molecular level, therefore, direct T-cell recognition of intact foreign MHC molecules is identical to the recognition of other foreign antigens presented in the context of self-MHC. Moreover, both scenarios arise from the inherent reactivity of T cells to MHC molecules (Yin et al. 2012). Raulet and colleagues showed more than 15 years ago that the high propensity of the mature T-cell repertoire to recognize cells expressing MHC molecules is already manifest in immature T cells that have undergone TCR gene rearrangement but have not yet been subject to thymic selection (the so-called preselection or germline repertoire) (Zerrahn et al. 1997). They found that reactivity to both self- and nonself-MHC molecules occurs at equally high frequencies in the preselection and postselection (mature) repertoires. More recently, the Marrack group provided evidence that TCR variable elements have indeed been selected by evolution to react with MHC proteins (Scott-Browne et al. 2011). The inherent bias of TCRs to "see" self-MHC is advantageous to the organism because it prepares T cells for positive selection in the thymus but, at the same time, is not conducive to the transplantation of allogeneic tissues because the same TCRs are cross-reactive with nonself-MHC molecules. Direct alloreactivity is therefore an unavoidable consequence of a TCR gene rearrangement process that creates a high frequency of T cells recognizing both self- and nonself-MHC molecules.

In summary, two forms of allorecognition of foreign MHC molecules exist. The first involves recognition by T cells of intact nonself-MHC molecules complexed to endogenous peptides. This is known as "direct" allorecognition and is characterized by uniquely high frequencies of responding T cells for the reasons outlined above. The second pathway involves the recognition of allogeneic MHC molecules as peptides presented by self-MHC molecules. This is referred to as "indirect" allorecognition, because it requires an intermediate antigen processing step; however, it could equally be referred to as "self-restricted" allorecognition because it is indistinguishable from the events that lead to recognition of all other protein antigens. The indirect pathway is in many regards similar to the recognition of mHAg; however, there is one factor that makes the indirect recognition of MHC alloantigens categorically distinct, and that is the role of allospecific B cells. One hallmark of mHAgs is that they are much less likely to provoke antibody responses, reflecting the fact that they are peptides derived from intracellular proteins. However, MHC alloantigens are readily bound by surface immunoglobulin on B cells. Indeed, when this occurs, and the allospecific B cell internalizes and processes the MHC molecule, this leads to indirect presentation by the B cell of allogeneic MHC peptides by the B cell's own MHC class II molecules. Given the efficiency of antigen-specific B cells as antigen-presenting cells (see "Role of Recipient B Cells"), this represents an important route of amplifying what is otherwise a relatively weak immune response. Both types of allorecognition owe their origins to the fundamental processes that generate a diverse mature T-cell repertoire. The first arises because TCR gene rearrangement is random and extensive, and the second because TCR gene rearrangement preferentially generates TCRs that bind MHC molecules, which enables T cells to undergo positive selection in the thymus. Thus, the elements of the adaptive immune response selected over evolutionary time to provide the organism with maximal protection against infection (TCR gene rearrangement, thymic T-cell selection, extreme polymorphism of the MHC, and self-restricted antigen recognition) are one and the same as those that underlie T-cell alloreactivity.

Are Innate Cells Alloreactive?

In a now classical monograph published in 1989 (Janeway 1989), Charles Janeway posited that nonself recognition extends beyond somatically rearranged receptors on T and B lymphocytes to germline-encoded pattern-recognition receptors on cells of the innate immune system. He postulated that these pattern recognition receptors, commonly referred to by the acronym PRR, detect conserved pathogen-associated molecular patterns (PAMPs) not shared by mammalian cells. Engaging PRRs by PAMPs, he predicted, not only triggers immediate innate defenses against microbes (e.g., phagocytosis and acute inflammation) but also induces the maturation of antigen-presenting cells (APCs). These, in turn, present microbial peptides to T cells and provide the costimulatory signals necessary for full T-cell activation. Janeway's predictions were rapidly borne out by the discovery of several families of PRR—the Toll-like receptor (TLR) family being a prime example—that detect specific classes of microbial products (e.g., lipopolysaccharide and viral nucleic acids) and trigger both inflammation and APC maturation (Beutler 2009; Palm and Medzhitov 2009). Therefore, the innate immune system as we know it now is not only a first-line defense system synonymous with inflammation but one that distinguishes between self and nonself and initiates the adaptive immune response. The new paradigm raises the fundamental question whether innate immune recognition extends beyond microbial detection to the detection of allogeneic nonself, and, if so, how is this advantageous to the host?

Two lines of evidence support the existence of innate allorecognition mechanisms. First, allograft rejection is not restricted to vertebrate animals endowed with adaptive immune systems but is common to many invertebrate organisms that predate the evolution of adaptive immunity (animals that lack T and B lymphocytes, NK cells, somatic gene rearrangement enzymes, and the MHC). Sessile, colonial marine invertebrates in at least four different phyla or subphyla (Porifera, sponges; Cnidaria, corals and hydroids; Bryozoa, sea mats; and Urochordata, tunicates or sea squirts) distinguish between their own tissues and those of unrelated members of the same species (Rosengarten and Nicotra 2011). Physical contact between two unrelated, genetically disparate colonies results in a rejection response, in which one colony induces the death of its opponent using specialized effector cells or establishes a barrier between itself and its opponent. If the interacting colonies are closely related (e.g., siblings), they fuse their tissues to become a physiologically integrated, chimeric individual. The genes or gene complexes responsible for allorecognition have been identified in two colonial marine invertebrates: the Cnidarian *Hydractinia symbiolongicarpus* and the Urochordate *Botryllus schlosseri* (De Tomaso et al. 2005; Nicotra et al. 2009; Rosa et al. 2010). In both organisms, overt rejection occurs if the interacting colonies do not share any allorecognition alleles and is inhibited if they share at least one haplotype (one set of alleles inherited from one parent) (Mokady and Buss 1996; McKitrick et al. 2011). These observations are reminiscent of NK responses in mice and humans, in which NK cells are inhibited by self-MHC molecules but are free to mount a cytotoxic attack if self-MHC is missing. Allorecognition in *Hydractinia* is determined by at least two highly polymorphic, allorecognition genes that reside in an immunoglobulin superfamily gene-rich complex and code for transmembrane proteins containing extracellular immunoglobulin-like domains, but neither the *Hydractinia* nor *Botryllus* allorecognition molecules have mammalian homologs (De Tomaso et al. 2005; Nicotra et al. 2009; Rosa et al. 2010). Therefore, the discovery of invertebrate allorecognition systems leaves us with the intriguing possibility that similar systems or remnants thereof have been conserved in vertebrates but does not provide direct clues to what they could be.

A second and perhaps more direct line of evidence that the mammalian innate immune system responds to allogeneic nonself derives from transplantation studies in mice devoid of lymphoid cells. Zecher et al. (2009) showed that injecting allogeneic splenocytes into the ear pinnae of $RAG^{-/-}$ mice, which lack T and B lymphocytes, elicits significantly greater swelling and infiltration of the skin with host myeloid cells than injecting syngeneic splenocytes. Maximal response to allogeneic cells required prior priming with alloantigen, showed crude specificity (donor-specific recall response > third-party response), occurred independent of NK cells and was mediated by monocytes. In this model, the monocyte response was to allodeterminants not linked to the MHC. A follow-up study showed that heart allografts transplanted to $\gamma c^{-/-}RAG^{-/-}$ mice, which lack T, B, and NK cells, are rapidly infiltrated by host monocytes that differentiate into mature, IL-12-expressing dendritic cells (DCs) (Zeng et al. 2012). The mature DCs presented antigen to T lymphocytes, driving their proliferation and differentiation to IFN-γ-producing cells. In contrast, syngeneic grafts elicited transient and limited monocytes to DC differentiation, and the DCs neither expressed IL-12 nor induced IFN-γ production by T cells. In vivo exposure of monocytes to allogeneic nonself precipitated T-cell-mediated rejection of single mHAg-mismatched grafts that are otherwise accepted by the host. These findings lend support to the hypothesis that, akin to the Janeway paradigm of innate microbial recognition, innate sensing of allogeneic nonself induces the maturation of APCs that then stimulate adaptive immunity. The allodeterminants that trigger APC maturation and the putative monocyte receptors that recognize them are not known yet.

In summary, there is mounting evidence that the mammalian innate immune system responds to allogeneic nonself independent of T, B, and NK cells. This form of allorecognition may initiate or amplify the responses of recipient T cells to MHC and mHAg alloantigens by inducing the maturation of APCs. It could also provide phagocytic cells with the means to kill allogeneic targets without inflicting damage on self-tissues (Liu et al. 2012). The ubiquity of allorecognition phenomena among sessile, colonial marine animals tempts one to speculate that mammalian innate allorecognition originated in invertebrate ancestors that routinely engaged in conspecific interactions in their natural habitats (Dishaw and Litman 2009). Allorecognition in these primitive ancestors not only conferred the ability to exclude unrelated individuals from one's living space—thus, enhancing survival, growth, and reproductive potential of the defending individual—but also provided safeguards against somatic and stem cell parasitism by preventing the fusion of genetically disparate individuals (Buss 1982; Laird et al. 2005; Poudyal et al. 2007). Fifty years ago, Sir Macfarlane Burnet (Burnet 1971) suggested that "a viviparously produced embryo in the wrong place might become a damaging parasite." Is it possible, then, that innate allorecognition systems were conserved in viviparous vertebrates to survey the maternal decidua for incipient invasion by fetal cells? Further studies are necessary to test this hypothesis.

MECHANISMS OF ALLORECOGNITION: *HOW* DOES THE IMMUNE SYSTEM RECOGNIZE A TRANSPLANTED ORGAN?

The accepted view is that the allogeneic response is initiated when host alloreactive T cells recognize alloantigens presented by mature APCs that originate from either the graft (donor APCs) or the recipient (recipient APCs) (Safinia et al. 2010). These distinct alloantigen presentation pathways have been dubbed "direct" and "indirect," respectively, because donor APCs, which express nonself-MHC molecules, interact with "directly" alloreactive T cells, whereas recipient APCs, which process and present allopeptides in the context self-MHC molecules, engage "indirectly" alloreactive T cells (Fig. 1) (Auchincloss and Sultan 1996). More recently, a third pathway of alloantigen presentation has been uncovered in which recipient APCs acquire intact allogeneic MHC–peptide complexes from donor cells and present them to directly reactive recipient T cells (Fig. 1) (Herrera et al. 2004). The latter pathway is referred to as the "semi-direct" or "cross-dressing" pathway to differentiate it from the "true" direct pathway performed by donor APCs and has also been shown to participate in initiating antiviral immunity (Wakim and Bevan 2011). Much research in

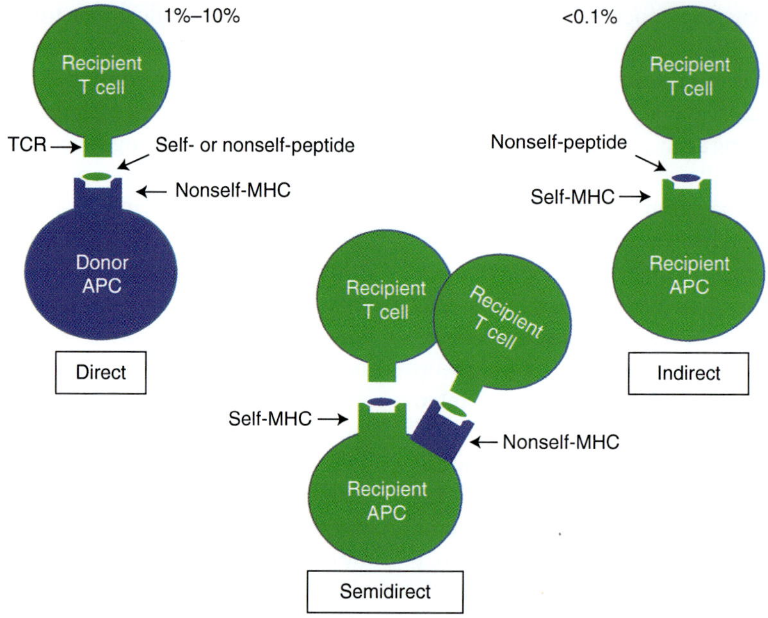

Figure 1. Pathways of alloantigen presentation. The three pathways of alloantigen presentation (direct, indirect, and semidirect) are shown. The frequency (%) of T cells that react to a specific pathway is indicated where known. APC, antigen-presenting cell; MHC, major histocompatibility complex molecule; TCR, T-cell receptor for antigen. (From Kaplan et al. 2012; reprinted, with permission, from John Wiley and Sons © 2012.)

transplantation has been dedicated to understanding these pathways and their relative contributions to acute and chronic rejection of solid organ allografts. Here we summarize salient observations and controversies in this area by asking which T cell and which APC are indeed instrumental to the recognition of transplanted organs.

Which T Cell?

T cells come in two general flavors: naïve and memory (Sallusto et al. 2010). Naïve T cells mount primary immune responses to foreign antigens not previously encountered by the host. They are present at a low precursor frequency for any given antigen, have a relatively high stimulation threshold, and can only be activated within secondary lymphoid tissues (Lakkis et al. 2000). Memory T cells, in contrast, mount secondary immune responses to antigens previously encountered by the individual, for example, after vaccination or infection. Memory T cells are long-lived lymphocytes that exist at a greater frequency for previously encountered antigens than their naïve counterparts, have a low stimulation threshold, a high proliferative capacity, and can be activated within secondary lymphoid tissues or at nonlymphoid sites—for example, within the allograft itself (Chalasani et al. 2002). Memory T-cell responses are therefore faster and more vigorous than naïve T-cell responses.

A key, albeit surprising, observation is that the allogeneic T-cell response in humans is not truly naïve but also includes memory T cells. Merkenschlager et al. (1988) and Lombardi et al. (1990) independently showed more than 20 years ago that the memory T-cell pool in individuals not previously exposed to alloantigens contains a high frequency of alloreactive T cells, equivalent to that observed in the naïve pool. The same observation was confirmed many years later in a larger group of healthy volunteers, and, as suspected, alloreactive memory T cells were shown to be cross-reactive memory

T cells that possess antimicrobial specificities (Macedo et al. 2009)—a phenomenon known as heterologous immunity (Ford and Larsen 2010). The participation of memory T cells in the allogeneic response, along with the polyclonality of alloreactive T cells (see "Origins of Allorecognition: *Why* Are We Alloreactive?"), contributes to the touted vigor of the "primary" allogeneic response both in vitro and in vivo. The memory feature of the allogeneic response also has important implications for how transplanted organs are recognized. Although the activation of naïve T cells is clearly dependent on encountering antigen presented by professional APCs (e.g., DCs) within secondary lymphoid tissues (Lakkis et al. 2000), memory T cells can be activated within the transplanted organ itself, and, importantly, their activation may or may not require professional APCs or the costimulatory stimuli that they provide (Chalasani et al. 2002). These issues are revisited in the next section in the context of which APC initiates allogeneic responses.

Which APC?

Antigen presentation, defined broadly as the ability to present either endogenous or exogenous antigens and activate T cells, can be shown in multiple cell types. First are the so-called professional APCs that include DCs (reportedly the most potent antigen-presenting cells) and phagocytic cells distributed throughout the body (e.g., macrophages). Second are endothelial cells, which when activated by inflammatory or immune stimuli, particularly in humans, acquire antigen-presenting function (Pober and Tellides 2012). Third are B lymphocytes, which although much less potent than DCs, proliferate in response to antigens and provide abundant total antigen-presenting capacity. So what roles do these different types of APCs have in the allogeneic response?

Role of Donor versus Recipient DCs

Historically, donor DCs that accompany the graft have been regarded as the dominant APC type responsible for initiating primary allogeneic responses. Evidence that grafts contain migratory, bone-marrow-derived, or "passenger" leukocytes that prime the recipient against graft antigens originated in the 1950s and was subsequently buttressed by seminal experiments in rodents (Snell 1957). Lafferty and coworkers showed that depletion of passenger leukocytes by in vitro culture prolongs the survival of thyroid and pancreatic islet grafts transplanted to allogeneic hosts (Lafferty et al. 1976; Talmage et al. 1976; Bowen et al. 1980). In 1982, Lechler and Batchelor (1982a,b) identified that DCs were the passenger leukocytes most likely responsible for priming the naïve host. They showed that renal allografts are accepted long-term by immunocompetent rats if first parked in an intermediate immunosuppressed host to deplete passenger leukocytes (Lechler and Batchelor 1982a). The transfer of donor-type DCs to the immunocompetent host resulted in T-cell priming and rejection of the grafts (Lechler and Batchelor 1982b), providing for the first time a cause–effect relationship between donor DCs and initiation of the allogeneic response. Proof that DCs, indeed, migrate out of grafts to the recipient's secondary lymphoid tissues shortly after transplantation provided further support for this hypothesis (Larsen et al. 1990a,b).

The primacy of donor DCs in initiating rejection was, however, called into question when it was observed that, in certain donor–recipient strain combinations, donor DC-depleted rat kidney grafts could provoke a rejection response, albeit more slowly, than DC-replete grafts. This led to the first postulation of the indirect pathway of allorecognition. This was subsequently shown using MHC gene knockout mice as graft donors to block antigen presentation by donor DCs or by introducing MHC-derived peptides rather than intact MHC molecules into the recipient; several groups showed that acute rejection, albeit sometimes of a slower nature, can be initiated by the indirect pathway alone (Auchincloss et al. 1993; Benham et al. 1995; Valujskikh et al. 1998). The indirect pathway of alloantigen presentation was subsequently shown to play a role in the pathogenesis of chronic allograft rejection (Safinia et al. 2010), a concept that is commensurate with the observation that direct presentation dwindles rapidly after transplantation because donor DCs that migrate to secondary lymphoid tissues are eliminated by recipient NK cells (Garrod et al. 2007). The view therefore emerged that direct alloantigen presentation is sufficient but not necessary for initiating acute rejection, whereas indirect alloantigen presentation contributes to acute rejection and may be the key driver of chronic rejection.

The recent development of transgenic mice in which DCs can be selectively deleted has provided the opportunity to test the roles of donor and recipient DCs in the allogeneic response directly. One early study has yielded surprising results. Using CD11c-DTR mice, in which the administration of diphtheria toxin (DT) deletes DCs because they express the transgene for the human diphtheria toxin receptor (DTR), Garrod et al. (2010) showed that donor DC depletion does not have an effect on acute cardiac allograft rejection, but recipient DC depletion prevents acute rejection and prolongs graft survival significantly. In the same study, the investigators showed that donor DCs are surrounded and killed by recipient NK cells very early after their migration to secondary lymphoid tissues, rendering direct alloantigen presentation a limited process at best. These findings suggest that recipient rather than donor DCs are necessary and sufficient for initiating acute rejection. It is possible that donor DCs deliver their antigenic cargo to recipient DCs in secondary lymphoid tissue either in the form of shed proteins that are then taken up and presented via the indirect pathway, or via exosomes that fuse to recipient DCs allowing for nonself-MHC presentation via the semidirect pathway (Smyth et al. 2007; Montecalvo et al. 2008). This model implies that a single recipient DC can stimulate indirectly reactive CD4 and directly reactive CD8 T cells at the same time, providing optimal CD4 help to CD8 T cells. The transfer of alloantigens from donor to recipient DCs may also be important for the induction of transplantation tolerance, for example, in experimental animals injected with tolerogenic donor DCs (Divito et al. 2010). Whether recipient DCs are indeed necessary for inducing alloimmunity or tolerance, however, should be validated in models other than the CD11c-DTR mouse because the CD11c promoter could potentially drive DTR expression on a variety of nondendritic cells (Satpathy et al. 2012).

Role of Graft Endothelium

Studies on the role of DCs in initiating allogeneic responses have been largely restricted to naïve laboratory rats and mice whose T-cell repertoires contain only a small percentage of memory cells. Humans, in contrast, possess much larger populations of memory T cells generated by repeated exposure to microbial infection. Because cross-reactive memory T cells contribute considerably to the primary alloimmune response in humans and have less stringent activation requirements than their naïve counterparts (see section, "Which T Cell?"), one is left wondering whether initiation of the human allogeneic response follows the same rules as in rodents? Pober, Tellides, and coworkers have long championed the hypothesis that the graft endothelium plays a central role in presenting antigen to and activating human alloreactive memory T cells (Pober and Tellides 2012). Their investigations, as well as those of others, have provided evidence that human endothelial cells express both MHC and costimulatory molecules, present antigens, and induce the activation and differentiation of memory CD4 and CD8 T cells (Shiao et al. 2007; Dengler and Pober 2008). Because the graft endothelium is presumably where circulating memory T cells first contact graft antigens, it is postulated that allogeneic responses could be initiated by endothelial cells, independent of DCs or other bone-marrow-derived APCs. Corroborating evidence that the graft endothelium is sufficient for activating directly alloreactive CD8 T cells exists in the mouse (Kreisel et al. 2002). An additional contribution of endothelial cell alloantigen presentation to transplant immunity has been referred to as the "shop window" effect. The vascular endothelium displays, as MHC-presented peptides, some of the contents within the tissue through which the vessel runs, and may accelerate, or in fact initiate, the transmigration of T cells specific for endothelial cell-displayed antigens. This has been shown to be true both in vitro and, using intravital microscopy, in vivo (Marelli-Berg et al. 2004; Walch et al. 2013). Gradual replacement of a small but significant proportion of donor endothelial cells with recipient cells over time suggests that the graft endothelium could also participate in indirect alloantigen presentation (Valujskikh et al. 2002). Additional studies, however, are needed to dissect which APC (DC or endothelial cell) is necessary and/or sufficient for the activation of alloreactive memory T cells after transplantation. Such studies are expected to have important clinical implications because of the conspicuous role that alloreactive memory T cells have in human allograft rejection in both naïve and sensitized recipients.

Role of Recipient B Cells

B-cell depletion in mice or humans does not prevent or delay acute T-cell-mediated rejection and, in some cases, could actually increase the risk of rejection (Tydén et al. 2009). There is mounting evidence, however, that B cells are required for maximal generation of memory T cells after either organ transplantation or microbial infection and that this effect is due to their antigen-presenting function (Lund and Randall 2010; Ng et al. 2010). Moreover, recipient B cells have been recently found to be instrumental for the pathogenesis of chronic heart allograft rejection in mice independent of alloantibody production (Sheriff et al. 2011). Therefore, B cells are not essential for initiating allogeneic responses but could have an important role in chronic rejection, possibly by providing long-term indirect alloantigen presentation to T cells.

CONCLUDING REMARKS

Allogeneic responses are potent immune responses that cause allograft rejection. They are characteristically polyclonal, involving many TCR specificities, and are brought about by the activation of both naïve and preexisting, cross-reactive memory T cells. Importantly, the alloreactive T-cell repertoire, whether "directly" or "indirectly" alloreactive, is one and the same as the "normal" T-cell repertoire devoted to the recognition of antigens in the context of self-MHC molecules. Both "directly" and "indirectly" alloreactive T cells bind to their target MHC–peptide complexes with the same molecular mechanisms. Mounting evidence suggests that allorecognition is not restricted to the adaptive immune system but that it may be an inherent property of the innate immune system as well. Cells or tissues transplanted from allogeneic donors induce the differentiation of host monocytes to mature DCs independent of T, B, and NK cells. These DCs play a key role in initiating T-cell-dependent rejection. Finally, despite long-standing emphasis on direct and indirect alloantigen presentation as independent processes competing for the attention of transplant immunologists, more recent evidence points to extensive antigen transfer between donor and recipient cells. The transfer of intact MHC molecules from graft cells, possibly donor DCs, to recipient DCs, for example, may have an important role in activating the allogeneic response. These and other unresolved questions highlighted in this review warrant further study, even if it entails exploring unorthodox ideas that lie outside the safe confines of our field.

REFERENCES

Auchincloss H Jr, Sultan H. 1996. Antigen processing and presentation in transplantation. *Curr Opin Immunol* **8:** 681–687.

Auchincloss H, Lee R, Shea S, Markowitz JS, Grusby MJ, Glimcher LH. 1993. The role of "indirect" recognition in initiating rejection of skin grafts from major histocompatibility complex class II–deficient mice. *Proc Natl Acad Sci* **90:** 3373–3377.

Benham AM, Sawyer GJ, Fabre JW. 1995. Indirect T cell allorecognition of donor antigens contributes to the rejection of vascularized kidney allografts. *Transplantation* **59:** 1028–1032.

Beutler BA. 2009. TLRs and innate immunity. *Blood* **113:** 1399–1407.

Bowen KM, Andrus L, Lafferty KJ. 1980. Successful allotransplantation of mouse pancreatic islets to nonimmunosuppressed recipients. *Diabetes* **29** (Suppl): 98–104.

Burnet FM. 1971. "Self-recognition" in colonial marine forms and flowering plants in relation to the evolution of immunity. *Nature* **232:** 230–235.

Buss L. 1982. Somatic cell parasitism and the evolution of somatic tissue compatibility. *Proc Natl Acad Sci* **79:** 5337–5341.

Chalasani G, Dai Z, Konieczny BT, Baddoura FK, Lakkis FG. 2002. Recall and propagation of allospecific memory T cells independent of secondary lymphoid organs. *Proc Natl Acad Sci* **99:** 6175–6180.

Dengler TJ, Pober JS. 2008. Human vascular endothelial cells stimulate memory but not naive $CD8^+$ T cells to differentiate into CTL retaining an early activation phenotype. *J Immunol* **164:** 5146–5155.

De Tomaso AW, Nyholm SV, Palmeri KJ, Ishizuka KJ, Ludington WB, Mitchel K, Weissman IL. 2005. Isolation and characterization of a protochordate histocompatibility locus. *Nature* **438:** 454–459.

Dierselhuis M, Goulmy E. 2009. The relevance of minor histocompatibility antigens in solid organ transplantation. *Curr Opin Organ Transplant* **14:** 419–425.

Dishaw LJ, Litman GW. 2009. Invertebrate allorecognition: The origins of histocompatibility. *Curr Biol* **19:** R286–R288.

Divito SJ, Wang Z, Shufesky WJ, Liu Q, Tkacheva OA, Montecalvo A, Erdos G, Larregina AT, Morelli AE. 2010. Endogenous dendritic cells mediate the effects of intravenously injected therapeutic immunosuppressive dendritic cells in transplantation. *Blood* **116:** 2694–2705.

Ford ML, Larsen CP. 2010. Overcoming the memory barrier in tolerance induction: Molecular mimicry and functional heterogeneity among pathogen-specific T-cell populations. *Curr Opin Organ Transplant* **15:** 405–410.

Garrod KR, Wei SH, Parker I, Cahalan MD. 2007. Natural killer cells actively patrol peripheral lymph nodes forming stable conjugates to eliminate MHC-mismatched targets. *Proc Natl Acad Sci* **104:** 12081–12086.

Garrod KR, Liu F-C, Forrest LE, Parker I, Kang S-M, Cahalan MD. 2010. NK cell patrolling and elimination of donor-derived dendritic cells favor indirect alloreactivity. *J Immunol* **184:** 2329–2336.

Hall BM, Dorsch S, Roser B. 1978a. The cellular basis of allograft rejection in vivo. I. The cellular requirements for first-set rejection of heart grafts. *J Exp Med* **148:** 878–889.

Hall BM, Dorsch S, Roser B. 1978b. The cellular basis of allograft rejection in vivo. II. The nature of memory cells mediating second set heart graft rejection. *J Exp Med* **148:** 890–902.

Herrera OB, Golshayan D, Tibbott R, Salcido Ochoa F, James MJ, Marelli-Berg FM, Lechler RI. 2004. A novel pathway of alloantigen presentation by dendritic cells. *J Immunol* **173:** 4828–4837.

Hildemann WH. 1974. Some new concepts in immunological phylogeny. *Nature* **250:** 116–129.

Janeway CA Jr. 1989. Approaching the asymptote? Evolution and revolution in immunology. *Cold Spring Harb Symp Quant Biol* **54** (Pt 1)**:** 1–13.

Janeway CJ. 2001. How the immune system works to protect the host from infection. *Proc Natl Acad Sci* **98:** 7461–7468.

Kaplan B, Burkhart GJ, Lakkis FG, eds. 2012. *Immunotherapy in transplantation: Principles and practice*, p. 13. Wiley, New York.

Kreisel D, Krupnick AS, Gelman AE, Engels FH, Popma SH, Krasinskas AM, Balsara KR, Szeto WY, Turka LA, Rosengard BR. 2002. Non-hematopoietic allograft cells directly activate $CD8^+$ T cells and trigger acute rejection: An alternative mechanism of allorecognition. *Nat Med* **8:** 233–239.

Lafferty KJ, Bootes A, Dart G, Talmage DW. 1976. Effect of organ culture on the survival of thyroid allografts in mice. *Transplantation* **22:** 138–149.

Laird D, De Tomaso A, Weissman I. 2005. Stem cells are unit of natural selection in a colonial ascidian. *Cell* **123:** 1351–1360.

Lakkis FG, Arakelov A, Konieczny BT, Inoue Y. 2000. Immunologic "ignorance" of vascularized organ transplants in the absence of secondary lymphoid tissue. *Nat Med* **6:** 686–688.

Larsen CP, Morris PJ, Austyn JM. 1990a. Migration of dendritic leukocytes from cardiac allografts into host spleens: A novel pathway for initiation of rejection. *J Exp Med* **171:** 307–314.

Larsen CP, Steinman RM, Witmer-Pack M, Hankins DF, Morris PJ, Austyn JM. 1990b. Migration and maturation of Langerhans cells in skin transplants and explants. *J Exp Med* **172:** 1483–1493.

Lechler RI, Batchelor JR. 1982a. Immunogenicity of retransplanted rat kidney allografts. Effect of inducing chimerism in the first recipient and quantitative studies on immunosuppression of the second recipient. *J Exp Med* **156:** 1835–1841.

Lechler RI, Batchelor JR. 1982b. Restoration of immunogenicity to passenger cell-depleted kidney allografts by the addition of donor strain dendritic cells. *J Exp Med* **155:** 31–41.

Liu W, Xiao X, Demirci G, Madsen J, Li XC. 2012. Innate NK cells and macrophages recognize and reject allogeneic nonself in vivo via different mechanisms. *J Immunol* **188:** 2703–2711.

Lombardi G, Sidhu S, Daly M, Batchelor JR, Makgoba W, Lechler RI. 1990. Are primary alloresponses truly primary? *Int Immunol* **2:** 9–13.

Lund FE, Randall TD. 2010. Effector and regulatory B cells: Modulators of $CD4^+$ T cell immunity. *Nat Rev Immunol* **10:** 236–247.

Macdonald WA, Chen Z, Gras S, Archbold JK, Tynan FE, Clements CS, Bharadwaj M, Kjer-Nielsen L, Saunders PM, Wilce MC, et al. 2009. T cell allorecognition via molecular mimicry. *Immunity* **31:** 897–908.

Macedo C, Orkis EA, Popescu I, Elinoff BD, Zeevi A, Shapiro R, Lakkis FG, Metes D. 2009. Contribution of naïve and memory T-cell populations to the human alloimmune response. *Am J Transplant* **9:** 2057–2066.

Marelli-Berg FM, James MJ, Dangerfield J, Dyson J, Millrain M, Scott D, Simpson E, Nourshargh S, Lechler RI. 2004. Cognate recognition of the endothelium induces HY-specific $CD8^+$ T-lymphocyte transendothelial migration (diapedesis) in vivo. *Blood* **103:** 3111–3116.

McKitrick TR, Muscat CC, Pierce JD, Bhattacharya D, De Tomaso AW. 2011. Allorecognition in a basal chordate consists of independent activating and inhibitory pathways. *Immunity* **34:** 616–626.

Merkenschlager M, Terry L, Edwards R, Beverley PC. 1988. Limiting dilution analysis of proliferative responses in human lymphocyte populations defined by the monoclonal antibody UCHL1: Implications for differential

CD45 expression in T cell memory formation. *Eur J Immunol* **18:** 1653–1661.

Miller JFAP. 1962. Effect of neonatal thymectomy on the immunological responsiveness of the mouse. *Proc R Soc Lond B* **156:** 415–428.

Mokady O, Buss LW. 1996. Transmission genetics of allorecognition in *Hydractinia symbiolongicarpus* (Cnidaria: Hydrozoa). *Genetics* **143:** 823–827.

Montecalvo A, Shufesky WJ, Stolz DB, Sullivan MG, Wang Z, Divito SJ, Papworth GD, Watkins SC, Robbins PD, Larregina AT, et al. 2008. Exosomes as a short-range mechanism to spread alloantigen between dendritic cells during T cell allorecognition. *J Immunol* **180:** 3081–3090.

Ng YH, Oberbarnscheidt MH, Chandramoorthy HC, Hoffman R, Chalasani G. 2010. B cells help alloreactive T cells differentiate into memory T cells. *Am J Transplant* **10:** 1970–1980.

Nicotra M, Powell A, Rosengarten R, Moreno M, Grimwood J, Lakkis F, Dellaporta S, Buss L. 2009. A hypervariable invertebrate allodeterminant. *Curr Biol* **19:** 583–589.

Palm NW, Medzhitov R. 2009. Pattern recognition receptors and control of adaptive immunity. *Immunol Rev* **227:** 221–233.

Peugh WN, Superina RA, Wood KJ, Morris PJ. 1986. The role of H-2 and non-H-2 antigens and genes in the rejection of murine cardiac allografts. *Immunogenetics* **23:** 30–37.

Pober JS, Tellides G. 2012. Participation of blood vessel cells in human adaptive immune responses. *Trends Immunol* **33:** 49–57.

Poudyal M, Rosa S, Powell AE, Moreno M, Dellaporta SL, Buss LW, Lakkis FG. 2007. Embryonic chimerism does not induce tolerance in an invertebrate model organism. *Proc Natl Acad Sci* **104:** 4559–4564.

Rosa SF, Powell AE, Rosengarten RD, Nicotra ML, Moreno MA, Grimwood J, Lakkis FG, Dellaporta SL, Buss LW. 2010. *Hydractinia* allodeterminant *alr1* resides in an immunoglobulin superfamily-like gene complex. *Curr Biol* **20:** 1122–1127.

Rosengarten RD, Nicotra ML. 2011. Model systems of invertebrate allorecognition. *Curr Biol* **21:** R82–R92.

Safinia N, Afzali B, Atalar K, Lombardi G, Lechler RI. 2010. T-cell alloimmunity and chronic allograft dysfunction. *Kidney Int Suppl* **2010:** S2–S12.

Sallusto F, Lanzavecchia A, Araki K, Ahmed R. 2010. From vaccines to memory and back. *Immunity* **33:** 451–463.

Satpathy AT, Wu X, Albring JC, Murphy KM. 2012. Re(de)fining the dendritic cell lineage. *Nat Immunol* **13:** 1145–1154.

Scott-Browne JP, Crawford F, Young MH, Kappler JW, Marrack P, Gapin L. 2011. Evolutionarily conserved features contribute to $\alpha\beta$ T cell receptor specificity. *Immunity* **35:** 526–535.

Sheriff KA, Zeng Q, Li Q, Ng YH, Marlowe A, Randhawa P, Hoffman R, Chalasani G. 2011. B cells and not antibodies are required for T cell activation and development of chronic rejection. *Am J Transplant* **11:** 40–41.

Sherman LA, Chattopadhyay S. 1993. The molecular basis of allorecognition. *Annu Rev Immunol* **11:** 385–402.

Shiao SL, Kirkiles-Smith NC, Shepherd BR, McNiff JM, Carr EJ, Pober JS. 2007. Human effector memory $CD4^+$ T cells directly recognize allogeneic endothelial cells in vitro and in vivo. *J Immunol* **179:** 4397–4404.

Smyth LA, Afzali B, Tsang J, Lombardi G, Lechler RI. 2007. Intercellular transfer of MHC and immunological molecules: Molecular mechanisms and biological significance. *Am J Transplant* **7:** 1442–1449.

Snell GD. 1957. The homograft reaction. *Annu Rev Microbiol* **11:** 439–458.

Takemoto SK, Terasaki PI, Gjertson DW, Cecka JM. 2000. Twelve years' experience with national sharing of HLA-matched cadaveric kidneys for transplantation. *N Engl J Med* **343:** 1078–1084.

Talmage DW, Dart G, Radovich J, Lafferty KJ. 1976. Activation of transplant immunity: Effect of donor leukocytes on thyroid allograft rejection. *Science* **191:** 385–388.

Tydén G, Genberg H, Tollemar J, Ekberg H, Persson NH, Tufveson G, Wadström J, Gäbel M, Mjörnstedt L. 2009. A randomized, doubleblind, placebo-controlled, study of single-dose rituximab as induction in renal transplantation. *Transplantation* **87:** 1325–1329.

Valujskikh A, Matesic D, Gilliam A, Anthony D, Haqqi TM, Heeger PS. 1998. T cells reactive to a single immunodominant self-restricted allopeptide induce skin graft rejection in mice. *J Clin Invest* **101:** 1398–1407.

Valujskikh A, Lantz O, Celli S, Matzinger P, Heeger PS. 2002. Cross-primed $CD8^+$ T cells mediate graft rejection via a distinct effector pathway. *Nat Immunol* **3:** 844–851.

Wakim LM, Bevan MJ. 2011. Cross-dressed dendritic cells drive memory $CD8^+$ T-cell activation after viral infection. *Nature* **471:** 629–632.

Walch JM, Zeng Q, Li Q, Oberbarnscheidt MH, Hoffman RA, Wiliams AL, Rothstein DM, Shlomchik WD, Kim JV, Camirand G, et al. 2013. Cognate antigen directs $CD8^+$ T cell migration to vascularized transplants. *J Clin Invest* **123:** 2663–2671.

Yin Y, Mariuzza RA. 2009. The multiple mechanisms of T cell receptor cross-reactivity. *Immunity* **31:** 849–851.

Yin L, Scott-Browne J, Kappler JW, Gapin L, Marrack P. 2012. T cells and their eons-old obsession with MHC. *Immunol Rev* **250:** 49–60.

Zecher D, van Rooijen N, Rothstein D, Shlomchik W, Lakkis F. 2009. An innate response to allogeneic nonself mediated by monocytes. *J Immunol* **183:** 7810–7816.

Zeng Q, Williams AL, Hoffman R, Rothstein DM, Oberbarnscheidt MH, Lakkis F. 2012. Innate recognition of allogeneic non-self induces monocyte differentiation to mature dendritic cells in vivo. *Am J Transplant* **12:** 148–148.

Zerrahn J, Held W, Raulet DH. 1997. The MHC reactivity of the T cell repertoire prior to positive and negative selection. *Cell* **88:** 627–636.

Effector Mechanisms of Rejection

Aurélie Moreau, Emilie Varey, Ignacio Anegon, and Maria-Cristina Cuturi

INSERM UMR 1064, Center for Research in Transplantation and Immunology-ITUN, CHU de Nantes 44093, France

Correspondence: ignacio.anegon@univ-nantes.fr

Organ transplantation appears today to be the best alternative to replace the loss of vital organs induced by various diseases. Transplants can, however, also be rejected by the recipient. In this review, we provide an overview of the mechanisms and the cells/molecules involved in acute and chronic rejections. T cells and B cells mainly control the antigen-specific rejection and act either as effector, regulatory, or memory cells. On the other hand, nonspecific cells such as endothelial cells, NK cells, macrophages, or polymorphonuclear cells are also crucial actors of transplant rejection. Last, beyond cells, the high contribution of antibodies, chemokines, and complement molecules in graft rejection is discussed in this article. The understanding of the different components involved in graft rejection is essential as some of them are used in the clinic as biomarkers to detect and quantify the level of rejection.

DIFFERENT TYPES OF REJECTION

Several types of rejection of vascularized organs can be defined according to their underlying mechanisms and tempos, the major types being hyperacute, acute, and chronic rejection. In allogeneic context and in the absence of preformed antidonor antibodies, cells and tissues are mainly rejected by acute cellular rejection mechanisms.

Hyperacute rejection appears in the first minutes following transplantation and occurs only in vascularized grafts. This very fast rejection is characterized by vessels thrombosis leading to graft necrosis. Hyperacute rejection is caused by the presence of antidonor antibodies existing in the recipient before transplantation. These antibodies induce both complement activation and stimulation of endothelial cells to secrete Von Willebrand procoagulant factor, resulting in platelet adhesion and aggregation. The result of these series of reactions is the generation of intravascular thrombosis leading to lesion formation and ultimately to graft loss. Today, this type of rejection is avoided in most cases by checking for ABO compatibility and by excluding the presence of antidonor human leukocyte antigen (HLA) antibodies by crossmatch techniques between donor graft cells and recipient sera. This type of rejection is also observed in models of xenotransplantation of vascularized organs between phylogenetically distant species when no immunosuppressive treatment is given to the recipients.

Acute rejection is caused by an immune response directed against the graft and occurs between 1 week and several months after transplantation. Acute rejection is diagnosed on histological analysis of a graft biopsy according to an international classification system, the Banff

Cite this article as *Cold Spring Harb Perspect Med* doi: 10.1101/cshperspect.a015461

classification for the kidney (Mengel et al. 2012). Acute rejection is thought to result from two immunological mechanisms that may act alone or in combination: (1) a T-cell-dependent process that corresponds to acute cellular rejection, and (2) a B-cell-dependent process that generates the acute humoral rejection. With current immunosuppressive treatment, acute rejection occurs in less than 15% of the transplants (Port et al. 2004) in nonsensitized patients.

Chronic rejection, on the other hand, is now the leading cause of graft rejection. Chronic rejection can be mediated by either humoral or cellular mechanisms linked to memory/plasma cells and antibodies. The presence of tertiary lymphoid organs in the graft is a characteristic of this form of rejection.

INNATE AND ADAPTIVE IMMUNE RESPONSES

Two major immunological mechanisms occur during allograft rejection: the nonspecific innate response that predominates in the early phase of the immune response, and the donor-specific adaptive response that results from alloantigen recognition by host T cells.

The Innate Response and Allograft Rejection

Although the adaptive response plays a central role in the mechanisms of allograft rejection, early proinflammatory signals (arising before the initiation of the T-cell response) are also considered as important factors of graft rejection. Inflammation is caused by the innate immune response induced independently of the adaptive response (Christopher et al. 2002; He et al. 2002, 2003; Land 2005). In fact, it was shown that 1 day after a heart transplant, the expression of genes coding for molecules linked to inflammation (proinflammatory cytokines, chemokines, components of the cellular infiltrate) was similar in normal mice and in mice deficient for T and B cells, but with normal NK and myeloid compartments (*Rag1* or *Rag2* knock-out mice) (He et al. 2003). These investigators also showed that the innate response is antigen independent, develops early after transplantation, and conditions the development of the adaptive response (He et al. 2003).

Innate immune responses are the consequence of several events associated with clinical transplantation, such as ischemia-reperfusion injury and infections, and lead to the release of damage-associated molecular patterns (DAMPs) and pathogen-associated molecular patterns (PAMPs) (Chong and Alegre 2012). DAMPs and PAMPs are recognized by so-called pattern-recognition receptors (PRRs) expressed by hematopoietic cells.

The specificity of PRRs is genetically determined and several subgroups can be classified based on their structure. The transmembrane group of PRRs includes several families of molecules such as toll-like receptors (TLRs), C-type lectins, RAGE (receptor for advanced glycation endproducts), complement receptors, scavenger receptors, and mannose receptors. The cytosolic PRR group includes retinoic acid-inducible gene-I-like receptors and nucleotide-binding domain and leucine-rich repeat-containing receptors (Iwasaki and Medzhitov 2010).

Immediately after transplantation, PRR-mediated danger signals activate dendritic cells (DCs) (LaRosa et al. 2007) leading to antigen-presenting cell (APC) maturation, up-regulation of costimulatory molecules, and secretion of proinflammatory cytokines. In this context, donor APCs migrate to the T cell areas of secondary lymphoid organs and induce the activation and differentiation of alloreactive naïve T cells into effector T helper cells. These effector cells migrate into the graft where they activate macrophages and granulocytes (neutrophils, eosinophils, and basophils) that have infiltrated the graft in response to inflammatory stimuli. The latter cells contribute to lesion formation, either directly or through the production of proinflammatory cytokines and chemokines. They also help boost and maintain the adaptive immune T-cell response. Another cell type involved in innate immunity is the NK cell. In a proinflammatory context, NK cells become activated and are able to kill target cells (Pratschke et al. 2009).

The complement system plays a central role in the effector mechanisms occurring during the innate response. The three complement pathways (classical, alternative, and mannose lectin) can be activated by DAMPs as well as by the inflammatory environment that develops during ischemia reperfusion injury (Castellano et al. 2010). Moreover, the intermediate product of complement, C5a, is able to bind to APC receptors, leading to Th1 activation and development of the adaptive immunity (Zhou et al. 2000).

Initiation of the Adaptive Response and Allograft Rejection

The adaptive immune response appears later than the innate response, its main characteristic being antigen specific. The initiation of the adaptive response is made possible by the presentation of alloantigens by APCs, mainly DCs and their allorecognition by recipient T cells. Three main pathways of allorecognition are described in the literature. The first is the direct pathway, in which donor DCs present in the graft act as passenger leukocytes. In the context of a proinflammatory environment, these cells mature and migrate to secondary lymphoid organs where they prime host T cells. The maturation of these cells is induced by proinflammatory signals such as IL-1β, TNF-α, and CD40. The second is the indirect allorecognition pathway in which recipient DCs capture, process, and present alloantigens as peptides on their host MHC molecules, and then prime T cells (Ochando et al. 2006). As dendritic cells express both MHC class I and class II, donor antigens can be presented to either $CD4^+$ or $CD8^+$ T cells. The third pathway is that of semidirect allorecognition, characterized by the dual ability of recipient DCs: (1) to present intact donor major histocompatibility complex (MHC) molecules acquired by cell-to-cell contact or fusion with donor exosomes, and (2) to internalize and process donor MHC as peptides on recipient MHC molecules. In this pathway, alloreactive $CD8^+$ and $CD4^+$ T cells are stimulated by direct and indirect allorecognition (Herrera et al. 2004; Smyth et al. 2006).

Using the two-photon microscopy technique, a recent elegant study by Bousso's group (Celli et al. 2011) provided a better understanding of the adaptive immune response by clarifying several different issues that until then had remained hypothetical. In a murine model of skin transplantation, they showed that donor dermal DCs disappeared rapidly from the graft and migrated to draining lymph nodes (LNs). However, these donor DCs were found to be dead in the secondary lymphoid organs. This is consistent with a previous study demonstrating that donor DCs are eliminated by NK cells (Garrod et al. 2010). It has been suggested that these dying cells could be a source of alloantigens for recipient APCs present in the draining LNs that stimulate T cells by the indirect pathway. Moreover, these experiments showed that host inflammatory monocytes and DCs act as antigen-transporting cells. These recipient graft-infiltrating cells indeed have the ability to reach the draining LNs and cross-prime $CD8^+$ T cells by the indirect pathway (Celli et al. 2011).

Initiation of the innate and adaptive immune response leads to acute and chronic graft rejection. Cellular and humoral responses, but also chemokines and innate cells, are present in both types of rejection and are described in this article.

ACUTE REJECTION

The Components of Acute Rejection

Chemokines

Chemokines are chemoattractant cytokines that influence immune cell migration. In transplantation settings, chemokines act during four key phases: (1) recruitment of mononuclear cells to the site of inflammation, (2) migration of APCs to draining LNs, (3) interaction between unprimed T cells and APCs in LNs, and (4) migration of primed alloreactive T cells to the graft (Belperio and Ardehali 2008). The roles of some chemokines in allograft rejection have been investigated directly by analyzing chemokine expression by immune cells as described in Figure 1. Furthermore, the function of these chemokine receptors and ligands has been shown using knockout (KO) mice or blocking antibodies.

One example is that of CCR5, which was shown to be expressed by mononuclear cells,

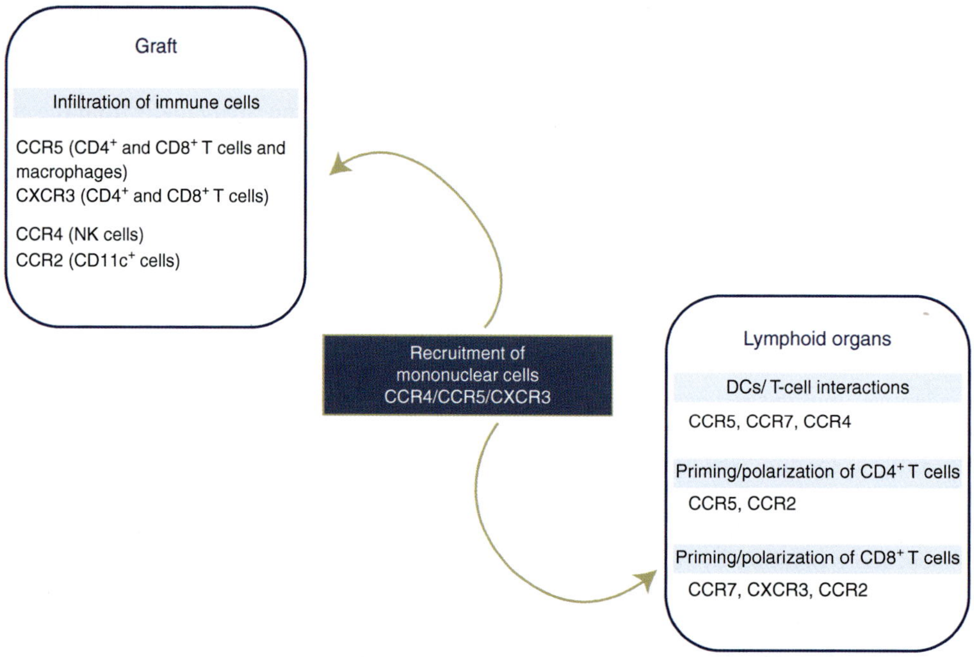

Figure 1. Chemokines: Homing of immune cells and role in acute rejection. Mononuclear cells are able to migrate into the graft or into the lymphoid organs in response to chemokine gradients. Experiments on animal models of transplantation and analysis of human samples from transplant recipients allow a better understanding of the involvement of chemokines in graft rejection.

suggesting that this chemokine helps their migration to inflammatory sites. Prolongation of cardiac allograft survival was obtained in $CCR5^{-/-}$ mice, demonstrating a major role for this chemokine (Gao et al. 2001). However, more recently, in a similar heart transplant model, Nozaki et al. (2007) only detected a modest effect of CCR5 on graft rejection using $CCR5^{-/-}$ mice. Involvement of CCR5 in graft rejection was also shown in murine models of pancreatic islet transplantation (Abdi et al. 2002). More recently, in a cardiac transplantation model in rhesus monkeys, cotreatment with a CCR5 antagonist and cyclosporine A was able to prolong heart survival. This prolongation was associated with a delay in alloantibody response, a lower graft infiltration, and a decrease in alternatively activated macrophages (Li et al. 2011). In addition to the recruitment of mononuclear cells to the graft, CCR5 polarizes T cells toward a type 1 response, and promotes the trafficking of T cells to LNs and favors DC/T-cell interaction within the LNs (Amano et al. 2005). Furthermore, Y-box protein-1, a molecule that acts as a transcription regulator of CCL5, one CCR5 ligand also called RANTES, expression in T cells, and monocytes/macrophages, was shown to be up-regulated during rejection in human kidney transplant biopsies (Raffetseder et al. 2009). Contrary to these studies, in a rat model of kidney transplant tolerance following CD28 blockade, CCL5 recruited $CD4^+$ Tregs to the graft via a graft-to-periphery CCL5 gradient (Dilek et al. 2012).

Another well-described chemokine involved in graft rejection is CXCR3. In a murine model, absence of CXCR3 in graft recipients was shown to prolong cardiac transplant survival (Hancock et al. 2000). However, as for CCR5, a more recent study obtained controversial results showing no delay in graft rejection between controls and CXCR3 KO mice (Kwun et al. 2008). Analysis of CXCR3 and its ligands, CXCL9 and CXCL10, was performed in biopsies from transplant patients undergoing acute rejection. Fahmy et al. (2003) showed an up-regulation of CXCL10 in endomyocardial biopsies, whereas an increase in CXCL10 was detected in macrophages of lung biopsies (Agostini et al. 2001). Furthermore, the number of CXCR3$^+$ cells present in human kidney transplant biopsies was reported to increase with graft rejection and to decrease with immunosuppressive treatment, suggesting that CXCR3 might be a good target to treat rejection (Hoffmann et al. 2006). Other recent studies have highlighted the role of chemokines as biomarkers of clinical rejection and are described in Table 1.

In mouse models of cardiac and pancreatic islet transplantation, expression of CXCR3 was detected in Tregs, memory CD4$^+$ and CD8$^+$ T cells, NKT, and NK cells (Uppaluri et al. 2008). Zhai et al. (2006) showed that CXCR3 is involved in the recruitment of CXCR3$^+$CD4$^+$ T cells following ischemia-reperfusion injury in a model of rat syngeneic orthotopic liver transplantation. Similarly, expression of CXCR3 by Tregs resulted in their recruitment to peripheral sites of inflammation, and a high number of circulating CXCR3$^+$ Tregs was detected in kidney transplant patients treated with an mTOR inhibitor as their immunosuppressive therapy (Hoerning et al. 2011). In addition to the role of CXCL9/CXCL10 in the recruitment of T cells to the allograft, Rosenblum et al. (2010) recently showed that these chemokines regulate donor-specific CD8$^+$ T-cell priming. As such, in a murine model of cardiac transplantation using KO mice for either the CXCL9 or CXCL10 gene, CXCL9 promoted, whereas CXCL10 inhibited the differentiation of IFN-γ-producing donor-specific CD8$^+$ T cells.

A new molecule called TAK-779 was identified as a blocking agent for both CCR5 and CXCR3. This molecule was shown to prevent cardiac allograft rejection in mice (Akashi et al. 2005) as well as kidney rejection in rats (Kakuta et al. 2012). In the latter study, prolongation of graft survival correlated with a decrease in macrophage infiltration (Kakuta et al. 2012). Last, coexpression of CXCL10 and fractalkine or CX3CL1 by endothelial cells was shown to be involved in effector memory T-cell recruitment (Manes et al. 2007; Manes and Pober 2008). Fractalkine acts as a chemoattractant and adhesion molecule for cells that express CX3CR1, including certain immune cells such as monocytes/macrophages, NK cells, and T cells (Imai et al. 1997; Fong et al. 1998; Foussat et al. 2000).

Some studies have also been performed to analyze other chemokine receptors. For example, CCR2 was shown to be an important chemoattractant receptor for mononuclear cells. This is because CCR2 induced mobilization of monocytes from bone marrow to blood and an accumulation of CD11c$^+$ cells in the allograft in a mouse model of lung allotransplantation (Gelman et al. 2010). Interaction of CCR2 with its ligand (CCL2) also caused T-cell clonal expansion and differentiation (Lee et al. 2003). In CCR2$^{-/-}$ mice, an increase in graft survival was observed after pancreatic islet transplantation (compared to control mice) and associated with the absence of CD8$^+$ effector T cells in these mice (Abdi et al. 2004).

CCR4 is an important chemokine for homing of memory T cells (Campbell et al. 1999) and was shown to be crucial for the recruitment of FoxP3$^+$ Tregs in a model of tolerance induced by donor-specific transfusion/CD154 therapy (46) as no tolerance was achieved in CCR4-deficient mice (Lee et al. 2005). Inhibition of CCR4 and its ligands led to an increase in heart graft survival associated with a decreased in DC/T-cell interactions and an inhibition of monocytes and NK cells present in the graft (Alferink et al. 2003; Huser et al. 2005). Last, CCR7 is known to be involved in the localization of both DCs and T cells in the T-cell-rich zones of the LNs, indicating a role for CCR7 in T-cell homing and priming (Hopken et al. 2004). Recently, Liu et al. (2011) showed that a lack of CCR7 prevented tolerance induction in a cardiac murine model of transplantation. This impairment of tolerance was partially rescued by adoptive transfer of wild-type plasmacytoid DCs to the CCR7-deficient mice (Liu et al. 2011).

As discussed before, some of these chemokines and chemokine receptors might be useful as markers of acute rejection. In the case of renal transplantation, rejection could be monitored via noninvasive urine sampling. The most well-known urine marker is IP-10 or CXCL10. An increase in IP-10 mRNA and protein in the

Table 1. Use of chemokines, cytokines, miRNA, and cells as biomarkers of graft rejection in clinical studies

Type of rejection	Molecules/cells	Transplant organ	Samples of analysis	Significance	References
Acute	CXCL10	Heart	Blood, serum	Higher serum level of CXCL10 in transplanted patients versus control ones. Among transplanted patients, high expression of CXCL10 is correlated with acute rejection.	Crescioli et al. 2009
	C reactive protein (CRP)	Heart	Blood, serum	Significant differential expression of CRP between rejected and nonrejected heart transplant recipient. In the first year postheart transplantation, CRP was the most useful parameter for noninvasive screening of acute cellular rejection.	Martinez-Dolz et al. 2009
	Granzyme B CTLA-4 IL-6 TNF-α TGF-β	Heart	Biopsies	Up-regulation in mRNA expression of all analyzed genes but the strongest increase is seen for granzyme B, TGF-β, and TNF-α genes. Correlation between the expression analysis of these three genes and acute rejection.	Ramsperger-Gleixner et al. 2011
	CD69	Heart	Blood, T lymphocytes	Overexpression of CD69 in T lymphocytes of patients with acute rejection. Relationship between increased expression by T lymphocytes and acute rejection is statistically more pronounced for $CD8^+$ T cells than for $CD4^+$ T cells. Presence of $CD69^+/CD8^+$ cells exceeding 15% in patients with acute rejection of all T lymphocytes is correlated with vigorous rejection.	Schowengerdt et al. 2000
	Functional markers of $CD4^+$ T-cell subsets: Th1 Th17 $FoxP3^+$ T cells	Heart	Blood and biopsies	Higher mRNA expression of the functional markers in patients with acute cardiac rejection in blood and biopsies samples (but not for Th2 mRNA cytokines). Higher percentage of Th1, Th17, and $FoxP3^+CD4^+$-circulating T cells. Graft infiltration by Th1 and Th17 cells. Kinetic changes of these $CD4^+$ T-cell subsets after cardiac allograft.	Wang et al. 2011
	CCL3L1	Liver	Blood, PBMC	Analysis of copy number variation in genes. Higher number of CCL3L1 gene in patients with acute rejection.	Li et al. 2012

Acute	CCL2 CXCL8 CCL5 CXCL9 CXCL10	Liver	Blood, serum	Chemokines screening using Luminex multiplex assay. Higher expression of molecules correlated with an up-regulation of NF-κB: CCL2, CCL5, and CXCL8. Higher expression of molecules correlated with T-cell immunity: CXCL9 and CXCL10. Biomarkers of early allograft dysfunction.	Friedman et al. 2012
	Granzyme B Perforin	Liver	Biopsies	Analysis by immunohistochemistry. Higher frequencies of perforin and granzyme B expression in acute rejection patients. In most of the biopsies, perforin and granzyme B are simultaneously expressed.	Ji et al. 2004
	Granzyme B	Liver	Biopsies	Higher granzyme B mRNA expression in the biopsies of the acute rejection group.	Krams et al. 1995
	Three subsets of T cells: $CD4^+CD7^+$ $CD8^+CD38^+$ $CD4^+CD25^+$	Liver	Biopsies and blood	Higher number of $CD4^+CD7^+$ and $CD8^+CD38^+$ cells in blood and liver of liver recipient with acute rejection. Decreased expansion of Treg cells ($CD4^+CD25^+$) in blood and liver in patients with acute rejection.	Perrella et al. 2006
	Hepatocyte-derived miRNAs (HDmiRs)	Liver	Blood, serum	Serum level of HDmiRs (miR-122, miR-148a, and miR194) are elevated up to 20-fold in patients with acute rejection. HDmiRs are promising candidates as biomarkers rejection, but also as hepatic injury, after liver transplantation.	Farid et al. 2012
	CCL4	Lung	Blood, PBMC	Analysis of copy number variation in genes. Higher number of CCL4 gene in patients with acute rejection.	Colobran et al. 2009
	Treg and cytotoxic markers	Lung	Bronchoalveolar lavage and blood	Higher mRNA expression for all of the markers in patients with acute rejection, but the expression is significantly increased only for CTLA-4, FoxP3, and granzyme B mRNA. But no difference have been observed in blood samples.	Madsen et al. 2010
	$CD4^+/CD8^+$ T cells NK NK-like T cells B cells Tregs Invariant receptor NK-T cells (iNKT)	Lung	Bronchoalveolar lavage fluid	Reduction of CD4:CD8 T-cell ratio in patients with acute lung rejection. Higher frequency of $CD4^+$ T cells expressing CD38 during rejection. Proportions of NK and Treg, as a frequency of total $CD3^+$ T cells, trended toward being reduced in rejection. Proportions of B cells, NK-like T cells, and the percentage of $CD4^-/CD8^-$ iNKT cells trended toward being increased in rejection.	Gregson et al. 2008

Continued

Table 1. *Continued*

Type of rejection	Molecules/cells	Transplant organ	Samples of analysis	Significance	References
Acute	$CD4^+/CD8^+$ T cells B cells Memory T cells	Lung	Bronchus-associated lymphoid tissues	Reduction of CD4:CD8 T-cell ratio is correlated with elevated acute rejection episode for patients in the first year of transplantation.	Shenoy et al. 2012
	CXCR1 CXCR2	Kidney	SNP genotype	Donor CXCR1-2668 GA/AA genotype is associated with acute rejection in kidney transplantation.	Ro et al. 2011
	CXCL10 CXCL13	Kidney	Blood, PBMC	Overexpression of CXCL10 and CXCL13 in PBMC of patients with acute rejection.	Mao et al. 2011
	Granzyme B Perforin FasL	Kidney	Blood, T lymphocytes	Significant higher perforin mRNA expression in patients with acute rejection. No consistent pattern of granzyme B and FasL expression.	Shin et al. 2005
	CD69 expression	Kidney	Blood, T lymphocytes	Higher percentages of $CD69^+$ cells in the $CD3^+$ and $CD8^+$ T-cell subsets of patients with acute rejection. Correlation between $CD69^+CD8^+$ T cells and acute rejection is the more clinically useful test (based on receiver-operator characteristics).	Posselt et al. 2003
	CXCL10	Kidney	Blood, serum	High serum level of CXCL10 is associated with acute risk of graft failure.	Rotondi et al. 2004
	CXCL9	Kidney	Blood, serum	High level of CXCL9 pretransplantation in recipient patients is a marker of high risk of acute rejection.	Rotondi et al. 2010
	Pattern of cytokines Granzyme B Perforin	Kidney	Blood, PBMC	High level of proinflammatory cytokines (IL-4, IL-5, IL-6, and IFN-γ) and cytotoxic molecules (perforin and granzyme B) are associated with acute rejection in kidney transplant recipient.	Dugre et al. 2000
	CXCL9 CXCL10	Kidney	Urine	Increased expression of CXCL9 and CXCL10 in urine of patients with acute rejection.	Jackson et al. 2011a
	Granzyme B Perforin	Kidney	Urine	Increased mRNA expression of granzyme B and perforin in urine of patients with acute rejection.	Li et al. 2001

Acute	Perforin Granzyme B FasL PI-9 FoxP3	Kidney	Urine	Increased mRNA signature expression in urine of patients with acute rejection. FoxP3 mRNA level is the most accurate and predictive for acute rejection with a sensibility and specificity of 100%.	Aquino-Dias et al. 2008
	miRNAs miR-210	Kidney	Urine	Down-regulation of miR-10b and miR-210 in patients with acute rejection (compared with controls). Up-regulation of miR-10a in patients with acute rejection (compared with controls). Expression of miR-210 is only one that changes before and after rejection episode. Low expression of miR-210 is correlated with acute rejection and prediction of long-term of kidney function.	Lorenzen et al. 2011
	CXCR3	Kidney	Biopsies	Increase number of CXCR3-positive cells in biopsies from patients with acute rejection and decrease with immunosuppressive treatment.	Hoffmann et al. 2006
	CX3CR1	Kidney	Biopsies	CX3CR1 intragraft staining is associated with worst outcome. CX3CR1 is expressed by macrophage and dendritic cells in acute rejection.	Hoffmann et al. 2010
	Pattern of chemokines	Kidney	Biopsies	Increase of inflammatory chemokines and chemokine receptors transcripts during acute rejection: CCL3/CCL5/CXCL9/CXCL10/CXCL11 and CCR5/CCR7/CXCR3. Rejection associated with Th1 polarization.	Lo et al. 2011
	miRNAs	Kidney	Biopsies	Up-regulation of seven miRNA and down-regulation of 10 miRNA samples of patients with acute rejection. Three samples (overexpressed) are predictive with a high accurancy of acute rejection: miR-142-5p, miR-155, and miR-223.	Anglicheau et al. 2009
Chronic	CCR4	Lung	Blood, PBMC	Reduction of $CCR4^+$ expression on $CD4^+$ memory T cells in patients with chronic rejection.	Paantjens et al. 2011
	CXCL9 CXCL10	Lung	Bronchoalveolar lavage fluid	Cumulative expression of CXCL9 and CXCL10 in bronchoalveolar liquid is a marker of graft failure.	Neujahr et al. 2012

Continued

Table 1 *Continued*

Type of rejection	Molecules/cells	Transplant organ	Samples of analysis	Significance	References
Chronic	HNPs (human neutrophil peptides 1–3)	Lung	Bronchoalveolar lavage fluid	High level of HNPs peptides by mass spectrometry in patients who have developed bronchiolitis obliterans syndrome (marker of chronic lung rejection). Validation by enzyme-linked immunosorbent assay (ELISA). Monitoring levels of HNPs can predate the clinical onset of disease up to 15 months.	Nelsestuen et al. 2005
	CCL2	Kidney	Urine	Early detection of CCL2 in urine is a marker of a development of IF/TA at 24 months.	Ho et al. 2010
	CXCL9 CXCL10	Kidney	Urine	Increased expression of CXCL9 and CXCL10 in urine of patients with chronic rejection.	Schaub et al. 2009
	CCR1 RANTES	Kidney	Biopsies	CCR1 and RANTES promote ongoing IF/TA and are linked to worst outcome.	Dikow et al. 2010
	Immunoproteasome β subunit 10	Kidney	Blood and biopsies	Immunoproteasome β subunit 10 mRNA expression is specifically up-regulated during chronic active antibody-mediated rejection.	Ashton-Chess et al. 2010
	Tribbles-1, -2, and -3	Kidney	Blood, biopsies and urine	Only mRNA expression of Tribbles-1 is significantly increased in blood and graft biopsies of kidney transplant recipients with chronic antibody-mediated rejection compared to patients with other histological and clinical diagnoses. In peripheral blood, mononuclear cell mRNA Tribbles-1 expression discriminates patients with chronic antibody-mediated rejection from those with other types of late allograft injury with high sensitivity and specificity. Tribbles-1 was found to be a potential blood and tissue biomarker of chronic antibody-mediated rejection.	Ashton-Chess et al. 2008

Acute and chronic	CXCL10	Kidney	Blood, serum	High serum levels of CXCL10 is associated with both acute rejection and chronic allograft nephropathy.	Lazzeri et al. 2005
	CXCL10 MIG MIP-1 OPG	Kidney	Urine	Chemokines screening using Luminex multiplex assay. CXCL10 and MIG: markers of acute rejection, but not chronic rejection and stable graft function. MIP-1 (Δ) and OPG: markers of both acute and chronic rejection, but not stable graft function.	Hu et al. 2009
	FoxP3 T-bet Granzyme B	Kidney	Blood, PBMC and biopsies	mRNA expression of FoxP3 and T-bet increase in biopsies, but not in PBMC samples of patients with acute rejection. mRNA expression of granzyme B increased in biopsy samples of patients with acute (cellular) and chronic (antibody-mediated) rejection. mRNA expression of granzyme B in PBMC is significantly reduced for patients with chronic rejection. Granzyme B, but not FoxP3 and T-bet, should be considered as a biomarker of chronic rejection.	Ashton-Chess et al. 2009

IF/TA, interstitial fibrosis and tubular atrophy.

urine of a kidney transplant recipient was shown to correlate with incidence of acute rejection episodes (Matz et al. 2006). Moreover, high levels of CXCL3 (Tatapudi et al. 2004) and CD103 mRNA (Ding et al. 2003) were present in a recipient undergoing acute rejection. Nevertheless, chemokines are not the only markers of rejection. In fact, other proteins, such as granzyme B and FoxP3, miRNAs, and cells could also be used as predictive biomarkers of rejection (Table 1).

Nonspecific Effector Cells (NK Cells, Macrophages, Monocytes, DCs, and Neutrophils)

Despite the finding that NK cell detection in the graft infiltrate was a negative prognostic factor (Sorrentino et al. 2006), this cell type does not appear to be necessary or sufficient to mediate allograft rejection. Some recent studies have even highlighted the involvement of NK cells in the phenomenon of tolerance (van der Touw et al. 2012). NK cells seem to act mainly as a bridge between innate and adaptive immune responses. This has been underlined by experiments using CD28-deficient mice. For example, by depleting NK1.1$^+$ cells using antibodies in CD28-deficient mice, some groups showed that absence of NK cells reduce cardiac allograft rejection in mice (Maier et al. 2001; McNerney et al. 2006). In this model, NK cells, but not NKT cells, were required for rejection by promoting T-cell expansion and effector function (McNerney et al. 2006). This effect was independent of the NK-cell activator NKG2D receptor.

In contrast, several studies have highlighted the role of the NKG2D receptor in rejection. In a model of cardiac transplantation in CD28-deficient mice, blocking NKG2D prevented graft rejection (Kim et al. 2007). Zhang and coworkers confirmed the involvement of NKG2D in graft rejection as they showed that the NKG2D receptor present on NK cells recognizes renal tubular epithelial cells, leading to their killing by a perforin-dependent pathway during ischemia reperfusion injury (Zhang et al. 2008). Promotion of CD4$^+$ T-cell activation by the indirect allorecognition pathway is another mechanism of action of NK cells investigated by NK-cell depletion and NKG2D blockade in a murine skin graft model (Ito et al. 2008). Two ligands of NKG2D, namely RAEL1 and RRLT, were found to be expressed in rat liver allograft and to be linked to acute rejection (Zhuo et al. 2010). Moreover, in human kidney transplant biopsies, an elevated level of NKG2D mRNA was also associated with acute rejection (Seiler et al. 2007). NK cells produce a variety of cytokines, including IFN-γ, TNF-α, and TGF-β, as well as colony-stimulating factors (Cuturi et al. 1987, 1989; Anegon et al. 1988). IFN-γ produced by NK cells is known to induce an up-regulation of MHC class I and class II molecules by endothelial cells leading to their killing by CD8$^+$ T cells (McDouall et al. 1997; Ayalon et al. 1998). More recently, a role for IL-15 in NK-cell-mediated mechanisms was shown, as in vivo stimulation of NK cells with IL-15 was found to induce skin graft rejection in RAG$^{-/-}$ mice (Kroemer et al. 2008). Last, an interesting effect of NK cells on donor DCs has been described. Coudert et al. (2002) showed that interaction of NK cells with donor DCs in the absence of CD8$^+$ T cells promoted CD4$^+$ T-cell priming and regulated the Th balance. In a more recent study performed in a model of CD4$^+$ T-cell-mediated allogeneic skin graft rejection, NK cells killed donor DCs in draining LNs by a perforin-dependent pathway that avoided T-cell activation (Laffont et al. 2008) thereby confirming that NK cells favor indirect allorecognition (Garrod et al. 2010).

Macrophages constitute another innate cell population that contributes to the tissue damage observed in rejected organs. In humans, macrophages were found to represent 40%–60% of the cellular infiltrate in renal allotransplants (Hancock et al. 1983). A large number of monocytes were also detected in infiltrates from the graft biopsies of kidney transplant recipients under T-cell-depleting therapy (anti-CD52 and FK506 monotherapy) (Salama et al. 2007). These cells were recruited to inflammatory sites by monocyte chemoattractant protein-1 (MCP-1). Blockade of MCP-1 was shown to increase pancreatic islet transplant survival in rodents (Lee et al. 2003). In addition, blockade of macrophage colony-stimulating factor-reduced macrophage proliferation and accumulation in the graft leading to a decrease in the severity of kidney transplant rejection in mice (Jose et al. 2003). Other studies have confirmed in macrophages a decreased infiltration and an increase in graft survival in murine models of both kidney (Qi et al. 2008) and heart (Takeiri et al. 2011) transplantation. Furthermore, in the context of human kidney transplantation, CX3CR1 expression by macrophages was associated with acute rejection, and was a negative prognostic factor in human kidney graft (Hoffmann et al. 2010).

There are several macrophage subsets, and specialized tissue-resident macrophages are widely distributed across the body including the liver (Kupffer cells) and bone (osteoclasts). Macrophages can also display heterogeneity in terms of their functional phenotype as well as their activation state. In the same terms, activated macrophages are classified as classically or alternatively activated macrophages, named M1 and M2 macrophages, respectively, referring to T-cell nomenclature (Mantovani et al. 2004).

M1 macrophages are activated by a combination of IFN-γ and TNF released by Th1 cells. M1 macrophages have the capacity to present antigen to T cells and express high levels of costimulatory molecules (CD86 and CD80) (Mantovani et al. 2004). These cells also express large amounts of the proinflammatory cytokine IL-12 and promote the development of the Th17 immune response by secreting IL-1, IL-6, and IL-23. M1 macrophages are thought to play a role in the defense against bacteria or viruses as well as tumor resistance. However, in several conditions, this macrophage subset can take part in chronic inflammation and autoimmune diseases (Mosser and Edwards 2008).

M2 macrophages, on the other hand, are associated with immunoregulation. Macrophage phenotype and function are thought to differ according to the chemokine environment at the time of their activation (Mantovani et al. 2004). In fact, macrophages develop a regulatory profile when activated by IL-10 or immune complexes associated with a second stimulus. These cells act as regulatory cells by producing high levels of IL-10 and other anti-inflammatory cytokines such as TGF-β, for example (Mosser and Edwards 2008). Alternatively, if resident macrophages are activated in the presence of IL-4 and IL-13, they are considered as M2a, or wound-healing macrophages, as they promote tissue repair. The latter cells are characterized by an up-regulation of mannose receptor and production of polyamine (Mantovani et al. 2004; Mosser and Edwards 2008).

However, depending on the context, activated macrophages are able to switch their phenotype and change their function too (Mosser and Edwards 2008). One example of this is the differentiation of M1 to hybrid macrophages that share the characteristics of both regulatory and wound-healing macrophages. Unfortunately, these cells are tumor-associated macrophages and inhibit antitumor immunity (Duluc et al. 2007). Another example is in the context of the autoimmune disease diabetes, in which adipose-tissue-associated macrophages with a wound-healing phenotype polarize toward a phenotype similar to classically activated macrophages (Lumeng et al. 2007).

A recent study performed in a rat model of lung transplantation showed that $CD68^{+}$ macrophages were the most abundant cell type observed during acute rejection, whereas $CD163^{+}$ cells were distributed around vessels and bronchioles (Jungraithmayr et al. 2010). Regarding their tissue localization, macrophage infiltration was significant in the interstitium, but also in the arterial intima of vascularized grafts (Sun et al. 2011). Infiltration was also observed in the glomeruli of kidney transplants.

In contrast to the arterial intima in which no correlation between antibody-mediated rejection (AMR) and macrophage infiltration was shown (Kozakowski et al. 2009), macrophages/monocytes were associated with AMR when present in the glomeruli. Moreover, presence of focal or diffuse C4d deposits was associated with macrophage/monocyte infiltrates, whereas T cells predominated when no C4d deposits were detected (Magil 2005; Magil and Tinckam 2006). These results were confirmed by Fahim et al. (2007), who showed an abundant recruitment of monocytes to peritubular and glomerular capillaries during humoral rejection. Moreover, in a murine model of hepatocellular allotransplantation in which rejection is mediated by a CD4-dependent alloantibody response, depletion of macrophages delayed hepatocyte rejection and inhibited in vivo allocytotoxicity (Horne et al. 2008).

Polymorphonuclear cells are also efficient producers of chemokines (Molesworth-Kenyon et al. 2005). As described in the previous section, chemokines act on T-cell activation, proliferation, and function. One type of polymorphonuclear cell, the neutrophil, was detected in large numbers among the cells infiltrating murine skin allografts during acute rejection (Celli et al. 2011). Neutrophils have also been shown to contribute to allograft rejection in models of costimulation blockade (El-Sawy et al. 2005), or in the absence of IFN-γ (Miura et al. 2003) and IL-4 (Surquin et al. 2005). There is also evidence of eosinophil involvement in models of rejection and their activation was shown to be dependent on a Th2 CD4$^+$ T-cell response (IL-4-dependent) (Braun et al. 2000; Surquin et al. 2005). Eosinophils were also abundant in intestinal graft infiltrates on T-cell elimination (Wu et al. 2006).

Recently, the role of platelets has been shown in allograft rejection, as these cells are able to induce immune stimulation. In this context, platelets are believed to act by recruiting mononuclear cells by secreting cytokines/chemokines and by stimulating monocytes, macrophages, and T cells by interaction with them via P-selectin/PSGL-1 or CD40/CD40L pathways (Xu et al. 2006; Kirk et al. 2009).

Last, endothelial cells are also able to induce allogeneic CD4$^+$ and CD8$^+$ T-cell activation and proliferation, favoring IFN-γ and IL-2 secretion and leading to graft rejection (Epperson and Pober 1994). In a recent review, Taflin et al. (2011, 2012) described how in the context of inflammation, such as observed in transplantation, endothelial cells are activated by proinflammatory cytokines and TLR ligands. This activation leads to memory CD4$^+$ T-cell proliferation and Th1/Th17 expansion by endothelial cell secretion of proinflammatory cytokines and chemokines (such as CXCL10, IL-6, and IL-1α) and overexpression of costimulatory molecules (such as LFA-3 or OX-40L) or adhesion molecules (such as ICAM-1) (Griffin et al. 2012). In humans, three nonclassical MHC class I antigens exist: HLA-E, HLA-F, and HLA-G. Coupel et al. (2007) showed that HLA-E is expressed only by leukocytes, such as NK cells, B and T lymphocytes, macrophages, and by endothelial cells. Moreover, they show that endothelial cells up-regulate HLA-E in inflammatory conditions, and endothelial cells produce a soluble form of HLA-E. Coupel's group additionally showed that increased expression of the membrane-bound form of HLA-E protected IFN-γ-activated endothelial cells from CD94/NKG2A-mediated cytolysis, whereas the soluble form of HLA-E protected other cell types. Endothelial cells expressed the MHC class-I related chain A, MICA, which is a ligand for NKG2D. NK cells and cytotoxic T cells express this activating receptor. Under inflammatory stimulation by TNF-α, human endothelial cells up-regulate the expression of MICA (Lin et al. 2012). These results are consistent with previous data suggesting that MICA is expressed by kidney macrovascular endothelial cells in kidney transplant recipients (Sumitran-Holgersson et al. 2002). Moreover, this study showed that MICA acts as an antibody target, which confirmed anti-MICA antibodies are associated with the reduction of graft survival (Zou et al. 2007).

Acute Cellular Response

T-Cell Activation and Effector T Cells

Effector T cells can mediate cell lysis either via cytokine or chemokine secretion, which induces necrosis of transplant tissue, or via direct contact with epithelial or endothelial cells and various mechanisms of cytotoxicity. However, T cells can also stimulate B cells to initiate humoral rejection, or other cells such as macrophages or neutrophils, whose role in transplant rejection is described above.

On activation by DCs, naïve alloreactive CD4$^+$ T cells can differentiate into T helper cells, including Th1, Th17, Th2, or into regulatory T cells (Tregs). In a proinflammatory environment, naïve CD4$^+$ T cells differentiate mainly into Th1 and Th17 cells. Th1 cells produce IFN-γ and IL-2 and are involved in cytotoxic T lymphocyte (CTL) priming, stimulation of the humoral response, and activation of other cell types such as NK cells. IL-17 produced by Th17 cells stimulates the production of inflammatory cytokines and chemokines leading to the recruitment of neutrophils and macrophages to the graft. In a Th1-depleted environment, Th17 cells (Yuan et al. 2008) or Th2 cells (Barbara et al. 2000) can efficiently promote graft rejection. Th2 cells release anti-inflammatory cytokines such as IL-10 and IL-4 T follicular helper cells and Th9 helper cells have not yet been described in detail, but they have been shown to induce B-cell maturation and mast cells recruitment, respectively.

In contrast, $CD4^+$ Tregs have been shown to be involved in the tolerance process and to prevent graft rejection in mice (Kingsley et al. 2002). These cells are able to suppress $CD4^+$ and $CD8^+$ effector T cells and can also target APCs, decreasing their capacity for antigen-presentation and costimulation (Shevach 2009). A recent study performed using humanized mice showed the superiority of donor alloantigen-specific $CD4^+$ Tregs over their polyclonal counterparts to suppress alloimmune responses (Sagoo et al. 2011). In tolerant liver transplant patients, an increase in peripheral blood $CD4^+$ Tregs has been observed by several groups (Li et al. 2004; Martinez Llordella et al. 2007; Pons et al. 2008; Nafady-Hego et al. 2010). Although this Treg signature does not seem to be present in the peripheral blood of kidney transplant patients, an increase in intragraft Tregs was detected in such patients (Bestard et al. 2007).

$CD8^+$ T cells can differentiate into CTLs, which themselves can be classified as Tc1, Tc2, or the more recently described Tc17 (Yuan et al. 2009). Although, naïve $CD8^+$ cells can only be activated by DCs, $CD4^+$ T cells have been shown to facilitate $CD8^+$ T-cell differentiation either by cell-to-cell contact or by secretion of IL-2/IFN-γ cytokines. In the latter case, both APCs and $CD4^+$ helper T cells have been shown to be involved in CTL differentiation (Ridge et al. 1998). CTLs subsequently migrate to the graft where they recognize target cells via their allogeneic MHC class I molecules. Killing by CTLs is mediated mainly by the secretion of perforin and granzyme B, or by the Fas/FasL pathway. These two pathways induce target cell apoptosis. CTLs are also able to secrete soluble mediators such as TNF-α. In terms of the dynamics of CTL action, in a model of skin graft rejection, an accumulation of CTLs surrounding the graft was detected, followed by an early phase of killing in small areas of the dermis/epidermis junction. The CTLs were shown to disseminate to the whole graft and induce tissue destruction (Celli et al. 2011).

Compared to their naïve counterparts, memory T cells induce a faster and more effective immune response with lower antigen stimulation (Rogers et al. 2000; Veiga-Fernandes et al. 2000). This is supported by data in humans showing that the pretransplant frequency of donor-specific memory cells correlates with the posttransplant risk of developing acute rejection episodes (Heeger et al. 1999). These results were confirmed in liver transplant patients in whom the presence of a high number of $CD8^+$ cells with a memory phenotype before transplantation was associated with reduced graft survival (Tanaka et al. 2006). More recently, in a model of kidney allotransplantation in nonhuman primates, Nadazdin et al. (2011) showed that the presence of memory T cells before transplantation was a barrier to transplantation tolerance.

Despite $CD8^+$ Tregs being of considerable interest in transplantation, their subsets and mechanisms of action are less understood in comparison to their $CD4^+$ Treg counterparts (Guillonneau et al. 2010). Understanding the function and mechanisms of action of the different $CD8^+$ Treg subsets is essential. In rodent models, several studies have shown the role of $CD8^+$ Treg subsets in the suppression of transplant rejection. For example, donor-antigen-specific $CD8^+CD45^{low}$ Tregs were shown to develop in a fully mismatched heart allograft model in rat on treatment with CD40Ig (Guillonneau et al. 2007). The therapeutic potential of these regulatory cells has also been highlighted in human studies; in liver and cardiac transplant recipients, expansion of $CD8^+CD28^-$ Tregs was shown to correlate with low exposure to immunosuppressive drugs and a decrease in acute and chronic allograft rejection (Guillonneau et al. 2010).

Memory T Cells

The presence of memory T cells able to react with alloantigens was shown in healthy individuals by Lombardi's group in 1990 (Lombardi et al. 1990). Memory T cells showed a faster proliferation, in accordance with observations in vivo during secondary-type immune responses (Akbar et al. 1990). Memory CD8$^+$ T cells act as CTLs as they are able to directly kill target cells, mainly via the granzyme/perforin pathway (Barber et al. 2003).

Two distinct subsets of memory T cells have been identified, and are termed central and effector memory cells (Brook et al. 2006). Central and effector memory cells are CD45RO$^+$ CD45RA$^-$, but differ by their expression of the CD62L/CCR7 markers. Contrary to central memory cells, effector memory cells express more β_1 and β_2 integrins. Moreover, the two subsets differ in their localization because central memory T cells express intermediate levels of CCR4 and CCR6, whereas effector memory T cells express high levels of CCR1, CCR3, and CCR5. Although central memory T cells require restimulation, this subset is more efficient in their killing function (Sallusto et al. 1999; Barber et al. 2003; Hengel et al. 2003).

In sensitized patients, the presence of memory T cells specific for alloantigens can be caused by prior blood transfusion, pregnancy, or transplant rejection. The presence of allospecific memory T cells is more difficult to explain in nonsensitized individuals, but may be the result of heterologous immunity or homeostatic proliferation.

Heterologous immunity is defined as the ability of alloreactive memory T cells specific for a microbial antigen to cross-react with allogeneic MHC molecules through direct allorecognition. This affinity of the T-cell receptor for allogeneic MHC molecules can be higher than that for self-antigens. Heterologous immunity has been shown both in mice and in humans (Smith et al. 2012). Homeostatic proliferation occurs in lymphopenic conditions (such as those caused by certain immunotherapies) and in the absence of an antigenic stimulus. In these conditions, peripheral T cells rapidly proliferate and acquire a memory phenotype that spontaneously skews toward a Th1 phenotype (Moxham et al. 2008). Certain cytokines, such as IL-7, have been shown to mediate this homeostatic proliferation (Schluns et al. 2000).

Memory T cells generated in the context of viral infections or by homeostatic proliferation may prevent tolerance induction, as shown using well-established rodent models of tolerance (Valujskikh et al. 2002; Zhai et al. 2002; Wu et al. 2004). These results show that memory T cells can act as barriers to tolerance induction. It is therefore necessary to develop therapies to prevent memory T-cell generation or to eliminate such cells. The effect of immunosuppressive drugs on memory T cells has started to be studied. So far, it is known that CD4$^+$ memory T cells are resistant to steroids, deoxyspergualin, and sirolimus. In contrast, calcineurin inhibitors tacrolimus and cyclosporine A inhibit their proliferation and in vitro activation (Pearl et al. 2005). Optimal activation of memory T cells requires costimulatory molecules, which are different from those acting on naïve cells, such as ICOS/ICOSL, OX40/OX40L, and CD27/CD70 (Croft 2003; Wu et al. 2004). Consequently, blockade of the CD27/CD70 pathway prolongs the survival of heart transplant in mice (Yamada et al. 2005). Moreover, IL-7 inhibition blocks allograft rejection mediated by memory T cells without affecting Tregs (Wang et al. 2006).

Acute Humoral Response

Although T cells have been considered for many years as playing a predominant role in graft rejection, it is now known that the acute humoral response is the main cause of acute graft loss (Terasaki 2003). In fact, acute humoral rejection accounts for 15% to 20% of graft rejection within the first posttransplant year, despite immunosuppressive therapies (Montgomery et al. 2004; Lucas et al. 2011).

The criteria for the diagnosis of acute humoral rejection of kidney transplants were initially established by the Banff working group in 2003 (Racusen et al. 2003). At least one of the following criteria must be present for a diagnosis of acute humoral rejection to be made: (1) morphological criteria: presence of neutrophils and/or monocytes/macrophages in peritubular capillaries (PTC) and/or glomeruli, fibrinous arterial necrosis, thrombosis in the glomerular capillaries, arterioles, and/or small arteries and acute tubular damage; (2) immunohistological criteria: C4d deposits in PTC, presence of immunoglobulins, or complement in the fibrinous necrosis of the arteries; (3) serological criteria: presence of circulating anti-donor (HLA and non-HLA) antibodies called donor-specific antibodies (DSA).

These criteria were optimized during the 9th Banff conference by including the quantification of C4d deposition (Solez et al. 2008).

Generally, detection of DSA and C4d deposits is associated with an increase in AMR and an inferior graft outcome (Kedainis et al. 2009). Nevertheless, although the presence of C4d deposits in kidney biopsies appears to be a good marker of AMR, during the last Banff meeting it was agreed that AMR can also be diagnosed in the absence of C4d detection (Mengel et al. 2012).

The new objectives of the Banff committee are therefore to characterize AMR by defining: (1) thresholds for microvascular injury and for DSA in AMR, noting that all injuries are specific and all patients with DSA develop AMR, (2) C4d-negative AMR before adding this category to the Banff classification, (3) acute versus chronic AMR, and (4) the significance of intimal arteritis in the absence of necrosis as a histological criteria (Mengel et al. 2012).

B Cells

The importance of B cells in allograft rejection was shown using B-cell-deficient mice. Several groups, including Brandle and coworkers, as well as Wasowska and colleagues showed that acute allograft rejection was delayed in B-cell-deficient mice and could be restored by the administration of hyperimmune sera (Brandle et al. 1998; Wasowska et al. 2001). In humans, the primary role of B cells as amplifiers of the alloresponse is suggested by the positive reports of B-cell depletion therapy (rituximab) as induction for ABO-incompatible renal transplant recipients (Tyden et al. 2009). However, rituximab-induction therapy was reported to lead to acute cellular rejection in some kidney transplant recipients (Clatworthy et al. 2009). It is therefore thought that B-cell depletion may be quicker at depleting the B cells that contribute to tolerance than those that promote rejection, thereby temporarily amplifying the alloresponse.

The ability of B cells to present antigen and to produce antibodies and cytokines means that these cells can take part in allograft rejection by different mechanisms. B cells have the ability to promote T-cell activation/differentiation through costimulatory pathways, or cytokine release (Constant 1999; Zarkhin et al. 2010). B cells have also been shown to promote alloreactive T-cell differentiation into memory T cells (Ng et al. 2010). Moreover, a deficiency in B-cell antigen presentation can prolong graft survival (Noorchashm et al. 2006).

More recently, several groups suggested that B cells might be involved in tolerance. For example, Deng and coworkers showed that anti-CD45RB-induced tolerance to heart transplants requires the presence of B cells (Deng et al. 2007). Furthermore, resting B cells expressing donor antigen indefinitely prolonged heterotopic heart transplant survival (Niimi et al. 1998). In rats, long-term allograft tolerance is characterized by an accumulation of B cells expressing genes such as BANK-1, which are associated with tolerance (Le Texier et al. 2011). In addition, anti-CD20-mediated depletion of B cells in mice was shown to accelerate skin graft rejection (DiLillo et al. 2011). A specific population of B cells expressing TIM-1 was identified as having regulatory properties in mouse models of islet transplantation (Ding et al. 2011). Furthermore, a B-cell signature was described in immunosuppressant-free transplant patients who were spontaneously tolerant to their HLA mismatched kidney transplant (Newell et al. 2010; Pallier et al. 2010; Sagoo et al. 2010; Silva et al. 2012). In contrast, spontaneously tolerant liver transplant recipients were reported to display an NK cell signature (Martinez-Llordella et al. 2008). These results were confirmed by the analysis of PBMC from kidney and liver transplant patients by the same laboratory, showing that the B-cell signature was only detectable in kidney in immunosuppressant-free kidney transplant patients (Lozano et al. 2011).

Antibodies

Halloran's group showed the role of alloantibodies in acute humoral rejection in the 1990s. They described a correlation between the prognosis of acute rejection and the production of antidonor antibodies after transplantation (Halloran et al. 1990). In presensitized patients, the presence of DSA at the time of transplantation was identified as a high risk factor for AMR and patients who develop anti-HLA DSA tend to have inferior long-term graft survival compared to those that do not (Lefaucheur et al. 2008, 2010). In support of these results, a recent study by Cooper et al. (2011) also showed that patients who develop de novo DSA after transplantation have an inferior graft outcome.

Alloantibodies that develop against the donor organ can recognize several types of antigens (for review, see Dragun 2008): HLA antigens class I and II (Terasaki and Ozawa 2004; Terasaki and Cai 2005), MICA and MICB antigens (MHC class I-related molecules A and B) (Zou et al. 2007; Li et al. 2010), minor histocompatibility antigens and non-HLA antigens including the angiotensin II type 1 receptor (Dragun et al. 2005), vimentin (Mahesh et al. 2007), myosin, the ABO blood group antigens (Montgomery et al. 2012), perlecan, type IV and VI collagen, agrin, unknown endothelial antigens (Jackson et al. 2011b), and ICAM-1 (Lawson et al. 2005).

Alloantibodies mainly induce AMR by complement-dependent mechanisms. In fact, the alloantibodies that fix C1q complement are associated with a worse outcome that does not fix complement (Yabu et al. 2011). A greater risk of acute humoral rejection was clearly shown when C4d deposits were associated with DSA and, more specifically, anti-HLA class I but not anti-HLA class II antibodies (Cosio et al. 2010). However, in the absence of C4d staining, association of DSA with altered expression of endothelial genes was also found to be a marker of AMR in kidney patients, as the alloantibodies modified the microcirculation (Sis et al. 2009). NK cells and macrophages were also involved in AMR in patients with DSA. On stimulation by alloantibodies, NK cells cause endothelial injury and act as effector cells by killing target cells through the antibody-dependent cellular cytotoxicity pathway (Hidalgo et al. 2010). The role of NK cells is particularly well-defined as NKG2D is a receptor for stress-inducible MICA and, as described previously, this molecule is also highly efficient at stimulating NK cells and certain T-cell subsets (Bauer et al. 1999).

Several desensitizing protocols have been tested to reduce the DSA that cause AMR, including plasmapheresis or IVIg, as well as treatments targeting complement C5 molecules, proteasome, or $CD20^{+}$ cells (Gueler et al. 2008; Raedler et al. 2011; Yoo et al. 2012). A combination of these treatments was shown to improve graft survival outcome (Lefaucheur et al. 2009; Montgomery et al. 2011).

CHRONIC REJECTION

The Origins and Mechanisms of Chronic Rejection

Since the early 1980s, 1-year graft survival of kidney allografts has increased significantly, reaching more than 90% (Hariharan et al. 2000; Pascual et al. 2002). Nevertheless, the graft survival in the long term has changed minimally, and the percentage of grafts lost annually after the first year has not changed (Meier-Kriesche et al. 2004). Death with a functioning graft and chronic allograft nephropathy (CAN) are the main causes of graft loss (Halloran et al. 1999; Ojo et al. 2000; Matas et al. 2002; Pascual et al. 2002). The predominance of CAN is 60%–70% in the first year posttransplant (Solez et al. 1998; Nankivell et al. 2003). CAN is an entity that encompasses several different mechanisms that include interstitial fibrosis and tubular atrophy. The natural history of CAN suggests that it can result from both immunological and nonimmu-

 Cite this article as *Cold Spring Harb Perspect Med* doi: 10.1101/cshperspect.a015461

nological phenomena, including calcineurin inhibitors (CNI) toxicity (Nankivell et al. 2003). Chronic dysfunction is also observed in other organ transplants and is often responsible for premature graft loss.

In recent years, significant efforts have been made to describe and classify CAN. In 1990, analyses based in histological studies were performed in the field of kidney transplantation, resulting in the description of a classification system for chronic allograft dysfunction (Classification "Banff 97") (Racusen et al. 1999). This classification is revisited regularly to incorporate the latest scientific data and to try to differentiate lesions associated with chronic rejection (CR) from those observed in response to cardiovascular risk factors or CNI toxicity. The 2005 Banff meeting differentiated the lesions suggestive of chronic allograft rejection (including antibody-dependent complement activation lesions) and cell arteritis from the less specific lesions of interstitial fibrosis/tubular atrophy (IF/TA) (Solez et al. 2007, 2008). IF/TA lesions can appear very early after transplantation. At 1 year posttransplant, >80% of kidneys have minimal lesions of IF/TA that will deteriorate over time, reaching >50% of kidneys with severe lesions at 5 years. The 2011 Banff meeting focused on refining the diagnostic criteria for AMR. As such, C4d-negative AMR was recognized, in which NK cells and endothelial activation were proposed to play a role (Mengel et al. 2012).

The Histological Lesions

Significant progress has been made in understanding the mechanisms involved in the onset of CR, with the description of arteritis lesions in animal models (Yuan et al. 2002). One predominant observation made during chronic allograft rejection is an increase in the thickness of the intima, resulting in a decrease in vessel caliber with destruction of the internal elastic lamina. This thickening is also because of an accumulation of extracellular matrix and proliferation of myofibroblasts (Pedagogos et al. 1997; Pilmore et al. 2000; Ramirez et al. 2006). An accumulation of macrophages and $CD4^+$ T cells has been observed at the periphery of the vessels (Thaunat and Nicoletti 2008; Thaunat et al. 2005, 2006, 2008), whereas $CD8^+$ T cells are rarely present. The involvement of T cells in the pathogenesis of CR has been investigated through the study of genetically invalidated mice. In a mouse model of cardiac allotransplantation in which the genes encoding CD40 or CD40L were invalidated, CR was not observed, suggesting that activated T lymphocytes are required to initiate the phenomenon of CR. Nevertheless, in a rat model of cardiac allotransplantation in which animals were treated with CD40Ig (a molecule interfering with CD40-CD40L interaction), acute rejection was successfully inhibited, but CR occurred in long-term-surviving animals (Guillot et al. 2002). In this same model, the development of CR was prevented by simultaneous blockade of other costimulation pathways, such as RANK and ICOS (Guillonneau et al. 2004, 2005).

In some cases, the graft infiltrate has been found to adopt an organization resembling that of ectopic lymphoid tissues, with the infiltrating B cells capable of producing antidonor antibodies. These lesions of tertiary lymphoid tissue have been identified in kidney transplants, and the process is referred to as lymphoid neogenesis, which is analogous to the ontogenic program triggered during embryo development, the ectopic germinal centers (eGCs) (Thaunat et al. 2010; Cheng et al. 2011). The eGCs participate in the mechanisms of CR maintaining local memory T and B cells capable of synthesizing cytokines and antibodies. Interestingly, there is minimal overlap between the intragraft humoral response, suggesting a local antibody repertoire. Recently, Thaunat and coworkers showed that a breakdown of B-cell tolerance occurs within the graft during CR (Thaunat et al. 2012). It is thought that the micrograft environment interferes with peripheral deletion of autoreactive immature B cells that, in turn, produce autoantibodies. Recent studies have shown an important role for autoimmunity in the pathogenesis of CR suggesting a cross talk between the alloimmune response and autoimmunity (Sarma et al. 2012). Although all types of Th polarization profiles can lead to terminal CR, Thaunat and coworkers (Deteix et al. 2010) reported on a correlation between shorter graft survival and the presence of Th17 cells that produce IL-17 and IL-21 in human transplant patients. The latter group hypothesized that IL-21 promotes intragraft lymphoid neogenesis, which in turn supports the development of a local humoral immune response. Moreover, Burlingham et al. (Burlingham et al. 2007; Keller and Burlingham 2011) showed that the alloimmune response is partly responsible for the development of a Th17 autoimmune response. In this context, Fukami et al. (2009) showed that an anti-IL-17 therapy decreases the production of autoantibodies and the development of CR in a murine model.

The Role of Antibodies and Endothelial Cells

The involvement of antibodies directed against the graft in chronic rejection has been suggested by several groups showing a negative correlation between the appearance of antidonor antibodies and graft survival (Mao et al. 2007a,b; Sis et al. 2007). The emergence of new techniques to identify antibodies by fluorescence-activated cell sorting, flow cytometry, ELISA (enzyme-linked immunosorbent assay), and the most sensitive Luminex technique (flow cytometry with microbeads coated with peptides) now enables a better analysis of the humoral component (Pei et al. 2003). Moreover, recent data showing the existence of deposits during CR suggest the involvement of anti-HLA antibodies with the ability to activate complement (Nickeleit et al. 2002; Regele et al. 2002). However, observations from ABO-incompatible transplants, a context in which complement activation occurs, suggest the existence of adaptation mechanisms, allowing endothelial cells to resist complement activation (Gloor et al. 2006).

However, the effects of antibodies are diverse as described in Figure 2, and not limited to complement activation. Antibodies can also bind to the surface molecules of target cells and recruit other cells by interactions via their constant domain (Rebellato et al. 2006; Won et al. 2006). Recently, Hirohashi et al. described a new role for NK cells in chronic allograft rejection in which donor-specific antibodies interact with NK cells in the absence of complement and trigger transplant vasculopathy (Hirohashi et al. 2012; Li and Baldwin 2012). Last, there is some in vitro data showing that the culture of endothelial cells with antidonor antibodies results in

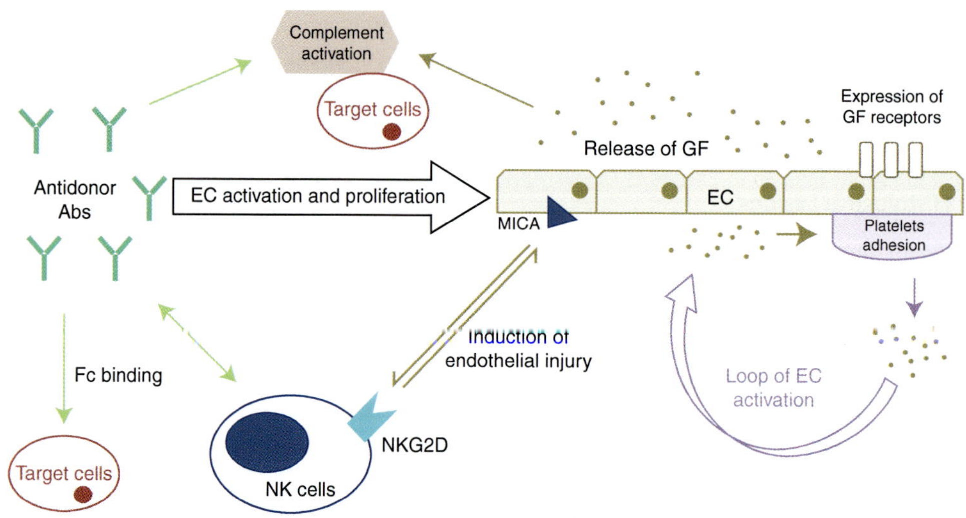

Figure 2. Induction of endothelial cells activation by antidonor antibodies. Antidonor antibodies (Abs) are known to induce chronic allograft rejection by several mechanisms of action involving their constant domain (Fc) or their capacity to induce activation of complement. Antidonor Abs also induce endothelial cell activation. Activated ECs secrete notably growth factors (GF) that induce the recruitment and activation/proliferation of several cells implicated in allograft rejection.

the activation and proliferation of endothelial cells (Bian and Reed 1999).

This activation step of endothelial cells is associated with the expression of various receptors on their cell surface (PDGF-R, EGF-R, FGF-R), as well as the synthesis of numerous growth factors (PDGF, EGF, FGF, VEGF, TGF-β, etc.) and synthesis of endothelin I (Bian and Reed 2001; Chen et al. 2001; Rossini et al. 2005). The local presence of growth factor is increased by the platelet adhesion that occurs during endothelial-cell activation, resulting in the generation of a local amplification loop with the release of many growth factors such as PDGF and TGF-β (MacDermott 1996; Yang et al. 2005). This phase of endothelial-cell activation promotes the stimulation of smooth muscle cells via release of endothelin I, and indirectly via the local synthesis of angiotensin II. It also causes local recruitment of inflammatory cells through the release of chemokines (MCP-1, IP-10), local activation of complement, local coagulation activation, promoting platelet adhesion, and releasing thromboxane A2. Finally, endothelial-cell activation stimulates the differentiation and proliferation of myofibroblast cells that synthesize the extracellular matrix involved in CR (Abbate et al. 2002; Cogan et al. 2002; Dewald et al. 2005; Li et al. 2007; Frangogiannis 2008; Haurani et al. 2008; Kennard et al. 2008; Wynn 2008). Myofibroblastic cells are important constituents of CR as they infiltrate the vessel wall and interstitial space. These cells express different cytoskeletal proteins (vimentin, α smooth muscle actin) as well as the myosin light chain, but do not express markers of lymphoid or epithelial cells (E-cadherin, ZO-1) (Badid et al. 2002). The origin of these cells is diverse. They originate from circulating stem cells capable of differentiating into endothelial cells or myocardiocytes (Direkze et al. 2003; Li et al. 2007). They can also originate from the transdifferentiation of endothelial cells or from renal tubular epithelial cells that transdifferentiate into myofibroblasts (Gressner 1996; Sommer et al. 2005; Hertig et al. 2006). The mechanisms involved remain poorly understood. However, the changes induced by ischemia/reperfusion or during CR appear to promote epithelial-mesenchymal transdifferentiation. TGF-β was also shown to be important for the promotion of epithelial-mesenchymal transdifferentiation (Fan et al. 1999; Mezzano et al. 2003; Lindert et al. 2005; Jiang et al. 2006; Meyer-ter-Vehn et al. 2006). Cell differentiation into myofibroblasts and expansion involves various growth factors including TGF-β, FGF, PDGF, IGF-1, angiotensin II, MCP-1, RANTES, TNF-α, IL-15, and connective tissue growth factor.

The Antigenic Targets of Antigens (Antidonor and Autoantibodies)

One of the independent risk factors for the development of chronic rejection is the presence of anti-HLA class I and especially class II antibodies (Ozawa et al. 2007). More than 80% of patients with transplant glomerulopathy have anti-HLA antibodies, 85% being directed against class I or class II antigens (Gloor et al. 2007; Sis et al. 2007; Issa et al. 2008). Less than half of biopsies (40%) display deposits of C4d, suggesting that mechanisms other than complement activation may be associated with CR (Solez et al. 2008). Apart from antibodies directed against the major histocompatibility complex HLA, other non-HLA antibodies may also contribute to the structural changes observed during chronic rejection. These include anti-MICA, anti-MICB (MHC class I-related molecules A and B), antiendothelial cells, antivimentin, and other autoantibodies directed against angiotensin II receptor (Dragun et al. 2005, 2008; Hilbrands et al. 2005; Zafar et al. 2006; Panigrahi et al. 2007; Zou et al. 2007; Baid-Agrawal and Frei 2008; Kamoun and Grossman 2008; Sumitran-Holgersson 2008).

Tissue Fibrosis

During the various immunological and nonimmunological rejection processes, several cell types, particularly endothelial cells, are responsible for the development of an extracellular matrix through the release of growth factors or cytokines such as endothelin I, angiotensin II, TNF, PDGF, TGF, etc. (Coupes et al. 1994; Lu et al. 2002; el-Agroudy et al. 2003; Baczkowska et al. 2005; Summers et al. 2005; Roos-van Groningen et al. 2006). Extracellular matrix accumulation is a lesion of fibrosis. In recent years, it has become apparent that tissue fibrosis is a dynamic equilibrium. Activation of different tissue protease, such as metalloproteinases, or the use of blocking molecule should prevent fibrosis and degrade extracellular matrix. These approaches must be validated in clinical models.

In conclusion, although significant progress has been made in developing immunosuppressive therapies that reduce the occurrence of acute rejection and improve allograft survival at 1 year, little progress has been made regarding graft survival in the longer term. Long-term survival is limited by complex and intricate mechanisms including chronic rejection, immunosuppressive drug toxicity, cardiovascular factors, dyslipidemia, diabetes, and so on. Understanding the mechanisms involved in the onset of chronic rejection should be promoted by approaches such as DNA chips or protein chips to better understand the different stages of development of chronic rejection and the different factors involved. This understanding should help develop new preventive and/or therapeutic approaches in organ transplantation.

REFERENCES

Abbate M, Zoja C, Rottoli D, Corna D, Tomasoni S, Remuzzi G. 2002. Proximal tubular cells promote fibrogenesis by TGF-β_1-mediated induction of peritubular myofibroblasts. *Kidney Int* **61:** 2066–2077.

Abdi R, Smith RN, Makhlouf L, Najafian N, Luster AD, Auchincloss H Jr, Sayegh MH. 2002. The role of CC chemokine receptor 5 (CCR5) in islet allograft rejection. *Diabetes* **51:** 2489–2495.

Abdi R, Means TK, Ito T, Smith RN, Najafian N, Jurewicz M, Tchipachvili V, Charo I, Auchincloss H Jr, Sayegh MH, et al. 2004. Differential role of CCR2 in islet and heart allograft rejection: Tissue specificity of chemokine/chemokine receptor function in vivo. *J Immunol* **172:** 767–775.

Agostini C, Calabrese F, Rea F, Facco M, Tosoni A, Loy M, Binotto G, Valente M, Trentin L, Semenzato G. 2001. Cxcr3 and its ligand CXCL10 are expressed by inflammatory cells infiltrating lung allografts and mediate chemotaxis of T cells at sites of rejection. *Am J Pathol* **158:** 1703–1711.

Akashi S, Sho M, Kashizuka H, Hamada K, Ikeda N, Kuzumoto Y, Tsurui Y, Nomi T, Mizuno T, Kanehiro H, et al. 2005. A novel small-molecule compound targeting CCR5 and CXCR3 prevents acute and chronic allograft rejection. *Transplantation* **80:** 378–384.

Akbar AN, Amlot PL, Timms A, Lombardi G, Lechler R, Janossy G. 1990. The development of primed/memory CD8$^+$ lymphocytes in vitro and in rejecting kidneys after transplantation. *Clin Exp Immunol* **81:** 225–231.

Alferink J, Lieberam I, Reindl W, Behrens A, Weiss S, Huser N, Gerauer K, Ross R, Reske-Kunz AB, Ahmad-Nejad P, et al. 2003. Compartmentalized production of CCL17 in vivo: Strong inducibility in peripheral dendritic cells contrasts selective absence from the spleen. *J Exp Med* **197:** 585–599.

Amano H, Bickerstaff A, Orosz CG, Novick AC, Toma H, Fairchild RL. 2005. Absence of recipient CCR5 promotes early and increased allospecific antibody responses to cardiac allografts. *J Immunol* **174:** 6499–6508.

Anegon I, Cuturi MC, Trinchieri G, Perussia B. 1988. Interaction of Fc receptor (CD16) ligands induces transcription of interleukin 2 receptor (CD25) and lymphokine genes and expression of their products in human natural killer cells. *J Exp Med* **167:** 452–472.

Anglicheau D, Sharma VK, Ding R, Hummel A, Snopkowski C, Dadhania D, Seshan SV, Suthanthiran M. 2009. MicroRNA expression profiles predictive of human renal allograft status. *Proc Natl Acad Sci* **106:** 5330–5335.

Aquino-Dias EC, Joelsons G, da Silva DM, Berdichevski RH, Ribeiro AR, Veronese FJ, Goncalves LF, Manfro RC. 2008. Non-invasive diagnosis of acute rejection in kidney transplants with delayed graft function. *Kidney Int* **73:** 877–884.

Ashton-Chess J, Giral M, Mengel M, Renaudin K, Foucher Y, Gwinner W, Braud C, Dugast E, Quillard T, Thebault P, et al. 2008. Tribbles-1 as a novel biomarker of chronic antibody-mediated rejection. *J Am Soc Nephrol* **19:** 1116–1127.

Ashton-Chess J, Dugast E, Colvin RB, Giral M, Foucher Y, Moreau A, Renaudin K, Braud C, Devys A, Brouard S, et al. 2009. Regulatory, effector, and cytotoxic T cell profiles in long-term kidney transplant patients. *J Am Soc Nephrol* **20:** 1113–1122.

Ashton-Chess J, Mai HL, Jovanovic V, Renaudin K, Foucher Y, Giral M, Moreau A, Dugast E, Mengel M, Racape M, et al. 2010. Immunoproteasome β subunit 10 is increased in chronic antibody-mediated rejection. *Kidney Int* **77:** 880–890.

Ayalon O, Hughes EA, Cresswell P, Lee J, O'Donnell L, Pardi R, Bender JR. 1998. Induction of transporter associated with antigen processing by interferon γ confers endothelial cell cytoprotection against natural killer-mediated lysis. *Proc Natl Acad Sci* **95:** 2435–2440.

Baczkowska T, Perkowska-Ptasinska A, Sadowska A, Lewandowski Z, Nowacka-Cieciura E, Cieciura T, Pazik J, Lewandowska D, Mroz A, Urbanowicz A, et al. 2005. Serum TGF-β_1 correlates with chronic histopathological lesions in protocol biopsies of kidney allograft recipients. *Transplant Proc* **37:** 773–775.

Badid C, Desmouliere A, Babici D, Hadj-Aissa A, McGregor B, Lefrancois N, Touraine JL, Laville M. 2002. Interstitial expression of α-SMA: An early marker of chronic renal allograft dysfunction. *Nephrol Dial Transplant* **17:** 1993–1998.

Baid-Agrawal S, Frei UA. 2008. Kidney-transplant rejection and anti-MICA antibodies. *N Engl J Med* **358:** 196.

Barbara JA, Turvey SE, Kingsley CI, Spriewald BM, Hara M, Witzke O, Morris PJ, Wood KJ. 2000. Islet allograft rejection can be mediated by $CD4^+$, alloantigen experienced, direct pathway T cells of TH1 and TH2 cytokine phenotype. *Transplantation* **70:** 1641–1649.

Barber DL, Wherry EJ, Ahmed R. 2003. Cutting edge: Rapid in vivo killing by memory CD8 T cells. *J Immunol* **171:** 27–31.

Bauer S, Groh V, Wu J, Steinle A, Phillips JH, Lanier LL, Spies T. 1999. Activation of NK cells and T cells by NKG2D, a receptor for stress-inducible MICA. *Science* **285:** 727–729.

Belperio JA, Ardehali A. 2008. Chemokines and transplant vasculopathy. *Circ Res* **103:** 454–466.

Bestard O, Cruzado JM, Mestre M, Caldes A, Bas J, Carrera M, Torras J, Rama I, Moreso F, Seron D, et al. 2007. Achieving donor-specific hyporesponsiveness is associated with $FoxP3^+$ regulatory T cell recruitment in human renal allograft infiltrates. *J Immunol* **179:** 4901–4909.

Bian H, Reed EF. 1999. Alloantibody-mediated class I signal transduction in endothelial cells and smooth muscle cells: Enhancement by IFN-γ and TNF-α. *J Immunol* **163:** 1010–1018.

Bian H, Reed EF. 2001. Anti-HLA class I antibodies transduce signals in endothelial cells resulting in FGF receptor translocation, down-regulation of ICAM-1 and cell proliferation. *Transplant Proc* **33:** 311.

Brandle D, Joergensen J, Zenke G, Burki K, Hof RP. 1998. Contribution of donor-specific antibodies to acute allograft rejection: Evidence from B cell-deficient mice. *Transplantation* **65:** 1489–1493.

Braun MY, Desalle F, Le Moine A, Pretolani M, Matthys P, Kiss R, Goldman M. 2000. IL-5 and eosinophils mediate the rejection of fully histoincompatible vascularized cardiac allografts: Regulatory role of alloreactive $CD8^+$ T lymphocytes and IFN-γ. *Eur J Immunol* **30:** 1290–1296.

Brook MO, Wood JK, Jones ND. 2006. The impact of memory T cells on rejection and the induction of tolerance. *Transplantation* **82:** 1–9.

Burlingham WJ, Love RB, Jankowska-Gan E, Haynes LD, Xu Q, Bobadilla JL, Meyer KC, Hayney MS, Braun RK, Greenspan DS, et al. 2007. IL-17-dependent cellular immunity to collagen type V predisposes to obliterative bronchiolitis in human lung transplants. *J Clin Invest* **117:** 3498–3506.

Campbell JJ, Haraldsen G, Pan J, Rottman J, Qin S, Ponath P, Andrew DP, Warnke R, Ruffing N, Kassam N, et al. 1999. The chemokine receptor CCR4 in vascular recognition by cutaneous but not intestinal memory T cells. *Nature* **400:** 776–780.

Castellano G, Melchiorre R, Loverre A, Ditonno P, Montinaro V, Rossini M, Divella C, Battaglia M, Lucarelli G, Annunziata GS, et al. 2010. Therapeutic targeting of classical and lectin pathways of complement protects from ischemia-reperfusion-induced renal damage. *Am J Pathol* **176:** 1648–1659.

Celli S, Albert ML, Bousso P. 2011. Visualizing the innate and adaptive immune responses underlying allograft rejection by two-photon microscopy. *Nat Med* **17:** 744–749.

Chen J, Fabry B, Schiffrin EL, Wang N. 2001. Twisting integrin receptors increases endothelin-1 gene expression in endothelial cells. *Am J Physiol Cell Physiol* **280:** C1475–C1484.

Cheng J, Torkamani A, Grover RK, Jones TM, Ruiz DI, Schork NJ, Quigley MM, Hall FW, Salomon DR, Lerner RA. 2011. Ectopic B-cell clusters that infiltrate transplanted human kidneys are clonal. *Proc Natl Acad Sci* **108:** 5560–5565.

Chong AS, Alegre ML. 2012. The impact of infection and tissue damage in solid-organ transplantation. *Nat Rev Immunol* **12:** 459–471.

Christopher K, Mueller TF, Ma C, Liang Y, Perkins DL. 2002. Analysis of the innate and adaptive phases of allograft rejection by cluster analysis of transcriptional profiles. *J Immunol* **169:** 522–530.

Clatworthy MR, Watson CJ, Plotnek G, Bardsley V, Chaudhry AN, Bradley JA, Smith KG. 2009. B-cell-depleting induction therapy and acute cellular rejection. *N Engl J Med* **360:** 2683–2685.

Cogan JG, Subramanian SV, Polikandriotis JA, Kelm RJ Jr, Strauch AR. 2002. Vascular smooth muscle α-actin gene transcription during myofibroblast differentiation requires Sp1/3 protein binding proximal to the MCAT enhancer. *J Biol Chem* **277:** 36433–36442.

Colobran R, Casamitjana N, Roman A, Faner R, Pedrosa E, Arostegui JI, Pujol-Borrell R, Juan M, Palou E. 2009. Copy number variation in the CCL4L gene is associated with susceptibility to acute rejection in lung transplantation. *Genes Immun* **10:** 254–259.

Constant SL. 1999. B lymphocytes as antigen-presenting cells for $CD4^+$ T cell priming in vivo. *J Immunol* **162:** 5695–5703.

Cooper JE, Gralla J, Cagle L, Goldberg R, Chan L, Wiseman AC. 2011. Inferior kidney allograft outcomes in patients with de novo donor-specific antibodies are due to acute rejection episodes. *Transplantation* **91:** 1103–1109.

Cosio FG, Lager DJ, Lorenz EC, Amer H, Gloor JM, Stegall MD. 2010. Significance and implications of capillaritis during acute rejection of kidney allografts. *Transplantation* **89:** 1088–1094

Coudert JD, Coureau C, Guery JC. 2002. Preventing NK cell activation by donor dendritic cells enhances allospecific CD4 T cell priming and promotes Th type 2 responses to transplantation antigens. *J Immunol* **169:** 2979–2987.

Coupel S, Moreau A, Hamidou M, Horejsi V, Soulillou JP, Charreau B. 2007. Expression and release of soluble HLA-E is an immunoregulatory feature of endothelial cell activation. *Blood* **109:** 2806–2814.

Coupes BM, Newstead CG, Short CD, Brenchley PE. 1994. Transforming growth factor β_1 in renal allograft recipients. *Transplantation* **57:** 1727–1731.

Crescioli C, Buonamano A, Scolletta S, Sottili M, Francalanci M, Giomarelli P, Biagioli B, Lisi G, Pradella F, Serio M, et al. 2009. Predictive role of pretransplant serum CXCL10 for cardiac acute rejection. *Transplantation* **87:** 249–255.

Croft M. 2003. Costimulation of T cells by OX40, 4-1BB, and CD27. *Cytokine Growth Factor Rev* **14:** 265–273.

Cuturi MC, Murphy M, Costa-Giomi MP, Weinmann R, Perussia B, Trinchieri G. 1987. Independent regulation of tumor necrosis factor and lymphotoxin production by human peripheral blood lymphocytes. *J Exp Med* **165:** 1581–1594.

Cuturi MC, Anegon I, Sherman F, Loudon R, Clark SC, Perussia B, Trinchieri G. 1989. Production of hematopoietic colony-stimulating factors by human natural killer cells. *J Exp Med* **169:** 569–583.

Deng S, Moore Dj, Huang X, Lian MM, Mohiuddin M, Velededeoglu E, Lee MKT, Sonawane S, Kim J, Wang J, et al. 2007. Cutting edge: Transplant tolerance induced by anti-CD45RB requires B lymphocytes. *J Immunol* **178:** 6028–6032.

Deteix C, Attuil-Audenis V, Duthey A, Patey N, McGregor B, Dubois V, Caligiuri G, Graff-Dubois S, Morelon E, Thaunat O. 2010. Intragraft Th17 infiltrate promotes lymphoid neogenesis and hastens clinical chronic rejection. *J Immunol* **184:** 5344–5351.

Dewald O, Zymek P, Winkelmann K, Koerting A, Ren G, Abou-Khamis T, Michael LH, Rollins BJ, Entman ML, Frangogiannis NG. 2005. CCL2/monocyte chemoattractant protein-1 regulates inflammatory responses critical to healing myocardial infarcts. *Circ Res* **96:** 881–889.

Dikow R, Becker LE, Schaier M, Waldherr R, Gross ML, Zeier M. 2010. In renal transplants with delayed graft function chemokines and chemokine receptor expression predict long-term allograft function. *Transplantation* **90:** 771–776.

Dilek N, Poirier N, Usal C, Martinet B, Blancho G, Vanhove B. 2012. Control of transplant tolerance and intragraft regulatory T cell localization by myeloid-derived suppressor cells and CCL5. *J Immunol* **188:** 4209–4216.

DiLillo DJ, Griffiths R, Seshan SV, Magro CM, Ruiz P, Coffman TM, Tedder TF. 2011. B lymphocytes differentially influence acute and chronic allograft rejection in mice. *J Immunol* **186:** 2643–2654.

Ding R, Li B, Muthukumar T, Dadhania D, Medeiros M, Hartono C, Serur D, Seshan SV, Sharma VK, Kapur S, et al. 2003. CD103 mRNA levels in urinary cells predict acute rejection of renal allografts. *Transplantation* **75:** 1307–1312.

Ding Q, Yeung M, Camirand G, Zeng Q, Akiba H, Yagita H, Chalasani G, Sayegh MH, Najafian N, Rothstein DM. 2011. Regulatory B cells are identified by expression of TIM-1 and can be induced through TIM-1 ligation to promote tolerance in mice. *J Clin Invest* **121:** 3645–3656.

Direkze NC, Forbes SJ, Brittan M, Hunt T, Jeffery R, Preston SL, Poulsom R, Hodivala-Dilke K, Alison MR, Wright NA. 2003. Multiple organ engraftment by bone-marrow-derived myofibroblasts and fibroblasts in bone-marrow-transplanted mice. *Stem Cells* **21:** 514–520.

Dragun D. 2008. Humoral responses directed against non-human leukocyte antigens in solid-organ transplantation. *Transplantation* **86:** 1019–1025.

Dragun D, Muller DN, Brasen JH, Fritsche L, Nieminen-Kelha M, Dechend R, Kintscher U, Rudolph B, Hoebeke J, Eckert D, et al. 2005. Angiotensin II type 1-receptor activating antibodies in renal-allograft rejection. *N Engl J Med* **352:** 558–569.

Dragun D, Scornik J, Meier-Kriesche HU. 2008. Kidney-transplant rejection and anti-MICA antibodies. *N Engl J Med* **358:** 196; author reply 196.

Dugre FJ, Gaudreau S, Belles-Isles M, Houde I, Roy R. 2000. Cytokine and cytotoxic molecule gene expression determined in peripheral blood mononuclear cells in the diagnosis of acute renal rejection. *Transplantation* **70:** 1074–1080.

Duluc D, Delneste Y, Tan F, Moles MP, Grimaud L, Lenoir J, Preisser L, Anegon I, Catala L, Ifrah N, et al. 2007. Tumor-associated leukemia inhibitory factor and IL-6 skew monocyte differentiation into tumor-associated macrophage-like cells. *Blood* **110:** 4319–4330.

el-Agroudy AE, Hassan NA, Foda MA, Ismail AM, el-Sawy EA, Mousa O, Ghoneim MA. 2003. Effect of angiotensin II receptor blocker on plasma levels of TGF-β_1 and interstitial fibrosis in hypertensive kidney transplant patients. *Am J Nephrol* **23:** 300–306.

El-Sawy T, Belperio JA, Strieter RM, Remick DG, Fairchild RL. 2005. Inhibition of polymorphonuclear leukocyte-mediated graft damage synergizes with short-term costimulatory blockade to prevent cardiac allograft rejection. *Circulation* **112:** 320–331.

Epperson DE, Pober JS. 1994. Antigen-presenting function of human endothelial cells. Direct activation of resting CD8 T cells. *J Immunol* **153:** 5402–5412.

Fahim T, Bohmig GA, Exner M, Huttary N, Kerschner H, Kandutsch S, Kerjaschki D, Brambock A, Nagy-Bojarszky K, Regele H. 2007. The cellular lesion of humoral rejection: Predominant recruitment of monocytes to peritubular and glomerular capillaries. *Am J Transplant* **7:** 385–393.

Fahmy NM, Yamani MH, Starling RC, Ratliff NB, Young JB, McCarthy PM, Feng J, Novick AC, Fairchild RL. 2003. Chemokine and receptor-gene expression during early and late acute rejection episodes in human cardiac allografts. *Transplantation* **75:** 2044–2047.

Fan JM, Ng YY, Hill PA, Nikolic-Paterson DJ, Mu W, Atkins RC, Lan HY. 1999. Transforming growth factor-β regulates tubular epithelial-myofibroblast transdifferentiation in vitro. *Kidney Int* **56:** 1455–1467.

Farid WR, Pan Q, van der Meer AJ, de Ruiter PE, Ramakrishnaiah V, de Jonge J, Kwekkeboom J, Janssen HL, Metselaar HJ, Tilanus HW, et al. 2012. Hepatocyte-derived microRNAs as serum biomarkers of hepatic injury and rejection after liver transplantation. *Liver Transpl* **18:** 290–297.

Fong AM, Robinson LA, Steeber DA, Tedder TF, Yoshie O, Imai T, Patel DD. 1998. Fractalkine and CX3CR1 mediate a novel mechanism of leukocyte capture, firm adhesion, and activation under physiologic flow. *J Exp Med* **188:** 1413–1419.

Foussat A, Coulomb-L'Hermine A, Gosling J, Krzysiek R, Durand-Gasselin I, Schall T, Balian A, Richard Y, Galanaud P, Emilie D. 2000. Fractalkine receptor expression by T lymphocyte subpopulations and in vivo production of fractalkine in human. *Eur J Immunol* **30:** 87–97.

Frangogiannis NG. 2008. The immune system and cardiac repair. *Pharmacol Res* **58:** 88–111.

Friedman BH, Wolf JH, Wang L, Putt ME, Shaked A, Christie JD, Hancock WW, Olthoff KM. 2012. Serum cytokine profiles associated with early allograft dysfunction in patients undergoing liver transplantation. *Liver Transpl* **18:** 166–176.

Fukami N, Ramachandran S, Saini D, Walter M, Chapman W, Patterson GA, Mohanakumar T. 2009. Antibodies to MHC class I induce autoimmunity: Role in the pathogenesis of chronic rejection. *J Immunol* **182:** 309–318.

Gao W, Faia KL, Csizmadia V, Smiley ST, Soler D, King JA, Danoff TM, Hancock WW. 2001. Beneficial effects of targeting CCR5 in allograft recipients. *Transplantation* **72:** 1199–1205.

Garrod KR, Liu FC, Forrest LE, Parker I, Kang SM, Cahalan MD. 2010. NK cell patrolling and elimination of donor-derived dendritic cells favor indirect alloreactivity. *J Immunol* **184:** 2329–2336.

Gelman AE, Okazaki M, Sugimoto S, Li W, Kornfeld CG, Lai J, Richardson SB, Kreisel FH, Huang HJ, Tietjens JR, et al. 2010. CCR2 regulates monocyte recruitment as well as CD4 T1 allorecognition after lung transplantation. *Am J Transplant* **10:** 1189–1199.

Gloor JM, Cosio FG, Rea DJ, Wadei HM, Winters JL, Moore SB, DeGoey SR, Lager DJ, Grande JP, Stegall MD. 2006. Histologic findings one year after positive crossmatch or ABO blood group incompatible living donor kidney transplantation. *Am J Transplant* **6:** 1841–1847.

Gloor JM, Sethi S, Stegall MD, Park WD, Moore SB, DeGoey S, Griffin MD, Larson TS, Cosio FG. 2007. Transplant glomerulopathy: Subclinical incidence and association with alloantibody. *Am J Transplant* **7:** 2124–2132.

Gregson AL, Hoji A, Saggar R, Ross DJ, Kubak BM, Jamieson BD, Weigt SS, Lynch JP III, Ardehali A, Belperio JA, et al. 2008. Bronchoalveolar immunologic profile of acute human lung transplant allograft rejection. *Transplantation* **85:** 1056–1059.

Gressner AM. 1996. Transdifferentiation of hepatic stellate cells (Ito cells) to myofibroblasts: A key event in hepatic fibrogenesis. *Kidney Int Suppl* **54:** S39–S45.

Griffin GK, Newton G, Tarrio ML, Bu DX, Maganto-Garcia E, Azcutia V, Alcaide P, Grabie N, Luscinskas FW, Croce KJ, et al. 2012. IL-17 and TNF-α sustain neutrophil recruitment during inflammation through synergistic effects on endothelial activation. *J Immunol* **188:** 6287–6299.

Gueler F, Rong S, Gwinner W, Mengel M, Brocker V, Schon S, Greten TF, Hawlisch H, Polakowski T, Schnatbaum K, et al. 2008. Complement 5a receptor inhibition improves renal allograft survival. *J Am Soc Nephrol* **19:** 2302–2312.

Guillonneau C, Louvet C, Renaudin K, Heslan JM, Heslan M, Tesson L, Vignes C, Guillot C, Choi Y, Turka LA, et al. 2004. The role of TNF-related activation-induced cytokine-receptor activating NF-κB interaction in acute allograft rejection and CD40L-independent chronic allograft rejection. *J Immunol* **172:**1619–1629.

Guillonneau C, Aubry V, Renaudin K, Seveno C, Usal C, Tezuka K, Anegon I. 2005. Inhibition of chronic rejection and development of tolerogenic T cells after ICOS-ICOSL and CD40-CD40L co-stimulation blockade. *Transplantation* **80:** 546–554.

Guillonneau C, Hill M, Hubert FX, Chiffoleau E, Herve C, Li XL, Heslan M, Usal C, Tesson L, Menoret S, et al. 2007. CD40Ig treatment results in allograft acceptance mediated by CD8CD45RC T cells, IFN-γ, and indoleamine 2,3-dioxygenase. *J Clin Invest* **117:** 1096–1106.

Guillonneau C, Picarda E, Anegon I. 2010. $CD8^+$ regulatory T cells in solid organ transplantation. *Curr Opin Organ Transplant* (to be published).

Guillot C, Guillonneau C, Mathieu P, Gerdes CA, Menoret S, Braudeau C, Tesson L, Renaudin K, Castro MG, Lowenstein PR, et al. 2002. Prolonged blockade of CD40-CD40 ligand interactions by gene transfer of CD40Ig results in long-term heart allograft survival and donor-specific hyporesponsiveness, but does not prevent chronic rejection. *J Immunol* **168:** 1600–1609.

Halloran PF, Wadgymar A, Ritchie S, Falk J, Solez K, Srinivasa NS. 1990. The significance of the anti-class I antibody response. I: Clinical and pathologic features of anti-class I-mediated rejection. *Transplantation* **49:** 85–91.

Halloran PF, Melk A, Barth C. 1999. Rethinking chronic allograft nephropathy: The concept of accelerated senescence. *J Am Soc Nephrol* **10:** 167–181.

Hancock WW, Thomson NM, Atkins RC. 1983. Composition of interstitial cellular infiltrate identified by monoclonal antibodies in renal biopsies of rejecting human renal allografts. *Transplantation* **35:** 458–463.

Hancock WW, Lu B, Gao W, Csizmadia V, Faia K, King JA, Smiley ST, Ling M, Gerard NP, Gerard C. 2000. Requirement of the chemokine receptor CXCR3 for acute allograft rejection. *J Exp Med* **192:** 1515–1520.

Hariharan S, Johnson CP, Bresnahan BA, Taranto SE, McIntosh MJ, Stablein D. 2000. Improved graft survival after renal transplantation in the United States, 1988 to 1996. *N Engl J Med* **342:** 605–612.

Haurani MJ, Cifuentes ME, Shepard AD, Pagano PJ. 2008. Nox4 oxidase overexpression specifically decreases endogenous Nox4 mRNA and inhibits angiotensin II-induced adventitial myofibroblast migration. *Hypertension* **52:** 143–149.

He H, Stone JR, Perkins DL. 2002. Analysis of robust innate immune response after transplantation in the absence of adaptive immunity. *Transplantation* **73:** 853–861.

He H, Stone JR, Perkins DL, Perkins. 2003. Analysis of differential immune responses induced by innate and adaptive immunity following transplantation. *Immunology* **109:** 185–196.

Heeger PS, Greenspan NS, Kuhlenschmidt S, Dejelo C, Hricik DE, Schulak JA, Tary-Lehmann M. 1999. Pretransplant frequency of donor-specific, IFN-γ-producing lymphocytes is a manifestation of immunologic memory and correlates with the risk of posttransplant rejection episodes. *J Immunol* **163:** 2267–2275.

Hengel RL, Thaker V, Pavlick MV, Metcalf JA, Dennis G Jr, Yang J, Lempicki RA, Sereti I, Lane HC. 2003. Cutting edge: L-selectin (CD62L) expression distinguishes small resting memory $CD4^+$ T cells that preferentially respond to recall antigen. *J Immunol* **170:** 28–32.

Herrera OB, Golshayan D, Tibbott R, Salcido Ochoa F, James MJ, Marelli-Berg FM, Lechler RI. 2004. A novel pathway of alloantigen presentation by dendritic cells. *J Immunol* **173:** 4828–4837.

Hertig A, Verine J, Mougenot B, Jouanneau C, Ouali N, Sebe P, Glotz D, Ancel PY, Rondeau E, Xu-Dubois YC. 2006. Risk factors for early epithelial to mesenchymal transition in renal grafts. *Am J Transplant* **6:** 2937–2946.

Hidalgo LG, Sis B, Sellares J, Campbell PM, Mengel M, Einecke G, Chang J, Halloran PF. 2010. NK cell transcripts and NK cells in kidney biopsies from patients with donor-specific antibodies: Evidence for NK cell involvement in antibody-mediated rejection. *Am J Transplant* **10:** 1812–1822.

Hilbrands L, Hoitsma A, Wetzels J. 2005. Angiotensin II type 1-receptor activating antibodies in renal-allograft rejection. *N Engl J Med* **352:** 2027–2028.

Hirohashi T, Chase CM, Della Pelle P, Sebastian D, Alessandrini A, Madsen JC, Russell PS, Colvin RB. 2012. A novel pathway of chronic allograft rejection mediated by NK cells and alloantibody. *Am J Transplant* **12:** 313–321.

Ho J, Rush DN, Gibson IW, Karpinski M, Storsley L, Bestland J, Stefura W, HayGlass KT, Nickerson PW. 2010. Early urinary CCL2 is associated with the later development of interstitial fibrosis and tubular atrophy in renal allografts. *Transplantation* **90:** 394–400.

Hoerning A, Koss K, Datta D, Boneschansker L, Jones CN, Wong IY, Irimia D, Calzadilla K, Benitez F, Hoyer PF, et al. 2011. Subsets of human $CD4^+$ regulatory T cells express the peripheral homing receptor CXCR3. *Eur J Immunol* **41:** 2291–2302.

Hoffmann U, Segerer S, Rummele P, Kruger B, Pietrzyk M, Hofstadter F, Banas B, Kramer BK. 2006. Expression of the chemokine receptor CXCR3 in human renal allografts—A prospective study. *Nephrol Dial Transplant* **21:** 1373–1381.

Hoffmann U, Bergler T, Segerer S, Rummele P, Kruger B, Banas MC, Reinhold S, Banas B, Kramer BK. 2010. Impact of chemokine receptor CX3CR1 in human renal allograft rejection. *Transpl Immunol* **23:** 204–208.

Hopken UE, Droese J, Li JP, Joergensen J, Breitfeld D, Zerwes HG, Lipp M. 2004. The chemokine receptor CCR7 controls lymph node-dependent cytotoxic T cell priming in alloimmune responses. *Eur J Immunol* **34:** 461–470.

Horne PH, Zimmerer JM, Fisher MG, Lunsford KE, Nadasdy G, Nadasdy T, van Rooijen N, Bumgardner GL. 2008. Critical role of effector macrophages in mediating CD4-dependent alloimmune injury of transplanted liver parenchymal cells. *J Immunol* **181:** 1224–1231.

Hu H, Kwun J, Aizenstein BD, Knechtle SJ. 2009. Noninvasive detection of acute and chronic injuries in human renal transplant by elevation of multiple cytokines/chemokines in urine. *Transplantation* **87:** 1814–1820.

Huser N, Tertilt C, Gerauer K, Maier S, Traeger T, Assfalg V, Reiter R, Heidecke CD, Pfeffer K. 2005. CCR4-deficient mice show prolonged graft survival in a chronic cardiac transplant rejection model. *Eur J Immunol* **35:** 128–138.

Imai T, Hieshima K, Haskell C, Baba M, Nagira M, Nishimura M, Kakizaki M, Takagi S, Nomiyama H, Schall TJ, et al. 1997. Identification and molecular characterization of fractalkine receptor CX3CR1, which mediates both leukocyte migration and adhesion. *Cell* **91:** 521–530.

Issa N, Cosio FG, Gloor JM, Sethi S, Dean PG, Moore SB, DeGoey S, Stegall MD. 2008. Transplant glomerulopathy: Risk and prognosis related to anti-human leukocyte antigen class II antibody levels. *Transplantation* **86:** 681–685.

Ito A, Shimura H, Nitahara A, Tomiyama K, Ito M, Kanekura T, Okumura K, Yagita H, Kawai K. 2008. NK cells contribute to the skin graft rejection promoted by $CD4^+$ T cells activated through the indirect allorecognition pathway. *Int Immunol* **20:** 1343–1349.

Iwasaki A, Medzhitov R. 2010. Regulation of adaptive immunity by the innate immune system. *Science* **327:** 291–295.

Jackson JA, Kim EJ, Begley B, Cheeseman J, Harden T, Perez SD, Thomas S, Warshaw B, Kirk AD. 2011a. Urinary chemokines CXCL9 and CXCL10 are noninvasive markers of renal allograft rejection and BK viral infection. *Am J Transplant* **11:** 2228–2234.

Jackson AM, Lucas DP, Melancon JK, Desai NM. 2011b. Clinical relevance and IgG subclass determination of non-HLA antibodies identified using endothelial cell precursors isolated from donor blood. *Transplantation* **92:** 54–60.

Ji J, Yu YY, Chen H, Deng XX, Shen BY, Zhou GW, Peng CW, Li HW. 2004. Perforin and granzyme B expression in liver allograft rejection. *Zhonghua Gan Zang Bing Za Zhi* **12:** 325–327.

Jiang YL, Dai AG, Li QF, Hu RC. 2006. Transforming growth factor-β1 induces transdifferentiation of fibroblasts into myofibroblasts in hypoxic pulmonary vascular remodeling. *Acta Biochim Biophys Sin (Shanghai)* **38:** 29–36.

Jose MD, Le Meur Y, Atkins RC, Chadban SJ. 2003. Blockade of macrophage colony-stimulating factor reduces macrophage proliferation and accumulation in renal allograft rejection. *Am J Transplant* **3:** 294–300.

Jungraithmayr W, Inci I, Bain M, Hillinger S, Korom S, Weder W. 2010. Distribution of macrophages and T cells in syngrafts and allografts after experimental rat lung transplantation. *Immunobiology* **215:** 206–214.

Kakuta Y, Okumi M, Miyagawa S, Tsutahara K, Abe T, Yazawa K, Matsunami K, Otsuka H, Takahara S, Nonomura N. 2012. Blocking of CCR5 and CXCR3 suppresses the infiltration of macrophages in acute renal allograft rejection. *Transplantation* **93:** 24–31.

Kamoun M, Grossman RA. 2008. Kidney-transplant rejection and anti-MICA antibodies. *N Engl J Med* **358:** 195.

Kedainis RL, Koch MJ, Brennan DC, Liapis H. 2009. Focal $C4d^+$ in renal allografts is associated with the presence of donor-specific antibodies and decreased allograft survival. *Am J Transplant* **9:** 812–819.

Keller MR, Burlingham WJ. 2011. Loss of tolerance to self after transplant. *Semin Immunopathol* **33:** 105–110.

Kennard S, Liu H, Lilly B. 2008. Transforming growth factor-β (TGF-1) down-regulates Notch3 in fibroblasts to promote smooth muscle gene expression. *J Biol Chem* **283:** 1324–1333.

Kim J, Chang CK, Hayden T, Liu FC, Benjamin J, Hamerman JA, Lanier LL, Kang SM. 2007. The activating immunoreceptor NKG2D and its ligands are involved in allograft transplant rejection. *J Immunol* **179:** 6416–6420.

Kingsley CI, Karim M, Bushell AR, Wood KJ. 2002. $CD25^+$ $CD4^+$ regulatory T cells prevent graft rejection: CTLA-4- and IL-10-dependent immunoregulation of alloresponses. *J Immunol* **168:** 1080–1086.

Kirk AD, Morrell CN, Baldwin WM III. 2009. Platelets influence vascularized organ transplants from start to finish. *Am J Transplant* **9:** 14–22.

Kozakowski N, Bohmig GA, Exner M, Soleiman A, Huttary N, Nagy-Bojarszky K, Ecker RC, Kikic Z, Regele H. 2009. Monocytes/macrophages in kidney allograft intimal arteritis: No association with markers of humoral rejection or with inferior outcome. *Nephrol Dial Transplant* **24:** 1979–1986.

Krams SM, Villanueva JC, Quinn MB, Martinez OM. 1995. Expression of the cytotoxic T cell mediator granzyme B during liver allograft rejection. *Transpl Immunol* **3:** 162–166.

Kroemer A, Xiao X, Degauque N, Edtinger K, Wei H, Demirci G, Li XC. 2008. The innate NK cells, allograft rejection, and a key role for IL-15. *J Immunol* **180:** 7818–7826.

Kwun J, Hazinedaroglu SM, Schadde E, Kayaoglu HA, Fechner J, Hu HZ, Roenneburg D, Torrealba J, Shiao L, Hong X, et al. 2008. Unaltered graft survival and intragraft lymphocytes infiltration in the cardiac allograft of $CXCR3^{-/-}$ mouse recipients. *Am J Transplant* **8:** 1593–1603.

Laffont S, Seillet C, Ortaldo J, Coudert JD, Guery JC. 2008. Natural killer cells recruited into lymph nodes inhibit alloreactive T-cell activation through perforin-mediated killing of donor allogeneic dendritic cells. *Blood* **112:** 661–671.

Land WG. 2005. The role of postischemic reperfusion injury and other nonantigen-dependent inflammatory pathways in transplantation. *Transplantation* **79:** 505–514.

LaRosa DF, Rahman AH, Turka LA. 2007. The innate immune system in allograft rejection and tolerance. *J Immunol* **178:**7503–7509.

Lawson C, Holder AL, Stanford RE, Smith J, Rose ML. 2005. Anti-intercellular adhesion molecule-1 antibodies in sera of heart transplant recipients: A role in endothelial cell activation. *Transplantation* **80:** 264–271.

Lazzeri E, Rotondi M, Mazzinghi B, Lasagni L, Buonamano A, Rosati A, Pradella F, Fossombroni V, La Villa G, Gacci M, et al. 2005. High CXCL10 expression in rejected kidneys and predictive role of pretransplant serum CXCL10 for acute rejection and chronic allograft nephropathy. *Transplantation* **79:** 1215–1220.

Lee I, Wang L, Wells AD, Ye Q, Han R, Dorf ME, Kuziel WA, Rollins BJ, Chen L, Hancock WW. 2003. Blocking the monocyte chemoattractant protein-1/CCR2 chemokine pathway induces permanent survival of islet allografts through a programmed death-1 ligand-1-dependent mechanism. *J Immunol* **171:** 6929–6935.

Lee I, Wang L, Wells AD, Dorf ME, Ozkaynak E, Hancock W. 2005. Recruitment of $FoxP3^{+}$ T regulatory cells mediating allograft tolerance depends on the CCR4 chemokine receptor. *J Exp Med* **201:** 1037–1044.

Lefaucheur C, Suberbielle-Boissel C, Hill GS, Nochy D, Andrade J, Antoine C, Gautreau C, Charron D, Glotz D. 2008. Clinical relevance of preformed HLA donor-specific antibodies in kidney transplantation. *Am J Transplant* **8:** 324–331.

Lefaucheur C, Nochy D, Andrade J, Verine J, Gautreau C, Charron D, Hill GS, Glotz D, Suberbielle-Boissel C. 2009. Comparison of combination plasmapheresis/IVIg/anti-CD20 versus high-dose IVIg in the treatment of antibody-mediated rejection. *Am J Transplant* **9:** 1099–1107.

Lefaucheur C, Loupy A, Hill GS, Andrade J, Nochy D, Antoine C, Gautreau C, Charron D, Glotz D, Suberbielle-Boissel C. 2010. Preexisting donor-specific HLA antibodies predict outcome in kidney transplantation. *J Am Soc Nephrol* **21:** 1398–1406.

Le Texier L, Thebault P, Lavault A, Usal C, Merieau E, Quillard T, Charreau B, Soulillou JP, Cuturi MC, Brouard S, et al. 2011. Long-term allograft tolerance is characterized by the accumulation of B cells exhibiting an inhibited profile. *Am J Transplant* **11:** 429–438.

Li XC, Baldwin WM III. 2012. NK cells: New partners in antibody-triggered chronic rejection. *Am J Transplant* **12:** 275–276.

Li B, Hartono C, Ding R, Sharma VK, Ramaswamy R, Qian B, Serur D, Mouradian J, Schwartz JE, Suthanthiran M. 2001. Noninvasive diagnosis of renal-allograft rejection by measurement of messenger RNA for perforin and granzyme B in urine. *N Engl J Med* **344:** 947–954.

Li Y, Koshiba T, Yoshizawa A, Yonekawa Y, Masuda K, Ito A, Ueda M, Mori T, Kawamoto H, Tanaka Y, et al. 2004. Analyses of peripheral blood mononuclear cells in operational tolerance after pediatric living donor liver transplantation. *Am J Transplant* **4:** 2118–2125.

Li J, Deane JA, Campanale NV, Bertram JF, Ricardo SD. 2007. The contribution of bone marrow-derived cells to the development of renal interstitial fibrosis. *Stem Cells* **25:** 697–706.

Li L, Chen A, Chaudhuri A, Kambham N, Sigdel T, Chen R, Sarwal MM. 2010. Compartmental localization and clinical relevance of MICA antibodies after renal transplantation. *Transplantation* **89:** 312–319.

Li J, Chen G, Ye P, Wang S, Zhang K, Chen W, Stepkowski SM, Li J, Zhong S, Xia J. 2011. CCR5 blockade in combination with cyclosporine increased cardiac graft survival and generated alternatively activated macrophages in primates. *J Immunol* **186:** 3753–3761.

Li H, Xie HY, Zhou L, Feng XW, Wang WL, Liang TB, Zhang M, Zheng SS. 2012. Copy number variation in CCL3L1 gene is associated with susceptibility to acute rejection in patients after liver transplantation. *Clin Transplant* **26:** 314–321.

Lin D, Lavender H, Soilleux EJ, O'Callaghan CA. 2012. NF-κB regulates MICA gene transcription in endothelial cell through a genetically inhibitable control site. *J Biol Chem* **287:** 4299–4310.

Lindert S, Wickert L, Sawitza I, Wiercinska E, Gressner AM, Dooley S, Breitkopf K. 2005. Transdifferentiation-dependent expression of α-SMA in hepatic stellate cells does not involve TGF-β pathways leading to coinduction of collagen type I and thrombospondin-2. *Matrix Biol* **24:** 198–207.

Liu X, Mishra P, Yu S, Beckmann J, Wendland M, Kocks J, Seth S, Hoffmann K, Hoffmann M, Kremmer E, et al. 2011. Tolerance induction towards cardiac allografts under costimulation blockade is impaired in CCR7-deficient animals but can be restored by adoptive transfer of syngeneic plasmacytoid dendritic cells. *Eur J Immunol* **41:** 611–623.

Lo DJ, Weaver TA, Kleiner DE, Mannon RB, Jacobson LM, Becker BN, Swanson SJ, Hale DA, Kirk AD. 2011. Chemokines and their receptors in human renal allotransplantation. *Transplantation* **91:** 70–77.

Lombardi G, Sidhu S, Daly M, Batchelor JR, Makgoba W, Lechler RI. 1990. Are primary alloresponses truly primary? *Int Immunol* **2:** 9–13.

Lorenzen JM, Volkmann I, Fiedler J, Schmidt M, Scheffner I, Haller H, Gwinner W, Thum T. 2011. Urinary miR-210 as a mediator of acute T-cell mediated rejection in renal allograft recipients. *Am J Transplant* **11:** 2221–2227.

Lozano JJ, Pallier A, Martinez-Llordella M, Danger R, Lopez M, Giral M, Londono MC, Rimola A, Soulillou JP, Brouard S, et al. 2011. Comparison of transcriptional and blood cell-phenotypic markers between operationally tolerant liver and kidney recipients. *Am J Transplant* **11:** 1916–1926.

Lu KC, Jaramillo A, Lecha RL, Schuessler RB, Aloush A, Trulock EP, Mendeloff EN, Huddleston CB, Alexander Patterson G, Mohanakumar T. 2002. Interleukin-6 and interferon-γ gene polymorphisms in the development of bronchiolitis obliterans syndrome after lung transplantation. *Transplantation* **74:** 1297–1302.

Lucas JG, Co JP, Nwaogwugwu UT, Dosani I, Sureshkumar KK. 2011. Antibody-mediated rejection in kidney transplantation: An update. *Expert Opin Pharmacother* **12:** 579–592.

Lumeng CN, Bodzin JL, Saltiel AR. 2007. Obesity induces a phenotypic switch in adipose tissue macrophage polarization. *J Clin Invest* **117:** 175–184.

MacDermott RP. 1996. Alterations of the mucosal immune system in inflammatory bowel disease. *J Gastroenterol* **31:** 907–916.

Madsen CB, Norgaard A, Iversen M, Ryder LP. 2010. Elevated mRNA levels of CTLA-4, FoxP3, and granzyme B in BAL, but not in blood, during acute rejection of lung allografts. *Transpl Immunol* **24:** 26–32.

Magil AB. 2005. Infiltrating cell types in transplant glomerulitis: Relationship to peritubular capillary C4d deposition. *Am J Kidney Dis* **45:** 1084–1089.

Magil AB, Tinckam KJ. 2006. Focal peritubular capillary C4d deposition in acute rejection. *Nephrol Dial Transplant* **21:** 1382–1388.

Mahesh B, Leong HS, McCormack A, Sarathchandra P, Holder A, Rose ML. 2007. Autoantibodies to vimentin cause accelerated rejection of cardiac allografts. *Am J Pathol* **170:** 1415–1427.

Maier S, Tertilt C, Chambron N, Gerauer K, Huser N, Heidecke CD, Pfeffer K. 2001. Inhibition of natural killer cells results in acceptance of cardiac allografts in $CD28^{-/-}$ mice. *Nat Med* **7:** 557–562.

Manes TD, Pober JS. 2008. Antigen presentation by human microvascular endothelial cells triggers ICAM-1-dependent transendothelial protrusion by, and fractalkine-dependent transendothelial migration of, effector memory $CD4^+$ T cells. *J Immunol* **180:** 8386–8392.

Manes TD, Shiao SL, Dengler TJ, Pober JS. 2007. TCR signaling antagonizes rapid IP-10-mediated transendothelial migration of effector memory $CD4^+$ T cells. *J Immunol* **178:** 3237–3243.

Mantovani A, Sica A, Sozzani S, Allavena P, Vecchi A, Locati M. 2004. The chemokine system in diverse forms of macrophage activation and polarization. *Trends Immunol* **25:** 677–686.

Mao Q, Terasaki PI, Cai J, Briley K, Catrou P, Haisch C, Rebellato L. 2007a. Extremely high association between appearance of HLA antibodies and failure of kidney grafts in a five-year longitudinal study. *Am J Transplant* **7:** 864–871.

Mao Q, Terasaki PI, Cai J, El-Awar N, Rebellato L. 2007b. Analysis of HLA class I specific antibodies in patients with failed allografts. *Transplantation* **83:** 54–61.

Mao Y, Wang M, Zhou Q, Jin J, Wang Y, Peng W, Wu J, Shou Z, Chen J. 2011. CXCL10 and CXCL13 Expression were highly up-regulated in peripheral blood mononuclear cells in acute rejection and poor response to anti-rejection therapy. *J Clin Immunol* **31:** 414–418.

Martinez-Dolz L, Almenar L, Reganon E, Vila V, Sanchez-Soriano R, Martinez-Sales V, Moro J, Aguero J, Sanchez-Lazaro I, Salvador A. 2009. What is the best biomarker for diagnosis of rejection in heart transplantation? *Clin Transplant* **23:** 672–680.

Martinez-Llordella M, Puig-Pey I, Orlando G, Ramoni M, Tisone G, Rimola A, Lerut J, Latinne D, Margarit C, Bilbao I, et al. 2007. Multiparameter immune profiling of operational tolerance in liver transplantation. *Am J Transplant* **7:** 309–319.

Martinez-Llordella M, Lozano JJ, Puig-Pey I, Orlando G, Tisone G, Lerut J, Benitez C, Pons JA, Parrilla P, Ramirez P, et al. 2008. Using transcriptional profiling to develop a diagnostic test of operational tolerance in liver transplant recipients. *J Clin Invest* **118:** 2845–2857.

Matas AJ, Humar A, Gillingham KJ, Payne WD, Gruessner RW, Kandaswamy R, Dunn DL, Najarian JS, Sutherland DE. 2002. Five preventable causes of kidney graft loss in the 1990s: A single-center analysis. *Kidney Int* **62:** 704–714.

Matz M, Beyer J, Wunsch D, Mashreghi MF, Seiler M, Pratschke J, Babel N, Volk HD, Reinke P, Kotsch K. 2006. Early post-transplant urinary IP-10 expression after kidney transplantation is predictive of short- and long-term graft function. *Kidney Int* **69:** 1683–1690.

McDouall RM, Batten P, McCormack A, Yacoub MH, Rose ML. 1997. MHC class II expression on human heart microvascular endothelial cells: Exquisite sensitivity to interferon-γ and natural killer cells. *Transplantation* **64:** 1175–1180.

McNerney ME, Lee KM, Zhou P, Molinero L, Mashayekhi M, Guzior D, Sattar H, Kuppireddi S, Wang CR, Kumar V, Alegre ML. 2006. Role of natural killer cell subsets in cardiac allograft rejection. *Am J Transplant* **6:** 505–513.

Meier-Kriesche HU, Schold JD, Kaplan B. 2004. Long-term renal allograft survival: Have we made significant progress or is it time to rethink our analytic and therapeutic strategies? *Am J Transplant* **4:** 1289–1295.

Mengel M, Sis B, Haas M, Colvin RB, Halloran PF, Racusen LC, Solez K, Cendales L, Demetris AJ, Drachenberg CB, et al. 2012. Banff 2011 meeting report: New concepts in antibody-mediated rejection. *Am J Transplant* **12:** 563–570.

Meyer-ter-Vehn T, Sieprath S, Katzenberger B, Gebhardt S, Grehn F, Schlunck G. 2006. Contractility as a prerequisite for TGF-β-induced myofibroblast transdifferentiation in human tenon fibroblasts. *Invest Ophthalmol Vis Sci* **47:** 4895–4904.

Mezzano SA, Aros CA, Droguett A, Burgos ME, Ardiles LG, Flores CA, Carpio D, Vio CP, Ruiz-Ortega M, Egido J. 2003. Renal angiotensin II up-regulation and myofibroblast activation in human membranous nephropathy. *Kidney Int Suppl* **86:** S39–S45.

Miura M, El-Sawy T, Fairchild FL. 2003. Neutrophils mediate parenchymal tissue necrosis and accelerate the rejection of complete major histocompatibility complex-disparate cardiac allografts in the absence of interferon-γ. *Am J Pathol* **162:** 509–519.

Molesworth-Kenyon SJ, Oakes JE, Lausch RN. 2005. A novel role for neutrophils as a source of T cell-recruiting chemokines IP-10 and Mig during the DTH response to HSV-1 antigen. *J Leukoc Biol* **77:** 552–559.

Montgomery RA, Hardy MA, Jordan SC, Racusen LC, Ratner LE, Tyan DB, Zachary AA. 2004. Consensus opinion from the antibody working group on the diagnosis, reporting, and risk assessment for antibody-mediated rejection and desensitization protocols. *Transplantation* **78:** 181–185.

Montgomery RA, Lonze BE, King KE, Kraus ES, Kucirka LM, Locke JE, Warren DS, Simpkins CE, Dagher NN, Singer AL, et al. 2011. Desensitization in HLA-incompatible kidney recipients and survival. *N Engl J Med* **365:** 318–326.

Montgomery JR, Berger JC, Warren DS, James NT, Montgomery RA, Segev DL. 2012. Outcomes of ABO-incompatible kidney transplantation in the United States. *Transplantation* **93:** 603–609.

Mosser DM, Edwards JP. 2008. Exploring the full spectrum of macrophage activation. *Nat Rev Immunol* **8:** 958–969.

Moxham VF, Karegli J, Phillips RE, Brown KL, Tapmeier TT, Hangartner R, Sacks SH, Wong W. 2008. Homeostatic proliferation of lymphocytes results in augmented memory-like function and accelerated allograft rejection. *J Immunol* **180:** 3910–3918.

Nadazdin O, Boskovic S, Murakami T, Tocco G, Smith RN, Colvin RB, Sachs DH, Allan J, Madsen JC, Kawai T, et al. 2011. Host alloreactive memory T cells influence tolerance to kidney allografts in nonhuman primates. *Sci Transl Med* **3:** 86ra51.

Nafady-Hego H, Li Y, Ohe H, Zhao X, Satoda N, Sakaguchi S, Wood K, Uemoto S, Koshiba T. 2010. The generation of donor-specific $CD4^+CD25^{2+}CD45RA^+$ naive regulatory T cells in operationally tolerant patients after pediatric living-donor liver transplantation. *Transplantation* **90:** 1547–1555.

Nankivell BJ, Borrows RJ, Fung CL, O'Connell PJ, Allen RD, Chapman JR. 2003. The natural history of chronic allograft nephropathy. *N Engl J Med* **349:** 2326–2333.

Nelsestuen GL, Martinez MB, Hertz MI, Savik K, Wendt CH. 2005. Proteomic identification of human neutrophil α-defensins in chronic lung allograft rejection. *Proteomics* **5:** 1705–1713.

Neujahr DC, Perez SD, Mohammed A, Ulukpo O, Lawrence EC, Fernandez F, Pickens A, Force SD, Song M, Larsen CP, et al. 2012. Cumulative exposure to γ interferon-dependent chemokines CXCL9 and CXCL10 correlates with worse outcome after lung transplant. *Am J Transplant* **12:** 438–446.

Newell KA, Asare A, Kirk AD, Gisler TD, Bourcier K, Suthanthiran M, Burlingham WJ, Marks WH, Sanz I, Lechler RI, et al. 2010. Identification of a B cell signature associated with renal transplant tolerance in humans. *J Clin Invest* **120:** 1836–1847.

Ng YH, Oberbarnscheidt MH, Chandramoorthy HC, Hoffman R, Chalasani G. 2010. B cells help alloreactive T cells differentiate into memory T cells. *Am J Transplant* **10:** 1970–1980.

Nickeleit V, Zeiler M, Gudat F, Thiel G, Mihatsch MJ. 2002. Detection of the complement degradation product C4d in renal allografts: Diagnostic and therapeutic implications. *J Am Soc Nephrol* **13:** 242–251.

Niimi M, Hara M, Witzke O, Morris PJ, Wood KJ. 1998. Donor resting B cells induce indefinite prolongation of fully allogeneic cardiac grafts when delivered with anti-immunoglobulin-D monoclonal antibody: Evidence for tolerogenicity of donor resting B cells in vivo. *Transplantation* **66:** 1786–1792.

Noorchashm H, Reed AJ, Rostami SY, Mozaffari R, Zekavat G, Koeberlein B, Caton AJ, Naji A. 2006. B cell-mediated antigen presentation is required for the pathogenesis of acute cardiac allograft rejection. *J Immunol* **177:** 7715–7722.

Nozaki T, Amano H, Bickerstaff A, Orosz CG, Novick AC, Tanabe K, Fairchild RL. 2007. Antibody-mediated rejection of cardiac allografts in CCR5-deficient recipients. *J Immunol* **179:** 5238–5245.

Ochando JC, Krieger NR, Bromberg JS. 2006. Direct versus indirect allorecognition: Visualization of dendritic cell distribution and interactions during rejection and tolerization. *Am J Transplant* **6:** 2488–2496.

Ojo AO, Hanson JA, Wolfe RA, Leichtman AB, Agodoa LY, Port FK. 2000. Long-term survival in renal transplant recipients with graft function. *Kidney Int* **57:** 307–313.

Ozawa M, Terasaki PI, Lee JH, Castro R, Alberu J, Alonso C, Alvarez I, Toledo R, Alvez H, Monterio M, et al. 2007. 14th International HLA and Immunogenetics Workshop: Report on the prospective chronic rejection project. *Tissue Antigens* **69:** 174–179.

Paantjens AW, van de Graaf EA, Kwakkel-van Erp JM, Hoefnagel T, van Kessel DA, van den Bosch JM, Otten HG. 2011. Lung transplantation affects expression of the chemokine receptor type 4 on specific T cell subsets. *Clin Exp Immunol* **166:** 103–109.

Pallier A, Hillion S, Danger R, Giral M, Racape M, Degauque N, Dugast E, Ashton-Chess J, Pettre S, Lozano JJ, et al. 2010. Patients with drug-free long-term graft function display increased numbers of peripheral B cells with a memory and inhibitory phenotype. *Kidney Int* **78:** 503–513.

Panigrahi A, Siddiqui JA, Rai A, Margoob A, Khaira A, Bhowmik D, Tiwari SC, Guleria S, Mehra MK. 2007. Allosensitization to HLA and MICA is an important measure of renal graft outcome. *Clin Transpl* **2007:** 211–217.

Pascual M, Theruvath T, Kawai T, Tolkoff-Rubin N, Cosimi AB. 2002. Strategies to improve long-term outcomes after renal transplantation. *N Engl J Med* **346:** 580–590.

Pearl JP, Parris J, Hale DA, Hoffmann SC, Bernstein WB, McCoy KL, Swanson SJ, Mannon RB, Roederer M, Kirk AD. 2005. Immunocompetent T-cells with a memory-like phenotype are the dominant cell type following antibody-mediated T-cell depletion. *Am J Transplant* **5:** 465–474.

Pedagogos E, Hewitson TD, Walker RG, Nicholis KM, Becker GJ. 1997. Myofibroblast involvement in chronic transplant rejection. *Transplantation* **64:** 1192–1197.

Pei R, Lee JH, Shih NJ, Chen M, Terasaki PI. 2003. Single human leukocyte antigen flow cytometry beads for accurate identification of human leukocyte antigen antibody specificities. *Transplantation* **75:** 43–49.

Perrella O, Sbreglia C, Arenga G, Perrella A, Ferrara A, D'Antonio A, Di Costanzo G, Atripaldi L, Alone C, Sciano D, et al. 2006. Acute rejection after liver transplantation: Is there a specific immunological pattern? *Transplant Proc* **38:** 3594–3596.

Pilmore HL, Painter DM, Bishop GA, McCaughan GW, Eris JM. 2000. Early up-regulation of macrophages and myofibroblasts: A new marker for development of chronic renal allograft rejection. *Transplantation* **69:** 2658–2662.

Pons JA, Revilla-Nuin B, Baroja-Mazo A, Ramirez P, Martinez-Alarcon L, Sanchez-Bueno F, Robles R, Rios A, Aparicio P, Parrilla P. 2008. FoxP3 in peripheral blood is associated with operational tolerance in liver transplant patients during immunosuppression withdrawal. *Transplantation* **86:** 1370–1378.

Port FK, Dykstra DM, Merion RM, Wolfe RA. 2004. Organ donation and transplantation trends in the USA, 2003. *Am J Transplant* **4:** 7–12.

Posselt AM, Vincenti F, Bedolli M, Lantz M, Roberts JP, Hirose R. 2003. CD69 expression on peripheral CD8 T cells correlates with acute rejection in renal transplant recipients. *Transplantation* **76:** 190–195.

Pratschke J, Stauch D, Kotsch K. 2009. Role of NK and NKT cells in solid organ transplantation. *Transpl Int* **22:** 859–868.

Qi F, Adair A, Ferenbach D, Vass DG, Mylonas KJ, Kipari T, Clay M, Kluth DC, Hughes J, Marson LP. 2008. Depletion of cells of monocyte lineage prevents loss of renal microvasculature in murine kidney transplantation. *Transplantation* **86:** 1267–1274.

Racusen LC, Solez K, Colvin RB, Bonsib SM, Castro MC, Cavallo T, Croker BP, Demetris AJ, Drachenberg CB, Fogo AB, et al. 1999. The Banff 97 working classification of renal allograft pathology. *Kidney Int* **55:** 713–723.

Racusen LC, Colvin RB, Solez K, Mihatsch MJ, Halloran PF, Campbell PM, Cecka MJ, Cosyns JP, Demetris AJ, Fishbein MC, et al. 2003. Antibody-mediated rejection criteria—An addition to the Banff 97 classification of renal allograft rejection. *Am J Transplant* **3:** 708–714.

Raedler H, Vieyra MB, Leisman S, Lakhani P, Kwan W, Yang M, Johnson K, Faas SJ, Tamburini P, Heeger PS. 2011. Anti-complement component C5 mAb synergizes with CTLA4Ig to inhibit alloreactive T cells and prolong cardiac allograft survival in mice. *Am J Transplant* **11:** 1397–1406.

Raffetseder U, Rauen T, Djudjaj S, Kretzler M, En-Nia A, Tacke F, Zimmermann HW, Nelson PJ, Frye BC, Floege J, et al. 2009. Differential regulation of chemokine CCL5 expression in monocytes/macrophages and renal cells by Y-box protein-1. *Kidney Int* **75:** 185–196.

Ramirez AM, Shen Z, Ritzenthaler JD, Roman J. 2006. Myofibroblast transdifferentiation in obliterative bronchiolitis: TGF-β signaling through smad3-dependent and independent pathways. *Am J Transplant* **6:** 2080–2088.

Ramsperger-Gleixner M, Spriewald BM, Tandler R, Kondruweit M, Amann K, Weyand M, Ensminger SM. 2011. Increased transcript levels of TNF-α, TGF-β, and granzyme B in endomyocardial biopsies correlate with allograft rejection. *Exp Clin Transplant* **9:** 387–392.

Rebellato LM, Ozawa M, Verbanac KM, Catrou P, Haisch CE, Terasaki PI. 2006. Clinical and anti-HLA antibody profile of nine renal transplant recipients with failed grafts: Donor-specific and non-donor-specific antibody development. *Clin Transpl* **2006:** 241–253.

Regele H, Bohmig GA, Habicht A, Gollowitzer D, Schillinger M, Rockenschaub S, Watschinger B, Kerjaschki D, Exner M. 2002. Capillary deposition of complement split product C4d in renal allografts is associated with basement membrane injury in peritubular and glomerular capillaries: A contribution of humoral immunity to chronic allograft rejection. *J Am Soc Nephrol* **13:** 2371–2380.

Ridge JP, Di Rosa F, Matzinger P. 1998. A conditioned dendritic cell can be a temporal bridge between a $CD4^+$ T-helper and a T-killer cell. *Nature* **393:** 474–478.

Ro H, Hwang YH, Kim H, Jeong JC, Lee H, Doh YS, Park HC, Oh KH, Park MH, Ha J, et al. 2011. Association of polymorphisms of interleukin-8, CXCR1, CXCR2, and selectin with allograft outcomes in kidney transplantation. *Transplantation* **91:** 57–64.

Rogers PR, Dubey C, Swain SL. 2000. Qualitative changes accompany memory T cell generation: Faster, more effective responses at lower doses of antigen. *J Immunol* **164:** 2338–2346.

Roos-van Groningen MC, Scholten EM, Lelieveld PM, Rowshani AT, Baelde HJ, Bajema IM, Florquin S, Bemelman FJ, de Heer E, de Fijter JW, et al. 2006. Molecular comparison of calcineurin inhibitor-induced fibrogenic responses in protocol renal transplant biopsies. *J Am Soc Nephrol* **17:** 881–888.

Rosenblum JM, Shimoda N, Schenk AD, Zhang H, Kish DD, Keslar K, Farber JM, Fairchild RL. 2010. CXC chemokine ligand CXCL9 and CXCL10 are antagonistic costimulation molecules during the priming of alloreactive T cell effectors. *J Immunol* **184:** 3450–3460.

Rossini M, Cheunsuchon B, Donnert E, Ma LJ, Thomas JW, Neilson EG, Fogo AB. 2005. Immunolocalization of fibroblast growth factor-1 (FGF-1), its receptor (FGFR-1), and fibroblast-specific protein-1 (FSP-1) in inflammatory renal disease. *Kidney Int* **68:** 2621–2628.

Rotondi M, Rosati A, Buonamano A, Lasagni L, Lazzeri E, Pradella F, Fossombroni V, Cirami C, Liotta F, La Villa G, et al. 2004. High pretransplant serum levels of CXCL10/IP-10 are related to increased risk of renal allograft failure. *Am J Transplant* **4:** 1466–1474.

Rotondi M, Netti GS, Lazzeri E, Stallone G, Bertoni E, Chiovato L, Grandaliano G, Gesualdo L, Salvadori M, Schena FP, et al. 2010. High pretransplant serum levels of CXCL9 are associated with increased risk of acute rejection and graft failure in kidney graft recipients. *Transpl Int* **23:** 465–475.

Sagoo P, Perucha E, Sawitzki B, Tomiuk S, Stephens DA, Miqueu P, Chapman S, Craciun L, Sergeant R, Brouard S, et al. 2010. Development of a cross-platform biomarker signature to detect renal transplant tolerance in humans. *J Clin Invest* **120:** 1848–1861.

Sagoo P, Ali N, Garg G, Nestle FO, Lechler RI, Lombardi G. 2011. Human regulatory T cells with alloantigen specificity are more potent inhibitors of alloimmune skin graft damage than polyclonal regulatory T cells. *Sci Transl Med* **3:** 83ra42.

Salama AD, Womer KL, Sayegh MH. 2007. Clinical transplantation tolerance: Many rivers to cross. *J Immunol* **178:** 5419–5423.

Sallusto F, Lenig D, Forster R, Lipp M, Lanzavecchia A. 1999. Two subsets of memory T lymphocytes with distinct homing potentials and effector functions. *Nature* **401:** 708–712.

Sarma NJ, Tiriveedhi V, Angaswamy N, Mohanakumar T. 2012. Role of antibodies to self-antigens in chronic allograft rejection: Potential mechanism and therapeutic implications. *Hum Immunol* **10:** 10.

Schaub S, Nickerson P, Rush D, Mayr M, Hess C, Golian M, Stefura W, Hayglass K. 2009. Urinary CXCL9 and CXCL10 levels correlate with the extent of subclinical tubulitis. *Am J Transplant* **9:** 1347–1353.

Schluns KS, Kieper WC, Jameson SC, Lefrancois L. 2000. Interleukin-7 mediates the homeostasis of naive and memory CD8 T cells in vivo. *Nat Immunol* **1:** 426–432.

Schowengerdt KO, Fricker FJ, Bahjat KS, Kuntz ST. 2000. Increased expression of the lymphocyte early activation marker CD69 in peripheral blood correlates with histologic evidence of cardiac allograft rejection. *Transplantation* **69:** 2102–2107.

Seiler M, Brabcova I, Viklicky O, Hribova P, Rosenberger C, Pratschke J, Lodererova A, Matz M, Schonemann C, Reinke P, et al. 2007. Heightened expression of the cytotoxicity receptor NKG2D correlates with acute and chronic nephropathy after kidney transplantation. *Am J Transplant* **7:** 423–433.

Shenoy KV, Solomides C, Cordova F, Rogers TJ, Ciccolella D, Criner GJ. 2012. Low CD4/CD8 ratio in bronchus-associated lymphoid tissue is associated with lung allograft rejection. *J Transplant* **2012:** 928081.

Shevach EM. 2009. Mechanisms of FoxP3$^+$ T regulatory cell-mediated suppression. *Immunity* **30:** 636–645.

Shin GT, Kim SJ, Lee TS, Oh CK, Kim H. 2005. Gene expression of perforin by peripheral blood lymphocytes as a marker of acute rejection. *Nephron Clin Pract* **100:** c63–c70.

Silva HM, Takenaka MC, Moraes-Vieira PM, Monteiro SM, Hernandez MO, Chaara W, Six A, Agena F, Sesterheim P, Barbe-Tuana FM, et al. 2012. Preserving the B-cell compartment favors operational tolerance in human renal transplantation. *Mol Med* **18:** 733–743.

Sis B, Campbell PM, Mueller T, Hunter C, Cockfield SM, Cruz J, Meng C, Wishart D, Solez K, Halloran PF. 2007. Transplant glomerulopathy, late antibody-mediated rejection and the ABCD tetrad in kidney allograft biopsies for cause. *Am J Transplant* **7:** 1743–1752.

Sis B, Jhangri GS, Bunnag S, Allanach K, Kaplan B, Halloran PF. 2009. Endothelial gene expression in kidney transplants with alloantibody indicates antibody-mediated damage despite lack of C4d staining. *Am J Transplant* **9:** 2312–2323.

Smith C, Miles JJ, Khanna R. 2012. Advances in direct T-cell alloreactivity: Function, avidity, biophysics and structure. *Am J Transplant* **12:** 15–26.

Smyth LA, Herrera OB, Golshayan D, Lombardi G, Lechler RI. 2006. A novel pathway of antigen presentation by dendritic and endothelial cells: Implications for allorecognition and infectious diseases. *Transplantation* **82:** S15–S18.

Solez K, Vincenti F, Filo RS. 1998. Histopathologic findings from 2-year protocol biopsies from a U.S. multicenter kidney transplant trial comparing tarolimus versus cyclosporine: A report of the FK506 Kidney Transplant Study Group. *Transplantation* **66:** 1736–1740.

Solez K, Colvin RB, Racusen LC, Sis B, Halloran PF, Birk PE, Campbell PM, Cascalho M, Collins AB, Demetris AJ, et al. 2007. Banff '05 meeting report: Differential diagnosis of chronic allograft injury and elimination of chronic allograft nephropathy ("CAN"). *Am J Transplant* **7:** 518–526.

Solez K, Colvin RB, Racusen LC, Haas M, Sis B, Mengel M, Halloran PF, Baldwin W, Banfi G, Collins AB, et al. 2008. Banff 07 classification of renal allograft pathology: Updates and future directions. *Am J Transplant* **8:** 753–760.

Sommer M, Gerth J, Stein G, Wolf G. 2005. Transdifferentiation of endothelial and renal tubular epithelial cells into myofibroblast-like cells under in vitro conditions: A morphological analysis. *Cells Tissues Organs* **180:** 204–214.

Sorrentino C, Scarinci A, D'Antuono T, Piccirilli M, Di Nicola M, Pasquale M, Di Iorio C, Di Carlo E. 2006. Endomyocardial infiltration by B and NK cells foreshadows the recurrence of cardiac allograft rejection. *J Pathol* **209:** 400–410.

Sumitran-Holgersson S. 2008. Relevance of MICA and other non-HLA antibodies in clinical transplantation. *Curr Opin Immunol* **20:** 607–613.

Sumitran-Holgersson S, Wilczek HE, Holgersson J, Soderstrom K. 2002. Identification of the nonclassical HLA molecules, mica, as targets for humoral immunity associated with irreversible rejection of kidney allografts. *Transplantation* **74:** 268–277.

Summers AM, Coupes BM, Brennan MF, Ralph SA, Short CD, Brenchley PE. 2005. VEGF-460 genotype plays an important role in progression to chronic kidney disease stage 5. *Nephrol Dial Transplant* **20:** 2427–2432.

Sun HJ, Zhou T, Wang Y, Fu YW, Jiang YP, Zhang LH, Zhang CB, Zhou HL, Gao BS, Shi YA, et al. 2011. Macrophages and T lymphocytes are the predominant cells in intimal arteritis of resected renal allografts undergoing acute rejection. *Transpl Immunol* **25:** 42–48.

Surquin M, Le Moine A, Flamand V, Rombaut K, Demoor FX, Salmon I, Goldman M, Abramowicz D. 2005. IL-4 deficiency prevents eosinophilic rejection and uncovers a role for neutrophils in the rejection of MHC class II disparate skin grafts. *Transplantation* **80:** 1485–1492.

Taflin C, Favier B, Baudhuin J, Savenay A, Hemon P, Bensussan A, Charron D, Glotz D, Mooney N. 2011. Human endothelial cells generate Th17 and regulatory T cells under inflammatory conditions. *Proc Natl Acad Sci* **108:** 2891–2896.

Taflin C, Charron D, Glotz D, Mooney N. 2012. Regulation of the CD4$^+$ T cell allo-immune response by endothelial cells. *Hum Immunol* **16:** 16.

Takeiri M, Tachibana M, Kaneda A, Ito A, Ishikawa Y, Nishiyama S, Goto R, Yamashita K, Shibasaki S, Hirokata G, et al. 2011. Inhibition of macrophage activation and suppression of graft rejection by DTCM-glutarimide, a novel piperidine derived from the antibiotic 9-methylstreptimidone. *Inflamm Res* **60:** 879–888.

Tanaka K, Ozawa K, Teramukai S, Takada Y, Egawa H, Kaihara S, Fujimoto Y, Ogura Y, Kasahara M, Ono M, et al. 2006. Classification of human liver transplant recipients by their preoperative $CD8^+$ T cell subpopulation and its relation to outcome. *Liver Transpl* **12:** 792–800.

Tatapudi RR, Muthukumar T, Dadhania D, Ding R, Li B, Sharma VK, Lozada-Pastorio E, Seetharamu N, Hartono C, Serur D, et al. 2004. Noninvasive detection of renal allograft inflammation by measurements of mRNA for IP-10 and CXCR3 in urine. *Kidney Int* **65:** 2390–2397.

Terasaki PI. 2003. Humoral theory of transplantation. *Am J Transplant* **3:** 665–673.

Terasaki PI, Cai J. 2005. Humoral theory of transplantation: Further evidence. *Curr Opin Immunol* **17:** 541–545.

Terasaki PI, Ozawa M. 2004. Predicting kidney graft failure by HLA antibodies: A prospective trial. *Am J Transplant* **4:** 438–443.

Thaunat O, Nicoletti A. 2008. Lymphoid neogenesis in chronic rejection. *Curr Opin Organ Transplant* **13:** 16–19.

Thaunat O, Field AC, Dai J, Louedec L, Patey N, Bloch MF, Mandet C, Belair MF, Bruneval P, Meilhac O, et al. 2005. Lymphoid neogenesis in chronic rejection: Evidence for a local humoral alloimmune response. *Proc Natl Acad Sci* **102:** 14723–14728.

Thaunat O, Patey N, Morelon E, Michel JB, Nicoletti A. 2006. Lymphoid neogenesis in chronic rejection: The murderer is in the house. *Curr Opin Immunol* **18:** 576–579.

Thaunat O, Patey N, Gautreau C, Lechaton S, Fremeaux-Bacchi V, Dieu-Nosjean MC, Cassuto-Viguier E, Legendre C, Delahousse M, Lang P, et al. 2008. B cell survival in intragraft tertiary lymphoid organs after rituximab therapy. *Transplantation* **85:** 1648–1653.

Thaunat O, Graff-Dubois S, Brouard S, Gautreau C, Varthaman A, Fabien N, Field AC, Louedec L, Dai J, Joly E, et al. 2010. Immune responses elicited in tertiary lymphoid tissues display distinctive features. *PLoS ONE* **5:** e11398.

Thaunat O, Graff-Dubois S, Fabien N, Duthey A, Attuil-Audenis V, Nicoletti A, Patey N, Morelon E. 2012. A stepwise breakdown of B-cell tolerance occurs within renal allografts during chronic rejection. *Kidney Int* **81:** 207–219.

Tyden G, Genberg H, Tollemar J, Ekberg H, Persson NH, Tufveson G, Wadstrom J, Gabel M, Mjornstedt L. 2009. A randomized, doubleblind, placebo-controlled, study of single-dose rituximab as induction in renal transplantation. *Transplantation* **87:** 1325–1329.

Uppaluri R, Sheehan KC, Wang L, Bui JD, Brotman JJ, Lu B, Gerard C, Hancock WW, Schreiber RD. 2008. Prolongation of cardiac and islet allograft survival by a blocking hamster anti-mouse CXCR3 monoclonal antibody. *Transplantation* **86:** 137–147.

Valujskikh A, Pantenburg B, Heeger PS. 2002. Primed allospecific T cells prevent the effects of costimulatory blockade on prolonged cardiac allograft survival in mice. *Am J Transplant* **2:** 501–509

Van der Touw W, Burrell B, Lal G, Bromberg JS. 2012. NK cells are required for costimulatory blockade induced tolerance to vascularized allografts. *Transplantation* **20:** 20.

Veiga-Fernandes H, Walter U, Bourgeois C, McLean A, Rocha B. 2000. Response of naive and memory $CD8^+$ T cells to antigen stimulation in vivo. *Nat Immunol* **1:** 47–53.

Wang Y, Dai H, Liu Z, Cheng X, Tellides G, Dai Z. 2006. Neutralizing IL-7 promotes long-term allograft survival induced by CD40/CD40L costimulatory blockade. *Am J Transplant* **6:** 2851–2860.

Wang S, Li J, Xie A, Wang G, Xia N, Ye P, Rui L, Xia J. 2011. Dynamic changes in Th1, Th17, and $FoxP3^+$ T cells in patients with acute cellular rejection after cardiac transplantation. *Clin Transplant* **25:** E177–E186.

Wasowska BA, Qian Z, Cangello DL, Behrens E, Van Tran K, Layton J, Sanfilippo F, Baldwin MW III. 2001. Passive transfer of alloantibodies restores acute cardiac rejection in IgKO mice. *Transplantation* **71:** 727–736.

Won DI, Jeong HD, Kim YL, Suh JS. 2006. Simultaneous detection of antibody binding and cytotoxicity in flow cytometry crossmatch for renal transplantation. *Cytometry B Clin Cytom* **70:** 82–90.

Wu Z, Bensinger SJ, Zhang J, Chen C, Yuan X, Huang X, Markmann JF, Kassaee A, Rosengard BR, Hancock WW, et al. 2004. Homeostatic proliferation is a barrier to transplantation tolerance. *Nat Med* **10:** 87–92.

Wu T, Bond G, Martin D, Nalesnik MA, Demetris AJ, Abu-Elmagd K. 2006. Histopathologic characteristics of human intestine allograft acute rejection in patients pretreated with thymoglobulin or alemtuzumab. *Am J Gastroenterol* **101:** 1617–1624.

Wynn TA. 2008. Cellular and molecular mechanisms of fibrosis. *J Pathol* **214:** 199–210.

Xu H, Zhang X, Mannon RB, Kirk AD. 2006. Platelet-derived or soluble CD154 induces vascularized allograft rejection independent of cell-bound CD154. *J Clin Invest* **116:** 769–774.

Yabu JM, Higgins JP, Chen G, Sequeira F, Busque S, Tyan DB. 2011. C1q-fixing human leukocyte antigen antibodies are specific for predicting transplant glomerulopathy and late graft failure after kidney transplantation. *Transplantation* **91:** 342–347.

Yamada A, Salama AD, Sho M, Najafian N, Ito T, Forman JP, Kewalramani R, Sandner S, Harada H, Clarkson MR, et al. 2005. CD70 signaling is critical for CD28-independent $CD8^+$ T cell-mediated alloimmune responses in vivo. *J Immunol* **174:** 1357–1364.

Yang WS, Han NJ, Kim CS, Ahn H, Lee SK, Lee KU, Park SK. 2005. STAT1-independent down-regulation of interferon-γ-induced class II transactivator and HLA-DR expression by transforming growth factor β-1 in human glomerular endothelial cells. *Nephron Exp Nephrol* **100:** e124–e131.

Yoo S, Lee EY, Huh KH, Kim MS, Kim YS, Kim HO. 2012. Role of plasma exchange in ABO-incompatible kidney transplantation. *Ann Lab Med* **32:** 283–288.

Yuan X, Dong VM, Coito AJ, Waaga AM, Salama AD, Benjamin CD, Sayegh MH, Chandraker A. 2002. A novel CD154 monoclonal antibody in acute and chronic rat vascularized cardiac allograft rejection. *Transplantation* **73:** 1736–1742.

Yuan X, Paez-Cortez J, Schmitt-Knosalla I, D'Addio F, Mfarrej B, Donnarumma M, Habicht A, Clarkson MR, Iacomini J, Glimcher LH, et al. 2008. A novel role of CD4 Th17 cells in mediating cardiac allograft rejection and vasculopathy. *J Exp Med* **205:** 3133–3144.

Yuan X, Ansari MJ, D'Addio F, Paez-Cortez J, Schmitt I, Donnarumma M, Boenisch O, Zhao X, Popoola J, Clarkson MR, et al. 2009. Targeting Tim-1 to overcome resistance to transplantation tolerance mediated by CD8 T17 cells. *Proc Natl Acad Sci* **106:** 10734–10739.

Zafar MN, Terasaki PI, Naqvi SA, Rizvi SA. 2006. Non-HLA antibodies after rejection of HLA identical kidney transplants. *Clin Transpl* **2006:** 421–426.

Zarkhin V, Chalasani G, Sarwal MM. 2010. The yin and yang of B cells in graft rejection and tolerance. *Transplant Rev (Orlando)* **24:** 67–78.

Zhai Y, Meng L, Gao F, Busuttil RW, Kupiec-Weglinski JW. 2002. Allograft rejection by primed/memory $CD8^+$ T cells is CD154 blockade resistant: Therapeutic implications for sensitized transplant recipients. *J Immunol* **169:** 4667–4673.

Zhai Y, Shen XD, Hancock WW, Gao F, Qiao B, Lassman C, Belperio JA, Strieter RM, Busuttil RW, Kupiec-Weglinski JW. 2006. $CXCR3^+CD4^+$ T cells mediate innate immune function in the pathophysiology of liver ischemia/reperfusion injury. *J Immunol* **176:** 6313–6322.

Zhang ZX, Wang S, Huang X, Min WP, Sun H, Liu W, Garcia B, Jevnikar AM. 2008. NK cells induce apoptosis in tubular epithelial cells and contribute to renal ischemia-reperfusion injury. *J Immunol* **181:** 7489–7498.

Zhou W, Farrar CA, Abe K, Pratt JR, Marsh JE, Wang Y, Stahl GL, Sacks SH. 2000. Predominant role for C5b-9 in renal ischemia/reperfusion injury. *J Clin Invest* **105:** 1363–1371.

Zhuo M, Fujiki M, Wang M, Piard-Ruster K, Wai LE, Wei L, Martinez OM, Krams SM. 2010. Identification of the rat NKG2D ligands, RAE1L and RRLT, and their role in allograft rejection. *Eur J Immunol* **40:** 1748–1757.

Zou Y, Stastny P, Susal C, Dohler B, Opelz G. 2007. Antibodies against MICA antigens and kidney-transplant rejection. *N Engl J Med* **357:** 1293–1300.

The Innate Immune System and Transplantation

Conrad A. Farrar[1], Jerzy W. Kupiec-Weglinski[2], and Steven H. Sacks[1]

[1]MRC Centre for Transplantation, Division of Transplantation Immunology and Mucosal Biology, King's College London School of Medicine at Guy's, King's College and St. Thomas' Hospitals, London SE1 9RT, United Kingdom

[2]The Dumont-UCLA Transplant Center, Division of Liver and Pancreas Transplantation, Department of Surgery, David Geffen School of Medicine at UCLA, Los Angeles, California 90095

Correspondence: steven.sacks@kcl.ac.uk

The sensitive and broadly reactive character of the innate immune system makes it liable to activation by stress factors other than infection. Thermal and metabolic stresses experienced during the transplantation procedure are sufficient to trigger the innate immune response and also augment adaptive immunity in the presence of foreign antigen on the donor organ. The resulting inflammatory and immune reactions combine to form a potent effector response that can lead to graft rejection. Here we examine the evidence that the complement and toll-like receptor systems are central to these pathways of injury and present a formidable barrier to transplantation. We review extensive information about the effector mechanisms that are mediated by these pathways, and bring together what is known about the damage-associated molecular patterns that initiate this sequence of events. Finally, we refer to two ongoing therapeutic trials that are evaluating the validity of these concepts in man.

The innate arm of the immune system is geared to rapidly react with broad groups of invasive pathogens but is also triggered by a variety of physical and metabolic insults. The response consists of soluble and cellular mediators of inflammation, which resolve once the threat has been eliminated. However, failure of this response to resolve can result in chronic inflammation, loss of tissue parenchyma and development of tissue fibrosis. The presence of foreign antigen stimulates a more specific pattern of clonal expansion and affinity maturation by cells of the adaptive immune system, the recruitment of which is enhanced by the innate response. Thus the ability to acquire long-lasting immunity mediated by high-affinity T- and B-cell receptors with exquisite sensitivity for antigenic peptides is an integrated response between innate and adaptive arms of host defense.

Whereas the adaptive system ever changes with lifetime experience of infection, the innate response is fixed by evolution. The innate system is directed against shared pathogenic sequences that often include carbohydrate and lipid moieties. A set of pattern-recognition receptors that engage with these ligands is found on tissue-resident and migratory cells. They are also present on antigen-presenting cells, T cells, and B cells. These innate receptors can detect pathogen- or damage-associated molecular patterns. Other innate receptors may detect second signals amplified by the complement and co-

Cite this article as *Cold Spring Harb Perspect Med* doi: 10.1101/cshperspect.a015479

agulation cascades. In effect, tissue cells and infiltrating leukocytes are "hard-wired" to sense potentially dangerous signals in the local environment and accordingly can initiate intracellular signaling pathways that mobilize the inflammatory and immune responses.

The complement and toll-like receptor (TLR) systems are well-characterized components of innate immunity. Their involvement in transplantation offers insight into the general principles outlined above. In addition, they illustrate how the innate immune response is layered into soluble and cellular components that integrate to provide immune surveillance at critical tissue sites. The purpose of this article is to understand how and to what extent these innate pathways can influence graft dysfunction and rejection. We shall also address whether manipulation of specific components at defined cellular locations can steer the response towards graft acceptance. A current perception of how these two pathways could interact is shown in Figure 1.

THE COMPLEMENT SYSTEM AND ITS ROLE IN ORGAN AND CELL TRANSPLANTATION

Complement proteins are expressed throughout the animal kingdom and function in antimicrobial defense. The pivotal protein C3 has a highly conserved thioester group, which enables covalent binding of the activated molecule on pathogen surfaces and elimination by phagocytic cells with receptors for activated C3. Covalently attached C3 also acts as a focal point for the conversion of C5 into an active form leading to the assembly of membrane attack complex, C5b-9 as lethal pores in the pathogen surface. These three basic mechanisms, namely the marking of cells for disposal, pore formation liberation of the proinflammatory fragments C3a and C5a, account for much of the injury to host tissue when the complement control mechanisms are overwhelmed or are defective. Injury of mammalian cells can therefore be related to direct membrane injury caused by C5b-

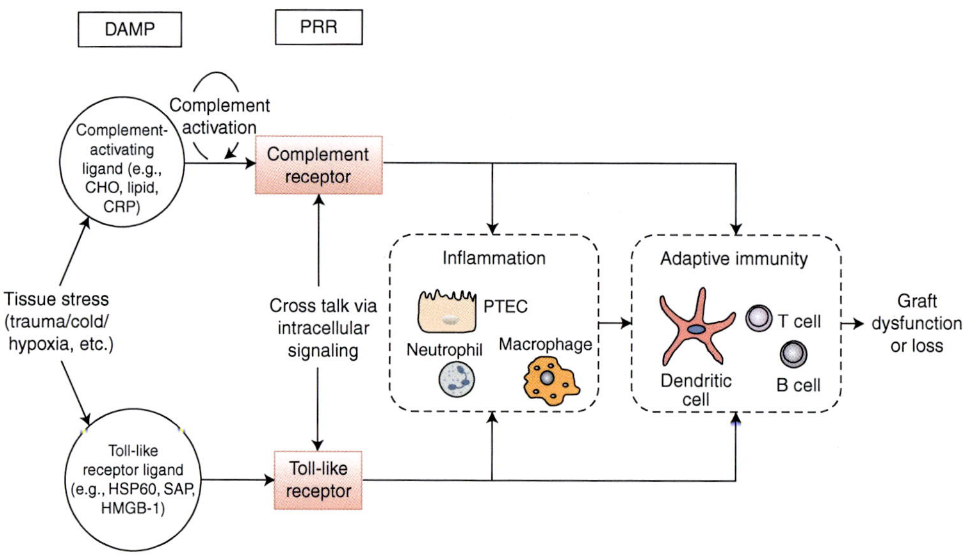

Figure 1. Pathway of injury mediated by innate immune receptors. Toll-like receptors (TLR) are pattern-recognition receptors (PRR) that sense damage-associated molecular patterns (DAMPS). Complement receptors (CR) are PRR that sense complement effector molecules (e.g., C3a, C5a, C3b, iC3b, C3d) generated by DAMP-mediated activation of complement. Stress-induced signaling through PRRs on resident tissue cells and infiltrating leukocytes mediate tissue injury, and on antigen-presenting cells and T cells promote the donor-specific immune response. Effector responses against donor antigen are also PRR-signal dependent. Cross talk between CR and TLR may alter the cellular response in a complex biological system.

 Cite this article as *Cold Spring Harb Perspect Med* doi: 10.1101/cshperspect.a015479

9, or to cell activation mediated by specific receptors for C3a and C5a (C3aR and C5aR), or to interaction with leukocytes that have receptors (CR1-4) for the fragment C3b and its inactive metabolites iC3b and C3d attached to the target cell surface (Fig. 2).

The conversion of C3 into an active form is achieved by enzyme complexes that are assembled by the classical (antibody-mediated), alternative (hydrolytically mediated), and lectin (carbohydrate-mediated) pathways. There is much overlap in the molecular signatures that initiate these pathways. For example, the classical pathway may be triggered by nonimmunoglobulin immune surveillance molecules such as C-reactive protein (Kaplan and Volanakis 1974) and serum amyloid P (Ying et al. 1993). Carbohydrate residues on IgA and IgM molecules may trigger the lectin pathway (Zhang et al. 2006; Shi et al. 2009). The alternative pathway is constantly "ticking over" and may serve to amplify the amount of C3 that is deposited after classical or lectin pathway activation. Precise identification of the relevant pathways may therefore be problematic and, in any case, it may be suboptimal to inhibit just one of these pathways for therapeutic purposes.

Regulators (e.g., CD35, CD46, and CD55) that disrupt the stability of the converting enzyme complexes and consequently limit the cleavage of C3 on cell surfaces are cell protective. Genetic or acquired defects in the function of these regulators can lead to inappropriate activation of complement. Such defects underpin common disorders in which inflammation and immunity play a role (Walport 2001a,b). Complement-mediated injury can therefore result from undue stimulation of the pathways that activate C3 or defective regulation of C3 cleavage, or both.

Local production of complement proteins by many types of tissue-resident cells and infiltrating leukocytes can enhance the functions of innate immunity in health and disease. As a measure of this local capacity, intrarenal synthesis of C3 largely by proximal tubular epithelial cells (PTECs) contributes up to 15% of the circulating pool, the remainder of which is mainly generated by the liver (Tang et al. 1999). Damage or pathogen-related stress is a potent stimulus to this tissue production, increasing the number of copies of local product by several hundredfold (Springall et al. 2001; Farrar et al. 2006). Other essential components (e.g., lectin pathway proteins), are exclusively manufactured in the liver or by infiltrating macrophages.

Complement-Mediated Inflammatory Injury in the Early Course of Transplantation

Complement activation within the donor organ can occur at any stage of the transplant procedure (e.g., activation as a result of severe brain injury in the donor). However, two peaks of activity are recognized, the first coinciding with ischemia-reperfusion (I/R) injury (Farrar et al. 2006) and the second during acute rejection (Pratt et al. 2000), as outlined in Figure 3. Complement activation in the pretransplant kidney biopsy is associated with poor midterm function of the transplanted organ, suggesting a causal relationship with innate immune expression.

Animal experiments have shown a clear role for complement in organ reperfusion injury. Complement-deficient or depleted mice or rats are protected from I/R injury of native or transplanted organs, including heart (Jordan et al. 2001), lung (Eppinger et al. 1997), liver (Strey et al. 2003; Fondevila et al. 2008), intestine (Hart et al. 2005), pancreas (Tjernberg et al. 2008), and kidney (Zhou et al. 2000; Farrar et al. 2006). Loss of renal function is reduced by up to 50% in protected mice (Zhou et al. 2000). Cardiac infarct size is reduced by almost one-half (Schwaeble et al. 2011). Regarding mechanism, C5b-9 is known to mediate parenchymal injury within the kidney, heart, and gut, increasing the cellular release of proinflammatory and fibrotic factors and contributing to tissue infarction. A second mechanism of toxicity involves a direct parenchymal action of C5a and an indirect effect mediated by leukocytes (Peng et al. 2012). In contrast, C3a appears to have no major impact, either in cardiac (Busche and Stahl 2010) or renal reperfusion damage (E. Asgari, pers. comm.).

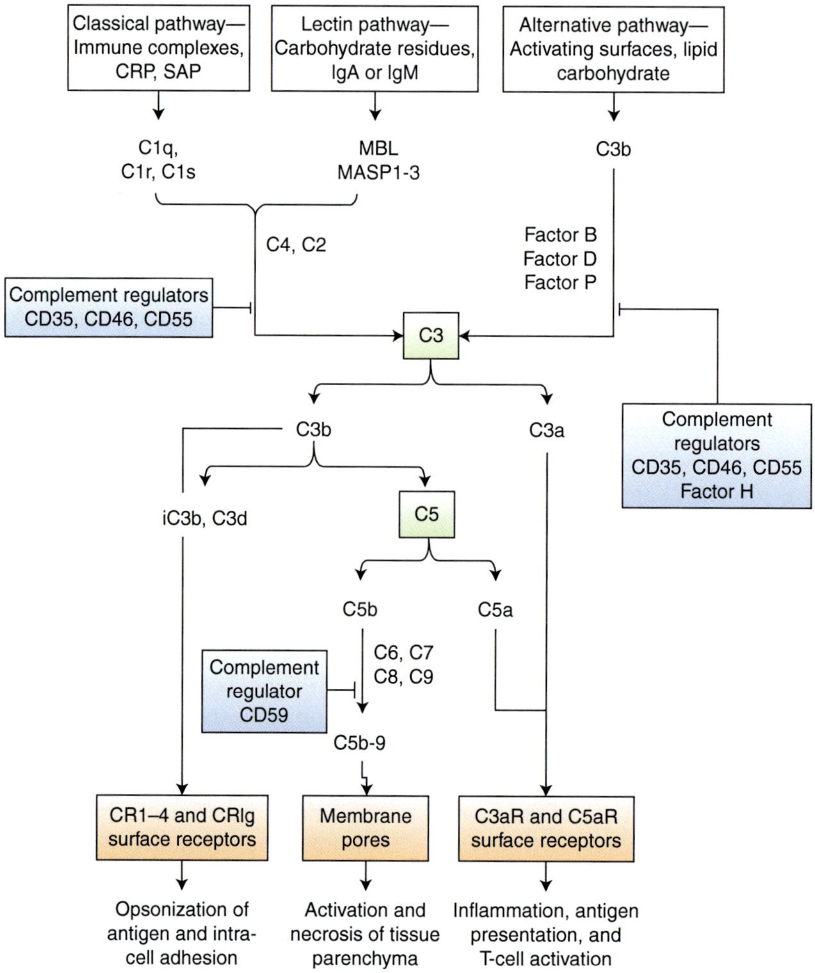

Figure 2. The complement cascade. Complement is activated by three major pathways: classical, lectin, and alternative. The classical pathway is triggered by the binding of C1 to immune surveillance molecules that are attached to the target sequence (e.g., immunoglobulin), C-reactive protein (CRP), and serum amyloid protein (SAP). The lectin pathway is initiated by mannose binding protein (MBL), which binds to carbohydrate residues on the pathogenic surface or IgA and IgM molecules. The alternative pathway is triggered by direct binding of C3b to activating surface. All three pathways progress to form enzyme complexes (classical or alternative pathway) that convert C3 and then C5 into active forms. This generates three groups of complement effectors. C3b is highly reactive and attaches to the activating surface. C3b and metabolites iC3b and C3d are ligands for receptors that are found on leukocytes and mediate inflammation, antigen uptake, and B-cell stimulation. C5b triggers the formation of C5b-9, a multimeric complex that creates a pore in the target cell membrane and induces cell activation and cell death. The small peptide fragments C3a and C5a interact with receptors on leukocytes and parenchymal cells to promote inflammation and, in the presence of foreign antigen, enhance T-cell stimulation. Regulators of complement activation are soluble (e.g., factor H) or membrane-associated, for example, CD35 (complement receptor 1, CR1), CD46 (membrane cofactor protein, MCP), and CD55 (decay-accelerating factor, DAF). The regulators bind C3b (and C4b) and increase its decay or proteolysis from the C3 and C5 convertases of the classical and alternative pathways. Factor H is unique to the alternative pathway. Other regulators inhibit the formation of C5b-9 (e.g., through binding of C3 by CD59).

 Cite this article as *Cold Spring Harb Perspect Med* doi: 10.1101/cshperspect.a015479

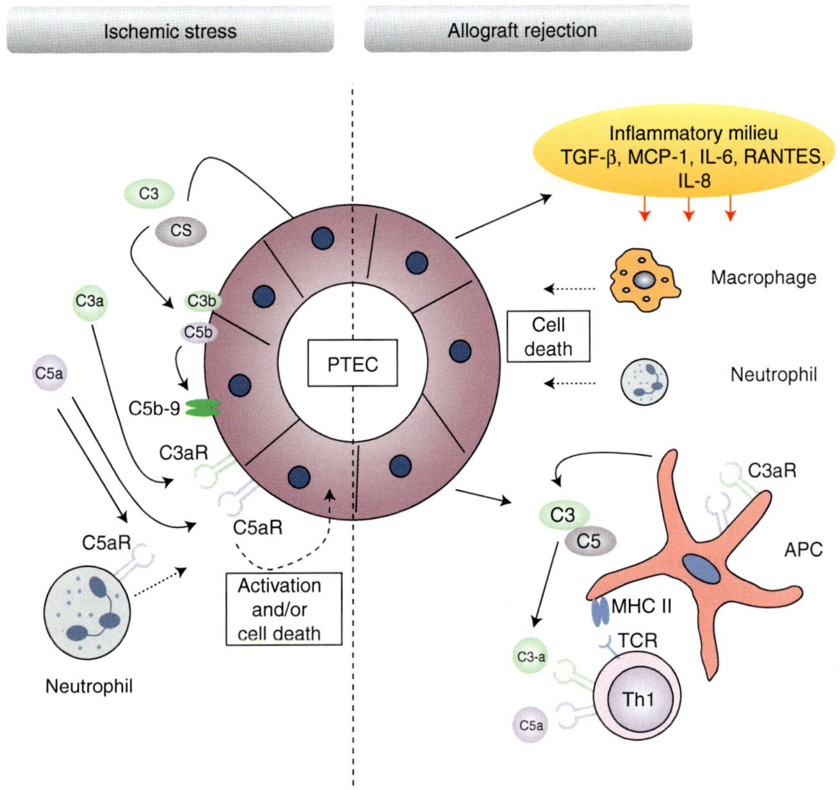

Figure 3. Kidney immune activation against ischemia-reperfusion and transplant injury. Ischemic stress to the kidney induces the local synthesis and secretion of complement proteins directly by proximal tubule epithelial cells. Subsequent activation of the complement cascade leads to C3b and C5b deposition and engagement of receptors by terminal pathway effector molecules such as C5a, which signals via C5aR. Deposition of membranes attack complex (C5b-9) and engagement of C5aR by C5a can lead to PTEC activation and/or cell death. During allograft rejection, chemokines secreted by perturbed PTECs, such as TGF-β, result in an inflammatory phenotype within the interstitium, where lymphocytes, neutrophils, macrophages, and antigen-presenting cells (APCs) enter. APCs express complement components and receptors following complement activation during the inflammatory response and can enhance DC-mediated T-cell priming and the generation of T helper 1 (Th1) cells. C3a and C5a recognition by naïve Th cells can also promote their differentiation to a Th1 phenotype. TCR, T-cell receptor; MHC II, major histocompatibility complex class II.

These pathological functions of complement in I/R injury are highly dependent on local production of C3. Mice that lack the ability to produce C3 within the donor organ were resistant to renal reperfusion damage, despite abundant C3 being produced by the recipient. Graft-derived complement is thus a potential target in strategies to limit the impact on reperfusion damage. More selective blockade may be possible by targeting the relevant activation pathway, but it is often uncertain which pathway is critical. Evidence suggests the lectin pathway plays a role in cardiac and renal models of postischemic injury, possibly triggered by carbohydrate motifs or binding of natural IgM to targets exposed as a result of tissue stress. This will need further resolution before committing to highly selective strategies for complement blockade.

Islet transplantation provides another example of the challenge posed by complement. Donor islets infused into the portal vein trigger the complement (and coagulation) systems (Ricordi and Strom 2004) and this is associated with marked reduction in cell mass. The protease inhibitor α1-antitrypsin (AAT), which includes activity against the complement and coagulation systems, had a protective effect, lowering the threshold number of islets required for successful engraftment (Koulmanda et al. 2012). At present, islets from more than one pancreas are often required for successful engraftment. Experiments are in progress to determine if islet sparing can be achieved by using more specific interventions directed at the complement and coagulation cascades.

So, complement provides a common target for modifying the biological responses of donor organs and cells used in transplantation on exposure to physical and metabolic stress. One therapeutic strategy has used C3 mRNA inhibition to limit production within the donor organ. Another strategy uses a therapeutic construct that is planted in the graft to inhibit the cleavage of C3 and prevent the generation of complement effectors. Having successfully completed preclinical and phase I studies, the construct is being evaluated in a phase II efficacy trial (Smith 2002). Our interest in complement has led to the design of a new imaging ligand based on the receptor CR2 (Badar et al. 2011); this detects organ-bound C3b and provides a means to quantify and monitor the effect of therapy on this specific pathway of injury.

Impact of Complement on Cell-Mediated Graft Rejection

Complement activation is a well-known feature of cell-mediated rejection, as detected by the measurement of activation products in the peripheral blood, urine, and the graft. Large increases in complement gene expression occur within the graft and correspond to cellular infiltration (Pratt et al. 2000, 2002) and release of T-cell cytokines that are known to regulate complement transcription (Gerritsma et al. 1996).

More than 10 years ago, mouse kidney transplant experiments confirmed that local production of complement was essential for allograft rejection. Most animals transplanted with donor kidney from C3-deficient mice were found to accept the grafts for > 100 days (Pratt et al. 2002). In contrast, recipients of wild-type organs rejected their kidney transplants rapidly, even when the recipient was complement-deficient. Heart transplant studies illustrated a similar principle using donor mice that were deficient in the complement regulator CD55. Here, uncontrolled activation of the complement cascade led to enhanced T-cell reactivity and promoted allograft rejection (Pavlov et al. 2008). These experiments highlight the requirement for complement to generate potent T-cell responses against alloantigen.

The mechanisms by which complement mediates T-cell alloreactivity have been investigated. To a large extent, C3a and C5a explain the effects of complement on the T-cell response. These small peptide fragments are usually associated with the inflammatory reaction in severe allergy or sepsis. However, C3a and C5a have also been shown to enhance the function of antigen-presenting cells (APC) (Peng et al. 2008, 2009) and T cells, providing essential costimulatory signals for donor-specific immune recognition (Lalli et al. 2008; Strainic et al. 2008). Receptor signaling induced by C3a and C5a increased the capacity of donor APC to generate Th1 response to alloantigen (Peng et al. 2008, 2009). In addition, C3a and C5a act directly on T cells increasing the differentiation of naïve T helper (Th0) cells to Th1, which mediate graft rejection. C3aR and C5aR signaling were also found to inhibit the development of T-regulatory cells, further directing the immune response down a pathway of rejection (Peng et al. 2006; Strainic et al. 2012; Kwan et al. 2013). Finally, donor parenchymal cells that are coated with complement were more strongly interactive than nonopsonized cells with donor-specific T cells, suggesting that intercell adhesion mediated by complement was an important factor for graft destruction by antigen-experienced T cells (Li et al. 2004). In principle, complement has multiple sites of interaction with the afferent and efferent limbs of the immune response that participate in graft rejection.

Further work has illuminated our understanding of which intracellular signaling pathways mediate the effect of C3a and C5a on antigen presentation. These not only include the expected effects on PI3-kinase and NF-κB signaling, which regulate MHC and costimulatory molecule expression on APC, but also negative regulation of cAMP, an intracellular second messenger with immunosuppressive actions (Li et al. 2008). Thus, both positive and negative signals within APC could transmit the effects of complement-induced signaling on the alloimmune response. In addition, it is possible that signaling via the mammalian target of rapamycin (mTOR) is induced by C5a, because the immunosuppressive effects of C5aR blockade were profoundly enhanced by rapamycin (Strainic et al. 2012). If verified, C5a blockade could provide a possible means to increase the immunosuppressive action of rapamycin.

Cell-autonomous production enhances these immunoregulatory functions of complement. For instance, APC derived from a variety of sources and species express complement components that lead to the generation of C3a and C5a (Li et al. 2011, 2012; Peng et al. 2006, 2009; Zhou et al. 2006). APC with specific complement defects were found to have low levels of MHC and costimulatory molecules on the cell surface and showed poor capacity for T-cell priming; instead these deficient APC promoted the development of Foxp3$^+$ T regulatory function (Peng et al. 2006). And as already noted, donor kidney epithelial cells that express complement components normally are much better targets for donor-specific T cells. None of this is surprising given the impact of donor derived complement on allograft rejection, at least in mice. It is as if cellular emission of complement marks out perturbed cells for the attention of the immune system.

Complement as a Marker and Activator of Humoral Rejection

The complement system was so named because it complements the ability of antibody and macrophages to clear pathogens from the blood. The effector role of complement in acute antibody-mediated rejection (AMR) is relatively well understood (Colvin and Smith 2005). Formed by classical pathway activation at the site of endothelium-bound anti-donor antibody, the terminal components C5a and C5b-9 mediate the acute inflammatory response. Inflammation of the vessel wall together with the pro-thrombotic effects of complement stimulates the coagulation cascade, and this result in vessel occlusion and distal tissue infarction.

Capillary wall C4d is a relatively stable marker of classical pathway activity. The detection of C4d is an integral part of the Banff classification of AMR and serves as a diagnostic and prognostic aid (Solez et al. 2008). The development of a C4d-based assay offers a further step toward improved recognition in serum of pathogenic antibodies against HLA antigens. Given the mounting significance of complement-fixing antibodies in allograft rejection, accurate markers and diagnostic tests with established specificity and sensitivity are likely to be a help in future trials of complement inhibitors. One caveat however is in the context of C5 blockade, which has already shown success in the prevention or reversal of acute AMR (Stegall et al. 2011). C4d will not be directly affected by anti-C5 treatment, because it is formed upstream of C5 in the complement cascade. Therefore C4d may not be a reliable marker in every setting.

Another important observation concerns the mechanism of endothelial resistance to complement attack. "Accommodation" refers to the acquired resistance of graft endothelial cells to injury, despite the presence of pathogenic antibody and complement, as first described in ABO-incompatible transplantation (Park et al. 2003). Although the molecular basis for this resistance remains uncertain, several observers have noted that expression of CD46, CD55, and CD59 is increased in "accommodated" cardiac and renal allografts, suggesting that endothelial stability is maintained through resistance to complement and inducible cell death (Gonzalez-Stawinski et al. 2008; Griesemer et al. 2009; Tan et al. 2009). If confirmed, this would support current attempts to induce longer periods of graft acceptance by increasing the number of regulatory molecules on the endothelial surface. Targeted delivery of therapeutic complement and coagulation regulators to the vessel wall provides a feasible means to achieve this (Smith et al. 2007).

It is also conceivable that prophylactic treatment to reduce complement activation on the vessel wall may help to reduce the sensitization of the recipient against donor HLA antigen. The initiation of IgG production against donor major histocompatibility complex antigen was found to be complement dependent, in common with other IgG responses to different antigens (Marsh et al. 2001). Opsonization with C3 is known to improve antigen retention in lymphoid tissue and increases the sensitivity of B cells to antigenic stimulation (Dempsey et al. 1996), giving a plausible explanation for the effect of complement on alloantibody production. Blockade of this afferent limb of the antibody response may have value in preventing sensitization against a range of alloantigens.

THE TOLL-LIKE RECEPTOR SYSTEM AND ITS ROLE IN ORGAN INJURY AND TRANSPLANTATION

Toll-like receptors were first discovered in 1998, in mice displaying endotoxin resistance but a high susceptibility to gram-negative bacterial infection (Poltorak et al. 1998). TLRs are an evolutionarily conserved group of trans-membrane proteins of which to date, 11 have been identified in humans and 13 in mice (Table 1). These innate receptors have a central role in immunity against invading pathogens by virtue of their ability to transduce signals in response to ligation of distinctive molecular motifs termed pathogen-associated molecular patterns (PAMPs). They are a major group of pattern-recognition receptors and are ubiquitous, being expressed on a host of both immune and non-immune cells (Arumugam et al. 2009). TLR-PAMP interactions lead to downstream effects such as cytokine and chemokine release and augmentation of costimulatory molecule expression (Sobek et al. 2004). All TLRs mediate signal transduction via the adapter molecule myeloid differentiation factor 88 (MyD88), apart from TLR3, which is dependent on the adapter molecule Toll/IL-1R domain-containing adapter-inducing IFN-β (Trif) and TLR4 through which signaling is dependent on both Trif and MyD88 (Robson 2009). The observation that TLRs are expressed on parenchymal cells suggests that they may have functions unrelated to immune-mediated destruction of pathogens. Indeed, it is now apparent that endogenous, cell-derived ligands (DAMPs) derived from both intracellular and extracellular sources during inflammation and tissue damage are capable of binding TLRs (Yu et al. 2010). DAMPs are released from cells and displayed on the cell surface following cellular injury such as hypoxia; therefore, under normal conditions, these molecules are not expressed and invisible to the immune system. A variety of endogenous DAMPs have been described that can engage TLRs, such as heat shock protein (Dempsey et al. 1996) (HSP60) (Ohashi et al. 2000), purines, heparan sulphate and degradation products of fibronectin, the EDA domain (Okamura et al. 2001). More recently, immune response to allografts has displayed an association of TLRs with heat shock proteins and high mobility group box 1 (HMGB1) (Kruger et al. 2010).

TLR-Mediated I/R Injury in Solid Organs

Organ procurement is often associated with a significant period of cold ischemia, which has a well-documented deleterious impact on graft survival. The period of ischemia and subsequent reperfusion leads to rapid complement activation as discussed earlier. Likewise, TLRs are rapidly upregulated during I/R injury. In animal models of renal I/R injury, in the absence of any alloimmune events, expression of both TLR2 (Shigeoka et al. 2007) and TLR4 has been characterized, with demonstrable upregulation of TLR4 in the outer medulla of ischemic kidney just 4 h postischemia, with extensive expression on proximal tubular epithelial cells (PTECs) 24 h after injury (Chen et al. 2011). MyD88-dependent signaling via TLR4 is required for full development of I/R injury as both TLR4 and MyD88 knockout mice are resistant to renal I/R injury (Wu et al. 2007). Conversely, TLR4 mediates hepatic I/R injury in an MyD88-independent fashion, showing the diversity and complexity of the signaling mechanisms in different organ models (Zhai et al. 2004). Indeed, the absence of TLR4 signal-

Table 1. TLRs—their microbial, endogenous ligands and cellular distribution

Receptor	Microbial ligand(s)	Endogenous ligands	Cellular expression
TLR1	Triacyl lipopeptides		B cells, monocytes, macrophages, and certain dendritic cells
TLR2	Peptodoglycan, zymosan, lipoteichoic acid, and glycolipids	HSP60, HSP70, hyaluronan, HMGB1	Monocytes and macrophages, mast cells and myeloid dendritic cells
TLR3	Double-stranded RNA, poly I:C	Messenger RNA (mRNA)	B cells, dendritic cells, and fibroblasts
TLR4	LPS	Fibrinogen, HSPs, surfactant protein A, b-defensin 2, hyaluronan, fibronectin extra domain A, heparin sulfate, HMGB-1	Monocytes and macrophages, mast cells, certain dendritic cells, B cells; intestinal epithelium and hepatocytes (low)
TLR5	Flagellin		Monocytes and macrophages, subset of dendritic cells; intestine
TLR6	Multiple diacyl lipopeptides on mycoplasma		B cells, mast cells, and macrophages
TLR7	Single-stranded RNA imidazoquinolines	RNA and protein complexes	Plasmacytoid dendritic cells, monocytes, and macrophages; B cells
TLR8	Single-stranded RNA imidazoquinolines and small synthetic compounds		Monocytes and macrophages; subset of dendritic cells; mast cells
TLR9	CpG oligodeoxynucleotide DNA		Monocytes, macrophages, and plasmacytoid dendritic cells
TLR10	Undefined		B cells, monocytes, and regulatory T cells
TLR11	Profilin		Kidney and urinary bladder epithelium
TLR12	Profilin		Macrophages, neurons, and dendritic cells
TLR13	Conserved bacterial 23S ribosomal RNA (rRNA) sequence		Monocytes, macrophages, and dendritic cells

Data based on Robson (2009).

ing in the donor organ is required to reduce I/R injury in a mouse liver transplantation model (Shen et al. 2007).

Mice that are either deficient for TLR2 or receive antisense TLR2 oligonucleotide therapy display significant protection from renal I/R injury (Leemans et al. 2005). This protective effect of therapeutic blockade of TLR2 is also observed after cardiac ischemia (Arslan et al. 2010). In the context of solid organ transplantation, both donor and recipient cells have the capacity to express TLR2. Notably, ablation of the recipient pool of TLR2 alone with a therapeutic agent directed at murine TLR2 conferred protection from transplantation-associated ischemic injury in an isograft model (Farrar et al. 2012), suggesting that recipient leukocyte expression of TLR2 is an important mediator of the observed injury and constitutes a viable therapeutic target. The acute kidney injury was reduced by up to 60% in terms of renal function. The mechanism of inflammatory damage resulting from TLR-ligand engagement is complex and not completely understood (Leventhal and Schroppel 2012). Evidence to date suggests a complex interaction between the induction of proinflammatory cytokines, complement activation, and TLR signaling. This potential cross talk between complement and TLR systems will be discussed in more detail later in this chapter.

T Cells Facilitate TLR-Mediated I/R Organ Damage

Although innate immune-driven I/R injury may develop in syngeneic grafts, ex vivo, or under sterile conditions, T cells, particularly of CD4 phenotype, are indispensable for the activation of local TLR-mediated proinflammatory immune sequel. The observation that systemic immunosuppression (CsA, FK506) attenuated peritransplant hepatocellular damage provided initial indirect evidence for T-cell involvement in the pathophysiology of organ I/R injury (Suzuki et al. 1993). Studies in T-cell-deficient (nude) and CD4-deficient mouse systems have proven the pivotal function of CD4 T cell in the process (Zwacka et al. 1997; Rabb et al. 2000; Burne et al. 2001; Shen et al. 2002). The question arises as to how T cells may function in the predominantly I/R-triggered innate response and in the absence of exogenous antigen stimulation?

The pathogenic role of T-cell costimulation was initially shown in a study in which CD28 blockade with CTLA-4Ig protected rat kidneys from local I/R damage (Takada et al. 1997). Consistent with the essential function of both CD28 and CD154 molecules to activate TLR-mediated inflammation cascade, livers in CD154 KO or CD28 KO mice and in WT mice treated with anti-CD154 or CTLA-4Ig are all protected from I/R damage (Shen et al. 2002). Indeed, Th1-type cells are critical in the process, as Stat4 KO (deficient in Th1 development) but not Stat6 KO, mice are I/R-injury resistant, whereas reconstitution of nude mice with T cells from Stat6KO, but not Stat4KO, mice restores cardinal features of IR damage (Shen et al. 2003).

Although the role of CD154 has been attributed to its costimulatory T-cell function, CD40 ligation on DC or macrophages by T-cell-derived CD154 represents, first of all, the critical activating signal to innate immune cells. It is plausible that endogenous ligands that trigger I/R injury may be insufficient to fully activate and sustain proinflammatory phenotype in I/R-stressed organ. For instance, Kupffer cells (KC) in the liver allograft may be less sensitive to TLR4 stimulation than peripheral macrophages because of the exposure to portal-drained and gut-derived endotoxin. Liver DC have also lower TLR4 expression levels and are less susceptible to LPS stimulation as compared with their spleen counterparts (De et al. 2005). In fact, conventional DC may exert immune-regulatory functions during I/R by producing IL-10 via a TLR9-mediated mechanism (Bamboat et al. 2010).

Interactions between the T-cell immunoglobulin mucin (TIM) family of costimulatory proteins constitute a novel molecular signaling pathway of T-cell—macrophage regulation at the innate—adaptive interface. Treatment of mice with anti-TIM-1 mAb ameliorates the hepatocellular damage, accompanied by decreased local neutrophil infiltration/activation, inhibition of T lymphocyte/macrophage sequestration and diminished homing of TIM-1 ligand-expressing TIM-4 cells in the ischemic liver (Uchida et al. 2010a). The induction of proinflammatory cytokine and chemokine programs was also blunted, data supported by findings from a renal I/R injury mouse model (Rong et al. 2011). The TIM-3–Gal-9, on the other hand, constitutes a “negative” T-cell costimulation signal, as TIM-3 blockade worsens tissue damage, along with increased IFN-γ and reciprocally depressed IL-10 expression in I/R-stressed organs (Uchida et al. 2010b). The PD-1 (B7)-PD-L1 (H1) “negative” T-cell pathway has been also shown to promote I/R cytoprotection (Ji et al. 2010; Ueki et al. 2011). Thus, multiple T-cell costimulatory pathways, both positive and negative, may function in a two-way traffic fashion to promote vs. inhibit TLR-dependent innate immune responses against I/R-insult (Fig. 4).

 Cite this article as *Cold Spring Harb Perspect Med* doi: 10.1101/cshperspect.a015479

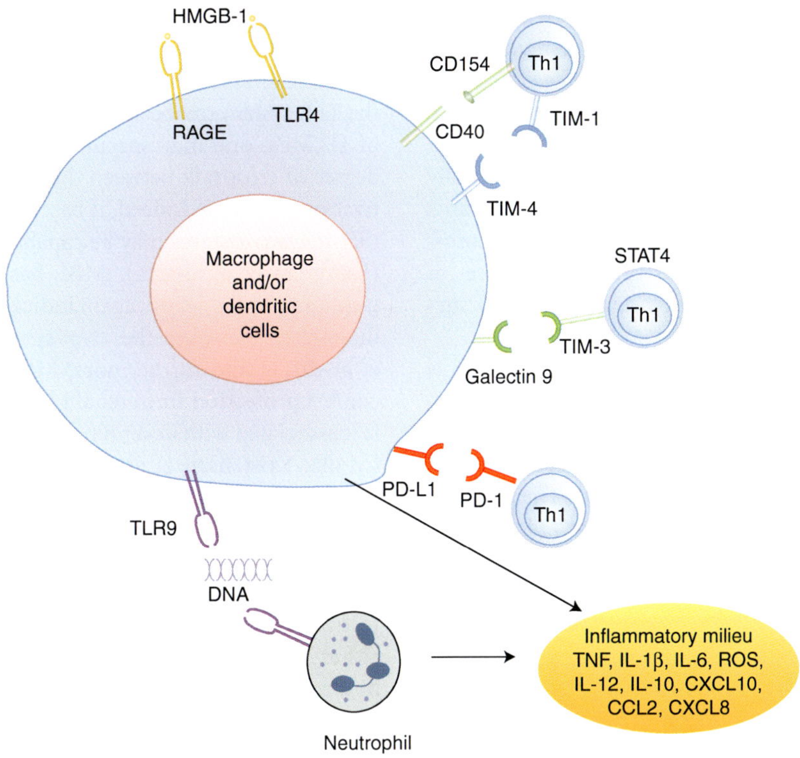

Figure 4. Liver immune activation against ischemia-reperfusion injury. The ischemia insult induces initial cell death, which results in diverse "danger" molecules, such as HMGB1, DNA fragments, and histones activating TLR4, RAGE, and TLR9 signaling on macrophages and/or dendritic cells and neutrophils. $CD4^+$ Th1 effectors might also facilitate and regulate local innate immune activation via CD154–CD40, TIM-1–TIM3, TIM-4–galectin 9, and PD-L1 pathways. The proinflammatory milieu, composed of TNF, IL-1β, IL-5, IL-12, CXCL10, CCL2, CXCL8, and ROS, further activates local immune cells and recruits circulating immune cells, culminating in inflammatory reperfusion injury. HMGB1, high-mobility group protein B1; PD-1, programmed cell death 1; PD-L1, programmed cell death ligand 1; RAGE, receptor for advanced glycation end products; STAT4, signal of transducer and activator of transcription 4; Th1, T-helper type 1 cell; TIM, T cell, immunoglobulin, mucin-containing molecules; TLR, Toll-like receptor. (Image adapted from data from a review by Zhai et al. 2013.)

TLR-Mediated Allograft Rejection

TLR-mediated I/R injury is likely to increase the immunogenicity of a solid organ graft by augmentation of the inflammatory response of innate myeloid cells and lymphocytes that mediate adaptive immune responses (Fig. 1). To this end, it could be argued that TLRs have no direct effect on antigen-specific responses to solid organ transplants but merely exacerbate rejection by increasing the damage associated purely with IR events, for example, through production and interaction of TLR4 and HGMB1 (Kruger et al. 2009). However, there is growing evidence suggesting TLR can directly affect the antigen-specific immune response. TLRs have high expression on DC and ligation of DC-expressed-TLR leads to increased expression of costimulatory molecules (Medzhitov 2001). Mice deficient for the TLR signal adapter molecule MyD88 have impaired $CD4^+$ T-cell function characterized by a skewing towards a Th2 phenotype following immunization with bacterial extracts (Jankovic et al. 2002), suggesting MyD88 is crucial for appropriate development of Th1 or Th2 T cells. Organ transplantation in the presence or absence of TLR and/or specific adapter molecules required for full signaling has yielded mixed results. The absence of MyD88 prolongs skin graft survival in the minor-histocompatibility (HY) model (Goldstein et al. 2003). However, in more robust models, fully MHC-mismatched skin and cardiac transplants show impaired Th1 responses with concomitant reduction in DCs in the absence of MyD88, but with no prolongation of graft survival (Tesar et al. 2004).

Islet transplantation has provided insight as to the role of TLR4 in the rejection process. TLR4 is upregulated after islet transplantation and recipient TLR4 deficiency prolonged allograft survival (Zhang et al. 2010). The islet isolation procedure leads to increased TLR4 expression, a process that can be attenuated by prior exposure of donor islets to carbon monoxide, resulting in prolonged islet allograft survival (Goldberg et al. 2007). In attempts to elucidate how TLR-induced production of endogenous DAMPs mediates allograft rejection, breaking tolerance induced by costimulatory blockade may be just one mechanism. Tolerance to skin grafts induced by donor-specific transfusion and anti-CD154 immunotherapy can be overcome by activation of TLR using TLR agonists, thereby preventing apoptotic clearance of $CD8^+$ effector T cells, which then reject the graft rapidly (Thornley et al. 2006). There is also a contributory T-regulatory element, highlighted by the observation that DAMP production during the alloimmune and concurrent TLR2 and TLR4 activation, leads to downregulation of $Foxp3^+$ regulatory T-cell production (Lal et al. 2011).

CROSS TALK BETWEEN COMPLEMENT AND TLR PATHWAYS

Acute renal failure (ARF) in mice can be circumvented by blocking TLR2 (Farrar et al. 2012). The study tentatively links TLR2-mediated renal injury with complement activation as decreased deposition of activated complement component C3d was observed in the protected kidneys. The lectin complement pathway can be activated during renal I/R injury (Farrar et al. 2009), mediated through an interaction of MBL-MASP-2 complexes bound to DAMPs that are expressed following transplantation. As the TLR system can be activated by engagement of DAMPs, one may speculate there will be a degree of cross talk between the two sets of activation pathways. Indeed, it has been suggested that the two systems may be capable of synergy (Damman et al. 2011a). MBL has been proposed as a TLR4 ligand, again indicative of close interaction between the two systems (Wang et al. 2011) and furthermore, MBL deficiency confers protection from renal I/R injury, an effect associated with lower renal expression of C3 (Moller-Kristensen et al. 2005). Mice that are deficient for both factor B and TLR2 develop severe acute renal failure even though knockout mice for the individual genes display curtailed injury (Amura et al. 2012), suggesting a protective mechanism may be mediated through either TLR2 or complement, or that there may be a regulatory link between the two pathways. Mice deficient in the membrane-bound complement regulator decay-accelerating factor (DAF, CD55) produce large amounts of proinflammatory cytokines in response to TLR4 agonist. In this model, regulation of TLR pathway activity by complement was completely abrogated in double-deficient mice (DAF and C3), an effect indicating strong dependence on complement activation (Zhang et al. 2007). Complement may regulate TLR4-mediated injury in the intestine where I/R injury-induced deposition of activated complement product occurs in a TLR4-dependent fashion (Pope et al. 2010).

Points at which there is cross talk between the two systems could prove attractive for designing targeted therapeutics. DAMPs, such as HGMB1 (Leventhal and Schroppel 2012) and MAP kinase (Zhang et al. 2007), may be key bridging points between the two systems. At present, no such studies in humans have analyzed the extent of cooperation between the two activation pathways (Damman et al. 2011a). However, in rodent models, targeting of DAMPs in islet allografts using a specific anti-HGMB1 or TLR has shown improved graft function and survival (Matsuoka et al. 2010).

 Cite this article as *Cold Spring Harb Perspect Med* doi: 10.1101/cshperspect.a015479

CLINICAL RELEVANCE OF COMPLEMENT AND TLR MECHANISMS OF I/R INJURY

Hyperexpression of the molecular components of the complement and TLR pathways in human transplantation (Tang et al. 1999; Kruger et al. 2009; Naesens et al. 2009; Damman et al. 2011b) provides circumstantial evidence of their pathological relevance and also support the experimental data derived in rodent studies. However, more significant proof of their clinical importance will require specific intervention by therapeutic trial. Reassuringly, at least two such trials are in progress. One is addressing the role of therapeutic complement regulator (mirococept) in the prevention of delayed graft function (DGF) in renal transplantation—an important manifestation of ischemia reperfusion injury with implication for acute rejection rate and long-term graft survival. Another study is assessing the benefit of TLR2 blockade with monoclonal antibody (OPN-305), also in the context of DGF. These two different studies, one aimed to protect the tubular epithelium against complement and the other targeting the function of migratory leukocytes in the evolution of postischemic damage, are primarily designed to assess patient benefit through control of innate immunity. But they are also much needed to establish the validity of two target pathways that are thought to be important in human disease.

ACKNOWLEDGMENTS

S.H.S. and C.A.F. are supported by the Medical Research Council (U.K.) Centre for Transplantation. Their research is also supported by the National Institute for Health Research (NIHR) Biomedical Research Centre based at Guy's and St. Thomas' NHS Foundation Trust and King's College London. The views expressed are those of the investigators and not necessarily those of the NHS, the NIHR, or the Department of Health.

REFERENCES

Amura CR, Renner B, Lyubchenko T, Faubel S, Simonian PL, Thurman JM. 2012. Complement activation and toll-like receptor-2 signaling contribute to cytokine production after renal ischemia/reperfusion. *Mol Immunol* **52:** 249–257.

Arslan F, Smeets MB, O'Neill LA, Keogh B, McGuirk P, Timmers L, Tersteeg C, Hoefer IE, Doevendans PA, Pasterkamp G, et al. 2010. Myocardial ischemia/reperfusion injury is mediated by leukocytic toll-like receptor-2 and reduced by systemic administration of a novel anti-toll-like receptor-2 antibody. *Circulation* **121:** 80–90.

Arumugam TV, Okun E, Tang SC, Thundyil J, Taylor SM, Woodruff TM. 2009. Toll-like receptors in ischemia-reperfusion injury. *Shock* **32:** 4–16.

Badar A, DeFreitas S, McDonnell JM, Yahya N, Thakor D, Razavi R, Smith R, Sacks S, Mullen GE. 2011. Recombinant complement receptor 2 radiolabeled with $[^{99m}Tc(CO)_3]^+$: A potential new radiopharmaceutical for imaging activated complement. *PLoS ONE* **6:** e18275.

Bamboat ZM, Ocuin LM, Balachandran VP, Obaid H, Plitas G, DeMatteo RP. 2010. Conventional DCs reduce liver ischemia/reperfusion injury in mice via IL-10 secretion. *J Clin Invest* **120:** 559–569.

Burne MJ, Daniels F, El GA, Mauiyyedi S, Colvin RB, O'Donnell MP, Rabb H. 2001. Identification of the $CD4^+$ T cell as a major pathogenic factor in ischemic acute renal failure. *J Clin Invest* **108:** 1283–1290.

Busche MN, Stahl GL. 2010. Role of the complement components C5 and C3a in a mouse model of myocardial ischemia and reperfusion injury. *Ger Med Sci* **8:** 1–11.

Chen J, John R, Richardson JA, Shelton JM, Zhou XJ, Wang Y, Wu QQ, Hartono JR, Winterberg PD, Lu CY. 2011. Toll-like receptor 4 regulates early endothelial activation during ischemic acute kidney injury. *Kidney Int* **79:** 288–299.

Colvin RB, Smith RN. 2005. Antibody-mediated organ-allograft rejection. *Nat Rev Immunol* **5:** 807–817.

Damman J, Daha MR, van Son WJ, Leuvenink HG, Ploeg RJ, Seelen MA. 2011a. Cross talk between complement and Toll-like receptor activation in relation to donor brain death and renal ischemia-reperfusion injury. *Am J Transplant* **11:** 660–669.

Damman J, Seelen MA, Moers C, Daha MR, Rahmel A, Leuvenink HG, Paul A, Pirenne J, Ploeg RJ. 2011b. Systemic complement activation in deceased donors is associated with acute rejection after renal transplantation in the recipient. *Transplantation* **92:** 163–169.

De CA, Abe M, Lau H, Hackstein H, Raimondi G, Thomson AW. 2005. Low TLR4 expression by liver dendritic cells correlates with reduced capacity to activate allogeneic T cells in response to endotoxin. *J Immunol* **174:** 2037–2045.

Dempsey PW, Allison ME, Akkaraju S, Goodnow CC, Fearon DT. 1996. C3d of complement as a molecular adjuvant: Bridging innate and acquired immunity. *Science* **271:** 348–350.

Eppinger MJ, Deeb GM, Bolling SF, Ward PA. 1997. Mediators of ischemia-reperfusion injury of rat lung. *Am J Pathol* **150:** 1773–1784.

Farrar CA, Zhou W, Lin T, Sacks SH. 2006. Local extravascular pool of C3 is a determinant of postischemic acute renal failure. *FASEB J* **20:** 217–226.

Farrar CA, Asgari E, Lynch N, Roscher S, Stover CM, Schwaeble WJ, Sacks SH. 2009. Mannan binding lectin associated serine protease-2 (MASP-2) is a critical player in the pathophysiology of renal ischaemia reperfusion (I/R) injury and mediates tissue injury in absence of complement C4. *Mol Immunol* **46:** 2832.

Farrar CA, Keogh B, McCormack W, O'Shaughnessy A, Parker A, Reilly M, Sacks SH. 2012. Inhibition of TLR2 promotes graft function in a murine model of renal transplant ischemia-reperfusion injury. *FASEB J* **26:** 799–807.

Fondevila C, Shen XD, Tsuchihashi S, Uchida Y, Freitas MC, Ke B, Busuttil RW, Kupiec-Weglinski JW. 2008. The membrane attack complex (C5b-9) in liver cold ischemia and reperfusion injury. *Liver Transpl* **14:** 1133–1141.

Gerritsma JS, Gerritsen AF, van KC, van Es LA, Daha MR. 1996. Interleukin-1α enhances the biosynthesis of complement C3 and factor B by human kidney proximal tubular epithelial cells in vitro. *Mol Immunol* **33:** 847–854.

Goldberg A, Parolini M, Chin BY, Czismadia E, Otterbein LE, Bach FH, Wang H. 2007. Toll-like receptor 4 suppression leads to islet allograft survival. *FASEB J* **21:** 2840–2848.

Goldstein DR, Tesar BM, Akira S, Lakkis FG. 2003. Critical role of the Toll-like receptor signal adaptor protein MyD88 in acute allograft rejection. *J Clin Invest* **111:** 1571–1578.

Gonzalez-Stawinski GV, Tan CD, Smedira NG, Starling RC, Rodriguez ER. 2008. Decay-accelerating factor expression may provide immunoprotection against antibody-mediated cardiac allograft rejection. *J Heart Lung Transplant* **27:** 357–361.

Griesemer AD, Okumi M, Shimizu A, Moran S, Ishikawa Y, Iorio J, Arn JS, Yamada K. 2009. Upregulation of CD59: Potential mechanism of accommodation in a large animal model. *Transplantation* **87:** 1308–1317.

Hart ML, Ceonzo KA, Shaffer LA, Takahashi K, Rother RP, Reenstra WR, Buras JA, Stahl GL. 2005. Gastrointestinal ischemia-reperfusion injury is lectin complement pathway dependent without involving C1q. *J Immunol* **174:** 6373–6380.

Jankovic D, Kullberg MC, Hieny S, Caspar P, Collazo CM, Sher A. 2002. In the absence of IL-12, $CD4^+$ T cell responses to intracellular pathogens fail to default to a Th2 pattern and are host protective in an IL-$10^{-/-}$ setting. *Immunity* **16:** 429–439.

Ji H, Shen X, Gao F, Ke B, Freitas MC, Uchida Y, Busuttil RW, Zhai Y, Kupiec-Weglinski JW. 2010. Programmed death-1/B7-H1 negative costimulation protects mouse liver against ischemia and reperfusion injury. *Hepatology* **52:** 1380–1389.

Jordan JE, Montalto MC, Stahl GL. 2001. Inhibition of mannose-binding lectin reduces postischemic myocardial reperfusion injury. *Circulation* **104:** 1413–1418.

Kaplan MH, Volanakis JE. 1974. Interaction of C-reactive protein complexes with the complement system. I: Consumption of human complement associated with the reaction of C-reactive protein with pneumococcal C-polysaccharide and with the choline phosphatides, lecithin and sphingomyelin. *J Immunol* **112:** 2135–2147.

Koulmanda M, Bhasin M, Fan Z, Hanidziar D, Goel N, Putheti P, Movahedi B, Libermann TA, Strom TB. 2012. α1-Antitrypsin reduces inflammation and enhances mouse pancreatic islet transplant survival. *Proc Natl Acad Sci* **109:** 15443–15448.

Kruger B, Krick S, Dhillon N, Lerner SM, Ames S, Bromberg JS, Lin M, Walsh L, Vella J, Fischereder M, et al. 2009. Donor Toll-like receptor 4 contributes to ischemia and reperfusion injury following human kidney transplantation. *Proc Natl Acad Sci* **106:** 3390–3395.

Kruger B, Yin N, Zhang N, Yadav A, Coward W, Lal G, Zang W, Heeger S, Bromberg JS, Murphy B, et al. 2010. Islet-expressed TLR2 and TLR4 sense injury and mediate early graft failure after transplantation. *Eur J Immunol* **40:** 2914–2924.

Kwan WH, van der TW, Paz-Artal E, Li MO, Heeger PS. 2013. Signaling through C5a receptor and C3a receptor diminishes function of murine natural regulatory T cells. *J Exp Med* **210:** 257–268.

Lal G, Yin N, Xu J, Lin M, Schroppel S, Ding Y, Marie I, Levy DE, Bromberg JS. 2011. Distinct inflammatory signals have physiologically divergent effects on epigenetic regulation of Foxp3 expression and Treg function. *Am J Transplant* **11:** 203–214.

Lalli PN, Strainic MG, Yang M, Lin F, Medof ME, Heeger PS. 2008. Locally produced C5a binds to T cell-expressed C5aR to enhance effector T-cell expansion by limiting antigen-induced apoptosis. *Blood* **112:** 1759–1766.

Leemans JC, Stokman G, Claessen N, Rouschop KM, Teske GJ, Kirschning CJ, Akira S, van der PF, Weening JJ, Florquin S. 2005. Renal-associated TLR2 mediates ischemia/reperfusion injury in the kidney. *J Clin Invest* **115:** 2894–2903.

Leventhal JS, Schroppel B. 2012. Toll-like receptors in transplantation: Sensing and reacting to injury. *Kidney Int* **81:** 826–832.

Li K, Patel H, Farrar CA, Hargreaves RE, Sacks SH, Zhou W. 2004. Complement activation regulates the capacity of proximal tubular epithelial cell to stimulate alloreactive T cell response. *J Am Soc Nephrol* **15:** 2414–2422.

Li K, Anderson KJ, Peng Q, Noble A, Lu B, Kelly AP, Wang N, Sacks SH, Zhou W. 2008. Cyclic AMP plays a critical role in C3a-receptor-mediated regulation of dendritic cells in antigen uptake and T-cell stimulation. *Blood* **112:** 5084–5094.

Li K, Fazekasova H, Wang N, Sagoo P, Peng Q, Khamri W, Gomes C, Sacks SH, Lombardi G, Zhou W. 2011. Expression of complement components, receptors and regulators by human dendritic cells. *Mol Immunol* **48:** 1121–1127.

Li K, Fazekasova H, Wang N, Peng Q, Sacks SH, Lombardi G, Zhou W. 2012. Functional modulation of human monocytes derived DCs by anaphylatoxins C3a and C5a. *Immunobiology* **217:** 65–73.

Marsh JE, Farmer CK, Jurcevic S, Wang Y, Carroll MC, Sacks SH. 2001. The allogeneic T and B cell response is strongly dependent on complement components C3 and C4. *Transplantation* **72:** 1310–1318.

Matsuoka N, Itoh T, Watarai H, Sekine-Kondo E, Nagata N, Okamoto K, Mera T, Yamamoto H, Yamada S, Maruyama I, et al. 2010. High-mobility group box 1 is involved in the initial events of early loss of transplanted islets in mice. *J Clin Invest* **120:** 735–743.

Medzhitov R. 2001. Toll-like receptors and innate immunity. *Nat Rev Immunol* **1:** 135–145.

Moller-Kristensen M, Wang W, Ruseva M, Thiel S, Nielsen S, Takahashi K, Shi L, Ezekowitz A, Jensenius JC, Gadjeva M. 2005. Mannan-binding lectin recognizes structures on ischaemic reperfused mouse kidneys and is implicated in tissue injury. *Scand J Immunol* **61:** 426–434.

Naesens M, Li L, Ying L, Sansanwal P, Sigdel TK, Hsieh SC, Kambham N, Lerut E, Salvatierra O, Butte AJ, et al. 2009. Expression of complement components differs between kidney allografts from living and deceased donors. *J Am Soc Nephrol* **20:** 1839–1851.

Ohashi K, Burkart V, Flohe S, Kolb H. 2000. Cutting edge: Heat shock protein 60 is a putative endogenous ligand of the toll-like receptor-4 complex. *J Immunol* **164:** 558–561.

Okamura Y, Watari M, Jerud ES, Young DW, Ishizaka ST, Rose J, Chow JC, Strauss JF III. 2001. The extra domain A of fibronectin activates Toll-like receptor 4. *J Biol Chem* **276:** 10229–10233.

Park WD, Grande JP, Ninova D, Nath KA, Platt JL, Gloor JM, Stegall MD. 2003. Accommodation in ABO-incompatible kidney allografts, a novel mechanism of self-protection against antibody-mediated injury. *Am J Transplant* **3:** 952–960.

Pavlov V, Raedler H, Yuan S, Leisman S, Kwan WH, Lalli PN, Medof ME, Heeger PS. 2008. Donor deficiency of decay-accelerating factor accelerates murine T cell-mediated cardiac allograft rejection. *J Immunol* **181:** 4580–4589.

Peng Q, Li K, Patel H, Sacks SH, Zhou W. 2006. Dendritic cell synthesis of C3 is required for full T cell activation and development of a Th1 phenotype. *J Immunol* **176:** 3330–3341.

Peng Q, Li K, Anderson K, Farrar CA, Lu B, Smith RA, Sacks SH, Zhou W. 2008. Local production and activation of complement up-regulates the allostimulatory function of dendritic cells through C3a-C3aR interaction. *Blood* **111:** 2452–2461.

Peng Q, Li K, Wang N, Li Q, Asgari E, Lu B, Woodruff TM, Sacks SH, Zhou W. 2009. Dendritic cell function in allostimulation is modulated by C5aR signaling. *J Immunol* **183:** 6058–6068.

Peng Q, Li K, Smyth LA, Xing G, Wang N, Meader L, Lu B, Sacks SH, Zhou W. 2012. C3a and C5a promote renal ischemia-reperfusion injury. *J Am Soc Nephrol* **23:** 1474–1485.

Poltorak A, He X, Smirnova I, Liu MY, Van HC, Du X, Birdwell D, Alejos E, Silva M, Galanos C, et al. 1998. Defective LPS signaling in C3H/HeJ and C57BL/10ScCr mice: Mutations in Tlr4 gene. *Science* **282:** 2085–2088.

Pope MR, Hoffman SM, Tomlinson S, Fleming SD. 2010. Complement regulates TLR4-mediated inflammatory responses during intestinal ischemia reperfusion. *Mol Immunol* **48:** 356–364.

Pratt JR, Abe K, Miyazaki M, Zhou W, Sacks SH. 2000. In situ localization of C3 synthesis in experimental acute renal allograft rejection. *Am J Pathol* **157:** 825–831.

Pratt JR, Basheer SA, Sacks SH. 2002. Local synthesis of complement component C3 regulates acute renal transplant rejection. *Nat Med* **8:** 582–587.

Rabb H, Daniels F, O'Donnell M, Haq M, Saba SR, Keane W, Tang WW. 2000. Pathophysiological role of T lymphocytes in renal ischemia-reperfusion injury in mice. *Am J Physiol Renal Physiol* **279:** F525–F531.

Ricordi C, Strom TB. 2004. Clinical islet transplantation: Advances and immunological challenges. *Nat Rev Immunol* **4:** 259–268.

Robson MG. 2009. Toll-like receptors and renal disease. *Nephron Exp Nephrol* **113:** e1–e7.

Rong S, Park JK, Kirsch T, Yagita H, Akiba H, Boenisch O, Haller H, Najafian N, Habicht A. 2011. The TIM-1:TIM-4 pathway enhances renal ischemia-reperfusion injury. *J Am Soc Nephrol* **22:** 484–495.

Schwaeble WJ, Lynch NJ, Clark JE, Marber M, Samani NJ, Ali YM, Dudler T, Parent B, Lhotta K, Wallis R, et al. 2011. Targeting of mannan-binding lectin-associated serine protease-2 confers protection from myocardial and gastrointestinal ischemia/reperfusion injury. *Proc Natl Acad Sci* **108:** 7523–7528.

Shen XD, Ke B, Zhai Y, Amersi F, Gao F, Anselmo DM, Busuttil RW, Kupiec-Weglinski JW. 2002. CD154-CD40 T-cell costimulation pathway is required in the mechanism of hepatic ischemia/reperfusion injury, and its blockade facilitates and depends on heme oxygenase-1 mediated cytoprotection. *Transplantation* **74:** 315–319.

Shen XD, Ke B, Zhai Y, Gao F, Anselmo D, Lassman CR, Busuttil RW, Kupiec-Weglinski JW. 2003. Stat4 and Stat6 signaling in hepatic ischemia/reperfusion injury in mice: HO-1 dependence of Stat4 disruption-mediated cytoprotection. *Hepatology* **37:** 296–303.

Shen XD, Ke B, Zhai Y, Gao F, Tsuchihashi S, Lassman CR, Busuttil RW, Kupiec-Weglinski JW. 2007. Absence of toll-like receptor 4 (TLR4) signaling in the donor organ reduces ischemia and reperfusion injury in a murine liver transplantation model. *Liver Transpl* **13:** 1435–1443.

Shi T, Moulton VR, Lapchak PH, Deng GM, le Lucca JJ, Tsokos GC. 2009. Ischemia-mediated aggregation of the actin cytoskeleton is one of the major initial events resulting in ischemia-reperfusion injury. *Am J Physiol Gastrointest Liver Physiol* **296:** G339–G347.

Shigeoka AA, Holscher TD, King AJ, Hall FW, Kiosses WB, Tobias PS, Mackman N, McKay DB. 2007. TLR2 is constitutively expressed within the kidney and participates in ischemic renal injury through both MyD88-dependent and -independent pathways. *J Immunol* **178:** 6252–6258.

Smith RA. 2002. Targeting anticomplement agents. *Biochem Soc Trans* **30:** 1037–1041.

Smith RAG, Koffman G, Chowdhury P, Smith KCG, Watson CJ, Nicholson ML, Zhou W, Sacks SH. 2007. Membrane-localising complement inhibitors—clinical progress. *Mol Immunol* **44:** 3915.

Sobek V, Birkner N, Falk I, Wurch A, Kirschning CJ, Wagner H, Wallich R, Lamers MC, Simon MM. 2004. Direct Toll-like receptor 2 mediated co-stimulation of T cells in the mouse system as a basis for chronic inflammatory joint disease. *Arthritis Res Ther* **6:** R433–R446.

Solez K, Colvin RB, Racusen LC, Haas M, Sis B, Mengel M, Halloran PF, Baldwin W, Banfi G, Collins AB, et al. 2008. Banff 07 classification of renal allograft pathology: Updates and future directions. *Am J Transplant* **8:** 753–760.

Springall T, Sheerin NS, Abe K, Holers VM, Wan H, Sacks SH. 2001. Epithelial secretion of C3 promotes colonization of the upper urinary tract by *Escherichia coli*. *Nat Med* **7:** 801–806.

Stegall MD, Diwan T, Raghavaiah S, Cornell LD, Burns J, Dean PG, Cosio FJ, Gandhi MJ, Kremers W, Gloor JM. 2011. Terminal complement inhibition decreases antibody-mediated rejection in sensitized renal transplant recipients. *Am J Transplant* **11:** 2405–2413.

Strainic MG, Liu J, Huang D, An F, Lalli PN, Muqim N, Shapiro VS, Dubyak GR, Heeger PS, Medof ME. 2008. Locally produced complement fragments C5a and C3a provide both costimulatory and survival signals to naive $CD4^+$ T cells. *Immunity* **28:** 425–435.

Strainic MG, Shevach EM, An F, Lin F, Medof ME. 2012. Absence of signaling into $CD4^+$ cells via C3aR and C5aR enables autoinductive TGF-β1 signaling and induction of $Foxp3^+$ regulatory T cells. *Nat Immunol* **14:** 162–171.

Strey CW, Markiewski M, Mastellos D, Tudoran R, Spruce LA, Greenbaum LE, Lambris JD. 2003. The proinflammatory mediators C3a and C5a are essential for liver regeneration. *J Exp Med* **198:** 913–923.

Suzuki S, Toledo-Pereyra LH, Rodriguez FJ, Cejalvo D. 1993. Neutrophil infiltration as an important factor in liver ischemia and reperfusion injury. Modulating effects of FK506 and cyclosporine. *Transplantation* **55:** 1265–1272.

Takada M, Chandraker A, Nadeau KC, Sayegh MH, Tilney NL. 1997. The role of the B7 costimulatory pathway in experimental cold ischemia/reperfusion injury. *J Clin Invest* **100:** 1199–1203.

Tan CD, Sokos GG, Pidwell DJ, Smedira NG, Gonzalez-Stawinski GV, Taylor DO, Starling RC, Rodriguez ER. 2009. Correlation of donor-specific antibodies, complement and its regulators with graft dysfunction in cardiac antibody-mediated rejection. *Am J Transplant* **9:** 2075–2084.

Tang S, Zhou W, Sheerin NS, Vaughan RW, Sacks SH. 1999. Contribution of renal secreted complement C3 to the circulating pool in humans. *J Immunol* **162:** 4336–4341.

Tesar BM, Zhang J, Li Q, Goldstein DR. 2004. TH1 immune responses to fully MHC mismatched allografts are diminished in the absence of MyD88, a toll-like receptor signal adaptor protein. *Am J Transplant* **4:** 1429–1439.

Thornley TB, Brehm MA, Markees TG, Shultz LD, Mordes JP, Welsh RM, Rossini A, Greiner DL. 2006. TLR agonists abrogate costimulation blockade-induced prolongation of skin allografts. *J Immunol* **176:** 1561–1570.

Tjernberg J, Ekdahl KN, Lambris JD, Korsgren O, Nilsson B. 2008. Acute antibody-mediated complement activation mediates lysis of pancreatic islets cells and may cause tissue loss in clinical islet transplantation. *Transplantation* **85:** 1193–1199.

Uchida Y, Ke B, Freitas MC, Ji H, Zhao D, Benjamin ER, Najafian N, Yagita H, Akiba H, Busuttil RW, et al. 2010a. The emerging role of T cell immunoglobulin mucin-1 in the mechanism of liver ischemia and reperfusion injury in the mouse. *Hepatology* **51:** 1363–1372.

Uchida Y, Ke B, Freitas MC, Yagita H, Akiba H, Busuttil RW, Najafian N, Kupiec-Weglinski JW. 2010b. T-cell immunoglobulin mucin-3 determines severity of liver ischemia/reperfusion injury in mice in a TLR4-dependent manner. *Gastroenterology* **139:** 2195–2206.

Ueki S, Castellaneta A, Yoshida O, Ozaki K, Zhang M, Kimura S, Isse K, Ross M, Shao L, Stolz DB, et al. 2011. Hepatic B7 homolog 1 expression is essential for controlling cold ischemia/reperfusion injury after mouse liver transplantation. *Hepatology* **54:** 216–228.

Walport MJ. 2001a. Complement. First of two parts. *N Engl J Med* **344:** 1058–1066.

Walport MJ. 2001b. Complement. Second of two parts. *N Engl J Med* **344:** 1140–1144.

Wang M, Chen Y, Zhang Y, Zhang L, Lu X, Chen Z. 2011. Mannan-binding lectin directly interacts with Toll-like receptor 4 and suppresses lipopolysaccharide-induced inflammatory cytokine secretion from THP-1 cells. *Cell Mol Immunol* **8:** 265–275.

Wu H, Chen G, Wyburn KR, Yin J, Bertolino P, Eris JM, Alexander SI, Sharland AF, Chadban SJ. 2007. TLR4 activation mediates kidney ischemia/reperfusion injury. *J Clin Invest* **117:** 2847–2859.

Ying SC, Gewurz AT, Jiang H, Gewurz H. 1993. Human serum amyloid P component oligomers bind and activate the classical complement pathway via residues 14–26 and 76–92 of the A chain collagen-like region of C1q. *J Immunol* **150:** 169–176.

Yu L, Wang L, Chen S. 2010. Endogenous toll-like receptor ligands and their biological significance. *J Cell Mol Med* **14:** 2592–2603.

Zhai Y, Shen XD, O'Connell R, Gao F, Lassman C, Busuttil RW, Cheng G, Kupiec-Weglinski JW. 2004. Cutting edge: TLR4 activation mediates liver ischemia/reperfusion inflammatory response via IFN regulatory factor 3-dependent MyD88-independent pathway. *J Immunol* **173:** 7115–7119.

Zhai Y, Petrowsky H, Hong JC, Busuttil RW, Kupiec-Weglinski JW. 2013. Ischaemia-reperfusion injury in liver transplantation—from bench to bedside. *Nat Rev Gastroenterol Hepatol* **10:** 79–89.

Zhang M, Alicot EM, Chiu I, Li J, Verna N, Vorup-Jensen T, Kessler B, Shimaoka M, Chan R, Friend D, et al. 2006. Identification of the target self-antigens in reperfusion injury. *J Exp Med* **203:** 141–152.

Zhang X, Kimura Y, Fang C, Zhou L, Sfyroera G, Lambris JD, Wetsel RA, Miwa T, Song WC. 2007. Regulation of Toll-like receptor-mediated inflammatory response by complement in vivo. *Blood* **110:** 228–236.

Zhang N, Kruger B, Lal G, Luan Y, Yadav A, Zang W, Grimm M, Waaga-Gasser AM, Murphy B, Bromberg JS, et al. 2010. Inhibition of TLR4 signaling prolongs Treg-dependent murine islet allograft survival. *Immunol Lett* **127:** 119–125.

Zhou W, Farrar CA, Abe K, Pratt JR, Marsh JE, Wang Y, Stahl GL, Sacks SH. 2000. Predominant role for C5b-9 in renal ischemia/reperfusion injury. *J Clin Invest* **105:** 1363–1371.

Zhou W, Patel H, Li K, Peng Q, Villiers MB, Sacks SH. 2006. Macrophages from C3-deficient mice have impaired potency to stimulate alloreactive T cells. *Blood* **107:** 2461–2469.

Zwacka RM, Zhang Y, Halldorson J, Schlossberg H, Dudus L, Engelhardt JF. 1997. $CD4^+$ T-lymphocytes mediate ischemia/reperfusion-induced inflammatory responses in mouse liver. *J Clin Invest* **100:** 279–289.

Immunosuppressive Drug Therapy

Choli Hartono, Thangamani Muthukumar, and Manikkam Suthanthiran

Division of Nephrology and Hypertension, Departments of Medicine and Transplantation Medicine, New York Presbyterian Hospital–Weill Cornell Medical College, New York, New York 10065

Correspondence: msuthan@med.cornell.edu

The first successful kidney transplantation between monozygotic identical twins did not require any immunosuppressive drugs. Clinical application of azathioprine and glucocorticosteroids allowed the transfer of organs between genetically disparate donors and recipients. Transplantation is now the standard of care, a lifesaving procedure for patients with failed organs. Progress in our understanding of the immunobiology of rejection has been translated to the development of immunosuppressive agents targeting T cells, B cells, plasma cells, costimulatory signals, complement products, and antidonor antibodies. Modern immunopharmacologic interventions have contributed to the clinical success observed following transplantation but challenges remain in personalizing immunosuppressive therapy.

Organ transplantation, in the absence of immunosuppressive drugs, can be performed between monozygotic identical twins, and the risk of immune rejection is invariant in the absence of drug therapy in almost all donor–recipient combinations. The goal of immunosuppressive therapy is to prevent the graft destructive immune response. Whether rejection prophylaxis strategies prevent the development of a tolerogenic response remains unresolved.

In the decades leading up to 1980, azathioprine and glucocorticosteroids were the primary immunosuppressive drugs. The introduction of calcineurin inhibitors (CNI), cyclosporine (CsA), and tacrolimus (Tac) in the 1980s ushered in an era of improved graft outcome. Small molecules and biologics became available as a result of advancements in drug design and use of recombinant DNA technology. Consequently, transplant clinicians/patients now have an array of agents such as mycophenolate mofetil (MMF), sirolimus, rabbit-antithymocyte globulin (rATG), alemtuzumab, and belatacept for clinical use. Treatment for steroid-resistant rejection is now feasible with novel agents such as rATG. Agents with direct efficacy against the humoral antiallograft response appeared to have improved the outcomes of patients with antibody-mediated rejection. However, we lack long-term data regarding efficacy and toxicity of the newer drugs. Moreover, adverse events such as polyomavirus infection and posttransplant EBV-associated lymphoma are directly related to the increased potency of newer agents. Importantly, the improvement in short-term outcome following their introduction has not extended substantively the life span of transplanted organs. Immunosuppressive agents are also increasingly used in novel protocols to induce transplant tolerance.

Cite this article as *Cold Spring Harb Perspect Med* doi: 10.1101/cshperspect.a015487

We briefly review the immunobiology of the antiallograft response to provide the conceptual framework for the clinical application of multidrug regimens to constrain the antiallograft repertory.

IMMUNOBIOLOGY OF REJECTION

Allograft rejection involves a highly orchestrated action of multiple cell types and mediators. Effective immunosuppression is achieved by targeting these cells and mediators at multiple levels (Fig. 1). Lymphocytes are the principal immune cells for the identification of the foreignness of the allograft and mediate graft damage (rejection) by cell-to-cell interactions and via their secretory products including antibodies that bind to antigens displayed by the allograft and recruit complement components (complement-dependent cytotoxicity) and/or Fc receptor-bearing cells (antibody-dependent cell-mediated cytotoxicity).

The α and β chains on the T cell that recognizes the peptide-major histocompatibility

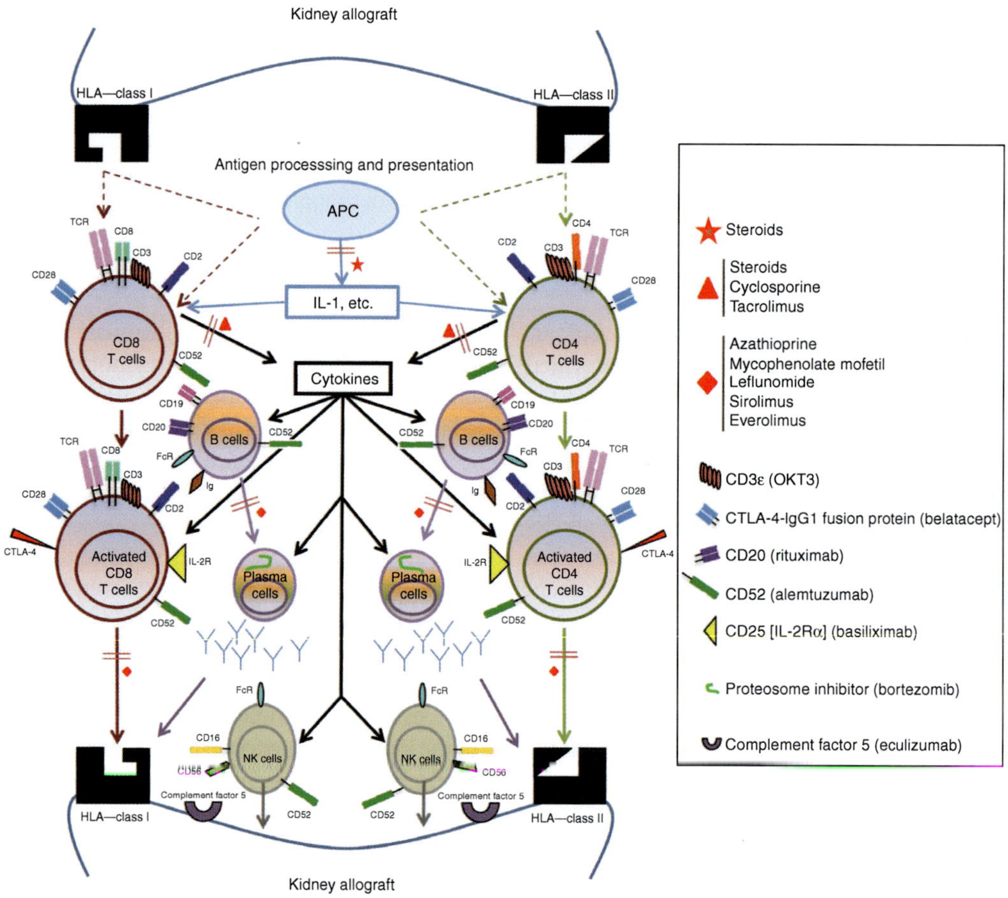

Figure 1. The antiallograft response and sites of action of common immunosuppressive drugs. Schematic representation of human leukocyte antigen (HLA), the primary stimulus for the initiation of the antiallograft response; cell surface proteins participating in antigenic recognition and signal transduction; contribution of the cytokines and multiple cell types to the immune response; and the potential sites for the action of commonly used immunosuppressive drugs. Figure 2 shows the cell surface proteins on antigen-presenting cells (APCs) interacting with T cells to generate costimulatory/coinhibitory signals. (Adapted from Suthanthiran and Strom 1994; reprinted, with permission, from the authors.)

Cite this article as *Cold Spring Harb Perspect Med* doi: 10.1101/cshperspect.a015487

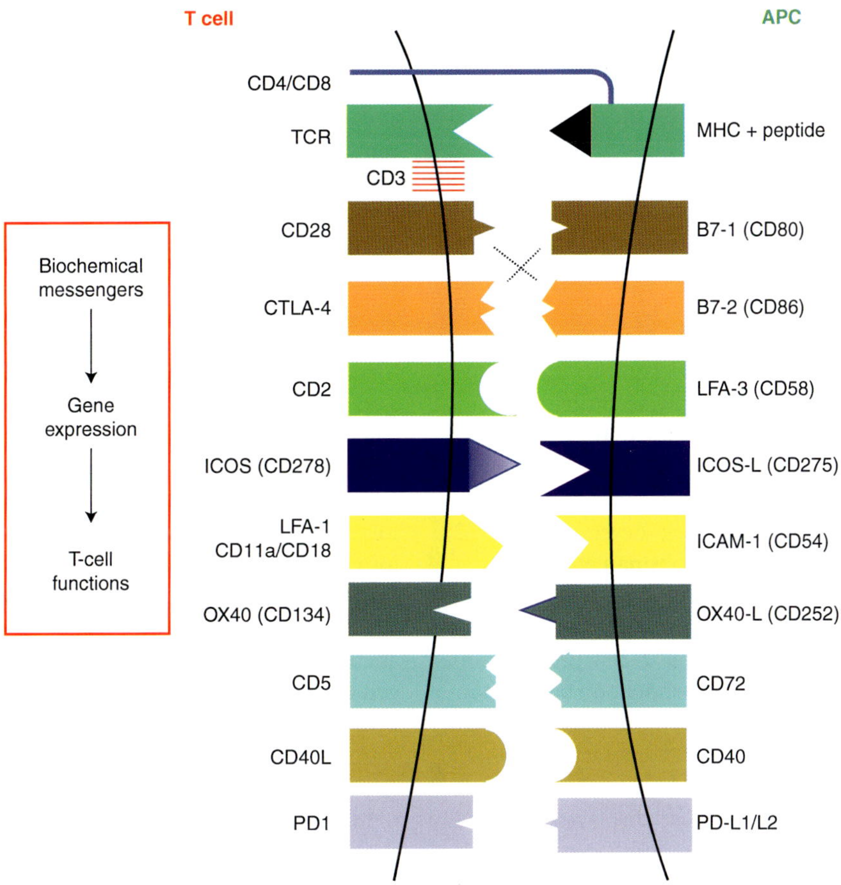

Figure 2. T-cell/APC contact sites and potential targets of immunosuppressive drugs. Schematic representation of T-lymphocyte and antigen-presenting cell contact sites. Signal transduction in T cells on recognition of antigen is not by the TCR itself, but proteins CD3 and ζ noncovalently linked to the TCR. Signaling of T cells via the TCR/CD3 complex (antigenic signal) is necessary, but insufficient in itself to induce maximal T-cell proliferation; plenary activation is dependent on both the antigenic signals and the costimulatory signals engendered by the physical interactions among the cell-surface proteins expressed on antigen-specific T cells and those displayed on APCs. Of all the APCs, mature dendritic cells express the highest level of costimulatory proteins. The best-characterized T-cell costimulation pathway is the interaction of CD28 protein on the T-cell surface with the B7-1 and B7-2 (CD80 and CD86) proteins expressed on activated APCs. In the absence of this second signal, T cells either remain unresponsive or become actively tolerant to antigens. CD28-mediated signals increase the production of cytokines as well as promote the survival of T cells by increasing the expression of antiapoptotic proteins. Although costimulatory pathways were discovered as mediators of T-cell activation, homologous molecules are involved in inhibiting T-cell activation. The key inhibitory receptor is the CTLA-4, a member of CD28 family. The higher the affinity of CTLA-4, as compared to CD28, to B7 may determine the differential binding of stimulatory CD28 and inhibitory CTLA-4 to the same B7 on APCs. Several molecules on the T cells or APCs are potential targets for immunosuppressive drug development. Monoclonal antibodies (anti-CD25, anti-CD2) and recombinant fusion proteins (belatacept) targeting specific cell-surface contact sites are available or undergoing clinical trials as immunosuppressive or tolerance-inducing drugs. (Adapted from Suthanthiran 1996; reprinted, with permission, from the author.)

complex on the surface of antigen-presenting cells (APCs) is the clonotypic T-cell receptor (TCR). Signal transduction in T cells on recognition of antigen is not by the TCR itself, but proteins CD3 and ζ chain noncovalently linked to the TCR. CD4 and CD8 proteins, coreceptors involved in T-cell activation, are expressed on reciprocal T-cell subsets and bind to nonpolymorphic domains of human leucocyte antigen (HLA) class II (DR, DP, DQ) and class I (A, B, C) molecules, respectively. Following activation by antigen, the TCR/CD3 complex and coclustered CD4 and CD8 activate protein tyrosine kinases that are associated with the cytoplasmic tail of CD4 or CD8 and result in activation of several downstream pathways (Brown et al. 1989; Suthanthiran 1990; Beyers et al. 1992; Lebedeva et al. 2004; Fooksman et al. 2010).

Antigenic signaling of T cells via the TCR/CD3 complex is necessary, but insufficient in itself to induce maximal T-cell proliferation; plenary activation is dependent on both the antigenic signals and the costimulatory signals engendered by the physical interactions among the cell-surface proteins expressed on antigen-specific T cells and those displayed on APCs (Fig. 2) (Suthanthiran and Garovoy 1983). Among multiple types of APCs, mature dendritic cells express the highest level of costimulatory proteins and are the most potent antigen-presenting cells. Although some of the costimulatory proteins are expressed in naïve T cells, several of them are expressed following activation of T cells. The best-characterized T-cell costimulation pathway is the interaction of CD28 protein on T cells with the B7-1 and B7-2 (CD80 and CD86) proteins on APCs. In the absence of this second signal, T cells either remain unresponsive or become tolerant to antigens. Although costimulatory pathways were discovered as mediators of T-cell activation, homologous molecules are involved in inhibiting T-cell activation. The key inhibitory receptor is the cytotoxic T-lymphocyte antigen 4 (CTLA-4), a member of CD28 family. The higher the affinity of CTLA-4, as compared to CD28, to B7 may determine the differential binding of stimulatory CD28 and inhibitory CTLA-4 to the same B7 on APCs.

Signal transduction by the TCR/CD3 complex culminates in the transcription of several genes, which includes T-cell growth, survival, and differentiation factor IL-2. Secreted IL-2 signals through IL-2 receptors are induced on activation. The IL-2 receptor signaling proceeds through multiple pathways: Shc/Ras/Raf-1/MAP kinase, JAK1/JAK3/STAT5, and PI 3-kinase/AKT/p70 S6 kinase pathway.

B cells recognize antigens and are activated in the lymphoid tissues. The B-cell antigen receptor complex is made of membrane IgM and IgD associated with the invariant Igα and Igβ molecules. The Igα and Igβ molecules in B cells function similar to CD3 and ζ proteins in the T cells. Complement components play an important role in B-cell activation. B cells express receptor for C3d, a degradation product of complement factor 3 (C3), called CR2 (CD21). The CD21-CD19-CD81 proteins on the B-cell membrane are termed the B-cell coreceptor complex. Antigen and C3d binding to the coreceptor complex activates several kinases resulting in B-cell activation (Batista and Harwood 2009). B-cell response to protein antigens requires recognition of the antigen by the T helper cells and antigen-specific T- and B-cell cooperation (Batista and Harwood 2009). Activated B cells differentiate into antibody-secreting plasma cells.

The net consequence of cytokine production and acquisition of cell-surface receptors for these transcellular molecules by the T cells is the emergence of antigen-specific and graft-destructive T cells. With help from T cells, the humoral arm of immunity is activated resulting in production of donor-specific antibodies. $CD8^+$ cytotoxic T-lymphocyte (CTL)-mediated killing of target cells is mainly accomplished by the directed release of perforin and granzyme, as well as by FasL–Fas interaction, all of which lead to the activation of several apoptotic pathways. Antibodies cause target cell destruction by complement-dependent or complement-independent mechanisms.

IMMUNOSUPPRESSIVE AGENTS

Immunosuppressive medications can be classified into induction or maintenance agents

 Cite this article as *Cold Spring Harb Perspect Med* doi: 10.1101/cshperspect.a015487

Table 1. Induction agents

Drug	Mechanisms of action	FDA approved	Off label
rATG	Polyclonal antibodies against CD2, CD3, CD4, CD8, CD11a, CD18, CD25, CD44, HLA-DR, HLA I heavy chain	FDA approved as antirejection drug Off label as induction agent	
Basiliximab	Chimeric human/murine monoclonal $IgG_{1\kappa}$ against CD25/Tac subunit	Yes	No
Alemtuzumab	Humanized monoclonal $IgG_{1\kappa}$ against CD52	No	Yes
Muromonab-CD3 (OKT3)	Murine monoclonal antibody against CD3 of T-cell receptor complex	FDA approved as antirejection drug Off label as induction agent Voluntarily discontinued	

FDA, Food and Drug Administration.

(Tables 1 and 2). Induction agents serve to eliminate alloreactive lymphocytes when the host's immune system is first exposed to alloantigens following transplantation. Several of the biologics such as rATG and OKT3 were approved for use as antirejection therapy rather than as induction agents. Maintenance agents provide continuous prophylaxis against rejection. Approved and off-label use of specialty drugs exert their effects against B cells and plasma cells or directly inhibit downstream complement activation following antibody binding (Table 3).

Induction Agents

Antithymocyte Globulin (ATG)

Antithymocyte globulin (ATG) is produced by immunizing rabbits or horses with human thy-

Table 2. Maintenance agents

Drugs	Mechanisms	FDA approved	Off label
Azathioprine	Converted to 6-mercaptopurine, inhibits purine biosynthesis and CD28 signaling	Yes	No
Glucocorticoids	Inhibits formation of free NF-κB and down-regulates expression of proinflammatory cytokines	NA	Yes
Cyclosporine	Binds cyclophilin and inhibits calcineurin, prevents activation of NFAT and expression of IL-2, stimulates TGF-β expression	Yes	No
Tacrolimus	Binds FK-binding protein, inhibits calcineurin, prevents activation of NFAT and expression of IL-2, stimulates TGF-β expression	Yes	No
Mycophenolate mofetil	Inhibits inosine monophosphate dehydrogenase and de novo purine biosynthetic pathway	Yes	No
Mycophenolic acid	Enteric coated, inhibits inosine monophosphate dehydrogenase and de novo purine biosynthetic pathway	Yes	No
Sirolimus	Binds FK-binding protein and inhibits mammalian target of rapamycin (mTOR) pathway by binding mTOR complex 1 leading to blockade of activation of 70-kDa S6 protein kinases, expression of bcl-2 proto-oncogene, Ca^{2+}-independent CD28-induced costimulatory pathway	Yes	No
Everolimus	Derivative of rapamycin	Yes	No
Belatacept	Fusion protein of modified CTLA-4-human Ig, blocks B7/CD28 costimulatory proteins	Yes	No
Leflunomide	Inhibits dihydroorotate dehydrogenase and pyrimidine biosynthesis	No	Yes

Table 3. Agents with activities against B cell, plasma cell, and complement components

Drug	Mechanisms	FDA approved	Off label
Bortezomib	Proteasome inhibitor	No	Yes
Eculizumab	Monoclonal antibody against complement C5	No	Yes
Rituximab	Monoclonal antibody against CD20 glycoprotein on B cells	No	Yes
IVIG	Suppress autoantibodies, cytokines, neutralize complements, up-regulate FcγRIIB, immunomodulation	No	Yes

IVIG, intravenous immune globulins.

mocytes generating polyclonal antithymocyte antibodies. The major mode of action for ATG is the depletion of T cells. ATG also targets a wide range of antigens displayed on B cells, dendritic cells, NK cells, and endothelial cells, and disrupts cell trafficking (Mueller 2007). The equine ATG (eATG) product ATGAM became available in the 1980s, whereas rabbit-ATG (rATG) or thymoglobulin was introduced commercially in 1999. rATG or eATG are used both as induction agents, especially in high-risk renal transplant recipients, and for the treatment of moderate to severe acute rejection. Head-to-head comparison study of rabbit versus equine ATG preparations in a randomized double-blinded fashion showed that the acute rejection rate was lower with rATG compared to eATG (4% vs. 25%, RR = 0.09, $P = 0.009$) (Brennan et al. 1999). At 10 yr, the acute rejection rate was 11% for rATG arm and 42% for eATG arm ($P = 0.004$) (Hardinger et al. 2008). Currently, rATG is the most common induction agent in kidney graft recipients.

Interleukin (IL)-2 Receptor Antagonists: Basiliximab and Daclizumab

Both basiliximab and daclizumab are monoclonal antibodies (mAbs) directed at the IL-2 receptor α (CD25 antigen); basiliximab is a chimeric mAb, whereas daclizumab is a humanized mAb. Both antibodies received Food and Drug Administration (FDA) approval as an induction agent for renal transplantation. A meta-analysis showed that when combined with a standard double or triple immunosuppressive regimen, these antibodies reduced the incidence of acute rejection by 34% and the incidence of steroid-resistant rejection by 49%, and that their efficacy in preventing acute rejection was similar to that of OKT3 and polyclonal antibody preparations and, importantly, with fewer side effects (Webster et al. 2004). A prospective randomized international trial tested the efficacy of a 5-day course of ATG versus two doses of basiliximab therapy in 278 deceased-donor renal transplants at risk for acute rejection or delayed graft function (DGF) (Brennan et al. 2006). Participants in the ATG group ($n = 141$) had a lower incidence of acute rejection (15.6% vs. 25.5%) at 12 mo compared to the basiliximab group ($n = 137$). The incidence of DGF, death, and graft loss was similar in the two groups. A Cochrane systematic review of 71 studies (10,537 participants) compared the effects of IL-2Ra to placebo in kidney graft recipients found reduced graft loss because of death with a functioning graft by 25% at 1 yr (RR 0.75, 95% CI, 0.62–0.90) (Webster et al. 2010). The incidence of biopsy-proven acute rejection at 1 yr was also reduced by 28% (RR 0.72, 95% CI, 0.64–0.81). Compared to ATG, induction with IL-2Ra resulted in similar rate of graft loss at any time point and an equivalent incidence of clinical rejection. However, ATG was superior to IL-2Ra when comparing biopsy-proven acute rejection at 1 yr (RR 1.30, 95% CI, 1.01–1.67), but at a cost of 75% increase in malignancy (RR 0.25, 95% CI, 0.07–0.87) and 32% increase in cytomegalovirus (CMV) disease (RR 0.68, 95% CI, 0.50–0.93).

Alemtuzumab

Alemtuzumab (Campath-1H), a humanized monoclonal $IgG_{1\kappa}$ directed against the CD52 glycoprotein and developed by the pathology department at Cambridge University, was approved by the FDA for B-cell chronic lymphocytic leukemia. The INTAC study group tested alemtuzumab against conventional induction agents in a randomized prospective multicenter trial of kidney graft recipients (Hanaway et al. 2011). All participants received Tac and MMF maintenance immunosuppression with rapid steroid discontinuation after 5 days of therapy. Compared to basiliximab, alemtuzumab induction in the low-risk group resulted in a lower rate of biopsy-proven acute rejection at 6 mo (2% vs. 18%, $P < 0.001$), 12 mo (3% vs. 20%, $P < 0.001$), and 36 mo (10% vs. 22%, $P = 0.003$). Compared to rATG, alemtuzumab induction in the high-risk group resulted in equivalent rates of biopsy-proven acute rejection at 6 mo (6% vs. 9% $P = 0.49$), 12 mo (10% vs. 13%, $P = 0.53$), and 36 mo (18% vs. 15%, $P = 0.63$). Patient survival at 3 yr was similar between alemtuzumab and basiliximab in the low-risk group (95% vs. 98%, $P = 0.19$) or between alemtuzumab and ATG in the high-risk group (99% vs. 91%, $P = 0.07$). The rates of late biopsy-proven acute rejection (between 12 and 36 mo) were higher in the alemtuzumab cohort when compared to participants in the conventional induction cohorts (8% vs. 3%, $P = 0.03$). After censoring for deaths, graft survival at 3 yr was also similar in the low-risk (97% vs. 94%, $P = 0.17$) or the high-risk group (91% vs. 84%, $P = 0.32$).

Muromonab—CD3 (OKT3)

OKT3, a mAb directed against the CD3 epsilon component of the CD3 complex and the first of its class to be applied in the clinic, received FDA approval as an antirejection treatment in 1986. A randomized, controlled trial showed that OKT3 reversed 94% of acute rejection episodes in 63 kidney graft recipients compared to 75% in the 60 recipients treated with corticosteroids ($P = 0.009$) (Ortho Multicenter Transplant Study Group 1985). Before its voluntary withdrawal from the U.S. market, OKT3 was used as an induction agent in renal transplantation and for the treatment of steroid-resistant acute rejection. OKT3-associated first-dose reaction as a result of cytokine release may be severe and includes fever, chills, respiratory symptoms, and headaches. A humanized preparation (HuM291) with lesser cytokine release reaction was investigated in a phase I study of kidney graft recipients (Norman et al. 2000) and in patients with new onset insulin-dependent diabetes mellitus (Herold et al. 2002).

Maintenance Drugs for Prophylaxis against Rejection

Azathioprine

Azathioprine (AZA) is a prodrug of 6-mercaptopurine and inhibits purine biosynthesis, and blocks CD28 costimulatory signaling via Rac 1 (Tiede et al. 2003). In a randomized conversion trial from MMF to AZA in 48 stable kidney transplant recipients at 6 mo following transplantation, acute rejection rates were comparable (4.5% vs. 3.8%) after a 6-mo observation period in the MMF ($n = 22$) or AZA ($n = 26$) arm. High-risk patients were, however, excluded in this trial (Wuthrich et al. 2000).

Glucocorticosteroids

Glucocorticosteroids serve as first-line agents for treating rejection. The mechanistic basis for their anti-inflammatory properties is diverse (Rhen and Cidlowski 2005). On binding the glucocorticoids receptor, the steroid-receptor complex inhibits the expression of proinflammatory genes/proteins dependent on NF-κB and activator protein 1. Glucocorticosteroid and its receptor complex may also block the transcription of IL-2 via disruption of key DNA-binding proteins (Vacca et al. 1992). High-dose glucocorticosteroids are given peri-transplantation and tapered to a lower dose for the life of the allograft. Serious side effects because of prolonged steroid exposure supported the implementation of a steroid-sparing regimen for kidney transplantation. At our center, following rATG induction, the cumulative incidence of acute kidney rejection with an early corticosteroid withdrawal regimen in kidney graft recipients was 12.0% over a 5-yr period and death-censored graft survival was 96.2%, 91.9%, and 87.6% at 1, 3, and 5 yr, respectively (Aull et al. 2012).

A Cochrane systematic review of 5949 participants in 30 randomized controlled trials showed that glucocorticosteroid-sparing regimens had no effect on patient mortality or graft loss including death with functioning graft (Pascual et al. 2009). However, patients on glucocorticosteroid-sparing regimens were at a higher risk of graft loss (excluding death) when compared to patients on conventional glucocorticosteroid regimen (RR 1.23, 95% CI, 1.00–1.52). The risk was higher in patients not concurrently on MMF/mycophenolic acid or everolimus (RR 1.70, 95% CI, 1.00–2.90). Compared with standard glucocorticosteroid-containing regimen and in the presence of CsA, the risk of acute rejection was higher without glucocorticosteroid maintenance therapy (RR 1.27, 95% CI, 1.14–1.40). Maintenance immunosuppression without glucocorticosteroid was beneficial in reducing antihypertensive drug need, cholesterol-lowering drug need, new-onset diabetes after transplantation, cardiovascular events, and the incidence of infection.

Calcineurin Inhibitors: Cyclosporine (CsA) and Tacrolimus (Tace)

CsA is a cyclic nonribosomal peptide of 11 amino acids isolated from the soil fungus Tolypocladium inflatum. CsA binds cyclophylin, an evolutionarily conserved peptidyl prolyl *cis–trans* isomerase, and the CsA and cyclophylin heterodimer targets and inactivates calcineurin, a serine-threonine phosphatase. Inhibition of calcineurin activity is central to the immunosuppressive effects of both CsA and Tac and results in lack of dephosphorylation of the nuclear factor of activated T-cells protein-1 (NFAT-1), nuclear import, and inhibition of NFAT-dependent transcription of a number of cytokine genes including IL-2. CsA as well as Tac enhances the expression of transforming growth factor β (TGF-β) (Sehajpal et al. 1993; Khanna et al. 1999). Because TGF-β is a potent inhibitor of T-cell proliferation and generation of antigen-specific CTL (Kehrl et al. 1986), heightened expression of TGF-β must contribute to the antiproliferative/immunosuppressive activity of CsA/Tac. This TGF-β-inducing effect of CsA/Tac suggests also a mechanism for renal fibrosis and tumor metastasis associated with CNIs because TGF-β is a fibrogenic and pro-angiogenic cytokine.

A Cochrane systematic review comparing Tac versus CsA regimen in 4102 participants from 30 studies (Webster et al. 2005) showed that Tac regimen reduced graft loss for up to 3 yr (RR 0.56, 95% CI, 0.36–0.86). Metaregression showed that this benefit was less at higher target trough levels of Tac ($P = 0.04$), after allowing for differences in CsA formulation ($P = 0.97$) and target trough level ($P = 0.38$). At 1 yr, Tac regimen resulted in less acute rejection (RR 0.69, 95% CI, 0.60–0.79) or steroid-resistant rejection (RR 0.49, 95% CI, 0.37–0.64). Tac-treated patients had more insulin-dependent diabetes mellitus (RR 1.86, 95% CI, 1.11–3.09), tremor, headache, and gastrointestinal symptoms such as diarrhea, dyspepsia, and vomiting. CsA-treated patients had more constipation and cosmetic side effects. The review revealed no significant differences in infection or malignancy between CsA and Tac regimens.

MMF and Enteric-Coated Mycophenolate Sodium (EC-MPS)

MMF is the prodrug of mycophenolic acid and reversibly inhibits inosine monophosphate dehydrogenase, an enzyme in the de novo pathway of guanosine biosynthesis. B and T lymphocytes are dependent on this biosynthetic pathway to satisfy their guanosine requirements. The FDA approved MMF for the prevention of kidney transplant rejection in 1995. Early clinical trials used MMF to replace azathioprine in the cyclosporine- and glucocorticosteroid-based immunosuppressive regimen. These controlled, prospective trials showed a diminished incidence of early acute rejection (Sollinger 1995; Lui and Halloran 1996; European Mycophenolate Mofetil Cooperative Study Group 1999). Although follow-up studies over a 3-yr period have indicated an advantage for MMF over azathioprine (European Mycophenolate Mofetil Cooperative Study Group 1999), a randomized trial comparing MMF with azathioprine in recipients of a first kidney transplant from a deceased donor found similar levels of acute rejection in the first 6 mo of transplantation (Remuzzi et al. 2004).

Enteric-coated mycophenolate sodium (EC-MPS) was developed to improve the gastrointestinal tolerability of MPA. An international phase III, randomized, double-blind, parallel group trial showed the therapeutic equivalence of MMF and EC-MPS (Salvadori et al. 2004). At 12 mo, the incidence of acute rejection, graft loss, and death was comparable for both groups and gastrointestinal complications were not different.

Sirolimus

Sirolimus was discovered in 1975 on the island of Rapa Nui (Easter Island) from *Streptomyces hygroscopicus* in the soil sample. Also known as rapamycin, it is a macrocyclic lactone that, like Tac, binds FKBP. However, sirolimus and Tac affect different and distinctive sites downstream in the signal transduction pathway. Whereas sirolimus blocks IL-2 and other growth factor-mediated signal transduction, Tac does not. In addition, the sirolimus–FKBP complex, unlike the Tac–FKBP complex, does not bind calcineurin. The antiproliferative activity of the sirolimus–FKBP complex is linked to the blockade of the activation of 70-kDa S6 protein kinases and blockade of expression of the bcl-2 protooncogene. Sirolimus also blocks the Ca^{2+}-independent CD28-induced costimulatory pathway. Sirolimus was approved in 1999 by the FDA and indicated for the prophylaxis against kidney rejection.

The Cochrane group analyzed the use of mTOR inhibitor for kidney transplant immunosuppression in 33 studies (7114 participants) including 27 sirolimus, five everolimus, and one head-to-head trial (Webster et al. 2006). Replacing CNI with mTOR inhibitor resulted in no difference in acute rejection, but lower serum creatinine (mean difference -18.31 µM/L, -30.96 to -5.67), and higher risk of bone marrow suppression (leukopenia: RR 2.02, 95% CI, 1.12–3.66; thrombocytopenia: RR 6.97, 95% CI, 2.97–16.36, and anemia: RR 1.67, 95% CI, 1.27–2.20). Replacing antimetabolites with mTOR inhibitor resulted in lower risk of acute rejection (RR 0.84, 95% CI, 0.71–0.99) and CMV infection (RR 0.49, 95% CI, 0.37–0.65), but higher risk of dyslipidemia (RR 1.65, 95% CI, 1.32–2.06).

Everolimus (RAD)

Everolimus is a derivative of sirolimus. The FDA approved everolimus in 2010 for prevention of kidney transplant rejection following its approval in 2009 for the treatment of advanced renal cell carcinoma in patients who have failed sunitinib or sofrafenib therapy. The use of everolimus in phase II clinical trials involving cyclosporine, steroids, and basiliximab induction resulted in excellent graft survival at 36 mo (Nashan et al. 2004). In a short-term phase III trial, everolimus was comparable to MMF with cyclosporine and steroids in preventing acute rejection (Vitko et al. 2004).

Belatacept

Belatacept is a recombinant protein created by fusion of modified CTLA-4 with an Fc portion of human immunoglobulin (CTLA-4Ig) and was designed to block the B7/CD28 costimulatory pathway. The FDA approved belatacept in 2011 for the prophylaxis of rejection in kidney transplant recipients. Phase III trials, BENEFIT, and BENEFIT-EXT have tested the effectiveness of belatacept as part of a CNI-free regimen (Durrbach et al. 2010; Vincenti et al 2010). The BENEFIT study was a 3-yr randomized, parallel group designed trial conducted at 100 transplant sites worldwide. Following basiliximab induction, participants were randomized to one of three groups consisting of a more intensive belatacept (MI) regimen, a less intensive belatacept regimen (LI), or cyclosporine with the addition of maintenance MMF and corticosteroids. The incidence of acute rejection at 12 mo was higher in belatacept groups (22% and 17%) compared to cyclosporine groups (7%). More participants also developed type IIa and IIb rejections in belatacept groups, but without an increase in donor-specific antibody production compared to cyclosporine-treated groups. The mean glomerular filtration rate (GFR) was better at 12 mo for belatacept groups compared to cyclosporine groups.

The BENEFIT-EXT trial was a 3-yr randomized, multicenter trial performed at 79 transplant sites worldwide to test the benefit of a CNI-free regimen containing belatacept in patients undergoing high risk transplant from expanded criteria donors. Following basiliximab induction, participants were randomized to one of three groups (MI, LI, cyclosporine) with the addition of maintenance MMF and corticosteroids. The incidence of acute rejection was not different among the three groups, but CNI-free belatacept regimens resulted in more type IIb rejections. The mean GFR was higher at 12 mo for MI groups (52.1 mL/min/1.73 m^2), but not for LI groups (49.5 mL/min/1.73 m^2) compared to cyclosporine groups (45.2 mL/min/1.73 m^2). Both the BENEFIT and BENEFIT-EXT trials showed that neither the MI nor LI belatacept regimen was noninferior to cyclosporine on patient and graft survival. At the 3-yr follow-up for the BENEFIT trial, the mean calculated GFR (cGFR) decreased by 2.0 mL/min/1.73 m^2 per year for cyclosporine groups, whereas mean cGFR increased in MI and LI groups (+1.0 mL/min/1.73 m^2 and +1.2 mL/min/1.73 m^2 per year, respectively) (Vincenti et al. 2012). The 3-yr follow-up study of BENEFIT-EXT showed similar patient survival in the three groups (MI 80%, LI 82%, and 80% in cyclosporine groups) (Pestana et al. 2012). The mean cGFR was 11 mL/min higher in belatacept-treated arms when compared to cyclosporine arms (MI 42.7 mL/min, LI 42.2 mL/min, cyclosporine 31.5 mL/min). Patients in the cyclosporine group were more likely to progress to a GFR of less than 30 mL/min when compared to MI and LI groups (44% vs. 27% and 30%, respectively).

Conversion from a CNI-based to a belatacept-based immunosuppression regimen was investigated in a randomized phase II study (Rostaing et al. 2011). Patients who were more than 6 mo but less than 3 yr after transplant were randomized to stay on CNI ($n = 89$) or undergo conversion to belatacept ($n = 84$). Post hoc analysis showed that conversion to belatacept resulted in improvement of mean cGFR at 1 yr when compared to baseline (7.0 mL/min belatacept vs. 2.1 mL/min for CNI group; $p = 0.0058$). Acute rejection occurred within 6 mo after conversion to a CNI-free regimen in six patients.

Leflunomide

Leflunomide is a disease-modifying antirheumatic drug approved by the FDA in 1998 for treatment of rheumatoid arthritis. It belongs to the family of drugs known as malonitrilamides and is a synthetic isoxazole derivative that inhibits dihydroorotate dehydrogenase—a key enzyme for de novo pyrimidine synthesis. Leflunomide has antiviral effects against CMV, herpes simplex virus type 1, and polyomavirus (Waldman et al. 1999; Knight et al. 2001; Farasati et al. 2005). A short-term, open-label, prospective crossover trial of leflunomide of 22 patients with chronic renal allograft dysfunction found 100% patient survival and 91% graft survival at 6-mo posttransplantation, and was well tolerated, with anemia being the most common adverse effect (Hardinger et al. 2002).

A systematic review of the treatment of polyomavirus infection in kidney transplant patients showed that the pooled death-censored graft loss rate was 8/100 patient-yr for reduction of immunosuppression and 13/100 patient-yr for the addition of leflunomide (Johnston et al. 2010). The utility of leflunomide for polyomavirus-associated nephropathy remains undefined.

Agents against B Cells, Plasma Cells, Complements

For agents discussed in this section, none are FDA approved for transplantation.

IVIG

Immune globulin therapy has been available since the 1950s for treatment of primary immunodeficiency diseases. The intravenous preparation is better tolerated and used to treat autoimmune diseases based on its immunomodulatory effects (Gelfand 2012). In renal transplantation, IVIG is utilized to reduce humoral immunity in two settings: (1) reduce the level of preexisting anti-HLA antibodies and convert a positive cross-match recipient to a negative cross-match recipient (Jordan et al. 2004), and (2) treat antibody-mediated rejection (White et al. 2004).

A combination of IVIG and rituximab has facilitated, in a safe manner, rapid transplantation of 16 of 20 highly sensitized patients (Vo et al. 2008). A protocol incorporating plasmapheresis and low-dose IVIG was successful in desensitizing 211 HLA-sensitized patients before kidney transplant (Montgomery et al. 2011). The study showed that compared with a matched control group of wait-listed patients who remained on dialysis, the patient survival rate of desensitized patients following kidney transplant was far superior after 8 yr of follow-up (30.5% vs. 80.6%, respectively; $P < 0.001$).

A desensitization regimen of IVIG (2 g/kg, maximum dose 120 g) and rituximab (375 mg/m^2) was tested on highly sensitized kidney transplant candidates with calculated panel reactive antibody (cPRA) of greater than 50% (Marfo et al. 2012). Two of 11 patients who completed the protocol received a kidney transplant compared to 14 of 27 patients in the nondesensitized cohort. Desensitization using one dose of rituximab and high-dose IVIG was not successful in reducing class I and class II cPRA levels.

Rituximab

Rituximab is a chimeric murine/human monoclonal $IgG_{1\kappa}$ directed against the CD20 antigen expressed on B cells. Rituximab is FDA approved for the treatment of CD20-positive, B-cell non-Hodgkin's lymphoma. Initial experience with rituximab shows promise in the treatment of steroid-resistant acute renal allograft rejection (Becker et al. 2004). Rituximab has been used as a component of a preconditioning regimen to prepare patients for renal transplantation from ABO incompatible donors (Tyden et al. 2005). Rituximab may exert a direct effect on podocyte actin cytoskeleton by interacting with sphingomyelin phosphodiesterase acid-like 3b protein (Fornoni et al. 2011). Pilot studies have yielded mixed outcomes when rituximab is added to a regimen of plasmapheresis to induce remission in recurrent focal segmental glomerulosclerosis (Yabu et al. 2008; Dello Strologo et al. 2009).

Bortezomib

Bortezomib is the first proteasome inhibitor approved by the FDA for the treatment of multiple myeloma and mantle cell lymphoma. Proteasomes are large cytosolic protease complexes and with ubiquitin they perform basic housekeeping protein degradation in all eukaryotic cells. The ubiquitin-proteasome pathway is essential for physiologic functions such as oncogenesis, inflammation, apoptosis, cell cycle progression, and immune activation. Plasma cells are professional antibody-secreting cells, and in the process of producing antibodies they are subjected to tremendous intracellular stress leading to proteasomal insufficiency and cell death if accumulation of polyubiquitinated proteins is left unchecked (Perry et al. 2009). Studies in kidney transplant recipients suggest that nonmalignant plasma cells are susceptible to proteasome inhibition (Perry et al. 2009). Preliminary results of bortezomib in antibody-mediated kidney rejection are promising (Everly et al. 2008).

Eculizumab

Eculizumab is a humanized monoclonal antibody against complement 5a molecule and approved by the FDA for the treatment of paroxysmal nocturnal hematuria. Several case reports described the use of eculizumab in renal transplant recipients with atypical hemolytic-uremic syndrome, as salvage therapy for antibody-mediated rejection, and renal transplant patients with catastrophic antiphospholipid antibody syndrome (Locke et al. 2009; Lonze et al. 2010; Zimmerhackl et al. 2010).

Novel Drugs in the Pipeline

Alefacept

Alefacept is a dimeric fusion protein made by linking the CD2-binding portion of the human lymphocyte function-associated antigen-3 to the Fc portion of human IgG1. It was marketed as Amevive and indicated for the treatment of psoriasis. A phase 2 randomized open-label parallel group multicenter trial enrolled 309 subjects and randomized kidney graft recipients to four arms: control (basiliximab induction with full-dose Tac, MMF, and steroids), alefacept/low-dose Tac/MMF/steroids (A), alefacept/full-dose Tac/steroids (B), and every other week alefacept/low-dose Tac/MMF/steroids (C) Bromberg et al. 2011). The incidence of BPAR was higher in group A compared to the control arm (26.3% vs. 12.7%, $P < 0.05$) whereas groups B and C had similar rates of biopsy-proven acute rejection compared to the controls (18.8% and 16.7%, respectively). At 6 mo, patient and graft survival as well as renal function were similar in all groups.

Siplizumab (MEDI-507)

Siplizumab is a humanized monoclonal antibody directed against CD2 antigen and has not received FDA approval. It has been investigated as an induction agent in a pioneering study of kidney transplant tolerance following bone marrow-induced mixed chimerism, and is the only monoclonal antibody shown to date to induce tolerance to MHC-incompatible kidney allografts in humans (Kawai et al. 2008).

Sotrastaurin (AEB-071)

AEB-071, a selective protein kinase C isoforms inhibitor, prevents T-cell activation independent of calcineurin inhibition and is being developed for the prophylaxis of kidney transplant rejection. Phase II studies in de novo kidney transplant recipients randomized to sotrastaurin ($n = 81$) showed a higher composite efficacy failure rate at 3 mo without CNI (Friman et al. 2011). Biopsy-proven acute rejection occurred in 17 patients in the sotrastaurin arm. The median eGFR at 3 mo was higher in patients who have not received Tac (59.0 $\pm$ 22.3 vs. 49.5 $\pm$ 17.7 mL/min/1.72 m^2, $P = 0.006$).

Janus Kinase (JAK)3 Inhibitor (CP-690550)

Tofacitinib inhibits the tyrosine kinase required for signal transduction downstream of cytokine receptors and is important for the activation and function of T cells as well as natural killer cells. It was approved in 2012 by the FDA for the treatment of rheumatoid arthritis. A phase IIA randomized, open-label multicenter trial was conducted in de novo kidney transplant recipients (Busque et al. 2009). Following induction with IL-2Ra mAbs, participants were randomized to lower-dose JAK3 inhibitor (CP15), higher-dose JAK3 inhibitor (CP30), or Tac and with maintenance MMF and corticosteroids. The trial was converted into an exploratory study without sufficient power to address its primary and secondary objectives because of four cases of BK virus nephropathy (BKVN) occurring in CP30 group. The CP15 group had a similar rejection rate when compared to the Tac group. The CP30 group had a higher rejection rate than the Tac group and the combination of JAK3 inhibition with MMF therapy yielded excessive viral opportunistic infections such as BKVN and CMV disease.

Voclosporin (ISA247)

Voclosporin is a modified CNI with an additional carbon molecule attached to the amino acid 1 residue of CsA resulting in more potency and less toxicities. A phase IIB multicenter study, conducted on de novo kidney transplant recipients, showed a lower incidence of new-onset diabetes after transplantation in patients on voclosporin and the incidence of biopsy-proven acute rejection was not inferior in voclosporin arms compared to Tac arm (Busque et al. 2011).

TOL101

TOL101, an orphan drug, is a murine mAb directed against the αβ TCR. Unlike anti-CD3 antibodies, which may activate intracellular immunoreceptor tyrosine-based activation motifs, anti-αβTCR antibodies do not, thus potentially limiting drug toxicity (Getts et al. 2010). Selectively targeting αβ T cells using TOL101 may also preserve the role of γδ T cells.

CONCLUSIONS

Sixty years after the pioneering tolerance studies of Medawar and his colleagues (Billingham et al. 1953), immune rejection continues to be a dreaded complication. The emergence of novel agents with efficacy not only against the cellular, but also the humoral arms of the immune response may improve the management of transplant recipients. A one-size-fits-all approach remains the clinical norm despite serious toxicities. A personalized immunosuppressive drug therapy based on mechanistic biomarkers remains an unmet objective and worthy of pursuit.

ACKNOWLEDGMENTS

We thank our patients, characterized as the true heroes of transplantation by the pioneering clinician-scientist Thomas E. Starzl, for their continued inspiration.

REFERENCES

Aull MJ, Dadhania D, Afaneh C, Leeser DB, Hartono C, Lee JB, Serur D, Del Pizzo JJ, Suthanthiran M, Kapur S. 2012. Early corticosteroid withdrawal in recipients of renal allografts: A single-center report of ethnically diverse recipients and recipients of marginal deceased-donor kidneys. *Transplantation* **94:** 837–844.

Batista FD, Harwood NE. 2009. The who, how and where of antigen presentation to B cells. *Nat Rev Immunol* **9:** 15–27.

Becker YT, Becker BN, Pirsch JD, Sollinger HW. 2004. Rituximab as treatment for refractory kidney transplant rejection. *Am J Transplant* **4:** 996–1001.

Beyers AD, Spruyt LL, Williams AF. 1992. Molecular associations between the T-lymphocyte antigen receptor complex and the surface antigens CD2, CD4, or CD8 and CD5. *Proc Natl Acad Sci* **89:** 2945–2949.

Billingham RE, Brent L, Medawar PB. 1953. Actively acquired tolerance of foreign cells. *Nature* **172:** 603–606.

Brennan DC, Flavin K, Lowell JA, Howard TK, Shenoy S, Burgess S, Dolan S, Kano JM, Mahon M, Schnitzler MA, et al. 1999. A randomized, double-blinded comparison of thymoglobulin versus ATGAM for induction immunosuppressive therapy in adult renal transplant recipients. *Transplantation* **67:** 1011–1018.

Brennan DC, Daller JA, Lake KD, Cibrik D, Del Castillo D, Thymoglobulin Induction Study Group. 2006. Rabbit antithymocyte globulin versus basiliximab in renal transplantation. *N Engl J Med* **355:** 1967–1977.

Bromberg J, Cibrik D, Steinberg S, West-Thielke P, Yang H, Erdman J, First R, Holman J. 2011. A phase 2 study to assess the safety and efficacy of alefacept (ALEF) in de novo kidney transplant recipients. *American Transplant Congress*, Abstract 533.

Brown MH, Cantrell DA, Brattsand G, Crumpton MJ, Gullberg M. 1989. The CD2 antigen associates with the T-cell antigen receptor CD3 antigen complex on the surface of human T lymphocytes. *Nature* **339:** 551–553.

Busque S, Leventhal J, Brennan DC, Steinberg S, Klintmalm G, Shah T, Mulgaonkar S, Bromberg JS, Vincenti F, Hariharan S, et al. 2009. Calcineurin-inhibitor-free immunosuppression based on the JAK inhibitor CP-690,550: A pilot study in de novo kidney allograft recipients. *Am J Transplant* **9:** 1936–1945.

Busque S, Cantarovich M, Mulgaonkar S, Gaston R, Gaber AO, Mayo PR, Ling S, Huizinga RB, Meier-Kriesche HU, PROMISE Investigators. 2011. The PROMISE study: A phase 2b multicenter study of voclosporin (ISA247) versus tacrolimus in de novo kidney transplantation. *Am J Transplant* **11:** 2675–2684.

Dello Strologo L, Guzzo I, Laurenzi C, Vivarelli M, Parodi A, Barbano G, Camilla R, Scozzola F, Amore A, Ginevri F, et al. 2009. Use of rituximab in focal glomerulosclerosis relapses after renal transplantation. *Transplantation* **88:** 417–420.

Durrbach A, Pestana JM, Pearson T, Vincenti F, Garcia VD, Campistol J, Rial Mdel C, Florman S, Block A, Di Russo G, et al. 2010. A phase III study of belatacept versus cyclosporine in kidney transplants from extended criteria donors (BENEFIT-EXT study). *Am J Transplant* **10:** 547–557.

European Mycophenolate Mofetil Cooperative Study Group. 1999. Mycophenolate mofetil in renal transplantation: 3-Year results from the placebo-controlled trial. *Transplantation* **68:** 391–396.

Everly MJ, Everly JJ, Susskind B, Brailey P, Arend LJ, Alloway RR, Roy-Chaudhury P, Govil A, Mogilishetty G, Rike AH, et al. 2008. Bortezomib provides effective therapy for antibody- and cell-mediated acute rejection. *Transplantation* **86:** 1754–1761.

Farasati NA, Shapiro R, Vats A, Randhawa P. 2005. Effect of leflunomide and cidofovir on replication of BK virus in an in vitro culture system. *Transplantation* **79:** 116–118.

Fooksman DR, Vardhana S, Vasiliver-Shamis G, Liese J, Blair DA, Waite J, Sacristan C, Victoria GD, Zanin-Zhorov A, Dustin ML. 2010. Functional anatomy of T cell activation and synapse formation. *Annu Rev Immunol* **28:** 79–105.

Fornoni A, Sageshima J, Wei C, Merscher-Gomez S, Aguillon-Prada R, Jauregui AN, Li J, Mattiazzi A, Ciancio G, Chen L, et al. 2011. Rituximab targets podocytes in recurrent focal segmental glomerulosclerosis. *Sci Transl Med* **3:** 85ra46.

Friman S, Arns W, Nashan B, Vincenti F, Banas B, Budde K, Cibrik D, Chan L, Klempnauer J, Mulgaonkar S, et al. 2011. Sotrastaurin, a novel small molecule inhibiting protein-kinase C: A randomized phase II study in renal transplant recipients. *Am J Transplant* **11:** 1444–1455.

Gelfand EW. 2012. Intravenous immune globulin in autoimmune and inflammatory diseases. *N Engl J Med* **367:** 2015–2025.

Getts DR, Getts MT, McCarthy DP, Chastain EM, Miller SD. 2010. Have we overestimated the benefits of human(ized) antibodies? *MAbs* **2:** 682–694.

Hanaway MJ, Woodle ES, Mulgaonkar S, Peddi VR, Kaufman DB, First MR, Croy R, Holman J, INTAC Study Group. 2011. Alemtuzumab induction in renal transplantation. *N Engl J Med* **364:** 1909–1919.

Hardinger KL, Wang CD, Schnitzler MA, Miller BW, Jendrisak MD, Shenoy S, Lowell JA, Brennan DC. 2002. Prospective, pilot, open-label, short-term study of conversion to leflunomide reverses chronic renal allograft dysfunction. *Am J Transplant* **2:** 867–871.

Hardinger KL, Rhee S, Buchanan P, Koch M, Miller B, Enkvetchakul D, Schuessler R, Schnitzler MA, Brennan DC. 2008. A prospective, randomized, double-blinded comparison of thymoglobulin versus ATGAM for induction immunosuppressive therapy: 10 Year results. *Transplantation* **86:** 947–952.

Herold KC, Hagopian W, Auger JA, Poumian-Ruiz E, Taylor L, Donaldson D, Gitelman SE, Harlan DM, Xu D, Zivin RA, et al. 2002. Anti-CD3 monoclonal antibody in new-onset type 1 diabetes mellitus. *N Engl J Med* **346:** 1692–1698.

Johnston O, Jaswal D, Gill JS, Doucette S, Fergusson DA, Knoll GA. 2010. Treatment of polyomavirus infection in kidney transplant recipients: A systemic review. *Transplantation* **89:** 1057–1070.

Jordan SC, Tyan D, Stablein D, McIntosh M, Rose S, Vo A, Toyoda M, Davis C, Shapiro R, Adey D, et al. 2004. Evaluation of intravenous immunoglobulin as an agent to lower allosensitization and improve transplantation in highly sensitized adult patients with end-stage renal disease: Report of the NIH IG02 trial. *J Am Soc Nephrol* **15:** 3256–3262.

Kawai T, Cosimi AB, Spitzer TR, Tolkoff-Rubin N, Suthanthiran M, Saidman SL, Shaffer J, Preffer FL, Ding R, Sharma V, et al. 2008. HLA-mismatched renal transplantation without maintenance immunosuppression. *N Engl J Med* **358:** 353–361.

Kehrl JH, Wakefield LM, Roberts AB, Jakowlew S, Alvarez-Mon M, Derynck R, Sporn MB, Fauci AS. 1986. Production of transforming growth factor β by human T lymphocytes and its potential role in the regulation of T cell growth. *J Exp Med* **163:** 1037–1050.

Khanna A, Cairns V, Hosenpud JD. 1999. Tacrolimus induces increased expression of transforming growth factor-β1 in mammalian lymphoid as well as nonlymphoid cells. *Transplantation* **67:** 614–619.

Knight DA, Hejmanowski AQ, Dierksheide JE, Williams JW, Chong AS, Waldman WJ. 2001. Inhibition of herpes simplex virus type 1 by the experimental immunosuppressive agent leflunomide. *Transplantation* **71:** 170–174.

Lebedeva TN, Anikeeva N, Kalams SA, Walker BD, Gaidarov I, Keen JH, Sykulev Y. 2004. Major histocompatibility complex class I-intercellular adhesion molecule-1 association on the surface of target cells: Implications for antigen presentation to cytotoxic T lymphocytes. *Immunology* **113:** 460–471.

Locke JE, Magro CM, Singer AL, Segev DL, Haas M, Hillel AT, King KE, Kraus E, Lees LM, Melancon JK, et al. 2009. The use of antibody to complement protein C5 for salvage treatment of severe antibody-mediated rejection. *Am J Transplant* **9:** 231–235.

Lonze BE, Singer AL, Montgomery RA. 2010. Eculizumab and renal transplantation in a patient with CAPS. *N Engl J Med* **362:** 1744–1745.

Lui SL, Halloran PF. 1996. Mycophenolate mofetil in kidney transplantation. Mycophenolate mofetil in kidney transplantation. *Curr Opin Nephrol Hypertens* **5:** 508–513.

Marfo K, Ling M, Bao Y, Calder B, Ye B, Hayde N, Greenstein S, Chapochnick-Friedman J, Glicklich D, de Boccardo G, et al. 2012. Lack of effect in desensitization with intravenous immunoglobulin and rituximab in highly sensitized patients. *Transplantation* **94:** 345–351.

Montgomery RA, Lonze BE, King KE, Kraus ES, Kucirka LM, Locke JE, Warren DS, Simpkins CE, Dagher NN, Singer AL, et al. 2011. Desensitization in HLA-incompatible kidney recipients and survival. *N Engl J Med* **365:** 318–326.

Mueller TF. 2007. Mechanisms of action of thymoglobulin. *Transplantation* **84:** S5–S10.

Nashan B, Curtis J, Ponticelli C, Mourad G, Jaffe J, Haas T, 156 Study Group. 2004. Everolimus and reduced-exposure cyclosporine in de novo renal-transplant recipients: A three-year phase II, randomized, multicenter, open-label study. *Transplantation* **78:** 1332–1340.

Norman DJ, Vincenti F, de Mattos AM, Barry JM, Levitt DJ, Wedel NI, Maia M, Light SE. 2000. Phase I trial of HuM291, a humanized anti-CD3 antibody, in patients receiving renal allografts from living donors. *Transplantation* **70:** 1707–1712.

Ortho Multicenter Transplant Study Group. 1985. A randomized clinical trial of OKT3 monoclonal antibody for acute rejection of cadaveric renal transplants. *N Engl J Med* **313:** 337–342.

Pascual J, Zamora J, Galeano C, Royuela A, Quereda C. 2009. Steroid avoidance or withdrawal for kidney transplant recipients. *Cochrane Database Syst Rev* **1:** CD005632.

Perry DK, Burns JM, Pollinger HS, Amiot BP, Gloor JM, Gores GJ, Stegall MD. 2009. Proteasome inhibition causes apoptosis of normal human plasma cells preventing alloantibody production. *Am J Transplant* **9:** 201–209.

Pestana JO, Grinyo JM, Vanrenterghem Y, Becker T, Campistol JM, Florman S, Garcia VD, Kamar N, Lang P, Manfro RC, et al. 2012. Three-year outcomes from BENEFIT-EXT: A phase III study of belatacept versus cyclosporine in recipients of extended criteria donor kidneys. *Am J Transplant* **12:** 630–639.

Remuzzi G, Lesti M, Gotti E, Ganeva M, Dimitrov BD, Ene-lordache B, Gherardi G, Donati D, Salvadori M, Sandrini S, et al. 2004. Mycophenolate mofetil versus azathioprine for prevention of acute rejection in renal transplantation (MYSS): A randomised trial. *Lancet* **364:** 503–512.

Rhen T, Cidlowski JA. 2005. Antiinflammatory action of glucocorticoids—New mechanisms for old drugs. *N Engl J Med* **353:** 1711–1723.

Rostaing L, Massari P, Garcia VD, Mancilla-Urrea E, Nainan G, del Carmen Rial M, Steinberg S, Vincenti F, Shi R, Di Russo G, et al. 2011. Switching from calcineurin inhibitor-based regimens to a belatacept-based regimen in renal transplant recipients: A randomized phase II study. *Clin J Am Soc Nephrol* **6:** 430–439.

Salvadori M, Holzer H, de Mattos A, Sollinger H, Arns W, Oppenheimer F, Maca J, Hall M, ERL B301 Study Groups. 2004. Enteric-coated mycophenolate sodium is therapeutically equivalent to mycophenolate mofetil in de novo renal transplant patients. *Am J Transplant* **4:** 231–236.

Sehajpal PK, Sharma VK, Ingulli E, Stenzel KH, Suthanthiran M. 1993. Synergism between the CD3 antigen- and CD2 antigen-derived signals. Exploration at the level of induction of DNA-binding proteins and characterization of the inhibitory activity of cyclosporine. *Transplantation* **55:** 1118–1124.

Sollinger HW. 1995. Mycophenolate mofetil for the prevention of acute rejection in primary cadaveric renal allograft recipients. *Transplantation* **60:** 225–232.

Suthanthiran M. 1990. A novel model for antigen-dependent activation of normal human T cells. Transmembrane signaling by crosslinkage of the CD3/T cell receptor-α/β complex with the cluster determinant 2 antigen. *J Exp Med* **171:** 1965–1979.

Suthanthiran M. 1996. Transplant tolerance: Fooling mother nature. *Proc Natl Acad Sci* **93:** 12072–12075.

Suthanthiran M, Garovoy MR. 1983. Immunologic monitoring of the renal transplant recipient. *Urol Clin North Am* **10:** 315–325.

Suthanthiran M, Strom TB. 1994. Renal transplantation. *N Engl J Med* **331:** 365–376.

Tiede I, Fritz G, Strand S, Poppe D, Dvorsky R, Strand D, Lehr HA, Wirtz S, Becker C, Atreya R, et al. 2003. CD28-dependent Rac1 activation is the molecular target of azathioprine in primary human $CD4^+$ T lymphocytes. *J Clin Invest* **111:** 1133–1145.

Tyden G, Kumlien G, Genberg H, Sandberg J, Lundgren T, Fehrman I. 2005. ABO incompatible kidney transplantations without splenectomy, using antigen-specific immunoadsorption and rituximab. *Am J Transplant* **5:** 145–148.

Vacca A, Felli MP, Farina AR, Martinotti S, Maroder M, Screpanti I, Meco D, Petrangeli E, Frati L, Gulino A. 1992. Glucocorticoid receptor-mediated suppression of the interleukin 2 gene expression through impairment of the cooperativity between nuclear factor of activated T cells and AP-1 enhancer elements. *J Exp Med* **175:** 637–646.

Vincenti F, Charpentier B, Vanrenterghem Y, Rostaing L, Bresnahan B, Darji P, Massari P, Mondragon-Ramirez GA, Agarwal M, Di Russo G, et al. 2010. A phase III study of belatacept-based immunosuppression regimens versus cyclosporine in renal transplant recipients (BENEFIT study). *Am J Transplant* **10:** 535–546.

Vincenti F, Larsen CP, Alberu J, Bresnahan B, Garcia VD, Kothari J, Lang P, Mancilla Urrea E, Massari P, Mondragon-Ramirez G, et al. 2012. Three-year outcomes from BENEFIT, a randomized, active-controlled, parallel-group study in adult kidney transplant recipients. *Am J Transplant* **12:** 210–217.

Vitko S, Margreiter R, Weimar W, Dantal J, Viljoen HG, Li Y, Jappe A, Cretin N, RADB 201 Study Group. 2004. Everolimus (Certican) 12-month safety and efficacy versus mycophenolate mofetil in de novo renal transplant recipients. *Transplantation* **78:** 1532–1540.

Vo AA, Lukovsky M, Toyoda M, Wang J, Reinsmoen NL, Lai CH, Peng A, Villicana R, Jordan SC. 2008. Rituximab and intravenous immune globulin for desensitization during renal transplantation. *N Engl J Med* **359:** 242–251.

Waldman WJ, Knight DA, Lurain NS, Miller DM, Sedmark DD, Williams JW, Chong AS. 1999. Novel mechanism of inhibition of cytomegalovirus by the experimental immunosuppressive agent leflunomide. *Transplantation* **68:** 814–825.

Webster AC, Playford EG, Higgins G, Chapman JR, Craig JC. 2004. Interleukin 2 receptor antagonists for renal transplant recipients: A meta-analysis of randomized trials. *Transplantation* **77:** 166–176.

Webster AC, Taylor RRS, Chapman JR, Craig JC. 2005. Tacrolimus versus cyclosporine as primary immunosuppression for kidney transplant recipients. *Cochrane Database Syst Rev* **4:** CD003961.

Webster AC, Lee VW, Chapman JR, Craig JC. 2006. Target of rapamycin inhibitors (TOR-I; sirolimus and everolimus) for primary immunosuppression in kidney transplant recipients. *Cochrane Database Syst Rev* **2:** CD004290.

Webster AC, Ruster LP, McGee R, Matheson SL, Higgins GY, Willis NS, Chapman JR, Craig JC. 2010. Interleukin 2 recpetor antagonists for kidney transplant recipients. *Cochrane Database Syst Rev* **1:** CD003897.

White NB, Greenstein SM, Cantafio AW, Schechner R, Glicklich D, McDonough P, Pullman J, Mohandas K, Boctor F, Uehlinger J, et al. 2004. Successful rescue therapy with plasmapheresis and intravenous immunoglobulin for acute humoral renal transplant rejection. *Transplantation* **78:** 772–774.

Wuthrich RP, Cicvara S, Ambuhl PM, Binswanger U. 2000. Randomized trial of conversion from mycophenolate mofetil to azathioprine 6 months after renal allograft transplantation. *Nephrol Dial Transplant* **15:** 1228–1231.

Yabu JM, Ho B, Scandling JD, Vincenti F. 2008. Rituximab failed to improve nephrotic syndrome in renal transplant patients with recurrent focal segmental glomerulosclerosis. *Am J Transplant* **8:** 222–227.

Zimmerhackl LB, Hofer J, Cortina G, Mark W, Wurzner R, Jungraithmayr TC, Khursigara G, Kliche KO, Radauer W. 2010. Prophylactic eculizumab after renal transplantation in atypical hemolytic-uremic syndrome. *N Engl J Med* **362:** 1746–1748.

Lessons and Limits of Mouse Models

Anita S. Chong[1], Maria-Luisa Alegre[2], Michelle L. Miller[2], and Robert L. Fairchild[3]

[1]Section of Transplantation, Department of Surgery, The University of Chicago, Chicago, Illinois 60637
[2]Section of Rheumatology, Department of Medicine, The University of Chicago, Chicago, Illinois 60637
[3]Department of Immunology, Lerner Research Institute, The Cleveland Clinic, Cleveland, Ohio 44195
Correspondence: achong@uchicago.edu

Seminal studies in rabbits and rodent transplantation models by Peter Medawar revealed that cellular processes, rather than humoral antibodies, are central to the acute rejection of transplanted organs, and much of basic transplantation research continues to be focused on the biology and control of these cells, which were subsequently shown to be T cells. However, the success of current immunosuppression at controlling T-cell-mediated rejection has resulted in an increasing awareness of antibody-mediated rejection in the clinic. This, in turn, has fueled an emerging interest in the biology of allospecific antibodies, the B cells that produce these antibodies, and the development of mouse models that allow their investigation. Here we summarize some of the more widely used mouse models that have been developed to study the immunobiology of alloreactivity, transplantation rejection and tolerance, and used to identify therapeutic strategies that modulate these events.

Rodent models in transplantation have been essential to developing a mechanistic understanding of the process of allograft rejection, as well as to the identification of novel therapeutic approaches that prevent rejection. Models of transplantation tolerance in mice and other rodents have paved the way to translational studies of tolerance induction in nonhuman primates and humans, whereas the failures in translating the successes in tolerance induction observed in mice into the clinic have led to a closer examination of the limitations of the mouse models and the identification of physiological barriers to tolerance induction.

ORGAN-SPECIFIC MODELS OF ACUTE REJECTION

Clinicians have long appreciated the importance of the organ type in shaping the alloreactive immune response, with lungs and small intestines having a higher propensity to being rejected compared with hearts, kidneys, or livers. Early models of organ transplantation were limited by microsurgical techniques, and the skin transplant model was extensively used. With technical advances, the heterotopic heart transplantation model is now the model of choice, although other organ transplantation models, such as kidney or liver, offer unique advantages (see Fig. 1).

Cite this article as *Cold Spring Harb Perspect Med* doi: 10.1101/cshperspect.a015495

Skin Transplantation

Following the unsuccessful efforts to use skin transplants to replace burned or damaged skin on injured British airmen and soldiers during World War II, Peter Medawar's experiments with rodent models of skin transplantation revealed fundamental rules of the acceptance of autologous and the rejection of heterologous skin grafts (Gibson and Medawar 1943; Medawar 1944, 1945; Billingham and Medawar 1951; Steinmuller 1984). The availability of genetically defined strains of mice has led to the understanding that complete major histocompatibility complex (MHC)-mismatched or even single class I or class II MHC mismatched skin allografts are quickly recognized by the recipient T cells and rejected. In contrast, rejection of MHC matched/multiple (e.g., BALB.B → BALB/c) or (male → female) minor histocompatibility antigen-mismatched allografts can take a considerable amount of time to be rejected and may not be rejected at all.

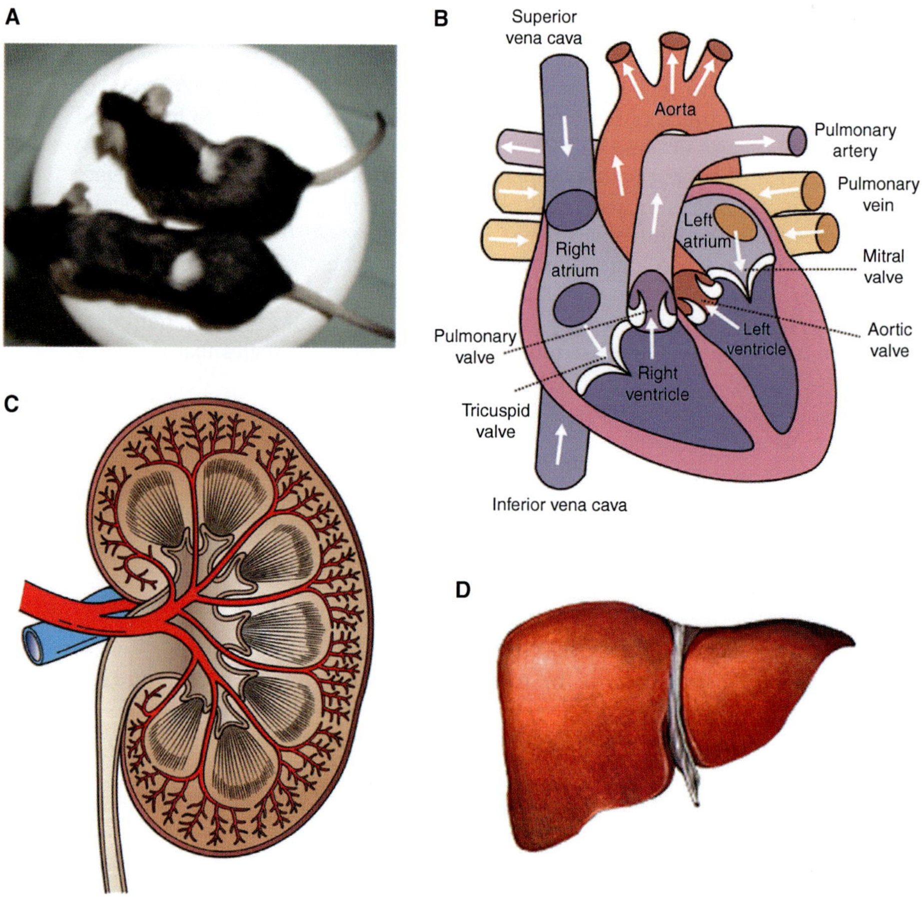

Figure 1. (*A*) Skin transplantation: Technically simple, a good model for acute T-cell-mediated rejection, and considered a stringent model of tolerance. (*B*) Heterotopic heart transplantation: Moderate technical difficulty, a good model for acute T-cell-mediated rejection, can model acute and chronic antibody-mediated rejection, notable differences in the pathology of chronic rejection from human chronic rejection, and most common model to investigate immunological tolerance. (*C*) Orthotopic kidney transplantation: Technically difficult, can be a life-sustaining graft, and can model spontaneous graft acceptance and chronic rejection. (*D*) Orthotopic liver transplantation: Technically very difficult, life-sustaining graft, and clinically relevant model of spontaneous graft acceptance.

The procedure for skin transplantation developed by Medawar and colleagues involves the removal of the abdominal skin from the donor and placement on the prepared graft bed in the recipient where the skin graft is wrapped. It takes 48–72 h for the graft to become revascularized by vascular endothelium growing from the recipient bed into the graft (Capla et al. 2006; Lindenblatt et al. 2008; Calcagni et al. 2011). This need for revascularization can impact the alloimmune response to the graft because the longer period of ischemia until revascularization occurs influences the inflammatory state within the skin allograft and therefore the function of the graft-derived antigen-presenting cells (APCs). These APCs have immediate access, through lymphatic drainage, to recipient secondary lymphoid organs where they stimulate donor-reactive T cells. Vascularization of skin allografts has also been shown to impact the early transmigration of donor T cells into the draining lymph nodes that then alter the specificity, tempo, and/or intensity of the alloreactive response by recipient T cells (Kwun et al. 2011). In contrast, host leukocytes including memory $CD8^+$ T cells may have early access to the allograft endothelium to promote early inflammation and subsequent rejection (Schenk et al. 2008, 2009). There is an additional nuance to the skin transplant model, particularly when only minor histocompatibility differences are involved and trunk skin allografts are eventually rejected, whereas tail skin allografts are not (McFarland and Rosenberg 2009). This differential susceptibility to rejection has been proposed to be due to the density of epidermal dendritic cells (DC) (i.e., Langerhans cells) and dermal dendritic cell populations, which is much higher in trunk skin versus tail skin (Bergstresser et al. 1980a,b; Chen and Silvers 1983). Thus, the trunk skin allograft model is considered a more robust model of rejection compared with the tail skin.

Heterotopic Heart Transplantation

The vascularized, intra-abdominal, heterotopic heart graft model in which the graft is transplanted into the recipient peritoneal cavity of a mouse was developed by Corry et al. (1973). The surgical procedure involves end-to-end anastomosis of the heart graft aorta to the recipient abdominal aorta and the graft pulmonary artery to the recipient inferior vena cava. Upon completion of the anastomosis, the heart graft quickly begins contracting. The advantages of this model are the immediate vascularization and the ease of monitoring rejection by palpating the heartbeat of the graft; thus, this is an excellent model to study alloreactivity that results in acute rejection. The shortcomings of the model are that it is a nonfunctional graft that does not pump blood despite heart graft contractions, and there is marked epicardial and endocardial inflammation due to chronic ischemia from abnormal blood flow. The absence of functional contraction results in progressive atrophy of the transplanted graft while the inflammatory response impinges on coronary arteries that are located near the epicardium and endocardium, especially on the right free wall and septum.

When complete MHC-disparate (e.g., both class I and class II MHC allelic differences) heart allografts are heterotopically transplanted into immunocompetent recipients, they are rejected (defined as the complete cessation of heartbeat) within 7–10 d (Corry et al. 1973; Schenk et al. 2008, 2009; Kwun et al. 2011). Characteristics of this rejection are similar to acute cell-mediated rejection in clinical transplantation, namely, an intense mononuclear cell infiltrate that includes macrophages and T cells, with $CD8^+$ T cells predominating over $CD4^+$ T cells. Transplant of single class I MHC– or class II MHC–mismatched heart allografts or MHC–matched/multiple minor mismatched allografts generally does not provoke acute rejection, and these grafts can survive for >60–80 d (Schenk et al. 2005; Hattori et al. 2012). A major reason for the lack of acute rejection is postulated to be the low frequency of donor-reactive T cells that permits the dominance of regulation over this response (Schenk et al. 2005). Studies by He et al. (2004) indicated that the size of the allograft is also an important factor in determining allograft outcome. For example, female recipients do not normally reject MHC-matched, minor antigen-disparate male heart allografts from adult, large-size donors but will reject smaller hearts from young (e.g., 3-wk-old) donors, suggesting that rejection can be affected by the graft itself, in addition to the immunological response it elicits.

Because the primary donor-specific lymphocytes infiltrating MHC-mismatched cardiac allografts are $CD8^+$ T cells, the role of $CD4^+$ T cells in mediating acute cell-mediated rejection has been investigated with allografts that have only class II MHC disparity. B6.H-2^{bm12} is a spontaneous mutation resulting in a three-amino-acid change in the third hypervariable region of the β-chain of the class II I-A^b molecule (McKenzie et al. 1979; Mengle-Gaw et al. 1984). B6.H-2^{bm12} skin grafts are acutely rejected by C57BL/6 recipients as a result of the priming of I-A^{bm12} reactive CD4 T cells (Yun et al. 2002; Schenk et al. 2005). The heart allografts are not acutely rejected but develop an occlusive vasculopathy that is also dependent on I-A^{bm12} reactive CD4 T cells. The explanation for heart allograft acceptance is that the skin is more immunogenic than the heart, and the anti-I-A^{bm12} response is highly regulated as a result of I-A^{bm12} being so similar to I-A^b (Lafuse et al. 1981). Indeed, the removal of $CD4^+$ $CD25^+$ T-regulatory cells results in an increased donor-reactive CD4 T-cell response and acute rejection of the I-A^{bm12} heart allografts (Schenk et al. 2005). Thus, several investigators have used the B6.H-2^{bm12} → C57BL/6 heart allograft model to study regulatory T-cell function as well as the development of allograft vasculopathy. Caveats of this model to study the pathogenesis of chronic rejection have been raised because the occlusive vasculopathy that develops in clinical heart transplants looks very different from that developing in the B6.H-2^{bm12} heart allografts.

There has been increasing discussion regarding the appropriate use of mouse heart transplant models to model clinical chronic rejection pathologies, including the anatomical and surgical differences with the human transplant model as discussed above, and the fact that immunosuppression in these models is often limited to subclinical levels of one agent, or that no immunosuppression is used when transplants are performed between weakly incompatible strains. As a result, most experimental heterotopic heart transplants incur acute rejection grades that are no longer observed in the clinic. In these murine models, acute interstitial inflammation resolves into a prominent interstitial fibrosis that is not typical of chronic rejection in human heart transplants. Suboptimal immunosuppression also results in a much greater acute vascular injury, resulting in the injured donor endothelial cells and smooth muscle cells being replaced by cells of recipient origin (Krasinskas et al. 2000); the most extreme example of donor cell replacement being observed with interpositional segmental aorta grafts in mice and rats that undergo complete intimal denudation and replacement by adjacent endothelium (Hagensen et al. 2011). Although the published data in human heart transplants are not in complete agreement, the more recent reports indicate that the majority of cells are of graft origin even years after transplantation (Atkinson et al. 2004; Minami et al. 2005). Another important difference between humans and mice is the anatomical location of the coronary arteries. Human coronaries are located on the epicardium, and chronic allograft vasculopathy in humans develops primarily in the epicardial portions of the coronary arteries with limited involvement of intramural branches (Lu et al. 2011). In contrast, rodent coronary arteries are largely intramural, and the pathological lesion found in intramural coronary branches is often characterized by medial necrosis and fibrosis. The most severe arterial pathology in murine cardiac allografts is usually found in the small epicardial segment of the coronary artery close to the aortic root (Wehner and Baldwin 2010). These differences may impact the clinical relevance of mechanistic studies of immune-mediated pathology and rejection of murine grafts, and thus caution has to be exercised.

Kidney Transplantation

An ideal solid organ transplant model is one that functions to support the life of the recipient—the *raison d'être* for clinical transplantation. Thus, in contrast to heterotopic heart transplants, renal transplantation can be performed in nephrectomized mice so that the viability of the recipient is dependent on the function of the graft. The kidney graft is harvested *en bloc* with the ureter so that the transplant surgery involves revascularization of the kidney graft and connection of the donor ureter to the recipient bladder, either through insertion of the ureter into the bladder or through generation of a bladder patch (Ge and Gong 2011). The function of the graft in nephrectomized recipients is assessed by the quantification of serum creatinine, similar to clinical renal transplantation.

Acute rejection of complete MHC-mismatched kidney allografts is observed in some mouse recipient→donor strain combinations, including C57BL/6→(C3H × DBA/2)F_1 (Skoskiewicz et al. 1973) and C57BL/6 (H-2^d)→ BALB/c (H-2^d) or→B10.BR (H-2^k) recipients (Pratt et al. 2002; Li et al. 2010). In other combinations, such as BALB/c→C57BL/6, renal allografts are slowly rejected, with 60% rejecting between day 12 and 50 posttransplant and the other 40% surviving long-term (>100 d) (Meng et al. 2008; Wu et al. 2012). Thus, the kidney transplant model provides a model for studying alloimmune responses and the pathology of acute and chronic graft rejection that is potentially more clinically accurate. However, it is considerably more technically demanding to perform the surgery and to monitor rejection compared with the skin and heterotopic heart model, thus limiting deeply mechanistic studies.

MODELS OF B-CELL AND ANTIBODY-MEDIATED REJECTION

Studies defining the role of B cells in transplantation have focused on the effects of alloantibodies secreted by plasma cells arising from the terminal differentiation of alloreactive B cells (summarized in Table 1). Antibodies can cause hyperacute (HAR), acute, and chronic antibody-mediated rejection (AMR); however, routine preoperative serological testing for preformed antihuman leucocyte antigen (HLA) and anti-ABO antibodies has reduced HAR to a rare clinical occurrence, whereas antibody-mediated acute or chronic rejection has emerged as a pressing problem in clinical transplantation (for review, see Mengel et al. 2012; Smith and

Table 1. Adoptive transfer of antibodies into B-cell or T- and B-cell (Rag)-deficient recipients to investigate antibody-mediated rejection and graft accommodation

Rejection type	Model	Passive transfer	References
Hyperacute	Xenografts in Rag-deficient recipients	High-titer antibodies or low-titer Abs + DAF deficiency	Shimizu et al. 2006; Ding et al. 2008a
	Allografts in Rag-deficient recipients	High-titer antibodies + xenogeneic complement	Koene et al. 1973
Acute, antibody mediated	Allografts in B-cell-deficient recipients	Delayed transfer of anti-MHC class I mAbs	Wasowska et al. 2001
	Allografts in Rag-deficient recipients	High-titer polyclonal antibodies from sensitized CCR5-deficient recipients	Nozaki et al. 2007
Chronic, antibody mediated	MHC-mismatched heart allograft	Repeated transfer of anti-MHC class I antibodies	Akiyoshi et al. 2012
	Native lung	Repeated intrabronchial transfer of anti-MHC class I Abs	Fukami et al. 2009, 2012a
Accommodation	Xenogeneic	Repeated transfer of anti-Gal antibodies	Ding et al. 2008b
	Single HLA-A2 mismatch heart allograft	Anti-HLA-A2 mAb transferred into donor 2 d before transplantation	Fukami et al. 2012b

Colvin 2012; Stegall et al. 2012). There is, therefore, a need to better predict outcomes based on circulating donor-specific antibody titers (DSA) and analysis of biopsies, and for treatments that prevent as well as resolve ongoing antibody production and antibody-mediated rejection.

Hyperacute Rejection

It is difficult to achieve HAR of vascularized allografts in mice even in the presence of high-titer DSA. HAR is classically dependent on complement activation, and some mouse strains have complete deficiencies of C5 (A/J, AKR, B10D2, and DBA/2), low levels of C3 (B10.D2, C57BL/10, and DBA.2), or C4 (CBA/J and C3H) that make them inappropriate for modeling HAR (for review, see Baldwin et al. 2010). In addition, a large family of complement-regulatory proteins (for review, see Morgan 1995) inhibits complement activation at different steps in the complement cascade and participates in the rapid clearance of active complement components from the cell surface, thereby protecting cells from damage. Indeed, Shimizu et al. (2006) reported that allografts lacking the complement regulator decay-accelerating factor (DAF) were susceptible to HAR in the presence of low-titer anti-α1,3Gal (Galα1-3Galβ1-[3]4GlcNAc [Gal]) antibodies, whereas wild-type grafts were not.

HAR in mice can be induced by the passive transfer of xenogeneic rabbit complement (Koene et al. 1973) or by the transplantation of xenogeneic grafts, the rationale for the latter being that the efficacy of complement-regulatory proteins controlling complement is reduced when they have to control the activation of complement from a different species. Using a xenogeneic model of rat hearts transplanted into recombination-activating gene (Rag)$^{-/-}$ Gal$^{-/-}$ mice, passive transfer of anti-Gal immunoglobulin G isotype (IgG3), an IgG subclass that strongly activates complement, elicited HAR in a complement-dependent manner (Ding et al. 2008a). However, monoclonal antibodies (mAbs) bearing identical antigen specificity but class switched to IgG1, IgG2a, and IgG2b subclasses that also engage IgC Fc receptors (FcγRs), HAR was shown to be dependent on complement activation as well as FcγRs and NK cells (Ding et al. 2008a). Those observations expanded our understanding of how HAR may be induced in vivo from a process that is only complement dependent to one in which FcγR-expressing cells also play essential roles.

Acute Antibody-Mediated Rejection

Seminal observations by Wasowska et al. (2001) showing that the adoptive transfer of allospecific mAbs in B-cell-deficient mice was able to restore acute cardiac rejection underscored the ability of alloantibodies to promote acute T-cell-mediated rejection. They went on to show that complement-fixing antibodies were most effective at promoting acute rejection, but that non-complement-fixing antibodies stimulated endothelial cell activation and synergized with complement-fixing antibodies to cause acute rejection (Rahimi et al. 2004). However, in those experiments, although antibodies were necessary, T cells were also important, resulting in a model of mixed antibody- and T-cell-mediated rejection.

The successful induction of pure acute antibody-mediated rejection in the absence of T cells and B cells was reported by Nozaki et al. (2007), following the adoptive transfer of sera containing extremely high-titer alloantibodies into Rag$^{-/-}$ recipients. For yet-to-be-defined reasons, the CCR5$^{-/-}$ mice transplanted with allogeneic allografts generate extremely high titers of donor-specific antibodies that were 20-fold higher than in wild-type recipients. Acute antibody-mediated rejection was observed within 12 d of passive transfer of sera from sensitized CCR5$^{-/-}$ but not wild-type recipients, with histological features of neutrophil and macrophage infiltration that are characteristic of acute antibody-mediated rejection. Their studies showed the importance of extremely high titers of antibodies for eliciting antibody-mediated rejection in Rag$^{-/-}$ mice.

Although much of the focus has been on antibody interaction with graft endothelium during acute rejection, antibodies can also play an important role in the generation of op-

Cite this article as *Cold Spring Harb Perspect Med* doi: 10.1101/cshperspect.a015495

sonins. Opsonins allow antigen to be presented simultaneously with bound antibodies that engage FcR and activated complement components that engage complement receptors on APCs. The engagement of these receptors promotes the up-regulation of costimulatory molecules on, and the production of cytokines by APCs, which are necessary for T-cell activation. Indeed, Burns and Chong (2011) provided experimental evidence for such opsonic activity of adoptively transferred anti-MHC class I monoclonal antibodies (mAbs). In that experimental model, recipients receiving anti-CD154 plus donor spleen cells accepted allogeneic skin long term, but in the presence of transferred antidonor K_d mAbs, the grafts were acutely rejected, even when only the donor spleen cells, but not the skin graft, expressed K_d. These observations provided support for an alternative way in which antibodies promote cellular rejection in the absence of C4d deposition in the graft.

Antibody-Mediated Chronic Rejection

Improved assessment of DSA and C4d deposition in the allograft has strengthened the association of alloantibodies with chronic rejection and poor graft survival in the clinic. Pathological markers of chronic AMR are specific for each organ—in renal allografts, it is characterized by transplant glomerulopathy (duplication and laminations of the glomerular basement membrane) and excess laminations of the peritubular capillaries (for review, see Smith and Colvin 2012); in cardiac allografts, the hallmark of transplantation-associated arteriosclerosis is concentric vascular intimal thickening seen in chronically rejecting cardiac allografts (for review, see Soleimani et al. 2006); and in lung allografts, obliterative bronchiolitis syndrome (BOS) arises from airway inflammation and the development of progressive narrowing and fibrosis in the small airways (for review, see Emtiazjoo and Wilkes 2013).

Passive antibody transfer models into T- and B-cell-deficient recipients provide the most definitive evidence for the sufficiency of DSA to induce chronic rejection. Colvin and colleagues developed a model of chronic cardiac allograft vasculopathy by the repeated transfer over 28 days of either complement-fixing IgG2a or non-complement-fixing anti-MHC class I mAb into B6.Rag1$^{-/-}$ mice recipients of B10.BR hearts (for review, see Akiyoshi et al. 2012). Importantly, antibody to class I MHC antigens did not require complement deposition in the graft to induce antibody-mediated rejection, whereas Fc- and NK-dependent events were necessary.

Mohanakumar and colleagues have developed an intriguing mouse model in which immunopathology of chronic rejection results from the repeated administration of anti-MHC class I antibodies intrabronchially into native lungs (Fukami et al. 2009). These mice develop marked CD4$^+$ T cell and CD11b$^+$ cell infiltration around the vessels and bronchioles of the lung by day 15, followed by epithelial hyperplasia, fibrosis, and occlusion of the distal airways by day 30, histopathological features similar to chronic rejection in human lung transplantation. De novo production of antibodies and increased frequencies of Th17 cells specific for self-antigens, including K-α1-tubulin and collagen V, contributed to lung pathology (Fukami et al. 2012a), similar to reports on obliterative bronchiolitis in human and rat lung transplantation (Burlingham et al. 2007). Although there are several significant caveats for this model of antibody-mediated injury of native lungs, it nonetheless provides a tractable model for an investigation into the proximal causes of lung remodeling following antibody-mediated injury.

Models involving the passive transfer of antibodies to induce chronic rejection do not provide opportunities to investigate the regulation of the antibody response that generates chronic rejection (vs. acute antibody-mediated rejection). Thus, the report by Kwun et al. (2012) of a chronic cardiac allograft rejection model in mice after alemtuzumab (antihuman CD52 mAb) is noteworthy. Induction therapy with alemtuzumab in human kidney transplant recipients has been associated with increased incidence of serum alloantibody, C4d deposition, and antibody-mediated rejection. In transgenic mice expressing the human CD52 only on T cells, alemtuzumab treatment resulted in profound peripheral T-cell depletion and prolonged survival of fully MHC-mismatched cardiac allografts. DSA was inhibited at 2 wk posttransplantation, but increasing titers of DSA were detected in 50% of recipients at week ≥7 posttransplantation. Hearts examined at ≥100 d posttransplantation show a striking dichotomy: those from recipients with DSA showing narrowing of vascular lumen due to neointimal thickening and fibrosis in the myocardium, whereas those from recipients without DSA having significantly reduced pathology. There was minimal T-cell infiltration of the grafts with cardiac allograft vasculopathy (CAV), and the frequency of allospecific T cells producing interferon γ (IFN-γ) was comparable in the recipients with or without DSA. Thus, this mouse transplant model of T-cell depletion with alemtuzumab mimics the clinical outcome of human patients, wherein continuous de novo emergence of allospecific B cells and donor-specific antibodies, but not alloreactive T cells, was correlated with the development of CAV and fibrosis.

Models Defining Antibody-Independent Roles for B Cells in Transplantation

Basic understanding of the multiple properties of B cells suggests that B cells may play a role in graft rejection independently of alloantibodies. Several models have attempted to address this issue. The first was through the generation of mixed bone marrow chimeras in which all the B cells were class II deficient, whereas 50% of all the other cell types were class II sufficient (Noorchashm et al. 2006). In those mixed bone marrow chimeras, fully mismatched MHC cardiac allograft survival was markedly prolonged compared with control counterparts reconstituted with B-cell-deficient and wild-type bone marrows (median survival time, >70 vs. 9.5 d). Mechanistically, deficiency in B-cell-mediated Ag presentation disrupted both alloantibody production and the progression of CD4 T-cell activation, suggesting that cognate T–B-cell interactions, alloantibodies, or both were necessary for acute rejection. More recently, Ng et al. (2010) used B-cell-deficient hosts and the adoptive transfer of activated T cells with or without B cells to investigate the role of B cells in the maintenance of T-cell memory. The presence of B cells promoted the optimal activation, differentiation, and survival of alloreactive CD4 and CD8 T cells into functional memory cells, assessed by an accelerated rejection of allograft rejection; however, the effects were modest and the necessity for antibodies unclear. Thus, the definitive demonstration of antibody-independent function of B cells in allograft rejection will require a model system that has defective B cells but normalized alloantibodies, or that has normal B cells but no alloantibodies.

Recent reports indicate that a subset of regulatory B cells (Bregs) plays an important role in modulating rejection and facilitating allograft tolerance in two models of allogeneic islet transplantation following treatment with anti-T-cell IG and Mucin 1 (TIM-1) (Ding et al. 2011) or with anti-CD45RB in combination with anti-TIM-1 (Lee et al. 2012). Bregs from these tolerant mice were enriched for expression of TIM-1, IL-10, and IL-4 and, upon adoptive transfer into B-cell-deficient recipients, were able to prolong allograft survival in a donor-specific and IL-10-dependent manner. Whether regulatory B cells play an important role in other organ transplant models, or models of tolerance, remains to be determined. Nonetheless, these models provide important insight into the divergent roles of B cells in allotransplantation and highlight the caveat of adoptive transfer of a heterogeneous population of B cells that may contain both effector and regulatory alloreactive subsets, as well as a potential unwanted consequence of eliminating all B cells with panreactive antibodies.

Antibodies and Graft Accommodation

One of the least understood but clinically important concepts in regard to graft response to DSA is "accommodation," which was originally defined as an acquired resistance to antibody-mediated injury in ABO-incompatible kidney allografts (Bach et al. 1991). A common histopathological hallmark of accommodation is C4d deposition without other signs of rejection—

 Cite this article as *Cold Spring Harb Perspect Med* doi: 10.1101/cshperspect.a015495

suggesting that antibody binding and early complement activation is intact but in the presence of regulatory pathways. In early hamster-to-rat heart transplantation models, inhibition of complement reproducibly induced accommodation associated with the expression of cytoprotective genes, A20 and B-cell lymphoma (Bcl)-2 (Bach et al. 1997), and the inhibition of the membrane attack complex of complement (Suhr et al. 2007). Ding and coworkers (Bobadilla et al. 2008) induced accommodation in xenogeneic rat hearts transplanted into $Rag^{-/-}$ $Gal^{-/-}$ mice by repeated low-dose transfusion of anti-Gal IgG1 mAbs. Accommodation was confirmed by the acquired resistance to HAR to doses of anti-Gal IgG1 or IgG3 that caused HAR in freshly transplanted grafts and was associated with the up-regulated expression of the complement-regulatory proteins DAF, complement receptor 1-related gene/protein-y (Crry), and, to a lesser extent, CD59. More recent studies suggest that accommodation may be a more frequent clinical occurrence and that anti-HLA antibodies may also induce graft accommodation via increased expression of cytoprotective genes in vitro (Atz and Reed 2008; Iwasaki et al. 2012) and in vivo (Fukami et al. 2012b). In a model of single HLA A2-mismatched heterotopic heart transplantation in presensitized mice, accommodation was induced by the transfer of anti-HLA framework (mouse IgG2a) mAb into the donor to precondition the graft before transplantation (Fukami et al. 2012b). Increased expression of Bcl-2, Bcl-x_L, and hemoxygenase-1 genes was observed in the accommodated hearts, and silencing of Bcl-2 prevented accommodation. Many mechanistic issues remain unresolved, including whether different states of accommodation are elicited by different antibody specificities or isotypes (or IgG subclasses) and whether graft accommodation is mechanistically and clinically distinct from chronic rejection in recipients with high-titer DSA.

MODELS OF TRANSPLANTATION TOLERANCE

Mechanisms of T-cell tolerance exist naturally to prevent the onset of autoimmunity in animals and humans, and efforts at inducing transplantation tolerance have attempted to harness these pathways. Mechanisms of self-tolerance include the deletion of autoreactive T cells in the thymus (central tolerance) or, for T cells that escape this negative selection, peripheral tolerance through deletion, intrinsic inactivation, or extrinsic suppression of effector T cells. In animals, transplantation tolerance can be confirmed by acceptance of donor but not third-party secondary grafts, a hallmark of donor specificity that correlates with low or undetectable T-cell responses upon stimulation with alloantigens in vitro (proliferation or IFN-γ production) and with the absence of alloantibodies in vivo. Here, we discuss three different approaches leading to tolerance to allogeneic grafts in mice (see also Table 2).

Induction of Central Tolerance with Mixed Chimerism

The only prospective strategy so far that has allowed the prospective induction of transplantation tolerance in the clinic is the generation of donor-recipient chimerism that enables donor APCs to populate the host thymus to induce negative selection of donor-reactive T cells (Manilay et al. 1998). Early successful mouse models involved lethal doses of irradiation and reconstitution of the immune system with a mixture of donor and host bone marrow. Such a regimen can achieve life-long coexistence of donor and host hematopoietic cells and enable donor-specific tolerance of fully mismatched skin grafts, the most stringent transplant model (Sharabi et al. 1992; Wekerle et al. 2000). However, the toxicity of lethal irradiation prompted conditioning strategies to evolve toward nonmyeloablative regimens that include partial depletion/inactivation of host T cells rather than complete elimination. More recent approaches include facilitating cells, a bone marrow cell type that can promote engraftment (Kaufman et al. 1994), or infusion of regulatory T cells (Tregs) to further reduce graft-versus-host disease (GVHD) (Pilat et al. 2010).

Successes in mouse models have led to attempts at inducing mixed chimerism in preclin-

Table 2. Animal models of central and peripheral tolerance

Type	Tolerance mechanism	Induction method	Animal models	References
Central	Negative selection	Mixed chimerism	Mice, cattle, dog, miniature swine, nonhuman primate	For review, see Pasquet et al. 2011
Peripheral	Deletion	Costimulation blockade	Mice, rats, nonhuman primates	Li et al. 2001; Preston et al. 2005
	Anergy	Costimulation blockade	Mice, rats, nonhuman primates	Lenschow et al. 1992; Lin et al. 1993; Wekerle et al. 2002; Preston et al. 2005
	Regulation	Costimulation blockade	Mice	Quezada et al. 2005

ical models, including the miniature swine, nonhuman primates, and ultimately, humans. In those studies, the recipients are preconditioned with nonmyeloablative regimens to facilitate allogeneic bone marrow engraftment. Although durable mixed chimerism was rarely achieved (Larsen et al. 2010), long-term acceptance of kidney allografts in a subset of the recipients was attained (Kawai et al. 1995). In a classic bench-to-bedside and then returning to the bench, efforts are now focused on identifying new ways to achieve stable mixed chimerism in the clinic, because this state resulted in the most robust form of tolerance in rodents, and also in understanding how transient chimerism favors the development of operational tolerance.

Induction of Peripheral Tolerance with Costimulation Blockade

Activation of T cells depends on engagement by the T-cell receptor (TCR) simultaneously with ligation of costimulatory receptors, of which CD28 is the predominant one on naïve T cells. Once stimulated, T cells up-regulate other costimulatory molecules that are important for the activation of effector and memory T cells, including CD154 (the ligand for CD40, a receptor expressed on APCs that promotes their activation), and CD152 (cytotoxic T-lymphocyte antigen 4 [CTLA-4], an inhibitory receptor that can terminate T-cell activation) (Alegre and Fallarino 2006). Blockade of activating costimulatory receptors during TCR engagement can result in T-cell death (Li et al. 2001) or anergy (durable hyporesponsiveness) (Wekerle et al. 2002), therefore predicting that costimulation blockade therapies administered at the time of transplantation (when only alloreactive T cells should be engaged) would induce transplantation tolerance. This prediction has been shown to be correct in multiple mouse transplantation models.

CD28 and cytotoxic T-lymphocyte antigen 4 (CTLA-4) share the same ligands, CD80 (B7-1) and CD86 (B7-2), but the affinity of CTLA-4 for its binding partners is greater than that of CD28, allowing CTLA-4 to sequester the ligands and thus prevent/minimize CD28 engagement (Alegre et al. 2001). The first-generation costimulation blockade reagent, CTLA-4Ig, was a fusion protein of the extracellular domain of CTLA-4 and the Fc portion of human IgG. This construct binds B7 molecules with high affinity and prevents activation of naïve T cells (Alegre and Fallarino 2006). Administration of CTLA-4Ig at the time of transplantation was initially shown to prevent rejection of xenogeneic human pancreatic islets transplanted into diabetic mice (Lenschow et al. 1992) and, when injected shortly after transplantation, of allogeneic cardiac transplants in rat (Lin et al. 1993). Although CTLA-4Ig can induce long-term allograft survival, it has failed to induce true immunological tolerance in some mouse models of allograft transplantation. This may be due to the ability of CTLA-4Ig to inhibit both the engagement of the positive costimulator CD28

 Cite this article as *Cold Spring Harb Perspect Med* doi: 10.1101/cshperspect.a015495

and the negative costimulator CTLA-4 on effector T cells and on Tregs. Specific inhibition of CTLA-4 has been shown to reduce the intrinsic restraint of effector T cells as well as diminish numbers and suppressive function of Tregs (Charbonnier et al. 2012; Riella et al. 2012). Thus, there is a need to understand the complex effects of CTLA-4Ig or higher-affinity variants, on different subsets of immune cells, because these insights may eventually lead to the optimal use of this reagent in the clinic, and possibly to the successful induction of tolerance.

Blockade of the CD40/CD154 pathway using anti-CD154 mAb is one of the most effective approaches in rodents at inducing donor-specific tolerance, especially if combined with an infusion of donor cells administered either before or on the day of solid organ transplantation. This regimen not only reduces activation and promotes death of alloreactive T cells, but also promotes the generation of Tregs, and these mechanisms collectively contribute to a robust state of tolerance (Quezada et al. 2005). Anti-CD154 was also successful at inducing long-term acceptance of renal allografts in nonhuman primates (Preston et al. 2005), leading to much optimism for achieving tolerance in the clinic. Anti-CD154-based models of tolerance permit basic investigations into transplantation tolerance; however, thromboembolic complications observed in humans treated with anti-CD154 mAb have stymied further clinical development. This was due to CD154 expression on human but not mouse platelets (Koyama et al. 2004). Nonetheless, the efficacy of anti-CD154 has prompted the development of an alternative approach of targeting CD40. The challenge with this approach is to identify reagents that lack agonistic activity when bound to CD40, while working comparably to anti-CD154 in preclinical models and humans (Adams et al. 2005; Aoyagi et al. 2009; Badell et al. 2012).

Because memory T cells up-regulate other costimulatory and adhesion molecules and do not rely as much on CD28 for their activation, they are resistant to costimulation blockade with CTLA-4Ig (Valujskikh and Li 2007). Animal models have been developed to test the ability of antibodies to such adhesion and costimulatory molecules to prevent activation of memory alloreactive T cells (Damle et al. 1992; Curry et al. 2004; Vu et al. 2006; Badell et al. 2010). It is unclear whether these approaches will be superior to conventional calcineurin inhibitors that can block TCR signaling in both naïve and memory T cells; nonetheless, these approaches of targeting adhesion molecules can significantly promote graft survival in nonhuman primates and point to the potential of controlling recall alloreactive responses with rationally designed biologics (Badell et al. 2010). The induction of tolerance in the face of a significant memory alloreactive T-cell repertoire represents a major area of investigation that has strong clinical relevance.

Spontaneous Transplant Models

In certain mouse strain combinations of donors and recipients, the kidney allografts are spontaneously accepted without immunosuppression or tolerogenic therapies despite MHC and minor antigen mismatches. Although this outcome does not recapitulate the vast majority of clinical data, they have been used to shed light into the immunological pathways that can achieve tolerance in a tissue-specific manner. The long-term survival (>60 d) of DBA/2 ($H2^d$) kidneys→C57BL/6 ($H\text{-}2^b$) recipients was originally reported by Russell and colleagues (1978). Spontaneous acceptance of complete MHC-mismatched renal allografts has also been reported in several other strain combinations including A/J ($H\text{-}2^a$)→C57BL/6 (Bickerstaff et al. 2008). Long-term survival was accompanied by an absence of renal allograft injury by histological analysis and without a rise in recipient serum creatinine levels. Essential roles for transforming growth factor β (TGF-β) signaling and regulatory T cells (Bickerstaff et al. 2001), as well as (idoleamine-2,3-dioxygenase [indoleamine-*pyrrole* 2,3-dioxygenase [IDO]) signaling, which increased over time in graft-resident dendritic cells, have been reported (Cook et al. 2008). Thirty percent of these kidney recipient mice with detectable IDO-producing dendritic cells accepted donor skin grafts without any immunosuppression, thus formally showing tolerance (Cook et al. 2008). In another strain combination, C57BL/6 kidney allografts into B10.BR recipients also showed spontaneous graft acceptance and tolerance to subsequent skin allografts (Wang et al. 2011). Increased Foxp3 mRNA expression was observed in those kidney allografts, suggesting a role for Tregs. Thus, these kidney models, although not reflective of the clinical situation, provide a unique opportunity to study the protolerogenic environment of the allograft.

Liver allografts are accepted in many rat and mouse strain combinations in the absence of immunosuppression. The acceptance of these allografts has been shown to be dependent upon $CD25^+$ regulatory cells (Li et al. 2008), invariant natural killer cells (Morita et al. 2007), Fas-FasL (Uchiyama et al. 2002), IFN-γ (Mele et al. 2003), CTLA-4 (Li et al. 2005), and PD-1 signaling (Morita et al. 2010), and low levels of NF-κB in liver dendritic cells (Ma et al. 2009). The broad spectrum of peripheral tolerance mechanisms, including deletion, anergy, and regulation, implicated in the liver allograft model underscores the critical role of the transplanted organ in facilitating the development of immunological tolerance, and also the complexity of the tolerant state. The tolerogenicity of the allogeneic liver is recapitulated in clinical transplantation, and thus further investigation with this model of tolerance is warranted, despite the rather enormous technical challenges of performing successful liver transplantation in mice.

Much work remains to be performed in order to reach an understanding of why certain organ types and certain histoincompatibilities promote graft acceptance, and to translate these insights into tolerogenic strategies for the clinic.

CONCLUDING REMARKS

It is clear that mouse models provide unique opportunities to gain insights into mechanisms of rejection and to test approaches that prevent rejection and induce transplantation tolerance. Nonetheless, there are limitations to the information generated from these models. The exclusive use of inbred mice with limited genetic diversity may mask pathways of rejection or tolerance that exist in humans. Indeed, the wide range of rejection tempos in transplantation with outbred mice, spanning HAR to chronic rejection, contrasting with the unimodal acute rejection of cardiac allografts observed when using inbred strains, underscore this concern (Reichenbach et al. 2013). Obesity, age, and the infectious history, as well as organ preservation times (Wehner et al. 2010; Oberhuber et al. 2012; Denecke et al. 2013), which are highly variable in human transplantation, can also have profound effects on immune responses and graft outcomes. Understanding these caveats and modifying the mouse transplantation models to capture these variables may introduce complexities into the investigation but may allow for a more accurate translation of findings from mice to man.

ACKNOWLEDGMENTS

We thank Dr. James W. Williams for his comments on the manuscript. This work is supported in part by grants NIAID P01AI-97113 to A.S.C. and M.-L.A.; RO1 AI071080 to M.-L.A.; and ROTRF 979162997 and NIAID R01 AI072630 to A.S.C. M.L.M. is funded by a Cardiovascular Pathophysiology and Biochemistry Training Grant (T32 HL07237) and an AHA Predoctoral Fellowship (13PRE14550022).

REFERENCES

Adams AB, Shirasugi N, Jones TR, Durham MM, Strobert EA, Cowan S, Rees P, Hendrix R, Price K, Kenyon NS, et al. 2005. Development of a chimeric anti-CD40 monoclonal antibody that synergizes with LEA29Y to prolong islet allograft survival. *J Immunol* **174:** 542–550.

Akiyoshi T, Hirohashi T, Alessandrini A, Chase CM, Farkash EA, Neal Smith R, Madsen JC, Russell PS, Colvin RB. 2012. Role of complement and NK cells in antibody mediated rejection. *Hum Immunol* **73:** 1226–1232.

Alegre ML, Fallarino F. 2006. Mechanisms of CTLA-4Ig in tolerance induction. *Curr Pharm Des* **12:** 149–160.

Alegre ML, Frauwirth KA, Thompson CB. 2001. T-cell regulation by CD28 and CTLA-4. *Nat Rev Immunol* **1:** 220–228.

Aoyagi T, Yamashita K, Suzuki T, Uno M, Goto R, Taniguchi M, Miura T, Okimura K, Itoh T, Shimizu A, et al. 2009. A human anti-CD40 monoclonal antibody, 4D11, for kidney transplantation in cynomolgus monkeys: Induction and maintenance therapy. *Am J Transplant* **9:** 1732–1741.

Atkinson C, Horsley J, Rhind-Tutt S, Charman S, Phillpotts CJ, Wallwork J, Goddard MJ. 2004. Neointimal smooth muscle cells in human cardiac allograft coronary artery vasculopathy are of donor origin. *J Heart Lung Transplant* **23:** 427–435.

Atz ME, Reed EF. 2008. Role of anti-MHC class I antibody in facilitating transplant accommodation. *Critical Rev Immunol* **28:** 485–511.

Bach FH, Turman MA, Vercellotti GM, Platt JL, Dalmasso AP. 1991. Accommodation: A working paradigm for progressing toward clinical discordant xenografting. *Transplant Proc* **23:** 205–207.

Bach FH, Ferran C, Hechenleitner P, Mark W, Koyamada N, Miyatake T, Winkler H, Badrichani A, Candinas D, Hancock WW. 1997. Accommodation of vascularized xenografts: Expression of "protective genes" by donor endothelial cells in a host Th2 cytokine environment. *Nat Med* **3:** 196–204.

Badell IR, Russell MC, Thompson PW, Turner AP, Weaver TA, Robertson JM, Avila JG, Cano JA, Johnson BE, Song M, et al. 2010. LFA-1-specific therapy prolongs allograft survival in rhesus macaques. *J Clin Invest* **120:** 4520–4531.

Badell IR, Thompson PW, Turner AP, Russell MC, Avila JG, Cano JA, Robertson JM, Leopardi FV, Strobert EA, Iwakoshi NN, et al. 2012. Nondepleting anti-CD40-based therapy prolongs allograft survival in nonhuman primates. *Am J Transplant* **12:** 126–135.

Baldwin WM III, Valujskikh A, Fairchild RL. 2010. Antibody-mediated rejection: Emergence of animal models to answer clinical questions. *Am J Transplant* **10:** 1135–1142.

Bergstresser PR, Toews GB, Gilliam JN, Streilein JW. 1980a. Unusual numbers and distributions of Langerhans cells in skin with unique immunologic properties. *J Invest Dermatol* **74:** 312–314.

Bergstresser PR, Fletcher CR, Streilein JW. 1980b. Surface densities of Langerhans cells in relation to rodent epidermal sites with special immunologic properties. *J Invest Dermatol* **74:** 77–80.

Bickerstaff AA, Wang JJ, Pelletier RP, Orosz CG. 2001. Murine renal allografts: Spontaneous acceptance is associated with regulated T cell-mediated immunity. *J Immunol* **167:** 4821–4827.

Bickerstaff A, Nozaki T, Wang J-J, Pelletier R, Hadley G, Nadasdy G, Fairchild RL. 2008. Acute humoral rejection of renal allografts in $CCR5^{-/-}$ recipients. *Am J Transplant* **8:** 557–566.

Billingham RE, Medawar PB. 1951. Technique of free skin grafting in mammals. *J Exp Biol* **28:** 385–391.

Bobadilla JL, Love RB, Jankowska-Gan E, Xu Q, Haynes LD, Braun RK, Hayney MS, Munoz del Rio A, Meyer K, Greenspan DS, et al. 2008. Th-17, monokines, collagen type V, and primary graft dysfunction in lung transplantation. *Am J Respir Crit Care Med* **177:** 660–668.

Burlingham WJ, Love RB, Jankowska-Gan E, Haynes LD, Xu Q, Bobadilla JL, Meyer KC, Hayney MS, Braun RK, Greenspan DS, et al. 2007. IL-17-dependent cellular immunity to collagen type V predisposes to obliterative bronchiolitis in human lung transplants. *J Clin Invest* **117:** 3498–3506.

Burns AM, Chong AS. 2011. Alloantibodies prevent the induction of transplantation tolerance by enhancing alloreactive T cell priming. *J Immunol* **186:** 214–221.

Calcagni M, Althaus MK, Knapik AD, Hegland N, Contaldo C, Giovanoli P, Lindenblatt N. 2011. In vivo visualization of the origination of skin graft vasculature in a wild-type/GFP crossover model. *Microvasc Res* **82:** 237–245.

Capla JM, Ceradini DJ, Tepper OM, Callaghan MJ, Bhatt KA, Galiano RD, Levine JP, Gurtner GC. 2006. Skin graft vascularization involves precisely regulated regression and replacement of endothelial cells through both angiogenesis and vasculogenesis. *Plast Reconstr Surg* **117:** 836–844.

Charbonnier LM, Vokaer B, Lemaître PH, Field KA, Leo O, Le Moine A. 2012. CTLA-4Ig restores rejection of MHC class-II mismatched allografts by disabling IL-2-expanded regulatory T cells. *Am J Transplant* **12:** 2313–2321.

Chen H-D, Silvers WK. 1983. Influence of Langerhans cells on the survial of H-Y incompatible skin grafts in rats. *J Invest Dermatol* **81:** 20–23.

Cook CH, Bickerstaff AA, Wang J-J, Nadasdy T, Della Pelle P, Colvin RB, Orosz CG. 2008. Spontaneous renal allograft acceptance associated with "regulatory" dendritic cells and IDO. *J Immunol* **180:** 3103–3112.

Corry RJ, Winn HJ, Russell PS. 1973. Primarily vascularized allografts of hearts in mice. The role of H-2D, H-2K, and non-H-2 antigens in rejection. *Transplantation* **16:** 343–350.

Curry AJ, Chikwe J, Smith XG, Cai M, Schwarz H, Bradley JA, Bolton EM. 2004. OX40 (CD134) blockade inhibits the co-stimulatory cascade and promotes heart allograft survival. *Transplantation* **78:** 807–814.

Damle NK, Klussman K, Linsley P, Aruffo A. 1992. Differential costimulatory effects of adhesion molecules B7, ICAM-1, LFA-3 and VCAM-1 on resting and antigen-primed $CD4^+$ T lymphocytes. *J Immunol* **148:** 1985–1989.

Denecke C, Yuan X, Ge X, Kim IK, Bedi D, Boenisch O, Weiland A, Jurisch A, Kotsch K, Pratschke J, et al. 2013. Synergistic effects of prolonged warm ischemia and donor age on the immune response following donation after cardiac death kidney transplantation. *Surgery* **153:** 249–261.

Ding JW, Zhou T, Zeng H, Ma L, Verbeek JS, Yin D, Shen J, Chong AS. 2008a. Hyperacute rejection by anti-Gal IgG1, IgG2a, and IgG2b is dependent on complement and Fc-γ receptors. *J Immunol* **180:** 261–268.

Ding JW, Zhou T, Ma L, Yin D, Shen J, Ding CP, Tang IY, Byrne GW, Chong AS. 2008b. Expression of complement regulatory proteins in accommodated xenografts induced by anti-α-Gal IgG1 in a rat-to-mouse model. *Am J Transplant* **8:** 32–40.

Ding Q, Yeung M, Camirand G, Zeng Q, Akiba H, Yagita H, Chalasani G, Sayegh MH, Najafian N, Rothstein DM. 2011. Regulatory B cells are identified by expression of TIM-1 and can be induced through TIM-1 ligation to promote tolerance in mice. *J Clin Invest* **121:** 3645–3656.

Emtiazjoo AM, Wilkes DS. 2013. Humoral immunity and the development of obliterative bronchiolitis post lung transplantation—Is there a link? *Am J Respir Cell Mol Biol* **48:** 145–149.

Fukami N, Ramachandran S, Saini D, Walter M, Chapman W, Patterson GA, Mohanakumar T. 2009. Antibodies to MHC class I induce autoimmunity: Role in the pathogenesis of chronic rejection. *J Immunol* **182:** 309–318.

Fukami N, Ramachandran S, Takenaka M, Weber J, Subramanian V, Mohanakumar T. 2012a. An obligatory role for lung infiltrating B cells in the immunopathogenesis of obliterative airway disease induced by antibodies to MHC class I molecules. *Am J Transplant* **12:** 867–876.

Fukami N, Ramachandran S, Narayanan K, Liu W, Nath DS, Jendrisak M, Chapman W, Mohanakumar T. 2012b. Mechanism of accommodation in a sensitized human leukocyte antigen transgenic murine cardiac transplant model. *Transplantation* **93:** 364–372.

Ge F, Gong W. 2011. Strategies for successfully establishing a kidney transplant in a mouse model. *Exp Clin Transplant* **9:** 287–294.

Gibson T, Medawar PB. 1943. The fate of skin homografts in man. *J Anat* **77:** 299–316.

Hagensen MK, Shim J, Falk E, Bentzon JF. 2011. Flanking recipient vasculature, not circulating progenitor cells, contributes to endothelium and smooth muscle in murine allograft vasculopathy. *Arterioscler Thromb Vasc Biol* **31:** 808–813.

Hattori Y, Bucy RP, Kubota Y, Baldwin WM III, Fairchild RL. 2012. Antibody-mediated rejection of single class I MHC-disparate cardiac allografts. *Am J Transplant* **12:** 2017–2028.

He C, Schenk S, Zhang Q, Valujskikh A, Bayer J, Fairchild RL, Heeger PS. 2004. Effects of T cell frequency and graft size on transplant outcome in mice. *J Immunol* **172:** 240–247.

Iwasaki K, Miwa Y, Ogawa H, Yazaki S, Iwamoto M, Furusawa T, Onishi A, Kuzuya T, Haneda M, Watarai Y, et al. 2012. Comparative study on signal transduction in endothelial cells after anti-a/b and human leukocyte antigen antibody reaction: Implication of accommodation. *Transplantation* **93:** 390–397.

Kaufman CL, Colson YL, Wren SM, Watkins S, Simmons RL, Ildstad ST. 1994. Phenotypic characterization of a novel bone marrow-derived cell that facilitates engraftment of allogeneic bone marrow stem cells. *Blood* **84:** 2436–2446.

Kawai T, Cosimi AB, Colvin RB, Powelson J, Eason J, Kozlowski T, Sykes M, Monroy R, Tanaka M, Sachs DH. 1995. Mixed allogeneic chimerism and renal allograft tolerance in cynomolgus monkeys. *Transplantation* **59:** 256–562.

Koene RA, Gerlag PG, Hagemann JF, van Haelst UJ, Wijdeveld PG. 1973. Hyperacute rejection of skin allografts in the mouse by the administration of alloantibody and rabbit complement. *J Immunol* **111:** 520–526.

Koyama I, Kawai T, Andrews D, Boskovic S, Nadazdin O, Wee SL, Sogawa H, Wu DL, Smith RN, Colvin RB, et al. 2004. Thrombophilia associated with anti-CD154 monoclonal antibody treatment and its prophylaxis in nonhuman primates. *Transplantation* **77:** 460–462.

Krasinskas AM, Eiref SD, McLean AD, Kreisel D, Gelman AE, Popma SH, Moore JS, Rosengard BR. 2000. Replacement of graft-resident donor-type antigen presenting cells alters the tempo and pathogenesis of murine cardiac allograft rejection. *Transplantation* **70:** 514–521.

Kwun J, Malarkannnan S, Burlingham WJ, Knechtle SJ. 2011. Primary vascularization of the graft determines the immunodominance of murine minor H antigens during organ transplantation. *J Immunol* **187:** 3997–4006.

Kwun J, Oh BC, Gibby AC, Ruhil R, Lu VT, Kim DW, Page EK, Bulut OP, Song MQ, Farris AB, et al. 2012. Patterns of de novo allo B cells and antibody formation in chronic cardiac allograft rejection after alemtuzumab treatment. *Am J Transplant* **12:** 2641–2651.

Lafuse WP, McCormick JF, Melvold RW, David CS. 1981. Serological and biochemical analysis of Ia molecules in the I-A mutant B6.C-H-2^{bm12}. *Transplantation* **31:** 434–438.

Larsen CP, Page A, Linzie KH, Russell M, Deane T, Stempora L, Strobert E, Penedo MC, Ward T, Wiseman R, et al. 2010. An MHC-defined primate model reveals significant rejection of bone marrow after mixed chimerism induction despite full MHC matching. *Am J Transplant* **10:** 2396–2409.

Lee KM, Kim JI, Stott R, Soohoo J, O'Connor MR, Yeh H, Zhao G, Eliades P, Fox C, Cheng N, et al. 2012. Anti-CD45RB/anti-TIM-1-induced tolerance requires regulatory B cells. *Am J Transplant* **12:** 2072–2078.

Lenschow DJ, Zeng Y, Thistlethwaite JR, Montag A, Brady W, Gibson MG, Linsley PS, Bluestone JA. 1992. Long-term survival of xenogeneic pancreatic islet grafts induced by CTLA-4Ig. *Science* **257:** 789–792.

Li W, Lu L, Wang Z, Wang L, Fung JJ, Thomson AW, Qian S. 2001. Costimulation blockade promotes the apoptotic death of graft-infiltrating T cells and prolongs survival of hepatic allografts from FLT3L-treated donors. *Transplantation* **72:** 1423–1432.

Li W, Zheng XX, Kuhr CS, Perkins JD. 2005. CTLA-4 engagement is required for induction of murine liver transplant spontaneous tolerance. *Am J Transplant* **5:** 978–986.

Li W, Kuhr CS, Zheng XX, Carper K, Thomson AW, Reyes JD, Perkins JD. 2008. New insights into mechanisms of spontaneous liver transplant tolerance: The role of Foxp3-expressing $CD25^+CD4^+$ regulatory T cells. *Am J Transplant* **8:** 1639–1651.

Li Q, Peng Q, Xing G, Li K, Wang N, Farrar CA, Meader L, Sacks SH, Zhou W. 2010. Deficiency of C5aR prolongs renal allograft survival. *J Am Soc Nephrol* **21:** 1344–1353.

Lin H, Bolling SF, Linsley PS, Wei RQ, Gordon D, Thompson CB, Turka LA. 1993. Long-term acceptance of major histocompatibility complex mismatched cardiac allografts induced by CTLA-4Ig plus donor-specific transfusion. *J Exp Med* **178:** 1801–1806.

Lindenblatt N, Calcagni M, Contaldo C, Menger MD, Giovanoli P, Vollmar B. 2008. A new model for studying the revascularization of skin grafts in vivo: The role of angiogenesis. *Plast Reconstr Surg* **122:** 1669–1680.

Lu WH, Palatnik K, Fishbein GA, Lai C, Levi DS, Perens G, Alejos J, Kobashigawa J, Fishbein MC. 2011. Diverse morphologic manifestations of cardiac allograft vasculopathy: A pathologic study of 64 allograft hearts. *J Heart Lung Transplant* **30:** 1044–1050.

Ma L-L, Gao X, Liu L, Xiang Z, Blackwell TS, Williams P, Chari RS, Yin DP. 2009. CpG oligodeoxynucleotide triggers the liver inflammatory reaction and abrogates spontaneous tolerance. *Liver Transpl* **15:** 915–923.

Manilay JO, Pearson DA, Sergio JJ, Swenson KG, Sykes M. 1998. Intrathymic deletion of alloreactive T cells in mixed bone marrow chimeras prepared with a nonmyeloablative conditioning regimen. *Transplantation* **66:** 96–102.

McFarland HI, Rosenberg AS. 2009. Skin allograft rejection. *Curr Protoc Immunol* **4:** 4.

McKenzie IFC, Morgan CM, Sanderin MS, Michaelides MM, Melvold RW, Kohn HL. 1979. B6.C-H-2^{bm12}, a new H-2 mutation in the I region in the mouse. *J Exp Med* **150:** 1323–1338.

Medawar PB. 1944. The behavior and fate of skin autografts and skin homografts in rabbits. *J Anat* **78:** 176–199.

Medawar PB. 1945. A second study of the behaviour and fate of skin homografts in rabbits. *J Anat* **79:** 157–177.

Mele TS, Kneteman NM, Zhu L-F, Ramassar V, Urmson J, Halloran B, Churchill TA, Jewell L, Kane K, Halloran PF. 2003. IFN-γ is an absolute requirement for spontaneous acceptance of liver allografts. *Am J Transplant* **3:** 942–951.

Meng L, Wu Z, Wang Y, Lassman C, Busuttil RW, Zhai Y, Kupiec-Weglinski JW, et al. 2008. Differential impact of CD154 costimulation blockade on alloreactive effector and regulatory T cells in murine renal transplant recipients. *Transplantation* **85:** 1332–1338.

Mengel M, Husain S, Hidalgo L, Sis B. 2012. Phenotypes of antibody-mediated rejection in organ transplants. *Transpl Int* **25:** 611–622.

Mengle-Gaw L, Conner S, McDevitt HO, Fathman CG. 1984. Gene conversion between murine class II major histocompatibility complex loci: Functional and molecular evidence from the *bm12* mutant. *J Exp Med* **160:** 1184–1194.

Minami E, Laflamme MA, Saffitz JE, Murry CE. 2005. Extracardiac progenitor cells repopulate most major cell types in the transplanted human heart. *Circulation* **112:** 2951–2958.

Morgan BP. 1995. Complement regulatory molecules: Application to therapy and transplantation. *Immunol Today* **16:** 257–259.

Morita M, Fujino M, Li X-K, Kimura H, Nakayama T, Taniguchi M, Sugioka A. 2007. Spontaneous tolerance involving natural killer T cells after hepatic grafting in mice. *Transplant Immunol* **18:** 142–145.

Morita M, Fujino M, Jiang G, Kitazawa Y, Xie L, Azuma M, Yagita H, Nagao S, Sugioka A, Kurosawa Y, et al. 2010. PD-1/B7-H1 interaction contribute to the spontaneous acceptance of mouse liver allograft. *Am J Transplant* **10:** 40–46.

Ng YH, Oberbarnscheidt MH, Chandramoorthy HC, Hoffman R, Chalasani G. 2010. B cells help alloreactive T cells differentiate into memory T cells. *Am J Transplant* **10:** 1970–1980.

Noorchashm H, Reed AJ, Rostami SY, Mozaffari R, Zekavat G, Koeberlein B, Caton AJ, Naji A. 2006. B cell-mediated antigen presentation is required for the pathogenesis of acute cardiac allograft rejection. *J Immunol* **177:** 7715–7722.

Nozaki T, Amano H, Bickerstaff A, Orosz CG, Novick AC, Tanabe K, Fairchild RL. 2007. Antibody-mediated rejection of cardiac allografts in CCR5-deficient recipients. *J Immunol* **179:** 5238–5245.

Oberhuber R, Ge X, Tullius SG. 2012. Donor age-specific injury and immune responses. *Am J Transplant* **12:** 38–42.

Pasquet L, Joffre O, Santolaria T, van Meerwijk JP. 2011. Hematopoietic chimerism and transplantation tolerance: A role for regulatory T cells. *Front Immunol* **2:** 80.

Pilat N, Baranyi U, Klaus C, Jaeckel E, Mpofu N, Wrba F, Golshayan D, Muehlbacher F, Wekerle T. 2010. Treg-therapy allows mixed chimerism and transplantation tolerance without cytoreductive conditioning. *Am J Transplant* **10:** 751–762.

Pratt JR, Basheer SA, Sacks SH. 2002. Local synthesis of complement component C3 regulates acute renal transplant rejection. *Nat Med* **8:** 582–587.

Preston EH, Xu H, Dhanireddy KK, Pearl JP, Leopardi FV, Starost MF, Hale DA, Kirk AD. 2005. IDEC-131 (anti-CD154), sirolimus and donor-specific transfusion facilitate operational tolerance in non-human primates. *Am J Transplant* **5:** 1032–1041.

Quezada SA, Bennett K, Blazar BR, Rudensky AY, Sakaguchi S, Noelle RJ. 2005. Analysis of the underlying cellular mechanisms of anti-CD154-induced graft tolerance: The interplay of clonal anergy and immune regulation. *J Immunol* **175:** 771–779.

Rahimi S, Qian Z, Layton J, Fox-Talbot K, Baldwin WM III, Wasowska BA. 2004. Non-complement- and complement-activating antibodies synergize to cause rejection of cardiac allografts. *Am J Transplant* **4:** 326–334.

Reichenbach DK, Li Q, Hoffman RA, Williams AL, Shlomchik WD, Rothstein DM, Demetris AJ, Lakkis FG. 2013. Allograft outcomes in outbred mice. *Am J Transplant* **13:** 580–588.

Riella LV, Liu T, Yang J, Chock S, Shimizu T, Mfarrej B, Batal I, Xiao X, Sayegh MH, Chandraker A. 2012. Deleterious effect of CTLA-4Ig on a Treg-dependent transplant model. *Am J Transplant* **12:** 846–855.

Russell PS, Chase CM, Colvin RB, Plate JMD. 1978. Kidney transplants in mice: An analysis of the immune status of mice bearing long-ter, H-2 incompatible transplants. *J Exp Med* **147:** 1449–1468.

Schenk S, Kish DD, He C, El-Sawy T, Chiffoleau E, Chen C, Wu Z, Sandner S, Gorbachev AV, Fukamachi K, et al. 2005. Alloreactive T cell responses and acute rejection of single class II MHC disparate heart allografts is under strict regulation by $CD4^{+}CD25^{+}$ T cells. *J Immunol* **174:** 3741–3748.

Schenk AD, Nozaki T, Rabant M, Valujskikh A, Fairchild RL. 2008. Donor-reactive CD8 memory T cells infiltrate cardiac allografts within 24-h posttransplant in naive recipients. *Am J Transplant* **8:** 1652–1661.

Schenk AD, Gorbacheva V, Rabant M, Fairchild RL, Valujskikh A. 2009. Effector functions of donor-reactive CD8 memory T cells are dependent on ICOS induced during division in cardiac grafts. *Am J Transplant* **9:** 64–73.

Sharabi Y, Abraham VS, Sykes M, Sachs DH. 1992. Mixed allogeneic chimeras prepared by a non-myeloablative regimen: Requirement for chimerism to maintain tolerance. *Bone Marrow Transplant* **9:** 191–197.

Shimizu I, Smith NR, Zhao G, Medof E, Sykes M. 2006. Decay-accelerating factor prevents acute humoral rejection induced by low levels of anti-αGal natural antibodies. *Transplantation* **81:** 95–100.

Skoskiewicz M, Chase C, Winn HJ, Russell PS. 1973. Kidney transplants between mice of graded immunogenetic diversity. *Transplant Proc* **5:** 721–725.

Smith RN, Colvin RB. 2012. Chronic alloantibody mediated rejection. *Semin Immunol* **24:** 115–121.

Soleimani B, Lechler RI, Hornick PI, George AJ. 2006. Role of alloantibodies in the pathogenesis of graft arteriosclerosis in cardiac transplantation. *Am J Transplant* **6:** 1781–1785.

Stegall MD, Chedid MF, Cornell LD. 2012. The role of complement in antibody-mediated rejection in kidney transplantation. *Nat Rev Nephrol* **8:** 670–678.

Steinmuller D. 1984. Skin grafting. *Methods Enzymol* **108:** 28–35.

Suhr BD, Black SM, Guzman-Paz M, Matas AJ, Dalmasso AP. 2007. Inhibition of the membrane attack complex of complement for induction of accommodation in the hamster-to-rat heart transplant model. *Xenotransplantation* **14:** 572–579.

Uchiyama H, Kishihara K, Minagawa R, Hashimoto K, Sugimachi K, Nomoto K. 2002. Crucial Fas–Fas ligand interaction in spontaneous acceptance of hepatic allografts in mice. *Immunology* **105:** 450–457.

Valujskikh A, Li XC. 2007. Frontiers in nephrology: T cell memory as a barrier to transplant tolerance. *J Am Soc Nephrol* **18:** 2252–2261.

Vu MD, Clarkson MR, Yagita H, Turka LA, Sayegh MH, Li XC. 2006. Critical, but conditional, role of OX40 in memory T cell-mediated rejection. *J Immunol* **176:** 1394–1401.

Wang C, Cordoba S, Hu M, Bertolino P, Bowen DG, Sharland AF, Allen RD, Alexander SI, McCaughan GW, Bishop GA. 2011. Spontaneous acceptance of mouse kidney allografts is associated with increased Foxp3 expression and differences in the B and T cell compartments. *Transplant Immunol* **24:** 149–156.

Wasowska BA, Qian Z, Cangello DL, Behrens E, Van Tran K, Layton J, Sanfilippo F, Baldwin WM III. 2001. Passive transfer of alloantibodies restores acute cardiac rejection in IgKO mice. *Transplantation* **71:** 727–736.

Wehner JR III, Baldwin WM. 2010. Cardiac allograft vasculopathy: Do adipocytes bridge alloimmune and metabolic risk. *Curr Opin Organ Transplant* **15:** 639–644.

Wekerle T, Kurtz J, Ito H, Ronquillo JV, Dong V, Zhao G, Shaffer J, Sayegh MH, Sykes M. 2000. Allogeneic bone marrow transplantation with co-stimulatory blockade induces macrochimerism and tolerance without cytoreductive host treatment. *Nat Med* **6:** 464–469.

Wekerle T, Kurtz J, Bigenzahn S, Takeuchi Y, Sykes M. 2002. Mechanisms of transplant tolerance induction using costimulatory blockade. *Curr Opin Immunol* **14:** 592–600.

Wu H, Noordmans GA, O'Brien MR, Ma J, Zhao CY, Zhang GY, Kwan TK, Alexander SI, Chadban SJ. 2012. Absence of MyD88 signaling induces donor-specific kidney allograft tolerance. *J Am Soc Nephrol* **23:** 1701–1716.

Yun JJ, Fischbein MP, Whiting D, Irie Y, Fishbein MC, Burdick MD, Belperio J, Strieter RM, Laks H, Berliner JA, et al. 2002. The role of MIG/CXCL9 in cardiac allograft vasculopathy. *Am J Pathol* **161:** 1307–1313.

Primate Models in Organ Transplantation

Douglas J. Anderson and Allan D. Kirk

Emory Transplant Center, Emory University School of Medicine, Atlanta, Georgia 30322

Correspondence: adkirk@emory.edu

Large animal models have long served as the proving grounds for advances in transplantation, bridging the gap between inbred mouse experimentation and human clinical trials. Although a variety of species have been and continue to be used, the emergence of highly targeted biologic- and antibody-based therapies has required models to have a high degree of homology with humans. Thus, the nonhuman primate has become the model of choice in many settings. This article will provide an overview of nonhuman primate models of transplantation. Issues of primate genetics and care will be introduced, and a brief overview of technical aspects for various transplant models will be discussed. Finally, several prominent immunosuppressive and tolerance strategies used in primates will be reviewed.

Animal experimentation has long provided a rational basis for the translation of treatments and techniques from the bench to the bedside. In general, all first-in-human trials require preparative animal experimentation to allow patients to make truly informed decisions about their participation. Properly designed animal studies in relevant species provide the necessary background experience with a novel approach to reasonably anticipate the efficacy or, at the very least, safety of a planned intervention. As such, they serve as a foundation on which human trials can be ethically designed, particularly in fields such as immunology, in which the complexity of the interactions involved has prevented the development of any sufficiently predictive in vitro model. Although animal models are far superior to in vitro models in projecting the potential of an approach, it must be recognized that they do not mimic clinical transplantation precisely, and thus cannot be expected to forecast the ultimate experience in humans.

The mouse model has formed the backbone of medical research and development for many years owing to the relative ease of breeding and genetic manipulation of the animals at a comparatively low cost. For immunology research, the mouse immune system offers sufficient homology for pathway determination and mechanistic studies, and indeed represents the ideal platform for this type of endeavor. In contrast, the large animal models (dog, pig, and primate) are significantly more expensive and, with the exception of inbred miniature swine (Sachs 1992; Mezrich et al. 2003), exhibit increased genetic diversity, making definitive mechanistic studies much more difficult, if not impossible. However, this complexity makes large animals suited to preclinical studies, in which the addition of often-unanticipated variables allows for the examination of practicality, safety, and generalized efficacy. In general, mice define pathways, and large animal models help establish whether a particular pathway's effect is suffi-

Cite this article as *Cold Spring Harb Perspect Med* doi: 10.1101/cshperspect.a015503

ciently robust to emerge as dominant in the midst of the numerous uncontrolled variables typical of heterogeneous human populations.

In specific regard to transplantation immunology, mice have several potential drawbacks. Laboratory mice bred in clean environments and studied between 4 and 8 weeks old have a largely naïve immune system (Blattman et al. 2002), a fact likely responsible for the success of therapies, including methods of tolerance induction, in mice, and their subsequent failure when translated to large animals (Kirk 2003; Sachs 2003), or mice exposed to pathogens (Adams et al. 2003). Additionally, mice do not constitutively express class II antigens on vascular endothelium, unlike other large animal models, which may explain the importance of class II matching in the large animals models (Pescovitz et al. 1984; Choo et al. 1997). Furthermore, the efficacy of any regimen may also be dependent on the strain of mice used (Williams et al. 2000). The relative genetic diversity and immunologic experience of large animals helps to avoid many of these shortcomings, and, indeed, experimentation in a large animal model, most frequently primates (for reasons discussed below), has become a de facto requirement before initiation of human trials in transplantation (Sachs 2003; 't Hart et al. 2004).

The complexities of the immune response often cause therapies to fail in transition to large animals, or to humans. This is most often owing to one or a few critical differences between species rather than a failure of the concept. Interspecies differences in drug pharmacokinetics may lead to apparent failure of a regimen that may have been successful if adjustments for distribution or metabolism had been considered. Furthermore, modern biologic and antibody-based therapies may be profoundly altered by minor differences in molecular structure of the target molecule. The immunologic diversity of large animals can lead to significant differences in outcome via heterologous immune interactions, an experimental parameter that is challenging to quantify and control for. Finally, the practicalities of animal husbandry during treatment can only approximate the care human patients receive. Indwelling catheters and wound care in surgical models are often difficult. Monitoring and vascular access is challenging in conscious animal models, often limiting the options for drug delivery and dosing schedules.

Another potential issue with large animal models is that of time, both in terms of the animal's age and survival of grafts. For practical reasons, adolescent animals are often used, and evidence suggests that these young animals, like young humans, have a predominately naïve immune system that will mature toward a memory phenotype as the animal ages (Nan et al. 1998a,b; Rodriquez-Carreno et al. 2002; Saalmüller et al. 2002). With regard to graft survival in animals, convention has established 100 days as indicative of acceptance in mouse models, whereas in larger animals, several years has been the norm. However, although this is practical in terms of the shorter life span and cost of maintaining the animals, no data support that graft survival should be interpreted proportionally to lifespan. Thus, data from these studies should be interpreted for what it is, and not extrapolated to indicate longer survival in longer-lived animals (Kirk 2003).

A full review of the ethical principles surrounding animal experimentation is beyond the scope of this article. However, it is worth noting that no animal experiment can be ethical unless it has a reasonable chance of answering a worthwhile question. Specific guidance for ethical treatment of animal subjects can be found in the *Guide for the Care and Use of Laboratory Animals* (National Research Council 2011) and in the publications of the American Association for Laboratory Animal Care (www.aaalac.org/resources). In general, all animal experiments should embrace the principles of reduction, replacement, and refinement. Animal models should be reserved for when no reasonable alternative exists, and any procedure should be maximally refined for the benefit of the animal's welfare. Any experiment should be designed to minimize the number of animals used without endangering the ability of the experiment to answer the question at hand. An experiment that fails to answer the question posed using fewer animals is less ethical than an experiment that succeeds using more animals, and is virtually useless.

 Cite this article as *Cold Spring Harb Perspect Med* doi: 10.1101/cshperspect.a015503

The choice of large animal model can often be daunting. Dogs, pigs, and nonhuman primates have all been used extensively in transplantation research, and each has sufficient anatomic and physiologic similarity to humans to be useful. In this article we will focus on primates, as that model has become the predominant model used owing to the need for highly conserved targets when testing biologic and antibody-based therapies. However, primates are not a model of convenience, and should be considered only when a therapy is being anticipated for translation to humans, and data are required to facilitate this transition in a safe and ethically viable fashion.

PRIMATE MODELS OF TRANSPLANTATION

Primate studies in transplantation have used a variety of species, including baboons, macaques, and, rarely, chimpanzees. Chimpanzees offer the highest degree of homology with humans, even serving (historically) as graft donors for humans (Reemtsma 1969); however, their endangered status and evolutionary stature have raised questions regarding the ethics of their use and led to increased restrictions, rendering them a historical model only. Indeed, the National Institutes of Health (NIH) has now eliminated the chimpanzee as a funded model. Baboons have commonly been used in xenotransplantation studies, as their relatively large size better accommodates the porcine organs. Macaques (cynomolgus, rhesus, and pigtail) are the most widely used of the nonhuman primates (NHPs) owing to their small size, defined major histocompatibility complex (MHC), and relative ease of care. Macaques are sufficiently homologous with humans such that most molecular targets will cross-react, with the notable exception of CD3. The remainder of this article will focus predominantly on the rhesus macaque. Numerous applications are described to provide the reader with a general knowledge of the procedures performed using NHPs. For specific methodological detail, the reader is referred to the primary literature (Kirk et al. 1997; Pierson et al. 1999; Hausen et al. 2000a; Elster et al. 2001a; Adams et al. 2002; Cendales et al. 2005).

Genetics in Primate Transplantation

In human studies of transplantation, the MHC matching of donor–recipient pairs is often known. Although the effect of MHC disparity on outcome varies by organ, generally, better-matched pairs have better results. The same can be said of primates; however, until recently, MHC typing of primates was difficult, based on incomplete mapping of the region, and frequently omitted. Although NHPs cannot be readily inbred to the same extent as mice, or even dogs or pigs, they are not truly outbred either. Most monkeys used in research are from dedicated colonies with limited interbreeding. The genetic disparity of potential transplant pairs was established using relatively limited techniques, usually mixed lymphocyte reactions and analysis of the DR region. Failure to adequately control for this experimental variable introduces a significant confounding factor in the interpretation of results.

Recent studies of the rhesus macaque paired with technological advances have begun to allow for easier assessment of MHC matching. The rhesus MHC is located on chromosome 4 and, owing to significant duplication of class I A and B genes (known as mamu-A and mamu-B) and, to a lesser extent, class II duplication, is significantly larger than the human MHC. Owing to the duplication of these genes, individual animals express up to four mamu-A genes and 14 mamu-B genes per chromosome (Kean et al. 2012). Furthermore, these genes are expressed in a hierarchical pattern rather than the codominant expression seen in humans (Otting et al. 2005). The combination of these two facts makes the rhesus MHC extremely complex, particularly for heterozygotes, and renders allele-specific DNA-based typing ineffective at predicting the expression of MHC molecules. Recently, several DNA microsatellite probes were identified that allowed for the determination of autosomal and sex chromosome inheritance between generations of macaques. This allowed for accurate pedigrees to be developed of rhesus breeding groups and determination of inheritance of complete MHC haplotypes (Andrade et al. 2004; Penedo et al. 2005). The development of massively parallel pyrosequencing allowed for comprehensive analysis of MHC expression (Wiseman et al. 2009) and confirmed the extreme complexity of the rhesus MHC. Pyrosequencing can be used to determine the complete MHC type for any individual animal. By combining these two technologies, investigators can design experiments with knowledge of the MHC disparity between potential transplant pairs, greatly increasing the validity of the results. To do so, however, requires substantial coordination with the primate breeding facility, and advanced planning to achieve the desired degree of match or mismatch while still minimizing the number of animals used.

Perioperative Care

The perioperative and follow-up care of NHPs can only approximate what is available for humans. Routine care and treatment of any complications should be performed in consultation with veterinary staff and be anticipated within the animal use protocols approved by the Institutional Animal Care and Use Committee (IACUC). The animal's clinical status should be monitored frequently, as changes in activity or appetite are often early signs of rejection, infection, or technical complications. Animals receiving nephrotoxic immunosuppression or those that receive renal grafts should be monitored for urine output. Laboratory values should be checked routinely postoperatively and with any clinical changes. Medications can be delivered by a number of routes; however, care should be taken when planning experiments to avoid overly frequent sedation for intravenous infusions. Medications requiring daily dosing should be given by mouth or by subcutaneous or intramuscular injection, if possible, unless specialized cages allowing for indwelling catheters in conscious animals are available. Anorexic or malnourished animals can receive supplemental nutrition via orogastric tube while sedated, and should be referred to veterinary staff for dietary modifications. Clinical medications analogous to the standard medications used for human transplantation, such as antibiotics or analgesics, are generally available through veterinary protocols, and although many require dose adjustments, most human agents work well in rhesus macaques. Finally, animals receiving immunosuppression should be monitored for the development of infections, and prophylaxis can be considered. Polymerase chain reaction (PCR)-based assays for rhesus macaque homologs of CMV and EBV, rhCMV (Yue and Barry 2008) and LCV (Rao et al. 2000; Rivailler et al. 2004), respectively, are available for monitoring. Regular monitoring of transplanted animals can provide insight into the extent of immunosuppression, and perhaps, indicate when a regimen may excessively compromise protective immunity (Lo et al. 2013; Lowe et al. 2013).

Renal Transplantation

Renal transplantation in primates mimics the intra-abdominal procedure used in small children. Following systemic heparinization of both the donor and recipient, the donor organ is excised and the vessels are anastamosed to the recipient's infrarenal aorta and vena cava (Fig. 1). Typically this is performed in a left-to-right fashion owing to the longer length of the left renal vessels. The donor ureter is tunneled through the retroperitoneum and a primary ureteroneocystostomy is formed, typically on the posterior wall of the bladder using a modified Leadbetter-Politano approach (Politano and Leadbetter 1958; Kayler et al. 2010). The recipient's native kidney(s) are removed following implantation (Kirk et al. 1997; Knechtle et al. 1997). When possible, the animals may be paired in a domino fashion, allowing each individual to act as a donor and recipient, reducing the number of animals needed and preventing the need for a simultaneous bilateral retroperitoneal dissection and native nephrectomy, which is particularly stressful for the animals. In general, renal transplantation is well tolerated by the animals and complications are limited, with most issues related to the ureter (Song et al. 2010).

Skin Transplantation

Like renal transplantation, skin transplantation in primates is relatively straightforward, and

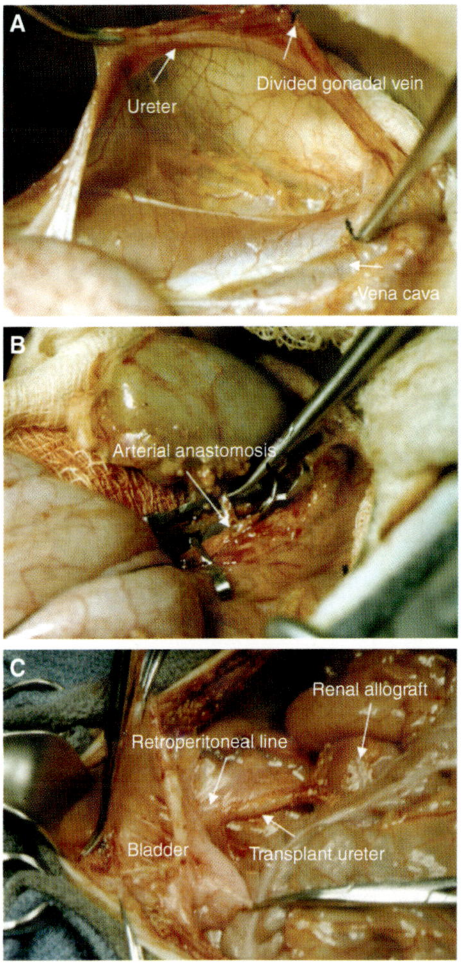

Figure 1. Primate renal transplantation. (*A*) A retroperitoneal pocket is created starting medial to the native ureter and moving laterally. (*B*) The donor vessels are anastomosed to the recipient vena cava and aorta. (*C*) Following reperfusion, the donor ureter is tunneled through the retroperitoneum to a posterior cystotomy and anchored within the bladder.

offers a rigorous model in which to test immunosuppressive or tolerance strategies (Elster et al. 2001b). Donor grafts are typically taken from the abdominal wall and implanted on the back of the recipient to reduce the likelihood of the animal picking at the graft. It is important to remove cutaneous fat from the skin graft before transplantation, and similarly important not to remove subcutaneous tissue and microvasculature from the recipient site. To further protect the grafts during healing, soft bolsters should be used to hold the graft against the underlying tissue (Fig. 2). These can be removed after approximately 1 week. The animals also need to wear protective jackets to protect the grafts and bandages from damage during grooming. The animals generally need to acclimatize to wearing the jacket for a period of time preoperatively, and this may take several weeks before the primate is comfortable with the jacket.

Islet Transplantation

Nonhuman primates have been used extensively in the study of pancreatic islet transplantation. Following a period of training the animals for frequent glucose monitoring, the animals can be rendered diabetic by pancreatectomy or injection of streptozotocin, and diabetes is confirmed by the absence of native c-peptide pro-

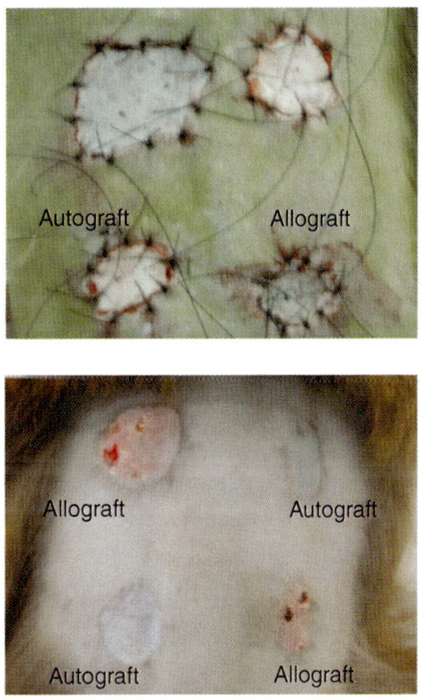

Figure 2. NHP skin grafts. In the *upper* panel, grafts have been sewn in place and anchor sutures placed. The sutures are left long so that they can be used to affix cotton bolsters over the grafts to stabilize the grafts for approximately 1 week. In the *lower* panel, the grafts are shown after having healed to the underlying tissue. Note that the allografts are undergoing early rejection and appear erythematous when compared with the autografts.

duction. Islet cells from the donor pancreas are isolated in the same manner in which human islets are isolated and then infused into the portal circulation by gravity (Adams et al. 2002; Badell et al. 2012a), although other sites for islet implantation are being investigated (Cantarelli and Piemonti 2011). Daily fasting and postprandial blood sugars can be easily monitored postoperatively to assess engraftment and survival of the islet cells. Histological implantation, rejection, and function all closely approximate human islet transplantation.

Heart and Lung Transplantation

The use of NHPs in studies of allogenic heart and lung transplantation is less common; however, examples do exist of their use (Hausen et al. 2000a,b; Azimzadeh et al. 2006). More frequently, primates are used as recipients of heart or lung grafts in studies of xenotransplantation, which will be discussed briefly below and more extensively in other portions of this text. Heart transplantation in primates is typically performed in a heterotopic fashion with the graft placed in the abdominal cavity (Pierson et al. 1999). The graft can then be followed by palpation, ultrasound, or even implanted EKG electrodes (Li et al. 2011). Larger NHPs, such as the baboon, or the cynomolgus macaque are often used in these procedures for technical reasons.

Vascularized Composite Allografts

Advances in immunosuppression and microsurgical technique has allowed for the transplantation of multiple tissues as a functional unit, such as hand and face transplantation. Multiple primate models of these vascularized composite allografts have been developed. Models involving either a portion of or whole hand transplantation in primates (Stark et al. 1987; Hovius et al. 1992) potentially negatively affect the function of the surviving recipient. Cendales et al. (2005) developed a model in which only a small portion of the radial forearm is transplanted, allowing for preserved function in both the donor and recipient (Fig. 3). A heterotopic model of facial transplantation has also been developed by a group at the University of Maryland involving part of the mandible, overlying soft tissue, skin, and vasculature (Silverman et al. 2008; Barth et al. 2009).

Xenotransplantation

Xenotransplantation is covered more extensively in another article in this collection, so we will mention it here only briefly. Studies of xenotransplantation in primates frequently use swine as the organ donor, owing to the ease of breeding and genetic manipulation, such as knockouts of the Gal antigen (Lai et al. 2002; Phelps et al. 2003; Kolber-Simonds et al. 2004; Nottle et al. 2007). Baboons are typically used for solid organ xenotransplant experiments, as their larger size better accommodates the porcine organs. Macaques are often used as recipients of xenoislet transplants. When planning these experiments, it is worth noting that cynomolgus macaques have lower resting blood glucose levels than humans (Casu et al. 2008). Although choosing an ideal donor age has provided some controversy in these experiments, both adult and neonatal islets have been successfully transplanted (Cardona et al. 2006, 2007; Hering and Walawalkar 2009; Thompson et al. 2011).

STANDARD IMMUNOSUPPRESSION

Virtually all clinically used standard immunosuppressants have been found to be successful in prolonging graft survival in primate models, particularly when adjustments for distribution and metabolism have been considered. To focus on the relatively newer immunosuppressants and biologic therapies, we will mention the conventional immunosuppressants, which have been reviewed extensively elsewhere (Kirk 2003), only briefly.

Both azathioprine and mycophenolate have been successfully used in primate models of transplantation, as have calcineurin inhibitors. The development of tacrolimus is worth particular mention, as it was tested in a canine model first with promising results but significant toxicity. Only after follow-up studies in primates that showed its safety was the drug translated

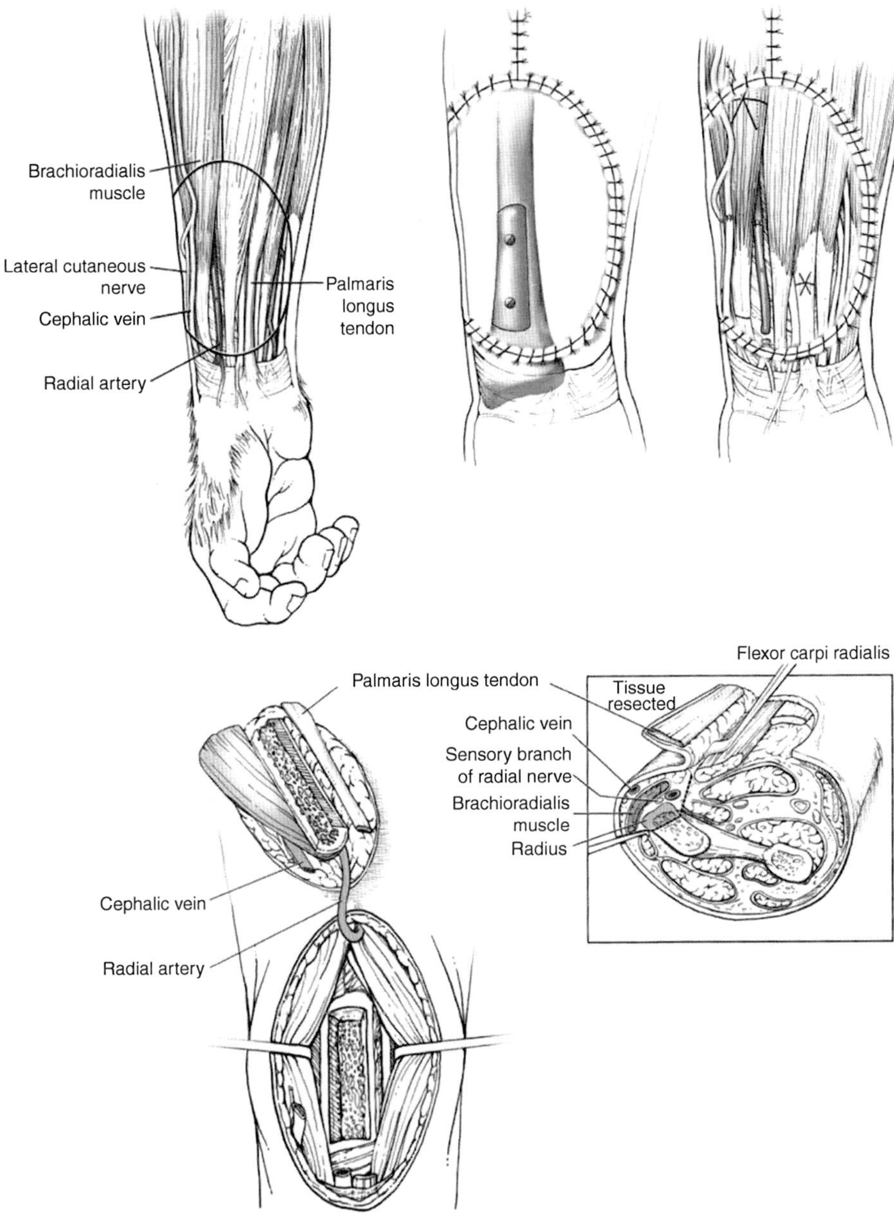

Figure 3. Radial forearm vascularized composite allotransplantation (VCA) containing bone, muscle, tendon, nerve, vasculature, subcutaneous fat, and skin. (From Cendales et al. 2005; reprinted, with express permission, from the author.)

to human use (Todo et al. 1988). A cyclosporine-based regimen has even been shown to induce tolerance in a partially MHC-matched rhesus model (Jonker et al. 1998). Given their success and almost universal use in human transplantation, the calcineurin inhibitors remain a frequently studied class of immunosuppressants, either in combination with new agents or as a control against which the new agent is tested. They are dosed at somewhat higher levels compared with humans dosing to achieve therapeutic targets. The mTOR inhibitor sirolimus has been frequently studied in primate models and been found to prolong graft survival alone (Weaver et al. 2009; Lo et al. 2013) or in combination with tacrolimus (Chen et al. 2000). However, it has been shown to have significant gastrointestinal toxicity in rhesus macaques (Montgomery et al. 2002a). Sirolimus has also been used in combination with T regulatory cell therapy to prolong renal graft survival in the rhesus model (Ma et al. 2011). Everolimus, another mTOR inhibitor, has also been successfully used in primate models of lung (Hausen et al. 2000b) and islet transplantation (Wijkstrom et al. 2004).

A bevy of newer immunosuppressants have also been tested in primate models. FTY720, which prevents the egress of lymphocytes out of lymph nodes, prolongs renal graft survival in macaques, either alone or in combination with other agents (Schuurman et al. 2002). Sotrastaurin, a selective protein kinase C inhibitor, also prolongs primate renal graft survival (Bigaud et al. 2012). It too was found to synergize with other agents, including the aforementioned FTY720 (Bigaud et al. 2006, 2012). The Janus kinase 3 inhibitor, tofacitinib (CP-690,550), was used in a cynomolgus renal transplant model and was shown to prolong graft survival and decrease circulating natural killer (NK) and T-cell populations, a result interpreted to be consistent with multiple cytokine blockade (Borie et al. 2005). Finally, a study of PG490-88, a prodrug form of triptolide, showed prolonged renal allograft survival in cynomolgus monkeys (Chen et al. 2006).

COSTIMULATION BLOCKADE

The blockade of costimulatory signals has repeatedly been used to induce tolerance in rodent models (Turka et al. 1992; Pearson et al. 1994; Larsen et al. 1996). However, the translation of these strategies to large animal models has been less successful. Two phenomena are believed to be primarily responsible for this discrepancy. First, more outbred animals will have a higher precursor frequency, and the alloreactive immune cells are capable of overcoming costimulation blockade by sheer numbers (Ford et al. 2007). Antigen recognition inefficiently generates some response in the absence of costimulation, and the high precursor frequency makes up for this inefficiency, allowing for an immunologically meaningful response (Viola and Lanzavecchia 1996; Germain 2001). Second, there are some cell populations, such as memory T cells, that appear to be less dependent on costimulation (London et al. 2000); thus, more experienced immune systems will be more apt to respond to antigen even during costimulatory blockade. In this review, we will focus on the CD28-B7 and CD40-CD154 pathways, which are the most extensively studied of the costimulatory pathways. However, as the role of costimulation has been increasingly studied, multiple receptor–ligand pairs have been discovered, and it is now generally accepted that multiple ancillary pathways exist to modulate the immune response (Ford and Larsen 2009; Pilat et al. 2011). Owing to the highly specific nature of the biologic agents used to block these signals, primate models have become the large animal model of choice for these studies (Kirk 1999).

Of existing costimulation blockade strategies, blockade of the CD28-B7 pathway has reached the furthest clinical development, with the FDA approval of belatacept for use in human kidney transplantation in July of 2011. Early work to study blockade of the CD28 pathway involved monoclonal antibodies against the B7 ligands CD80 and CD86. Three groups showed prolongation of renal allograft survival using the antibodies 1F1 and 3D1 either alone or in combination with other immunosuppressants (Hausen et al. 2001; Kirk et al. 2001; Montgomery et al. 2002b; Birsan et al. 2003). Another group used different clones, B7-24 and 1G10, in rhesus skin and renal transplant models with modest success (Ossevoort et al. 1998a,b). Despite these successes, further development of these antibodies was abandoned, as it was shown that both B7 molecules needed to be blocked to have efficacy (Kirk et al. 2001), and the B7-specific fusion protein approach accomplished this with a single agent, thus simplifying development plans.

A second strategy to block the CD28 pathway is via the soluble form of the CTLA4 receptor, the fusion protein CTLA4-Ig, which binds both B7 ligands. Following success in several rodent models (Pearson et al. 1994; Akalin et al. 1996; Bolling et al. 1996), early studies in multiple primate models showed only modest prolongation of graft survival (Kirk et al. 1997; Levisetti et al. 1997; Krieger et al. 1998). Persistent interest in this pathway led to the development of belatacept, which differs from the base CTLA4-Ig protein by only two amino acids, but shows significantly higher inhibition of CD28 signaling (Larsen et al. 2005). Belatacept was shown to be effective as maintenance therapy in primate models of islet and renal transplantation (Cardona et al. 2006; Lo et al. 2013; Lowe et al. 2013), and subsequent clinical trials led to its successful translation to human use.

Blockade of the CD28 pathway via targeting of the B7 ligands has the potential drawback of blockade of inhibitory signals through CTLA4, and significant interest exists in developing therapeutics that will simultaneously block CD28 signaling while leaving CTLA4 signaling intact. The use of full monoclonal antibodies directed against CD28 was set back by the disastrous results of the TGN1412 trial, in which an anti-CD28 antibody with superagonistic qualities caused life-threatening inflammatory responses in all six human volunteers who received the drug (Suntharalingam et al. 2006). Although several recommendations have been made for the design of future trials to avoid a repeat of the TGN141 trial (Stebbings et al. 2007; Findlay et al. 2010), there has been, understandably, extreme caution in developing antibodies against CD28. Since TGN1412, two trials using modified CD28 antibodies have been reported: the monovalent fusion antibody sc28AT has been shown to prolong primate graft survival (Poirier et al. 2010), and a pegylated monovalent Fab' antibody, FR104, has been promising in rodent models, but thus far only pharmacokinetic studies have been reported in primates (Poirier et al. 2012).

The CD40-CD154 pathway has also been extensively studied in primate models. The monoclonal antibody hu5C8, directed against CD154, has been shown to be efficacious in multiple primate transplant models involving several organs (Kenyon et al. 1999a,b; Kirk et al. 1999; Pierson et al. 1999; Elster et al. 2001b; Xu et al. 2001). Similarly, two other antibodies against CD154, IDEC-131 and ABI793, have been shown to prolong graft survival in primate models (Pfeiffer et al. 2001; Pierson et al. 2001; Xu et al. 2001, 2003; Kanmaz et al. 2004; Schuler et al. 2004; Preston et al. 2005; Azimzadeh et al. 2006; Pearl et al. 2007). Somewhat fortuitously, hu5C8 was one of the first antibodies to be tested at the relatively high dose of 20 mg/kg. This was important, as it was later discovered that CD154 was present on multiple cell types throughout the body, offering large quantities of the ligand to compete for binding, and, in some cases, initiate rejection (Xu et al. 2006). It is possible that the efficacy of CD154 blockade would not have been appreciated had a lower dose been used. Unfortunately, however, the widespread distribution of CD154, particularly on platelets, has led to concerns regarding the thromboembolic potential of CD154-directed therapies, and this has limited the clinical development of these therapies (Kawai et al. 2000).

The success of the CD154 antibodies prompted continued investigation of the pathway, focusing on CD40. Multiple antibodies against CD40 have been tested, with efficacy on graft survival and effect on B-cell populations depending on the clone used: Chi220 (Pearson et al. 2002; Adams et al. 2005), 3A8 (Badell et al. 2012a,b; Page et al. 2012a), 4D11 (Imai et al. 2007; Aoyagi et al. 2009), ch5D12 (de Vos et al. 2004), or 2C10 (Lowe et al. 2012).

The combination of CD28-B7 and CD40-CD154 pathway blockade has been studied on multiple occasions. Depending on the agents used, the combination has only sometimes provided a graft survival advantage over either agent alone. However, even in cases in which no synergy was seen in terms of graft survival, there does appear to be an advantage of combination therapy in abating the humoral response (Ossevoort et al. 1998b; Montgomery et al. 2002b). Clearly, as our understanding of these and other costimulatory pathways improves, there will be plenty of opportunity to further develop strategies to take advantage of these molecules for immune modulation.

ADHESION BLOCKADE

A number of surface molecules on immune cells mediate the interaction between these cells, their environment, and each other. Many of these adhesion molecules are important for cell trafficking and homing, whereas some have been found to stabilize the immune synapse, and even provide costimulatory signals. This section will focus on two prominent adhesion molecules in transplantation experiments.

The integrin LFA-1 is the most extensively studied of the adhesion molecules. It has been shown to be important in cell trafficking and homing (Hamann et al. 1988; Kavanaugh et al. 1991; Warnock et al. 1998), antigen presentation (Moy and Brian 1992; Pribila et al. 2004), and costimulation (Van Seventer et al. 1990; Nicolls and Gill 2006). The apparent importance of LFA-1 in multiple areas of the immune response prompted significant interest as a strategy for immunosuppression. Anti-LFA-1-based regimens have been successful in prolonging graft survival in primate models of skin (Berlin et al. 1992), cardiac (Poston et al. 2000), and islet transplantation (Badell et al. 2010). Translation of these strategies to human use has been challenging. Studies of the humanized anti-LFA-1 agent, efalizumab, in kidney (Vincenti et al. 2007) and islet (Turgeon et al. 2010) transplantation were promising. However, the drug, which was widely used for psoriasis, was found to be the cause of four cases of progressive multifocal leukoencephalopathy in that population, and was removed from the market. Any future anti-LFA-1-based regimens will likely need to be more selective in their blockade to avoid deleterious effects on protective immunity.

The interaction of CD2 and LFA-3 primarily strengthens the interaction of T cells and antigen-presenting cells (APCs), but also, like LFA-1, is capable of transducing costimulatory signals directly (Bromgerg 1993). Multiple agents to block this interaction have been developed with modest results in primate models (Kaplon et al. 1996; Dehoux et al. 2000). Weaver et al. (2009) used the fusion protein alefacept (LFA3-Ig) in addition to CTLA4-Ig in a rhesus renal transplant model with significant prolongation of graft survival. However, a subsequent study by the same group, which replaced CTLA4-Ig with belatacept, showed no benefit with the addition of alefacept and an increase in viral infections in both renal and islet models (Lo et al. 2013; Lowe et al. 2013). Clearly, further experiments will be needed to clarify the role of CD2 blockade in the future.

Much like the exploding number of costimulatory receptor–ligand pairs that have been discovered over recent years, increasing numbers of adhesion molecules are being discovered and studied. As our understanding of these molecules improves, it is likely we may be able to exploit these pathways to modulate the alloimmune response. One such candidate is the integrin VLA-4, for which an agent, natalizumab, is already approved for use in multiple sclerosis, making it an ideal candidate for translation for use in transplantation.

DEPLETION

One strategy to deal with the aforementioned issue of relatively high precursor frequency and sensitization is to remove those cells from the body before transplantation. Thus, lymphocyte depletion is seen as an adjuvant to other therapies in the development of tolerance regimens. However, at present there are no readily available NHP-specific depletional agents; thus, human specific preparations are used. These have efficacy in NHPs, but they do not mimic the effects seen in humans. Thus, NHP models are not as satisfactory for studying depletion as they are for the more specific pathway blockade described above.

Lymphocyte depletion with polyclonal preparations was one of the first strategies developed. These preparations include both antilymphocytic antibodies as well as antibodies against a variety of other cell-surface molecules, particularly adhesion molecules, which may be of use in cases in which ischemic injury leads to activated endothelium providing an interface for leukocyte adhesion and migration (Hammer and Thein 2002). Rabbit antithymocyte globulin (ATG) must be given in much higher doses in primates than humans, likely owing to decreased specificity for NHP epitopes, notably CD3 (Preville et al. 2001). ATG has been shown to prolong primate graft survival (Thomas et al. 1978); however, its dosing requirements, half-life, and specificity differ so significantly from that seen in humans (the species to which ATG is raised) that it is unlikely that ATG's behavior in NHPs is a satisfactory surrogate for its use in the clinic. ATG fails to induce tolerance in most models, and is thus used primarily as induction therapy followed by some form of maintenance therapy, a variety of which have been successful (Hirshberg et al. 2003; Liu et al. 2007). However, a study by Haanstra et al. (2006) found that ATG combined with costimulation blockade shortened the time to rejection, possibly owing to decreased intragraft inhibitory molecules. The lymphocyte repopulation following depletion with ATG relatives enriches memory populations (Pearl et al. 2007), potentially with lower costimulation requirements.

A variety of monoclonal antibodies have been developed to eliminate or incapacitate reactive effector T cells. Depletion with CD3 and CD4 antibodies conjugated to idarubicin allowed for operational tolerance in half of baboon kidney grafts when paired with TLI, but only when given preoperatively (Myburgh et al. 2001). However, use of a nondepleting antibody, OKT4A, only modestly prolonged graft survival (Cosimi et al. 1990; Wee et al. 1992; Mourad et al. 1998). Depletion based on another potential target, CD45RB, has been shown to prolong graft survival in cynomolgus macaques (Chen et al. 2007). B cells have also been targeted in depletional studies, with the agent rituximab showing prolongation of graft survival in primate islet (Liu et al. 2007) and cardiac models (Kelishadi et al. 2010).

The most promising results in NHP lymphocyte depletion have been achieved using immunotoxin (IT), a fusion of a modified diptheria toxin with a monoclonal antibody against CD3 (Neville et al. 1996; Ma et al. 1997). Knechtle has been able to show pretransplant depletion with IT promotes tolerance in renal transplant models (Knechtle et al. 1997, 1998a). Despite the ability of IT to reverse rejection in a rhesus renal transplant model (Knechtle et al. 1998b), it has been somewhat less effective at prolonging graft survival when given at the time of transplant rather than beforehand (Armstrong et al. 1998; Jonker et al. 2002). The development of a humoral response leading to graft loss in these models using chemically linked IT, or the somewhat less effective recombinant IT (Ma et al. 1997; Kim et al. 2007), has been proposed as a useful model of chronic allograft rejection (Torrealba et al. 2003; Page et al. 2012b).

CHIMERISM

Chimerism has long been studied as a method of inducing allograft tolerance, dating to the studies of Owen (1945) in freemartin cattle, and later Medawar and colleagues (Billingham et al. 1956). Hematopoietic chimerism can be classified based on the relative number of donor cells in the general cell population. Infusion of hematopoietic stem cells into an unconditioned recipient leads to rapid rejection of the cells and failure to engraft. Microchimerism, in which $<1\%$ of circulating cells are of donor origin, and these are only detectable by very sensitive techniques such as PCR, is the most this model can produce. Establishment of mixed chimerism ($>1\%$ but $<100\%$) or especially full chimerism ($\sim 100\%$), requires more intensely ablative conditioning of the recipient. The potential toxicity of these conditioning regimens has been a major barrier to clinical translation of chimerism-based approaches, given the relative safety of modern immunosuppression.

Donor-specific marrow infusions without ablative preconditioning have been studied in attempts to prolong primate allograft survival. However, it is worth noting that all transplant recipients are likely microchimeric to some degree (Starzl et al. 1992), and this does not lead to widespread tolerance. It is possible the establishment of microchimerism is thus simply a marker for sufficient immunosuppression (Wood and Sachs 1996).

Full chimerism, in which virtually all of the hematopoietic cells are of donor origin, is generally only achievable following intense conditioning in HLA identical siblings, if graft-versus-host-disease (GvHD) is to be avoided.

However, there are multiple reports of full chimerism inducing donor-specific tolerance of subsequent renal grafts (Sayegh et al. 1991; Helg et al. 1994; Jacobsen et al. 1994) in humans. Full chimerism does have some drawbacks, however, both in the risk of GvHD and an apparent immune incompetence likely owing to MHC disparity between donor-derived immune cells and recipient-derived APCs (Sykes 2001).

Based on the studies of microchimerism, it is generally accepted that macrochimerism is necessary for durable tolerance to develop, and the drawbacks and impracticalities of full chimerism approaches have made mixed chimerism an area of intense study. A group in Boston has induced mixed chimerism in cynomolgus macaques using irradiation, bone marrow infusion, T-cell depletion, and splenectomy across MHC barriers without GvHD (Kawai et al. 1995; Kimikawa et al. 1997). These animals accepted renal allografts from their marrow donors and, in most cases, also did not produce donor-specific antibodies (Kawai et al. 1999). However, results in cardiac transplantation were somewhat less promising (Kawai et al. 2002), and this regimen also induced increased rates of posttransplant lymphoproliferative disorder (PTLD), likely related to lymphocryptovirus (LCV) in monkeys (Schmidtko et al. 2002). Further refinement of the regimen replaced splenectomy with CD154 blockade (Kawai et al. 2004), and these regimens formed the basis for the first human trial of mixed chimerism (Kawai et al. 2008).

Another group has attempted to induce mixed chimerism in rhesus macaques. In this model, busulfan is the primary conditioning agent, followed by basiliximab, combined costimulation blockade, and sirolimus (Kean et al. 2007). Although the regimen was successful in inducing long-lasting mixed chimerism, the marrow was found to be rapidly rejected once the maintenance immunosuppression was stopped (Larsen et al. 2010), and viral infections continued to be an issue (Kean et al. 2007). Replacing CD154 blockade with CD40 blockade did not prolong marrow graft acceptance beyond the cessation of immunosuppression (Page et al. 2012a). To date, the effect of these mixed chimerism regimens on solid organ transplantation has not been reported. Despite the challenges of chimerism-based approaches to tolerance, they represent one of the few strategies that have been successfully translated to humans, and the promising results of preclinical studies warrant further research into the field.

SUMMARY

In this article, we have reviewed the use of primates in preclinical studies of transplantation, both in the technical aspects of working with primates, and also a number of strategies to prolong graft survival currently being studied. Some of these approaches have already found their way into the clinic, whereas others are still a far way off, in need of significant refinement before they can become a clinical reality. Primate models of transplantation offer an ideal arena in which to study these therapies and hone their application so they may be of the greatest benefit for our patients.

REFERENCES

Adams AB, Shirasugi N, Durham MM, Strobert E, Anderson D, Rees P, Cowan S, Xu H, Blinder Y, Cheung M, et al. 2002. Calcineurin inhibitor-free CD28 blockade-based protocol protects allogeneic islets in nonhuman primates. *Diabetes* **51:** 265–270.

Adams AB, Pearson TC, Larsen CP. 2003. Heterologous immunity: An overlooked barrier to tolerance. *Immunol Rev* **196:** 147–160.

Adams AB, Shirasugi N, Jones TR, Durham MM, Strobert EA, Cowan S, Rees P, Hendrix R, Price K, Kenyon NS, et al. 2005. Development of a chimeric anti-CD40 monoclonal antibody that synergizes with LEA29Y to prolong iset allograft survival. *J Immunol* **174:** 542–550.

Akalin E, Chandraker A, Russell ME, Turka LA, Hancock WW, Sayegh MH. 1996. CD28-B7 T cell costimulatory blockade by CTLA4Ig in the rat renal allograft model: Inhibition of cell-mediated and humoral immune responses in vivo. *Transplantation* **62:** 1942–1945.

Andrade MCR, Penedo MCT, Ward T, Silva VF, Bertolini LR, Roberts JA, Leite JP, Cabello PH. 2004. Determination of genetic status in a closed colony of rhesus monkeys (Macaca mulatta). *Primates* **45:** 183–186.

Aoyagi T, Yamashita K, Suzuki T, Uno M, Goto R, Taniguchi M, Shimamura T, Takahashi N, Miura T, Okimura K, et al. 2009. A human anti-CD40 monoclonal antibody, 4D11, for kidney transplantation in cynomolgus monkeys: Induction and maintenance therapy. *Am J Transplant* **9:** 1732–1741.

Armstrong N, Buckley P, Oberley T, Fechner J, Dong YC, Hong XN, Kirk A, Neville D Jr, Knechtle S. 1998. Analysis of primate renal allografts after T-cell depletion with anti-CD3-CRM9. *Transplantation* **66:** 5–13.

Azimzadeh AM, Pfeiffer S, Wu G, Schroder C, Zorn GL III, Kelishadi SS, Ozkaynak E, Kehry M, Atkinson JB, Miller GG, et al. 2006. Alloimmunity in primate heart recipients with CD154 blockade: Evidence for alternative costimulation mechanisms. *Transplantation* **81:** 255–264.

Badell IR, Russell MC, Thompson PW, Turner AP, Weaver TA, Robertson JM, Avila JG, Cano JA, Johnson BE, Song M, et al. 2010. LFA-1-specific therapy prolongs allograft survival in rhesus macaques. *J Clin Invest* **120:** 4520–4531.

Badell IR, Russell MC, Cardona K, Shaffer VO, Turner AP, Avila JG, Cano JA, Leopardi FV, Song M, Strobert EA, et al. 2012a. CTLA4Ig prevents alloantibody formation following nonhuman primate islet transplantation using the CD40-specific antibody 3A8. *Am J Transplant* **12:** 1918–1923.

Badell IR, Thompson PW, Turner AP, Russell MC, Avila JG, Cano JA, Robertson JM, Leopardi FV, Strobert EA, Iwakoshi NN, et al. 2012b. Nondepleting anti-CD40-based therapy prolongs allograft survival in nonhuman primates. *Am J Transplant* **12:** 126–135.

Barth RN, Bluebond-Langner R, Nam A, Stanwix M, Shipley S, Bartlett ST. 2009. Facial subunit composite tissue allografts in nonhuman primates: I. Technical and immunosuppressive requirements for prolonged graft survival. *Plast Reconstr Surg* **123:** 493–501.

Berlin PJ, Bacher JD, Sharrow SO, Gonzalez C, Gress RE. 1992. Monoclonal antibodies against human T cell adhesion molecules—Modulation of immune function in nonhuman primates. *Transplantation* **53:** 840–849.

Bigaud M, Wieczorek G, Preussing E, Riesen S, Cordoba F, Audet M, Blancher A, Heusser C, Bruns C, Wagner J, et al. 2006. NVP-AEB071 (AEB), the novel oral inhibitor of protein kinase C (PKC) and early T-cell activation, prolongs non-human primates (NHP) kidney allograft survival when combined with everolimus (RAD), ERL080 (ERL) or FTY720 (FTY) without calcineurin inhibitor (CNI). *Am J Transplant* **6:** 251–252.

Bigaud M, Wieczorek G, Beerli C, Audet M, Blancher A, Heusser C, Morris RE, Wagner J. 2012. Sotrastaurin (AEB071) alone and in combination with cyclosporine A prolongs survival times of non-human primate recipients of life-supporting kidney allografts. *Transplantation* **93:** 156–164.

Billingham RE, Brent L, Medawar PB. 1956. Quantitative studies on tissue transplantation immunity. III: Actively acquired tolerance. *Philos Trans R Soc Lond B Biol Sci* **239:** 357–412.

Birsan T, Hausen B, Higgins JP, Hubble RW, Klupp J, Stalder M, Celniker A, Friedrich S, O'Hara RM, Morris RE. 2003. Treatment with humanized monoclonal antibodies against CD80 and CD86 combined with sirolimus prolongs renal allograft survival in cynomolgus monkeys. *Transplantation* **75:** 2106–2113.

Blattman JN, Antia R, Sourdive DJD, Wang X, Kaech SM, Murali-Krishna K, Altman JD, Ahmed R. 2002. Estimating the precursor frequency of naive antigen-specific CD8 T cells. *J Exp Med* **195:** 657–664.

Bolling SF, Lin H, Wei RQ, Turka LA. 1996. Preventing allograft rejection with CTLA4IG: effect of donor-specific transfusion route or timing. *J Heart Lung Transplant* **15:** 928–935.

Borie DC, Changelian PS, Larson MJ, Si MS, Paniagua R, Higgins JP, Holm B, Campbell A, Lau M, Zhang S, et al. 2005. Immunosuppression by the JAK3 inhibitor CP-690,550 delays rejection and significantly prolongs kidney allograft survival in nonhuman primates. *Transplantation* **79:** 791–801.

Bromberg JS. 1993. The biology of CD2: Adhesion, transmembrane signal, and regulatory receptor of immunity. *J Surg Res* **54:** 258–267.

Cantarelli E, Piemonti L. 2011. Alternative transplantation sites for pancreatic islet grafts. *Curr Diab Rep* **11:** 364–374.

Cardona K, Korbutt GS, Milas Z, Lyon J, Cano J, Jiang W, Bello-Laborn H, Hacquoll B, Strobert E, Gangappa S, et al. 2006. Long-term survival of neonatal porcine islets in nonhuman primates by targeting costimulation pathways. *Nat Med* **12:** 304–306.

Cardona K, Milas Z, Strobert E, Cano J, Jiang W, Safley SA, Gangappa S, Hering BJ, Weber CJ, Pearson TC, et al. 2007. Engraftment of adult porcine islet xenografts in diabetic nonhuman primates through targeting of co-stimulation pathways. *Am J Transplant* **7:** 2260–2268.

Casu A, Bottino R, Balamurugan AN, Hara H, van der Windt DJ, Campanile N, Smetanka C, Cooper DK, Trucco M. 2008. Metabolic aspects of pig-to-monkey (*Macaca fascicularis*) islet transplantation: Implications for translation into clinical practice. *Diabetologia* **51:** 120–129.

Cendales LC, Xu H, Bacher J, Eckhaus MA, Kleiner DE, Kirk AD. 2005. Composite tissue allotransplantation: Development of a preclinical model in nonhuman primates. *Transplantation* **80:** 1447–1454.

Chen H, Peng J, Luo H, Loubeau M, Wan X, Xu D, Daloze P, Fitzsimmons WE, Bekersky I, Peets J, et al. 2000. Compromised kidney graft rejection response in Vervet monkeys after withdrawal of immunosuppressants tacrolimus and sirolimus. *Transplantation* **69:** 1555–1561.

Chen G, Sun H, Arp J, Garcia B, Wang X, Wise Y, Liu W, Ramcharran S, Huang X, Xiang Y, et al. 2006. A synergistic effect between PG490-88 and tacrolimus prolongs renal allograft survival in monkeys. *Am J Transplant* **6:** 714–723.

Chen G, Luke PPW, Yang H, Visser L, Sun H, Garcia B, Qian H, Xiang Y, Huang X, Liu W, et al. 2007. Anti-CD45RB monoclonal antibody prolongs renal allograft survival in cynomolgus monkeys. *Am J Transplant* **7:** 27–37.

Choo JK, Seebach JD, Nickeleit V, Shimizu A, Lei H, Sachs DH, Madsen JC. 1997. Species differences in the expression of major histocompatibility complex class II antigens on coronary artery endothelium: Implications for cell-mediated xenoreactivity. *Transplantation* **64:** 1315–1322.

Cosimi AB, Delmonico FL, Wright JK, Wee SL, Preffer FI, Jolliffe LK, Colvin RB. 1990. Prolonged survival of nonhuman primate renal allograft recipients treated only with anti-CD4 monoclonal antibody. *Surgery* **108:** 406–413.

Dehoux JP, Talpe S, Dewolf N, Otsuka M, Oike F, Jamar F, de la Parra B, Latinne D, Bazin H, Gianello P. 2000. Effects on human and nonhuman primate immune response of a new rat anti-CD2 monoclonal antibody. *Transplantation* **69:** 2622–2633.

de Vos AF, Melief MJ, van Riel D, Boon L, van Eijk M, de Boer M, Laman JD. 2004. Antagonist anti-human CD40 antibody inhibits germinal center formation in cynomolgus monkeys. *Eur J Immunol* **34:** 3446–3455.

Elster EA, Xu H, Tadaki DK, Burkly LC, Berning JD, Baumgartner RE, Cruzata F, Patterson NB, Harlan DM, Kirk AD. 2001a. Primate skin allotransplantation with anti-CD154 monotherapy. *Transplant Proc* **33:** 675–676.

Elster EA, Xu H, Tadaki DK, Montgomery S, Burkly LC, Berning JD, Baumgartner RE, Cruzata F, Marx R, Harlan DM, et al. 2001b. Treatment with the humanized CD154-specific monoclonal antibody, hu5C8, prevents acute rejection of primary skin allografts in nonhuman primates. *Transplantation* **72:** 1473–1478.

Findlay L, Eastwood D, Stebbings R, Sharp G, Mistry Y, Ball C, Hood J, Thorpe R, Poole S. 2010. Improved in vitro methods to predict the in vivo toxicity in man of therapeutic monoclonal antibodies including TGN1412. *J Immunol Methods* **352:** 1–12.

Ford ML, Larsen CP. 2009. Translating costimulation blockade to the clinic: Lessons learned from three pathways. *Immunol Rev* **229:** 294–306.

Ford ML, Koehn BH, Wagener ME, Jiang W, Gangappa S, Pearson TC, Larsen CP. 2007. Antigen-specific precursor frequency impacts T cell proliferation, differentiation, and requirement for costimulation. *J Exp Med* **204:** 299–309.

Germain RN. 2001. The art of the probable: System control in the adaptive immune system. *Science* **293:** 240–245.

Haanstra KG, Sick EA, Ringers J, Wubben JAM, Kuhn EM, 't Hart BA, Boon L, Jonker M. 2006. No synergy between ATG induction and costimulation blockade induced kidney allograft survival in rhesus monkeys. *Transplantation* **82:** 1194–1201.

Hamann A, Jablonski-Westrich D, Duijvestijn A, Butcher EC, Baisch H, Harder R, Thiele HG. 1988. Evidence for an accessory role of LFA-1 in lymphocyte-high endothelium interaction during homing. *J Immunol* **140:** 693–699.

Hammer C, Thein E. 2002. Visualization of the effect of polyclonal antithymocyte globulins on adhesion of leukocytes. *Transplant Proc* **34:** 2486–2487.

Hausen B, Gummert J, Berry GJ, Christians U, Serkova N, Ikonen T, Hook L, Legay F, Schuler W, Schreier MH, et al. 2000a. Prevention of acute allograft rejection in nonhuman primate lung transplant recipients: Induction with chimeric anti-interleukin-2 receptor monoclonal antibody improves the tolerability and potentiates the immunosuppressive activity of a regimen using low doses of both microemulsion cyclosporine and 40-*O*-(2-hydroxyethyl)-rapamycin. *Transplantation* **69:** 488–496.

Hausen B, Ikonen T, Briffa N, Berry GJ, Christians U, Robbins RC, Hook L, Serkova N, Benet LZ, Schuler W, et al. 2000b. Combined immunosuppression with cyclosporine (neoral) and SDZ RAD in non-human primate lung transplantation: Systematic pharmacokinetic-based trials to improve efficacy and tolerability. *Transplantation* **69:** 76–86.

Hausen B, Klupp J, Christians U, Higgins JP, Baumgartner RE, Hook LE, Friedrich S, Celnicker A, Morris RE. 2001. Coadministration of either cyclosporine or steroids with humanized monoclonal antibodies against CD80 and CD86 successfully prolong allograft survival after life supporting renal transplantation in cynomolgus monkeys. *Transplantation* **72:** 1128–1137.

Helg C, Chapuis B, Bolle JF, Morel P, Salomon D, Roux E, Antonioli V, Jeannet M, Leski M. 1994. Renal transplantation without immunosuppression in a host with tolerance induced by allogeneic bone marrow transplantation. *Transplantation* **58:** 1420–1422.

Hering BJ, Walawalkar N. 2009. Pig-to-nonhuman primate islet xenotransplantation. *Transpl Immunol* **21:** 81–86.

Hirshberg B, Preston EH, Xu H, Tal MG, Neeman Z, Bunnell D, Soleimanpour S, Hale DA, Kirk AD, Harlan DM. 2003. Rabbit antithymocyte globulin induction and sirolimus monotherapy supports prolonged islet allograft function in a nonhuman primate islet transplantation model. *Transplantation* **76:** 55–60.

Hovius SE, Stevens HP, Van Nierop PW, Godschalk M, Kusuma A, Deelen G, Van de Bergh M, Van der Meulen JC. 1992. Replantation of the radial side of the hand in the rhesus monkey: Anatomical and functional aspects. A preliminary study to composite tissue allografting. *J Hand Surg Br* **17:** 651–656.

Imai A, Suzuki T, Sugitani A, Itoh T, Ueki S, Aoyagi T, Yamashita K, Taniguchi M, Takahashi N, Miura T, et al. 2007. A novel fully human anti-CD40 monoclonal antibody, 4D11, for kidney transplantation in cynomolgus monkeys. *Transplantation* **84:** 1020–1028.

Jacobsen N, Taaning E, Ladefoged J, Kristensen JK, Pedersen FK. 1994. Tolerance to an HLA-B,DR disparate kidney allograft after bone-marrow transplantation from same donor. *Lancet* **343:** 800.

Jonker M, van de Hout Y, Neuhaus P, Ringers J, Kuhn EM, Bruijn JA, Noort R, Doxiadis G, Otting N, Bontrop RE, et al. 1998. Complete withdrawal of immunosuppression in kidney allograft recipients: A prospective study in rhesus monkeys. *Transplantation* **66:** 925–927.

Jonker M, Ringers J, Ossevoort MA, Slingerland W, van den Hout Y, Haanstra K, Wubben J, Kuhn E, Friend P, Calne R. 2002. Long-term kidney graft survival by delayed T cell ablative treatment in rhesus monkeys. *Transplantation* **73:** 874–880.

Kanmaz T, Fechner JJH, Torrealba J, Kim HT, Dong Y, Oberley TD, Schultz JM, Bloom DD, Katayama M, Dar W, et al. 2004. Monotherapy with the novel human anti-CD154 monoclonal antibody ABI793 in rhesus monkey renal transplantation model. *Transplantation* **77:** 914–920.

Kaplon RJ, Hochman PS, Michler RE, Kwiatkowski PA, Edwards NM, Berger CL, Xu H, Meier W, Wallner BP, Chisholm P, et al. 1996. Short course single agent therapy with an LFA-3-IgG1 fusion protein prolongs primate cardiac allograft survival. *Transplantation* **61:** 356–363.

Kavanaugh AF, Lightfoot E, Lipsky PE, Oppenheimer-Marks N. 1991. Role of CD11/CD18 in adhesion and transendothelial migration of T cells. Analysis utilizing CD18-deficient T cell clones. *J Immunol* **146:** 4149–4156.

Kawai T, Cosimi AB, Colvin RB, Powelson J, Eason J, Kozlowski T, Sykes M, Monroy R, Tanaka M, Sachs DH. 1995. Mixed allogeneic chimerism and renal allograft tolerance in cynomolgus monkeys. *Transplantation* **59:** 256–262.

Kawai T, Poncelet A, Sachs DH, Mauiyyedi S, Boskovic S, Wee SL, Ko DS, Bartholomew A, Kimikawa M, Hong HZ, et al. 1999. Long-term outcome and alloantibody production in a non-myeloablative regimen for induction of renal allograft tolerance. *Transplantation* **68:** 1767–1775.

Kawai T, Andrews D, Colvin R, Sachs D, Cosimi A. 2000. Thromboembolic complications after treatment with monoclonal antibody against CD40 ligand. *Nat Med* **6:** 114.

Kawai T, Cosimi AB, Wee SL, Houser S, Andrews D, Sogawa H, Phelan J, Boskovic S, Nadazdin O, Abrahamian G, et al. 2002. Effect of mixed hematopoietic chimerism on cardiac allograft survival in cynomolgus monkeys. *Transplantation* **73:** 1757–1764.

Kawai T, Sogawa H, Boskovic S, Abrahamian G, Smith RN, Wee SL, Andrews D, Nadazdin O, Koyama I, Sykes M, et al. 2004. CD154 blockade for induction of mixed chimerism and prolonged renal allograft survival in nonhuman primates. *Am J Transplant* **4:** 1391–1398.

Kawai T, Cosimi AB, Spitzer TR, Tolkoff-Rubin N, Suthanthiran M, Saidman SL, Shaffer J, Preffer FI, Ding R, Sharma V, et al. 2008. HLA-mismatched renal transplantation without maintenance immunosuppression. *N Engl J Med* **358:** 353–361.

Kayler L, Kang D, Molmenti E, Howard R. 2010. Kidney transplant ureteroneocystostomy techniques and complications: Review of the literature. *Transplant Proc* **42:** 1413–1420.

Kean LS, Adams AB, Strobert E, Hendrix R, Gangappa S, Jones TR, Shirasugi N, Ridby MR, Hamby K, Jian J, et al. 2007. Induction of chimerism in rhesus macaques through stem cell transplant and costimulation blockade-based immunosuppression. *Am J Transplant* **7:** 320–335.

Kean LS, Singh K, Blazar BR, Larsen CP. 2012. Nonhuman primate transplant models finally evolve: Detailed immunogenetic analysis creates new models and strengthens the old. *Am J Transplant* **12:** 812–819.

Kelishadi SS, Azimzadeh AM, Zhang T, Stoddard T, Welty E, Avon C, Higuchi M, Laaris A, Cheng XF, McMahon C, et al. 2010. Preemptive $CD20^+$ B cell depletion attenuates cardiac allograft vasculopathy in cyclosporine-treated monkeys. *J Clin Invest* **120:** 1275.

Kenyon NS, Fernandez LA, Lehmann R, Masetti M, Ranuncoli A, Chatzipetrou M, Iaria G, Han D, Wagner JL, Ruiz P, et al. 1999a. Long-term survival and function of intrahepatic islet allografts in baboons treated with humanized anti-CD154. *Diabetes* **48:** 1473–1481.

Kenyon NS, Chatzipetrou M, Masetti M, Ranuncoli A, Oliveira M, Wagner JL, Kirk AD, Harlan DM, Burkly LC, Ricordi C. 1999b. Long-term survival and function of intrahepatic islet allografts in rhesus monkeys treated with humanized anti-CD154. *Proc Natl Acad Sci* **96:** 8132–8137.

Kim GB, Wang Z, Liu YY, Stavrou S, Mathias A, Goodwin KJ, Thomas JM, Neville DM. 2007. A fold-back single-chain diabody format enhances the bioactivity of an anti-monkey CD3 recombinant diphtheria toxin-based immunotoxin. *Protein Eng Des Sel* **20:** 425–432.

Kimikawa M, Sachs DH, Colvin RB, Bartholomew A, Kawai T, Cosimi AB. 1997. Modifications of the conditioning regimen for achieving mixed chimerism and donor-specific tolerance in cynomolgus monkeys. *Transplantation* **64:** 709–716.

Kirk AD. 1999. Transplantation tolerance: A look at the nonhuman primate literature in the light of modern tolerance theories. *Crit Rev Immunol* **19:** 349–388.

Kirk AD. 2003. Crossing the bridge: Large animal models in translational transplantation research. *Immunol Rev* **196:** 176–196.

Kirk AD, Harlan DM, Armstrong NN, Davis TA, Dong Y, Gray GS, Hong X, Thomas D, Fechner JH Jr, Knechtle SJ. 1997. CTLA4-Ig and anti-CD40 ligand prevent renal allograft rejection in primates. *Proc Natl Acad Sci* **94:** 8789–8794.

Kirk AD, Burkly LC, Batty DS, Baumgartner RE, Berning JD, Buchanan K, Fechner JH Jr, Germond RL, Kampen RL, Patterson NB, et al. 1999. Treatment with humanized monoclonal antibody against CD154 prevents acute renal allograft rejection in nonhuman primates. *Nat Med* **5:** 686–693.

Kirk AD, Tadaki DK, Celniker A, Batty DS, Berning JD, Colonna JO, Cruzata F, Elster EA, Gray GS, Kampen RL, et al. 2001. Induction therapy with monoclonal antibodies specific for CD80 and CD86 delays the onset of acute renal allograft rejection in non-human primates. *Transplantation* **72:** 377–384.

Knechtle SJ, Vargo D, Fechner J, Zhai Y, Wang J, Hanaway MJ, Scharff J, Hu H, Knapp L, Watkins D, et al. 1997. FN18-CRM9 immunotoxin promotes tolerance in primate renal allografts. *Transplantation* **63:** 1–6.

Knechtle SJ, Fechner JH, Dong YC, Stavrou S, Neville DM, Oberley T, Buckley P, Armstrong N, Rusterholz K, Hong X, et al. 1998a. Primate renal transplants using immunotoxin. *Surgery* **124:** 438–446.

Knechtle SJ, Fechner JH, Stavrou S, Neville DM, Dong Y, Hong X, Tsuchida M, Hamawy MM. 1998b. Reversal of acute allograft rejection using immunotoxin. *Transplant Proc* **30:** 2150–2151.

Kolber-Simonds D, Lai L, Watt SR, Denaro M, Arn S, Augenstein ML, Betthauser J, Carter DB, Greenstein JL, Hao Y, et al. 2004. Production of α-1,3-galactosyltransferase null pigs by means of nuclear transfer with fibroblasts bearing loss of heterozygosity mutations. *Proc Natl Acad Sci* **101:** 7335–7340.

Krieger NR, Yuh D, McIntyre WB, Flavin TF, Yin D, Robbins R, Fathman CG. 1998. Prolongation of cardiac graft survival with anti-CD4Ig plus hCTLA4Ig in primates. *J Surg Res* **76:** 174–178.

Lai L, Kolber-Simonds D, Park KW, Cheong HT, Greenstein JL, Im GS, Samuel M, Bonk A, Rieke A, Day BN, et al. 2002. Production of α-1,3-galactosyltransferase knockout pigs by nuclear transfer cloning. *Science* **295:** 1089–1092.

Larsen CP, Elwood ET, Alexander DZ, Ritchie SC, Hendrix R, Tucker-Burden C, Cho HR, Aruffo A, Hollenbaugh D, Linsley PS, et al. 1996. Long-term acceptance of skin and cardiac allografts after blocking CD40 and CD28 pathways. *Nature* **381:** 434–438.

Larsen CP, Pearson TC, Adams AB, Tso P, Shirasugi N, Strobert E, Anderson D, Cowan S, Price K, Naemura J, et al. 2005. Rational development of LEA29Y (belatacept), a high-affinity variant of CTLA4-Ig with potent immunosuppressive properties. *Am J Transplant* **5:** 443–453.

Larsen CP, Page A, Linzie KH, Russell M, Deane T, Stempora L, Strobert E, Penedo MC, Ward T, Wiseman R, et al. 2010. An MHC-defined primate model reveals significant rejection of bone marrow after mixed chimerism induction despite full MHC matching. *Am J Transplant* **10:** 2396–2409.

Levisetti MG, Padrid PA, Szot GL, Mittal N, Meehan SM, Wardrip CL, Gray GS, Bruce DS, Thistlethwaite JR Jr, Bluestone JA. 1997. Immunosuppressive effects of human CTLA4Ig in a non-human primate model of allogeneic pancreatic islet transplantation. *J Immunol* **159:** 5187–5191.

Li J, Chen G, Ye P, Wang S, Zhang K, Chen W, Stepkowski SM, Li J, Zhong S, Xia J. 2011. CCR5 blockade in combination with cyclosporine increased cardiac graft survival and generated alternatively activated macrophages in primates. *J Immunol* **186:** 3753–3761.

Liu C, Noorchashm H, Sutter JA, Naji M, Prak EL, Boyer J, Green T, Rickels MR, Tomaszewski JE, Koeberlein B, et al. 2007. B lymphocyte–directed immunotherapy promotes long-term islet allograft survival in nonhuman primates. *Nat Med* **13:** 1295–1298.

Lo DJ, Anderson DJ, Weaver TA, Leopardi F, Song M, Farris AB, Strobert EA, Jenkins J, Turgeon NA, Mehta AK, et al. 2013. Belatacept and sirolimus prolong nonhuman primate renal allograft survival without a requirement for memory T cell depletion. *Am J Transplant* **13:** 320–328.

London CA, Lodge MP, Abbas AK. 2000. Functional responses and costimulator dependence of memory $CD4^+$ T cells. *J Immunol* **164:** 265–272.

Lowe M, Badell IR, Thompson P, Martin B, Leopardi F, Strobert E, Price AA, Abdulkerim HS, Wang R, Iwakoshi NN, et al. 2012. A novel monoclonal antibody to CD40 prolongs islet allograft survival. *Am J Transplant* **12:** 2079–2087.

Lowe MC, Badell IR, Turner AP, Thompson PW, Leopardi FV, Strobert EA, Larsen CP, Kirk AD. 2013. Belatacept and sirolimus prolong nonhuman primate islet allograft survival: Adverse consequences of concomitant alefacept therapy. *Am J Transplant* **13:** 312–319.

Ma SL, Hu HZ, Thompson J, Stavrou S, Scharff J, Neville DM. 1997. Genetic construction and characterization of an anti-monkey CD3 single-chain immunotoxin with a truncated diphtheria toxin. *Bioconjug Chem* **8:** 695–701.

Ma A, Qi S, Song L, Hu Y, Dun H, Massicotte E, Dupuis M, Daloze P, Chen H. 2011. Adoptive transfer of $CD4^+$ $CD25^+$ regulatory cells combined with low-dose sirolimus and anti-thymocyte globulin delays acute rejection of renal allografts in Cynomolgus monkeys. *Int Immunopharmacol* **11:** 618–629.

Mezrich JD, Haller GW, Arn JS, Houser SL, Madsen JC, Sachs DH. 2003. Histocompatible miniature swine: An inbred large-animal model. *Transplantation* **75:** 904–907.

Montgomery SP, Mog SR, Xu H, Tadaki DK, Hirshberg B, Berning JD, Leconte J, Harlan DM, Hale D, Kirk AD. 2002a. Efficacy and toxicity of a protocol using sirolimus, tacrolimus and daclizumab in a nonhuman primate renal allotransplant model. *Am J Transplant* **2:** 381–385.

Montgomery SP, Xu H, Tadaki DK, Celniker A, Burkly LC, Berning JD, Cruzata F, Elster EA, Gray G, Kampen RL, et al. 2002b. Combination induction therapy with monoclonal antibodies specific for CD80, CD86, and CD154 in nonhuman primate renal transplantation. *Transplantation* **74:** 1365–1369.

Mourad GJ, Preffer FI, Wee SL, Powelson JA, Kawai T, Delmonico FL, Knowles RW, Cosimi AB, Colvin RB. 1998. Humanized IgG1 and IgG4 anti-CD4 monoclonal antibodies: Effects on lymphocytes in the blood, lymph nodes, and renal allografts in cynomolgus monkeys. *Transplantation* **65:** 632–641.

Moy VT, Brian AA. 1992. Signaling by lymphocyte function-associated antigen 1 (LFA-1) in B cells: Enhanced antigen presentation after stimulation through LFA-1. *J Exp Med* **175:** 1–7.

Myburgh JA, Smit JA, Pieterz GA, Ling S, Sandrin MS, McKenzie IFC. 2001. Tolerance in baboon kidney transplantation with total lymphoid irradiation (TLI) and anti-CD3/CD4-Idarubicin conjugates. *Transplantation* **72:** 1150–1152.

Nam KH, Akari H, Terao K, Ohto H, Itagaki S, Yoshikawa Y. 1998a. Age-dependent remodeling of peripheral blood $CD4^+$ $CD8^+$ T lymphocytes in cynomolgus monkeys. *Dev Comp Immunol* **22:** 239–248.

Nam KH, Akari H, Terao K, Itagaki S, Yoshikawa Y. 1998b. Age-related changes in major lymphocyte subsets in cynomolgus monkeys. *Exp Anim* **47:** 159–166.

National Research Council. 2011. *Guide for the care and use of laboratory animals*, 8th ed. The National Academies Press, Washington, DC.

Neville DM, Scharff J, Hu HZ, Rigaut K, Shiloach J, Slingerland W, Jonker M. 1996. A new reagent for the induction of T-cell depletion, anti-CD3-CRM9. *J Immunother* **19:** 85–92.

Nicolls MR, Gill RG. 2006. LFA-1 (CD11a) as a therapeutic target. *Am J Transplant* **6:** 27–36.

Nottle MB, Beebe LFS, Harrison SJ, McIlfatrick SM, Ashman RJ, O'Connell PJ, Salvaris EJ, Fisicaro N, Pommey S, Cowan PJ, et al. 2007. Production of homozygous α-1,3-galactosyltransferase knockout pigs by breeding and somatic cell nuclear transfer. *Xenotransplantation* **14:** 339–344.

Ossevoort MA, de Boer M, Lorre K, Van de Voorde A, Jonker M. 1998a. Blocking of costimulatory pathways using monoclonal antibodies as a new strategy to prevent transplant rejection in a non-human primate model. *Transplant Proc* **30:** 1061–1062.

Ossevoort MA, Ringers J, Boon L, Lorre K, van den Hout Y, Kuhn EM, de Boer M, Jonker M, de Waele P. 1998b. Blocking of costimulation prevents kidney graft rejection in rhesus monkeys. *Transplant Proc* **30:** 2165–2166.

Otting N, Heijmans CMC, Noort RC, de Groot NG, Doxiadis GGM, van Rood JJ, Watkins DI, Bontrop RE. 2005. Unparalleled complexity of the MHC class I region in rhesus macaques. *Proc Natl Acad Sci* **102:** 1626–1631.

Owen RD. 1945. Immunogenetic consequences of vascular anastomoses between bovine twins. *Science* **102:** 400–401.

Page A, Srinivasan S, Singh K, Russell M, Hamby K, Deane T, Sen S, Stemora L, Leopardi F, Price AA, et al. 2012a. CD40 blockade combines with CTLA4Ig and sirolimus to produce mixed chimerism in an MHC-defined rhesus macaque transplant model. *Am J Transplant* **12:** 115–125.

Page EK, Page AJ, Kwun J, Gibby AC, Leopardi F, Jenkins JB, Strobert EA, Song M, Hennigar RA, Iwakoshi N, et al. 2012b. Enhanced de novo alloantibody and antibody-mediated injury in rhesus macaques. *Am J Transplant* **12:** 2395–2405.

Pearl JP, Xu H, Leopardi F, Preston E, Kirk AD. 2007. CD154 blockade, sirolimus, and donor-specific transfusion prevents renal allograft rejection in cynomolgus monkeys despite homeostatic T-cell activation. *Transplantation* **83:** 1219–1225.

Pearson TC, Alexander DZ, Winn KJ, Linsley PS, Lowry RP, Larsen CP. 1994. Transplantation tolerance induced by CTLA4-Ig. *Transplantation* **57:** 1701–1706.

Pearson TC, Trambley J, Odom K, Anderson DC, Cowan SR, Bray R, Lin A, Hollenbaugh D, Aruffo A, Siadak AW, et al. 2002. Anti-CD40 therapy extends renal allograft survival in rhesus macaques. *Transplantation* **74:** 933–940.

Penedo MCT, Bontrop RE, Heijmans CMC, Otting N, Noort R, Rouweler AJ, de Groot N, de Groot NG, Ward T, Doxiadis GG. 2005. Microsatellite typing of the rhesus macaque MHC region. *Immunogenetics* **57:** 198–209.

Pescovitz MD, Thistlethwaite JR, Auchincloss H, Ildstad ST, Sharp TG, Terrill R, Sachs DH. 1984. Effect of class II antigen matching on renal allograft survival in miniature swine. *J Exp Med* **160:** 1495–1508.

Pfeiffer S, Iii GLZ, Azimzadeh AM, Atkinson J, Newman R, Pierson RN. 2001. Monotherapy with anti-CD40 ligand antibody (IDEC 131) for non-human primate allograft heart transplantation. *J Heart Lung Transplant* **20:** 250.

Phelps CJ, Koike C, Vaught TD, Boone J, Wells KD, Chen SH, Ball S, Specht SM, Polejaeva IA, Monahan JA, et al. 2003. Production of α1,3-galactosyltransferase-deficient pigs. *Science* **299:** 411–414.

Pierson RN III, Chang AC, Blum MG, Blair KS, Scott MA, Atkinson JB, Collins BJ, Zhang JP, Thomas DW, Burkly LC, et al. 1999. Prolongation of primate cardiac allograft survival by treatment with ANTI-CD40 ligand (CD154) antibody. *Transplantation* **68:** 1800–1805.

Pierson RN, Crowe JE, Pfeiffer S, Atkinson J, Azimzadeh A, Miller GG. 2001. CD40-ligand in primate cardiac allograft and viral immunity. *Immunol Res* **23:** 253–262.

Pilat N, Sayegh MH, Wekerle T. 2011. Costimulatory pathways in transplantation. *Semin Immunol* **23:** 293–303.

Poirier N, Azimzadeh AM, zhang T, Dilek N, Mary C, Nguyen B, Tillou X, Wu G, Reneaudin K, Hervouet J, et al. 2010. Inducing CTLA-4-dependent immune regulation by selective CD28 blockade promotes regulatory T cells in organ transplantation. *Sci Transl Med* **2:** p17ra10.

Poirier N, Mary C, Dilek N, Hervouet J, Minault D, Blancho G, Vanhove B. 2012. Preclinical efficacy and immunological safety of FR104, an antagonist anti-CD28 monovalent Fab' antibody. *Am J Transplant* **12:** 2630–2640.

Politano VA, Leadbetter WF. 1958. An operative technique for the correction of vesicoureteral reflux. *J Urol* **79:** 932–941.

Poston RS, Robbins RC, Chan B, Simms P, Presta L, Jardieu P, Morris RE. 2000. Effects of humanized monoclonal antibody to rhesus CD11a in rhesus monkey cardiac allograft recipients. *Transplantation* **69:** 2005–2013.

Preston EH, Xu H, Dhanireddy KK, Pearl JP, Leopardi FV, Starost MF, Hale DA, Kirk AD. 2005. IDEC-131 (anti-CD154), sirolimus and donor-specific transfusion facilitate operational tolerance in non-human primates. *Am J Transplant* **5:** 1032–1041.

Préville X, Flacher M, LeMauff B, Beauchard S, Davelu P, Tiollier J, Revillard JP. 2001. Mechanisms involved in antithymocyte globulin immunosuppressive activity in a nonhuman primate model. *Transplantation* **71:** 460–468.

Pribila JT, Quale AC, Mueller KL, Shimizu Y. 2004. Integrins and T cell-mediated immunity. *Annu Rev Immunol* **22:** 157–180.

Qi S, Xu D, Peng J, Vu MD, Wu J, Bekersky I, Fitzsimmons WE, Peets J, Sehgal S, Daloze P, et al. 2000. Effect of tacrolimus (FK506) and sirolimus (rapamycin) mono- and combination therapy in prolongation of renal allograft survival in the monkey. *Transplantation* **69:** 1275–1283.

Rao P, Jiang H, Wang F. 2000. Cloning of the rhesus lymphocryptovirus viral capsid antigen and Epstein–Barr virus-encoded small RNA homologues and use in diagnosis of acute and persistent infections. *J Clin Microbiol* **38:** 3219–3225.

Reemtsma K. 1969. Heterotransplantation. *Transplant Proc* **1:** 251–255.

Rivailler P, Carville A, Kaur A, Rao P, Quink C, Kutok JL, Westmoreland S, Klumpp S, Simon A, Aster JC, et al. 2004. Experimental rhesus lymphocryptovirus infection in immunosuppressed macaques: An animal model for Epstein–Barr virus pathogenesis in the immunosuppressed host. *Blood* **104:** 1482–1489.

Rodriguez-Carreno MP, Lopez-Fuertes L, Revilla C, Ezquerra A, Alonso F, Dominguez J. 2002. Phenotypic characterization of porcine IFN-γ-producing lymphocytes by flow cytometry. *J Immunol Methods* **259:** 171–179.

Saalmüller A, Werner T, Fachinger V. 2002. T-helper cells from naive to committed. *Vet Immunol Immunopathol* **87:** 137–145.

Sachs DH. 1992. MHC homozygous miniature swine. In *Swine as models in biomedical research* (ed. Swindle MM, Moody DC, Phillips LD), pp. 3–15. Iowa State University Press, Ames, Iowa.

Sachs DH. 2003. Tolerance: Of mice and men. *J Clin Invest* **111:** 1819–1821.

Sayegh MH, Fine NA, Smith JL, Rennke HG, Milford EL, Tilney NL. 1991. Immunologic tolerance to renal allografts after bone marrow transplants from the same donors. *Ann Intern Med* **114:** 954–955.

Schmidtko J, Wang R, Wu CL, Mauiyyedi S, Harris NL, Pelle Della P, Brousaides N, Zagachin L, Ferry JA, Wang F, et al. 2002. Posttransplant lymphoproliferative disorder associated with an Epstein–Barr-related virus in cynomolgus monkeys. *Transplantation* **73:** 1431–1439.

Schuler W, Bigaud M, Brinkmann V, Di Padova F, Geisse S, Gram H, Hungerford V, Kleuser B, Kristofic C, Menninger K, et al. 2004. Efficacy and safety of ABI793, a novel human anti-human CD154 monoclonal antibody, in cynomolgus monkey renal allotransplantation. *Transplantation* **77:** 717–726.

Schuurman HJ, Menninger K, Audet M, Kunkler A, Maurer C, Vedrine C, Bernhard M, Gaschen L, Brinkmann V, Quesniaux V. 2002. Oral efficacy of the new immunomodulator FTY720 in cynomolgus monkey kidney allotransplantation, given alone or in combination with cyclosporine or RAD. *Transplantation* **74:** 951–960.

Silverman RP, Banks ND, Detolla LJ, Shipley ST, Panda A, Sanchez RA, Azimzadeh AM, Pierson RN III, Wang D, Rodriguez ED, et al. 2008. A heterotopic primate model for facial composite tissue transplantation. *Ann Plast Surg* **60:** 209–216.

Song L, Qi S, Dun H, Hu Y, Ma A, Yu G, Xiong Z, Zhu S, Wang X, Xu D, et al. 2010. Surgical complications in kidney transplantation in nonhuman primates. *Microsurgery* **30:** 327–331.

Stark GB, Swartz WM, Narayanan K, Møller AR. 1987. Hand transplantation in baboons. *Transplant Proc* **19:** 3968–3971.

Starzl TE, Demetris AJ, Trucco M, Ramos H, Zeevi A, Rudert WA, Kocova M, Ricordi C, Iistad S, Murase N. 1992. Systemic chimerism in human female recipients of male livers. *Lancet* **340:** 876–877.

Stebbings R, Findlay L, Edwards C, Eastwood D, Bird C, North D, Mistry Y, Dilger P, Liefooghe E, Cludts I, et al. 2007. "Cytokine storm" in the phase I trial of monoclonal antibody TGN1412: Better understanding the causes to improve preclinical testing of immunotherapeutics. *J Immunol* **179:** 3325–3331.

Suntharalingam G, Perry MR, Ward S, Brett SJ, Castello-Cortes A, Brunner MD, Panoskaltsis N. 2006. Cytokine storm in a phase 1 trial of the anti-CD28 monoclonal antibody TGN1412. *N Engl J Med* **7:** 1018–1028.

Sykes M. 2001. Mixed chimerism and transplant tolerance. *Immunity* **14:** 417–424.

't Hart BA, Amor S, Jonker M. 2004. Evaluating the validity of animal models for research into therapies for immune-based disorders. *Drug Discov Today* **9:** 517–524.

Thomas JM, Williams EA, Thomas FT, Millington GM. 1978. T cell-specific activity in rabbit anti-human thymocyte globulin. Correlation with immunosuppresive activity in nonhuman primates. *Transplantation* **25:** 97–102.

Thompson P, Badell IR, Lowe M, Cano J, Song M, Leopardi F, Avila J, Ruhil R, Strobert E, Korbutt G, et al. 2011. Islet xenotransplantation using gal-deficient neonatal donors improves engraftment and function. *Am J Transplant* **11:** 2593–2602.

Todo S, Ueda Y, Demetris JA, Imventarza O, Nalesnik M, Venkataramanan R, Makowka L, Starzl TE. 1988. Immunosuppression of canine, monkey, and baboon allografts by FK 506: With special reference to synergism with other drugs and to tolerance induction. *Surgery* **104:** 239–249.

Torrealba JR, Fernandez LA, Kanmaz T, Oberley TD, Schultz JM, Brunner KG, Peters D, Fechner JH Jr, Dong Y, Hu H, et al. 2003. Immunotoxin-treated rhesus monkeys: A model for renal allograft chronic rejection. *Transplantation* **76:** 524–530.

Turgeon NA, Avila JG, Cano JA, Hutchinson JJ, Badell IR, Page AJ, Adams AB, Sears MH, Bowen PH, Kirk AD, et al. 2010. Experience with a novel efalizumab-based immunosuppressive regimen to facilitate single donor islet cell transplantation. *Am J Transplant* **10:** 2082–2091.

Turka LA, Linsley PS, Lin H, Brady W, Leiden JM, Wei RQ, Gibson ML, Zheng XG, Myrdal S, Gordon D, et al. 1992. T-cell activation by the CD28 ligand B7 is required for cardiac allograft rejection in vivo. *Proc Natl Acad Sci* **89:** 11102–11105.

Van Seventer GA, Shimizu Y, Horgan KJ, Shaw S. 1990. The LFA-1 ligand ICAM-1 provides an important costimulatory signal for T cell receptor-mediated activation of resting T cells. *J Immunol* **144:** 4579–4586.

Vincenti F, Mendez R, Pescovitz M, Rajagopalan PR, Wilkinson AH, Butt K, Laskow D, Slakey DP, Lorder MI, Garg JP, et al. 2007. A phase I/II randomized open-label multicenter trial of efalizumab, a humanized anti-CD11a, anti-LFA-1 in renal transplantation. *Am J Transplant* **7:** 1770–1777.

Viola A, Lanzavecchia A. 1996. T cell activation determined by T cell receptor number and tunable thresholds. *Science* **273:** 104–106.

Warnock RA, Askari S, Butcher EC, Andrian von UH. 1998. Molecular mechanisms of lymphocyte homing to peripheral lymph nodes. *J Exp Med* **187:** 205–216.

Weaver TA, Charafeddine AH, Agarwal A, Turner AP, Russell M, Leopardi FV, Kampen RL, Stempora L, Song M, Larsen CP, et al. 2009. Alefacept promotes co-stimulation blockade based allograft survival in nonhuman primates. *Nat Med* **15:** 746–749.

Wee SL, Stroka DM, Preffer FI, Jolliffe LK, Colvin RB, Cosimi AB. 1992. The effects of OKT4A monoclonal antibody on cellular immunity of nonhuman primate renal allograft recipients. *Transplantation* **53:** 501–507.

Wijkstrom M, Kenyon NS, Kirchhof N, Kenyon NM, Mullon C, Lake P, Cottens S, Ricordi C, Hering BJ. 2004. Islet allograft survival in nonhuman primates immunosuppressed with basiliximab, RAD, and FTY720. *Transplantation* **77:** 827–835.

Williams MA, Trambley J, Ha J, Adams AB, Durham MM, Rees P, Cowan SR, Pearson TC, Larsen CP. 2000. Genetic characterization of strain differences in the ability to mediate CD40/CD28-independent rejection of skin allografts. *J Immunol* **165:** 6849–6857.

Wiseman RW, Karl JA, Bimber BN, O'Leary CE, Lank SM, Tuscher JJ, Detmer AM, Bouffard P, Levenkova N, Turcotte CL, et al. 2009. Major histocompatibility complex genotyping with massively parallel pyrosequencing. *Nat Med* **15:** 1322–1326.

Wood K, Sachs DH. 1996. Chimerism and transplantation tolerance: Cause and effect. *Immunol Today* **17:** 584–587.

Xu H, Elster E, Batty D, Berning J, Burkly L, Kampen R, Swanson S, Tadaki D, Harlan D, Kirk A. 2001. Effects of dose and duration of anti-CD154 antibody therapy in preventing renal allograft rejection in a nonhuman primate model. *Transplant Proc* **33:** 223–224.

Xu H, Montgomery SP, Preston EH, Tadaki DK, Hale DA, Harlan DM, Kirk AD. 2003. Studies investigating pretransplant donor-specific blood transfusion, rapamycin, and the CD154-specific antibody IDEC-131 in a nonhuman primate model of skin allotransplantation. *J Immunol* **170:** 2776–2782.

Xu H, Zhang X, Mannon RB, Kirk AD. 2006. Platelet-derived or soluble CD154 induces vascularized allograft rejection independent of cell-bound CD154. *J Clin Invest* **116:** 769–774.

Yue Y, Barry PA. 2008. Rhesus cytomegalovirus a nonhuman primate model for the study of human cytomegalovirus. *Adv Virus Res* **72:** 207–226.

Lymphodepletional Strategies in Transplantation

Eugenia Page, Jean Kwun, Byoungchol Oh, and Stuart Knechtle

Department of Surgery, Division of Transplantation, Emory University School of Medicine, Atlanta, Georgia 30322

Correspondence: stuart.knechtle@emoryhealthcare.org

Because lymphocytes were shown to mediate transplant rejection, their depletion has been studied as a mechanism of preventing rejection and perhaps inducing immunologic tolerance. Agents that profoundly deplete lymphocytes have included monoclonal antibodies, cytotoxic drugs, and radiation. We have studied several such agents but focused on antibodies that deplete not only peripheral blood lymphocytes, but also lymph node lymphocytes. Depletion of lymph node T lymphocytes appears to permit peripheral tolerance at least for T cells in animal models. Nevertheless, B-cell responses may be resistant to such approaches, and T memory cells are likewise relatively resistant to depleting antibodies. We review the experimental and clinical approaches to depletion strategies and outline some of the pitfalls of depletion, such as limitations of currently available agents, duration of tolerance, infection, and malignancy. It is notable that most tolerogenic strategies that have been attempted experimentally and clinically include depleting agents even when they are not named as the underlying strategy. Thus, there is an implicitly acknowledged role for reducing the precursor frequency of donor antigen-specific lymphocytes when approaching the daunting goal of transplant tolerance.

Intense immunosuppressive treatment at the time of transplantation is categorized as an induction regimen in organ transplantation. The goals of induction therapy in transplantation have evolved from preventing acute rejection to allowing lower doses of conventional immunosuppression and eventually inducing T-cell nonresponsiveness, also known as operational tolerance (Orlando et al. 2010). Induction therapy has been widely used in organ transplantation, involving 83% of renal transplant and 45% of heart transplant recipients in the United States in 2011 (Annual Report of the U.S. Organ Procurement and Transplantation Network and the Scientific Registry of Transplant Recipients 2011). The use of depletional agents as an induction therapy also has been growing; 59% of adult kidney transplant recipients and 18% of adult heart transplant recipients now receive lymphodepletion. Generally, depleting antibodies activate the classical complement cascade upon binding to the target antigen and induce complement-mediated cell lysis of cells expressing target antigen. Furthermore, phagocytic cells with Fc receptor (FcR) preferentially engulf antibody-coated cells through ADCC (antibody-dependent cell cytotoxicity). However, other modes of action could also induce

Cite this article as *Cold Spring Harb Perspect Med* doi: 10.1101/cshperspect.a015511

lymphocytic depletion, such as limiting survival factors of target cells.

Lymphocyte depletion prior to or beginning at the time of transplantation is beneficial in reducing maintenance immunosuppression (Calne et al. 1998, 1999; Swanson et al. 2002; Kirk et al. 2003; Starzl et al. 2003; Torrealba et al. 2003). Many depleting agents have been studied in animal models and clinical trials, and have been proven efficacious in reducing the rate of acute rejection when combined with maintenance regimens. Indeed, near-tolerance states were induced in many animal models. We showed, for example, that lymphodepletion by anti-CD3 immunotoxin (FN18-CRM9) prolongs renal allograft survival in the nonhuman primate renal transplantation model, but when combined with other immunosuppressive regimens it can induce long-term metastable tolerance (Torrealba et al. 2003, 2004). In human patients, toxicities of chronic calcineurin inhibitor use support the clinical need for depletional strategies (Torrealba et al. 2006). Depleting agents applied in conjunction with CNIs have shown fewer incidents of CNI-related side effects with similar outcomes in preventing acute rejection (Alexander et al. 2006).

This article provides an overview of lymphodepletion in organ transplantation (Fig. 1). We will discuss small and large animal transplantation models using depletional approaches, agents used in the clinic, and challenges to lymphodepletion, including protective immunity, homeostatic proliferation of recalcitrant memory populations, and humoral responses.

LYMPHODEPLETION IN MOUSE MODELS

The large precursor frequency of allospecific T cells among host T cells poses a significant challenge for transplantation. For that reason, lymphodepletion has become a common immunosuppressive strategy at the time of solid organ transplantation (Kwun et al. 2012a). Although this strategy has helped improve early graft survival, long-term outcomes after depletion are still afflicted with challenges. Animal models have been of great importance in exploring transplant immunology. The rodent model in particular provides essential and critical insight on the basic mechanisms of lymphocyte depletion and homeostasis for field of transplantation, although there are gaps between clinical conditions and animal models (Kwun et al.

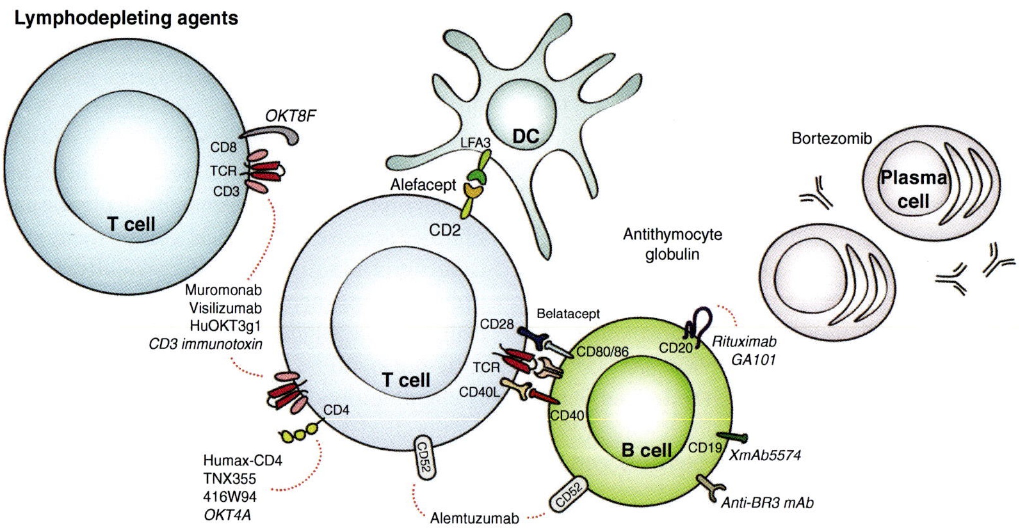

Figure 1. Lymphodepleting agents. Portrayed in this figure are preclinical and clinical agents found to have depleting properties on T cells, B cells, and plasma cells. The dotted lines indicate target specificities for the agents.

2012a). Here we will discuss depletion of T and B cells, plasma cells, and natural killer (NK) T cells in rodent models.

T-Cell Depletion

The first antibodies used since the 1960s, antithymocyte globulin (ATG) induces a rapid and profound lymphodepletion. ATG-induced lymphocyte depletion not only resulted in transplant tolerance in several rodent transplant models but also enhanced regulatory T-cell (Treg) number and function (Feng et al. 2008; Joseph et al. 2012; Shi et al. 2012). Hongo et al. further described that in a combined bone marrow and heart transplant murine model, Treg production of IL-10 and up-regulation of PD-1 was enhanced by IL-4 produced by host natural killer cells (Hongo et al. 2012). Despite the development of newer antibodies, ATG continues to be investigated in murine models today.

The first monoclonal antibody, OKT3, was produced in the late 1970s (Kung et al. 1979), and many researchers have used anti-CD3 mAb for inducing transplant tolerance. The administration of anti-CD3 mAb led to the rapid depletion of T cells from peripheral blood and suppression of heart and islet graft rejection in the rodent model (Nicolls et al. 1993; Tang et al. 2003). However, depletion of T cells from the spleen and lymph nodes was both delayed and incomplete.

As helper CD4 T lymphocytes control humoral and cellular responses, targeting subpopulations using anti-CD4 antibodies appeared to be the best strategy to achieve tolerance to heart allografts in the rodent model (Pearson et al. 1993). Treatment with this monoclonal antibody rapidly depletes the recipient of circulating $CD4^+$ T cells, and repopulation takes about 40–50 days. Their median survival time was over 60 days after transplant (Orosz et al. 1996). Anti-CD4 antibodies depleted Tregs but not as efficiently as $CD4^+CD25^-$ cells, resulting in an enhanced peripheral $CD4^+CD25^+/CD4^+CD25^-$ ratio and thus promoting tolerance (Yi et al. 2008). Cytotoxic CD8 T cells are highly aggressive toward allografts, but unlike CD4 T-cell depletion, the depletion of these cells by the anti-CD8 antibody was not efficient in preventing acute rejection in rodent models.

Mouse T-cell depletion with anti-CD52 antibody alemtuzumab requires the use of human CD52 transgenic mice, in which hCD52 is expressed under the direction of the mouse CD2 promotor and thus only on T cells. Our murine experience with alemtuzumab showed profound peripheral T-cell depletion with the repopulation of $CD44^{hi}CD62L^{lo}$ memory T cells. Despite complete T-cell repopulation by 10 weeks, treated animals showed no evidence of acute rejection, and showed excellent graft survival and function for more than 200 days. Half of the alemtuzumab recipients, however, showed alloantibody production and antibody-mediated chronic rejection (Kwun et al. 2012b). Furthermore, alloantibody-producing animals had increased allospecific B cells found in their spleens, as well as ongoing germinal center reactions represented by CD38-alloreactive B cells.

Other T-cell depleting agents targeting CD2 (Sido et al. 1998) and TCRαβ (Heidecke et al. 1995) have resulted in long-term allograft survival. Sido et al. evaluated three anti-CD2 monoclonal and found that graft survival correlated with dose-dependent down-modulation of CD2, and that humoral responses were preserved. Heidecke et al. observed that, despite incomplete depletion of $CD5^+$ cells, the α/β T-cell receptor blockade was able to induce hyporesponsiveness in a murine heart transplant model after skin graft sensitization.

B-Cell Depletion

Agents that deplete B cells in human recipients, such as alemtuzumab and ATG, have limited efficacy at B-cell depletion in rodents. CD20 is expressed early in B-cell development, downregulates through maturation, and is then absent on mature plasma cells. Anti-CD20 mAB depletes mature, germinal center, and memory B cells but does not affect plasma cell or serum IgG levels (DiLillo et al. 2008). In rodent models, mature B-cell depletion using anti-CD20 mAb had no effect on acute skin allograft rejection and chronic renal rejection but inhibited the development of allospecific IgG (DiLillo et al. 2011). CD19 is expressed on memory B cells and short-lived plasma cells. Anti-CD19 antibody treatment has the effect of depleting B-cell subsets and some plasma cells resulted in enhanced allograft survival and a reduction in circulating allo-specific IgG in human CD19 transgenic mouse models (DiLillo et al. 2011).

Plasma Cell Depletion

Bortezomib is a proteasome inhibitor that selectively targets the 26s proteasome complex. Plasma cells produce large quantities of protein, a proportion of which will be misfolded and will require degradation and disposal by the proteasome. Inhibition of proteasome activity results in the accumulation of misfolded proteins and consequent apotosis of the plasma cell. In rodent transplant models, proteasome inhibition suppresses ongoing rejection and increases mean survival time after heterotropic heart transplantation, which suggests possible efficacy in transplantation (Luo et al. 2001).

Recent clinical data have shown acute cellular rejection after B-cell depletion with rituximab, suggesting positive or negative effects on graft rejection, depending on the nature of the allograft and response (Clatworthy et al. 2009). Because mature B-cell depletion with CD20 mAb will not be sufficient to prevent acute and chronic allograft rejection, new therapies such as CD19 mAb and the next generation of proteasome inhibitors may offer a new approach for depleting both B cells and plasma cells. For testing these agents to investigate their mechanisms, proper animal models will be needed due to constraints in strain and species specifics (Kwun et al. 2012b).

NK Depletion

NK cells are a subset of lymphocytes with cytokine-secreting and cytotoxic effector functions that respond to viral infection, cell stress, and tumors (Vivier et al. 2008). Defining the role of NK cells in transplantation is less clear because of evidence suggesting their involvement both in rejection and in tolerance. During acute rejection, graft infiltrating NK cells secrete chemokines that amplify local inflammation and IFN-γ, which exacerbates allograft rejection and enhances T-cell-mediated immunity. Also, in murine heart transplant models, NK cell depletion has been shown to reduce alloantibody-mediated chronic allograft vasculopathy (Hirohashi et al. 2012). Others have observed that NK cell-depleted recipients showed monocyte and macrophages infiltration in grafts (van der Touw et al. 2012).

LYMPHODEPLETION IN NONHUMAN PRIMATES

En route to clinical translation, lymphodepleting agents have been widely explored in nonhuman primate models of transplantation since the 1980s. Some depleting agents were used as primary induction agents whereas others were incorporated into a combination of therapeutic modalities for transplant tolerance induction. This section will focus on lymphodepletion by pharmacologic and biologic agents and not by virus- or retrovirus-mediated methods.

Polyclonal Antibody Preparations

Polyclonal antilymphocyte globulins, initially prepared from horses immunized with human spleens, thymus, and lymph nodes, have been used for immunosuppression in renal transplantation since the 1960s (Starzl et al. 1963, 1967; Najarian and Simmons 1971). A technology discovered in the 1890s by Mechnikov (Gaber et al. 2010), the mechanisms of action of the modern preparations—rabbit and equine ATG—are vastly diverse due to target antigens that range from surface immune response antigens, costimulation markers, adhesion and trafficking molecules, and many more (Bonnefoy-Berard et al. 1991; Rebellato et al. 1994; Bourdage and Hamlin 1995; Mohty 2007). Preville et al. found that rabbit ATG-administered cynomolgus macaques showed a dose-dependent lymphodepletion in peripheral blood, axillary lymph nodes, and spleens but not in thymuses (Preville et al. 2001). At very high doses (20 mg/kg), peripheral blood T counts dropped to zero within 2 h of the first ATG infusion.

Early nonhuman primate studies showed that combining ATG with other modalities could significantly increase allograft survival and immunologic unresponsiveness. Thomas et al. observed that kidney transplanted rhesus macaques receiving 50 mg/kg of rATG for 5 days followed by a single infusion of unfractionated donor bone marrow on day 12 experienced long-term survival past 248 days without evidence of graft versus host disease, infection, or rejection (compared to 35.8 days for rATG alone, $p < 0.001$); a second donor bone marrow infusion on day 24 reversed the benefit, however, as the animals rejected with a mean graft survival time of 76.6 days (Thomas et al. 1983). In revisiting the ATG + dBM model, the group observed a reduced median survival of 70 days, still a considerable feat in the absence of chronic immunosuppression (Thomas et al. 1989).

Several groups have observed long-term graft acceptance when combining ATG with other modalities. Pearl et al. studied thymectomized cynomologus macaques depleted with rATG (20 mg/kg × 7 days), and found that anti-CD154, sirolimus, and donor specific whole blood transfusion successfully inhibit the homeostatically proliferating memory T-cell response in the depleted animals to yield rejection-free survivals of >300 days (Pearl et al. 2007). Kawai et al. observed renal allograft tolerance in cynomolgus macaques achieving mixed chimerism after treatment with equine ATG (50 mg/kg × 3 days), nonmyeloablative total-body irradiation, thymic irradiation, and donor bone marrow infusion (Kawai et al. 1995). When anti-CD154 mAb was added to this regimen (+CsA × 28 days), donor bone marrow engraftment was enhanced, leading to long-term renal allograft survival >688 days (Kawai et al. 2004). The further addition of humanized anti-CD8 mAb (cM-T807) depleted remaining CD8 memory T cells and allowed for animals to achieve renal allograft tolerance (Koyama et al. 2007). Other groups have combined rATG with anti-CD20 mAb rituximab (Liu et al. 2007) or with sirolimus (Hirshberg et al. 2003) to prolong islet allograft survival in nonhuman primates.

CD3 Immunotoxin

In contrast to the wide heterogeneity of ATG's target specificities, the single antigen-specific antirhesus CD3 immunotoxin developed by Neville et al. (Neville et al. 1992) was pursued with much interest by our group and others (Knechtle 2001). The antirhesus monoclonal antibody FN18 conjugated with mutated diphtheria toxin CRM9 was able to deplete CD3 T cells by 2–3 logs in peripheral blood and by 2 logs in lymph nodes, when we administered a three-day course to rhesus macaques (Knechtle et al. 1998). We observed long-term renal allograft acceptance (mean survival time, 614 days) with donor-specific tolerance confirmed by skin grafting in our long-term survivors; however, a high incidence of donor-specific alloantibodies (73%) and chronic rejection (100%) was also observed (Knechtle et al. 1997, 1998; Torrealba et al. 2003).

We ventured to investigate the activation of the humoral response and eventual chronic rejection in the FN19-CRM9 immunotoxin-treated animals, but the drug was replaced with a new recombinant immunotoxin (A-dmDT390-scfbDb[C207]) because of production yield difficulties and potential immunogenicity of the former chemically conjugated construct (Ma et al. 1997; Kim et al. 2007; Wang et al. 2011). Although the use of the new immunotoxin provided profound peripheral T-cell depletion, the rapid proliferation of memory T cells prevented sustained graft survival. Tacrolimus and alefacept were added to target the memory T cells, which resulted in early and consistent alloantibody production and antibody-mediated graft injury in all treated macaques (Page et al. 2012). Nishimura et al. doubled the immunotoxin dose to 50 µg/kg × 8 doses and combined it with an anti-CD2 antibody in a GalT-KO pig-to-baboon thymokidney model, which led to the preservation of graft function but death due to pneumonia or heart failure at 15 and 23 days (Nishimura et al. 2011).

Alemtuzumab

Alemtuzumab (Campath-1H), a humanized IgG1 monoclonal antibody, targets the membrane glycoprotein CD52, resulting in profound and sustained T-cell depletion as well as transient depletion of B cells, macrophages, and natural killer cells (Hale 2001; Magliocca and Knechtle 2006). Its use in nonhuman primates has been limited as most Old World monkeys express CD52 also on red blood cells, potentially leading to lethal hemolytic anemia if administered the drug. Recently, van der Windt et al. evaluated an agglutination assay and identified cynomolgus macaques of Indonesian but not Chinese origin to be RBC-CD52 negative (van der Windt et al. 2010). As they achieved greater than 99.5% depletion of T and B cells with 20 mg/kg of alemtuzumab, this model provides opportunities to further investigate the drug in preclinical studies.

Other T-Cell Depleting Agents

The advent of monoclonal antibodies provoked much interest in the transplant community, with several studies applying depleting antibodies to T-cell targets (OKT3, OKT4, OKT8, and OKT11 to CD3, CD4, CD8, and E rosettes, respectively) in nonhuman primate models (Jonker et al. 1983; Letvin et al. 1985; Cosimi et al. 1990; Rao et al. 1991). These early monoclonal antibodies, namely OKT3, although effective at treating acute rejection, were associated with cytokine release syndrome, increased risk of posttransplant lymphoproliferative disease, and xenosensitization to its murine component (Chatenoud et al. 1990; Schroeder et al. 1990; Cherikh et al. 2003). To diminish the antimurine response, humanized chimeric preparations of several of the early monoclonal antibodies were developed and are still used in nonhuman primate studies today (Woodle et al. 1992; Delmonico et al. 1993; Engram et al. 2010).

Other depleting agents no longer used in human organ transplantation include alefacept (LFA3-Ig) and fingolimod (FTY720). The CD2-specific fusion protein alefacept (LFA3-Ig), which selectively depletes memory T cells, has been used as adjunct immunosuppression in nonhuman primate models of cardiac, renal, and xenoislet transplantation (Kaplon et al. 1996; Weaver et al. 2009; Page et al. 2012; Thompson et al. 2012). Fingolimod, a sphingoside-1-phosphate receptor agonist administered orally, was found to effectively deplete T and B cells in baboons and cynomolgus macaques (Quesniaux et al. 1999, 2000), and prevent acute rejection in nonhuman primate renal, allo-islet, and xeno-islet transplantation (Schuurman et al. 2002; Wijkstrom et al. 2004; Hering et al. 2006).

B-Cell Depletion

Several B-cell-targeting therapeutics have emerged over the last decade, as our understanding of the impact of B cells on allograft rejection as well as transplant tolerance has increased (Redfield et al. 2011; Kwun et al. 2012a). These strategies may be of particular interest in the setting of T-cell depletion in which humoral immunity is often up-regulated. Their applications may also be extended to sensitized patients pre- and posttransplant.

The following agents targeting the B-cell surface molecules CD19 and CD20, costimulatory target CD40, and B-cell survival factors in the BAFF pathway have resulted in effective B-cell depletion. Rituximab (IDEC-C2B8), originally developed for B-cell lymphomas, has been used in several preclinical models, including islet cell transplantation (in combination with rATG), renal transplantation (with rATG, sirolimus, and donor bone marrow infusion), cardiac transplantation (with cyclosporine A), and pig-to-baboon cardiac xenotransplantation (with rATG, splenectomy, and maintenance immunosuppression) (McGregor et al. 2005; Liu et al. 2007; Maki et al. 2008; Kelishadi et al. 2010). GA101, a humanized IgG1 to CD20, induced further B-cell depletion particularly in lymphoid tissues in cynomolgus macaques (Mossner et al. 2010). XmAb5574, an Fc-engineered antibody to CD19, showed sustained B-cell depletion to $<10\%$ of baseline as well as transient NK cell depletion in cynomolguc macaques (Zalevsky et al. 2009). Although some anti-CD40 antibodies are nondepleting, Chi220 administered to rhesus macaques showed near complete peripheral B-cell depletion (Pearson et al. 2002). BR3 is one of three receptors for the B-cell activating factor BAFF; humanized anti-BR3 antibodies reduced B-cell populations in peripheral blood as well as lymphoid tissues, but not as profoundly as anti-CD20 treatment (Lin et al. 2007).

CURRENT LYMPHOCYTIC DEPLETION IN CLINIC

Currently used antibodies to deplete immune cells in solid organ transplant patients include thymoglobulin (rabbit ATG, Genzyme), ATGAM (equine ATG, Pfizer), and alemtuzumab (Campath-1H). Another rabbit ATG is marketed in Europe by Fresenius. Alemtuzumab was bought by Sanofi in 2011 and is being remarketed at a higher price under the name Lemtrada. Thymoglobulin is approved by the U.S. Federal Drug Administration for treatment of acute renal allograft rejection but not as an induction agent nor for other organs, although the drug is commonly used for these other purposes. Alemtuzumab is approved for treatment of lymphoma but not for organ transplantation, and its ownership by more than four drug companies over the past 10 years and the economics of its development make it unlikely to receive regulatory approval in the near future for organ transplantation. Nevertheless, these two agents are commonly used, and there is considerable experience, especially in kidney transplantation, with both. The published data on both agents inform our use and practice, but they leave unclear the ideal dose, duration, and indications for depletional therapy despite extensive experience. According to the 2011 Annual Report of the Scientific Registry of Transplant Recipients (SRTR/OPTN), about 60% of renal transplant patients in the United States received a depleting agent at the time of transplantation in 2010.

As discussed with regard to nonhuman primate (NHP) depletional therapy, CD3-immunotoxin (CD3-IT), developed by Neville for NHP use, potently depletes not only peripheral blood $CD3^+$ T cells, but also lymph node cells, within days. This agent ($FN18^-CRM9$) was in development by Novartis for human use with the intention of developing a genetically derived recombinant fusion protein, thus avoiding chemical conjugation of CD3 mAb and mutated diphtheria toxin (CRM9) and inevitable unconjugated products that would contaminate such a compound. However, this development project was finally aborted after the new constructs, such as a monovalent antibody fused to toxin, failed to have similar potency, binding properties, or stability compared to the NHP original construct, as discussed above. However, taking a cue from the remarkable ability of the CD3-IT depletion to promote tolerance, Calne revisited clinical use of Campath-1H, a potent anti-CD52 mAb produced by Herman Waldmann in the Cambridge, England pathology department (hence the name, with H for humanized). Calne wisely combined Campath-1H with low dose cyclosporine, aiming for near or "prope" tolerance in humans rather than a completely drug-free state. The data from the Cambridge kidney transplant experience with Campath-1H have been published with both early and long-term follow up, and indicate the safety of profound depletion from an infection and malignancy perspective and the ability of depletion to replace some degree of maintenance immunosuppression (Calne et al. 1998; Watson ct al. 2005).

Studies by our group supported by the Immune Tolerance Network (ITN) in human kidney transplantation evaluated the ability of high dose alemtuzumab to achieve tolerance in 10 haploidentical living donors or other well-matched renal transplant recipients. Although four of 10 patients achieved stable renal function for >3 years on rapamycin 1 mg daily with levels <4 ng/ml, none of these patients totally withdrew immunosuppression. Another observation of this study was that most other patients (5/10) developed detectable donor-specific alloantibodies after the withdrawal of tacrolimus that occurred at 60 days in the protocol. This high incidence of alloantibodies was shown by our group in other studies to be driven by tacrolimus withdrawal and to correlate with high

BAFF levels (Shapiro et al. 2008; Bloom et al. 2009; Ruhil et al. 2012). Nevertheless, this study suggests that well-matched kidney transplant patients treated with profound depletion in the setting of low dose tacrolimus, later weaned from tacrolimus, and then weaned to very low dose rapamycin may become unresponsive. Mechanistic assays are in the process of confirming the nature of the donor-specific unresponsiveness of these selected patients.

An ongoing study by Kirk reflects a modification of the ITN study above and substitutes belatacept for tacrolimus, perhaps an ideal combination that avoids the high alloantibody incidence associated with tacrolimus withdrawal and takes advantage of the known benefit of belatacept to prevent alloantibody. This study has been reported in abstract form (Kirk et al. 2011) and follow-up studies are in progress. Although it is too early to compare such a study to conventional therapy, results have been excellent; however, once again they show that complete drug withdrawal may not be advisable. The longer term outcomes of this study remain of high interest.

Perhaps the best and largest prospective, randomized study of alemtuzumab induction in renal transplantation was by Hanaway et al. (2011). This study compared alemtuzumab to anti-CD25 (Simulect) in low risk recipients and to thymoglobulin in high-risk recipients. All groups received maintenance immunosuppression with tacrolimus, mycophenolate, and steroids. The main findings were that alemtuzumab was associated with significantly less acute rejection compared to Simulect in low risk patients, but that there was no significant difference in acute rejection between the two high-risk groups (alemtuzumab vs. thymoglobulin). There were no significant differences in infection or side effects among the four groups.

Thymoglobulin induction in kidney transplantation continues to be used particularly for patients considered to be at higher than average risk of rejection. According to the SRTR 2011 annual report on kidney transplantation, abuse was driven by a publication from Brennan et al. (2006) comparing anti-CD25 induction to thymoglobulin, with slightly better results seen with depletion. Nevertheless, a recent trend toward less depletion and more reliance on tacrolimus and anti-CD25 reflects the relatively thin layer of objective data that show superiority of depletion for even the higher risk patients. Higher risk refers to retransplants, African-American recipients, and sensitized or previously sensitized but current cross-match negative recipients of renal allografts.

One of the lessons learned in the past 15 years about depletion induction therapy in renal transplantation is that whereas it clearly protects from acute rejection in the early posttransplant period, maintenance therapy is necessary during cell repopulation to particularly prevent T memory cells (Tmem) from driving rejection and alloantibody formation. As Pearl nicely showed, Tmem are particularly resistant to depletion and perhaps best controlled by calcineurin inhibitors (CNI) (Pearl et al. 2005). Thus, depletional strategies that have been most successful, such as the alemtuzumab/tacrolimus protocol described by Shapiro et al. (2007), rely on at least a low dose of CNI maintenance therapy. Similarly, thymoglobulin protocols typically use triple drug maintenance therapy with tacrolimus, mycophenolate, and steroids for higher risk patients (Gurk-Turner et al. 2008) with good results.

The long-term safety of depletion with respect to opportunistic infection (cytomegalovirus [CMV], other viral and fungal infections) and malignancy (posttransplantation lymphoproliferative disease [PTLD]) depends on the agent used and the intensity and duration of therapy. Given the age-related senescence of the lymphocyte compartment, it would stand to reason that older patients would tolerate depletion less well than younger patients. Kirk et al. (2007) reported that in contrast to thymoglobulin and OKT3 that are associated with increased risk of PTLD, the cumulative experience with alemtuzumab did not suggest an increased risk of PTLD. However, repeated use of alemtuzumab in transplant patients to maintain a low white blood cell (WBC) count is not recommended and was associated with unacceptable side effects in a Minnesota experience with the drug (Nath et al. 2005).

Use of depleting antibodies in liver transplantation continues at some centers although the majority of U.S. centers do not use any routine induction and only about 15% of recipients receive any induction antibody. Many centers add anti-CD25 induction for patients with renal dysfunction with the notion of delaying CNI introduction until renal function improves. Data from the University of Pittsburgh suggest that using alemtuzumab induction in hepatitis C-positive liver transplant recipients is associated with unacceptable outcomes and is ill-advised (Marcos et al. 2004). Other experiences with thymoglobulin induction in liver transplantation suggest that it does not reduce acute rejection incidence but increases the incidence of leukopenia (Boillot et al. 2009). A recent randomized controlled trial suggests that rather than depleting antibodies, lower dose tacrolimus in combination with everolimus spares renal function and effectively prevents rejection (De Simone et al. 2012). This strategy may be more effective than increasing the potency of induction immunosuppression by depletion in liver transplant patients who are often immunocompromised already from their illness.

CHALLENGES IN LYMPHODEPLETION

Although depletion is highly effective and increasingly used as an induction regimen, lymphodepletional agents are associated with several problems. As mentioned above, polyclonal preparations bring undesirable complications of transient cytokine release, viral reactivation, and antianimal antibody production. Although monoclonal OKT-3 solved some of these issues with polyclonal preparations, it was still subject to a significant antimouse response, as OKT3 is an entirely mouse-derived antibody (Webster et al. 2006). These xenogeneic antibodies can dramatically reduce the efficacy of these agents and extended/repeated use may lead to anaphylaxis and serum sickness (Prin Mathieu et al. 1997; Regan et al. 1997, 2001). The first dose of these antibodies can be followed by fevers, chills, and gastointestinal, respiratory, and cardiac complications, mostly due to the release of pyrogenic cytokines such as IL-6 and TNF-α from target cell lysis (Debets et al. 1989; Chatenoud et al. 1990; Vallhonrat et al. 1999). Alternatively, these cytokines could be released by a transient activation of T cells before they undergo cell lysis (Hardinger 2006; Webster et al. 2006). This cytokine release syndrome can be limited by methylprednisolone. These adverse effects were much improved with the development of monoclonal antibodies and fusion proteins. To avoid xenogeneic antibody production, the gene fragment encoding xenogenic epitopes in the mAb may be exchanged with equivalent portions of human antibodies (Boulianne et al. 1984; Morrison et al. 1984). Other early complications, including cardiovascular and infectious deaths, correlate with lymphodepletion, but the interpretation of this relationship is confounded by the preferential use of antibodies in high-risk patients (Opelz 1995).

Infectious Complications

Patients with prolonged lymphopenia are more likely to develop serious infections, especially when depletion is combined with other immunosuppressants before or after transplantation. Polyclonal antibodies have been shown to increase reactivation and development of primary viral disease with cytomegalovirus, herpes simplex virus, Epstein–Barr virus (EBV), and varicella, which more likely represent intensive immunosuppression (Abbott et al. 2002; Gourishankar et al. 2004). More recent reports have shown no increased risk of cytomegalovirus (CMV) infection, possibly due to the proper usage of antiviral prophylaxis (Webster et al. 2006). Although higher infection rates were observed in patients receiving thymoglobulin compared to nondepleting anti-CD25 mAb treated patients (85.8% vs. 75.2%, $P = 0.03$), a lower incidence of cytomegalovirus disease was documented from thymoglobulin treated patients (7.8% vs. 17.5%, $P = 0.02$) (Brennan and Schnitzler 2008). Alemtuzumab did not show a significant risk of infection compared to thymoglobulin or basiliximab (Hanaway et al. 2011). Owing to its intensity on immunosuppression in general, lymphodepletion should be accompanied by broad prophylaxis to prevent opportunistic viral (cytomegalovirus and BK virus) and bacterial infections (urinary tract infection, pneumonia, and bacteremia) (Hibberd et al. 1992; Batiuk et al. 2002).

Malignant Complications

Early depletion approaches, such as polyclonal antibodies and OKT3, have been linked with the risk of developing posttransplantation lymphoproliferative disease (PTLD) when combined with maintenance immunosuppressants (Penn 1990; Malatack et al. 1991; Meier-Kriesche et al. 2002; Cherikh et al. 2003). In an exemplary analysis of 157 consecutive cardiac transplant recipients, a ninefold higher incidence of PTLD was observed in OKT-3 treated cardiac allograft patients. Furthermore, a dose-related risk was also observed, with more frequent PTLD observed in patients receiving a cumulative dose of OKT3 exceeding 75 mg than in those who received a lower total dose (35.7% vs. 6.2%; $p < 0.001$) (Swinnen et al. 1990). On the other hand, analysis of 207 patients with thymoglobulin did not reveal an association with an additional risk of malignancies (El-Hamamsy et al. 2005). Interestingly, alemtuzumab induction has also shown a low PTLD incidence that approximates that of the nondepletional CD25-specific mAbs (Kirk et al. 2007). This could relate to B-cell depleting properties of alemtuzumab because B cells are the dominant reservoir of EBV. The impact of induction therapy on all-cause mortality remains unclear, especially when combined with newer immunosuppressants for maintenance, as does the risk for increased rejection and malignancy.

Homeostatic Repopulation after Lymphodepletion

Despite its efficacy in controlling T-cell-mediated acute rejection with low dose maintenance immunosuppression, lymphodepletion alone does not prevent acute rejection in the human patient (Trzonkowski et al. 2006). Turka's group elegantly showed that repopulating T cells and their phenotypic switching to memory type under homeostatic pressure is a great barrier to promoting tolerance (Wu et al. 2004; Neujahr et al. 2006). Lymphocyte reconstitution involves a dynamic balance between the peripheral expansion of residual lymphocytes and the generation of new lymphocytes from progenitors. Following lymphocytic depletion, homeostatic mechanisms drive the immediate reconstitution of lymphocyte populations whereas lymphopoiesis may not fully impact reconstitution for 1–2 years. Two major components for homeostatic maintenance of lymphocyte numbers and diversity include cytokines/chemokines and survival factors (Marrack et al. 2000; Ploix et al. 2001; Jameson 2002). Continuous engagement of T-cell receptors with self-peptide/MHC-complex is also required for the T-cell homeostatic proliferation (Viret et al. 1999).

T-cell subsets have been increasingly studied in the postdepletion setting, particularly after observing increased Treg populations with alemtuzumab or ATG induction (Ciancio et al. 2005; Bloom et al. 2008; Morelon et al. 2010). The exact mechanism associated with relative or absolute expansion of $CD4^{+}CD25^{+}$ T cells remains unclear; however, it is suggested that the increase of Treg is not completely dependent on the sparing of this subset after depletion, but by active conversion of non-Treg to Treg cells (Bloom et al. 2008). This relative increase in Tregs may counter the postdepletional increase in memory-type T cells and their function. Th17 cells also play a role in transplant rejection, as IL-17 has been found in biopsies of rejecting kidney, lung, and liver allografts (Loong et al. 2002; Vanaudenaerde et al. 2008; Fabrega et al. 2009). One study documented an increased number of Th17 with sirolimus monotherapy after alemtuzumab induction (Hester et al. 2011), although its role in rejection after lymphodepletion is unclear. These observations suggest that T-cell subsets have different susceptibilities to depletion and repopulation dynamics.

Follicular helper T cells (Tfh) play a unique role in helping the B-cell response (Crotty 2011). Although little is known on the direct role of T-cell depletion on Tfh cells, our preclinical and clinical studies have reported increased humoral

 Cite this article as *Cold Spring Harb Perspect Med* doi: 10.1101/cshperspect.a015511

responses and de novo alloantibody production after T-cell depletion (Barth et al. 2006; Page et al. 2012; Kwun et al. 2012b). Tfh cells are known to express high levels of IL-21, which is required for cognate B-cell help and germinal center (GC) formation (Silver and Hunter 2008). Interestingly, increased serum IL-21 levels have been found in a significant subset of multiple sclerosis patients treated with alemtuzumab, which also correlated to their development of autoimmune thyroiditis (Jones et al. 2009). Incomplete depletion in lymph nodes has been previously reported (Kirk et al. 2007; Page et al. 2012); it is possible that lymph node structures provide more resistance to these depletional agents, which may favor an enhanced Tfh cell and germinal center response. Owing to the increasing interest on antibody-mediated rejection in organ transplantation, follicular helper T-cell subsets in lymphodepletion need further careful evaluation.

CONCLUSION

Lymphodepletion has been an effective strategy for addressing the precursor frequency of alloreactive T cells at the time of organ transplantation and for preventing acute allograft rejection. Several preclinical studies have found that combining lymphodepletion with other modalities may be tolerogenic, although this has yet to be fully achieved in humans. Depletion has allowed, however, for clinicians to decrease maintenance immunosuppression and reduce the side effects of long-term steroids and calcineurin inhibition. For these reasons, it has been an increasingly used method of induction immunosuppression in the United States in recent years. Future studies should evaluate ways to inhibit postdepletional humoral responses and memory T-cell repopulation, while taking care not to over-immunosuppress and subject the patient to infectious and malignant complications.

REFERENCES

Abbott KC, Hypolite IO, Viola R, Poropatich RK, Hshieh P, Cruess D, Hawkes CA, Agodoa LY. 2002. Hospitalizations for cytomegalovirus disease after renal transplantation in the United States. *Ann Epidemiol* **12:** 402–409.

Alexander JW, Goodman HR, Cardi M, Austin J, Goel S, Safdar S, Huang S, Munda R, Fidler JP, Buell JF, et al. 2006. Simultaneous corticosteroid avoidance and calcineurin inhibitor minimization in renal transplantation. *Transpl Int* **19:** 295–302.

Annual Report of the U.S. Organ Procurement and Transplantation Network and the Scientific Registry of Transplant Recipients: Transplant Data 1998–2011. 2011. Department of Health and Human Services, Health Resources and Services Administration, Healthcare Systems Bureau, Division of Transplantation, Rockville, MD; United Network for Organ Sharing, Richmond, VA; University Renal Research and Education Association, Ann Arbor, MI.

Barth RN, Janus CA, Lillesand CA, Radke NA, Pirsch JD, Becker BN, Fernandez LA, Thomas Chin L, Becker YT, Odorico JS, et al. 2006. *Transpl Int* **19:** 885–892.

Batiuk TD, Bodziak KA, Goldman M. 2002. Infectious disease prophylaxis in renal transplant patients: A survey of U.S. transplant centers. *Clin Transplant* **16:** 1–8.

Bloom DD, Chang Z, Fechner JH, Dar W, Polster SP, Pascual J, Turka LA, Knechtle SJ. 2008. $CD4^+$ $CD25^+$ $FOXP3^+$ regulatory T cells increase de novo in kidney transplant patients after immunodepletion with Campath-1H. *Am J Transplant* **8:** 793–802.

Bloom D, Chang Z, Pauly K, Kwun J, Fechner J, Hayes C, Samaniego M, Knechtle S. 2009. BAFF is increased in renal transplant patients following treatment with alemtuzumab. *Am J Transplant* **9:** 1835–1845.

Boillot O, Seket B, Dumortier J, Pittau G, Boucaud C, Bouffard Y, Scoazec JY. 2009. Thymoglobulin induction in liver transplant recipients with a tacrolimus, mycophenolate mofetil, and steroid immunosuppressive regimen: A five-year randomized prospective study. *Liver Transpl* **15:** 1426–1434.

Bonnefoy-Berard N, Vincent C, Revillard JP. 1991. Antibodies against functional leukocyte surface molecules in polyclonal antilymphocyte and antithymocyte globulins. *Transplantation* **51:** 669–673.

Boulianne GL, Hozumi N, Shulman MJ. 1984. Production of functional chimaeric mouse/human antibody. *Nature* **312:** 643–646.

Bourdage JS, Hamlin DM. 1995. Comparative polyclonal antithymocyte globulin and antilymphocyte/antilymphoblast globulin anti-CD antigen analysis by flow cytometry. *Transplantation* **59:** 1194–1200.

Brennan DC, Schnitzler MA. 2008. Long-term results of rabbit antithymocyte globulin and basiliximab induction. *N Engl J Med* **359:** 1736–1738.

Brennan DC, Daller JA, Lake KD, Cibrik D, Del Castillo D. 2006. Rabbit antithymocyte globulin versus basiliximab in renal transplantation. *N Engl J Med* **355:** 1967–1977.

Calne R, Friend P, Moffatt S, Bradley A, Hale G, Firth J, Bradley J, Smith K, Waldmann H. 1998. Prope tolerance, perioperative campath 1H, and low-dose cyclosporin monotherapy in renal allograft recipients. *Lancet* **351:** 1701–1702.

Calne R, Moffatt SD, Friend PJ, Jamieson NV, Bradley JA, Hale G, Firth J, Bradley J, Smith KG, Waldmann H. 1999. Campath IH allows low-dose cyclosporine monotherapy in 31 cadaveric renal allograft recipients. *Transplantation* **68:** 1613–1616.

Chatenoud L, Ferran C, Legendre C, Thouard I, Merite S, Reuter A, Gevaert Y, Kreis H, Franchimont P, Bach JF. 1990. In vivo cell activation following OKT3 administration. Systemic cytokine release and modulation by corticosteroids. *Transplantation* **49:** 697–702.

Cherikh WS, Kauffman HM, McBride MA, Maghirang J, Swinnen LJ, Hanto DW. 2003. Association of the type of induction immunosuppression with posttransplant lymphoproliferative disorder, graft survival, and patient survival after primary kidney transplantation. *Transplantation* **76:** 1289–1293.

Ciancio G, Burke GW, Gaynor JJ, Carreno MR, Cirocco RE, Mathew JM, Mattiazzi A, Cordovilla T, Roth D, Kupin W, et al. 2005. A randomized trial of three renal transplant induction antibodies: Early comparison of tacrolimus, mycophenolate mofetil, and steroid dosing, and newer immune-monitoring. *Transplantation* **80:** 457–465.

Clatworthy MR, Watson CJ, Plotnek G, Bardsley V, Chaudhry AN, Bradley JA, Smith KG. 2009. B-cell-depleting induction therapy and acute cellular rejection. *N Engl J Med* **360:** 2683–2685.

Cosimi AB, Delmonico FL, Wright JK, Wee SL, Preffer FI, Jolliffe LK, Colvin RB. 1990. Prolonged survival of nonhuman primate renal allograft recipients treated only with anti-CD4 monoclonal antibody. *Surgery* **108:** 406–413; discussion 413–404.

Crotty S. 2011. Follicular helper CD4 T cells (TFH). *Annu Rev Immunol* **29:** 621–663.

Debets JM, Leunissen KM, van Hooff HJ, van der Linden CJ, Buurman WA. 1989. Evidence of involvement of tumor necrosis factor in adverse reactions during treatment of kidney allograft rejection with antithymocyte globulin. *Transplantation* **47:** 487–492.

Delmonico FL, Cosimi AB, Kawai T, Cavender D, Lee WH, Jolliffe LK, Knowles RW. 1993. Nonhuman primate responses to murine and humanized OKT4A. *Transplantation* **55:** 722–728.

De Simone P, Nevens F, De Carlis L, Metselaar HJ, Beckebaum S, Saliba F, Jonas S, Sudan D, Fung J, Fischer L, et al. 2012. Everolimus with reduced tacrolimus improves renal function in de novo liver transplant recipients: A randomized controlled trial. *Am J Transplant* **12:** 3008–3020.

DiLillo DJ, Hamaguchi Y, Ueda Y, Yang K, Uchida J, Haas KM, Kelsoe G, Tedder TF. 2008. Maintenance of long-lived plasma cells and serological memory despite mature and memory B cell depletion during CD20 immunotherapy in mice. *J Immunol* **180:** 361–371.

DiLillo DJ, Griffiths R, Seshan SV, Magro CM, Ruiz P, Coffman TM, Tedder TF. 2011. B lymphocytes differentially influence acute and chronic allograft rejection in mice. *J Immunol* **186:** 2643–2654.

El-Hamamsy I, Stevens LM, Carrier M, Pelletier G, White M, Tremblay F, Perrault LP. 2005. Incidence and prognosis of cancer following heart transplantation using RATG induction therapy. *Transpl Int* **18:** 1280–1285.

Engram JC, Cervasi B, Borghans JA, Klatt NR, Gordon SN, Chahroudi A, Else JG, Mittler RS, Sodora DL, de Boer RJ, et al. 2010. Lineage-specific T-cell reconstitution following in vivo $CD4^+$ and $CD8^+$ lymphocyte depletion in nonhuman primates. *Blood* **116:** 748–758.

Fabrega E, Lopez-Hoyos M, San Segundo D, Casafont F, Pons-Romero F. 2009. Changes in the serum levels of interleukin-17/interleukin-23 during acute rejection in liver transplantation. *Liver Transpl* **15:** 629–633.

Feng X, Kajigaya S, Solomou EE, Keyvanfar K, Xu X, Raghavachari N, Munson PJ, Herndon TM, Chen J, Young NS. 2008. Rabbit ATG but not horse ATG promotes expansion of functional $CD4^+CD25^{high}FOXP3^+$ regulatory T cells in vitro. *Blood* **111:** 3675–3683.

Gaber AO, Monaco AP, Russell JA, Lebranchu Y, Mohty M. 2010. Rabbit antithymocyte globulin (thymoglobulin): 25 Years and new frontiers in solid organ transplantation and haematology. *Drugs* **70:** 691–732.

Gourishankar S, McDermid JC, Jhangri GS, Preiksaitis JK. 2004. Herpes zoster infection following solid organ transplantation: Incidence, risk factors and outcomes in the current immunosuppressive era. *Am J Transplant* **4:** 108–115.

Gurk-Turner C, Airee R, Philosophe B, Kukuruga D, Drachenberg C, Haririan A. 2008. Thymoglobulin dose optimization for induction therapy in high risk kidney transplant recipients. *Transplantation* **85:** 1425–1430.

Hale G. 2001. The CD52 antigen and development of the CAMPATH antibodies. *Cytotherapy* **3:** 137–143.

Hanaway MJ, Woodle ES, Mulgaonkar S, Peddi VR, Kaufman DB, First MR, Croy R, Holman J. 2011. Alemtuzumab induction in renal transplantation. *N Engl J Med* **364:** 1909–1919.

Hardinger KL. 2006. Rabbit antithymocyte globulin induction therapy in adult renal transplantation. *Pharmacotherapy* **26:** 1771–1783.

Heidecke CD, Hancock WW, Jakobs F, Zantl N, Kurrle R, Westerholt S, Sewczik T, Deusch K, Kupiec-Weglinski J. 1995. α/β–T cell receptor-directed therapy in rat cardiac allograft recipients. Treatment before alloantigen exposure prevents sensitization and abrogates accelerated rejection. *Transplantation* **59:** 78–84.

Hering BJ, Wijkstrom M, Graham ML, Hardstedt M, Aasheim TC, Jie T, Ansite JD, Nakano M, Cheng J, Li W, et al. 2006. Prolonged diabetes reversal after intraportal xenotransplantation of wild-type porcine islets in immunosuppressed nonhuman primates. *Nat Med* **12:** 301–303.

Hester J, Mills N, Shankar S, Carvalho-Gaspar M, Friend P, Wood KJ. 2011. Th17 cells in alemtuzumab-treated patients: The effect of long-term maintenance immunosuppressive therapy. *Transplantation* **91:** 744–750.

Hibberd PL, Tolkoff-Rubin NE, Cosimi AB, Schooley RT, Isaacson D, Doran M, Delvecchio A, Delmonico FL, Auchincloss H Jr, Rubin RH. 1992. Symptomatic cytomegalovirus disease in the cytomegalovirus antibody seropositive renal transplant recipient treated with OKT3. *Transplantation* **53:** 68–72.

Hirohashi T, Chase CM, Della Pelle P, Sebastian D, Alessandrini A, Madsen JC, Russell PS, Colvin RB. 2012. A novel pathway of chronic allograft rejection mediated by NK cells and alloantibody. *Am J Transplant* **12:** 313–321.

Hirshberg B, Preston EH, Xu H, Tal MG, Neeman Z, Bunnell D, Soleimanpour S, Hale DA, Kirk AD, Harlan DM. 2003. Rabbit antithymocyte globulin induction and sirolimus monotherapy supports prolonged islet allograft function in a nonhuman primate islet transplantation model. *Transplantation* **76:** 55–60.

Hongo D, Tang X, Dutt S, Nador RG, Strober S. 2012. Interactions between NKT cells and Tregs are required for tolerance to combined bone marrow and organ transplants. *Blood* **119:** 1581–1589.

Jameson SC. 2002. Maintaining the norm: T-cell homeostasis. *Nat Rev Immunol* **2:** 547–556.

Jones JL, Phuah CL, Cox AL, Thompson SA, Ban M, Shawcross J, Walton A, Sawcer SJ, Compston A, Coles AJ. 2009. IL-21 drives secondary autoimmunity in patients with multiple sclerosis, following therapeutic lymphocyte depletion with alemtuzumab (Campath-1H). *J Clin Invest* **119:** 2052–2061.

Jonker M, Goldstein G, Balner H. 1983. Effects of in vivo administration of monoclonal antibodies specific for human T cell subpopulations on the immune system in a rhesus monkey model. *Transplantation* **35:** 521–526.

Joseph A, Neff K, Richard J, Gao L, Bangari D, Joly M, Culm-Merdek K, Garman R, Williams J, Richards S, et al. 2012. Transient low-dose methotrexate induces tolerance to murine anti-thymocyte globulin and together they promote long-term allograft survival. *J Immunol* **189:** 732–743.

Kaplon RJ, Hochman PS, Michler RE, Kwiatkowski PA, Edwards NM, Berger CL, Xu H, Meier W, Wallner BP, Chisholm P, et al. 1996. Short course single agent therapy with an LFA-3-IgG1 fusion protein prolongs primate cardiac allograft survival. *Transplantation* **61:** 356–363.

Kawai T, Cosimi AB, Colvin RB, Powelson J, Eason J, Kozlowski T, Sykes M, Monroy R, Tanaka M, Sachs DH. 1995. Mixed allogeneic chimerism and renal allograft tolerance in cynomolgus monkeys. *Transplantation* **59:** 256–262.

Kawai T, Sogawa H, Boskovic S, Abrahamian G, Smith RN, Wee SL, Andrews D, Nadazdin O, Koyama I, Sykes M, et al. 2004. CD154 blockade for induction of mixed chimerism and prolonged renal allograft survival in nonhuman primates. *Am J Transplant* **4:** 1391–1398.

Kelishadi SS, Azimzadeh AM, Zhang T, Stoddard T, Welty E, Avon C, Higuchi M, Laaris A, Cheng XF, McMahon C, et al. 2010. Preemptive CD20^{+} B cell depletion attenuates cardiac allograft vasculopathy in cyclosporine-treated monkeys. *J Clin Invest* **120:** 1275–1284.

Kim GB, Wang Z, Liu YY, Stavrou S, Mathias A, Goodwin KJ, Thomas JM, Neville DM. 2007. A fold-back single-chain diabody format enhances the bioactivity of an anti-monkey CD3 recombinant diphtheria toxin-based immunotoxin. *Protein Eng Des Sel* **20:** 425–432.

Kirk AD, Hale DA, Mannon RB, Kleiner DE, Hoffmann SC, Kampen RL, Cendales LK, Tadaki DK, Harlan DM, Swanson SJ. 2003. Results from a human renal allograft tolerance trial evaluating the humanized CD52-specific monoclonal antibody alemtuzumab (CAMPATH-1H). *Transplantation* **76:** 120–129.

Kirk AD, Cherikh WS, Ring M, Burke G, Kaufman D, Knechtle SJ, Potdar S, Shapiro R, Dharnidharka VR, Kauffman HM. 2007. Dissociation of depletional induction and posttransplant lymphoproliferative disease in kidney recipients treated with alemtuzumab. *Am J Transplant* **7:** 2619–2625.

Kirk AD, Mead S, Xu H, Mehta A, Guasch A, Cheeseman J, Joseph J, Horan J, Kean L, Larsen CP, et al. 2011. Kidney transplantation using alemtuzumab induction and belatacept/sirolimus maintenance therapy. *Am J Transplant* **11:** 45.

Knechtle SJ. 2001. Treatment with immunotoxin. *Philos Trans R Soc London B Biol Sci* **356:** 681–689.

Knechtle SJ, Vargo D, Fechner J, Zhai Y, Wang J, Hanaway MJ, Scharff J, Hu H, Knapp L, Watkins D, et al. 1997. FN18-CRM9 immunotoxin promotes tolerance in primate renal allografts. *Transplantation* **63:** 1–6.

Knechtle SJ, Fechner JH Jr, Dong Y, Stavrou S, Neville DM Jr, Oberley T, Buckley P, Armstrong N, Rusterholz K, Hong X, et al. 1998. Primate renal transplants using immunotoxin. *Surgery* **124:** 438–446; discussion 446–437.

Koyama I, Nadazdin O, Boskovic S, Ochiai T, Smith RN, Sykes M, Sogawa H, Murakami T, Strom TB, Colvin RB, et al. 2007. Depletion of CD8 memory T cells for induction of tolerance of a previously transplanted kidney allograft. *Am J Transplant* **7:** 1055–1061.

Kung P, Goldstein G, Reinherz EL, Schlossman SF. 1979. Monoclonal antibodies defining distinctive human T cell surface antigens. *Science* **206:** 347–349.

Kwun J, Bulut P, Kim E, Dar W, Oh B, Ruhil R, Iwakoshi N, Knechtle SJ. 2012a. The role of B cells in solid organ transplantation. *Semin Immunol* **24:** 96–108.

Kwun J, Oh BC, Gibby AC, Ruhil R, Lu VT, Kim DW, Page EK, Bulut OP, Song MQ, Farris AB, et al. 2012b. Patterns of de novo allo B cells and antibody formation in chronic cardiac allograft rejection after alemtuzumab treatment. *Am J Transplant* **12:** 2641–2651.

Letvin NL, Ritz J, Guida LJ, Yetz JM, Lambert JM, Reinherz EL, Schlossman SF. 1985. In vivo administration of lymphocyte-specific monoclonal antibodies in nonhuman primates: I. Effects of anti-T11 antibodies on the circulating T cell pool. *Blood* **66:** 961–966.

Lin WY, Gong Q, Seshasayee D, Lin Z, Ou Q, Ye S, Suto E, Shu J, Lee WP, Lee CW, et al. 2007. Anti-BR3 antibodies: A new class of B-cell immunotherapy combining cellular depletion and survival blockade. *Blood* **110:** 3959–3967.

Liu C, Noorchashm H, Sutter JA, Naji M, Prak EL, Boyer J, Green T, Rickels MR, Tomaszewski JE, Koeberlein B, et al. 2007. B lymphocyte-directed immunotherapy promotes long-term islet allograft survival in nonhuman primates. *Nat Med* **13:** 1295–1298.

Loong CC, Hsieh HG, Lui WY, Chen A, Lin CY. 2002. Evidence for the early involvement of interleukin 17 in human and experimental renal allograft rejection. *J Pathol* **197:** 322–332.

Luo H, Wu Y, Qi S, Wan X, Chen H, Wu J. 2001. A proteasome inhibitor effectively prevents mouse heart allograft rejection. *Transplantation* **72:** 196–202.

Ma S, Hu H, Thompson J, Stavrou S, Scharff J, Neville DM Jr. 1997. Genetic construction and characterization of an anti-monkey CD3 single-chain immunotoxin with a truncated diphtheria toxin. *Bioconjug Chem* **8:** 695–701.

Magliocca JF, Knechtle SJ. 2006. The evolving role of alemtuzumab (Campath-1H) for immunosuppressive therapy in organ transplantation. *Transpl Int* **19:** 705–714.

Maki T, Carville A, Stillman IE, Sato K, Kodaka T, Minamimura K, Ogawa N, Kanamoto A, Gottschalk R, Monaco

AP, et al. 2008. SV40 infection associated with rituximab treatment after kidney transplantation in nonhuman primates. *Transplantation* **85:** 893–902.

Malatack JF, Gartner JC Jr, Urbach AH, Zitelli BJ. 1991. Orthotopic liver transplantation, Epstein-Barr virus, cyclosporine, and lymphoproliferative disease: A growing concern. *J Pediatr* **118:** 667–675.

Marcos A, Eghtesad B, Fung JJ, Fontes P, Patel K, Devera M, Marsh W, Gayowski T, Demetris AJ, Gray EA, et al. 2004. Use of alemtuzumab and tacrolimus monotherapy for cadaveric liver transplantation: With particular reference to hepatitis C virus. *Transplantation* **78:** 966–971.

Marrack P, Bender J, Hildeman D, Jordan M, Mitchell T, Murakami M, Sakamoto A, Schaefer BC, Swanson B, Kappler J. 2000. Homeostasis of $\alpha\beta TCR^+$ T cells. *Nat Immunol* **1:** 107–111.

McGregor CG, Davies WR, Oi K, Teotia SS, Schirmer JM, Risdahl JM, Tazelaar HD, Kremers WK, Walker RC, Byrne GW, et al. 2005. Cardiac xenotransplantation: Recent preclinical progress with 3-month median survival. *J Thoracic Cardio Surgery* **130:** 844–851.

Meier-Kriesche HU, Arndorfer JA, Kaplan B. 2002. Association of antibody induction with short- and long-term cause-specific mortality in renal transplant recipients. *J Am Soc Nephrol* **13:** 769–772.

Mohty M. 2007. Mechanisms of action of antithymocyte globulin: T-cell depletion and beyond. *Leukemia* **21:** 1387–1394.

Morelon E, Lefrancois N, Besson C, Prevautel J, Brunet M, Touraine JL, Badet L, Touraine-Moulin F, Thaunat O, Malcus C. 2010. Preferential increase in memory and regulatory subsets during T-lymphocyte immune reconstitution after thymoglobulin induction therapy with maintenance sirolimus vs cyclosporine. *Transplant Immunol* **23:** 53–58.

Morrison SL, Johnson MJ, Herzenberg LA, Oi VT. 1984. Chimeric human antibody molecules: mouse antigen-binding domains with human constant region domains. *Proc Natl Acad Sci* **81:** 6851–6855.

Mossner E, Brunker P, Moser S, Puntener U, Schmidt C, Herter S, Grau R, Gerdes C, Nopora A, van Puijenbroek E, et al. 2010. Increasing the efficacy of CD20 antibody therapy through the engineering of a new type II anti-CD20 antibody with enhanced direct and immune effector cell-mediated B-cell cytotoxicity. *Blood* **115:** 4393–4402.

Najarian JS, Simmons RL. 1971. The clinical use of antilymphocyte globulin. *N Engl J Med* **285:** 158–166.

Nath DS, Kandaswamy R, Gruessner R, Sutherland DE, Dunn DL, Humar A. 2005. Fungal infections in transplant recipients receiving alemtuzumab. *Transplant Proc* **37:** 934–936.

Neujahr DC, Chen C, Huang X, Markmann JF, Cobbold S, Waldmann H, Sayegh MH, Hancock WW, Turka LA. 2006. Accelerated memory cell homeostasis during T cell depletion and approaches to overcome it. *J Immunol* **176:** 4632–4639.

Neville DM Jr, Scharff J, Srinivasachar K. 1992. In vivo T-cell ablation by a holo-immunotoxin directed at human CD3. *Proc Natl Acad Sci* **89:** 2585–2589.

Nicolls MR, Aversa GG, Pearce NW, Spinelli A, Berger MF, Gurley KE, Hall BM. 1993. Induction of long-term specific tolerance to allografts in rats by therapy with an anti-CD3-like monoclonal antibody. *Transplantation* **55:** 459–468.

Nishimura H, Scalea J, Wang Z, Shimizu A, Moran S, Gillon B, Sachs DH, Yamada K. 2011. First experience with the use of a recombinant CD3 immunotoxin as induction therapy in pig-to-primate xenotransplantation: The effect of T-cell depletion on outcome. *Transplantation* **92:** 641–647.

Opelz G. 1995. Efficacy of rejection prophylaxis with OKT3 in renal transplantation. Collaborative Transplant Study. *Transplantation* **60:** 1220–1224.

Orlando G, Hematti P, Stratta RJ, Burke GW 3rd, Di Cocco P, Pisani F, Soker S, Wood K. 2010. Clinical operational tolerance after renal transplantation: Current status and future challenges. *Ann Surg* **252:** 915–928.

Orosz CG, Wakely E, Bergese SD, VanBuskirk AM, Ferguson RM, Mullet D, Apseloff G, Gerber N. 1996. Prevention of murine cardiac allograft rejection with gallium nitrate. Comparison with anti-CD4 monoclonal antibody. *Transplantation* **61:** 783–791.

Page EK, Page AJ, Kwun J, Gibby AC, Leopardi F, Jenkins JB, Strobert EA, Song M, Hennigar RA, Iwakoshi N, et al. 2012. Enhanced de novo alloantibody and antibody-mediated injury in rhesus macaques. *Am J Transplant* **12:** 2395–2405.

Pearl JP, Parris J, Hale DA, Hoffmann SC, Bernstein WB, McCoy KL, Swanson SJ, Mannon RB, Roederer M, Kirk AD. 2005. Immunocompetent T-cells with a memory-like phenotype are the dominant cell type following antibody-mediated T-cell depletion. *Am J Transplant* **5:** 465–474.

Pearl JP, Xu H, Leopardi F, Preston E, Kirk AD. 2007. CD154 blockade, sirolimus, and donor-specific transfusion prevents renal allograft rejection in cynomolgus monkeys despite homeostatic T-cell activation. *Transplantation* **83:** 1219–1225.

Pearson TC, Bushell AR, Darby CR, West LJ, Morris PJ, Wood KJ. 1993. Lymphocyte changes associated with prolongation of cardiac allograft survival in adult mice using anti-CD4 monoclonal antibody. *Clin Exp Immunol* **92:** 211–217.

Pearson TC, Trambley J, Odom K, Anderson DC, Cowan S, Bray R, Lin A, Hollenbaugh D, Aruffo A, Siadak AW, et al. 2002. Anti-CD40 therapy extends renal allograft survival in rhesus macaques. *Transplantation* **74:** 933–940.

Penn I. 1990. Cancers complicating organ transplantation. *N Engl J Med* **323:** 1767–1769.

Ploix C, Lo D, Carson MJ. 2001. A ligand for the chemokine receptor CCR7 can influence the homeostatic proliferation of CD4 T cells and progression of autoimmunity. *J Immunol* **167:** 6724–6730.

Preville X, Flacher M, LeMauff B, Beauchard S, Davelu P, Tiollier J, Revillard JP. 2001. Mechanisms involved in antithymocyte globulin immunosuppressive activity in a nonhuman primate model. *Transplantation* **71:** 460–468.

Prin Mathieu C, Renoult E, Kennel De March A, Bene MC, Kessler M, Faure GC. 1997. Serum anti-rabbit and anti-horse IgG, IgA, and IgM in kidney transplant recipients. *Nephrol Dial Transplant* **12:** 2133–2139.

Quesniaux V, Fullard L, Arendse H, Davison G, Markgraaff N, Auer R, Ehrhart F, Kraus G, Schuurman HJ. 1999. A novel immunosuppressant, FTY720, induces peripheral lymphodepletion of both T and B cells and immunosuppression in baboons. *Transplant Immunol* **7:** 149–157.

Quesniaux VF, Menninger K, Kunkler A, Vedrine C, Bernhard M, Hedinger R, Kraus G, Schuurman HJ. 2000. The novel immunosuppressant FTY720 induces peripheral lymphodepletion of both T and B cells in cynomolgus monkeys when given alone, with Cyclosporine Neoral or with RAD. *Transplant Immunol* **8:** 177–187.

Rao PE, Olini G, Kille J, Muchmore E, Talle MA, Brake G, Rudnick SA. 1991. OKT3E, an anti-CD3 antibody that does not elicit side effects or antiidiotype responses in chimpanzees. *Transplantation* **52:** 691–697.

Rebellato LM, Gross U, Verbanac KM, Thomas JM. 1994. A comprehensive definition of the major antibody specificities in polyclonal rabbit antithymocyte globulin. *Transplantation* **57:** 685–694.

Redfield RR II, Rodriguez E, Parsons R, Vivek K, Mustafa MM, Noorchashm H, Naji A. 2011. Essential role for B cells in transplantation tolerance. *Curr Opin Immunol* **23:** 685–691.

Regan J, Campbell K, van Smith L, Pouletty P, Schroeder TJ, Guttmann RD, Buelow R. 1997. Characterization of anti-thymoglobulin, anti-Atgam and anti-OKT3 IgG antibodies in human serum with an 11-min ELISA. *Transplant Immunol* **5:** 49–56.

Regan JF, Lyonnais C, Campbell K, Smith LV, Buelow R. 2001. Total and active thymoglobulin levels: Effects of dose and sensitization on serum concentrations. *Transplant Immunol* **9:** 29–36.

Ruhil R, Oh B, Lu VT, Bulut P, Kwun J, Talebagha S, Knechtle S, Iwakoshi N. 2012. Tacrolimus inhibits alloantibody and plasma cell development but permits germinal center reaction and accumulation of memory B cells: Implications for humoral rejection. *Am J Transplant* **12:** 53.

Schroeder TJ, First MR, Mansour ME, Hurtubise PE, Hariharan S, Ryckman FC, Munda R, Melvin DB, Penn I, Ballistreri WF, et al. 1990. Antimurine antibody formation following OKT3 therapy. *Transplantation* **49:** 48–51.

Schuurman HJ, Menninger K, Audet M, Kunkler A, Maurer C, Vedrine C, Bernhard M, Gaschen L, Brinkmann V, Quesniaux V. 2002. Oral efficacy of the new immunomodulator FTY720 in cynomolgus monkey kidney allotransplantation, given alone or in combination with cyclosporine or RAD. *Transplantation* **74:** 951–960.

Shapiro R, Ellis D, Tan HP, Moritz ML, Basu A, Vats AN, Kayler LK, Erkan E, McFeaters CG, James G, et al. 2007. Alemtuzumab pre-conditioning with tacrolimus monotherapy in pediatric renal transplantation. *Am J Transplant* **7:** 2736–2738.

Shapiro R, Zeevi A, Basu A, Tan HP, Kayler LK, Blisard DM, Thai NL, Girnita AL, Randhawa PS, Gray EA, et al. 2008. Alemtuzumab preconditioning with tacrolimus monotherapy—The impact of serial monitoring for donor-specific antibody. *Transplantation* **85:** 1125–1132.

Shi Q, Lees JR, Scott DW, Farber DL, Bartlett ST. 2012. Endogenous expansion of regulatory T cells leads to long-term islet graft survival in diabetic NOD mice. *Am J Transplant* **12:** 1124–1132.

Sido B, Dengler TJ, Otto G, Zimmermann R, Muller P, Meuer SC. 1998. Differential immunosuppressive activity of monoclonal CD2 antibodies on allograft rejection versus specific antibody production. *Eur J Immunol* **28:** 1347–1357.

Silver JS, Hunter CA. 2008. With a little help from their friends: interleukin-21, T cells, and B cells. *Immunity* **29:** 7–9.

Starzl TE, Marchioro TL, Waddell WR. 1963. The reversal of rejection in human renal homografts with subsequent development of homograft tolerance. *Surg Gynecol Obstet* **117:** 385–395.

Starzl TE, Marchioro TL, Hutchinson DE, Porter KA, Cerilli GJ, Brettschneider L. 1967. The clinical use of antilymphocyte globulin in renal homotransplantation. *Transplantation* **5:** 1100–1105.

Starzl TE, Murase N, Abu-Elmagd K, Gray EA, Shapiro R, Eghtesad B, Corry RJ, Jordan ML, Fontes P, Gayowski T, et al. 2003. Tolerogenic immunosuppression for organ transplantation. *Lancet* **361:** 1502–1510.

Swanson SJ, Hale DA, Mannon RB, Kleiner DE, Cendales LC, Chamberlain CE, Polly SM, Harlan DM, Kirk AD. 2002. Kidney transplantation with rabbit antithymocyte globulin induction and sirolimus monotherapy. *Lancet* **360:** 1662–1664.

Swinnen LJ, Costanzo-Nordin MR, Fisher SG, O'Sullivan EJ, Johnson MR, Heroux AL, Dizikes GJ, Pifarre R, Fisher RI. 1990. Increased incidence of lymphoproliferative disorder after immunosuppression with the monoclonal antibody OKT3 in cardiac-transplant recipients. *N Engl J Med* **323:** 1723–1728.

Tang Q, Smith JA, Szot GL, Zhou P, Alegre ML, Henriksen KJ, Thompson CB, Bluestone JA. 2003. CD28/B7 regulation of anti-CD3-mediated immunosuppression in vivo. *J Immunol* **170:** 1510–1516.

Thomas FT, Carver FM, Foil MB, Pryor WH, Larkin EW, Hall WR, Haisch CE, Thomas JM. 1983. Long-term incompatible kidney survival in outbred higher primates without chronic immunosuppression. *Ann Surg* **198:** 370–378.

Thomas JM, Carver M, Cunningham P, Sash C, Park K, Thomas F. 1989. Promotion of incompatible allograft acceptance in rhesus monkeys given posttransplant antithymocyte globulin and donor bone marrow. II. Effects of adjuvant immunosuppressive drugs. *Transplantation* **47:** 209–215.

Thompson P, Badell IR, Lowe M, Turner A, Cano J, Avila J, Azimzadeh A, Cheng X, Pierson RN 3rd, Johnson B, et al. 2012. Alternative immunomodulatory strategies for xenotransplantation: CD40/154 pathway-sparing regimens promote xenograft survival. *Am J Transplant* **12:** 1765–1775.

Torrealba JR, Fernandez LA, Kanmaz T, Oberley TD, Schultz JM, Brunner KG, Peters D, Fechner JH Jr, Dong Y, Hu H, et al. 2003. Immunotoxin-treated rhesus monkeys: A model for renal allograft chronic rejection. *Transplantation* **76:** 524–530.

Torrealba JR, Katayama M, Fechner JH Jr, Jankowska-Gan E, Kusaka S, Xu Q, Schultz JM, Oberley TD, Hu H, Hamawy MM, et al. 2004. Metastable tolerance to rhesus monkey renal transplants is correlated with allograft TGF-$\beta 1^+$

$CD4^+$ T regulatory cell infiltrates. *J Immunol* **172:** 5753–5764.

Trzonkowski P, Zilvetti M, Friend P, Wood KJ. 2006. Recipient memory-like lymphocytes remain unresponsive to graft antigens after CAMPATH-1H induction with reduced maintenance immunosuppression. *Transplantation* **82:** 1342–1351.

Vallhonrat H, Williams WW, Cosimi AB, Tolkoff-Rubin N, Ginns LC, Wain JC, Preffer F, Olszak I, Wee S, Delmonico FL, et al. 1999. In vivo generation of C4d, Bb, iC3b, and SC5b-9 after OKT3 administration in kidney and lung transplant recipients. *Transplantation* **67:** 253–258.

Vanaudenaerde BM, De Vleeschauwer SI, Vos R, Meyts I, Bullens DM, Reynders V, Wuyts WA, Van Raemdonck DE, Dupont LJ, Verleden GM. 2008. The role of the IL23/IL17 axis in bronchiolitis obliterans syndrome after lung transplantation. *Am J Transplant* **8:** 1911–1920.

van der Touw W, Burrell B, Lal G, Bromberg JS. 2012. NK Cells are required for costimulatory blockade induced tolerance to vascularized allografts. *Transplantation* **94:** 575–584.

van der Windt DJ, Smetanka C, Macedo C, He J, Lakomy R, Bottino R, Ekser B, Echeverri GJ, Metes D, Ijzermans JN, et al. 2010. Investigation of lymphocyte depletion and repopulation using alemtuzumab (Campath-1H) in cynomolgus monkeys. *Am J Transplant* **10:** 773–783.

Viret C, Wong FS, Janeway CA Jr, 1999. Designing and maintaining the mature TCR repertoire: The continuum of self-peptide:self-MHC complex recognition. *Immunity* **10:** 559–568.

Vivier E, Tomasello E, Baratin M, Walzer T, Ugolini S. 2008. Functions of natural killer cells. *Nat Immunol* **9:** 503–510.

Wang Z, Duran-Struuck R, Crepeau R, Matar A, Hanekamp I, Srinivasan S, Neville DM Jr, Sachs DH, Huang CA. 2011. Development of a diphtheria toxin based antiporcine CD3 recombinant immunotoxin. *Bioconjug Chem* **22:** 2014–2020.

Watson CJ, Bradley JA, Friend PJ, Firth J, Taylor CJ, Bradley JR, Smith KG, Thiru S, Jamieson NV, Hale G, et al. 2005. Alemtuzumab (CAMPATH 1H) induction therapy in cadaveric kidney transplantation—efficacy and safety at five years. *Am J Transplant* **5:** 1347–1353.

Weaver TA, Charafeddine AH, Agarwal A, Turner AP, Russell M, Leopardi FV, Kampen RL, Stempora L, Song M, Larsen CP, et al. 2009. Alefacept promotes co-stimulation blockade based allograft survival in nonhuman primates. *Nat Med* **15:** 746–749.

Webster AC, Pankhurst T, Rinaldi F, Chapman JR, Craig JC. 2006. Monoclonal and polyclonal antibody therapy for treating acute rejection in kidney transplant recipients: A systematic review of randomized trial data. *Transplantation* **81:** 953–965.

Wijkstrom M, Kenyon NS, Kirchhof N, Kenyon NM, Mullon C, Lake P, Cottens S, Ricordi C, Hering BJ. 2004. Islet allograft survival in nonhuman primates immunosuppressed with basiliximab, RAD, and FTY720. *Transplantation* **77:** 827–835.

Woodle ES, Thistlethwaite JR, Jolliffe LK, Zivin RA, Collins A, Adair JR, Bodmer M, Athwal D, Alegre ML, Bluestone JA. 1992. Humanized OKT3 antibodies: Successful transfer of immune modulating properties and idiotype expression. *J Immunol* **148:** 2756–2763.

Wu Z, Bensinger SJ, Zhang J, Chen C, Yuan X, Huang X, Markmann JF, Kassaee A, Rosengard BR, Hancock WW, et al. 2004. Homeostatic proliferation is a barrier to transplantation tolerance. *Nat Med* **10:** 87–92.

Yi H, Zhen Y, Zeng C, Zhang L, Zhao Y. 2008. Depleting anti-CD4 monoclonal antibody (GK1.5) treatment: Influence on regulatory $CD4^+CD25^+Foxp3^+$ T cells in mice. *Transplantation* **85:** 1167–1174.

Zalevsky J, Leung IW, Karki S, Chu SY, Zhukovsky EA, Desjarlais JR, Carmichael DF, Lawrence CE. 2009. The impact of Fc engineering on an anti-CD19 antibody: Increased Fcγ receptor affinity enhances B-cell clearing in nonhuman primates. *Blood* **113:** 3735–3743.

Induction of Tolerance through Mixed Chimerism

David H. Sachs[1], Tatsuo Kawai[2], and Megan Sykes[3]

[1]Transplantation Biology Research Center, Massachusetts General Hospital, Harvard Medical School, Boston, Massachusetts 02129

[2]Transplant Center, Massachusetts General Hospital, Harvard Medical School, Boston, Massachusetts 02115

[3]Columbia Center for Translational Immunology, Columbia University Medical Center, New York, New York 10032

Correspondence: sachs@helix.mgh.harvard.edu

"Mixed chimerism" refers to a state in which the lymphohematopoietic system of the recipient of allogeneic hematopoietic stem cells comprises a mixture of host and donor cells. This state is usually attained through either bone marrow or mobilized peripheral blood stem cell transplantation. Although numerous treatment regimens have led to transplantation tolerance in mice, the induction of mixed chimerism is currently the only treatment modality that has been successfully extended to large animals and to the clinic. Here we describe and compare the use of mixed chimerism to establish transplantation tolerance in mice, pigs, and monkeys and in the clinic. We also attempt to correlate the mechanisms involved in achieving tolerance with the nature of the tolerance that has resulted in each case.

It has been known for many years that the hematopoietic chimerism that occurs after bone marrow transplantation carries with it transplantation tolerance for any other tissue or organ from the same donor. This phenomenon has been observed not only in numerous animal models, but also in humans. Thus, patients treated during childhood for leukemia by bone marrow transplantation from an HLA-identical sibling, who have subsequently developed renal failure and received kidney transplants from the same donors, have not required immunosuppression (Sayegh et al. 1991; Helg et al. 1994; Jacobsen et al. 1994). These cases, however, represent a very specialized situation, in which the transplants were HLA identical and in which complete replacement of the bone marrow was desirable in order to ensure ablation of the hematopoietic malignancy.

In the more common situation for organ transplantation, donors and recipients are not HLA identical, and it would also not be desirable to completely replace all bone marrow elements, because immunocompetence depends on interactions between T cells educated in the thymus and bone-marrow-derived antigen-presenting cells in the periphery (Zinkernagel et al. 1980). Fortunately, for the induction of transplantation tolerance through hematopoietic chimerism, complete ablation and re-

Cite this article as *Cold Spring Harb Perspect Med* doi: 10.1101/cshperspect.a015529

constitution are not required. The survival of even a small percentage of donor bone-marrow-derived elements, even transiently, has been found to be sufficient to allow tolerance of a simultaneously transplanted organ to be achieved (Fuchimoto et al. 1999). This phenomenon, in which both host and donor bone-marrow-derived elements coexist in the recipient has been termed "mixed chimerism" (Sykes and Sachs 1988). Unlike many other treatment regimens that have been successful in the induction of tolerance in rodents but not in large animals (Sachs 2003; Cosimi and Sachs 2004), induction of tolerance through mixed chimerism has been successful in mice, in large animals, and, most recently, in the clinic.

Nevertheless, this transition from mice to the clinic has required numerous changes in treatment protocols, and the tolerance obtained has likely been achieved through more than one mechanism. In this review, we describe and compare the establishment of transplantation tolerance in mice, pigs, monkeys, and in the clinic, with an attempt to correlate the mechanisms involved in achieving tolerance with the nature of the tolerance that has resulted in each case.

STUDIES IN MOUSE MODELS

Proof of principle that mixed chimerism was associated with donor-specific skin graft tolerance was obtained in adult mice receiving lethal total body irradiation (TBI) and reconstitution with a mixture of T-cell-depleted host plus donor bone marrow. Mixed lymphohematopoietic chimerism was observed, in which all lymphoid and hematopoietic cells are derived from a mixture of donor- and host-derived hematopoietic stem cells (HSCs). The tolerant state was systemic, as evidenced by in vitro measures of alloreactivity. Graft-versus-host disease (GVHD) was avoided by T-cell depletion of the donor marrow component, and depletion of the recipient marrow component was required to prevent rejection of the donor marrow (Ildstad and Sachs 1984; Ildstad et al. 1985, 1986). Using a different approach, mixed chimerism and tolerance without GVHD were also shown in mice receiving fractionated high-dose total lymphoid irradiation (TLI) conditioning with shielding of marrow-containing bones before allogeneic BMT (Slavin et al. 1977).

GVHD is the major complication of clinical hematopoietic cell transplantation (HCT) that has precluded the routine use of extensively HLA-mismatched donors even in patients with malignancies. Although a low level of GVHD is associated with improved outcomes owing to reduced leukemic relapse rates in patients receiving BMT for this more standard indication (Weiden et al. 1981), its severity is unacceptable when HLA barriers are transgressed, mandating a search for closely HLA-matched donors in this patient group. Even a low level of GVHD would be an unacceptable complication if HCT were to be used expressly for the purpose of organ allograft tolerance induction. Although GVHD can be avoided by T-cell-depleting the donor product, rejection of HLA mismatched hematopoietic cells then becomes common, and very heavy host conditioning is needed to prevent rejection (Martin et al. 1985, 1988; Kernan et al. 1987; Bordignon et al. 1989; Fleischhauer et al. 1990; Aversa et al. 1998). Because organ transplantation is routinely performed across extensive HLA barriers, avoiding GVHD while crossing HLA barriers is a major challenge for the induction of mixed chimerism and tolerance.

The challenge is further augmented by the requirement that relatively nontoxic, low-intensity conditioning is mandated in using HCT for this purpose. The high-dose chemo-/radiotherapy regimens used for patients with malignant disease would not be appropriate in patients receiving HCT for tolerance induction. To reduce the toxicity of recipient conditioning regimens, it is necessary to more specifically target the host immune elements that resist allogeneic marrow engraftment. Recipient T cells are largely responsible for rejection of MHC mismatched marrow in animals and humans (Sharabi et al. 1992; Vallera et al. 1994; Hayashi et al. 1996), and a low-toxicity host conditioning regimen must either eliminate mature host T cells or permit preexisting T cells to be rendered tolerant. This requirement applies to both peripheral and intrathymic alloreactive T cells. When peripheral T cells are globally depleted with mAbs and thymocytes are globally depleted with local irradiation before allogeneic BMT with low-dose (3 Gy) TBI (Sharabi and Sachs 1989), donor- and host-reactive T cells arising de novo in the thymus are specifically deleted (Tomita et al. 1994, 1996a,b; Khan et al. 1996; Manilay et al. 1998). Coexistence of donor and recipient HSCs in the marrow generates lifelong thymic APCs of each type to ensure central, deletional T-cell tolerance (Tomita et al. 1994, 1996a,b; Khan et al. 1996; Manilay et al. 1998). This model built on an earlier regimen requiring 6 Gy of TBI combined with in vivo depletion of host T cells to achieve MHC-mismatched allogeneic marrow engraftment and donor-specific tolerance (Cobbold et al. 1990). Because T-cell-depleting mAbs almost completely eliminate peripheral but not intrathymic T cells (Sharabi and Sachs 1989), the addition of thymic irradiation, which eliminates mature alloreactive thymocytes (Nikolic et al. 2001), was needed to reduce the TBI requirement to only 3 Gy. Intrathymic clonal deletion maintains long-term donor-specific tolerance in association with the presence of donor class II^{high} cells (Tomita et al. 1994). Several lines of evidence (Khan et al. 1996) proved that intrathymic chimerism is necessary and sufficient for the maintenance of tolerance. These studies showed that peripheral chimerism and suppressive mechanisms did not play a meaningful role in the maintenance of tolerance in these mixed chimeras that were created in the environment of a completely T-cell-depleted "clean slate."

We subsequently developed conditioning regimens that are less T-cell-depleting than that described above and that involve more complex mechanisms in order to achieve initial mixed chimerism. Both thymic irradiation and T-cell-depleting mAbs can be avoided if an initial course of costimulatory blockade is used (Wekerle et al. 1998), and TBI can be eliminated if a high enough marrow dose is administered (Wekerle et al. 2000). If applied clinically, these approaches would obviate concerns about delayed T-cell recovery in older individuals with limited thymic function (Haynes et al. 2000). Once mixed chimerism is achieved with costimulatory blockade-based regimens, long-term tolerance is maintained by intrathymic deletion (Wekerle et al. 1998, 2000; Ito et al. 2001). Preexisting peripheral T cells that recognize the donor in recipients of anti-CD154 plus allogeneic BMT are specifically deleted, but the remainder of the T-cell repertoire is spared (Kurtz et al. 2004; Fehr et al. 2005). Mechanisms of tolerance of peripheral CD4 and CD8 cells differ in this context, although both culminate in deletion only of donor-reactive cells. For CD8 cells, initial tolerance requires CD4 cells that do not appear to be classical Tregs. Once peripheral deletion of donor-specific CD8 cells is complete (at ~2 wk), CD4 cells are no longer needed for maintenance of CD8 tolerance (Fehr et al. 2005). Tolerance of CD4 cells, in contrast, does not involve any regulatory mechanisms (Kurtz et al. 2004). The expressions of MHC class II on recipient APCs, as well as recipient dendritic cells and B cells, all are also required to tolerize preexisting CD8 cells, but not CD4 cells, in this model (Fehr et al. 2008; Mollov et al. 2010). Moreover, PD-1/PD-L1 interactions are required to tolerize CD8 cells but not CD4 cells (Haspot et al. 2008), and the CD8 cells themselves must express PD-1 in order to be tolerized (Lucas et al. 2011). NFAT-1, also in a CD8-cell-intrinsic manner, is required for CD8 tolerance and plays no role in tolerizing the CD4 compartment (Fehr et al. 2010). TGF-β and LAG-3 also play a critical role in tolerizing preexisting CD8 but not CD4 cells in mice receiving BMT with anti-CD40L (Lucas et al. 2011). Tolerance of both subsets is independent of IFN-γ and IL-10 (Takeuchi et al. 2004; Fehr et al. 2005; Mollov et al. 2010). Deletion of peripheral donor-reactive CD4 and CD8 cells is preceded by specific unresponsiveness to the donor (Kurtz et al. 2001, 2004; Haspot et al. 2008).

Long-term tolerance involves central deletion of donor-reactive thymocytes. Regulatory cells do not play a significant role in maintaining long-term tolerance in this model of BMT with anti-CD40L (Ito et al. 2001; Kurtz et al. 2004; Takeuchi et al. 2004; Fehr et al. 2005). As a general rule, deletional tolerance seems to preclude the expansion and requirement for specific regulatory cells in maintenance of tolerance. In contrast, some models using CTLA-4Ig and anti-CD154 as conditioning for allogeneic BMT are associated with less complete deletion of preexisting donor-reactive T cells and do appear to rely on long-term regulatory mechanisms (Bemelman et al. 1998; Bigenzahn et al. 2005; Domenig et al. 2005).

Translation into patients of an approach that involved initial mixed chimerism induction in the treatment of hematologic malignancies in mouse models played an important role in the development of clinical mixed chimerism/tolerance protocols. We have attempted to avoid GVHD to permit HLA-mismatched HCT for the treatment of hematologic malignancies in a series of clinical protocols by using nonmyeloablative conditioning with recipient and donor graft T-cell depletion to achieve initial mixed chimerism. Because conditioning-induced tissue inflammation is a critical checkpoint in the development of GVHD (Chakraverty et al. 2006a), MHC-directed GVH alloreactivity can be confined to the lymphohematopoietic system when nontolerant donor T cells are given to mixed chimeras after host recovery from the initial conditioning regimen has occurred (Sykes et al. 1988; Mapara et al. 2002, 2003; Chakraverty et al. 2006a,b). GVH reactions of nontolerant T cells in donor leukocyte infusions (DLI) are not opposed by a host-versus-graft response in tolerant mixed chimeras. The DLI convert mixed hematopoietic chimerism to full donor chimerism and mediate strong graft-versus-leukemia/lymphoma (GVL) effects (Sykes et al. 1988; Mapara et al. 2002) without inducing GVHD (Sykes et al. 1988; Pelot et al. 1999). Recipient professional APCs expressing both class I (Mapara et al. 2002) and class II (Chakraverty et al. 2006b) MHC are required to induce this antihost reactivity and maximal GVL. These DLI-derived GVH-reactive T cells do not migrate to the epithelial GVHD target tissues (skin, intestines, and liver) in the absence of inflammatory signals in those tissues (Chakraverty et al. 2006a; Li et al. 2012). Clinical trials based on this approach to separating GVHD and GVL provided proof-of-principle that GVH responses can be confined to the lymphohematopoietic system following delayed DLI and thereby fail to induce GVHD in patients who received nonmyeloablative HCT with an initially T-cell-depleted product, even across extensive HLA barriers (Spitzer et al. 2003).

Translational studies aiming to apply this approach to separating GVHD and GVL in humans led to a clinical regimen with sufficient safety and antitumor efficacy to be used for renal allograft tolerance induction in multiple myeloma patients with HLA-identical donors (Fudaba et al. 2006). Attempts to apply this approach to separating GVHD and GVL in the HLA-mismatched related donor setting led to the development of a nonmyeloablative regimen that achieved transient chimerism without any GVHD in patients with hematologic malignancies (Spitzer et al. 2003). The freedom from GVHD in these mismatched recipients ultimately permitted the trials of allograft tolerance induction in patients without malignant disease that are described below.

STUDIES IN SWINE

Studies of mixed chimerism in swine have been performed in a herd of miniature swine that have been bred selectively for the past 40 years as a large-animal model for studies of transplantation (Sachs 1992; Hanekamp et al. 2012). These animals are similar to humans with regard to size and many physiological parameters (Sachs 1992, 1994). Because of their breeding characteristics, it has been possible to develop lines of swine homozygous for the MHC (called SLA in swine), as well as several recombinant lines separating the genes encoding class I and class II antigens (Pennington et al. 1981; Lunney and Sachs 1995). Figure 1 shows the three inbred haplotypes and several of the recombinant haplotypes currently available. These animals therefore represent a unique resource in which transplants across reproducible, defined genetic combinations can be performed in a large animal model

Early studies of bone marrow transplantation in miniature swine involved lethal irradiation as a preparative regimen, with full donor

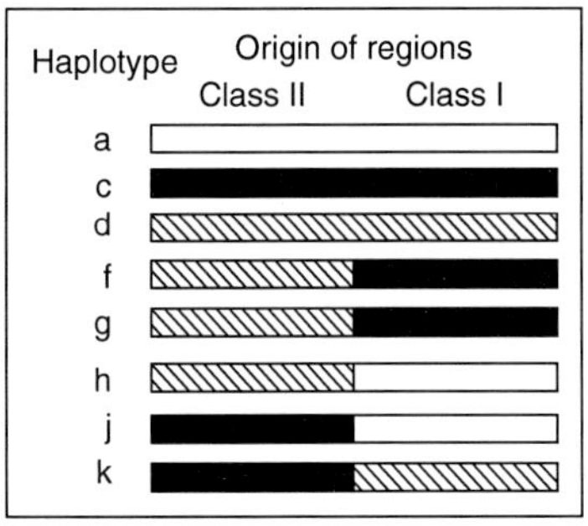

Figure 1. Origin of haplotypes of inbred miniature swine.

hematopoietic reconstitution as the result (Pennington et al. 1988; Popitz-Bergez et al. 1988; Sakamoto et al. 1988). Use of F_1 into parental and parental into F_1 combinations permitted the study of graft-versus-host (GVH) and host-versus-graft (HVG) reactions in this species, as well as an appreciation of the role of T cells in each (Sakamoto et al. 1987). Initial attempts to induce mixed chimerism across full two-haplotype SLA barriers used a lethal preparative regimen of (650 + 650) cGy TBI on two successive days followed by infusion of 7.5 × 108 bone marrow cells/kg + 1.0 × 10^8 T-cell-depleted autologous bone marrow (BM) cells. Although there were a few long-term survivors (one animal survived 154 d), most of the animals died earlier, either from the aplasia resulting from bone marrow rejection or from GVHD (Smith et al. 1993).

Better rates of engraftment and survival were obtained using a clinically relevant genetic combination of one-haplotype mismatched F_1 into F_1 (i.e., AC into AD, simulating the clinical situation of haploidentical related) and a preparative regimen of either (650 + 650) cGy TBI on two successive days or (500 + 650) cGy TBI, and 50 mg/kg cyclophosphamide. Once some of these animals had attained prolong survival and immunocompetence, renal transplants from donors SLA-matched to the bone marrow donors were performed, and all animals were found to be tolerant of the kidneys (Sundt et al. 1988; Guzzetta et al. 1991).

T-cell depletion in these early experiments was achieved by treating the donor inoculum in vitro with available anti-CD4 and anti-CD8 monoclonal antibodies (Lunney and Pescovitz 1987; Pescovitz et al. 1990) and lethally ablating the recipients. Achievement of mixed chimerism with a nonmyeloablative preparative regimen awaited the development of a T-cell-depleting reagent effective in vivo, because most of the anti-pig T-cell monoclonal antibodies available were capable of depleting in vitro in the presence of rabbit complement, but were poorly depleting in vivo. For this purpose, a swine CD3 immunotoxin, pCD3-CRM9, was developed by conjugating an available anti-pig CD3 monoclonal (Pescovitz et al. 1998; Huang et al. 1999a) with a mutant diphtheria toxin, developed by Neville et al. (1996). This reagent was found to be very effective for depleting mature T cells from the peripheral blood, lymph nodes, and thymus of miniature swine (Huang et al. 1999b). In addition, another source of hematopoietic stem cells was obtained through cytokine mobilization and apheresis, allowing collection of enormous numbers of peripheral blood stem cells (PBSC), which were capable of reconstituting ablated miniature swine just as effectively as bone marrow (Nash et al. 1999).

Huang et al. first reported the induction of stable mixed chimerism and donor-specific tolerance following either BM or PBSCs, using this CD3 immunotoxin for T-cell depletion in an SLA-matched combination (Huang et al. 1999b, 2000). Marrow was harvested from the long bones and/or vertebra following exsanguination of the donors, and 2 × 10^8 cells/kg administered by intravenous injection (i.v.) PBSC were collected following daily mobilization with recombinant porcine stem cell factor (pSCF) and IL-3 (pIL-3), with or without recombinant human G-CSF (rhuG-CSF; Amgen), and 20 × 10^8 cells/kg were administered. Irradiation consisted of two sequential daily doses of 150 cGy midline TBI and 700 cGy thymic irradiation on d −2. T-cell depletion was achieved by administering pCD3-CRM9 i.v. on d −2, and immunosuppression consisted of oral cyclosporine from d −1 to d 30. Excellent T-cell depletion was achieved by this regimen, with T-cell counts falling to <0.2% of their pretransplant values, ~30-fold to 150-fold greater depletion than the same preparative regimen without immunotoxin. Five of the six animals treated in this study developed stable mixed-lineage hematopoietic chimerism as well as thymic chimerism, and four of these animals that were subsequently tested with skin grafts showed markedly prolonged survival of donor versus third-party skin (Huang et al. 2000).

In subsequent studies, Fuchimoto et al. (2000) showed that a similar preparative regimen could also achieve stable mixed chimerism across a one-haplotype full SLA mismatch, and that by using a much larger dose of PBSC (100 × 108 cells/kg), the requirement for TBI could be avoided. The resulting animals showed evidence for specific unresponsiveness to donor by MLR and CML reactions in vitro and by specific prolongation of donor skin graft survival in vivo. In addition, four of these animals subsequently received kidney allografts SLA-matched to the donors without immunosuppression. Three of these animals accepted these renal allografts long term. The fourth animal rejected its allograft but remained unresponsive in vitro to the donor SLA. A second donor-matched kidney was therefore placed and survived for >100 d, suggesting that rejection of the first SLA-matched kidney might have been due to minor antigen disparities. All of these animals maintained peripheral mixed chimerism long term and showed chimerism of dendritic cells in the thymus by immunohistochemistry.

Because of an increased incidence of host-derived PTLD using this protocol, it was subsequently decided to add a low dose (100 cGy) of TBI back into the preparative regimen (Cina et al. 2006). It was also shown that, unlike the mouse (Sharabi and Sachs 1989) and the monkey (Kawai et al. 1995), in both of which thymic irradiation (TI) has been shown to be essential for the induction of long-term, stable mixed chimerism, TI could be omitted in the swine model and mixed chimerism could still be achieved, albeit at a lower level (Cina et al. 2006). The reason for this discrepancy may be related to the fact that antibody depletion of T cells is ineffective in the thymus when this depletion is dependent on the usual antibody-dependent mechanisms, including complement-mediated killing and antibody-dependent cell-mediated cytotoxicity (ADCC). However, in the swine model, depletion using the CD3 immunotoxin would not be expected to depend on anything other than the toxin itself. Therefore, depletion of mature T cells in the thymus may be sufficient using this protocol, even without TI.

By examining many of the parameters of chimerism detectable in the miniature-swine large-animal model, Horner et al. (2006) have attempted to determine which parameters are most indicative of induction of transplantation tolerance by this chimerism. The presence of donor cells in bone marrow, thymus, and various lineages of peripheral blood were examined in a series of 22 HCT recipients that also received subsequent donor renal allografts without immunosuppression. Bone marrow was assayed for the presence of donor colony-forming units (CFUs) by PCR, thymus by both fluorescence-activated cell sorting (FACS) analysis on thymic cell suspensions and by Southern blots, and peripheral blood by FACS and by in vitro responsiveness to donor MHC. These comparisons showed that engraftment, as indicated by the presence of donor-derived CFU in the bone marrow and detectable thymic chimerism, was the most reliable predictor of subsequent acceptance of a donor-matched renal allograft, whereas peripheral blood chimerism and in vitro assays of responsiveness were less-accurate predictors, although they were also highly correlated with tolerance (Horner et al. 2006).

STUDIES IN NONHUMAN PRIMATES

There are significant biological differences between rodents and large animals that are attributable to both genetic and developmental differences. The discrepancy in the ease of tolerance induction between rodents and primates may involve, for example, differential expression of MHC antigens, especially class II (Pescovitz et al. 1984a,b), as well as differences in levels of heterologous memory T cells depending on whether or not recipients are housed in environmentally controlled conditions (Adams et al. 2003a; Wu et al. 2004; Nadazdin et al. 2010). Elucidating these biological differences between species has proved pivotal to the successful extension of basic observations in mice to large animals and humans.

With regard to the induction of mixed chimerism with a nonmyeloablative regimen, earlier murine studies (see above) have shown that stable mixed chimerism and skin allograft tolerance can be consistently achieved across full MHC barriers with a nonmyeloablative conditioning regimen that consisted of low-dose total body irradiation (TBI, 3 Gy), thymic irradiation (7 Gy), and anti-T-cell antibody (Sharabi and Sachs 1989; Sykes and Sachs 2001). The recipient's mature T cells had to be severely depleted so that newly maturing T cells would be clonally deleted as they repopulated the T-cell repertoire. However, similar nonmyeloablative conditioning regimens failed to induce stable mixed chimerism in nonhuman primates (NHPs), likely because of the inability to completely deplete mature T cells. By adding splenectomy and a 1-mo course of cyclosporine to the conditioning regimen, mixed chimerism was found inducible, but only transiently (Kawai et al. 1995; Kimikawa et al. 1997). In subsequent experiments, mixed chimerism was significantly improved by adding CD154 blockade in place of splenectomy, but it still failed to induce stable mixed chimerism (Kawai et al. 2004). Kean and colleagues have also reported difficulty inducing stable chimerism in NHPs with a nonmyeloablative conditioning regimen. In their study, NHP recipients were treated with a busulfan-based conditioning regimen with CD40/CD40L and CD28/B7 blockade but failed to achieve stable mixed chimerism even in MHC-matched pairs (Pescovitz et al. 1984b; Larsen et al. 2010). This difference in chimerism induction between rodents and NHPs is presumably due to the presence of abundant memory T cells (Tmem) in NHP, which are not present in laboratory mice (Adams et al. 2003a,b). These Tmem are resistant to costimulatory blockade (Valujskikh et al. 2002) and to conventional anti-T-cell antibody treatment (Yamada et al. 2012), which may result in failure to achieve stable mixed chimerism.

Despite an inability to induce stable mixed chimerism in NHPs, we found that renal allograft tolerance was inducible across full MHC barriers even with transient mixed chimerism, as long as the kidneys were transplanted before the loss of peripheral chimerism (Kawai et al. 1995, 1999, 2004; Kimikawa et al. 1997). Continued survival of the renal allografts after the loss of chimerism suggested that peripheral mechanisms were also operative in tolerant recipients. As detailed below, the renal allograft itself has been shown to play a critical role in the maintenance of tolerance (Seidl et al. 1992).

For induction of renal allograft tolerance, splenectomy was initially thought to be a prerequisite in our conditioning regimen, because recipients consistently developed acute humoral rejection without splenectomy (Kawai et al. 1999, 2004). Because the spleen is the largest single site of lymphoid tissue where close collaboration between antigen-driven T and B cells occurs (Davis and Bjorkman 1990), splenectomy prevents such T/B-cell collaboration, especially when T-cell depletion by antibody treatment is not sufficient. Because of the pivotal role of the CD154 (i.e., the CD40 L-CD40) pathway in T- and B-cell responses, it was hypothesized that agents reacting with those molecules might also be able to prevent humoral rejection. By adding a short course of CD154 blockade in place of splenectomy, eight of eight recipients developed improved chimerism, and most recipients survived long term (Kawai et al. 2004). Unfortunately, anti-CD154 mAb is not likely to be available clinically because of its thrombogenic side effects (Kawai et al. 2000); therefore, applications of other costimulatory blockades, such as CTLA-4Ig, are also under current investigation for this purpose.

The conditioning regimen originally used in our laboratories for tolerance induction in NHP required treatment of recipients beginning 6 d before organ transplantation, which would limit the applicability of this procedure to only living donor allograft recipients. Therefore, we have recently developed a "delayed tolerance" protocol, in which kidney transplantation is performed first, with conventional immunosuppressive therapy, followed by conditioning and donor bone marrow transplantation (DBMT) at a subsequent time. With this approach, any stable recipient of either living or deceased donor kidney transplantation (KTx) could be a potential candidate, if either fresh (living donor) or cryopreserved (deceased donor) donor bone marrow (DBM) cells were available. In the delayed tolerance protocol, however, induction of mixed chimerism was found to be more difficult, likely because the prior transplanted kidney allograft may have sensitized the recipients before DBMT. We speculated that the faster homeostatic recovery of $CD8^+$ Tmem observed in the recipients of the delayed protocol in comparison to the standard protocol may have prevented induction of mixed chimerism. Following the addition of a humanized anti-CD8 mAb, cM-T807 (Schmitz et al. 1999) to the treatment regimen, recipients consistently developed mixed chimerism and achieved renal allograft tolerance (Koyama et al. 2007; Yamada et al. 2012). The timing of DBMT was also found to be critical. Recipients acquired allograft tolerance when DBMT was performed at 4 mo but consistently failed when DBMT was performed at 1 mo after kidney transplantation. Significantly higher inflammatory cytokine levels were detected at 1 mo than at 4 mo, suggesting that higher inflammatory responses caused by DBMT conditioning prevented tolerance induction (Yamada et al. 2012).

Studies attempting to extend the mixed chimerism approach to organ allografts other than kidneys (heart, lung, and islet) in cynomolgus monkeys have also been performed in this laboratory. In a heterotopic cardiac allograft model, three of five recipients developed multilineage chimerism, with allograft survivals in these recipients prolonged to 138, 428, and 509 d. Although in vitro assays showed donor-specific hyporesponsiveness, long-term heart allograft recipients eventually developed humoral and cellular immunity against the donor and rejected the grafts, indicating failure to induce full tolerance (Kawai et al. 2002).

Encouraged by our studies in MHC inbred miniature swine (Mezrich et al. 2004), we subsequently tested the effects of heart and kidney cotransplantation in NHP recipients using our mixed chimerism approach. We found that the presence of a kidney allograft consistently conferred stable and long-term tolerance to recipients of cotransplanted heart grafts from the same donors, even with transient chimerism (Tonsho 2012). Critical involvement of the kidney allograft in the maintenance of tolerance was shown by the fact that when the donor kidneys were removed, the heart allografts were rejected soon thereafter. Similar to the heart allograft, islet allograft tolerance was not achieved by induction of transient mixed chimerism. However, an islet allograft was accepted without immunosuppression when islets from the original BM and kidney donor were transplanted under the renal capsule of the allograft recipient that had previously been rendered tolerant by DBMT (Kawai et al. 2001). These results suggest that the presence of the renal allograft is a prerequisite for the maintenance of tolerance of cotransplanted tissues and organs after the loss of mixed chimerism. Our ultimate goal is to incorporate this yet-to-be-defined "kidney factor" into a novel and clinically relevant tolerance protocol that could induce unresponsiveness to other organ allografts without the requirement for a kidney transplant.

CLINICAL TOLERANCE INDUCTION THROUGH MIXED CHIMERISM

Although numerous tolerance induction strategies have been reported in rodent models and several have been reported in large animals, clinical renal allograft tolerance has been achieved to date only by induction of mixed or full hematopoietic chimerism. We review here three clinical studies in which such tolerance has been reported.

The Stanford group has developed a total lymphoid irradiation (TLI)–based conditioning regimen to induce renal allograft tolerance with either mixed or full chimerism in HLA-identical kidney transplantation (Millan 2005; Scandling et al. 2008, 2012). This protocol was based on decades of research in rodents, large animals, and clinical studies, attempting to improve the outcomes of HCT for the treatment of

 Cite this article as *Cold Spring Harb Perspect Med* doi: 10.1101/cshperspect.a015529

hematologic malignancies (Slavin et al. 1979; Strober et al. 1989; Millan et al. 2002; Lowsky et al. 2005; Kohrt et al. 2009). Fifteen of 16 patients receiving the Stanford regimen developed persistent mixed chimerism, and eight patients with chimerism lasting >6 mo were reported to have been completely withdrawn from immunosuppressive medications, with the longest allograft survival exceeding 3 yr. Four patients continued to receive immunosuppressive drugs because of recurrence of focal segmental glomerulosclerosis in one and rejection episodes during the tapering of cyclosporine in the others. The Stanford group also attempted induction of renal allograft tolerance with a similar regimen in HLA-mismatched kidney transplantation (Millan et al. 2002). However, all recipients failed to achieve tolerance, and it was concluded that this protocol, in its current form, did not induce allograft tolerance in HLA-mismatched recipients.

At Massachusetts General Hospital (MGH), a total of seven patients with renal failure secondary to refractory multiple myeloma have received HLA-identical combined kidney and bone marrow transplantation (CKBMT). The first six of these patients were treated on a study funded by the NIAID's Immune Tolerance Network (ITN). These patients received cyclophosphamide (CYP)-based nonmyeloablative conditioning that was based on a mouse model for separating GVHD and GVL (see above). The mouse protocol involved CYP and recipient T-cell depletion plus thymic irradiation in a nonmyeloablative regimen for induction of mixed chimerism, which was followed by delayed DLI (Pelot et al. 1999; Mapara et al. 2002, 2003). Clinical trials in patients with hematologic malignancies showed sufficient safety and anti-tumor efficacy to allow use of the same protocol for renal allograft tolerance induction in multiple myeloma patients with HLA-identical donors (Spitzer et al. 1999, 2011; Buhler et al. 2002; Fudaba et al. 2006). Either transient or durable (mixed or full) chimerism was achieved, and five of seven patients remain alive at the time of this writing. Remission of the myeloma was observed even in some patients with transient chimerism, suggesting a GVL effect despite loss of chimerism, as was subsequently shown in a mouse model (see above) (Rubio et al. 2003, 2005, 2006; Saito et al. 2006). Immunosuppression was successfully withdrawn in all three recipients who had transient chimerism and in one recipient with stable mixed chimerism, with the longest kidney allograft survival now exceeding 14 yr. One patient with transient chimerism subsequently received a second stem cell transplantation from the same donor because of progressive myeloma. She developed stable full chimerism, and, like all other patients with full donor chimerism, she remained thereafter on immunosuppression for GVHD prophylaxis. These observations show that CKBMT with a nonmyeloablative regimen from an HLA-matched donor can be an excellent option for renal failure secondary to myeloma, where no effective treatment option has previously been available.

Shortly after it became clear that renal allograft tolerance could be achieved by this mixed chimerism approach in an HLA-matched setting, a study to extend this approach to HLA-mismatched renal transplantation was begun at MGH, also sponsored by the ITN (Kawai et al. 2008). Ten patients received HLA-mismatched CKBMT after conditioning with CYP-based nonmyeloablative conditioning regimens. After acute humoral rejection was observed in patient 3, the original regimen (NKD03) was modified by adding two doses of pretransplant rituximab (modified NKD03 regimen). Patients 6 to 10 received a further modified conditioning regimen, with four peritransplant doses of rituximab (ITN036). All 10 subjects developed transient chimerism, which became undetectable by Day 21, and no GVHD occurred, meeting a critical safety criterion in patients who had no malignant disease. Immunosuppression was successfully discontinued 8–14 mo posttransplant in seven of 10 patients. Patient 1 has remained well, without rejection for >10 yr. Patient 2 has also remained rejection free for >9 yr, although mycophenolate mofetil (MMF) was added after 7 yr for recurrence of membranoproliferative glomerulonephritis (MPGN), his original disease. Kidney allograft function has remained stable for >7 yr in patient 4, although MMF was instituted at 5 yr because of a histological diagnosis of chronic humoral rejection. Patient 5 remained rejection free for 6 yr despite the development of low levels of DSA after discontinuing immunosuppression (IS). His most recent protocol biopsy at 6.8 yr, however, shows minor transplant glomerulopathy (C4d negative), which could indicate the incipient onset of chronic rejection. ITN036-treated patients 6, 7, and 9 successfully discontinued IS and remain stable, without evidence of rejection or DSA for periods of 3–4 yr. Patient 8 returned to dialysis after losing kidney function at 6 mo posttransplantation because of thrombotic microangiopathy. IS was reinstituted in patient 10 when cellular rejection was diagnosed 2 mo after IS withdrawal. His renal function became compromised and he eventually underwent a second kidney transplantation at 3 yr. In contrast to the development of DSA observed in recipients of the modified NKD03 regimen, no DSA has been detected in recipients of the ITN036 regimen, which suggests that prolonged depletion of $CD20^+$ B cells adequately controls DSA. "Engraftment syndrome," which causes transient renal dysfunction (Spitzer 2001), was observed in nine of these 10 patients and remains a problem to be solved. Several protocol modifications are planned to overcome this complication, which has not been observed in nonhuman primates treated with a similar protocol (Kawai et al. 1995, 2004). Possible mechanisms of this phenomenon are discussed below.

A third clinical approach has recently been reported by Leventhal et al. (2012) at Northwestern University. These investigators used a conditioning strategy consisting of total body irradiation (200 Gy), fludarabine, and high-dose cyclophosphamide, both pre- and post-BMT, together with administration of donor hematopoietic stem cells for treatment of eight HLA-mismatched kidney transplant recipients. The conditioning regimen was more intensive than that used in the MGH protocol, and full donor chimerism was achieved in several of these patients, reportedly without GVHD. These results suggest either that the preparative regimen was myeloablative or that donor T cells in the graft destroyed recipient hematopoiesis via a GVH reaction. Although their protocol has not been fully disclosed, they claim that the administration of novel "tolerogenic $CD8^+$ /TCR-facilitating cells (FC)" (Williams 1988; Kaufman et al. 1994) prevented rejection. The criterion for weaning of all maintenance immunosuppression was stated as persistent donor chimerism without GVHD for 1 yr. This criterion has apparently been reached in eight of 15 patients in a follow-up study (Leventhal et al. 2013). The highly myelosuppressive nature of the conditioning and/or the potential GVHR associated with the transplant is highlighted by the prolonged neutropenia and thrombocytopenia in these patients, which necessitated support with G-CSF and platelet transfusions (Leventhal et al. 2013).

These results contrast markedly with those reported in patients with hematologic malignancies who received HCT from less extensively mismatched (haploidentical related) donors with a similar regimen and in whom a 34% incidence of acute GVHD as well as significant chronic GVHD was reported (Luznik et al. 2008; Kasamon et al. 2010). Thus, the Northwestern results, if validated with longer follow-up and in larger numbers of patients, are unexpected and remarkable. However, significant morbidity due to putative infections occurred in the study (Leventhal et al. 2012), and there are reasons to be concerned that these patients may not regain normal immune incompetence (Strober et al. 2011; Kean and Blazar 2012; Li and Sykes 2012). In this regard, it is noteworthy that the patients in the HLA-mismatched HCT trials at MGH have not suffered from any significant opportunistic infections.

MECHANISTIC STUDIES

As is discussed above, the durable chimerism achieved in mouse models involving depletion or tolerization with specific deletion of preexisting donor-reactive T cells, ultimately involves central deletion as a major mechanism maintaining long-term tolerance to the donor. In these cases, the tolerance achieved is long-lasting and independent of the presence of other donor organs or tissues. On the other hand, in both the NHP models and the clinical studies described above, chimerism has generally been transient. Transient chimerism achieved in the face of only partial depletion of the preexisting T-cell repertoire is unlikely to result in the pure central, deletional form of tolerance achieved in the murine models. As is discussed below, the kidney also seems to play an important role in the achievement of lasting tolerance in primates and humans.

There are several surprising aspects of the clinical outcomes in the myeloma patients that raise important mechanistic questions. Myeloma outcomes were very good overall, and several of the patients achieved prolonged, complete remissions of their myelomas despite achieving only transient chimerism (Fudaba et al. 2006; Spitzer et al. 2011). This, and observations in other patients achieving lymphoma remissions despite transient chimerism with similar regimens (Dey et al. 2005), raised the possibility that initial marrow engraftment followed by rejection, which was evidenced by sensitized anti-donor T-cell responses in some patients (Fudaba et al. 2006), could lead to antitumor responses. Studies in an animal model confirmed this hypothesis (Rubio et al. 2003, 2005, 2006; Saito et al. 2006). The other surprising aspect in these patients was the long-term acceptance of donor kidneys after disappearance of chimerism. In at least some patients, this loss of chimerism was clearly due to rejection, because sensitization to minor histocompatibility antigens (HAs) expressed on hematopoietic cells was evident in vitro (Fudaba et al. 2006). Thus, renal allograft tolerance might be specific for kidney and include a role for the kidney graft itself. Indeed, T cells from these patients were unresponsive to donor renal tubular epithelial cells (Fudaba et al. 2006), suggesting that tolerance is specific for minor antigens expressed on the kidney graft itself.

Even more evanescent (lasting only a few weeks) chimerism was also observed in all of the recipients (who did not have a malignant disease) of HLA-mismatched CKBMT with the anti-CD2-based regimen (Kawai et al. 2008; Locascio et al. 2010). Loss of chimerism was associated with robust recipient hematopoiesis, showing that the regimen was truly nonmyeloablative. Because the donors were HLA mismatched, most of these patients had clearly measurable anti-donor alloresponses in vitro before transplant. Remarkably, despite the loss of chimerism, all four of the tolerant patients in the first study showed the progressive development of complete donor-specific unresponsiveness in both mixed lymphocyte response (MLR) and cell-mediated lympholysis (CML) assays. In vitro unresponsiveness to cultured donor renal tubular epithelial cells also developed. Normal anti-third-party alloresponses recovered in all four patients (Kawai et al. 2008; Andreola et al. 2011), suggesting that the state of donor-specific tolerance was systemic. Limiting dilution analyses (LDAs) showed a long-term absence of donor-reactive IL-2-producing cells or CTL precursors, suggesting that eventual tolerance was deletional (Andreola et al. 2011). There is an apparent contradiction between these observations and those in recipients of HLA-identical combined kidney and bone marrow transplants, who sometimes showed sensitization to donor hematopoietic antigens in association with loss of chimerism (Fudaba et al. 2006). Although it might be thought that this indicates that the mechanisms of tolerance differ in the HLA-identical versus the mismatched setting, a more unifying hypothesis is that in both groups, tolerance is restricted to antigens expressed by the kidney. In the HLA-mismatched transplant setting, most of the many different donor HLA/peptide complexes eliciting strong pretransplant immune responses are likely expressed both on hematopoietic cells and the kidney. Thus, tolerance of the donor kidney would extend to hematopoietic cells, as observed in MLR and CML assays. In the HLA-identical CKBMT recipients, in contrast, complete tolerance to the small number of minor HAs expressed by the kidney could still leave a significant pool of T cells recognizing minor HAs expressed only on hematopoietic cells, resulting in rejection of the marrow and sensitized in vitro responses to donor alloantigens, as observed (Kraus et al. 2003; Fudaba et al. 2006), whereas the kidney graft is accepted. This "unifying" explanation implies tolerance to alloantigens of the kidney in both the HLA-identical and HLA-mismatched CKBMT groups. In the HLA-mismatched setting, the donor-specific unresponsiveness seen in vitro in CKBMT recipients contrasted strikingly with results in patients with hematological malignancies who received a similar haploidentical BMT regimen without a kidney transplant. The latter group tended to have stronger anti-donor than anti-third-party responses following the loss of chimerism (Shaffer et al. 2007), suggesting that the kidney played an active role in the systemic (as evidenced by the MLR, CML, and LDA results) tolerance achieved in the CKBMT recipients.

Indeed, the transient chimerism achieved in the HLA-mismatched CKBMT recipients makes it difficult to imagine that the long-term tolerance achieved is due to a pure central deletional mechanism. These observations have led us to consider alternative hypotheses. Intragraft levels of the Treg-associated transcription factor FoxP3 relative to Granzyme B mRNA were increased in tolerant patients compared with conventional transplant patients, raising the possibility that regulatory T cells might play a role in tolerance (Kawai et al. 2008). Tregs are enriched among the peripheral blood T cells initially present in recipients of this regimen for BMT alone (Shaffer et al. 2007) and in the CKBMT recipients (Fudaba et al. 2006; Andreola et al. 2011). Depletion of Tregs revealed anti-donor responses that were not apparent in unfractionated PBMCs in the first year posttransplant in some, but not all of the patients (Andreola et al. 2011). Although the inconsistency of this observation is unsatisfying, the uniform inability to reveal anti-donor responses with Treg-depleted PBMCs >1 yr posttransplant is consistent with a long-term deletional process induced by the "quiescent" kidney graft itself. Alternatively, the relevant Tregs (and possibly donor-reactive effector T cells) may all reside in the graft. We are currently addressing this possibility by expanding T cells from protocol biopsies from the tolerant CKBMT patients, and preliminary phenotypical data suggest the presence of both Tregs and effector T cells in the lines we have cultured (B Sprangers, T Morokata, and M Sykes, unpubl.). Studies are in progress to identify the function and specificity of these T cells.

In the second cohort of CKBMT patients, we have performed more detailed phenotypical and kinetic studies on the Tregs that are enriched in the $CD4^+$ PBMC early posttransplant and on the conventional T cells. Preliminary studies suggest that the early Treg enrichment may reflect a combination of lymphopenia-driven expansion and a possible early wave of thymic Treg emigration (B Sprangers, T Morokata, and M Sykes, unpubl.). Because peripheral APCs recirculate to the thymus and promote tolerance there by both deletional and Treg-mediated mechanisms (Donskoy and Goldschneider 2003; Bonasio et al. 2006; Proietto et al. 2008; Hadeiba et al. 2012), it is possible that this early wave of Treg emigration contributes to the tolerance observed.

Several patients in the first, but not the second, cohort of HLA-mismatched CKBMT patients developed alloantibodies to their donors. In light of the donor-specific unresponsiveness detected in vitro, the development of DSA may reflect dysregulated, T-cell-independent B-cell activation as suggested (Porcheray et al. 2009), or, alternatively, may reflect failure to tolerize the indirect T-cell alloresponse. Because the assays discussed above measure predominantly direct alloresponses, studies are in progress to specifically assess anti-donor indirect alloresponses in these patients.

There is less mechanistic data available for tolerance induced by durable chimerism. As mentioned above, a TLI/ATG-based HCT protocol, also supported by safety data obtained in patients with hematological malignancies (Lowsky et al. 2005), has recently been used to achieve renal allograft tolerance in eight patients receiving HLA-identical related donor combined hematopoietic and kidney transplantation (Scandling et al. 2008, 2012). The mechanisms of tolerance in these patients may involve Th2 cytokines secreted by NKT cells, which, along with Tregs, are enriched in these patients. In contrast to patients with transient chimerism, these chimeric recipients do not show sensitization and actually show reduced in vitro

 Cite this article as *Cold Spring Harb Perspect Med* doi: 10.1101/cshperspect.a015529

responses to donor minor HAs (Scandling et al. 2012). Unfortunately, a similar protocol was unsuccessful in the HLA-mismatched setting (Millan et al. 2002; Strober et al. 2004).

CONCLUSIONS

It has become clear over the past two decades that the usual definition of immunological tolerance as "the specific absence of immune response to an antigen in the absence of immunosuppression" requires modification in the field of transplantation, where it is now clear that tolerance can also result from the presence of a down-regulatory immune response to a transplantation antigen. We can therefore more comprehensively define transplantation tolerance as "the specific absence of a destructive immune response to a transplanted tissue or organ in the absence of immunosuppression." This definition incorporates both central, deletional tolerance mechanisms and peripheral, down-regulatory tolerance mechanisms. Interestingly, both sets of mechanisms appear to be relevant to the induction of tolerance through mixed chimerism, as summarized in this review.

We have attempted here to provide a brief overview of the current status of tolerance induction through mixed chimerism in small- and large-animal models as well as in the clinic. Although it is clear that this modality has been successful in inducing tolerance in both small and large animals, it is also clear that both the methodology for induction of chimerism and the mechanism by which tolerance is induced have been different in different species. Probably the most important element required for achieving mixed chimerism through the induction of purely central, deletional tolerance is complete depletion of mature, donor-reactive T cells as a consequence of the preparative regimen. Thus, even in mice, where antibody reagents capable of depleting all mature T cells peripherally are available, the addition of thymic irradiation was required to achieve long-lasting, deletional tolerance (Sharabi and Sachs 1989), because mechanisms for antibody-mediated depletion appear not to be effective in the thymus (Nikolic et al. 2001). Consistent with this hypothesis are the data in miniature swine, in which durable mixed chimerism was possible without thymic irradiation only when an immunotoxin was used for T-cell depletion, presumably allowing depletion of thymic T cells directly by the toxin rather than by the usual antibody-mediated mechanisms (Huang et al. 1999b, 2000). Models for achieving durable mixed chimerism without complete depletion of mature, donor-reactive T cells have generally relied on the addition of costimulatory blockade to the preparative regimen. Costimulatory blockade with BMT promotes anergy followed by deletion of pre-existing donor-reactive T cells, both in the periphery (Kurtz et al. 2004; Fehr et al. 2005) and the thymus (J Kurtz and M Sykes, unpubl.).

In the primate models we have discussed, complete depletion of mature T cells has not been achieved, and mixed chimerism has been transient (Kawai et al. 2011). Nevertheless, tolerance has been established, undoubtedly through the generation of regulatory T cells capable of down-regulating the alloaggressive response that might otherwise have resulted from activation of the residual T cells. To achieve tolerance of a donor kidney, it had to be transplanted while mixed chimerism was still detectable in the recipient animal, and continued presence of the donor kidney was found to be essential to maintenance of the Treg and to maintenance of tolerance. With the possible exception of recent studies from Northwestern (Leventhal et al. 2012), in which durable full chimerism across HLA barriers was reported, induction of tolerance of mismatched organ allografts in the clinic has, as in NHP models, involved transient mixed chimerism.

Because the ultimate goal of using mixed chimerism for tolerance induction is long-term survival of organ allografts without the need for chronic immunosuppression, it is not clear whether attaining this goal through central versus peripheral mechanisms is as important practically as it may seem from a theoretical viewpoint. Indeed, although tolerance due to the absence of alloreactive T cells achieved centrally might appear to have an advantage in being more complete and less susceptible to disruption by inflammatory stimuli, it could have disadvantages over peripheral tolerance because of the absence of mechanisms to down-regulate residual alloreactive T cells that, like T cells causing autoimmunity, may become activated and expand in the peripheral environment. In addition, Tregs are thought to be responsible for the phenomenon of linked suppression, which could permit down-regulation of responses to tissue-specific antigens present on cells of organ allografts but not on the hematopoietic stem cells used to produce mixed chimerism. In any case, it appears that mixed chimerism can be intimately associated with both forms of tolerance and that a combination of both may be the best way of achieving our long-term goals.

ACKNOWLEDGMENTS

We gratefully acknowledge Rebecca A. Wark for expert editorial assistance.

REFERENCES

Adams AB, Pearson TC, Larsen CP. 2003a. Heterologous immunity: An overlooked barrier to tolerance. *Immunol Rev* **196:** 147–160.

Adams AB, Williams MA, Jones TR, Shirasugi N, Durham MM, Kaech SM, Wherry EJ, Onami T, Lanier JG, Kokko KE, et al. 2003b. Heterologous immunity provides a potent barrier to transplantation tolerance. *J Clin Invest* **111:** 1887–1895.

Andreola G, Chittenden M, Shaffer J, Cosimi AB, Kawai T, Cotter P, Dey B, Tolkoff-Rubin N, Preffer F, Bonnefoix T, et al. 2011. Mechanisms of donor-specific tolerance in recipients of haploidentical combined bone marrow/kidney transplantation. *Am J Transplant* **11:** 1236–1247.

Aversa F, Tabilio A, Velardi A, Cunningham I, Terenzi A, Falzetti F, Ruggeri L, Barbabietoloa G, Aristei C, Latini P, et al. 1998. Treatment of high-risk acute leukemia with T cell-depleted stem cells from related donors with one fully mismatched haplotype. *New Engl J Med* **339:** 1186–1193.

Bemelman F, Honey K, Adams E, Cobbold S, Waldmann H. 1998. Bone marrow transplantation induces either clonal deletion or infectious tolerance depending on the dose. *J Immunol* **160:** 2645–2648.

Bigenzahn S, Blaha P, Koporc Z, Pree I, Selzer E, Bergmeister H, Wrba F, Heusser C, Wagner K, Muehlbacher F, et al. 2005. The role of non-deletional tolerance mechanisms in a murine model of mixed chimerism with costimulation blockade. *Am J Transplant* **5:** 1237–1247.

Bonasio R, Scimone ML, Schaerli P, Grabie N, Lichtman AH, von Andrian UH. 2006. Clonal deletion of thymocytes by circulating dendritic cells homing to the thymus. *Nat Immunol* **7:** 1092–1100.

Bordignon C, Keever CA, Small TN, Flomenberg N, Dupont B, O'Reilly RJ, Kernan NA. 1989. Graft failure after T-cell-depleted human leukocyte antigen identical marrow transplants for leukemia: II. In vitro analyses of host effector mechanisms. *Blood* **74:** 2237–2243.

Buhler LH, Spitzer TR, Sykes M, Sachs DH, Delmonico FL, Tolkoff-Rubin N, Saidman SL, Sackstein R, McAfee S, Dey B, et al. 2002. Induction of kidney allograft tolerance after transient lymphohematopoietic chimerism in patients with multiple myeloma and end-stage renal disease. *Transplantation* **74:** 1405–1409.

Chakraverty R, Cote D, Buchli J, Cotter P, Hsu R, Zhao G, Sachs T, Psitilides C, Bronson R, Means T, et al. 2006a. An inflammatory checkpoint regulates recruitment of graft-versus-host-reactive T cells to peripheral tissues. *J Exp Med* **203:** 2021–2031.

Chakraverty R, Eom HS, Sachs J, Buchli J, Cotter P, Hsu R, Zhao G, Sykes M. 2006b. Host MHC class II^+ antigen-presenting cells and CD4 cells are required for CD8-mediated graft-versus-leukemia responses following delayed donor leukocyte infusions. *Blood* **108:** 2106–2113.

Cina RA, Wikiel KJ, Lee PW, Cameron AM, Hettiarachy S, Rowland H, Goodrich J, Colby C, Spitzer TR, Neville DM Jr, et al. 2006. Stable multilineage chimerism without graft versus host disease following nonmyeloablative haploidentical hematopoietic cell transplantation. *Transplantation* **81:** 1677–1685.

Cobbold SP, Qin S, Waldmann H. 1990. Reprogramming the immune system for tolerance with monoclonal antibodies. *Sem Immunol* **2:** 377–387.

Cosimi AB, Sachs DH. 2004. Mixed chimerism and transplantation tolerance. *Transplantation* **77:** 943–946.

Davis MM, Bjorkman PJ. 1990. A model for T cell receptor and MHC/peptide interaction. *Adv Exp Med Biol* **254:** 13–16.

Dey BR, MCafee S, Colby C, Cieply K, Caron M, Saidman S, Preffer F, Shaffer J, Tarbell N, Sackstein R, et al. 2005. Anti-tumor response despite loss of donor chimerism in patients treated with nonmyeloablative conditioning and allogeneic stem cell transplantation. *Br J Haematol* **128:** 351–359.

Domenig C, Sanchez-Fueyo A, Kurtz J, Alexopoulos AP, Mariat C, Sykes M, Strom TB, Zheng XX. 2005. Roles of deletion and regulation in creating mixed chimerism and allograft tolerance using a nonlymphoablative irradiation-free protocol. *J Immunol* **175:** 51–60.

Donskoy E, Goldschneider I. 2003. Two developmentally distinct populations of dendritic cells inhabit the adult mouse thymus: Demonstration by differential importation of hematogenous precursors under steady state conditions. *J Immunol* **170:** 3514–3521.

Fehr T, Takeuchi Y, Kurtz J, Sykes M. 2005. Early regulation of CD8 T cell alloreactivity by $CD4^+CD25^-$ T cells in recipients of anti-CD154 antibody and allogeneic BMT is followed by rapid peripheral deletion of donor-reactive $CD8^+$ T cells, precluding a role for sustained regulation. *Eur J Immunol* **35:** 2679–2690.

Fehr T, Haspot F, Mollov J, Chittenden M, Hogan T, Sykes M. 2008. Alloreactive CD8 T cell tolerance requires recipient B cells, dendritic cells and MHC class II. *J Immunol* **181:** 165–173.

Fehr T, Lucas CL, Kurtz J, Onoe T, Zhao G, Hogan T, Vallot C, Rao A, Sykes M. 2010. A CD8 T cell-intrinsic role for the calcineurin-NFAT pathway for tolerance induction in vivo. *Blood* **115:** 1280–1287.

Fleischhauer K, Kernan NA, O'Reilly RJ, Dupont B, Yang SY. 1990. Bone marrow-allograft rejection by T lymphocytes recognizing a single amino acid difference in HLA-B44. *New Engl J Med* **323:** 1818–1822.

Fuchimoto Y, Yamada K, Shimizu A, Yasumoto A, Sawada T, Huang CH, Sachs DH. 1999. Relationship between chimerism and tolerance in a kidney transplantation model. *J Immunol* **162:** 5704–5711.

Fuchimoto Y, Huang CA, Yamada K, Shimizu A, Kitamura H, Colvin RB, Ferrara V, Murphy MC, Sykes M, White-Scharf M, et al. 2000. Mixed chimerism and tolerance without whole body irradiation in a large animal model. *J Clin Invest* **105:** 1779–1789.

Fudaba Y, Spitzer TR, Shaffer J, Kawai T, Fehr T, Delmonico F, Preffer F, Tolkoff-Rubin N, Dey BR, Saidman SL, et al. 2006. Myeloma responses and tolerance following combined kidney and nonmyeloablative marrow transplantation: In vivo and in vitro analyses. *Am J Transplant* **6:** 2121–2133.

Guzzetta PC, Sundt TM, Suzuki T, Mixon A, Rosengard BR, Sachs DH. 1991. Induction of kidney transplantation tolerance across MHC barriers by bone marrow transplantation in miniature swine. *Transplantation* **51:** 862–866.

Hadeiba H, Lahl K, Edalati A, Oderup C, Habtezion A, Pachynski R, Nguyen L, Ghodsi A, Adler S, Butcher EC. 2012. Plasmacytoid dendritic cells transport peripheral antigens to the thymus to promote central tolerance. *Immunity* **36:** 438–450.

Hanekamp JS, Duran-Struuck R, Sachs DH. 2011. Transplantation in miniature swine. In *The minipig in biomedical research* (ed. McAnulty PA), pp. 357–372. Taylor & Francis, London, 2011.

Haspot F, Fehr T, Gibbons C, Zhao G, Hogan T, Honjo T, Freeman GJ, Sykes M. 2008. Peripheral deletional tolerance of alloreactive CD8 but not CD4 cells is dependent on the PD-1/PD-L1 pathway. *Blood* **112:** 2149–2155.

Hayashi H, LeGuern C, Sachs DH, Sykes M. 1996. Alloresistance to K locus mismatched bone marrow engraftment is mediated entirely by $CD4^+$ and $CD8^+$ T cells. *Bone Marrow Transplant* **18:** 285–292.

Haynes BF, Markert ML, Sempowski GD, Patel DD, Hale LP. 2000. The role of the thymus in immune reconstitution in aging, bone marrow transplantation, and HIV-1 infection. *Annu Rev Immunol* **18:** 529–560.

Helg C, Chapuis B, Bolle JF, Morel P, Salomon D, Roux E, Antonioli V, Jeannet M, Leski M. 1994. Renal transplantation without immunosuppression in a host with tolerance induced by allogeneic bone marrow transplantation. *Transplantation* **58:** 1420–1422.

Horner BM, Cina RA, Wikiel KJ, Lima B, Ghazi A, Lo DP, Yamada K, Sachs DH, Huang CA. 2006. Predictors of organ allograft tolerance following hematopoietic cell transplantation. *Am J Transplant* **6:** 2894–2902.

Huang CA, Lorf T, Arn JS, Koo GC, Blake T, Sachs DH. 1999a. Characterization of a monoclonal anti-porcine CD3 antibody. *Xenotransplantation* **6:** 201–212.

Huang CA, Yamada K, Murphy MC, Shimizu A, Colvin RB, Neville DMJ, Sachs DH. 1999b. In vivo T cell depletion in miniature swine using the swine CD3 immunotoxin, pCD3-CRM9. *Transplantation* **68:** 855–860.

Huang CA, Fuchimoto Y, Scheier-Dolberg R, Murphy MC, Neville DMJ, Sachs DH. 2000. Stable mixed chimerism and tolerance using a nonmyeloablative preparative regimen in a large-animal model. *J Clin Invest* **105:** 173–181.

Ildstad ST, Sachs DH. 1984. Reconstitution with syngeneic plus allogeneic or xenogeneic bone marrow leads to specific acceptance of allografts or xenografts. *Nature* **307:** 168–170.

Ildstad ST, Wren SM, Bluestone JA, Barbieri SA, Sachs DH. 1985. Characterization of mixed allogeneic chimeras. Immunocompetence, in vitro reactivity, and genetic specificity of tolerance. *J Exp Med* **162:** 231–244.

Ildstad ST, Wren SM, Bluestone JA, Barbieri SA, Stephany D, Sachs DH. 1986. Effect of selective T cell depletion of host and/or donor bone marrow on lymphopoietic repopulation, tolerance, and graft-vs-host disease in mixed allogeneic chimeras (B10 + B10.D2 → B10). *J Immunol* **136:** 28–33.

Ito H, Kurtz J, Shaffer J, Sykes M. 2001. CD4 T cell-mediated alloresistance to fully MHC-mismatched allogeneic bone marrow engraftment is dependent on CD40–CD40 L interactions, and lasting T cell tolerance is induced by bone marrow transplantation with initial blockade of this pathway. *J Immunol* **166:** 2970–2981.

Jacobsen N, Taaning E, Ladefoged J, Kristensen JK, Pedersen FK. 1994. Tolerance to an HLA-B,DR disparate kidney allograft after bone-marrow transplantation from same donor. *Lancet* **343:** 800.

Kasamon YL, Luznik L, Leffell MS, Kowalski J, Tsai HL, Bolanos-Meade J, Morris LE, Crilley PA, O'Donnell PV, Rossiter N, et al. 2010. Nonmyeloablative HLA-haploidentical bone marrow transplantation with high-dose posttransplantation cyclophosphamide: Effect of HLA disparity on outcome. *Biol Blood Marrow Transplant* **16:** 482–489.

Kaufman CL, Colson YL, Wren SM, Watkins S, Simmons RL, Ildstad ST. 1994. Phenotypic characterization of a novel bone marrow-derived cell that facilitates engraftment of allogeneic bone marrow stem cells. *Blood* **84:** 2436–2446.

Kawai T, Cosimi AB, Colvin RB, Powelson J, Eason J, Kozlowski T, Sykes M, Monroy R, Tanaka M, Sachs DH. 1995. Mixed allogeneic chimerism and renal allograft tolerance in cynomolgus monkeys. *Transplantation* **59:** 256–262.

Kawai T, Poncelet A, Sachs DH, Mauiyyedi S, Boskovic S, Wee SL, Ko DS, Bartholomew A, Kimikawa M, Hong HZ, et al. 1999. Long-term outcome and alloantibody production in a non-myeloablative regimen for induction of renal allograft tolerance. *Transplantation* **68:** 1767–1775.

Kawai T, Andrews D, Colvin RB, Sachs DH, Cosimi AB. 2000. Thromboembolic complications after treatment with monoclonal antibody against CD40 ligand [letter]. *Nat Med* **6:** 114.

Kawai T, Sogawa H, Koulmanda M, Smith RN, O'Neil JJ, Wee SL, Boskovic S, Sykes M, Colvin RB, Sachs DH, et al. 2001. Long-term islet allograft function in the absence of chronic immunosuppression: A case report of a nonhuman primate previously made tolerant to a renal allograft from the same donor. *Transplantation* **72:** 351–354.

Kawai T, Cosimi AB, Wee SL, Houser S, Andrews D, Sogawa H, Phelan J, Boskovic S, Nadazdin O, Abrahamian G, et al. 2002. Effect of mixed hematopoietic chimerism on cardiac allograft survival in cynomolgus monkeys. *Transplantation* **73:** 1757–1764.

Kawai T, Sogawa H, Boskovic S, Abrahamian G, Smith RN, Wee SL, Andrews D, Nadazdin O, Koyama I, Sykes M, et al. 2004. CD154 blockade for induction of mixed chimerism and prolonged renal allograft survival in nonhuman primates. *Am J Transplant* **4:** 1391–1398.

Kawai T, Cosimi AB, Spitzer TR, Tolkoff-Rubin N, Suthanthiran M, Saidman SL, Shaffer J, Preffer FI, Ding R, Sharma V, et al. 2008. HLA-mismatched renal transplantation without maintenance immunosuppression. *New Engl J Med* **358:** 353–361.

Kawai T, Cosimi AB, Sachs DH. 2011. Preclinical and clinical studies on the induction of renal allograft tolerance through transient mixed chimerism. *Curr Opin Organ Transplant* **16:** 366–371.

Kean LS, Blazar BR. 2012. Cooking up tolerance: Has a new recipe been created? *Am J Transplant* **12:** 1667–1669.

Kernan NA, Flomenberg N, Dupont B, O'Reilly RJ. 1987. Graft rejection in recipients of T-cell-depleted HLA-nonidentical marrow transplants for leukemia. *Transplantation* **43:** 842–847.

Khan A, Tomita Y, Sykes M. 1996. Thymic dependence of loss of tolerance in mixed allogeneic bone marrow chimeras after depletion of donor antigen. Peripheral mechanisms do not contribute to maintenance of tolerance. *Transplantation* **62:** 380–387.

Kimikawa M, Sachs DH, Colvin RB, Bartholomew A, Kawai T, Cosimi AB. 1997. Modifications of the conditioning regimen for achieving mixed chimerism and donor-specific tolerance in cynomolgus monkeys. *Transplantation* **64:** 709–716.

Kohrt HE, Turnbull BB, Heydari K, Shizuru JA, Laport GG, Miklos DB, Johnston LJ, Arai S, Weng WK, Hoppe RT, et al. 2009. TLI and ATG conditioning with low risk of graft-versus-host disease retains antitumor reactions after allogeneic hematopoietic cell transplantation from related and unrelated donors. *Blood* **114:** 1099–1109.

Koyama I, Nadazdin O, Boskovic S, Ochiai T, Smith RN, Sykes M, Sogawa H, Murakami T, Strom TB, Colvin RB, et al. 2007. Depletion of CD8 memory T cells for induction of tolerance of a previously transplanted kidney allograft. *Am J Transplant* **7:** 1055–1061.

Kraus AB, Shaffer J, Toh HC, Preffer F, Dombkowski D, Saidman S, Colby C, George R, MCafee S, Sackstein R, et al. 2003. Early host CD8 T-cell recovery and sensitized anti-donor IL-2-producing and cytolytic T-cell responses associated with marrow graft rejection following nonmyeloablative bone marrow transplantation. *Exp Hematol* **31:** 609–621.

Kurtz J, Ito H, Wekerle T, Shaffer J, Sykes M. 2001. Mechanisms involved in the establishment of tolerance through costimulatory blockade and BMT: Lack of requirement for CD40 L-mediated signaling for tolerance or deletion of donor-reactive $CD4^+$ cells. *Am J Transplant* **1:** 339–349.

Kurtz J, Shaffer J, Anosova N, Benichou G, Sykes M. 2004. Mechanisms of early peripheral CD4 T cell tolerance induction by anti-CD154 monoclonal antibody and allogeneic bone marrow transplantation: Evidence for anergy and deletion, but not regulatory cells. *Blood* **103:** 4336–4343.

Larsen CP, Page A, Linzie KH, Russell M, Deane T, Stempora L, Strobert E, Penedo MC, Ward T, Wiseman R, et al. 2010. An MHC-defined primate model reveals significant rejection of bone marrow after mixed chimerism induction despite full MHC matching. *Am J Transplant* **10:** 2396–2409.

Leventhal J, Abecassis M, Miller J, Gallon L, Ravindra K, Tollerud DJ, King B, Elliott MJ, Herzig G, Herzig R, et al. 2012. Chimerism and tolerance without GVHD or engraftment syndrome in HLA-mismatched combined kidney and hematopoietic stem cell transplantation. *Sci Transl Med* **4:** 124ra28.

Leventhal J, Abecassis M, Miller J, Gallon L, Tollerud D, Elliott MJ, Bozulic LD, Houston C, Sustento-Reodica N, Ildstad ST. 2013. Tolerance induction in HLA disparate living donor kidney transplantation by donor stem cell infusion: Durable chimerism predicts outcome. *Transplantation* **95:** 169–176.

Li HW, Sykes M. 2012. Emerging concepts in haematopoietic cell transplantation. *Nat Rev Immunol* **12:** 403–416.

Li HW, Sachs J, Pichardo C, Bronson R, Zhao G, Sykes M. 2012. Nonalloreactive T cells prevent donor lymphocyte infusion-induced graft-versus-host disease by controlling microbial stimuli. *J Immunol* **189:** 5572–5581.

Locascio SA, Morokata T, Chittenden M, Preffer FI, Dombkowski DM, Andreola G, Crisalli K, Kawai T, Saidman SL, Spitzer TR, et al. 2010. Mixed chimerism, lymphocyte recovery, and evidence for early donor-specific unresponsiveness in patients receiving combined kidney and bone marrow transplantation to induce tolerance. *Transplantation* **90:** 1607–1615.

Lowsky R, Takahashi T, Liu YP, Dejbakhsh-Jones S, Grumet FC, Shizuru JA, Laport GG, Stockerl-Goldstein KE, Johnston LJ, Hoppe RT, et al. 2005. Protective conditioning for acute graft-versus-host disease. *N Engl J Med* **353:** 1321–1331.

Lucas CL, Workman CJ, Beyaz S, Locascio S, Zhao G, Vignali DA, Sykes M. 2011. LAG-3, TGF-β, and cell-intrinsic PD-1 inhibitory pathways contribute to CD8 but not CD4 T-cell tolerance induced by allogeneic BMT with anti-CD40 L. *Blood* **117:** 5532–5540.

Lunney JK, Pescovitz MD. 1987. Phenotypic and functional characterization of pig lymphocyte populations. *Vet Immunol Immunopathol* **17:** 135–144.

Lunney JK, Sachs DH. 1995. The swine leukocyte antigen (SLA) complex. In *Transplantation immunology* (ed. Bach FH, Auchincloss Jr H), pp. 339–345. Wiley, New York.

Luznik L, O'Donnell PV, Symons HJ, Chen AR, Leffell MS, Zahurak M, Gooley TA, Piantadosi S, Kaup M, Ambinder RF, et al. 2008. HLA-haploidentical bone marrow transplantation for hematologic malignancies using nonmyeloablative conditioning and high-dose, posttransplantation cyclophosphamide. *Biol Blood Marrow Transplant* **14:** 641–650.

Manilay JO, Pearson DA, Sergio JJ, Swenson KG, Sykes M. 1998. Intrathymic deletion of alloreactive T cells in mixed bone marrow chimeras prepared with a nonmyeloablative conditioning regimen. *Transplantation* **66:** 96–102.

Mapara MY, Kim YM, Wang SP, Bronson R, Sachs DH, Sykes M. 2002. Donor lymphocyte infusions mediate superior graft-versus-leukemia effects in mixed compared to fully allogeneic chimeras: A critical role for host antigen-presenting cells. *Blood* **100:** 1903–1909.

Mapara MY, Kim YM, Marx J, Sykes M. 2003. DLI-mediated GVL effects in mixed chimeras established with a nonmyeloablative conditioning regimen: Extinction of GVL effects coincides with loss of alloreactive cells following conversion to full donor chimerism. *Transplantation* **76:** 297–305.

Martin PJ, Hansen JA, Buckner CD, Sanders JE, Deeg HJ, Stewart P, Appelbaum FR, Clift R, Fefer A, Witherspoon RP, et al. 1985. Effects of in vitro depletion of T cells in HLA-identical allogeneic marrow grafts. *Blood* **66:** 664–672.

Martin PJ, Hansen JA, Torok-Storb B, Durnam D, Przepiorka D, O'Quigley J, Sanders J, Sullivan KM, Witherspoon RP, Deeg HJ, et al. 1988. Graft failure in patients receiving T cell-depleted HLA-identical allogeneic marrow transplants. *Bone Marrow Transplant* **3:** 445–456.

Mezrich J, Yamada K, Sachs DH, Madsen JC. 2004. Regulatory T cells generated by the kidney may mediate the beneficial immune effects of combining kidney with heart transplantation. *Surgery* **135:** 473–478.

Millan MT. 2005. Studies of tolerance and chimerism after combined blood stem cell and kidney transplantation in humans. *Am J Transplant* **5**(Suppl): 544.

Millan MT, Shizuru JA, Hoffmann P, Dejbakhsh-Jones S, Scandling JD, Grumet FC, Tan JC, Salvatierra O, Hoppe RT, Strober S. 2002. Mixed chimerism and immunosuppressive drug withdrawal after HLA-mismatched kidney and hematopoietic progenitor transplantation. *Transplantation* **73:** 1386–1391.

Mollov JL, Lucas CL, Haspot F, Kurtz J, Gaspar C, Guzman A, Sykes M. 2010. Recipient dendritic cells, but not B cells, are required antigen-presenting cells for peripheral alloreactive $CD8^+$ T cell tolerance. *Am J Transplant* **10:** 518–526.

Nadazdin O, Boskovic S, Murakami T, O'Connor DH, Wiseman RW, Karl JA, Tuscher JJ, Sachs DH, Madsen JC, Tocco G, et al. 2010. Phenotype, distribution and alloreactive properties of memory T cells from cynomolgus monkeys. *Am J Transplant* **10:** 1375–1384.

Nash K, Chang Q, Watts A, Treter S, Oravec G, Ferrara V, Buhler L, Basker M, Gojo S, Sachs DH, et al. 1999. Peripheral blood progenitor cell mobilization and leukapheresis in pigs. *Lab Anim Sci* **49:** 645–649.

Neville DMJ, Scharff J, Hu HZ, Rigaut K, Shiloach J, Slingerland W, Jonker M. 1996. A new reagent for the induction of T-cell depletion, anti-CD3-CRM9. *J Immunother Emphasis Tumor Immunol* **19:** 85–92.

Nikolic B, Khan A, Sykes M. 2001. Induction of tolerance by mixed chimerism with nonmyeloblative host conditioning: The importance of overcoming intrathymic alloresistance. *Biol Blood Marrow Transplant* **7:** 144–153.

Pelot MR, Pearson DA, Swenson K, Zhao G, Sachs J, Yang YG, Sykes M. 1999. Lymphohematopoietic graft-vs-host reactions can be induced without graft-vs-host disease in murine mixed chimeras established with a cyclophosphamide-based non-myeloablative conditioning regimen. *Biol Blood Marrow Transplant* **5:** 133–143.

Pennington LR, Lunney JK, Sachs DH. 1981. Transplantation in miniature swine. VIII. Recombination within the major histocompatibility complex of miniature swine. *Transplantation* **31:** 66–71.

Pennington LR, Sakamoto K, Popitz-Bergez FA, Pescovitz MD, McDonough MA, MacVittie TJ, Gress RE, Sachs DH. 1988. Bone marrow transplantation in miniature swine. I: Development of the model. *Transplantation* **45:** 21–26.

Pescovitz MD, Sachs DH, Lunney JK, Hsu SM. 1984a. Localization of class II MHC antigens on porcine renal vascular endothelium. *Transplantation* **37:** 627–630.

Pescovitz MD, Thistlethwaite JR Jr, Auchincloss H Jr, Ildstad ST, Sharp TG, Terrill R, Sachs DH. 1984b. Effect of class II antigen matching on renal allograft survival in miniature swine. *J Exp Med* **160:** 1495–1508.

Pescovitz MD, Hsu SM, Katz SI, Lunney JK, Shimada S, Sachs Dh. 1990. Characterization of a porcine CD1-specific mAb that distinguishes CD4/CD8 double-positive thymic from peripheral T lymphocytes. *Tissue Antigens* **35:** 151–156.

Pescovitz MD, Book BK, Aasted B, Dominguez J, Ezguerra A, Trebichavsky I, Novikov B, Valpotic I, Neilson J, Arn S, et al. 1998. Analyses of monoclonal antibodies reacting with porcine CD3: Results from the Second International Swine CD Workshop. *Vet Immunol Immunopathol* **60:** 261–268.

Popitz-Bergez FA, Sakamoto K, Pennington LR, Pescovitz MD, McDonough MA, MacVittie TJ, Gress RE, Sachs DH. 1988. Bone marrow transplantation in miniature swine. II. Effect of selective genetic differences on marrow engraftment and recipient survival. *Transplantation* **45:** 27–31.

Porcheray F, Wong W, Saidman SL, De Vito J, Girouard TC, Chittenden M, Shaffer J, Tolkoff-Rubin N, Dey BR, Spitzer TR, et al. 2009. B-cell immunity in the context of T-cell tolerance after combined kidney and bone marrow transplantation in humans. *Am J Transplant* **9:** 2126–2135.

Proietto AI, van Dommelen S, Zhou P, Rizzitelli A, D'Amico A, Steptoe RJ, Naik SH, Lahoud MH, Liu Y, Zheng P, et al. 2008. Dendritic cells in the thymus contribute to T-regulatory cell induction. *Proc Natl Acad Sci* **105:** 19869–19874.

Rubio MT, Kim YM, Sachs T, Mapara M, Zhao G, Sykes M. 2003. Anti-tumor effect of donor marrow graft rejection induced by recipient leukocyte infusions in mixed chimeras prepared with nonmyeloablative conditioning: Critical role for recipient-derived IFN-γ. *Blood* **102:** 2300–2307.

Rubio MT, Saito TI, Kattelman K, Zhao G, Buchli J, Sykes M. 2005. Mechanisms of the anti-tumor responses and host-versus graft reactions induced by recipient leukocyte infusions in mixed chimeras prepared with nonmyeloablative conditioning: A critical role for recipient $CD4^+$ T cells and recipient leukocyte infusion-derived IFN-γ-producing $CD8^+$ T cells. *J Immunol* **175:** 665–676.

Rubio MT, Zhao G, Buchli J, Chittenden M, Sykes M. 2006. Role of indirect allo- and autoreactivity in anti-tumor responses induced by recipient leukocyte infusions (RLI) in mixed chimeras prepared with nonmyeloablative conditioning. *Clin Immunol* **120:** 33–44.

Sachs DH. 1992. MHC homozygous miniature swine. In *Swine as models in biomedical research* (ed Swindle MM, et al.), pp. 3–15. Iowa State University Press, Ames, IA.

Sachs DH. 1994. The pig as a potential xenograft donor. *Vet Immunol Immunopathol* **43:** 185–191.

Sachs DH. 2003. Tolerance: Of mice and men. *J Clin Invest* **111:** 1819–1821.

Saito TI, Rubio MT, Sykes M. 2006. Clinical relevance of recipient leukocyte infusion as antitumor therapy following nonmyeloablative allogeneic hematopoietic cell transplantation. *Exp Hematol* **34:** 1271–1277.

Sakamoto K, Pennington LR, Popitz-Bergez FA, Pescovitz MD, Gress RE, McDonough MA, Shimada S, Katz SI, Sachs DH. 1987. Swine GVHD model and the effect of T cell depletion of marrow by monoclonal antibodies. *Prog Bone Marrow Transplant* 449–453.

Sakamoto K, Sachs DH, Shimada S, Popitz-Bergez FA, Pennington LR, Pescovitz MD, McDonough MA, MacVittie TJ, Katz SI, Gress RE. 1988. Bone marrow transplantation in miniature swine. III. Graft-versus-host disease and the effect of T cell depletion of marrow. *Transplantation* **45:** 869–875.

Sayegh MH, Fine NA, Smith JL, Rennke HG, Milford EL, Tilney NL. 1991. Immunologic tolerance to renal allografts after bone marrow transplants from the same donors. *Ann Intern Med* **114:** 954–955.

Scandling JD, Busque S, Dejbakhsh-Jones S, Benike C, Millan MT, Shizuru JA, Hoppe RT, Lowsky R, Engleman EG, Strober S. 2008. Tolerance and chimerism after renal and hematopoietic-cell transplantation. *N Engl J Med* **358:** 362–368.

Scandling JD, Busque S, Dejbakhsh-Jones S, Benike C, Sarwal M, Millan MT, Shizuru JA, Lowsky R, Engleman EG, Strober S. 2012. Tolerance and withdrawal of immunosuppressive drugs in patients given kidney and hematopoietic cell transplants. *Am J Transplant* **12:** 1133–1145.

Schmitz JE, Simon MA, Kuroda MJ, Lifton MA, Ollert MW, Vogel CW, Racz P, Tenner-Racz K, Scallon BJ, Dalesandro M, et al. 1999. A nonhuman primate model for the selective elimination of $CD8^+$ lymphocytes using a mouse-human chimeric monoclonal antibody. *Am J Pathol* **154:** 1923–1932.

Seidl C, Saraiya C, Osterweil Z, Fu YP, Lee JS. 1992. Genetic complexity of regulatory mutants defective for HLA class II gene expression. *J Immunol* **148:** 1576–1584.

Shaffer J, Villard J, Means TK, Alexander S, Dombkowski D, Dey BR, McAffee S, Ballen KK, Saidman S, Preffer FI, et al. 2007. Regulatory T-cell recovery in recipients of haploidentical nonmyeloablative hematopoietic cell transplantation with a humanized anti-CD2 mAb, MEDI-507, with or without fludarabine. *Exp Hematol* **35:** 1140–1152.

Sharabi Y, Sachs DH. 1989. Mixed chimerism and permanent specific transplantation tolerance induced by a nonlethal preparative regimen. *J Exp Med* **169:** 493–502.

Sharabi Y, Sachs DH, Sykes M. 1992. T cell subsets resisting induction of mixed chimerism across various histocompatibility barriers. In *Progress in immunology. VIII. Proceedings of the Eighth International Congress of Immunology, Budapest* (ed. Gergely J, et al.), pp. 801–805. Springer-Verlag, Heidelberg.

Slavin S, Strober S, Fuks Z, Kaplan HS. 1977. Induction of specific tissue transplantation tolerance using fractionated total lymphoid irradiation in adult mice: Long-term survival of allogeneic bone marrow and skin grafts. *J Exp Med* **146:** 34.

Slavin S, Gottlieb M, Strober S, Bieber C, Hoppe R, Kaplan HS, Grumet FC. 1979. Transplantation of bone marrow in outbred dogs without graft-versus-host disease using total lymphoid irradiation. *Transplantation* **27:** 139–142.

Smith CV, Suzuki T, Guzzetta PC, Nakajima K, Sundt TM, Mixon A, Spitzer TR, Eckhaus M, Sachs DH. 1993. Bone marrow transplantation in miniature swine: IV. Development of myeloablative regimens that allow engraftment across major histocompatibility barriers. *Transplantation* **56:** 541–549.

Spitzer TR. 2001. Engraftment syndrome following hematopoietic stem cell transplantation. *Bone Marrow Transplant* **27:** 893–898.

Spitzer TR, Delmonico F, Tolkoff-Rubin N, McAfee S, Sackstein R, Saidman S, Colby C, Sykes M, Sachs DH, Cosimi AB. 1999. Combined histocompatibility leukocyte antigen-matched donor bone marrow and renal transplantation for multiple myeloma with end stage renal disease: The induction of allograft tolerance through mixed lymphohematopoietic chimerism. *Transplantation* **68:** 480–484.

Spitzer TR, MCafee S, Dey BR, Colby C, Hope J, Grossberg H, Preffer F, Shaffer J, Alexander SI, Sachs DH, et al. 2003. Non-myeloablative haploidentical stem cell transplantation using anti-CD2 monoclonal antibody (MEDI-507)–based conditioning for refractory hematologic malignancies. *Transplantation* **75:** 1748–1751.

Spitzer TR, Sykes M, Tolkoff-Rubin N, Kawai T, McAfee SL, Dey BR, Ballen K, Delmonico F, Saidman S, Sachs DH, et al. 2011. Long-term follow-up of recipients of combined human leukocyte antigen-matched bone marrow and kidney transplantation for multiple myeloma with end-stage renal disease. *Transplantation* **91:** 672–676.

Strober S, Dhillon M, Schubert M, Holm B, Engleman E, Benike C, Hoppe R, Sibley R, Myburgh JA, Collins G, et al. 1989. Acquired immune tolerance to cadaveric renal allografts: A study of three patients treated with total lymphoid irradiation. *N Engl J Med* **321:** 28–33.

Strober S, Lowsky RJ, Shizuru JA, Scandling JD, Millan MT. 2004. Approaches to transplantation tolerance in humans. *Transplantation* **77:** 932–936.

Strober S, Spitzer TR, Lowsky R, Sykes M. 2011. Translational studies in hematopoietic cell transplantation: Treatment of hematologic malignancies as a stepping stone to tolerance induction. *Semin Immunol* **23:** 273–281.

Sundt TM III, Suzuki T, Kortz O, Eckhaus ME, Gress RE, Sachs DH. 1988. Induction of specific transplant tolerance in a large animal model by bone marrow transplantation. *Surgical Forum* **39:** 365–367.

Sykes M, Sachs DH. 1988. Mixed allogeneic chimerism as an approach to transplantation tolerance. *Immunol Today* **9:** 23–27.

Sykes M, Sachs DH. 2001. Mixed chimerism. *Philos Trans R Soc Lond B Biol Sci* **356:** 707–726.

Sykes M, Sheard MA, Sachs DH. 1988. Graft-versus-host-related immunosuppression is induced in mixed chimeras by alloresponses against either host or donor lymphohematopoietic cells. *J Exp Med* **168:** 2391–2396.

Takeuchi Y, Ito H, Kurtz J, Wekerle T, Ho L, Sykes M. 2004. Earlier low-dose TBI or DST overcomes $CD8^+$ T-cell-mediated alloresistance to allogeneic marrow in recipients of anti-CD40 L. *Am J Transplant* **4:** 31–40.

Tomita Y, Khan A, Sykes M. 1994. Role of intrathymic clonal deletion and peripheral anergy in transplantation tolerance induced by bone marrow transplantion in mice conditioned with a non-myeloablative regimen. *J Immunol* **153:** 1087–1098.

Tomita Y, Khan A, Sykes M. 1996a. Mechanism by which additional monoclonal antibody injections overcome the requirement for thymic irradiation to achieve mixed chimerism in mice receiving bone marrow transplantation after conditioning with anti-T cell mAbs and 3 Gy whole body irradiation. *Transplantation* **61:** 477–485.

Tomita Y, Sachs DH, Khan A, Sykes M. 1996b. Additional mAb injections can replace thymic irradiation to allow induction of mixed chimerism and tolerance in mice receiving bone marrow transplantation after conditioning with anti-T cell mAbs and 3 Gy whole body irradiation. *Transplantation* **61:** 469–477.

Tonsho M. 2012. Induction of tolerance to heart allografts by donor kidney co-transplantation in nonhuman primates (NHPs). *American Transplant Congress 2012 Conference Proceedings*, Boston, p. 157.

Vallera DA, Taylor PA, Sprent J, Blazar BR. 1994. The role of host T cell subsets in bone marrow rejection directed to isolated major histocompatibility complex class I versus class II differences of *bm1* and *bm12* mutant mice. *Transplantation* **57:** 249–256.

Valujskikh A, Fedoseyeva E, Benichou G, Heeger PS. 2002. Development of autoimmunity after skin graft rejection via an indirect alloresponse. *Transplantation* **73:** 1130–1137.

Weiden PL, Sullivan KM, Flournoy N, Storb R, Thomas ED. 1981. Antileukemic effect of chronic graft-versus-host disease. *New Engl J Med* **304:** 1529–1533.

Wekerle T, Sayegh MH, Hill J, Zhao Y, Chandraker A, Swenson KG, Zhao G, Sykes M. 1998. Extrathymic T cell deletion and allogeneic stem cell engraftment induced with costimulatory blockade is followed by central T cell tolerance. *J Exp Med* **187:** 2037–2044.

Wekerle T, Kurtz J, Ito H, Ronquillo JV, Dong V, Zhao G, Shaffer J, Sayegh MH, Sykes M. 2000. Allogeneic bone marrow translantation with costimulatory blockade induces macrochimerism and tolerance without cytoreductive host treatment. *Nat Med* **6:** 464–469.

Williams RC Jr. 1988. Hypothesis: Rheumatoid factors are antiidiotypes related to bacterial or viral Fc receptors. *Arthritis Rheum* **31:** 1204–1207.

Wu Z, Bensinger SJ, Zhang J, Chen C, Yuan X, Huang X, Markmann JF, Kassaee A, Rosengard BR, Hancock WW, et al. 2004. Homeostatic proliferation is a barrier to transplantation tolerance. *Nat Med* **10:** 87–92.

Yamada Y, Boskovic S, Aoyama A, Murakami T, Putheti P, Smith RN, Ochiai T, Nadazdin O, Koyama I, Boenisch O, et al. 2012. Overcoming memory T-cell responses for induction of delayed tolerance in nonhuman primates. *Am J Transplant* **12:** 330–340.

Zinkernagel RM, Althage A, Callahan G, Welsh RM Jr. 1980. On the immunocompetence of H-2 incompatible irradiation bone marrow chimeras. *J Immunol* **124:** 2356–2365.

T-Cell Costimulatory Blockade in Organ Transplantation

Jonathan S. Maltzman[1] and Laurence A. Turka[2]

[1]Department of Medicine, Perelman School of Medicine at the University of Pennsylvania, Philadelphia, Pennsylvania 19104

[2]Transplantation Biology Research Center, Department of Surgery, Massachusetts General Hospital and Harvard Medical School, Boston, Massachusetts 02129

Correspondence: lturka@partners.org

Before it became possible to derive T-cell lines and clones, initial experimentation on the activation requirements of T lymphocytes was performed on transformed cell lines, such as Jurkat. These studies, although technically correct, proved misleading as most transformed T cells can be activated by stimulation of the clonotypic T-cell receptor (TCR) alone. In contrast, once it became possible to study nontransformed T cells, it quickly became clear that TCR stimulation by itself is insufficient for optimal activation of naïve T cells, but in fact, induces a state of anergy. It then became clear that functional activation of T cells requires not only recognition of major histocompatibility complex (MHC) and peptide by the TCR, but also requires ligation of costimulatory receptors expressed on the cell surface.

Costimulatory molecules can be divided based on structural similarities, expression patterns, and functional outcome of stimulation. Costimulatory molecules are structurally categorized into immunoglobulin (Ig) superfamily, tumor necrosis factor receptor (TNFR) family, and the T-cell Ig and mucin (TIM) family. This review will focus on costimulatory receptor structure and signaling and review experimental, preclinical, and clinical data regarding their efficacy in transplantation.

CD28/CTLA-4

CD28 is both the first identified and the prototypical costimulatory receptor expressed by T cells (June et al. 1987; Lindsten et al. 1989). Costimulation via CD28 is critical for activation of naïve T lymphocytes and avoidance of anergy (Jenkins et al. 1987a,b; DeSilva et al. 1991; Harding et al. 1992). CD28 is expressed as a homodimer and binds the ligands CD80 (B7-1) and CD86 (B7-2) (June et al. 1994). In mice, CD28 is constitutively expressed by all T cells and by a small subset of myeloid-derived cells. In humans, CD28 is similarly restricted to T cells; all CD4 T cells constitutively express CD28, but expression is absent on a subset of CD8 memory T cells (Lenschow et al. 1996).

CD28 is a type I transmembrane receptor with a single immunoglobulin domain in the extracellular portion and a relatively short cy-

Cite this article as *Cold Spring Harb Perspect Med* doi: 10.1101/cshperspect.a015537

toplasmic tail (Fig. 1). The cytoplasmic region (41 aa in human and 38 aa in mouse) has no intrinsic enzymatic activity, but contains several protein interaction motifs (Rudd and Schneider 2003). There are four tyrosines that can be phosphorylated, presumably by the src-family kinases LCK and FYN. When phosphorylated, these tyrosines can act as a binding site for proteins containing src-homology (SH)2 domains. There are also two PxxP motifs that mediate interactions with SH3 domains. Furthermore, there is a single YMNM motif allowing for interaction with the p85 subunit of PI3K as well as the Grb2/GADS adaptors. These initial phosphorylation and adaptor binding events lead to changes in downstream signal transduction and eventually changes in gene expression. Whether these proximal events mediate pathways distinct from the TCR or amplify TCR signals remains controversial. What is clear is that coligation of TCR/CD3 and CD28 leads directly to up-regulation of genes involved in T-cell activation including IL-2 and the IL-2 receptor α (CD25) genes. CD28 costimulation also results in transactivation of the antiapoptotic factors BCL2 and BCL-XL (Boise et al. 1995).

CTLA-4 also binds CD80/CD86 but differs in expression pattern, signaling, and functional outcome (Linsley et al. 1991; Krummel and Allison 1995). CTLA-4 is expressed as a homodimer on activated but not naïve T cells and constitutively on Foxp3$^+$ regulatory T cells. There are multiple splice variants of CTLA-4 that have been correlated with susceptibility to autoimmune disease (Ueda et al. 2003). The extracellular portion of CTLA-4 is highly homologous to that of CD28 but has a higher binding affinity for CD80/CD86. The higher affinity has been exploited to generate a fusion protein (CTLA-4Ig) that interferes with CD28 activation and can alter in vivo immune responses (see below). The intracellular regions differ substantially resulting in alternate function. The CTLA-4 cytoplasmic tail contains a YxxM motif that binds PI3K but also the PP2A and SHP-2 phosphatases. Given the binding

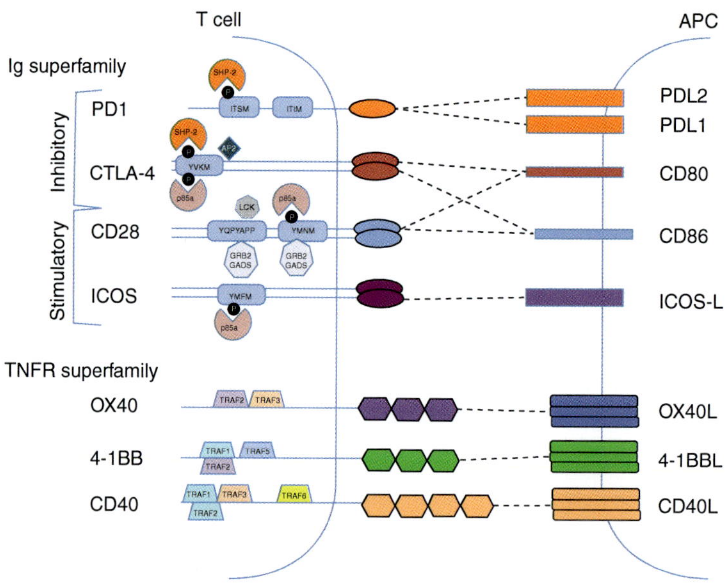

Figure 1. Costimulatory molecules and proximal signaling components. Immunoglobulin superfamily (*top*) and TNF superfamily receptors (*bottom*) expressed on T cells are shown on the *left*. Ligands are depicted in the antigen-presenting cell (APC) on the *right*. Dotted lines represent receptor–ligand pairs. Signaling motifs within the cytoplasmic tails are indicated in blue rectangles. Black circles represent phosphorylatable tyrosines. Proteins that interact directly with the cytoplasmic tails of costimulatory molecules are shown. Details of the signal transduction proteins are described in the text.

 Cite this article as *Cold Spring Harb Perspect Med* doi: 10.1101/cshperspect.a015537

of negative regulators, it is not surprising that CTLA-4 ligation sends a "negative" signal involving phosphatase-mediated inactivation of a number of proximal signaling pathways.

Until very recently, the major mechanism of CTLA-4-mediated function was believed to be cell intrinsic owing to increased affinity to CD80/CD86 and intracellular signal dampening, but there is a growing realization that CTLA-4 also functions in a cell-extrinsic manner (Walker and Sansom 2011; Corse and Allison 2012; Wang et al. 2012). Interestingly, a majority of CTLA-4 expressed in Tregs is localized to membranes of intracellular compartments rather than the cell surface (Leung et al. 1995; Valk et al. 2006). CTLA-4 is cycled to the surface of these cells following a TCR stimulation energy-dependent process likely involving TRIM (TCR interacting molecule) and the clatherin adaptor AP2 (Shiratori et al. 1997; Valk et al. 2006). Following binding of CTLA-4 to B7 expressed on antigen-presenting cells, CTLA-4 internalizes and via continued interaction with CD80/CD86, removes those critical molecules from antigen-presenting cells through a process termed *trans*-endocytosis or trogocytosis (Qureshi et al. 2011). Thus, CTLA-4 can negatively regulate an immune response through cell intrinsic signal dampening and cell-extrinsic alteration of antigen-presenting cells.

Immunosuppressive agents to block the pathway by targeting the CD28 "positive" signal have been in development for more than two decades. Among the first approaches used, and one which turned out to be successful as well, was to use the extracellular portion of CTLA-4 to generate a fusion protein with the Fc portion of IgG1, called CTLA-4Ig (Linsley et al. 1992). Administration of CTLA-4Ig prolongs allograft and xenograft survival in rodent models of islet, heart, and kidney transplantation and can induce tolerance under some conditions (Lenschow et al. 1992; Turka et al. 1992; Lin et al. 1993; Pearson et al. 1994; Azuma et al. 1996). CTLA-4Ig has been used more effectively in combination with additional agents and is a cornerstone for most regimens of costimulation blockade in rodents (see below). Interestingly, depending on the details of the timing of administration and degree of mismatch, CTLA-4Ig administration can accelerate rejection, likely owing to the effects on regulatory T cells (Charbonnier et al. 2012; Riella et al. 2012).

ICOS

Inducible T-cell costimulator (ICOS; CD278) is an Ig superfamily member expressed as a disulfide-linked homodimer (Hutloff et al. 1999; Simpson et al. 2010). ICOS transcription is regulated by calcineurin-dependent signals downstream from TCR and CD28 (Tan et al. 2006). Original descriptions had expression limited to activated T lymphocytes, but ICOS has subsequently been detected on NK, NKT, and T regulatory cells.

ICOS activity is initiated by interaction with its ligand ICOSL (also known as B7RP-1, B7h, CD275) leading to increased survival and proliferation with reduced IL-2 production when compared with CD28 costimulation. The intracellular region of ICOS contains an SH2 binding YMFM motif that recruits both the p85 and p50α regulatory isoforms of PI3K (Fos et al. 2008). The p50α isoform imparts a stronger signal and greater downstream AKT activity when compared with CD28-mediated activation via p85. ICOS/ICOSL may be sufficient for AKT phosphorylation-independent TCR cross-linking (Fos et al. 2008). ICOS/ICOSL also activates the MAPK pathway with resulting phosphorylation of ERK and p38, but interestingly not JNK (Parry et al. 2003).

ICOS expression is critical for T-cell differentiation of Th1, Th2, and Th17 cells and their effector function (Gonzalo et al. 2001; Rottman et al. 2001; Banquet et al. 2009). There is conflicting data as to whether ICOS predominantly regulates Th1 versus Th2 responses. ICOS is critical for T-cell-dependent B-cell activation and function. ICOS-deficient mice display defective germinal center formation and immunoglobulin isotype switching (McAdam et al. 2001). These defects result from reduced T_{FH} and Th17 generation and a reduction in CD40 L/CD154 expression on T cells (McAdam et al. 2001; Bauquet et al. 2009). Allograft survival of MHC-mismatched hearts is prolonged from ~7 to 21 days in both $ICOS^{-/-}$ recipient mice and in the setting of anti-ICOS antibody treatment (Ozkaynak et al. 2001). Interestingly, the combination of anti-ICOS plus cyclosporine was synergistic leading to >100 day survival without histological evidence of chronic rejection (Ozkaynak et al. 2001).

PD1

The programmed death 1 (PD1/CD279) receptor is a type 1 transmembrane protein with a single extracellular immunoglobulin domain that confers specificity to its ligands PDL1 and PDL2. PD1 is constitutively expressed on NK, NK-T, a subpopulation of dendritic cells (DCs), and activated monocytes. PD1 is not expressed on naïve T or B cells, but is up-regulated following activation, and is maintained by continuous TCR stimulation. High levels of PD1 are found on "exhausted" CD8 T cells during chronic infection (Barber et al. 2006). Transcription of PD1 is in part regulated by NFATc1; calcineurin inhibitor (CNI) treatment decreases PD1 levels (Oestreich et al. 2008). PD1 is expressed constitutively on Treg and TFH subsets of $CD4^+$ T cells.

The intracellular region of PD1 contains immunoreceptor tyrosine-based inhibitory and switch motifs (ITIM and ITSM). Phosphorylation of these motifs leads to recruitment of SHP1 and SHP2 phosphatases. SHP1 and SHP2 dephosphorylate a number of key signaling molecules including CD3, ZAP70, and PI3K (Parry et al. 2005), thus inhibiting activating signals and costimulation. Signaling through PD1 can alter differentiation of T cells through decreasing GATA3, Tbet, and eomesodermin expression. PD1 ligation decreases TCR-mediated generation of effector cytokines such as IFN-γ, TNF-α, and IL-2. Interestingly, blockade of PD1/PDL1 has been exploited to reverse T-cell exhaustion in mouse models of chronic infection and cancer (Iwai et al. 2002; Barber et al. 2006).

Experimental manipulation of the PD1/PDL1 axis has shown that this inhibitory pathway is critical for allograft tolerance and rejection. In general, limiting PD1/PDL1 accelerates rejection. For example, PD1 KO or PDL1 KO recipients overcome costimulation blockade (CD154 + CTLA-4Ig)-induced tolerance and reject BALB/c vascularized heart allografts within 1–2 weeks (Wang et al. 2007). Conversely, blockade with a PDL1 fusion protein (PDL1-Ig) in combination with cyclosporine or rapamycin significantly prolongs fully MHC-mismatched cardiac allograft survival (Ozkaynak et al. 2002).

CD40/CD40L (CD154)

CD40 and its ligand CD154 (CD40L/CD40 ligand) are costimulatory molecules critical for B–T interactions required for generation of high affinity, isotype switched antibodies (Elgueta et al. 2009). CD40 is a type II transmembrane protein that is a member of the TNFR superfamily. The extracellular region contains a β-sheet, α-helix loop, β-sheet sequence that allows for trimerization of the receptor. CD40 is expressed on antigen-presenting cells including dendritic cells, monocytes, and importantly on B lymphocytes. CD40 expression has also been shown on multiple nonhematopoietic cell types including fibroblasts, epithelial, and endothelial cells. CD154 is expressed primarily on activated T cells, but has been detected on activated B cells and platelets as well as monocytes, NK cells, mast cells, and basophils. Thrombosis as a result of CD154 expressed on platelets has limited its applicability as a therapeutic target (see below).

CD40 signaling depends on interaction of its cytoplasmic tail with TNFR-associated factors (TRAFs). There are six members of the TRAF family, designated TRAF1 through TRAF6. TRAF1, TRAF2, and TRAF 3 are recruited to CD40 via a PxQxT motif (Pullen et al. 1999); TRAF6 is recruited to the more membrane proximal QxPxEx sequence (Bishop et al. 2007). Activation of CD40 and recruitment of TRAFs leads to activation of the NF-κB, p38, Jnk, and Akt pathways. The specific pathways regulated by different TRAF proteins is beyond the scope of this review and can be found in detail elsewhere (Elgueta et al. 2009).

CD40/CD40L interactions have roles in both T-cell- and B-cell-mediated immunity through direct effects on lymphocytes and via generation and stabilization of the germinal center response (Foy et al. 1993, 1994; Van den Eertwegh et al. 1993). CD40 expressing DCs and B cells can activate T cells in lymph nodes during a normal immune response generating germinal centers. CD40/CD40L interactions are critical for maturation and survival of DCs via activation of the NF-κB pathway and up-regulation of the antiapoptotic factor Bcl-xL (Ouaaz et al. 2002). Survival and proliferation of B cells and activation of antigen-specific T cells also depend on CD40/CD40L interaction.

Interest in the CD40/CD40L pathway is second only to CD28/B7 in transplantation (reviewed in Quezada et al. 2004). A single dose of anti-CD40L (MR1 clone) is sufficient to induce long-term survival of fully mismatched cardiac allografts in a majority of mice (Hancock et al. 1996). Anti-CD40 treatment is synergistic with donor-specific transfusion (DST) of splenocytes resulting in long-term survival in 100% of recipients (Hancock et al. 1996). Combining MR1 with CTLA-4Ig leads to long-term acceptance of heart or skin Balb/c to C3H allografts (Larsen et al. 1996). Conversely, CD40/CD28 double knockout mice reject DBA/2 to C57BL/6 skin with only slightly delayed kinetics (Demirci et al. 2004).

OX40

OX40 (CD134; TNFRSF4) is a 50 kDa type 1 transmembrane protein that is a member of the TNFR superfamily. OX40 is expressed on most CD4 and CD8 T lymphocytes including naïve, Th1, Th2, Th17, and Treg. It is not expressed on resting memory T cells but can be rapidly reexpressed following TCR stimulation. Expression on natural Tregs begins during development in the thymus and is constitutive. OX40 is also expressed by NK and NKT cells. OX40L, the only known ligand for OX40, is expressed as a trimer cell surface of APCs including B cells.

The OX40 cytoplasmic tail contains a single QEE motif that acts as a TRAF-binding site (Ye et al. 1999; Croft 2010). Ligation by OX40L leads trimerizes OX40 recruiting TRAF2, TRAF3, and TRAF5, which activate the canonical and noncannonical NF-κB pathways (Arch and Thompson 1998). NF-κB activity results in up-regulation of a number of prosurvival and proliferative factors such as AKT, Bcl-2, Bcl-xL, and surviving (Song et al. 2004, 2005).

OX40 functions to promote immune responses through increased survival and proliferation of effector T cells. $OX40^{-/-}$ and $OX40L^{-/-}$ mice have intact $CD8^{+}$ but diminished $CD4^{+}$ effector responses to in vivo challenge (Chen et al. 1999; Kopf et al. 1999). The role of OX40 in Tregs has been more controversial. Genetic deletion of OX40 has no effect on Treg development or function (Vu et al. 2007). Conversely, ligation of OX40 and TCR on Tregs decreases Foxp3 transcription, per cell Foxp3 protein levels, and suppressive activity in vitro and in vivo (Vu et al. 2007).

Blockade of this pathway alone has little effect on allograft outcomes. In contrast, OX40/OX40L interactions play a nonredundant role in the absence of CD28 and CD40. As mentioned above, $CD28^{-/-}CD40^{-/-}$ mice reject DBA/2 skin. In these double knockouts or in the setting of CTLA-4Ig/anti-CD40L treatment, addition of OX40 blockade (but not anti-ICOS anti-4-1BB) induced long-term skin survival (Demirci et al. 2004).

4-1BB

4-1BB (CD137, tnfrsf9) is a transmembrane protein expressed on lymphocytes monocytes, dendritic cells, and B cells. Similar to many other costimulatory molecules discussed here, 4-1BB is not expressed on naïve T cells but rather on activated effector (Pollok et al. 1993; Dawicki and Watts 2004). The level of expression on Tregs is even higher and when combined with the absence of CD40L may allow for separation of human $Foxp3^{+}$ Tregs from $Foxp3^{-}$ effectors (Schoenbrunn et al. 2012). The ligand for 4-1BB, 4-1BBL, is expressed on APCs including mature DCs, macrophages, and activated B cells (Pollok et al. 1994).

Similar to other TNF superfamily receptors, ligand binding leads to trimerization and recruitment of adaptors. In this case, it is TRAF1 and TRAF2 that transduce signals leading to activation of the NF-κB and MAPK pathways. The cytoplasmic tail of murine 4-1BB also has a CxCP sequence that mediates binding of the src-family kinase LCK (Kim et al. 1993). The relevance of the 4-1BB–LCK association remains unclear.

Depending on the experimental system, 4-1BB has been shown to have a role in $CD4^+$ and/or $CD8^+$ T-cell responses. Specifically, 4-1BB-deficient mice have an intact CD4 response and impaired CD8 expansion and secondary responses in response to acute infection or vaccination (Tan et al. 1999, 2000). In contrast, the predominant effect of 4-1BBL deficiency in the setting of chronic infection is on the CD4 response (Fuse et al. 2007). 4-1BB-deficient mice reject allogeneic skin transplants with normal kinetics; mice deficient in both 4-1BB and CD28 show slightly delayed rejection (DeBenedette et al. 1999). Similarly, $CD8^+$ T-cell-mediated rejection of intestinal allografts is accelerated by treatment with 4-1BB agonistic antibody and delayed by 4-1BB/4-1BB blockade (Wang et al. 2003). Furthermore, antibody-mediated blockade leads to prolongation of fully MHC-mismatched heterotopic cardiac allografts and decreases allograft vasculopathy in a B6 into bm12 model (Cho et al. 2004; Saiki et al. 2008). Taken together, 4-1BB likely has a role in both CD4- and CD8-mediated aspects of transplant rejection.

MOVING INTO THE CLINIC—FROM LARGE ANIMALS TO MAN

Nonhuman Primate Studies

It is difficult to overestimate the excitement with which scientists and clinicians greeted the results from the initial mouse and rat studies of costimulatory blockade in transplantation. The findings that indefinite graft survival and tolerance could be induced by short-term administration of CTLA-4Ig and/or anti-CD154 mAb held the promise that tolerance might really be within reach, and led to the initiation of nonhuman primate studies of both agents.

CTLA-4Ig

The earliest reports using CTLA-4Ig in nonhuman primates were in islet allografts and renal allografts. In the islet allograft system, Levisetti et al. (1997) conducted a study in which five cynomologous monkeys were rendered diabetic with streptozotocin, transplanted with 12,000–20,000 islet equivalents, and treated with humanized CTLA-4Ig (hCTLA-4Ig) at a dose of 12.5 mg/kg daily on days 1–7 followed by 12.5 mg twice a day on days 8–14. Although three of the five animals rapidly rejected their grafts (day 6), two of the five animals achieved prolonged graft survival, with insulin independence 30 days in one, and 50 days in another. Of note in this study, CTLA-4Ig prevented alloantibody formation in all monkeys, even those that rapidly rejected their grafts. Contemporaneous with that report, Kirk et al. (1997) performed renal allografts in rhesus monkeys. Two animals were treated with CTLA-4Ig. One received a dose of 10 mg/kg daily for 5 days and rejected its graft on day 20. The other was dosed every other day for 12 days and rejected its graft on day 30.

Anti-CD154

Although these results with CTLA-4Ig were certainly disappointing, as noted above, studies in mice had revealed that antibodies to CD154 (at that time termed CD40L) were similarly effective as CTLA-4Ig for the induction of long-term graft survival, and that the two together had additive if not synergistic effects. Not surprisingly then, experiments using anti-CD154 mAb, either alone or in combination with CTLA-4Ig in nonhuman primates, were under way as well. It was first observed that a short course of anti-CD154 mAb alone was more effective than CTLA-4Ig alone and that the two together prolonged allograft survival up to 9 months, but did not lead to indefinite drug-free survival (Kirk et al. 1997, 1999).

Subsequently, Kirk and colleagues, rather than attempting to induce long-term graft survival with very short courses of one or both agents, treated animals with longer courses of drug, focusing on anti-CD154 as this seemed the more promising of the two. Using hu5C8, a humanized anti-CD154 mAb, they tested a clinically applicable protocol in which drug was given six times during the first 4 weeks, and at monthly intervals thereafter. Remarkably, eight of nine animals had long-term rejection-free survival, including periods exceeding 500 days. In five of these animals, drug had been discontinued for as long as a year. A second critical aspect of this study was that the addition of conventional immunosuppressive drugs (MMF, steroids, and/or tacrolimus) antagonized the salutary effects of anti-CD154, provoking rejection and graft loss. This finding paralleled similar results in mice, in which it was shown that a key mechanism by which costimulatory blockade worked in transplantation was through the deletion of alloreactive cells, and that calcineurin inhibitors prevented tolerance induction by blocking deletion (Li et al. 1999; Wells et al. 1999).

Other notable findings of Kirk et al. were that anti-CD154 did not act by global T-cell depletion and did not prevent donor-specific antibody formation. In retrospect, this latter point is interesting in light of accumulating reports of tolerant renal and liver transplant recipients who have donor-specific antibody but excellent stable graft function.

Into Man with Anti-CD40L

Unfortunately, efforts to translate anti-CD154 into humans revealed unacceptable toxicity of the drug. This manifested as elevated incidences of thromboembolic events in early clinical trials of two anti-CD154 mAbs, hu5c8, and toralizumab (IDEC 131). It was subsequently shown that activated platelets express CD154 (Henn et al. 1998) and that hu5C8 induced platelet expression of adhesion molecules and chemokines, providing a mechanistic explanation for the side effects observed in humans, and demonstrating that this was actually an "on-target" effect.

Back to CTLA-4Ig

In the wake of the problems with anti-CD154, the community refocused its interests on CTLA-4Ig, with the notion that rather than being used to induce tolerance, perhaps CTLA-4Ig could be used as part of an improved (either because of enhanced efficacy or reduced toxicity) immunosuppressive regimen. Consistent with this, a phase I study of CTLA-4Ig in 43 patients with psoriasis showed that the drug was safe, and compared with historical controls, induced improvement in disease activity, although this effect was transient and reverted after the drug was discontinued (Abrams et al. 1999). Building on this experience, CTLA-4Ig was developed for use in rheumatoid arthritis (Kremer et al. 2003), and is currently an approved drug for this indication under the generic name abatacept.

The relatively disappointing results with CTLA-4Ig in nonhuman primate transplantation studies led to a reevaluation of its binding interactions. These experiments revealed that although CTLA-4Ig does have a higher affinity for CD80 and CD86 than did CD28, it is markedly less effective in blocking CD86-mediated costimulation than CD80, perhaps because of a relatively lower avidity for CD86 (Larsen et al. 2005). By performing mutagenesis and screening, a higher avidity "version" of CTLA-4Ig with two amino acid substitutions was created, and was termed LEA29Y or belatacept (Larsen et al. 2005). The first set of studies of belatacept showed a synergistic effect with the anti-CD25 mAb basiliximab, with five of six recipients having stable creatinines for >100 days, and no rejection episodes occurring during the period of belatacept administration (which was given biweekly through day 70). These encouraging results paved the way for the development of belatacept for renal transplantation. As will be discussed in detail in another article of this monograph, belatacept gained FDA approval via registration trials designed to show its noninferiority to calcineurin ihibitors.

As noted above, in murine systems costimulatory blockade is able to induce transplant tolerance via deletion of alloreactive cells. Concomitant with this, and perhaps owing to the suppression of inflammation and an effector immune response, Foxp3^{+} regulatory T cells accumulate (owing to expansion of preexisting nTregs and/or induction of adaptive Tregs). Moreover, Tregs are required for maintenance of tolerance in mice. In humans, costimulatory blockade does not induce tolerance per se, but there is evidence to suggest that it may promote Tregs. This may appear surprising, as data from murine systems shows that Tregs require CD28 for development and maintenance (Zhang et al. 2013), and that CTLA-4Ig may cause a loss of Tregs from blood and lymphoid tissues (Salomon et al. 2000; Bour-Jordan and Bluestone 2009; Yang et al. 2009). Nonetheless, Bluestone et al. (2008) in a series of 8–10 patients, have reported that a basiliximab/belatacept regimen does not have a long-term effect on circulating Foxp3$^+$ Tregs, perhaps because CD86 is not fully saturated at these doses, and thus allows for CD28 costimulation. Furthermore, in one small series of patients, a belatacept regimen that did not contain anti-CD25 therapy also augmented Treg percentages in the peripheral blood of transplant recipients (Bestard et al. 2011).

Additional insight may be gained by findings in a related clinical setting, bone marrow transplantation. Beginning more than a decade ago, Guinan and colleagues performed a series of clinical studies in which patients undergoing MHC-mismatched myeloablative bone marrow transplants for hematologic malignancies received donor bone marrow that had been "anergized" against recipient alloantigens by brief culture with irradiated recipient APCs plus CTLA-4Ig before infusion. These studies (Guinan et al. 1999) revealed that the procedure was safe and did not result in either engraftment failure or increased susceptibility to infections. Moreover, graft versus host disease was less than what would have been predicted based on historical controls. Most importantly, they found that in vitro culture with CTLA-4Ig promoted the accumulation of donor-specific Foxp3$^+$ Tregs, while not affecting responses to third-party antigens or pathogens (Davies et al. 2008, 2009).

Based on this data, this group is currently initiating studies to exploit this approach in solid organ transplantation. Here, using one-haplotype mismatched live-donor renal transplants, the plan is to produce/expand recipient antidonor antigen-specific Tregs by culture of unfractionated recipient T cells with irradiated donor APCs plus belatacept, followed by magnetic bead separation of Tregs (identified as CD25^{hi}CD127^{lo}). These cells could then be used for infusion posttransplantation. Particularly attractive is a clinical protocol involving induction thymoglobulin and maintenance rapamycin, as this would "create space" and use an immunosuppressive milieu that would promote the expansion of cells postinfusion once the thymoglobulin is cleared.

What's Next?

Selective CD28 Blockade

Although still unproven, in theory, agents that bind to B7 molecules may have undesired counteracting effects by preventing them from engaging negative signaling molecules on T cells such as CTLA-4 and PD-1. Although this concern remains hypothetical, in part because of the continued uncertainty regarding the mechanism of action of CTLA-4 and whether or not negative signaling is part of how it functions to inhibit immune responses, it has nonetheless been the impetus to develop agents that would selectively bind to and block CD28 itself, rather than its ligands.

Most anti-CD28 Abs are stimulatory, but a few monovalent fusion antibodies have been developed that, most likely because they are incapable of cross-linking CD28, are blocking and inhibitory. These were initially used successfully in mouse and rat transplant experiments, and more recently have been applied to nonhuman primate renal and cardiac transplantation (Poirier et al. 2010). In this setting, long-term survival of both renal and cardiac allografts was achieved by the combination of a selective anti-CD28 inhibitor in combination with either tacrolimus or cyclosporine. Notably, this regimen also inhibited alloantibody production and cardiac allograft vasculopathy.

Further studies in primates and mice suggest that Tregs and negative signaling via CTLA-4 may be an important component of long-term engraftment (Poirier et al. 2010; Zhang et al. 2011). It cannot necessarily be concluded, however, that these features are owing to "sparing" of a B7:CTLA-4 interaction, as long-term engraftment induced by a brief course of CTLA-4Ig similarly depends on Tregs and CTLA-4 (Tran et al. 1997). More likely is the fact that the establishment of an allograft in a noninflammatory environment promotes the development of regulatory T cells. However, the synergy of CD28-blockade with calcineurin inhibitors (CNIs) is particularly noteworthy because, as discussed above, CNIs antagonize the effects of anti-CD154 in nonhuman primate transplantation (Kirk et al. 1997, 1999).

Agents that selectively block CD28 may still have the theoretical drawback of blocking regulatory T-cell pathways because, as noted above, CD28 is needed for optimal development and survival of Tregs. However, very early clinical data suggests that belatacept as currently dosed may not block the maintenance or expansion of Tregs, and thus it is possible that a CD28-specific inhibitor might be developed that would show the same pharmacodynamic feature. Perhaps the biggest barrier to the application of selective anti-CD28 mAbs in human transplant trials is the catastrophic immune activation that occurred in the trial of the superagonist anti-CD28 Ab TGN1412 (Suntharalingam et al. 2006). Although there is no evidence that anti-CD28 blocking agents would have a similar effect, caution is likely to be the byword in this area, and extensive preclinical testing that takes into account species-specific differences in the types of T cells that express CD28 will be required (Eastwood et al. 2010; Hunig 2012).

Activating Inhibition

With the discovery that T cells express "co-inhibitory" as well as costimulatory receptors, agents that were capable of activating these negative-regulatory pathways became an obvious approach to blocking immune responses. The best such candidate molecules to target at present would appear to be CTLA-4 and PD-1. Unfortunately, application of this concept is hindered by the absence of reagents that "turn-on" these inhibitory pathways; all reagents available at present are blocking antibodies, which thus lead to immunoenhancement. Nonetheless, recent studies showing that blockade of CTLA-4 or PD-1 can be a therapeutically useful maneuver in cancer immunotherapy (Hodi et al. 2010; Topalian et al. 2012) provide evidence for the importance of these pathways in humans and support the notion that activating negative-regulatory molecules might be an effective way to turn off pathogenic immune responses.

CD154:CD40

Despite the problems noted above associated with the anti-CD154 mAb hu5c8, interest in blocking this pathway has continued owing to its impressive efficacy in preclinical studies. One idea has been to develop anti-CD154-targeted agents that will not induce thromboembolic events. A number of companies are currently working on this area, although preclinical data are not yet available. An alternative approach is to block CD40, and a number of such Abs have been studied in nonhuman primate models. The anti-CD40 mAb 4D11 can induce relatively long-term renal allograft survival in cynomolgus monkeys, although animals eventually rejected even when given continued maintenance treatment (Aoyagi et al. 2009). Another anti-CD40 mAb, 3A8, was shown to be effective in rhesus islet transplantation, and the combination of 3A8 and anti-CD25 mAb plus sirolimus induced long-term survival (Badell et al. 2012). Similar results were seen with the anti-CD40 mAb 2C10 (Lowe et al. 2012). Notably, in each of the anti-CD40 studies, no thromboembolic complications were observed. Based on the promise of these studies, it seems likely that we will see clinical development of one or more of these agents.

LFA-3 (CD58)

The precise role of the CD58–CD2 interaction in human T-cell activation remains poorly understood. In part this stems from the fact that in mice, CD48, rather than CD58, is the CD2 ligand. Nonetheless, ample evidence for the importance of this pathway in humans led to the development of alefacept, an LFA-3/Fc fusion protein that would bind to, and block, CD2. Alefacept is currently FDA approved for the treatment of psoriasis and until 2011 was marketed by Astellas under the trade name Amevive.

Astellas has recently decided to discontinue marketing of the drug for business reasons, and thus the future of agents that target CD58 are uncertain. This is unfortunate, as alefacept was recently shown to selectively target memory T cells and, in combination with CTLA-4Ig plus sirolimus, lead to long-term renal allograft survival in rhesus macaques (Weaver et al. 2009).

CONCLUSIONS

More than two decades ago, it seemed to many that blocking T-cell costimulation held the promise to achieve transplantation tolerance. Although that has yet to come to fruition, the concept of costimulatory blockade has been translated into clinical practice in both autoimmunity and transplantation. For the near to medium term future, it is likely the use of these agents will be in the context of strategies to minimize immunosuppression and immunosuppressive side effects, with tolerance studies being appropriately restricted to carefully monitored clinical trials.

REFERENCES

Abrams JR, Lebwohl MG, Guzzo CA, Jegasothy BV, Goldfarb MT, Goffe BS, Menter A, Lowe NJ, Krueger G, Brown MJ, et al. 1999. CTLA4Ig-mediated blockade of T-cell costimulation in patients with psoriasis vulgaris. *J Clin Invest* **103:** 1243–1252.

Aoyagi T, Yamashita K, Suzuki T, Uno M, Goto R, Taniguchi M, Shimamura T, Takahashi N, Miura T, Okimura K, et al. 2009. A human anti-CD40 monoclonal antibody, 4D11, for kidney transplantation in cynomolgus monkeys: Induction and maintenance therapy. *Am J Transplant* **9:** 1732–1741.

Arch RH, Thompson CB. 1998. 4-1BB and Ox40 are members of a tumor necrosis factor (TNF)-nerve growth factor receptor subfamily that bind TNF receptor-associated factors and activate nuclear factor κB. *Mol Cell Biol* **18:** 558–565.

Azuma H, Chandraker A, Nadeau K, Hancock WW, Carpenter CB, Tilney NL, Sayegh MH. 1996. Blockade of T-cell costimulation prevents development of experimental chronic renal allograft rejection. *Proc Natl Acad Sci* **93:** 12439–12444.

Badell IR, Thompson PW, Turner AP, Russell MC, Avila JG, Cano JA, Robertson JM, Leopardi FV, Strobert EA, Iwakoshi NN, et al. 2012. Nondepleting anti-CD40-based therapy prolongs allograft survival in nonhuman primates. *Am J Transplant* **12:** 126–135.

Barber DL, Wherry EJ, Masopust D, Zhu B, Allison JP, Sharpe AH, Freeman GJ, Ahmed R. 2006. Restoring function in exhausted CD8 T cells during chronic viral infection. *Nature* **439:** 682–687.

Bauquet AT, Jin H, Paterson AM, Mitsdoerffer M, Ho IC, Sharpe AH, Kuchroo VK. 2009. The costimulatory molecule ICOS regulates the expression of c-Maf and IL-21 in the development of follicular T helper cells and TH-17 cells. *Nat Immunol* **10:** 167–175.

Bestard O, Cassis L, Cruzado JM, Torras J, Franquesa M, Gil-Vernet S, Lucia M, Grinyó JM. 2011. Costimulatory blockade with mTor inhibition abrogates effector T-cell responses allowing regulatory T-cell survival in renal transplantation. *Transpl Int* **24:** 451–460.

Bishop GA, Moore CR, Xie P, Stunz LL, Kraus ZJ. 2007. TRAF proteins in CD40 signaling. *Adv Exp Med Biol* **597:** 131–151.

Bluestone JA, Liu W, Yabu JM, Laszik ZG, Putnam A, Belingheri M, Gross DM, Townsend RM, Vincenti F. 2008. The effect of costimulatory and interleukin 2 receptor blockade on regulatory T cells in renal transplantation. *Am J Transplant* **8:** 2086–2096.

Boise LH, Minn AJ, Noel PJ, June CH, Accavitti MA, Lindsten T, Thompson CB. 1995. CD28 costimulation can promote T cell survival by enhancing the expression of Bcl-xL. *Immunity* **3:** 87–98.

Bour-Jordan H, Bluestone JA. 2009. Regulating the regulators: Costimulatory signals control the homeostasis and function of regulatory T cells. *Immunol Rev* **229:** 41–66.

Charbonnier LM, Vokaer B, Lemaître PH, Field KA, Leo O, Le Moine A. 2012. CTLA4-Ig restores rejection of MHC class-II mismatched allografts by disabling IL-2-expanded regulatory T cells. *Am J Transplant* **12:** 2313–2321.

Chen AI, McAdam AJ, Buhlmann JE, Scott S, Lupher ML Jr, Greenfield EA, Baum PR, Fanslow WC, Calderhead DM, Freeman GJ, et al. 1999. Ox40-ligand has a critical costimulatory role in dendritic cell: T cell interactions. *Immunity* **11:** 689–698.

Cho HR, Kwon B, Yagita H, La S, Lee EA, Kim JE, Akiba H, Kim J, Suh JH, Vinay DS, et al. 2004. Blockade of 4-1BB (CD137)/4-1BB ligand interactions increases allograft survival. *Transpl Int* **17:** 351–361.

Corse E, Allison JP. 2012. Cutting edge: CTLA-4 on effector T cells inhibits in *trans*. *J Immunol* **189:** 1123–1127.

Croft M. 2010. Control of immunity by the TNFR-related molecule OX40 (CD134). *Annu Rev Immunol* **28:** 57–78.

Davies JK, Gribben JG, Brennan LL, Yuk D, Nadler LM, Guinan EC. 2008. Outcome of alloanergized haploidentical bone marrow transplantation after ex vivo costimulatory blockade: Results of 2 phase 1 studies. *Blood* **112:** 2232–2241.

Davies JK, Nadler LM, Guinan EC. 2009. Expansion of allospecific regulatory T cells after anergized, mismatched bone marrow transplantation. *Sci Transl Med* **1:** p1ra3.

Dawicki W, Watts TH. 2004. Expression and function of 4-1BB during CD4 versus CD8 T cell responses in vivo. *Eur J Immunol* **34:** 743–751.

DeBenedette MA, Wen T, Bachmann MF, Ohashi PS, Barber BH, Stocking KL, Peschon JJ, Watts TH. 1999. Analysis of 4-1BB ligand (4-1BBL)-deficient mice and of mice lacking both 4-1BBL and CD28 reveals a role for 4-1BBL in skin allograft rejection and in the cytotoxic T cell response to influenza virus. *J Immunol* **163:** 4833–4841.

Demirci G, Amanullah F, Kewalaramani R, Yagita H, Strom TB, Sayegh MH, Li XC. 2004. Critical role of OX40 in CD28 and CD154-independent rejection. *J Immunol* **172:** 1691–1698.

DeSilva DR, Urdahl KB, Jenkins MK. 1991. Clonal anergy is induced in vitro by T cell receptor occupancy in the absence of proliferation. *J Immunol* **147:** 3261–3267.

Eastwood D, Findlay L, Poole S, Bird C, Wadhwa M, Moore M, Burns C, Thorpe R, Stebbings R. 2010. Monoclonal antibody TGN1412 trial failure explained by species differences in CD28 expression on $CD4^+$ effector memory T-cells. *Br J Pharmacol* **161:** 512–526.

Elgueta R, Benson MJ, de Vries VC, Wasiuk A, Guo Y, Noelle RJ. 2009. Molecular mechanism and function of CD40/CD40 L engagement in the immune system. *Immunol Rev* **229:** 152–172.

Fos C, Salles A, Lang V, Carrette F, Audebert S, Pastor S, Ghiotto M, Olive D, Bismuth G, Nunès JA. 2008. ICOS ligation recruits the pp50α PI3K regulatory subunit to the immunological synapse. *J Immunol* **181:** 1969–1977.

Foy TM, Shepherd DM, Durie FH, Aruffo A, Ledbetter JA, Noelle RJ. 1993. In vivo CD40-gp39 interactions are essential for thymus-dependent humoral immunity. II: Prolonged suppression of the humoral immune response by an antibody to the ligand for CD40, gp39. *J Exp Med* **178:** 1567–1575.

Foy TM, Laman JD, Ledbetter JA, Aruffo A, Claassen E, Noelle RJ. 1994. gp39-CD40 interactions are essential for germinal center formation and the development of B cell memory. *J Exp Med* **180:** 157–163.

Fuse S, Bellfy S, Yagita H, Usherwood EJ. 2007. $CD8^+$ T cell dysfunction and increase in murine gammaherpesvirus latent viral burden in the absence of 4-1BB ligand. *J Immunol* **178:** 5227–5236.

Gonzalo JA, Tian J, Delaney T, Corcoran J, Rottman JB, Lora J, Al-garawi A, Kroczek R, Gutierrez-Ramos JC, Coyle AJ. 2001. ICOS is critical for T helper cell-mediated lung mucosal inflammatory responses. *Nat Immunol* **2:** 597–604.

Guinan EC, Boussiotis VA, Neuberg D, Brennan LL, Hirano N, Nadler LM, Gribben JG. 1999. Transplantation of anergic histoincompatible bone marrow allografts. *N Engl J Med* **340:** 1704–1714.

Hancock WW, Sayegh MH, Zheng XG, Peach R, Linsley PS, Turka LA. 1996. Costimulatory function and expression of CD40 ligand, CD80, and CD86 in vascularized murine cardiac allograft rejection. *Proc Natl Acad Sci* **93:** 13967–13972.

Harding FA, McArthur JG, Gross JA, Raulet DH, Allison JP. 1992. CD28-mediated signalling co-stimulates murine T cells and prevents induction of anergy in T-cell clones. *Nature* **356:** 607–609.

Henn V, Slupsky JR, Gräfe M, Anagnostopoulos I, Förster R, Müller-Berghaus G, Kroczek RA. 1998. CD40 ligand on activated platelets triggers an inflammatory reaction of endothelial cells. *Nature* **391:** 591–594.

Hodi FS, O'Day SJ, McDermott DF, Weber RW, Sosman JA, Haanen JB, Gonzalez R, Robert C, Schadendorf D, Hassel JC, et al. 2010. Improved survival with ipilimumab in patients with metastatic melanoma. *N Engl J Med* **363:** 711–723.

Hunig T. 2012. The storm has cleared: Lessons from the CD28 superagonist TGN1412 trial. *Nat Rev Immunol* **12:** 317–318.

Hutloff A, Dittrich AM, Beier KC, Eljaschewitsch B, Kraft R, Anagnostopoulos I, Kroczek RA. 1999. ICOS is an inducible T-cell co-stimulator structurally and functionally related to CD28. *Nature* **397:** 263–266.

Iwai Y, Ishida M, Tanaka Y, Okazaki T, Honjo T, Minato N. 2002. Involvement of PD-L1 on tumor cells in the escape from host immune system and tumor immunotherapy by PD-L1 blockade. *Proc Natl Acad Sci* **99:** 12293–12297.

Jenkins MK, Pardoll DM, Mizuguchi J, Chused TM, Schwartz RH. 1987a. Molecular events in the induction of a nonresponsive state in interleukin 2-producing helper T-lymphocyte clones. *Proc Natl Acad Sci* **84:** 5409.

Jenkins MK, Ashwell JD, Schwartz RH. 1987b. Allogeneic non-T spleen cells restore the responsiveness of normal T cell clones stimulated with antigen and chemically modified antigen-presenting cells. *J Immunol* **140:** 3324–3330.

June CH, Ledbetter JA, Gillespie MM, Lindsten T, Thompson CB. 1987. T cell proliferation involving the CD28 pathway is associated with cyclosporine-resistant interleukin 2 gene expression. *Mol Cell Biol* **7:** 4472–4481.

June CH, Bluestone JA, Nadler LM, Thompson CB. 1994. The B7 and CD28 receptor families. *Immunol Today* **15:** 321–331.

Kim YJ, Pollok KE, Zhou Z, Shaw A, Bohlen JB, Fraser M, Kwon BS. 1993. Novel T cell antigen 4-1BB associates with the protein tyrosine kinase pp56lck1. *J Immunol* **151:** 1255–1262.

Kirk AD, Harlan DM, Armstrong NN, Davis TA, Dong Y, Gray GS, Hong X, Thomas D, Fechner JH Jr, Knechtle SJ. 1997. CTLA4-Ig and anti-CD40 ligand prevent renal allograft rejection in primates. *Proc Natl Acad Sci* **94:** 8789–8794.

Kirk AD, Burkly LC, Batty DS, Baumgartner RE, Berning JD, Buchanan K, Fechner JH Jr, Germond RL, Kampen RL, Patterson NB, et al. 1999. Treatment with humanized monoclonal antibody against CD154 prevents acute renal allograft rejection in nonhuman primates. *Nat Med* **5:** 686–693.

Kopf M, Ruedl C, Schmitz N, Gallimore A, Lefrang K, Ecabert B, Odermatt B, Bachmann MF. 1999. OX40-deficient mice are defective in Th cell proliferation but are competent in generating B cell and CTL Responses after virus infection. *Immunity* **11:** 699–708.

Kremer JM, Westhovens R, Leon M, Di Giorgio E, Alten R, Steinfeld S, Russell A, Dougados M, Emery P, Nuamah IF, et al. 2003. Treatment of rheumatoid arthritis by selective inhibition of T-cell activation with fusion protein CTLA4Ig. *N Engl J Med* **349:** 1907–1915.

Krummel MF, Allison JP. 1995. CD28 and CTLA-4 have opposing effects on the response of T cells to stimulation. *J Exp Med* **182:** 459–465.

Larsen CP, Elwood ET, Alexander DZ, Ritchie SC, Hendrix R, Tucker-Burden C, Cho HR, Aruffo A, Hollenbaugh D, Linsley PS, et al. 1996. Long-term acceptance of skin and cardiac allografts after blocking CD40 and CD28 pathways. *Nature* **381:** 434–438.

Larsen CP, Pearson TC, Adams AB, Tso P, Shirasugi N, Strobert E, Anderson D, Cowan S, Price K, Naemura J, et al. 2005. Rational development of LEA29Y, a high affinity variant of CTLA4Ig with potent immunosuppressive properties. *Am J Transpl* **5:** 443–453.

Lenschow DJ, Zeng Y, Thistlethwaite JR, Montag A, Brady W, Gibson MG, Linsley PS, Bluestone JA. 1992. Long-term survival of xenogeneic pancreatic islet grafts induced by CTLA4Ig. *Science* **257:** 789–792.

Lenschow DJ, Walunas TL, Bluestone JA. 1996. CD28/B7 system of T cell costimulation. *Ann Rev Immunol* **14:** 233–258.

Leung HT, Bradshaw J, Cleaveland JS, Linsley PS. 1995. Cytotoxic T lymphocyte-associated molecule-4, a high-avidity receptor for CD80 and CD86, contains an intracellular localization motif in its cytoplasmic tail. *J Biol Chem* **270:** 25107–25114.

Levisetti MG, Padrid PA, Szot GLMittal N, Meehan M, Wardrip CL, Gray GS, Bruce DS, Thistlethwaite JR, Bluestone JA. 1997. Immunosuppressive effects of human CTLA4Ig in a non-human primate model of allogeneic pancreatic islet transplantation. *J Immunol* **159:** 5187–5191.

Li Y, Li XC, Wells AD, Zheng XX, Turka LA, Strom TB. 1999. Blocking both signal 1 and signal 2 of T cell activation prevents apoptosis of alloreactive T cells and induction of peripheral allograft tolerance. *Nat Med* **5:** 1298–1302.

Lin H, Bolling SF, Linsley PS, Wei RQ, Gordon D, Thompson CB, Turka LA. 1993. Long-term acceptance of major histocompatibility complex mismatched cardiac allografts induced by CTLA4Ig plus donor-specific transfusion. *J Exp Med* **178:** 1801–1806.

Lindsten T, June CH, Ledbetter JA, Stella G, Thompson CB. 1989. Regulation of lyphokine messenger RNA stability by surface-mediated T cell activation pathway. *Science* **244:** 339.

Linsley PS, Brady W, Urnes M, Grosmaire LS, Damle NK, Ledbetter JA. 1991. CTLA-4 is a second receptor for the B cell activation antigen B7. *J Exp Med* **174:** 561–569.

Linsley PS, Wallace PM, Johnson J, Gibson MG, Greene JL, Ledbetter JA, Singh C, Tepper MA. 1992. Immunosuppression in vivo by a soluble form of the CTLA-4 T cell activation molecule. *Science* **257:** 792–795.

Lowe M, Badell IR, Thompson P, Martin B, Leopardi F, Strobert E, Price AA, Abdulkerim HS, Wang R, Iwakoshi NN, et al. 2012. A novel monoclonal antibody to CD40 prolongs islet allograft survival. *Am J Transplant* **12:** 2079–2087.

McAdam AJ, Greenwald RJ, Levin MA, Chernova T, Malenkovich N, Ling V, Freeman GJ, Sharpe AH. 2001. ICOS is critical for CD40-mediated antibody class switching. *Nature* **409:** 102–105.

Oestreich KJ, Yoon H, Ahmed R, Boss JM. 2008. NFATc1 regulates PD-1 expression upon T cell activation. *J Immunol* **181:** 4832–4839.

Ouaaz F, Arron J, Zheng Y, Choi Y, Beg AA. 2002. Dendritic cell development and survival require distinct NF-κB subunits. *Immunity* **16:** 257–270.

Ozkaynak E, Gao W, Shemmeri N, Wang C, Gutierrez-Ramos JC, Amaral J, Qin S, Rottman JB, Coyle AJ, Hancock WW. 2001. Importance of ICOS-B7RP-1 costimulation in acute and chronic allograft rejection. *Nat Immunol* **2:** 591–596.

Ozkaynak E, Wang L, Goodearl A, McDonald K, Qin S, O'Keefe T, Duong T, Smith T, Gutierrez-Ramos JC, Rottman JB, et al. 2002. Programmed death-1 targeting can promote allograft survival. *J Immunol* **169:** 6546–6553.

Parry RV, Rumbley CA, Vandenberghe LH, June CH, Riley JL. 2003. CD28 and inducible costimulatory protein Src homology 2 binding domains show distinct regulation of phosphatidylinositol 3-kinase, Bcl-xL, and IL-2 expression in primary human CD4 T lymphocytes. *J Immunol* **171:** 166–174.

Parry RV, Chemnitz JM, Frauwirth KA, Lanfranco AR, Braunstein I, Kobayashi SV, Linsley PS, Thompson CB, Riley JL. 2005. CTLA-4 and PD-1 receptors inhibit T-cell activation by distinct mechanisms. *Mol Cell Biol* **25:** 9543–9553.

Pearson TC, Alexander DZ, Winn KJ, Linsley PS, Lowry RP, Larsen CP. 1994. Transplantation tolerance induced by CTLA4-Ig. *Transplantation* **57:** 1701–1706.

Poirier N, Azimzadeh AM, Zhang T, Dilek N, Mary C, Nguyen B, Tillou X, Wu G, Reneaudin K, Hervouet J, et al. 2010. Inducing CTLA-4-dependent immune regulation by selective CD28 blockade promotes regulatory T cells in organ transplantation. *Sci Transl Med* **2:** 17ra10.

Pollok KE, Kim YJ, Zhou Z, Hurtado J, Kim KK, Pickard RT, Kwon BS. 1993. Inducible T cell antigen 4-1BB. Analysis of expression and function. *J Immunol* **150:** 771–781.

Pollok KE, Kim YJ, Hurtado J, Zhou Z, Kim KK, Kwon BS. 1994. 4-1BB T-cell antigen binds to mature B cells and macrophages, and costimulates anti-μ-primed splenic B cells. *Eur J Immunol* **24:** 367–374.

Pullen SS, Labadia ME, Ingraham RH, McWhirter SM, Everdeen DS, Alber T, Crute JJ, Kehry MR. 1999. High-affinity interactions of tumor necrosis factor receptor-associated factors (TRAFs) and CD40 require TRAF trimerization and CD40 multimerization. *Biochemistry* **38:** 10168–10177.

Quezada SA, Jarvinen LZ, Lind EF, Noelle RJ. 2004. CD40/CD154 interactions at the interface of tolerance and immunity. *Annu Rev Immunol* **22:** 307–328.

Qureshi OS, Zheng Y, Nakamura K, Attridge K, Manzotti C, Schmidt EM, Baker J, Jeffery LE, Kaur S, Briggs Z, et al. 2011. Trans-endocytosis of CD80 and CD86: A molecular basis for the cell-extrinsic function of CTLA-4. *Science* **332:** 600–603.

Riella LV, Liu T, Yang J, Chock S, Shimizu T, Mfarrej B, Batal I, Xiao X, Sayegh MH, Chandraker A. 2012. Deleterious effect of CTLA4-Ig on a Treg-dependent transplant model. *Am J Transplant* **12:** 846–855.

Rottman JB, Smith T, Tonra JR, Ganley K, Bloom T, Silva R, Pierce B, Gutierrez-Ramos JC, Ozkaynak E, Coyle AJ. 2001. The costimulatory molecule ICOS plays an important role in the immunopathogenesis of EAE. *Nat Immunol* **2:** 605–611.

Rudd CE, Schneider H. 2003. Unifying concepts in CD28, ICOS and CTLA4 co-receptor signalling. *Nat Rev Immunol* **3:** 544–556.

Saiki H, Suzuki J, Kosuge H, Haraguchi G, Ishihara T, Haga T, Maejima Y, Isobe M, Uede T. 2008. Blockade of the 4-1BB pathway attenuates graft arterial disease in cardiac allografts. *Int Heart J* **49:** 105–118.

Salomon B, Lenschow DJ, Rhee L, Ashourian N, Singh B, Sharpe A, Bluestone JA. 2000. B7/CD28 costimulation is essential for the homeostasis of the CD4$^+$CD25$^+$ immunoregulatory T cells that control autoimmune diabetes. *Immunity* **12:** 431–440.

Schoenbrunn A, Frentsch M, Kohler S, Keye J, Dooms H, Moewes B, Dong J, Loddenkemper C, Sieper J, Wu P, et al. 2012. A converse 4-1BB and CD40 ligand expression pattern delineates activated regulatory T cells (Treg) and conventional T cells enabling direct isolation of alloantigen-reactive natural Foxp3$^+$ Treg. *J Immunol* **189:** 5985–5994.

Shiratori T, Miyatake S, Ohno H, Nakaseko C, Isono K, Bonifacino JS, Saito T. 1997. Tyrosine phosphorylation controls internalization of CTLA-4 by regulating its interaction with clathrin-associated adaptor complex AP-2. *Immunity* **6:** 583–589.

Simpson TR, Quezada SA, Allison JP. 2010. Regulation of CD4 T cell activation and effector function by inducible costimulator (ICOS). *Curr Opin Immunol* **22:** 326–332.

Song J, Salek-Ardakani S, Rogers PR, Cheng M, Van Parijs L, Croft M. 2004. The costimulation-regulated duration of PKB activation controls T cell longevity. *Nat Immunol* **5:** 150–158.

Song J, So T, Cheng M, Tang X, Croft M. 2005. Sustained survivin expression from OX40 costimulatory signals drives T cell clonal expansion. *Immunity* **22:** 621–631.

Suntharalingam G, Perry MR, Ward S, Brett SJ, Castello-Cortes A, Brunner MD, Panoskaltsis N. 2006. Cytokine storm in a phase 1 trial of the anti-CD28 monoclonal antibody TGN1412. *N Engl J Med* **355:** 1018–1028.

Tan JT, Whitmire JK, Ahmed R, Pearson TC, Larsen CP. 1999. 4-1BB ligand, a member of the TNF family, is important for the generation of antiviral CD8 T cell responses. *J Immunol* **163:** 4859–4868.

Tan JT, Whitmire JK, Murali-Krishna K, Ahmed R, Altman JD, Mittler RS, Sette A, Pearson TC, Larsen CP. 2000. 4-1BB costimulation is required for protective anti-viral immunity after peptide vaccination. *J Immunol* **164:** 2320–2325.

Tan AH, Wong SC, Lam KP. 2006. Regulation of mouse inducible costimulator (ICOS) expression by Fyn-NFATc2 and ERK signaling in T cells. *J Biol Chem* **281:** 28666–28678.

Topalian SL, Hodi FS, Brahmer JR, Gettinger SN, Smith DC, McDermott DF, Powderly JD, Carvajal RD, Sosman JA, Atkins MB, et al. 2012. Safety, activity, and immune correlates of anti-PD-1 antibody in cancer. *N Engl J Med* **366:** 2443–2454.

Tran HM, Nickerson PW, Restifo AC, Ivis-Woodward MA, Patel A, Allen RD, Strom TB, O'Connell PJ. 1997. Distinct mechanisms for the induction and maintenance of allograft tolerance with CTLA4-Fc treatment. *J Immunol* **159:** 2232–2239.

Turka LA, Linsley PS, Lin H, Brady W, Leiden JM, Wei RQ, Gibson ML, Zheng XG, Myrdal S, Gordon D, et al. 1992. T-cell activation by the CD28 ligand B7 is required for cardiac allograft rejection in vivo. *Proc Natl Acad Sci* **89:** 11102–11105.

Ueda H, Howson JM, Esposito L, Heward J, Snook H, Chamberlain G, Rainbow DB, Hunter KM, Smith AN, Di Genova G, et al. 2003. Association of the T-cell regulatory gene CTLA4 with susceptibility to autoimmune disease. *Nature* **423:** 506–511.

Valk E, Leung R, Kang H, Kaneko K, Rudd CE, Schneider H. 2006. T cell receptor-interacting molecule acts as a chaperone to modulate surface expression of the CTLA-4 coreceptor. *Immunity* **25:** 807–821.

Van den Eertwegh AJ, Noelle RJ, Roy M, Shepherd DM, Aruffo A, Ledbetter JA, Boersma WJ, Claassen E. 1993. In vivo CD40-gp39 interactions are essential for thymus-dependent humoral immunity. I: In vivo expression of CD40 ligand, cytokines, and antibody production delineates sites of cognate T-B cell interactions. *J Exp Med* **178:** 1555–1565.

Vu MD, Xiao X, Gao W, Degauque N, Chen M, Kroemer A, Killeen N, Ishii N, Li XC. 2007. OX40 costimulation turns off Foxp3$^+$ Tregs. *Blood* **110:** 2501–2510.

Walker LS, Sansom DM. 2011. The emerging role of CTLA4 as a cell-extrinsic regulator of T cell responses. *Nat Rev Immunol* **11:** 852–863.

Wang J, Guo Z, Dong Y, Kim O, Hart J, Adams A, Larsen CP, Mittler RS, Newell KA. 2003. Role of 4-1BB in allograft rejection mediated by CD8$^+$ T cells. *Am J Transplant* **3:** 543–551.

Wang L, Han R, Hancock WW. 2007. Programmed cell death 1 (PD-1) and its ligand PD-L1 are required for allograft tolerance. *Eur J Immunol* **37:** 2983–2990.

Wang CJ, Kenefeck R, Wardzinski L, Attridge K, Manzotti C, Schmidt EM, Qureshi OS, Sansom DM, Walker LS. 2012. Cutting edge: Cell-extrinsic immune regulation by CTLA-4 expressed on conventional T cells. *J Immunol* **189:** 1118–1122.

Weaver TA, Charafeddine AH, Agarwal A, Turner AP, Russell M, Leopardi FV, Kampen RL, Stempora L, Song M, Larsen CP, et al. 2009. Alefacept promotes co-stimulation blockade based allograft survival in nonhuman primates. *Nat Med* **15:** 746–749.

Wells AD, Li XC, Li Y, Walsh M, Zheng XX, Wu Z, Nuñez G, Tang A, Sayegh M, Hancock WW, Strom TB, Turka LA. 1999. Requirement for T cell apoptosis in the induction of peripheral transplantation tolerance. *Nat Med* **5:** 1303–1307.

Yang J, Riella LV, Boenisch O, Popoola J, Robles S, Watanabe T, Vanguri V, Yuan X, Guleria I, Turka LA, et al. 2009. Paradoxical functions of B7: CD28 costimulation in a MHC class II-mismatched cardiac transplant model. *Am J Transplant* **9:** 2837–2844.

Ye H, Park YC, Kreishman M, Kieff E, Wu H. 1999. The structural basis for the recognition of diverse receptor sequences by TRAF2. *Mol Cell* **4:** 321–330.

Zhang T, Fresnay S, Welty E, Sangrampurkar N, Rybak E, Zhou H, Cheng XF, Feng Q, Avon C, Laaris A, et al. 2011. Selective CD28 blockade attenuates acute and chronic rejection of murine cardiac allografts in a CTLA-4-dependent manner. *Am J Transplant* **11:** 1599–1609.

Zhang R, Huynh A, Whitcher G, Chang J, Maltzman JS, Turka LA. 2013. An obligate cell-intrinsic function for CD28 in Tregs. *J Clin Invest* **123:** 580–593.

Regulatory Cells and Transplantation Tolerance

Stephen P. Cobbold and Herman Waldmann

Sir William Dunn School of Pathology, University of Oxford, South Parks Road, Oxford OX1 3RE, United Kingdom

Correspondence: stephen.cobbold@path.ox.ac.uk

Transplantation tolerance is a continuing therapeutic goal, and it is now clear that a subpopulation of T cells with regulatory activity (Treg) that express the transcription factor foxp3 are crucial to this aspiration. Although reprogramming of the immune system to donor-specific transplantation tolerance can be readily achieved in adult mouse models, it has yet to be successfully translated in human clinical practice. This requires that we understand the fundamental mechanisms by which donor antigen-specific Treg are induced and function to maintain tolerance, so that we can target therapies to enhance rather than impede these regulatory processes. Our current understanding is that Treg act via numerous molecular mechanisms, and critical underlying components such as mTOR inhibition, are only now emerging.

Evidence for immune regulation and suppressive cells has pervaded the field of transplantation ever since the classic experiments that showed the feasibility of inducing allogeneic chimerism and tolerance in the neonate (Billingham et al. 1953, 1954). Throughout the latter half of the twentieth century, while adoptively transferrable suppression of graft rejection remained a consistent finding (Streilein and Gruchalla 1981; Hall 1985), various lymphocyte subsets with supposedly suppressive properties came and went, as the technologies used to characterize them were found to be inadequate or flawed. Only in the last 10 years, with the discovery that foxp3 is a genetic determinant in autoimmunity and "master transcription factor" for a subset of $CD4^+$ regulatory T cells (Fontenot et al. 2003; Hori et al. 2003; Khattri et al. 2003), has the study and exploitation of regulatory cells become a defining pursuit for the transplant community.

It was only with the development of monoclonal antibodies (mAbs) that targeted functional molecules on the surface of T cells that transplantation tolerance could be achieved in the adult mouse (Cobbold et al. 1986; Qin et al. 1989). Initially these mAbs were used to deplete T cells, allowing bone marrow to be given to achieve donor chimerism and tolerance to skin grafts in a manner analogous to the neonatal model. Intriguingly, however, where antibodies were used to block T-cell function, antigen-specific T cells were not always deleted, but rather seemed to survive in a state of relative unresponsiveness known as anergy (Qin et al. 1989; Leong et al. 1992; Scully et al. 1994). These anergic T

Cite this article as *Cold Spring Harb Perspect Med* doi: 10.1101/cshperspect.a015545

cells failed to proliferate to their specific antigen in vitro, but were hyperesponsive to IL-2, which we now know is a feature of Treg.

The concept of reprogramming the peripheral immune system toward a state of transplantation tolerance was firmly established (Cobbold et al. 1992) with the discovery that a combination of functionally blocking mAbs against the T-cell coreceptors CD4 and CD8 could induce tolerance directly to allogeneic skin grafts (Qin et al. 1990). Tolerance was induced in the adult mouse without the need for T-cell depletion or stable chimerism and was found to be independent of the thymus. The tolerant state was also dominant—able to resist the infusion of a large number of naïve T cells, which was a strong pointer to the presence of immune regulation rather than any deletion of donor reactive T cells.

LINKED SUPPRESSION AND INFECTIOUS TOLERANCE

With the demonstration that the tolerance induced to allogeneic skin grafts in the periphery must involve some form of immune regulation, it became clear that we had to revisit the concept of suppression, despite considerable resistance from the immunological community at that time. The ability to adoptively transfer tolerance with T cells to secondary recipients had remained a consistent but mostly ignored theme even through the antisuppression era of the 1980s. But the availability of better reagents clearly associated suppression with the $CD4^+$ subpopulation of T cells (Qin et al. 1993) and not the $CD8^+$ "cytotoxic/suppressor" cells of the past. These $CD4^+$ "regulatory" T cells could not only transfer tolerance to secondary recipients, but they could also confer a state of tolerance onto new cohorts of naïve T cells—a process that merited the term "infectious tolerance" originally coined many years earlier (Gershon and Kondo 1971) to simply describe the phenomenon of suppression. In a similar fashion, tolerance could "spread" from one set of allogeneic antigens (from strain A) on a tolerated graft to the second set of antigens (from strain B) on a graft from a (strain A × strain B) F1 cross (Davies et al. 1996). This process was termed "linked suppression" to indicate that the two sets of antigens needed to be expressed, or linked, on the same APC.

REGULATORY T CELLS AND AUTOIMMUNITY

Around the same time that suppression was being revisited in transplantation tolerance, others working in certain rodent models of autoimmune disease were being faced with a similar challenge. In particular, it was found that $CD4^+$ T cells could be separated into two subsets on the basis of their expression of different isoforms of CD45—transfer of one of these subsets into immunodeficient secondary recipients caused a severe lymphoproliferative wasting disease, whereas addition of the other subset could suppress it (Powrie et al. 1994). This was interpreted as the former subset containing autoreactive T cells that required the presence of Treg in the latter subset to maintain immune homeostasis. A different model of autoimmunity had been pursued, particularly by Sakaguchi's group, where mice were thymectomized on day 3 after birth (Asano et al. 1996). The inflammatory gastritis that developed in this model could be prevented by the infusion of mature $CD4^+$ T cells, and, once again, these could be separated into two subsets—in this case those that expressed CD25 were the regulatory subpopulation. Although there were strong similarities between the $CD4^+$ regulatory subsets in the different autoimmune and transplantation models, there were also some apparent discrepancies (Alyanakian et al. 2003), and it was not until the discovery of foxp3 as the causative gene of autoimmune disease in the scurfy mouse and the human IPEX syndrome that the field of regulatory T cells began to focus on a unique and unifying T-cell subset (Fontenot et al. 2003; Hori et al. 2003; Khattri et al. 2003).

REGULATORY T CELLS AND Foxp3

Foxp3 was identified as a transcription factor of the forkhead box family that, in both mice and humans, is required for the generation of the

 Cite this article as *Cold Spring Harb Perspect Med* doi: 10.1101/cshperspect.a015545

thymic $CD4^+CD25^+$ regulatory T-cell subset often termed "natural" Treg (nTreg). Transduction of the foxp3 gene was shown to convert naïve T cells into regulatory T cells (Fontenot et al. 2003; Hori et al. 2003) in a manner that recapitulated the suppressive functions previously ascribed to $CD4^+CD25^+$ cells in a range of autoimmune and transplantation models. Naïve T cells could be induced to express foxp3 and acquire at least some of the functions of regulatory T cells by activating them in the presence of TGF-β in vitro (Chen et al. 2003). TGF-β-dependent conversion of naïve, peripheral $CD4^+$ T cells into iTreg was also observed in vivo, particularly in TCR transgenic mice on a $RAG^{-/-}$ background (Cobbold et al. 2004; Kretschmer et al. 2005). Treatment of female mice expressing a TCR specific for the male transplantation antigen DBY, as presented by MHC-II, with a brief course of nondepleting anti-CD4 antibody, could induce lifelong tolerance to male skin grafts (Cobbold et al. 2004). This tolerance was associated with a peripheral induction of foxp3 expression with the Treg concentrated within the tolerated skin graft (Graca et al. 2002; Cobbold et al. 2006). Positive selection in the thymus for the male-specific TCR in $RAG^{-/-}$ background mice generated only naïve T cells, and foxp3 expression could not be detected in the thymus, even by sensitive RT-PCR, indicating that tolerance in this model depended entirely on peripheral iTreg induction. Recently, the use of a mouse that expresses hCD2 on the cell surface as a reporter gene for foxp3 expression has allowed proof that these $foxp3^+$ Treg maintain transplantation tolerance (Kendal et al. 2011; Regateiro et al. 2012): administering a depleting hCD2 mAb caused rejection of the graft, either in the original recipients or in secondary $RAG^{-/-}$ recipients of the tolerated graft, proving that $foxp3^+$ Treg were actively regulating effector cells within the graft itself. Similar experiments in these hCD2 reporter mice also showed that the $foxp3^+$ Treg were necessary for linked suppression and infectious tolerance in mice that had a normal polyclonal TCR repertoire, and when crossed onto DBY-specific TCR transgenic mice on a $RAG^{-/-}$ background gave confirmation that tolerance to male skin grafts depended on the peripheral induction of $foxp3^+$ iTreg.

POSITIVE SELECTION OF Treg, THEIR SPECIFICITY, EPIGENETICS, AND STABILITY

Although the importance of nTreg in controlling autoimmunity and immune homeostasis is now generally accepted, the role of $foxp3^+$ iTreg has remained more controversial. Despite the publication of a number of candidate markers distinguishing nTreg and iTreg, such as helios (Thornton et al. 2010) and neuropilin-1 (Yadav et al. 2012), these have not proven reliable, and may instead relate to different activation states of either cell type. The key feature that distinguishes nTreg from iTreg is the specificity of their TCRs (Lee et al. 2012). nTreg are positively selected with a moderate to high affinity for self-antigens, including peripheral, tissue-specific antigen peptides driven by AIRE on thymic epithelium (Aschenbrenner et al. 2007). This means that nTreg are always likely to meet a constant source of their specific "self"-antigen in the periphery, which may be an important factor in their constitutive expression of molecules considered as markers of activation in non-Treg cells such as CD25, GITR, and CTLA4, and may also be a factor in their longevity and stability. In contrast, iTreg are derived from naïve T cells originally selected in the thymus with low affinities for self-antigen peptides, but likely to have higher affinities for the "foreign" antigen eliciting them in the periphery. Once the foreign antigen is eliminated, for example, if an allograft is rejected, the iTreg may lack sufficient TCR stimulus for continued maintenance or stability.

One important factor in the stability of Treg is the epigenetic status of the foxp3 gene (Floess et al. 2007; Lal et al. 2009; Zheng et al. 2010), specifically the extent to which the intronic enhancer element known as the "Treg cell-specific demethylation region" (TSDR) is demethylated (Floess et al. 2007) to allow the access of a number of important positive transcription factors such as STAT5 and NF-κB as well as foxp3 itself (Fig. 1). This TSDR starts to be demethylated early in the differentiation of nTreg in the thymus, continues until nTreg in the periphery are fully demethylated, and is associated with stable foxp3 expression. Surprisingly, the

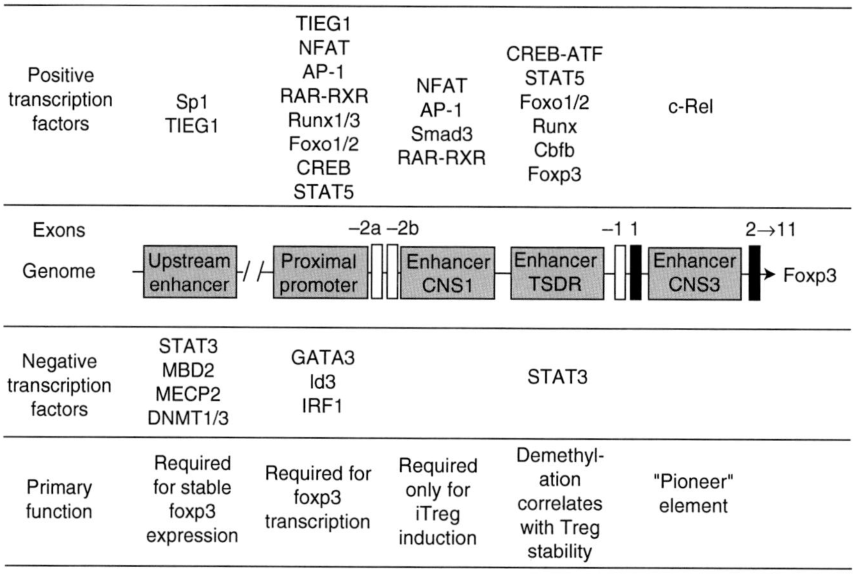

Figure 1. Regulation of the foxp3 locus. The genomic organization of the foxp3 locus is depicted (not to scale) with the position of the most important regulatory elements shown, together with the positive and negative regulatory transcription factors that have been shown to bind to them and the primary function of each element.

demethylation process in nTreg has recently been shown to be independent of any actual foxp3 expression (Ohkura et al. 2012). In contrast, iTreg recently generated in vitro by TCR activation in the presence of TGF-β have very little demethylation of the TSDR, and also rapidly lose stable foxp3 expression, particularly if TGF-β is withheld from the culture. The epigenetic status and stability of iTreg induced in vivo as a result of tolerance induction remains unclear, although historical transplantation experiments suggested that regulatory T-cell activity was maintained as long as the source of antigen (i.e., the allograft) was present (Scully et al. 1994).

It has been claimed that the addition of all transrentinoic acid (ATRA) or rapamycin during Treg expansion improves their stability in vitro (Zhou et al. 2010; Takahashi et al. 2012; Yurchenko et al. 2012), whereas others have attempted to provide IL-2/anti-IL-2 complexes with a long half-life in vivo to promote Treg activity (Daniel et al. 2010; Letourneau et al. 2010). Although we tend to think of T-cell subsets as fixed lineages proceeding to terminal differentiation, there are a number of examples showing "plasticity" allowing Treg cells to "transdifferentiate" into effector cells, although this remains controversial (Zhou et al. 2009; Rubtsov et al. 2010). One possibility is that plasticity and lineage diversification are generated by asymmetry during cell division, as has been observed during stem cell (Neumuller and Knoblich 2009) and $CD8^+$ T-cell differentiation (Chang et al. 2007), whereas symmetric divisions expand and mature in cell populations without further diversification. The stability of in vitro induced or expanded Treg is particularly important for their potential use in proposed cell-based therapies, where Treg are given to transplant recipients to suppress alloreactivity and enforce tolerance to the graft. Identification of the mechanisms that stabilize Treg is a clear target for new strategies to enhance tolerogenic therapies.

THERAPEUTIC APPLICATION OF Treg CELLS: WHICH Treg AND WHICH ANTIGENS?

One of the first therapeutic applications considered for Treg was to limit graft versus host disease (GVHD) in allogeneic bone marrow and stem cell transplantation. Although GVHD can be eliminated by depleting any contaminating donor T cells from the marrow/stem cell inoculum, this increases the risk the graft may be rejected, and also removes any potential for a curative graft versus leukemia effect (Hale and Waldmann 1996). It has been repeatedly shown that the administration of an excess of purified or in vitro expanded nTreg from the donor inoculum can reduce GVHD and enhance tolerance in mouse models of bone marrow transplantation, without losing the GVL effect (Edinger et al. 2003; Sato et al. 2003; Trenado et al. 2003). An important safety issue for clinical transplantation is the purity and stability of the nTreg cells infused—they should not contain any regulation-resistant memory T cells, nor should they be able to revert to effector cells, either of which could potentially exacerbate GVHD rather than suppress it. As more markers are combined (e.g., $CD4^+$, $CD25^{high}$, $CD45RA^+$, $CD127^{low}$) (Liu et al. 2006) and the stringency of sorting is increased, then so will the yield of Treg decrease. This means that in vitro expansion (Earle et al. 2005) is likely to be needed for routine application, and the long-term stability of such cells remains questionable (McClymont et al. 2011).

A large excess of nTreg have been shown in mouse models to be able to induce tolerance to allogeneic skin grafts given to lymphopenic recipients, although they seemed to do so without any particular specificity for the donor antigens (Graca et al. 2004). This is in contrast to the Treg that have been induced in mice made tolerant to allogeneic skin grafts, by treatment with nondepleting mAbs, where regulation requires donor antigen (although this can then lead to linked suppression of additional antigens, as discussed earlier). This raises a number of questions as to whether therapeutic Treg cells should be derived from recipient nTreg or iTreg, should they be expanded or induced on antigen, and in the context of transplantation, whether antigen be presented in the context of donor (direct presentation) or recipient (indirect presentation) MHC (Jiang et al. 2004; Sanchez-Fueyo et al. 2007)? It would seem that Treg with specificity for both directly and indirectly presented donor antigens are the most effective at inducing tolerance to allografts (Tsang et al. 2009), and are considerably more potent than unselected nTreg (Chen et al. 2009; Veerapathran et al. 2011).

It remains a challenge as to how sufficiently large numbers of antigen-specific Treg, stably expressing foxp3, and uncontaminated by effector T cells can be generated for clinical therapy. Human T cells appear to have a more promiscuous expression of foxp3 than mouse Treg (Wang et al. 2007), such that demonstrable expression of foxp3 alone may not be sufficient to ensure that the population will guarantee suppressive function in vivo. In addition, the usual in vitro readout for Treg function is the suppression of naïve T-cell proliferation, but as we do not yet know the molecular mechanisms of immune regulation, this assay may not faithfully reflect the in vivo efficacy. It is beyond the scope of this particular review to detail how attempts to develop clinical Treg therapy are progressing, as this topic is considered in more detail elsewhere in this collection.

MOLECULAR MECHANISMS OF Treg FUNCTION

One of the first identified properties of regulatory T cells was that they failed to proliferate in response to antigen or TCR stimulation in vitro—so called anergy (Qin et al. 1989; Schwartz 1990; Chen et al. 2004; Fu et al. 2004; Park et al. 2004). This is primarily associated with an inability to produce IL-2 for the autocrine stimulation of proliferation, although anergic cells may also overexpress the high-affinity IL-2 receptor (CD25) and be hyper-responsive to exogenous sources of IL-2 (Beverly et al. 1992; Setoguchi et al. 2005). This has led some to suggest that Treg function, at least in the in vitro suppression of proliferation assay, by preferentially consuming IL-2 (Shevach 2009). However, IL-2 is generally required for the survival of Treg, and mice deficient in either IL-2 or its receptor experience autoimmunity rather than immunosuppression (Horak et al. 1995; Kramer et al. 1995; Caudy et al. 2007).

Another mechanism proposed for Treg function is the expression of granzymes that kill APCs (Shevach et al. 2006), thereby suppressing antigen presentation. The phenomenon of linked suppression strongly suggests that Treg act via modulating the APC is some way (Davies et al. 1996); but if all APCs were killed, then one has to question how antigen is presented to allow the induction of Treg. Specific killing of donor APC, however, could be a means to eliminate directly presented donor antigen to allow only indirect presentation by recipient APCs, which in the presence of TGF-β could still induce Treg for tolerance.

TGF-β1-deficient mice and mice expressing a transgenic dominant negative TGF-BRII on their T cells are both prone to develop autoimmunity (Shull et al. 1992; Shah et al. 2002). Under some conditions, Treg are themselves able to secrete TGF-β, or may have TGF-β–LAP complexes expressed on their surface (Nakamura et al. 2001), tethered via a molecule associated with anergic cells called GARP (Tran et al. 2009). TGF-β is known to be broadly anti-inflammatory, and can also induce foxp3 expression and peripheral conversion of naïve T cells to iTreg, and has therefore been implicated as a mechanism of the cell contact-mediated suppression observed in vitro and infectious tolerance (Nakamura et al. 2004). Although we know TGF-β signaling to T cells in vivo is required for the induction of tolerance and iTreg, there are many sources of TGF-β other than Treg in vivo, and any TGF-β generated requires activation by appropriate proteases or α5 integrins (Paidassi et al. 2010, 2011) not normally expressed on Treg. In addition, exposure of naïve T cells to TGF-β, while inducing some of the markers such as CD103 associated with Treg, only induces foxp3 in a proportion of exposed cells (Regateiro et al. 2012). If TGF-β exposed T cells are sorted using a reporter gene (e.g., hCD2, as above) then only the cells expressing foxp3 can transfer transplantation tolerance while the foxp3 negative (but TGF-β exposed T cells still cause graft rejection) (Regateiro et al. 2012). All this suggests that TGF-β secretion, at best, only explains part of the functional capabilities of Treg.

Two other cytokines that can be secreted by Treg are IL-10 and IL-35 (Collison et al. 2007). Both of these cytokines are considered to have anti-inflammatory functions and have been implicated in the mechanism by which Treg suppress in inflammatory bowel disease (Uhlig et al. 2006), experimental allergic encephalomyelitis (Sundstedt et al. 2003; Yu et al. 2005; Fitzgerald et al. 2007), collagen-induced arthritis (Mauri et al. 2003), and allergic airway inflammation (Whitehead et al. 2012). Like TGF-β, there are many sources of IL-10 other than Treg cells, and the importance of Treg-derived IL-10 in transplantation tolerance remains unclear. The more recently identified cytokine IL-35 is currently thought to be more specific to Treg, although there is very little data concerning other possible sources, partly because it consists of a specific combination of EBI-3 and p35 chains, both of which can pair with other members of the IL-12 family of heterodimeric cytokines. Whereas IL-10 is thought to limit the induction of inflammation, IL-35 seems to act much later to dampen down ongoing inflammatory processes (Whitehead et al. 2012). There is, as yet, very little information on any possible role of IL-35 in transplantation tolerance.

CTLA4-deficient mice also develop an autoimmune proliferative disease similar to foxp3-deficient mice (Tivol et al. 1995). CTLA4 is transiently expressed on activated T cells, is constitutive on some foxp3^{+} Treg cells, and is thought to compete with the costimulatory binding of CD28 to B7 ligands on APCs. The high affinity of CTLA4 has even been observed to deplete B7 ligands from APCs by capturing and internalizing them on the T cells by a process termed "trogocytosis," thereby forcing the APCs into a less inflammatory and more tolerogenic mode of antigen presentation (Qureshi et al. 2011; Zhang et al. 2011). It is not yet clear how important this process is for Treg function in the context of transplantation, but mice with a conditionally inducible CTLA4 knockout allele in their foxp3^{+} T cells have shown that CTLA4 expression on Treg is continuously required to avoid the development of autoimmune disease (Sojka et al. 2009).

 Cite this article as *Cold Spring Harb Perspect Med* doi: 10.1101/cshperspect.a015545

TOLEROGENIC MICROENVIRONMENTS AND THE ROLE OF LOCAL, INDUCED IMMUNE PRIVILEGE

Although defects in Treg tend to cause a generalized lymphoproliferative disease, most autoimmune diseases are restricted to certain organs or tissues, suggesting a more localized loss of regulation. In TCR transgenic models of transplantation tolerance, the Treg seem to concentrate to the transplanted tissue and can be shown to act locally (Cobbold et al. 2006; Kendal et al. 2011). Evidence is now accumulating that peripheral tolerance is in part dependent on the induction of a form of immune privilege within the microenvironment of, and dependent on an active participation by, the tolerated tissue. Within the tissue, and potentially modified by the local microenvironment, are a number of different cell types that can present antigen on MHC-II antigens to Treg; some are migratory such as dendritic cells, macrophages, and B cells, but others are mostly resident, including endothelial cells and mast cells (Lu et al. 2006). Locally active TGF-β, whether derived from Treg or the tissue, may represent an important component of a tolerogenic microenvironment that can, for example, induce the expression of CD39 and CD73 on T cells, macrophages, and dendritic cells (Regateiro et al. 2011, 2013). CD39 and CD73 are two ectoenzymes expressed on the cell surface that are able to convert inflammatory extracellular ATP (released during cell damage) into AMP and then adenosine, which has anti-inflammatory properties by binding to adenosine receptors on T cells and dendritic cells and inducing the intracellular second messenger cAMP. Raised levels of cAMP have been associated with the anergy of Treg (Cone et al. 1996; Powell et al. 1999), and it has been suggested that infectious tolerance may be explained by cAMP, in essence, being passed from Treg to induce an anergic state in nearby naïve T cells, either via the CD39/CD73-mediated adenosine pathway, or perhaps directly via the formation of tight junctions between adjacent T-cell membranes (Bopp et al. 2007).

Treg can themselves also contribute to a state of immune privilege by inducing the local expression of enzymes that catabolize essential amino acids (EAAs) (Cobbold et al. 2009). IDO (indoleamine dioxygenase), that catabolizes tryptophan, was the first such enzyme to be shown as important in maintaining a form of transplantation tolerance in mice, as the specific inhibitor 1-MT was able to cause the rejection of fetuses expressing allogeneic, but not syngeneic, paternal antigens (Munn et al. 1998). It has recently been suggested that maintenance of tolerance to the allogeneic fetus may be the main physiological role for peripherally induced iTreg cells, particularly as the TGF-β responsive elements (Zheng et al. 2010) of the foxp3 enhancer appeared in evolutionary time consistent with the development of the placenta (Andersen et al. 2012; Samstein et al. 2012b).

An in vitro model of linked suppression, where a $CD4^+$ regulatory T-cell clone expressing constitutive CTLA4 was able to suppress the proliferation of a $CD8^+$ CTL clone was also found to depend on the induction of IDO in a subset of dendritic APC (Mellor et al. 2004). In this case, it was shown that IDO acted specifically by depleting tryptophan from the medium, rather than generating potentially inhibitory kynurenines (Mezrich et al. 2010), as adding excess tryptophan could overcome the suppression. In a similar manner, it has been shown that Treg cells can induce in APC many different catabolic or synthetic enzymes that can deplete EAAs, including IDO, arginase, and iNOS (which can both use arginine), and IL-4i1 (which depletes phenylalanine) (Cobbold et al. 2009). Histidine decarboxylase and tryptophan hydroxylase can be induced particularly in mast cells, and depletion of tryptophan by the latter enzyme (rather than its synthetic function of producing serotonin) has recently been shown to explain why mast cells seem to be required in some models of transplantation tolerance (Nowak et al. 2012). The healthy or healing tissue itself can also contribute to this EAA-depleted tolerogenic environment by expressing further catabolic enzymes for the branched chain amino acids and threonine (Cobbold et al. 2009).

The depletion of EAAs is sensed by T cells through two different pathways. One is known as the "integrated stress response," which relies on GCN2 detecting an excess of free tRNAs within the cell, and which is thought to be particularly important for detecting tryptophan depletion (Munn et al. 2005). Although it was initially thought that GCN2 signaling could inhibit proliferation and induce foxp3 and regulatory T-cell activity directly in the T cell, most evidence now suggests that it acts indirectly by modifying differentiation of the APC to a more tolerogenic phenotype (Sharma et al. 2007). The second amino acid-sensing pathway depends on the "ragulator" complex (Sancak et al. 2010; Zoncu et al. 2011), which acts via the RAG family of GTPases, and in the absence of amino acids this fails to recruit and activate the TORC1 complex via Rheb on the lysosomal membrane. This mode of sensing amino acid depletion can also be mimicked by the mTOR inhibitor rapamycin, which also acts to disrupt the TORC1 complex (Hara et al. 2002). Whereas rapamycin-mediated mTOR inhibition is also able to induce tolerogenic properties in dendritic cells (Taner et al. 2005), it can act directly on T cells to inhibit proliferation and to induce foxp3. Treg cells are also relatively resistant to mTOR inhibition, possibly as a result of the foxp3-mediated expression of PIM2 kinase that can promote T-cell survival in a manner independent of the PI3k/mTOR pathway (Basu et al. 2008).

Experiments in mice with specific defects in either TORC1 or TORC2 have suggested that both of these mTOR complexes need to be inhibited to enforce effective Treg differentiation (Delgoffe et al. 2009). Although rapamycin has usually been thought of as a specific inhibitor of TORC1, it is increasingly recognized that longer-term exposure to this inhibitor can also reduce TORC2 and downstream signaling through AKT (Barquilla et al. 2008). Even more recently, it has been shown that amino acids can actually activate both of the mTOR complexes (Tato et al. 2011). Although it is unclear how mTOR inhibition is linked to foxp3 expression, there is evidence that it may act, in part, via HIF1α and the hypoxia-sensing pathway (Ben-Shoshan et al. 2008; Shi et al. 2011; Clambey et al. 2012).

It is intriguing that immunosuppressive drugs that can inhibit the mTOR pathway directly (e.g., rapamycin/sirolimus) or indirectly (e.g., fingolimod), and which should therefore promote Treg differentiation, are already being used in clinical practice. However, the proportion of patients who can be weaned off of continuous immunosuppression remains extremely low (Roussey-Kesler et al. 2006). It remains to be seen whether this situation can be improved as new induction protocols, such as T-cell depletion (Morgan et al. 2012), allow for minimization of other immunosuppressive agents that may counteract tolerance.

In summary, there may not be a single, unique mechanism of action by which Treg induce and maintain tolerance, but rather a complex set of interactions between Treg, APCs, and the tissue that generate a microenvironment (Fig. 2) that not only suppresses potential inflammatory responses but can also promote further Treg differentiation via linked suppression and infectious tolerance (Cobbold et al. 2006). One key feature of the tolerogenic microenvironment may be that it is compartmentalized from the inflammatory signals such as IL-6 (Yang et al. 2008) and complement activation (Kwan et al. 2012; Le Friec et al. 2012) that would otherwise compromise the stability of Treg and their further generation. Although infectious tolerance has mainly been considered a mechanism for foxp3$^+$ Treg to generate a second foxp3$^+$ iTreg cohort (Kendal et al. 2011), it remains unclear whether this is necessary or sufficient to maintain the tolerant state long term. In particular, it remains possible that IL-10-producing Tr1 cells may also play a role (Cobbold et al. 2003; Stassen et al. 2004).

T-CELL-SUBSET-SPECIFIC REGULATION AND COEXPRESSION OF TRANSCRIPTION FACTORS

Although Treg appear to act via many apparently parallel and potentially redundant mechanisms to inhibit almost the entire range of innate and adaptive immune responses, evidence of some specialization within subsets of foxp3$^+$ Treg is beginning to emerge. It seems that for

 Cite this article as *Cold Spring Harb Perspect Med* doi: 10.1101/cshperspect.a015545

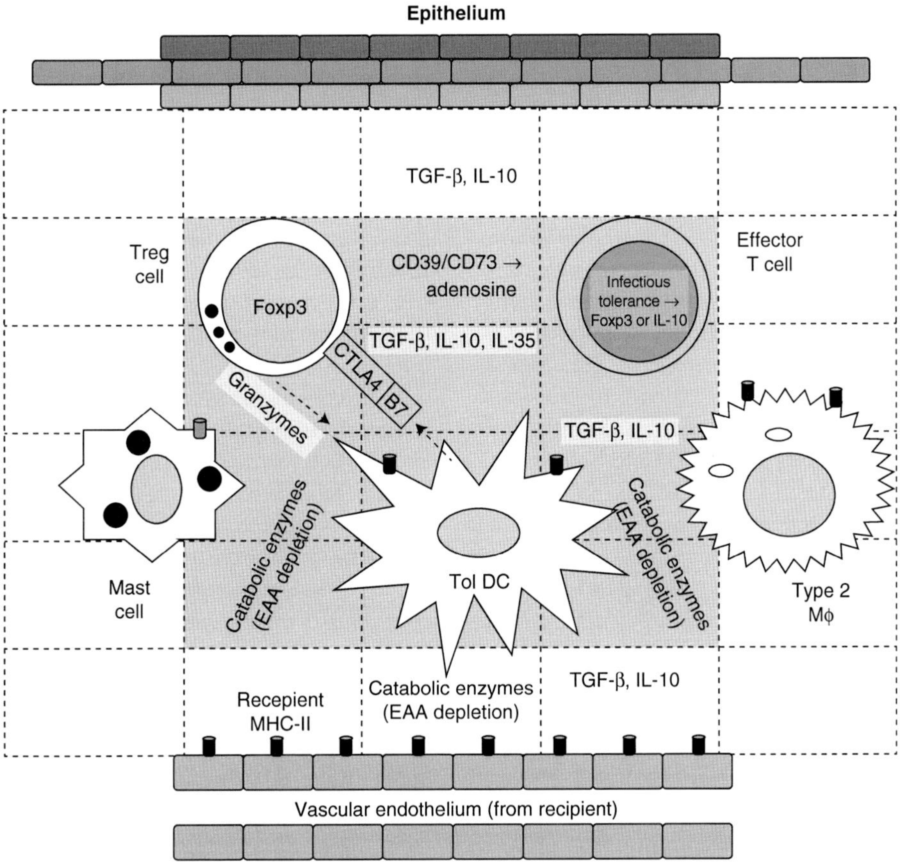

Figure 2. Treg cells act within tissue to create a state of induced immune privilege. A localized tolerogenic microenvironment is established in tolerated tissues by the interaction of foxp3$^+$ Treg cells with cells capable of presenting antigen in the context of MHC-II. These MHC-II$^+$-presenting cells, including dendritic cells (modulated to "TolDC"), macrophages (modulated to Type 2 Mϕ), and mast cells both contribute to, and are modulated by, this tolerogenic microenvironment, which is maintained by the cytokines TGF-β, IL10, and IL-35, the CD39- and CD73-mediated conversion of ATP to adenosine, and the expression of multiple enzymes that deplete essential amino acids (EAAs). In addition, Treg cells may down-regulate antigen presentation by CTLA4-mediated trogocytosis of the costimulatory B7 ligands, or by granzyme-mediated killing of inflammatory/donor APCs. In this tolerogenic microenvironment, effector T cells may be "reprogrammed" to differentiate into Treg cells expressing foxp3 and/or IL-10 in a form of infectious tolerance.

Treg to suppress different functional T-cell subsets they need to co-express the appropriate associated transcription factors such as Tbet for suppression of Th1 (Koch et al. 2009) or GATA3 (Wang et al. 2011) and IRF4 to inhibit Th2 (Cretney et al. 2011). It is thought that direct molecular interactions between foxp3 and these Th subset transcription factors in some way modifies the expression pattern of foxp3 target genes (Samstein et al. 2012a), perhaps to ensure that the appropriate Treg cell subset is engaged and will express appropriate chemokine and homing receptors to be able to localize to the same site as the effector cells, so ensuring efficient regulation at the site of inflammation (Wing and Sakaguchi 2012). There also seem to be a subset of foxp3 positive follicular Treg cells (Tfr cells) that inhibit antibody responses (Wollenberg et al. 2011), which like their foxp3 negative Tfh helper cell equivalents, express high levels of BCL6 and CXCR5, and are restricted to the B cell follicle (Chung et al. 2011; Linterman et al. 2011).

MEMORY RESPONSES AND HETEROLOGOUS IMMUNITY

Although foxp3-expressing Treg are able to control the initiation of all arms of the immune response, are they effective in the face of a memory response? This is particularly important in attempts to translate tolerogenic therapies to humans, where previous blood transfusions, pregnancy, and heterologous immunity from infections (Adams et al. 2003) and homeostatic expansion after T-cell depletion (Wu et al. 2004) can all generate memory to donor alloantigens. In addition, memory T cells have a different requirement for costimulation to naïve T cells (Markovic-Plese et al. 2001; Fuse et al. 2008), so therapeutic approaches using a costimulation blockade may be less effective (Wu et al. 2004). We know, however, that one can induce tolerance in primed recipients (Cobbold et al. 1990; Marshall et al. 1996), and it seems that this is more a quantitative issue, where more Treg or additional immunosuppression are required to control the higher frequency of effector memory cells (Neujahr et al. 2006; Siepert et al. 2012).

OTHER CELLS WITH REGULATORY ACTIVITY

As we have seen, foxp3$^+$ Treg cells function via the combined expression of many different molecular pathways, many of which can also be expressed on other, non-foxp3-expressing T-cell subsets. It is therefore not surprising that a variety of non-foxp3-expressing T cells possess some regulatory activity. Activated T cells, for example, can express CTLA4 and, just like foxp3$^+$ Treg, are able to capture costimulatory ligands by trogocytosis and down-regulate APC (Zhang et al. 2011). Similarly, as has been claimed for granzyme-expressing foxp3$^+$ Treg cells, any cell capable of killing APC, including conventional CTL and NK cells (Cassell and Forman 1990), may be able to suppress the immune response, particular to directly presented alloantigens (He et al. 2007). TGF-β is particularly produced by non-foxp3-expressing Th3 cells implicated in some models of oral tolerance (Fukaura et al. 1996), while IL-10 secretion is a characteristic feature of the foxp3 negative Tr1 cell subset (Groux et al. 1997). And, like Treg, they can constitutively express CTLA4 (Zelenika et al. 2002; Mellor et al. 2004). Adoptive transfer of purified or cloned antigen-specific Tr1 cells can block the induction of inflammatory bowel disease in mice and seem to be as effective as foxp3$^+$ Treg in suppressing allogeneic skin graft rejection (Zelenika et al. 2002). IL-10-secreting Tr1 cells also seem to play an important role in regulating allergic responses in humans (Akdis et al. 2004; Mobs et al. 2010)—with some of the best evidence coming from the studies of seasonal and reciprocal changes in cytokine-secreting T-cell subsets and in response to immunotherapy (McHugh et al. 1996; Fellrath et al. 2003).

Not all cells with regulatory activity are even T cells. As previously discussed, within tolerated grafts and healing self-tissues, many different migratory and tissue cells can respond by contributing components toward the local state of induced immune privilege. Dendritic cells and macrophages, for example, can differentiate, particularly in the presence of anti-inflammatory cytokines such as TGF-β and IL-10, or in the presence of retinoic acid or vitamin D3 (Farquhar et al. 2010), into alternately activated states that preferentially present antigen for Treg differentiation (Gregori et al. 2001). In some transplantation models, it is possible to promote tolerance induction by the adoptive transfer of these syngenic "TolDC" to recipients of skin allografts, although this usually requires some additional immunosuppressive conditioning (Peche et al. 2005). Most recently, there has also been a surge in interest in a population of IL-10-producing "regulatory" B cells (Breg), which were shown to suppress inflammatory responses in various autoimmune disease models (Mauri and Ehrenstein 2008). B-cell-deficient μMT mice, for example, develop a chronic form of experimental autoimmune encephalomyelitis (EAE) compared with the normally self-resolving acute inflammation seen in wild-type mice (Gonnella et al. 2001). Breg seem to be related to the $CD1d^{+}CD5^{+}$ Type 2 and marginal zone B cells often associated with autoreactivity (Lemoine et al. 2009), but their distinguishing feature seems to be the high levels of IL-10 secretion (Carter et al. 2012). In humans, increased frequencies of IL-10-expressing B cells are associated with a positive outcome in renal transplantation patients (Ranjbar et al. 2012). Although Breg cells may be able to induce FasL-mediated apoptosis of effector cells (Lundy 2009), it is thought that they act similarly to TolDC in presenting antigen, in this case in the context of IL-10, for the preferential induction of $foxp3^{+}$ Treg cells (Kessel et al. 2012).

REGULATORY GENES AS POTENTIAL BIOMARKERS OF TOLERANCE?

A major problem for translating tolerogenic- and Treg-based therapies to clinical transplantation is that there is no simple means to identify when tolerance to the donor graft has actually been achieved. This would facilitate any attempt to wean patients off of long-term immunosuppression. As $foxp3^{+}$ Treg play such an important part in the induction and maintenance of transplantation tolerance in mouse models, one might expect that biomarkers of Treg and their activity might be useful surrogate indicators of the development of a tolerant state. This has proven not to be the case for a number of reasons (Cobbold et al. 2011). First, the most effective Treg have specificity for the donor antigen, so the overall frequency of $foxp3^{+}$ Treg is not necessarily correlated with tolerance in mice with a normal TCR repertoire. In addition, antigen-specific Treg may be concentrated in the transplanted organ and poorly correlated with circulating Treg cell frequencies. Whereas graft biopsies might be useful to identify infiltrating Treg cells or the expression of genes associated with a tolerogenic microenvironment, these same processes appear to a large extent to overlap with the normal resolution of inflammation and healing and may therefore not be prognostic for the development of donor-specific tolerance. The only potential biomarker assays we currently have available are derived from gene expression studies in the very small number of patients who have managed to be weaned off immunosuppression and are therefore considered operationally tolerant of their graft (Newell et al. 2010; Sagoo et al. 2010; Brouard et al. 2011). These patterns of gene expression seem to be specific to the organ grafted, or perhaps the immunosuppressive protocol the patients were exposed to, as there seems to be no similarity between, for example, recipients of livers when compared with kidney transplant patients (Londono et al. 2012). Clinical trials to use these biomarkers as a guide to select patients for weaning are at too early a stage to evaluate.

CONCLUDING REMARKS

We are now beginning to understand from studies in animal models the importance of regulatory cells and their interactions with tissues in determining whether a graft is tolerated or rejected, but we are still struggling to successfully translate this knowledge to clinical transplantation. This is, in part, attributable to the remarkable success in developing combinations of immunosuppressive drugs that allow us to effectively control acute graft rejection, yet make it ethically difficult to move to alternative, as yet unproven, tolerogenic strategies. If Treg therapies are ever to be clinically applicable, their logistics must be amenable for commercial development, and this may limit their potential. If it were possible to use drugs and antibodies to potentiate Treg within the patient, then this might be within the realm of current pharmaceutical practice. However, until we more fully understand how Treg function to ensure tolerance in vivo, it will remain a considerable challenge to exploit all their complex properties for therapeutic benefit.

REFERENCES

Adams AB, Pearson TC, Larsen CP. 2003. Heterologous immunity: An overlooked barrier to tolerance. *Immunol Rev* **196:** 147–160.

Akdis M, Verhagen J, Taylor A, Karamloo F, Karagiannidis C, Crameri R, Thunberg S, Deniz G, Valenta R, Fiebig H, et al. 2004. Immune responses in healthy and allergic individuals are characterized by a fine balance between allergen-specific T regulatory 1 and T helper 2 cells. *J Exp Med* **199:** 1567–1575.

Alyanakian MA, You S, Damotte D, Gouarin C, Esling A, Garcia C, Havouis S, Chatenoud L, Bach JF. 2003. Diversity of regulatory $CD4^+$ T cells controlling distinct organ-specific autoimmune diseases. *Proc Natl Acad Sci* **100:** 15806–15811.

Andersen KG, Nissen JK, Betz AG. 2012. Comparative genomics reveals key gain-of-function events in Foxp3 during regulatory T cell evolution. *Front Immunol* **3:** 113.

Asano M, Toda M, Sakaguchi N, Sakaguchi S. 1996. Autoimmune disease as a consequence of developmental abnormality of a T cell subpopulation. *J Exp Med* **184:** 387–396.

Aschenbrenner K, D'Cruz LM, Vollmann EH, Hinterberger M, Emmerich J, Swee LK, Rolink A, Klein L. 2007. Selection of $Foxp3^+$ regulatory T cells specific for self antigen expressed and presented by $Aire^+$ medullary thymic epithelial cells. *Nat Immunol* **8:** 351–358.

Barquilla A, Crespo JL, Navarro M. 2008. Rapamycin inhibits trypanosome cell growth by preventing TOR complex 2 formation. *Proc Natl Acad Sci* **105:** 14579–14584.

Basu S, Golovina T, Mikheeva T, June CH, Riley JL. 2008. Cutting edge: Foxp3-mediated induction of pim 2 allows human T regulatory cells to preferentially expand in rapamycin. *J Immunol* **180:** 5794–5798.

Ben-Shoshan J, Maysel-Auslender S, Mor A, Keren G, George J. 2008. Hypoxia controls $CD4^+CD25^+$ regulatory T-cell homeostasis via hypoxia-inducible factor-1α. *Eur J Immunol* **38:** 2412–2418.

Beverly B, Kang SM, Lenardo MJ, Schwartz RH. 1992. Reversal of in vitro T cell clonal anergy by IL-2 stimulation. *Int Immunol* **4:** 661–671.

Billingham RE, Brent L, Medawar PB. 1953. Actively acquired tolerance of foreign cells. *Nature* **172:** 603–606.

Billingham RE, Brent L, Medawar PB. 1954. Quantitative studies on tissue transplantation immunity. II. The origin, strength and duration of actively and adoptively acquired immunity. *Proc R Soc Lond B Biol Sci* **143:** 58–80.

Bopp T, Becker C, Klein M, Klein-Hessling S, Palmetshofer A, Serfling E, Heib V, Becker M, Kubach J, Schmitt S, et al. 2007. Cyclic adenosine monophosphate is a key component of regulatory T cell-mediated suppression. *J Exp Med* **204:** 1303–1310.

Brouard S, Le Bars A, Dufay A, Gosselin M, Foucher Y, Guillet M, Cesbron-Gautier A, Thervet E, Legendre C, Dugast E, et al. 2011. Identification of a gene expression profile associated with operational tolerance among a selected group of stable kidney transplant patients. *Transpl Int* **24:** 536–547.

Carter NA, Rosser EC, Mauri C. 2012. Interleukin-10 produced by B cells is crucial for the suppression of Th17/Th1 responses, induction of T regulatory type 1 cells and reduction of collagen-induced arthritis. *Arthritis Res Ther* **14:** R32.

Cassell DJ, Forman J. 1990. Regulation of the cytotoxic T lymphocyte response against Qa-1 alloantigens. *J Immunol* **144:** 4075–4081.

Caudy AA, Reddy ST, Chatila T, Atkinson JP, Verbsky JW. 2007. CD25 deficiency causes an immune dysregulation, polyendocrinopathy, enteropathy, X-linked-like syndrome, and defective IL-10 expression from CD4 lymphocytes. *J Allergy Clin Immunol* **119:** 482–487.

Chang JT, Palanivel VR, Kinjyo I, Schambach F, Intlekofer AM, Banerjee A, Longworth SA, Vinup KE, Mrass P, Oliaro J, et al. 2007. Asymmetric T lymphocyte division in the initiation of adaptive immune responses. *Science* **315:** 1687–1691.

Chen W, Jin W, Hardegen N, Lei KJ, Li L, Marinos N, McGrady G, Wahl SM. 2003. Conversion of peripheral $CD4^+CD25-$ naive T cells to $CD4^+CD25^+$ regulatory T cells by TGF-β induction of transcription factor Foxp3. *J Exp Med* **198:** 1875–1886.

Chen TC, Cobbold SP, Fairchild PJ, Waldmann H. 2004. Generation of anergic and regulatory T cells following prolonged exposure to a harmless antigen. *J Immunol* **172:** 5900–5907.

Chen LC, Delgado JC, Jensen PE, Chen X. 2009. Direct expansion of human allospecific $FoxP3^+CD4^+$ regulatory T cells with allogeneic B cells for therapeutic application. *J Immunol* **183:** 4094–4102.

Chung Y, Tanaka S, Chu F, Nurieva RI, Martinez GJ, Rawal S, Wang YH, Lim H, Reynolds JM, Zhou XH, et al. 2011. Follicular regulatory T cells expressing Foxp3 and Bcl-6 suppress germinal center reactions. *Nat Med* **17:** 983–988.

Clambey ET, McNamee EN, Westrich JA, Glover LE, Campbell EL, Jedlicka P, de Zoeten EF, Cambier JC, Stenmark KR, Colgan SP, et al. 2012. Hypoxia-inducible factor-1 α-dependent induction of FoxP3 drives regulatory T-cell abundance and function during inflammatory hypoxia of the mucosa. *Proc Natl Acad Sci* **109:** E2784–E2793.

Cobbold SP, Martin G, Qin S, Waldmann H. 1986. Monoclonal-antibodies to promote marrow engraftment and tissue graft tolerance. *Nature* **323:** 164–166.

Cobbold SP, Martin G, Waldmann H. 1990. The induction of skin graft tolerance in major histocompatibility complex-mismatched or primed recipients: Primed T cells can be tolerized in the periphery with anti-CD4 and anti-CD8 antibodies. *Eur J Immunol* **20:** 2747–2755.

Cobbold SP, Qin SX, Leong LYW, Martin G, Waldmann H. 1992. Reprogramming the immune-system for peripheral tolerance with Cd4 and Cd8 monoclonal-antibodies. *Immunol Rev* **129:** 165–201.

Cobbold SP, Nolan KF, Graca L, Castejon R, Le Moine A, Frewin M, Humm S, Adams E, Thompson S, Zelenika D, et al. 2003. Regulatory T cells and dendritic cells in transplantation tolerance: molecular markers and mechanisms. *Immunol Rev* **196:** 109–124.

Cobbold SP, Castejon R, Adams E, Zelenika D, Graca L, Humm S, Waldmann H. 2004. Induction of $foxP3^+$ regulatory T cells in the periphery of T cell receptor transgenic mice tolerized to transplants. *J Immunol* **172:** 6003–6010.

Cobbold SP, Adams E, Graca L, Daley S, Yates S, Paterson A, Robertson NJ, Nolan KF, Fairchild PJ, Waldmann H. 2006. Immune privilege induced by regulatory T cells in transplantation tolerance. *Immunol Rev* **213:** 239–255.

Cobbold SP, Adams E, Farquhar CA, Nolan KF, Howie D, Lui KO, Fairchild PJ, Mellor AL, Ron D, Waldmann H. 2009. Infectious tolerance via the consumption of essential amino acids and mTOR signaling. *Proc Natl Acad Sci* **106:** 12055–12060.

Cobbold SP, Adams E, Waldmann H. 2011. Biomarkers of transplantation tolerance: More hopeful than helpful? *Front Immunol* **2:** 9.

Collison LW, Workman CJ, Kuo TT, Boyd K, Wang Y, Vignali KM, Cross R, Sehy D, Blumberg RS, Vignali DA. 2007. The inhibitory cytokine IL-35 contributes to regulatory T-cell function. *Nature* **450:** 566–569.

Cone RE, Cochrane R, Lingenheld EG, Clark RB. 1996. Elevation of intracellular cyclic AMP induces an anergic-like state in Th1 clones. *Cell Immunol* **173:** 246–251.

Cretney E, Xin A, Shi W, Minnich M, Masson F, Miasari M, Belz GT, Smyth GK, Busslinger M, Nutt SL, et al. 2011. The transcription factors Blimp-1 and IRF4 jointly control the differentiation and function of effector regulatory T cells. *Nat Immunol* **12:** 304–311.

Daniel C, Wennhold K, Kim HJ, von Boehmer H. 2010. Enhancement of antigen-specific Treg vaccination in vivo. *Proc Natl Acad Sci* **107:** 16246–16251.

Davies JD, Leong LYW, Mellor A, Cobbold SP, Waldmann H. 1996. T cell suppression in transplantation tolerance through linked recognition. *J Immunol* **156:** 3602–3607.

Delgoffe GM, Kole TP, Zheng Y, Zarek PE, Matthews KL, Xiao B, Worley PF, Kozma SC, Powell JD. 2009. The mTOR kinase differentially regulates effector and regulatory T cell lineage commitment. *Immunity* **30:** 832–844.

Earle KE, Tang Q, Zhou X, Liu W, Zhu S, Bonyhadi ML, Bluestone JA. 2005. In vitro expanded human $CD4^+$ $CD25^+$ regulatory T cells suppress effector T cell proliferation. *Clin Immunol* **115:** 3–9.

Edinger M, Hoffmann P, Ermann J, Drago K, Fathman CG, Strober S, Negrin RS. 2003. $CD4^+CD25^+$ regulatory T cells preserve graft-versus-tumor activity while inhibiting graft-versus-host disease after bone marrow transplantation. *Nat Med* **9:** 1144–1150.

Farquhar CA, Paterson AM, Cobbold SP, Garcia Rueda H, Fairchild PJ, Yates SF, Adams E, Saunders NJ, Waldmann H, Nolan KF. 2010. Tolerogenicity is not an absolute property of a dendritic cell. *Eur J Immunol* **40:** 1728–1737.

Fellrath JM, Kettner A, Dufour N, Frigerio C, Schneeberger D, Leimgruber A, Corradin G, Spertini F. 2003. Allergen-specific T-cell tolerance induction with allergen-derived long synthetic peptides: Results of a phase I trial. *J Allergy Clin Immunol* **111:** 854–861.

Fitzgerald DC, Zhang GX, El-Behi M, Fonseca-Kelly Z, Li H, Yu S, Saris CJ, Gran B, Ciric B, Rostami A. 2007. Suppression of autoimmune inflammation of the central nervous system by interleukin 10 secreted by interleukin 27-stimulated T cells. *Nat Immunol* **8:** 1372–1379.

Floess S, Freyer J, Siewert C, Baron U, Olek S, Polansky J, Schlawe K, Chang HD, Bopp T, Schmitt E, et al. 2007. Epigenetic control of the foxp3 locus in regulatory T cells. *PLoS Biol* **5:** e38.

Fontenot JD, Gavin MA, Rudensky AY. 2003. Foxp3 programs the development and function of $CD4^+CD25^+$ regulatory T cells. *Nat Immunol* **4:** 330–336.

Fu S, Yopp AC, Mao X, Chen D, Zhang N, Chen D, Mao M, Ding Y, Bromberg JS. 2004. $CD4^+CD25^+CD62^+$ T-regulatory cell subset has optimal suppressive and proliferative potential. *Am J Transplant* **4:** 65–78.

Fukaura H, Kent SC, Pietrusewicz MJ, Khoury SJ, Weiner HL, Hafler DA. 1996. Induction of circulating myelin basic protein and proteolipid protein-specific transforming growth factor-β1-secreting Th3 T cells by oral administration of myelin in multiple sclerosis patients. *J Clin Invest* **98:** 70–77.

Fuse S, Zhang W, Usherwood EJ. 2008. Control of memory $CD8^+$ T cell differentiation by CD80/CD86-CD28 costimulation and restoration by IL-2 during the recall response. *J Immunol* **180:** 1148–1157.

Gershon RK, Kondo K. 1971. Infectious immunological tolerance. *Immunology* **21:** 903–914.

Gonnella PA, Waldner HP, Weiner HL. 2001. B cell-deficient (μMT) mice have alterations in the cytokine microenvironment of the gut-associated lymphoid tissue (GALT) and a defect in the low dose mechanism of oral tolerance. *J Immunol* **166:** 4456–4464.

Graca L, Cobbold SP, Waldmann H. 2002. Identification of regulatory T cells in tolerated allografts. *J Exp Med* **195:** 1641–1646.

Graca L, Le Moine A, Lin CY, Fairchild PJ, Cobbold SP, Waldmann H. 2004. Donor-specific transplantation tolerance: The paradoxical behavior of $CD4^+CD25^+$ T cells. *Proc Natl Acad Sci* **101:** 10122–10126.

Gregori S, Casorati M, Amuchastegui S, Smiroldo S, Davalli AM, Adorini L. 2001. Regulatory T cells induced by 1 α,25-dihydroxyvitamin D3 and mycophenolate mofetil treatment mediate transplantation tolerance. *J Immunol* **167:** 1945–1953.

Groux H, O'Garra A, Bigler M, Rouleau M, Antonenko S, de Vries JE, Roncarolo MG. 1997. A $CD4^+$ T-cell subset inhibits antigen-specific T-cell responses and prevents colitis. *Nature* **389:** 737–742.

Hale G, Waldmann H. 1996. Recent results using CAMPATH-1 antibodies to control GVHD and graft rejection. *Bone Marrow Transplant* **17:** 305–308.

Hall BM. 1985. Mechanisms maintaining enhancement of allografts. I. Demonstration of a specific suppressor cell. *J Exp Med* **161:** 123–133.

Hara K, Maruki Y, Long X, Yoshino K, Oshiro N, Hidayat S, Tokunaga C, Avruch J, Yonezawa K. 2002. Raptor, a binding partner of target of rapamycin (TOR), mediates TOR action. *Cell* **110:** 177–189.

He KM, Ma Y, Wang S, Min WP, Zhong R, Jevnikar A, Zhang ZX. 2007. Donor double-negative Treg promote allogeneic mixed chimerism and tolerance. *Eur J Immunol* **37:** 3455–3466.

Horak I, Lohler J, Ma A, Smith KA. 1995. Interleukin-2 deficient mice: A new model to study autoimmunity and self-tolerance. *Immunol Rev* **148:** 35–44.

Hori S, Nomura T, Sakaguchi S. 2003. Control of regulatory T cell development by the transcription factor Foxp3. *Science* **299:** 1057–1061.

Jiang S, Herrera O, Lechler RI. 2004. New spectrum of allorecognition pathways: Implications for graft rejection and transplantation tolerance. *Curr Opin Immunol* **16:** 550–557.

Kendal AR, Chen Y, Regateiro FS, Ma JB, Adams E, Cobbold SP, Hori S, Waldmann H. 2011. Sustained suppression by Foxp3$^+$ regulatory T cells is vital for infectious transplantation tolerance. *J Exp Med* **208:** 2043–2053.

Kessel A, Haj T, Peri R, Snir A, Melamed D, Sabo E, Toubi E. 2012. Human CD19$^+$CD25^{high} B regulatory cells suppress proliferation of CD4$^+$ T cells and enhance Foxp3 and CTLA-4 expression in T-regulatory cells. *Autoimmun Rev* **11:** 670–677.

Khattri R, Cox T, Yasayko SA, Ramsdell F. 2003. An essential role for Scurfin in CD4$^+$CD25$^+$ T regulatory cells. *Nat Immunol* **4:** 337–342.

Koch MA, Tucker-Heard G, Perdue NR, Killebrew JR, Urdahl KB, Campbell DJ. 2009. The transcription factor T-bet controls regulatory T cell homeostasis and function during type 1 inflammation. *Nat Immunol* **10:** 595–602.

Kramer S, Schimpl A, Hunig T. 1995. Immunopathology of interleukin IL-2-deficient mice: Thymus dependence and suppression by thymus-dependent cells with an intact IL-2 gene. *J Exp Med* **182:** 1769–1776.

Kretschmer K, Apostolou I, Hawiger D, Khazaie K, Nussenzweig MC, von Boehmer H. 2005. Inducing and expanding regulatory T cell populations by foreign antigen. *Nat Immunol* **6:** 1219–1227.

Kwan WH, van der Touw W, Heeger PS. 2012. Complement regulation of T cell immunity. *Immunol Res* **54:** 247–253.

Lal G, Zhang N, van der Touw W, Ding Y, Ju W, Bottinger EP, Reid SP, Levy DE, Bromberg JS. 2009. Epigenetic regulation of Foxp3 expression in regulatory T cells by DNA methylation. *J Immunol* **182:** 259–273.

Le Friec G, Sheppard D, Whiteman P, Karsten CM, Shamoun SA, Laing A, Bugeon L, Dallman MJ, Melchionna T, Chillakuri C, et al. 2012. The CD46-Jagged1 interaction is critical for human TH1 immunity. *Nat Immunol* **13:** 1213–1221.

Lee HM, Bautista JL, Scott-Browne J, Mohan JF, Hsieh CS. 2012. A broad range of self-reactivity drives thymic regulatory T cell selection to limit responses to self. *Immunity* **37:** 475–486.

Lemoine S, Morva A, Youinou P, Jamin C. 2009. Regulatory B cells in autoimmune diseases: How do they work? *Ann NY Acad Sci* **1173:** 260–267.

Leong LYW, Qin SX, Cobbold SP, Waldmann H. 1992. Classical transplantation tolerance in the adult—the interaction between myeloablation and immunosuppression. *Eur J Immunol* **22:** 2825–2830.

Letourneau S, van Leeuwen EM, Krieg C, Martin C, Pantaleo G, Sprent J, Surh CD, Boyman O. 2010. IL-2/anti-IL-2 antibody complexes show strong biological activity by avoiding interaction with IL-2 receptor α subunit CD25. *Proc Natl Acad Sci* **107:** 2171–2176.

Linterman MA, Pierson W, Lee SK, Kallies A, Kawamoto S, Rayner TF, Srivastava M, Divekar DP, Beaton L, Hogan JJ, et al. 2011. Foxp3$^+$ follicular regulatory T cells control the germinal center response. *Nat Med* **17:** 975–982.

Liu WH, Putnam AL, Xu-Yu Z, Szot GL, Lee MR, Zhu S, Gottlieb PA, Kapranov P, Gingeras TR, Fazekas de St Groth B, et al. 2006. CD127 expression inversely correlates with FoxP3 and suppressive function of human CD4$^+$ T reg cells. *J Exp Med* **203:** 1701–1711.

Londono MC, Danger R, Giral M, Soulillou JP, Sanchez-Fueyo A, Brouard S. 2012. A need for biomarkers of operational tolerance in liver and kidney transplantation. *Am J Transplant* **12:** 1370–1377.

Lu LF, Lind EF, Gondek DC, Bennett KA, Gleeson MW, Pino-Lagos K, Scott ZA, Coyle AJ, Reed JL, Van Snick J, et al. 2006. Mast cells are essential intermediaries in regulatory T-cell tolerance. *Nature* **442:** 997–1002.

Lundy SK. 2009. Killer B lymphocytes: The evidence and the potential. *Inflam Res* **58:** 345–357.

Markovic-Plese S, Cortese I, Wandinger KP, McFarland HF, Martin R. 2001. CD4$^+$CD28$^-$ costimulation-independent T cells in multiple sclerosis. *J Clin Invest* **108:** 1185–1194.

Marshall SE, Cobbold SP, Davies JD, Martin GM, Phillips JM, Waldmann H. 1996. Tolerance and suppression in a primed immune system. *Transplantation* **62:** 1614–1621.

Mauri C, Ehrenstein MR. 2008. The "short" history of regulatory B cells. *Trends Immunol* **29:** 34–40.

Mauri C, Gray D, Mushtaq N, Londei M. 2003. Prevention of arthritis by interleukin 10-producing B cells. *J Exp Med* **197:** 489–501.

McClymont SA, Putnam AL, Lee MR, Esensten JH, Liu WH, Hulme MA, Hoffmuller U, Baron U, Olek S, Bluestone JA, et al. 2011. Plasticity of human regulatory t cells in healthy subjects and patients with type 1 diabetes. *J Immunol* **186:** 3918–3926.

McHugh S, Deighton J, Rifkin I, Ewan P. 1996. Kinetics and functional implications of Th1 and Th2 cytokine production following activation of peripheral blood mononuclear cells in primary culture. *Eur J Immunol* **26:** 1260–1265.

Mellor AL, Chandler P, Baban B, Hansen AM, Marshall B, Pihkala J, Waldmann H, Cobbold S, Adams E, Munn DH. 2004. Specific subsets of murine dendritic cells acquire potent T cell regulatory functions following CTLA4-mediated induction of indoleamine 2,3 dioxygenase. *Int Immunol* **16:** 1391–1401.

Mezrich JD, Fechner JH, Zhang X, Johnson BP, Burlingham WJ, Bradfield CA. 2010. An interaction between kynurenine and the aryl hydrocarbon receptor can generate regulatory T cells. *J Immunol* **185:** 3190–3198.

Mobs C, Slotosch C, Loffler H, Jakob T, Hertl M, Pfutzner W. 2010. Birch pollen immunotherapy leads to differential induction of regulatory T cells and delayed helper T cell immune deviation. *J Immunol* **184:** 2194–2203.

Morgan RD, O'Callaghan JM, Knight SR, Morris PJ. 2012. Alemtuzumab induction therapy in kidney transplantation: A systematic review and meta-analysis. *Transplantation* **93:** 1179–1188.

Munn DH, Zhou M, Attwood JT, Bondarev I, Conway SJ, Marshall B, Brown C, Mellor AL. 1998. Prevention of allogeneic fetal rejection by tryptophan catabolism. *Science* **281:** 1191–1193.

Munn DH, Sharma MD, Baban B, Harding HP, Zhang Y, Ron D, Mellor AL. 2005. GCN2 kinase in T cells mediates proliferative arrest and anergy induction in response to indoleamine 2,3-dioxygenase. *Immunity* **22:** 633–642.

Nakamura K, Kitani A, Strober W. 2001. Cell contact-dependent immunosuppression by CD4$^+$CD25$^+$ regulatory T cells is mediated by cell surface-bound transforming growth factor β. *J Exp Med* **194:** 629–644.

Nakamura K, Kitani A, Fuss I, Pedersen A, Harada N, Nawata H, Strober W. 2004. TGF-β 1 plays an important role in the mechanism of $CD4^+CD25^+$ regulatory T cell activity in both humans and mice. *J Immunol* **172:** 834–842.

Neujahr DC, Chen CQ, Huang X, Markmann JF, Cobbold S, Waldmann H, Sayegh MH, Hancock WW, Turka LA. 2006. Accelerated memory cell homeostasis during T cell depletion and approaches to overcome it. *J Immunol* **176:** 4632–4639.

Neumuller RA, Knoblich JA. 2009. Dividing cellular asymmetry: Asymmetric cell division and its implications for stem cells and cancer. *Genes Dev* **23:** 2675–2699.

Newell KA, Asare A, Kirk AD, Gisler TD, Bourcier K, Suthanthiran M, Burlingham WJ, Marks WH, Sanz I, Lechler RI, et al. 2010. Identification of a B cell signature associated with renal transplant tolerance in humans. *J Clin Invest* **120:** 1836–1847.

Nowak EC, de Vries VC, Wasiuk A, Ahonen C, Bennett KA, Le Mercier I, Ha DG, Noelle RJ. 2012. Tryptophan hydroxylase-1 regulates immune tolerance and inflammation. *J Exp Med* **209:** 2127–2135.

Ohkura N, Hamaguchi M, Morikawa H, Sugimura K, Tanaka A, Ito Y, Osaki M, Tanaka Y, Yamashita R, Nakano N, et al. 2012. T cell receptor stimulation-induced epigenetic changes and foxp3 expression are independent and complementary events required for treg cell development. *Immunity* **37:** 785–799.

Paidassi H, Acharya M, Lacy-Hulbert A. 2010. α (v) Integrins license regulatory T cells to apoptotic cells and self-associated antigens. *Ann NY Acad Sci* **1209:** 68–76.

Paidassi H, Acharya M, Zhang A, Mukhopadhyay S, Kwon M, Chow C, Stuart LM, Savill J, Lacy-Hulbert A. 2011. Preferential expression of integrin αvβ8 promotes generation of regulatory T cells by mouse $CD103^+$ dendritic cells. *Gastroenterology* **141:** 1813–1820.

Park HB, Paik DJ, Jang E, Hong S, Youn J. 2004. Acquisition of anergic and suppressive activities in transforming growth factor-β-costimulated $CD4^+CD25^-$ T cells. *Int Immunol* **16:** 1203–1213.

Peche H, Trinite B, Martinet B, Cuturi MC. 2005. Prolongation of heart allograft survival by immature dendritic cells generated from recipient type bone marrow progenitors. *Am J Transplant* **5:** 255–267.

Powell JD, Lerner CG, Ewoldt GR, Schwartz RH. 1999. The -180 site of the IL-2 promoter is the target of CREB/CREM binding in T cell anergy. *J Immunol* **163:** 6631–6639.

Powrie F, Correa-Oliveira R, Mauze S, Coffman RL. 1994. Regulatory interactions between $CD45RB^{high}$ and $CD45RB^{low}$ $CD4^+$ T cells are important for the balance between protective and pathogenic cell-mediated immunity. *J Exp Med* **179:** 589–600.

Qin SX, Cobbold S, Benjamin R, Waldmann H. 1989. Induction of classical transplantation tolerance in the adult. *J Exp Med* **169:** 779–794.

Qin SX, Wise M, Cobbold SP, Leong L, Kong YCM, Parnes JR, Waldmann H. 1990. Induction of tolerance in peripheral T-cells with monoclonal-antibodies. *Eur J Immunol* **20:** 2737–2745.

Qin SX, Cobbold SP, Pope H, Elliott J, Kioussis D, Davies J, Waldmann H. 1993. Infectious transplantation tolerance. *Science* **259:** 974–977.

Qureshi OS, Zheng Y, Nakamura K, Attridge K, Manzotti C, Schmidt EM, Baker J, Jeffery LE, Kaur S, Briggs Z, et al. 2011. Trans-endocytosis of CD80 and CD86: A molecular basis for the cell-extrinsic function of CTLA-4. *Science* **332:** 600–603.

Ranjbar M, Solgi G, Mohammadnia M, Nikbin B, Pourmand G, Ansaripour B, Amirzargar A. 2012. Regulatory T-cell subset analysis and profile of interleukin (IL)-10, IL-17 and interferon-γ cytokine-producing cells in kidney allograft recipients with donor cells infusion. *Clin Exp Nephrol* **16:** 636–646.

Regateiro FS, Howie D, Nolan KF, Agorogiannis EI, Greaves DR, Cobbold SP, Waldmann H. 2011. Generation of anti-inflammatory adenosine by leukocytes is regulated by TGF-β. *Eur J Immunol* **41:** 2955–2965.

Regateiro FS, Chen Y, Kendal AR, Hilbrands R, Adams E, Cobbold SP, Ma JB, Andersen KG, Betz AG, Zhang M, et al. 2012. Foxp3 expression is required for the induction of therapeutic tissue tolerance. *J Immunol* **189:** 3947–3956.

Regateiro FS, Cobbold SP, Waldmann H. 2013. CD73 and adenosine generation in the creation of regulatory microenvironments. *Clin Exp Immunol* **171:** 1–7.

Roussey-Kesler G, Giral M, Moreau A, Subra JF, Legendre C, Noel C, Pillebout E, Brouard S, Soulillou JP. 2006. Clinical operational tolerance after kidney transplantation. *Am J Transplant* **6:** 736–746.

Rubtsov YP, Niec RE, Josefowicz S, Li L, Darce J, Mathis D, Benoist C, Rudensky AY. 2010. Stability of the regulatory T cell lineage in vivo. *Science* **329:** 1667–1671.

Sagoo P, Perucha E, Sawitzki B, Tomiuk S, Stephens DA, Miqueu P, Chapman S, Craciun L, Sergeant R, Brouard S, et al. 2010. Development of a cross-platform biomarker signature to detect renal transplant tolerance in humans. *J Clin Invest* **120:** 1848–1861.

Samstein RM, Arvey A, Josefowicz SZ, Peng X, Reynolds A, Sandstrom R, Neph S, Sabo P, Kim JM, Liao W, et al. 2012a. Foxp3 exploits a pre existent enhancer landscape for regulatory T cell lineage specification. *Cell* **151:** 153–166.

Samstein RM, Josefowicz SZ, Arvey A, Treuting PM, Rudensky AY. 2012b. Extrathymic generation of regulatory T cells in placental mammals mitigates maternal-fetal conflict. *Cell* **150:** 29–38.

Sancak Y, Bar-Peled L, Zoncu R, Markhard AL, Nada S, Sabatini DM. 2010. Ragulator-Rag complex targets mTORC1 to the lysosomal surface and is necessary for its activation by amino acids. *Cell* **141:** 290–303.

Sanchez-Fueyo A, Domenig CM, Mariat C, Alexopoulos S, Zheng XX, Strom TB. 2007. Influence of direct and indirect allorecognition pathways on $CD4^+CD25^+$ regulatory T-cell function in transplantation. *Transpl Int* **20:** 534–541.

Sato K, Yamashita N, Yamashita N, Baba M, Matsuyama T. 2003. Regulatory dendritic cells protect mice from murine acute graft-versus-host disease and leukemia relapse. *Immunity* **18:** 367–379.

Schwartz RH. 1990. A cell culture model for T lymphocyte clonal anergy. *Science* **248:** 1349–1356.

Scully R, Qin SX, Cobbold S, Waldmann H. 1994. Mechanisms in Cd4 antibody-mediated transplantation tolerance—Kinetics of induction, antigen dependency and role of regulatory T-cells. *Eur J Immunol* **24:** 2383–2392.

Setoguchi R, Hori S, Takahashi T, Sakaguchi S. 2005. Homeostatic maintenance of natural Foxp3$^+$ CD25$^+$ CD4$^+$ regulatory T cells by interleukin IL-2 and induction of autoimmune disease by IL-2 neutralization. *J Exp Med* **201:** 723–735.

Shah AH, Tabayoyong WB, Kimm SY, Kim SJ, Van Parijs L, Lee C. 2002. Reconstitution of lethally irradiated adult mice with dominant negative TGF-β type II receptor-transduced bone marrow leads to myeloid expansion and inflammatory disease. *J Immunol* **169:** 3485–3491.

Sharma MD, Baban B, Chandler P, Hou DY, Singh N, Yagita H, Azuma M, Blazar BR, Mellor AL, Munn DH. 2007. Plasmacytoid dendritic cells from mouse tumor-draining lymph nodes directly activate mature Tregs via indoleamine 2,3-dioxygenase. *J Clin Invest* **117:** 2570–2582.

Shevach EM. 2009. Mechanisms of foxp3$^+$ T regulatory cell-mediated suppression. *Immunity* **30:** 636–645.

Shevach EM, DiPaolo RA, Andersson J, Zhao DM, Stephens GL, Thornton AM. 2006. The lifestyle of naturally occurring CD4$^+$CD25$^+$ Foxp3$^+$ regulatory T cells. *Immunol Rev* **212:** 60–73.

Shi LZ, Wang R, Huang G, Vogel P, Neale G, Green DR, Chi H. 2011. HIF1α-dependent glycolytic pathway orchestrates a metabolic checkpoint for the differentiation of TH17 and Treg cells. *J Exp Med* **208:** 1367–1376.

Shull MM, Ormsby I, Kier AB, Pawlowski S, Diebold RJ, Yin M, Allen R, Sidman C, Proetzel G, Calvin D, et al. 1992. Targeted disruption of the mouse transforming growth factor-β 1 gene results in multifocal inflammatory disease. *Nature* **359:** 693–699.

Siepert A, Ahrlich S, Vogt K, Appelt C, Stanko K, Kuhl A, van den Brandt J, Reichardt HM, Nizze H, Lehmann M, et al. 2012. Permanent CNI treatment for prevention of renal allograft rejection in sensitized hosts can be replaced by regulatory T cells. *Am J Transplant* **12:** 2384–2394.

Sojka DK, Hughson A, Fowell DJ. 2009. CTLA-4 is required by CD4$^+$CD25$^+$ Treg to control CD4$^+$ T-cell lymphopenia-induced proliferation. *Eur J Immunol* **39:** 1544–1551.

Stassen M, Fondel S, Bopp T, Richter C, Muller C, Kubach J, Becker C, Knop J, Enk AH, Schmitt S, et al. 2004. Human CD25$^+$ regulatory T cells: Two subsets defined by the integrins α4β7 or α4β1 confer distinct suppressive properties upon CD4$^+$ T helper cells. *Eur J Immunol* **34:** 1303–1311.

Streilein JW, Gruchalla RS. 1981. Analysis of neonatally induced tolerance of H-2 alloantigens. I. Adoptive transfer indicates that tolerance of class I and class II antigens is maintained by distinct mechanisms. *Immunogenetics* **12:** 161–173.

Sundstedt A, O'Neill EJ, Nicolson KS, Wraith DC. 2003. Role for IL-10 in suppression mediated by peptide-induced regulatory T cells in vivo. *J Immunol* **170:** 1240–1248.

Takahashi H, Kanno T, Nakayamada S, Hirahara K, Sciume G, Muljo SA, Kuchen S, Casellas R, Wei L, Kanno Y, et al. 2012. TGF-β and retinoic acid induce the microRNA miR-10a, which targets Bcl-6 and constrains the plasticity of helper T cells. *Nat Immunol* **13:** 587–595.

Taner T, Hackstein H, Wang Z, Morelli AE, Thomson AW. 2005. Rapamycin-treated, alloantigen-pulsed host dendritic cells induce ag-specific T cell regulation and prolong graft survival. *Am J Transplant* **5:** 228–236.

Tato I, Bartrons R, Ventura F, Rosa JL. 2011. Amino acids activate mammalian target of rapamycin complex 2 (mTORC2) via PI3K/Akt signaling. *J Biol Chem* **286:** 6128–6142.

Thornton AM, Korty PE, Tran DQ, Wohlfert EA, Murray PE, Belkaid Y, Shevach EM. 2010. Expression of Helios, an Ikaros transcription factor family member, differentiates thymic-derived from peripherally induced Foxp3$^+$ T regulatory cells. *J Immunol* **184:** 3433–3441.

Tivol EA, Borriello F, Schweitzer AN, Lynch WP, Bluestone JA, Sharpe AH. 1995. Loss of CTLA-4 leads to massive lymphoproliferation and fatal multiorgan tissue destruction, revealing a critical negative regulatory role of CTLA-4. *Immunity* **3:** 541–547.

Tran DQ, Andersson J, Wang R, Ramsey H, Unutmaz D, Shevach EM. 2009. GARP (LRRC32) is essential for the surface expression of latent TGF-β on platelets and activated FOXP3$^+$ regulatory T cells. *Proc Natl Acad Sci* **106:** 13445–13450.

Trenado A, Charlotte F, Fisson S, Yagello M, Klatzmann D, Salomon BL, Cohen JL. 2003. Recipient-type specific CD4$^+$CD25$^+$ regulatory T cells favor immune reconstitution and control graft-versus-host disease while maintaining graft-versus-leukemia. *J Clin Invest* **112:** 1688–1696.

Tsang JY, Tanriver Y, Jiang S, Leung E, Ratnasothy K, Lombardi G, Lechler R. 2009. Indefinite mouse heart allograft survival in recipient treated with CD4$^+$CD25$^+$ regulatory T cells with indirect allospecificity and short term immunosuppression. *Transpl Immunol* **21:** 203–209.

Uhlig HH, Coombes J, Mottet C, Izcue A, Thompson C, Fanger A, Tannapfel A, Fontenot JD, Ramsdell F, Powrie F. 2006. Characterization of Foxp3$^+$CD4$^+$CD25$^+$ and IL-10-secreting CD4$^+$CD25$^+$ T cells during cure of colitis. *J Immunol* **177:** 5852–5860.

Veerapathran A, Pidala J, Beato F, Yu XZ, Anasetti C. 2011. Ex vivo expansion of human Tregs specific for alloantigens presented directly or indirectly. *Blood* **118:** 5671–5680.

Wang J, Ioan-Facsinay A, van der Voort EI, Huizinga TW, Toes RE. 2007. Transient expression of FOXP3 in human activated nonregulatory CD4$^+$ T cells. *Eur J Immunol* **37:** 129–138.

Wang Y, Su MA, Wan YY. 2011. An essential role of the transcription factor GATA-3 for the function of regulatory T cells. *Immunity* **35:** 337–348.

Whitehead GS, Wilson RH, Nakano K, Burch LH, Nakano H, Cook DN. 2012. IL-35 production by inducible costimulator (ICOS)-positive regulatory T cells reverses established IL-17-dependent allergic airways disease. *J Allergy Clin Immunol* **129:** 207–215.

Wing JB, Sakaguchi S. 2012. Multiple treg suppressive modules and their adaptability. *Front Immunol* **3:** 178.

Wollenberg I, Agua-Doce A, Hernandez A, Almeida C, Oliveira VG, Faro J, Graca L. 2011. Regulation of the germinal center reaction by Foxp3$^+$ follicular regulatory T cells. *J Immunol* **187:** 4553–4560.

Wu Z, Bensinger SJ, Zhang J, Chen C, Yuan X, Huang X, Markmann JF, Kassaee A, Rosengard BR, Hancock WW, et al. 2004. Homeostatic proliferation is a barrier to transplantation tolerance. *Nat Med* **10:** 87–92.

Yadav M, Louvet C, Davini D, Gardner JM, Martinez-Llordella M, Bailey-Bucktrout S, Anthony BA, Sverdrup FM, Head R, Kuster DJ, et al. 2012. Neuropilin-1 distinguishes natural and inducible regulatory T cells among regulatory T cell subsets in vivo. *J Exp Med* **209:** 1713–1722.

Yang XO, Nurieva R, Martinez GJ, Kang HS, Chung Y, Pappu BP, Shah B, Chang SH, Schluns KS, Watowich SS, et al. 2008. Molecular antagonism and plasticity of regulatory and inflammatory T cell programs. *Immunity* **29:** 44–56.

Yu P, Gregg RK, Bell JJ, Ellis JS, Divekar R, Lee HH, Jain R, Waldner H, Hardaway JC, Collins M, et al. 2005. Specific T regulatory cells display broad suppressive functions against experimental allergic encephalomyelitis upon activation with cognate antigen. *J Immunol* **174:** 6772–6780.

Yurchenko E, Shio MT, Huang TC, Da Silva Martins M, Szyf M, Levings MK, Olivier M, Piccirillo CA. 2012. Inflammation-driven reprogramming of CD4$^+$ Foxp3$^+$ regulatory T cells into pathogenic Th1/Th17 T effectors is abrogated by mTOR inhibition in vivo. *PloS ONE* **7:** e35572.

Zelenika D, Adams E, Humm S, Graca L, Thompson S, Cobbold SP, Waldmann H. 2002. Regulatory T cells overexpress a subset of Th2 gene transcripts. *J Immunol* **168:** 1069–1079.

Zhang L, Gu P, Gao JF, Kowalczyk A, D'Souza CA, Chou KY. 2011. Trogocytosis of CD80 and CD86 by induced regulatory T cells. *Am J Transplant* **11:** 187–188.

Zheng Y, Josefowicz S, Chaudhry A, Peng XP, Forbush K, Rudensky AY. 2010. Role of conserved non-coding DNA elements in the Foxp3 gene in regulatory T-cell fate. *Nature* **463:** 808–812.

Zhou X, Bailey-Bucktrout S, Jeker LT, Bluestone JA. 2009. Plasticity of CD4$^+$ FoxP3$^+$ T cells. *Curr Opin Immunol* **21:** 281–285.

Zhou X, Kong N, Wang J, Fan H, Zou H, Horwitz D, Brand D, Liu Z, Zheng SG. 2010. Cutting edge: All-*trans* retinoic acid sustains the stability and function of natural regulatory T cells in an inflammatory milieu. *J Immunol* **185:** 2675–2679.

Zoncu R, Bar-Peled L, Efeyan A, Wang S, Sancak Y, Sabatini DM. 2011. mTORC1 senses lysosomal amino acids through an inside-out mechanism that requires the vacuolar H$^+$-ATPase. *Science* **334:** 678–683.

Regulatory T-Cell Therapy in Transplantation: Moving to the Clinic

Qizhi Tang[1] and Jeffrey A. Bluestone[2]

[1]Department of Surgery, University of California, San Francisco, San Francisco, California 94143

[2]UCSF Diabetes Center, University of California, San Francisco, San Francisco, California 94143

Correspondence: qizhi.tang@ucsfmedctr.org

Regulatory T cells (Tregs) are essential to transplantation tolerance and their therapeutic efficacy is well documented in animal models. Moreover, human Tregs can be identified, isolated, and expanded in short-term ex vivo cultures so that a therapeutic product can be manufactured at relevant doses. Treg therapy is being planned at multiple transplant centers around the world. In this article, we review topics critical to effective implementation of Treg therapy in transplantation. We will address issues such as Treg dose, antigen specificity, and adjunct therapies required for transplant tolerance induction. We will summarize technical advances in Treg manufacturing and provide guidelines for identity and purity assurance of Treg products. Clinical trial designs and Treg manufacturing plans that incorporate the most up-to-date scientific understanding in Treg biology will be essential for harnessing the tolerogenic potential of Treg therapy in transplantation.

One of the major challenges facing the field of transplantation is the management of immunosuppression. Although immunosuppression is necessary to prevent immune attacks of the transplanted organ, it also imposes substantial morbidity and mortality risks for transplant recipients. Chronic global immunosuppression impairs immune responses to microbial pathogens and hinders tumor immunosurveillance. Often, infections and posttransplant cancers, rather than allograft rejection, are the major contributors of transplant-related mortality, especially beyond the first year after transplant (Penn 1990; Euvrard et al. 2003; Soltys et al. 2007). In addition to these immunological complications, immunosuppressive drugs are often causes of morbidity owing to their off-target effects such as nephrotoxicity, diabetes, hyperlipidemia, hypertension, cardiovascular diseases, and obesity (Berenson et al. 1992; Textor et al. 2000; Nair et al. 2002; Ojo et al. 2003). All these complications may necessitate reduction or even withdrawal of immunosuppression that leads to graft rejection and graft loss. Therefore, the key to improving immunosuppression after transplantation is to selectively block the immune responses against the graft without impeding other protective immune functions or causing nonspecific toxicities.

In this article, we propose that such goal can be accomplished by harnessing the natural immune regulatory mechanisms using cell-

Cite this article as *Cold Spring Harb Perspect Med* doi: 10.1101/cshperspect.a015552

based therapies. Various types of T cells have been shown to contribute to transplant tolerance. These include the $CD4^+CD25^+$ regulatory T cells (Tregs) that express the transcription factor FOXP3, IL-10-producing Tr1 cells, $CD8^+28^-$ T cells, and anergic T cells. In this article, we will focus on the FOXP3-expressing Tregs. We summarize parameters that are important for effective application of Treg therapy to prevent graft rejection in experimental models and review advances in translating these preclinical experiences to the clinic.

ADVANTAGES OF Treg THERAPY IN TRANSPLANTATION

None of the current immunosuppressive drugs can suppress immune responses to transplant antigens without potentially altering immune surveillance toward tumor antigens and microbial pathogens. This is because immunosuppressive drugs target common pathways of immune activation, such as calcium signaling (cyclosporin A, FK506), purine biosynthesis (Mycophenolate), and T-cell costimulation (CTLA4Ig). Other immunosuppressive drugs such as antithymocyte globulin, Campath-1, and anti-CD20, massively and nonspecifically delete immune cells. In contrast, T cells have an extraordinary ability to distinguish minute differences among different antigens. Aside from the specificity conferred by the T-cell receptors, specificity of therapeutic T cells is also amplified by their ability to seek their targets throughout the body and deliver effector functions locally where they are most effective and specific. These properties underlie the remarkable efficacy of T-cell therapy in treating drug-resistant recurrent cancers (Restifo et al. 2012; Scholler et al. 2012). For example, using autologous cytotoxic T cells engineered to express a chimeric receptor that recognizes CD19, 19 out 20 patients with end-stage therapy-refractory B-cell lymphoma have been successfully treated (Porter et al. 2011; Scholler et al. 2012). Thus, T cells can be used to deliver highly specific and targeted therapies.

Similarly, immune tolerance mediated by Tregs is also highly antigen specific. Immunosuppressive functions of Tregs are activated by the engagement of their T-cell receptors locally at the site of antigen deposition. In a mouse model of autoimmune diabetes and autoimmune pancreatitis, we have observed complete protection against islet destruction using islet antigen-specific Tregs, whereas autoimmune attack of the surrounding exocrine pancreas progressed without hindrance (Meagher et al. 2008). In the transplant setting, tolerance maintained by Tregs is specific to the graft donor, whereas unrelated grafts are rejected (Joffre et al. 2008). Tolerance can spread to a new transplant antigen, and this "linked suppression" is mediated by Tregs in the graft. The establishment of linked suppression requires that the new antigen and the antigen the host is already tolerant to are present in the same graft. Thus, immune tolerance to transplant antigens mediated by Tregs is highly specific and highly localized. Despite their high antigen specificity, Tregs are highly versatile and can control responses of various immune cells including conventional $CD4^+$ T cells, $CD8^+$ T cells, natural killer (NK) cells, natural killer T (NKT) cells, B cells, and various antigen-presenting cells. Moreover, Tregs have a collection of more than a dozen different immunosuppressive mechanisms and can deploy different strategies depending on the tissue microenvironment (Tang and Bluestone 2008; Yamaguchi et al. 2011). In comparison to standard immunosuppressive drugs used in transplantation today, Tregs are "smart" therapeutic agents that are highly antigen specific and highly adaptable, capable of selectively targeting graft tissues by tuning their activities in response to the tissue microenvironment.

Besides the problems with long-term global immunosuppression and off-target toxicities, most of the current immunosuppressive drugs prevent tolerance induction and create a dependence on continuous immunosuppression. Immunosuppressive drugs used in the clinic today are selected based on their ability to prevent immune activation. Research from the past three decades shows that acquisition of immune tolerance to self-antigens and transplantation antigens is an active process that requires antigen exposure. A tolerogenic antigen exposure leads to inactivation of antigen-reactive T

 Cite this article as *Cold Spring Harb Perspect Med* doi: 10.1101/cshperspect.a015552

cells through apoptosis, anergy, and induction and expansion of immune regulatory mechanisms that maintain tolerance. The current approach of immunosuppressing transplant recipients blindfolds the immune system to prevent rejection, but also impedes tolerance induction. This may explain why spontaneous transplant tolerance is rare and the best predictor of tolerance is time after transplant (Sanchez-Fueyo 2011), likely through the cumulative effects of a low level of donor antigen exposure over a long period of time. To induce tolerance and freedom from chronic immunosuppression, transplant recipients should receive immunoregulatory regimens that block rejection while permitting donor antigen recognition. In this regard, Tregs can induce other T cells to acquire regulatory functions in vitro (Dieckmann et al. 2002; Jonuleit et al. 2002; Oliveira et al. 2011). In a mouse model of transplantation, Treg therapy not only prevents rejection, but also allows the induction of new Tregs with broader specificities through a process of infectious tolerance (Waldmann 2008). Therefore, Treg therapy can potentially turn the graft tissue from a target of immune attack into a tolerogenic organ that promotes its own long-term survival.

CRITICAL PARAMETERS FOR EFFECTIVE Treg THERAPY IN TRANSPLANTATION

Dosing

Tregs ensure normal immune homeostasis by providing a counterbalance for the effector arm of the immune system. In an immunologically quiescent state, Tregs represent ~5%–10% of $CD4^+$ T cells in lymphoid tissues. At this ratio, Tregs prevents unwanted immune activation by reducing expression of costimulatory molecules CD80 and CD86 via CTLA-4-mediated trogocytosis (Muthukumar et al. 2005; Dijke et al. 2007) and by sopping up IL-2 and other common γ-chain-binding cytokines (Pandiyan et al. 2007; O'Gorman et al. 2009). During an active immune response, Tregs proliferate, traffic, and accumulate at the site of inflammation, particularly at the later phase of the response, to restore normal immune homeostasis using a wider array of effector mechanisms including immunosuppressive cytokines IL-10, IL-35, TGF-β, cell-surface ATPases, granzyme-dependent killing of antigen-presenting cells, IL-9-mediated recruitment of mast cell, etc. (Tang and Bluestone 2008; Yamaguchi et al. 2011). Consequently, the number of Tregs often increases with inflammation and graft rejection (Muthukumar et al. 2005; Dijke et al. 2007). In the transplant setting, such increases are usually not sufficient and too late to prevent graft damage; therefore, effective Treg therapy should be given at a dose sufficient to tip the balance in favor of Tregs before the rise of effector responses.

Two approaches can be applied to estimate the effective dose of Tregs in humans to prevent graft rejection: one based on achieving a percentage of Tregs required to tip the balance to tolerance and the other based on allometric scaling from mouse transplant models. Early proof-of-principle experiments in mouse models relied on the use of adoptive transfer of a mixture of effector T cells and Tregs to lymphopenic hosts to precisely control and balance between the two populations (Hara et al. 2001; Graca et al. 2002). In these settings, a high ratio of at least one Treg per two effector T cells, and sometimes as high as five Tregs per one effector cell, is needed to prevent rejection (Nishimura et al. 2004; Golshayan et al. 2007). This suggests that a minimum of 33% Tregs is required to prevent rejection. Experiments in lymphoreplete mice also find that 30% Tregs in the grafts is associated with prevention of graft rejection (Fan et al. 2010). Interestingly, Tregs are also found to accumulate to 30% in tumor tissues and are thought to contribute to the suppression of antitumor immunity. Thus, 30% Tregs is likely the tipping point between productive immunity and tolerance. We have estimated that an average adult human has 166×10^9 $CD4^+$ T cells in the body, including lymphoid and nonlymphoid organs, and among which 13×10^9 are Tregs (Tang and Lee 2012). To increase the percentage of Tregs to 30% by using Treg therapy alone, 53×10^9 Treg would be needed. Aside from the technical challenge of producing such a high dose of Tregs (see below), it is not clear if Tregs can engraft when infused at such a high bolus dose. However, 10 times less (5×10^9)

Treg would be sufficient if Treg therapy is combined with prior depletion of 90% of the T cells using an immune T-cell-depleting agent such as antithymocyte immunoglobulin.

Scaling for immune cell therapy may be less complicated and unpredictable than that for small molecule drugs because pharmacokinetic and pharmacodynamic properties of immune cells (i.e., rules for cell trafficking, turnover, and dose response) are relatively insensitive to the differences in body mass and metabolic rate of different species (Wiegel and Perelson 2004; Perelson and Wiegel 2009). For example, clinical data obtained from hematopoietic stem cell transplant shows that the minimal effective dose for neutrophil and platelet reconstitution in humans is consistent with that predicted by allometric scaling of data from animal models. Thus, the number of Tregs needed to achieve efficacy in humans may be extrapolated using numbers from mouse experiments multiplied by the ratio between the sizes of the human and the mouse lymphoid compartments. One way to approximate the relative size of the mouse and human lymphoid compartments is by comparing the lymphocyte cellularity of the spleens. Adult human spleens contain 70×10^9 lymphocytes (Westermann and Pabst 1992; Ganusov and De Boer 2007; Nylen et al. 2007), 1000 times that in adult mouse spleens. Therefore effective Treg dose in humans is likely 1000 times that found in mouse models. There is very little experimental evidence for effective Treg dose in lymphoreplete mouse models. In a mouse islet transplant model, 1×10^6 unmanipulated polyclonal Tregs were able to prolong graft survival by 2 wk (Zhang et al. 2009). Others used 2 to 5×10^6 Tregs enriched for donor antigen reactivity and also observed limited prolongation of graft survival unless combined with other treatments (Golshayan et al. 2007; Joffre et al. 2008; Tsang et al. 2008). We have found that 30×10^6 polyclonal Tregs was able to induce long-term graft survival only when combined with substantial deletion of donor-reactive T cells from the hosts (K Lee and Q Tang, unpubl.). Thus, using allomeric scaling, effective dose of polyclonal Tregs for preventing rejection in humans is estimated to be $>30 \times 10^9$.

The total number of Tregs in an adult human is estimated to be 13×10^9 and most of the Tregs reside in lymphoid organs (Tang and Lee 2012). The number of Tregs circulating in the blood is 0.25×10^9, which is the most that can be isolated from an autologous source for therapeutic use. It is important to point out that there is limited information about the percentage of Tregs in the body that actually circulates. Tregs in the skin and gut have been suggested to be largely resident with little ability to circulate. This is especially true of memory Tregs (Rosenblum et al. 2011). In a recent clinical trial in type 1 diabetes conducted at UCSF, deuterium-labeled ex vivo expanded Tregs were tracked in vivo. Our preliminary results showed that a dose of $\sim 350 \times 10^6$ led to a 3%–5% increase of the circulating Tregs. It is clear that ex vivo expansion is needed to dramatically increase the number of Tregs (JA Bluestone, MK Hellerstein, SE Gitelman et al., unpubl.).

Specificity

In mouse models of Treg therapy in transplantation, a consistent feature is that Tregs purified from tolerant hosts are more effective in transferring tolerance to new hosts than Tregs from naïve hosts. The improved efficacy is likely owing to an increased frequency of donor alloantigen-reactive Tregs induced by various tolerance-inducing protocols (Bushell et al. 1995; Cobbold et al. 2004; Karim et al. 2004; Ochando et al. 2006; Yates et al. 2007; Verginis et al. 2008; Francis et al. 2011). There are two types of alloantigen-reactive T cells. The "direct" alloantigen-reactive T cells recognize intact alloantigen expressed on donor cells, and the "indirect" alloantigen-reactive T cells recognize processed donor alloantigens presented by host antigen-presenting cells (Gould and Auchincloss 1999; Rogers and Lechler 2001). Interestingly, allograft tolerance is primarily mediated through the indirect pathway (Hara et al. 2001; Yamada et al. 2001; Callaghan et al. 2007; Sanchez-Fueyo et al. 2007; Gokmen et al. 2008), and many tolerance-inducing protocols expand Tregs with the indirect specificity (Wise et al. 1998; Ochando et al. 2006; Verginis et al. 2008). Two independent studies compared

 Cite this article as *Cold Spring Harb Perspect Med* doi: 10.1101/cshperspect.a015552

the efficacy of direct Tregs with Tregs of mixed direct and indirect specificities, and both found that additional indirect specificity improved efficacy (Joffre et al. 2008; Tsang et al. 2008). Indirect Tregs alone confer some protection against rejection, but the protection is very limited (Golshayan et al. 2009; Tsang et al. 2009), not much better than protection conferred by direct Tregs with a single specificity from T-cell receptor (TCR) transgenic mice (Brennan et al. 2011). Because current Treg manufacturing technology is only capable of large-scale production of polyclonal and direct alloreactive Tregs (see below for more details), we have compared the relative potency of direct and polyclonal Tregs. We found that when combined with deletion of 80% donor-reactive T cells, Tregs of direct alloreactivity could induce long-term islet allograft survival. Polyclonal Tregs were also capable of inducing long-term graft acceptance, but five times more cells were needed (K Lee and Q Tang, unpubl.). It has been thought that the impact of direct Tregs would be short lived because their activation and function depends on short-lived donor-derived antigen-presenting cells. However, there is some evidence that host antigen-presenting cells can acquire intact donor human leukocyte antigen (HLA) antigens expressed by the grafts and present alloantigens in a "semidirect" fashion, thus professional direct alloantigen presentation, and hence direct Treg function, may persist long after transplantation (Sagoo et al. 2012). Alternatively, direct Treg may create a tolerogenic milieu in the graft tissue to promote indirect Tregs through infectious tolerance, thus achieving long-term protection. Taken together, we believe combining direct and indirect alloreactive Tregs may be optimal in inducing graft survival when limited Treg numbers and/or minimal immunosuppression are used. When combined with adequate preconditioning and given in sufficient numbers, both direct and polyclonal Tregs may be able to induce long-term graft survival.

Adjunct Immunosuppression

Experimental data show that Tregs cannot prevent rejection as a stand-alone therapy. Treg induction of long-term graft survival requires short-term adjunct immunosuppression to create a therapeutic window. Most immunosuppressive drugs used in the transplant clinics today were selected based on their ability to block immune responses before the existence of Tregs and their role in transplant tolerance was established. Some of these immunosuppressive drugs antagonize Treg function and survival, whereas others are less harmful to Tregs or even beneficial (Table 1). Therefore, selection of adjunct immunosuppression will have a significant impact on the efficacy of Treg therapy. Currently planned clinical trials have to rely on available drugs to combine with Treg therapy, and developing new "Treg-supportive" immune-modulatory drugs will be instrumental for improving efficacy of Treg therapy. For example, activation of antigen-presenting cells via CD40 can abrogate Treg-mediated suppression (Serra et al. 2003), and blocking CD40 and CD40L interaction in mouse and nonhuman primate models of transplantation has shown promising results (Li et al. 2008). However, anti-CD40L antibodies had unexpected thrombotic complications in humans owing to platelet activation. Using nonthrombogenic alternatives or targeting CD40 may solve the problem. Given the differential impact of PI3 kinase-AKT-mTOR signaling in Tregs and conventional T cells, it is also possible to preferentially inhibit conventional T-cell activation by antagonizing the PI3 kinase-signaling pathway (Han et al. 2012). In addition, directly boosting Tregs may be an effective strategy to combine with Treg therapy. IL-2 therapy induces significant expansion of Tregs, but can also expand $CD8^+$ T cells and NK cells, particularly when higher doses of IL-2 are used (Bluestone 2011; Koreth et al. 2011; Saadoun et al. 2011; Long et al. 2012). In mouse models, it was possible to preferentially deliver IL-2 to Tregs using anti-IL-2 antibodies and repair defects of Tregs in a mouse model of autoimmune diabetes (Boyman et al. 2006; Tang et al. 2008). The efficacy of such strategy in humans remains to be determined. Last, histone deacetylases destabilize Tregs by promoting FOXP3 degradation, and histone deacetylase inhibitors have been shown to improve Treg homeostasis and

Table 1. Impact of immunosuppressive drugs on Tregs

Drug	Mechanism of immunosuppression	Impact on Tregs	References
Corticosteroids	Binds to nuclear receptor to inhibit AP1 and NF-κB and expression of proinflammatory cytokines	May support Tregs by reducing inflammation	Karagiannidis et al. 2004; Xu et al. 2009
CNI	Inhibits calcineurin, calcium signaling pathway, NFAT activation, IL-2 production	Detrimental to Treg function and survival	Baan et al. 2005; Pascual et al. 2008; Zeiser et al. 2008; Demirkiran et al. 2009
Rapamycin	Inhibits mTOR, protein synthesis, proliferation	Spares Tregs, thus increases percentage of Tregs	Baan et al. 2005; Pascual et al. 2008; Zeiser et al. 2008; Demirkiran et al. 2009
MMF	Inhibits purine biosynthesis, T- and B-cell proliferation	Likely neutral	Baan et al. 2005; Pascual et al. 2008; Zeiser et al. 2008; Demirkiran et al. 2009
ATG	Deletes T cells, NK cells, B cells	Deletes Tregs less efficiently, thus increases percentage of Tregs	Lopez et al. 2006; Morelon et al. 2010
Anti-CD25	Deletes CD25-expressing cells	Deletes Tregs	Bluestone et al. 2008; Toso et al. 2009
CTLA4-Ig	Blocks CD80 and CD86 costimulation of CD28 and T-cell clonal expansion	Spares Tregs when used at subsaturating dose	Bluestone et al. 2008
Anti-LFA-1	Blocks LFA-1 and ICAM-1 interaction and T-cell activation, blocks leukocyte trafficking	Dramatically increases circulating Treg percentage	Posselt et al. 2010

CNI, calcineurin inhibitor; MMF, mycophenolate mofetil; ATG, antithymocyte globulin.

induce allograft tolerance (Tao and Hancock 2007; Tao et al. 2007). New discoveries and understanding in Treg biology and transplantation tolerance have led to ever expanding therapeutic opportunities. However, it is extremely unlikely that any single agent will be effective against transplant rejection or capable of inducing tolerance. The challenge is to design an optimal combinational regimen and delivery program to steer the immune system away from rejection and into tolerance.

Treg MANUFACTURING

Treg Isolation

The first step in Treg expansion is the isolation of Tregs. A two-step magnetic activated cell sorting (MACS) protocol has been proposed for good manufacturing practice (GMP)-compliant isolation of human Tregs (Hoffmann et al. 2006; Peters et al. 2008b). However, Tregs isolated using this protocol are often contaminated with conventional T cells. This is because CD25 expression on human $CD4^+$ cells is not restricted to Tregs and there is no clear separation between Tregs and non-Tregs based on the level of CD25 expression. Selecting only the top 2% $CD25^{hi}$ Tregs increases purity at the expense of low yield. The additional use of CD127 allowed high-yield and high-purity recovery human Tregs based on the cell-surface phenotype of $CD4^+CD25^+$ $CD127^{lo/-}$ (Liu et al. 2006), and on average 1×10^6 Tregs can be isolated from 100 mL of blood using this approach. Tregs selected based on $CD4^+CD25^+CD127^{lo/-}$ markers have been found to be more effective than $CD4^+CD25^+$ Tregs in controlling alloimmune-mediated arterial atherosclerosis in a humanized mouse model (Nadig et al. 2010). It has also been reported that human Tregs can be identified on the basis of CD4, CD25, and CD45RA expression with naïve Tregs having the $CD4^+CD25^{mid}CD45RA^+$ phenotype and antigen-experienced effector Tregs having a $CD4^+CD25^{hi}CD45RA^-$ phenotype (Miyara et al. 2009). It has been shown that $CD45RA^+$Tregs are more stable during repeated in vitro stimulation with anti-CD3 and anti-CD28, whereas $CD45RA^-$ Tregs progressively lost FOXP3 expression after each round of stimulation (Hoffmann et al. 2009). We have found that $CD4^+CD25^+CD127^{lo/-}$ Tregs contain both $CD45RA^+$ and $CD45RA^-$ populations and the $CD45RA^+$ subset preferentially expands in culture (Putnam et al. 2009), as others have reported (Miyara et al. 2009), and the presence of the $CD45RA^-$ subset does not negatively impact the purity of the Tregs at the end of the expansion. Selecting Tregs with three or more cell-surface markers improves purity and yield, but makes it challenging to isolate the cells using MACS. Fluorescence-activated cell sorting (FACS) is capable of identifying cell subsets based a panel of three or more markers, however, currently, there is no GMP-compliant FACS and the process is slow, which poses a limit on the number of Tregs that can be purified and manufactured. An instrument that combines the bulk processing capability of MACS with the high precision multiparameter-based FACS would be ideal for fast isolation of highly pure Tregs.

Polyclonal Treg Expansion

Tregs can be readily expanded using anti-CD3 and anti-CD28-coated beads supplemented with IL-2 (Levings et al. 2001; Herold et al. 2002; Hoffmann et al. 2004; Earle et al. 2005; Putnam et al. 2009). Conventional $CD4^+$ T cells and $CD8^+$ T cells expand better than Tregs using this protocol, therefore high purity of the starting population is essential for producing highly pure expanded Tregs. This is particularly problematic for MACS-purified $CD4^+CD25^+$ Tregs, as they are often contaminated with $CD25^+$ $FOXP3^-$ conventional cells. The addition of rapamycin improves the purity of the culture because rapamycin suppresses the proliferation of conventional T cells and Treg growth is less affected (Battaglia et al. 2005). Despite these advances, large-scale manufacturing of Tregs remains challenging because even highly pure Tregs lose FOXP3 expression with repeated stimulation even in the presence of rapamycin (Hoffmann et al. 2009; Hippen et al. 2011b). The loss of FOXP3 is likely owing to destabilization of FOXP3 expression in Tregs instead of outgrowth of a few contaminating conventional T cells (Hoffmann et al. 2009). The cellular and molecular basis for Treg destabilization during in vitro stimulation is presently unclear. Nonetheless, it will be important to determine an optimized expansion protocol to maximize yield without compromising purity.

Alloantigen-Reactive Treg Expansion

The frequency of direct alloreactive Tregs has been estimated to be between 1% and 10% (Lin et al. 2008; Veerapathran et al. 2011). Proof-of-principle experiments have shown that alloantigen-reactive Tregs can be expanded using donor antigen-presenting cells such as dendritic cells, B cells, and unfractionated peripheral blood mononuclear cells (Peters et al. 2008a; Chen et al. 2009; Sagoo et al. 2011; Veerapathran et al. 2011; Tran et al. 2012). Collectively, these studies reported that alloantigen-expanded Tregs were 5–32 times more potent at suppressing alloantigen-stimulated proliferation in vitro than polyclonal Tregs. These results suggest that 5–32 times less alloantigen-expanded Tregs may be sufficient to achieve the same therapeutic efficacy as polyclonal Tregs. Based on our estimate that 5×10^9 polyclonal Tregs would be sufficient to induce tolerance when combined with 90% deletion of endogenous T cells, 150×10^6 to 1×10^9 alloantigen-reactive Tregs would be needed to achieve similar efficacy. Thus, clinical translation of alloantigen-reactive Treg therapy will require reliable manufacturing of more than hundreds of millions cells under GMP conditions. In this regard, we have developed a process using CD40L-activated allogeneic B cells to selectively stimulate the expansion of alloantigen-reactive Tregs (Putnam et al. 2013). After primary expansion, the cells are restimulated with anti-CD3 and anti-CD28-coated beads to increase cell yield. Using this approach, we are able to achieve 200- to 4000-fold expansion in 16 d. The cells are highly donor reactive and have demethylated the Treg-specific demethylation region (TSDR). Clinical trials applying Tregs manufactured with this protocol in kidney and liver transplantations are planned to start in 2014.

Although most successes in expanding human alloantigen-reactive Tregs have been in generating direct Tregs, efforts have been made to expand human indirect Tregs with less spectacular results (Jiang et al. 2003; Veerapathran et al. 2011). This is likely because of the 100 times lower frequency of indirect alloreactive Tregs in the Treg pool (Veerapathran et al. 2011). Given the experimental evidence of improved efficacy in combining direct and indirect Tregs to induce tolerance, developing protocols to manufacture indirect alloreactive Tregs is an important future direction. Alternative to selective expansion of indirect Tregs from the existing repertoire, forced expression of TCRs with indirect alloreactivity during expansion of direct alloreactive Tregs can generate dual specificity Tregs with improved ability to protect grafts in mouse models (Tsang et al. 2008). Genetic engineering Treg specificity has been successfully shown with human Tregs (Brusko et al. 2010). In addition to TCRs, other desirable features can be introduced into the Tregs such as traceable markers, tunable TCRs, chemotactic receptors to synthetic ligands, and drug-inducible suicidal enzymes (Lim 2010). These "designer" features would allow monitoring of the infused Tregs, controlling their activities and trafficking patterns, and eliminating them when needed.

Ex Vivo Induction of Tregs

Alternative to isolating preexisting Tregs for ex vivo expansion, it has been proposed that conventional $CD4^+$ T cells can be converted to Tregs during ex vivo expansion with the addition of TGF-β together with rapamycin or all-*trans* retinoic acid (Lu et al. 2010; Hippen et al. 2011a). These in vitro-induced Tregs acquire some features of Tregs such as expression of CD25 and FOXP3, reduced expression of effector cytokines, and ability to suppress in vitro and in vivo in a humanized mouse model of xenogeneic graft-versus-host disease (GvHD). Activated human T cells transiently express FOXP3; therefore FOXP3 expression alone does not distinguish activated T cells from Tregs. One distinction between ex vivo-isolated Tregs and in vitro induced Tregs is the DNA methylation status of the TSDR, which is important for Treg commitment and stability (more details in the paragraph below). Ex vivo-isolated Tregs have a demethylated TSDR, whereas in vitro-induced Tregs have fully methylated TSDR suggesting that they are not committed Tregs (Baron et al. 2007; Hippen et al. 2011a). Inhibiting or knocking down the DNA methyltransferase promotes demethylation of the FOXP3 locus and may drive commitment and stabilization of induced Tregs (Kim and Leonard 2007; Lal et al. 2009; Sanchez-Abarca et al. 2010). Further experimental evidence on the commitment and stability of ex vivo induced Tregs is needed before they can be considered as a viable source of therapeutic Tregs for humans.

Assessing Treg Identity and Purity after Expansion

Tregs are typically $CD3^+CD4^+$ and express the transcription factor FOXP3. Although CD3 and CD4 are clearly not unique for Tregs, conventional T cells transiently express FOXP3 after activation; therefore FOXP3 expression does not distinguish Tregs from conventional T cells either (Allan et al. 2007). It has been proposed that Tregs are best defined by their ability to suppress conventional T-cell proliferation in the "classical" in vitro suppression assay. It is important to remember that activated human conventional $CD4^+$ T cells show dose-dependent inhibition of T-cell proliferation similar to the suppression curve obtained using Tregs (Walker et al. 2003, 2005). On the other hand, there are numerous examples in mouse models that cells with normal suppressive activities measured in the in vitro suppression assay fail to function in vivo (Zhou et al. 2008; Wohlfert et al. 2011; Ouyang et al. 2012). In addition, in vitro suppression assays cannot detect contaminating Tconv cells. Typically, Tregs suppress maximally when used at 1:2 to 1:4 Treg to conventional T-cell ratios, which means this assay cannot detect the presence of as many as 50% contaminating T cells. Collectively, the in vitro suppression assay can misidentify cells as Tregs and is not an appropriate approach for assessing the identity or the purity of a Treg product.

To properly identify Tregs, it is helpful to know how Tregs develop and what defines their lineage. Treg development in the thymus is a multistep process controlled by cytokines and signaling through T-cell receptor and CD28, which culminates in FOXP3 expression and epigenetic modification of the FOXP3 locus (Burchill et al. 2008; Lio and Hsieh 2008; Long et al. 2009; Lio et al. 2010; Ohkura et al. 2012). The TSDR of the FOXP3 locus is demethylated in cells that have committed to the Treg lineage to ensure stable inheritance of FOXP3 expression in dividing cells and the lineage stability of Tregs (Floess et al. 2007; Huehn et al. 2009; Josefowicz and Rudensky 2009; Zheng et al. 2010; Ohkura et al. 2012). Although FOXP3 expression is necessary for Treg development and function, Treg lineage specification and stabilization also depends on coordinated activities of up to 300 proteins to ensure persistent high expression of FOXP3 (Hill et al. 2007; Fu et al. 2012; Rudra et al. 2012; Samstein et al. 2012). This suggests that coexpression of FOXP3 and a collection of proteins essential for Treg lineage specification may define a Treg. However, the list of these essential proteins is not defined currently and there is evidence that many of these proteins have redundant roles in defining the Treg lineage (Fu et al. 2012) and there may never be an unambiguous definition for Tregs based on patterns of protein expression. Because the function of these proteins is reflected in the epigenome of a Treg to ensure inheritable high expression of FOXP3 and simultaneous repression of IL-2 locus and other genes characteristic of effector cells (Ohkura et al. 2012), an epigenetic profile most accurately defines a bona fide Treg (Hori 2008). Until a comprehensive epigenomic fingerprint of a Treg is defined, demethylated TSDR can be used currently to determine the purity and identity of Treg products (Wieczorek et al. 2009).

EARLY EXPERIENCE OF REGULATORY T-CELL THERAPY IN HUMANS

As of April 2013, there have been four reported clinical trials of Treg therapy in humans, three in GvHD and one in type 1 diabetes (Trzonkowski et al. 2009; Brunstein et al. 2010; Di Ianni et al. 2011; Marek-Trzonkowska et al. 2012). In addition, there are seven Treg trials registered in www.clinicaltrials.org, four in GvHD, one in type 1 diabetes, one in liver transplantation (NCT01624077), and one in kidney transplantation (NCT01446484). Some features of the published trials are summarized in Table 2 with emphasis on the Treg product infused. All four reported trials showed acceptable safety and promising efficacy of the treatment. Although none of these trials are in solid organ transplantation, information from these trials is instrumental for the design of future solid organ trials. For example, in the Brunstein et al. (2010) trial, because of the mismatch of HLA of the Treg cord blood donors from the host and the conventional T-cell donors, persistence of Tregs can be tracked. The infused Tregs could be detected between 4 hr and 7 d after infusion. The increase in infused Tregs in circulation was dramatically less after a second infusion of cryopreserved Treg products, suggesting cryopreservation may compromise the viability or stability of the Tregs, although other factors may also contribute. In addition, a clear trend of more consistent overall increase in Tregs was observed in rapamycin-treated patients when compared with cyclosporine-treated patients. In the Marek-Trzonkowska et al. trial, we learn that infusion of 20×10^6/kg body weight polyclonally expanded Tregs in type 1 diabetic children between the ages of 8 and 16 is safe and the treatment is associated with immediate doubling of the percentage of circulating Tregs and a trend of increase at 2 wk. In the Di Ianni trial, therapeutic Tregs enabled the infusion of a higher dose of conventional T cells and

Table 2. Summary of published Treg trials in humans

References	Indication (No. of patients)	Tregs			
		Isolation	Expansion	% FOXP3$^+$	Dosing/kg
Trzonkowski et al. 2009	GvHD ($n = 2$)	FACS	2–4 Wk, weekly stimulations	40%–90%	$0.1–3 \times 10^6$
Brunstein et al. 2010	GvHD ($n = 23$)	MACS	18 ± 1 d, Anti-CD3/CD28 on day 0	31%–96%	$0.1–6 \times 10^6$
Di Ianni et al. 2011	GvHD ($n = 28$)	MACS	None	$69\% \pm 14\%$	$2–4 \times 10^6$
Marek-Trzonkowska et al. 2012	Type 1 diabetes in children ($n = 10$)	FACS	14 d, Anti-CD3/CD28 on days 0 and 7	90%–97%	$10–20 \times 10^6$

GvHD, graft-versus-host disease; FACS, fluorescence-activated cell sorting; MACS, magnetic-activated cell sorting.

better immunity against opportunistic infections. Particularly, the investigators observed markedly improved prevention against cytomegalovirus (CMV) disease. Five out of seven patients who received influenza vaccine after Treg infusion achieved protective antibody titers. All these results suggest that Treg infusion did not impede protective immunity against infections or lead to global immunosuppression. Overall, experience in clinical application of Treg therapy thus far shows that it is feasible, safe, and potentially effective.

It is worthwhile to review the quality of the Treg products in these trials closely, so that lessons learned from these early experiences in humans can be applied to the planned trials in solid organ transplantations. The investigator reported a wide range of Treg purity ranging from 31% to 97% FOXP3$^+$. Two factors contributed to the variability of Treg purity. In the Di Ianni trial when no ex vivo expansion was used, purity below 50% was reported. In the Brunstein trial, the percentage of FOXP3$^+$ cells after MACS isolation varied between 20% and 80% despite the use of cord blood as source materials. These results illustrate the problem with the current MACS technology. Additionally, loss of FOXP3 during repeated stimulation of Tregs was evident in the Trzonkowski trial that reported the percentage of FOXP3$^+$ decreasing from 90% to 70% and 40% after successive weekly stimulations. It is encouraging that by combining FACS purification and limiting to two rounds of stimulations in the Marek-Trzonkowska trial, the purity of the expanded Tregs was consistently above 90%, however, the authors noted an insufficient Treg yield in four out of 10 patients. We have completed infusion of 10 type 1 diabetic patients of up to 320×10^6 expanded Tregs at UCSF. By using FACS purification, and two rounds of anti-CD3 and anti-CD28 stimulations, we are able to produce on average 2×10^9 Tregs (range $0.2–3.1 \times 10^9$) with an average of 92% FOXP3$^+$ (range 76%–97%) from 1 unit of blood donation (J Bluestone, unpubl.). In the GvHD setting, up to 70% of FOXP3$^-$ conventional T cells are tolerated because conventional T cells are needed for improving hematopoietic stem cell engraftment, restoring immunity, and mediating graft-versus-leukemia effect. In the solid organ transplant setting, high-purity Tregs will be needed to ensure safety, potency, and consistency of results.

CONCLUDING REMARKS

Taming a powerful multifaceted immune response against transplanted organ requires equally potent and versatile therapy without compromising the overall immune competence of the patient. Tregs have the desired specificity, versatility, and adaptability and decades of research has shown their therapeutic efficacy in transplantation. However, Tregs do not have sufficient potency as a stand-alone therapy for transplantation, and factors critical to the efficacy of Treg therapy in transplantation are dose, specificity, and adjunct immunosuppression.

Although early clinical trials will be mainly focused on safety, it is important to design these trials with efficacy end points in mind so that the safety of the right dose with the desired specificity administered with optimized adjunct therapy can be evaluated. Central to the successful implementation of Treg therapy is the reliable production of consistently high-quality Tregs and the challenge is in balancing the yield and the purity of the products. The future of Treg therapy depends on effective and informative clinical trial designs, technological advancement in Treg manufacture, and better mechanistic understanding of Treg biology and transplantation tolerance in humans.

ACKNOWLEDGMENTS

The authors acknowledge generous support by the Nicholas family (Q.T.), UCSF Department of Surgery (Q.T.), and research grants from Roche Organ Transplantation Research Foundation (Q.T.), Immune Tolerance Network (N01 AI15416; J.A.B.), Juvenile Diabetes Research Foundation 4-2011-248 (J.A.B.), European Commission (ONE Study) Project Number 260687 (Q.T. and J.A.B.), and National Institutes of Health U19 AI056388 (J.A.B.).

REFERENCES

Allan SE, Crome SQ, Crellin NK, Passerini L, Steiner TS, Bacchetta R, Roncarolo MG, Levings MK. 2007. Activation-induced FOXP3 in human T effector cells does not suppress proliferation or cytokine production. *Int Immunol* **19:** 345–354.

Baan CC, van der Mast BJ, Klepper M, Mol WM, Peeters AM, Korevaar SS, Balk AH, Weimar W. 2005. Differential effect of calcineurin inhibitors, anti-CD25 antibodies and rapamycin on the induction of FOXP3 in human T cells. *Transplantation* **80:** 110–117.

Baron U, Floess S, Wieczorek G, Baumann K, Grutzkau A, Dong J, Thiel A, Boeld TJ, Hoffmann P, Edinger M, et al. 2007. DNA demethylation in the human FOXP3 locus discriminates regulatory T cells from activated FOXP3$^+$ conventional T cells. *Eur J Immunol* **37:** 2378–2389.

Battaglia M, Stabilini A, Roncarolo MG. 2005. Rapamycin selectively expands CD4$^+$CD25$^+$FoxP3$^+$ regulatory T cells. *Blood* **105:** 4743–4748.

Berenson GS, Wattigney WA, Tracy RE, Newman WP 3rd, Srinivasan SR, Webber LS, Dalferes ER Jr, Strong JP. 1992. Atherosclerosis of the aorta and coronary arteries and cardiovascular risk factors in persons aged 6 to 30 years and studied at necropsy (The Bogalusa Heart Study). *Am J Cardiol* **70:** 851–858.

Bluestone JA. 2011. The yin and yang of interleukin-2-mediated immunotherapy. *N Engl J Med* **365:** 2129–2131.

Bluestone JA, Liu W, Yabu JM, Laszik ZG, Putnam A, Belingheri M, Gross DM, Townsend RM, Vincenti F. 2008. The effect of costimulatory and interleukin 2 receptor blockade on regulatory T cells in renal transplantation. *Am J Transplant* **8:** 2086–2096.

Boyman O, Kovar M, Rubinstein MP, Surh CD, Sprent J. 2006. Selective stimulation of T cell subsets with antibody-cytokine immune complexes. *Science* **311:** 1924–1927.

Brennan TV, Tang Q, Liu FC, Hoang V, Bi M, Bluestone JA, Kang SM. 2011. Requirements for prolongation of allograft survival with regulatory T cell infusion in lymphosufficient hosts. *J Surg Res* **169:** e69–e75.

Brunstein CG, Miller JS, Cao Q, McKenna DH, Hippen KL, Curtsinger J, Defor T, Levine BL, June CH, Rubinstein P, et al. 2010. Infusion of ex vivo expanded T regulatory cells in adults transplanted with umbilical cord blood: Safety profile and detection kinetics. *Blood* **117:** 1061–1070.

Brusko TM, Koya RC, Zhu S, Lee MR, Putnam AL, McClymont SA, Nishimura MI, Han S, Chang LJ, Atkinson MA, et al. 2010. Human antigen-specific regulatory T cells generated by T cell receptor gene transfer. *PLoS ONE* **5:** e11726.

Burchill MA, Yang J, Vang KB, Moon JJ, Chu HH, Lio CW, Vegoe AL, Hsieh CS, Jenkins MK, Farrar MA. 2008. Linked T cell receptor and cytokine signaling govern the development of the regulatory T cell repertoire. *Immunity* **28:** 112–121.

Bushell A, Morris PJ, Wood KJ. 1995. Transplantation tolerance induced by antigen pretreatment and depleting anti-CD4 antibody depends on CD4$^+$ T cell regulation during the induction phase of the response. *Eur J Immunol* **25:** 2643–2649.

Callaghan CJ, Rouhani FJ, Negus MC, Curry AJ, Bolton EM, Bradley JA, Pettigrew GJ. 2007. Abrogation of antibody-mediated allograft rejection by regulatory CD4 T cells with indirect allospecificity. *J Immunol* **178:** 2221–2228.

Chen LC, Delgado JC, Jensen PE, Chen X. 2009. Direct expansion of human allospecific FoxP3$^+$CD4$^+$ regulatory T cells with allogeneic B cells for therapeutic application. *J Immunol* **183:** 4094–4102.

Cobbold SP, Castejon R, Adams E, Zelenika D, Graca L, Humm S, Waldmann H. 2004. Induction of *foxP3*$^+$ regulatory T cells in the periphery of T cell receptor transgenic mice tolerized to transplants. *J Immunol* **172:** 6003–6010.

Demirkiran A, Sewgobind VD, van der Weijde J, Kok A, Baan CC, Kwekkeboom J, Tilanus HW, Metselaar HJ, van der Laan LJ. 2009. Conversion from calcineurin inhibitor to mycophenolate mofetil-based immunosuppression changes the frequency and phenotype of CD4$^+$ FOXP3$^+$ regulatory T cells. *Transplantation* **87:** 1062–1068.

Di Ianni M, Falzetti F, Carotti A, Terenzi A, Castellino F, Bonifacio E, Del Papa B, Zei T, Ostini RI, Cecchini D, et al. 2011. Tregs prevent GVHD and promote immune reconstitution in HLA-haploidentical transplantation. *Blood* **117:** 3921–3928.

Dieckmann D, Bruett CH, Ploettner H, Lutz MB, Schuler G. 2002. Human $CD4^+CD25^+$ regulatory, contact-dependent T cells induce interleukin 10-producing, contact-independent type 1-like regulatory T cells [corrected]. *J Exp Med* **196:** 247–253.

Dijke IE, Velthuis JH, Caliskan K, Korevaar SS, Maat AP, Zondervan PE, Balk AH, Weimar W, Baan CC. 2007. Intragraft FOXP3 mRNA expression reflects antidonor immune reactivity in cardiac allograft patients. *Transplantation* **83:** 1477–1484.

Earle KE, Tang Q, Zhou X, Liu W, Zhu S, Bonyhadi ML, Bluestone JA. 2005. In vitro expanded human $CD4^+$ $CD25^+$ regulatory T cells suppress effector T cell proliferation. *Clin Immunol* **115:** 3–9.

Euvrard S, Kanitakis J, Claudy A. 2003. Skin cancers after organ transplantation. *N Engl J Med* **348:** 1681–1691.

Fan Z, Spencer JA, Lu Y, Pitsillides CM, Singh G, Kim P, Yun SH, Toxavidis V, Strom TB, Lin CP, et al. 2010. In vivo tracking of "color-coded" effector, natural and induced regulatory T cells in the allograft response. *Nat Med* **16:** 718–722.

Floess S, Freyer J, Siewert C, Baron U, Olek S, Polansky J, Schlawe K, Chang HD, Bopp T, Schmitt E, et al. 2007. Epigenetic control of the *foxp3* locus in regulatory T cells. *PLoS Biol* **5:** e38.

Francis RS, Feng G, Tha-In T, Lyons IS, Wood KJ, Bushell A. 2011. Induction of transplantation tolerance converts potential effector T cells into graft-protective regulatory T cells. *Eur J Immunol* **41:** 726–738.

Fu W, Ergun A, Lu T, Hill JA, Haxhinasto S, Fassett MS, Gazit R, Adoro S, Glimcher L, Chan S, et al. 2012. A multiply redundant genetic switch "locks in" the transcriptional signature of regulatory T cells. *Nat Immunol* **13:** 972–980.

Ganusov VV, De Boer RJ. 2007. Do most lymphocytes in humans really reside in the gut? *Trends Immunol* **28:** 514–518.

Gokmen MR, Lombardi G, Lechler RI. 2008. The importance of the indirect pathway of allorecognition in clinical transplantation. *Curr Opin Immunol* **20:** 568–574.

Golshayan D, Jiang S, Tsang J, Garin MI, Mottet C, Lechler RI. 2007. In vitro-expanded donor alloantigen-specific $CD4^+CD25^+$ regulatory T cells promote experimental transplantation tolerance. *Blood* **109:** 827–835.

Golshayan D, Wyss JC, Abulker CW, Schaefer SC, Lechler RI, Lehr HA, Pascual M. 2009. Transplantation tolerance induced by regulatory T cells: In vivo mechanisms and sites of action. *Int Immunopharmacol* **9:** 683–688.

Gould DS, Auchincloss H Jr, 1999. Direct and indirect recognition: The role of MHC antigens in graft rejection. *Immunol Today* **20:** 77–82.

Graca L, Thompson S, Lin CY, Adams E, Cobbold SP, Waldmann H. 2002. Both $CD4^+CD25^+$ and $CD4^+CD25^-$ regulatory cells mediate dominant transplantation tolerance. *J Immunol* **168:** 5558–5565.

Han JM, Patterson SJ, Levings MK. 2012. The role of the PI3K signaling pathway in $CD4^+$ T cell differentiation and function. *Front Immunol* **3:** 245.

Hara M, Kingsley CI, Niimi M, Read S, Turvey SE, Bushell AR, Morris PJ, Powrie F, Wood KJ. 2001. IL-10 is required for regulatory T cells to mediate tolerance to alloantigens in vivo. *J Immunol* **166:** 3789–3796.

Herold KC, Hagopian W, Auger JA, Poumian-Ruiz E, Taylor L, Donaldson D, Gitelman SE, Harlan DM, Xu D, Zivin RA, et al. 2002. Anti-CD3 monoclonal antibody in new-onset type 1 diabetes mellitus. *N Engl J Med* **346:** 1692–1698.

Hill JA, Feuerer M, Tash K, Haxhinasto S, Perez J, Melamed R, Mathis D, Benoist C. 2007. Foxp3 transcription-factor-dependent and -independent regulation of the regulatory T cell transcriptional signature. *Immunity* **27:** 786–800.

Hippen KL, Merkel SC, Schirm DK, Nelson C, Tennis NC, Riley JL, June CH, Miller JS, Wagner JE, Blazar BR. 2011a. Generation and large-scale expansion of human inducible regulatory T cells that suppress graft-versus-host disease. *Am J Transplant* **11:** 1148–1157.

Hippen KL, Merkel SC, Schirm DK, Sieben CM, Sumstad D, Kadidlo DM, McKenna DH, Bromberg JS, Levine BL, Riley JL, et al. 2011b. Massive ex vivo expansion of human natural regulatory T cells (T_{regs}) with minimal loss of in vivo functional activity. *Sci Transl Med* **3:** 83ra41.

Hoffmann P, Eder R, Kunz-Schughart LA, Andreesen R, Edinger M. 2004. Large-scale in vitro expansion of polyclonal human $CD4^+CD25^{high}$ regulatory T cells. *Blood* **104:** 895–903.

Hoffmann P, Boeld TJ, Eder R, Albrecht J, Doser K, Piseshka B, Dada A, Niemand C, Assenmacher M, Orso E, et al. 2006. Isolation of $CD4^+CD25^+$ regulatory T cells for clinical trials. *Biol Blood Marrow Transplant* **12:** 267–274.

Hoffmann P, Boeld TJ, Eder R, Huehn J, Floess S, Wieczorek G, Olek S, Dietmaier W, Andreesen R, Edinger M. 2009. Loss of FOXP3 expression in natural human $CD4^+$ $CD25^+$ regulatory T cells upon repetitive in vitro stimulation. *Eur J Immunol* **39:** 1088–1097.

Hori S. 2008. Rethinking the molecular definition of regulatory T cells. *Eur J Immunol* **38:** 928–930.

Huehn J, Polansky JK, Hamann A. 2009. Epigenetic control of FOXP3 expression: The key to a stable regulatory T-cell lineage? *Nat Rev Immunol* **9:** 83–89.

Jiang S, Camara N, Lombardi G, Lechler RI. 2003. Induction of allopeptide-specific human $CD4^+CD25^+$ regulatory T cells ex vivo. *Blood* **102:** 2180–2186.

Joffre O, Santolaria T, Calise D, Al Saati T, Hudrisier D, Romagnoli P, van Meerwijk JP. 2008. Prevention of acute and chronic allograft rejection with $CD4^+CD25^+Foxp3^+$ regulatory T lymphocytes. *Nat Med* **14:** 88–92.

Jonuleit H, Schmitt E, Kakirman H, Stassen M, Knop J, Enk AH. 2002. Infectious tolerance: Human $CD25^+$ regulatory T cells convey suppressor activity to conventional $CD4^+$ T helper cells. *J Exp Med* **196:** 255–260.

Josefowicz SZ, Rudensky A. 2009. Control of regulatory T cell lineage commitment and maintenance. *Immunity* **30:** 616–625.

Karagiannidis C, Akdis M, Holopainen P, Woolley NJ, Hense G, Ruckert B, Mantel PY, Menz G, Akdis CA, Blaser K, et al. 2004. Glucocorticoids upregulate FOXP3 expression and regulatory T cells in asthma. *J Allergy Clin Immunol* **114:** 1425–1433.

Karim M, Kingsley CI, Bushell AR, Sawitzki BS, Wood KJ. 2004. Alloantigen-induced $CD25^+CD4^+$ regulatory T cells can develop in vivo from $CD25^-CD4^+$ precursors in a thymus-independent process. *J Immunol* **172:** 923–928.

Kim HP, Leonard WJ. 2007. CREB/ATF-dependent T cell receptor-induced *FoxP3* gene expression: A role for DNA methylation. *J Exp Med* **204:** 1543–1551.

Koreth J, Matsuoka K, Kim HT, McDonough SM, Bindra B, Alyea EP 3rd, Armand P, Cutler C, Ho VT, Treister NS, et al. 2011. Interleukin-2 and regulatory T cells in graft-versus-host disease. *N Engl J Med* **365:** 2055–2066.

Lal G, Zhang N, van der Touw W, Ding Y, Ju W, Bottinger EP, Reid SP, Levy DE, Bromberg JS. 2009. Epigenetic regulation of Foxp3 expression in regulatory T cells by DNA methylation. *J Immunol* **182:** 259–273.

Levings MK, Sangregorio R, Roncarolo MG. 2001. Human $CD25^+CD4^+$ T regulatory cells suppress naive and memory T cell proliferation and can be expanded in vitro without loss of function. *J Exp Med* **193:** 1295–1302.

Li XL, Menoret S, Le Mauff B, Angin M, Anegon I. 2008. Promises and obstacles for the blockade of CD40-CD40L interactions in allotransplantation. *Transplantation* **86:** 10–15.

Lim WA. 2010. Designing customized cell signalling circuits. *Nat Rev Mol Cell Biol* **11:** 393–403.

Lin YJ, Hara H, Tai HC, Long C, Tokita D, Yeh P, Ayares D, Morelli AE, Cooper DK. 2008. Suppressive efficacy and proliferative capacity of human regulatory T cells in allogeneic and xenogeneic responses. *Transplantation* **86:** 1452–1462.

Lio CW, Hsieh CS. 2008. A two-step process for thymic regulatory T cell development. *Immunity* **28:** 100–111.

Lio CW, Dodson LF, Deppong CM, Hsieh CS, Green JM. 2010. CD28 facilitates the generation of $Foxp3^-$ cytokine responsive regulatory T cell precursors. *J Immunol* **184:** 6007–6013.

Liu W, Putnam AL, Xu-Yu Z, Szot GL, Lee MR, Zhu S, Gottlieb PA, Kapranov P, Gingeras TR, Fazekas de St Groth B, et al. 2006. CD127 expression inversely correlates with FoxP3 and suppressive function of human $CD4^+$ T reg cells. *J Exp Med* **203:** 1701–1711.

Long M, Park SG, Strickland I, Hayden MS, Ghosh S. 2009. Nuclear factor-κB modulates regulatory T cell development by directly regulating expression of Foxp3 transcription factor. *Immunity* **31:** 921–931.

Long SA, Rieck M, Sanda S, Bollyky JB, Samuels PL, Goland R, Ahmann A, Rabinovitch A, Aggarwal S, Phippard D, et al. 2012. Rapamycin/IL-2 combination therapy in patients with type 1 diabetes augments Tregs yet transiently impairs β-cell function. *Diabetes* **61:** 2340–2348.

Lopez M, Clarkson MR, Albin M, Sayegh MH, Najafian N. 2006. A novel mechanism of action for anti-thymocyte globulin: Induction of $CD4^+CD25^+Foxp3^+$ regulatory T cells. *J Am Soc Nephrol* **17:** 2844–2853.

Lu L, Zhou X, Wang J, Zheng SG, Horwitz DA. 2010. Characterization of protective human CD4CD25 FOXP3 regulatory T cells generated with IL-2, TGF-β and retinoic acid. *PLoS ONE* **5:** e15150.

Marek-Trzonkowska N, Mysliwiec M, Dobyszuk A, Grabowska M, Techmanska I, Juscinska J, Wujtewicz MA, Witkowski P, Mlynarski W, Balcerska A, et al. 2012. Administration of $CD4^+CD25^{high}CD127^-$ regulatory T cells preserves β-cell function in type 1 diabetes in children. *Diabetes Care* **35:** 1817–1820.

Meagher C, Tang Q, Fife BT, Bour-Jordan H, Wu J, Pardoux C, Bi M, Melli K, Bluestone JA. 2008. Spontaneous development of a pancreatic exocrine disease in CD28-deficient NOD mice. *J Immunol* **180:** 7793–7803.

Miyara M, Yoshioka Y, Kitoh A, Shima T, Wing K, Niwa A, Parizot C, Taflin C, Heike T, Valeyre D, et al. 2009. Functional delineation and differentiation dynamics of human $CD4^+$ T cells expressing the FoxP3 transcription factor. *Immunity* **30:** 899–911.

Morelon E, Lefrancois N, Besson C, Prevautel J, Brunet M, Touraine JL, Badet L, Touraine-Moulin F, Thaunat O, Malcus C. 2010. Preferential increase in memory and regulatory subsets during T-lymphocyte immune reconstitution after Thymoglobulin induction therapy with maintenance sirolimus vs cyclosporine. *Transpl Immunol* **23:** 53–58.

Muthukumar T, Dadhania D, Ding R, Snopkowski C, Naqvi R, Lee JB, Hartono C, Li B, Sharma VK, Seshan SV, et al. 2005. Messenger RNA for FOXP3 in the urine of renal-allograft recipients. *N Engl J Med* **353:** 2342–2351.

Nadig SN, Wieckiewicz J, Wu DC, Warnecke G, Zhang W, Luo S, Schiopu A, Taggart DP, Wood KJ. 2010. In vivo prevention of transplant arteriosclerosis by ex vivo-expanded human regulatory T cells. *Nat Med* **16:** 809–813.

Nair S, Verma S, Thuluvath PJ. 2002. Obesity and its effect on survival in patients undergoing orthotopic liver transplantation in the United States. *Hepatology* **35:** 105–109.

Nishimura E, Sakihama T, Setoguchi R, Tanaka K, Sakaguchi S. 2004. Induction of antigen-specific immunologic tolerance by in vivo and in vitro antigen-specific expansion of naturally arising $Foxp3^+CD25^+CD4^+$ regulatory T cells. *Int Immunol* **16:** 1189–1201.

Nylen S, Maurya R, Eidsmo L, Manandhar KD, Sundar S, Sacks D. 2007. Splenic accumulation of IL-10 mRNA in T cells distinct from $CD4^+CD25^+$ (Foxp3) regulatory T cells in human visceral leishmaniasis. *J Exp Med* **204:** 805–817.

O'Gorman WE, Dooms H, Thorne SH, Kuswanto WF, Simonds EF, Krutzik PO, Nolan GP, Abbas AK. 2009. The initial phase of an immune response functions to activate regulatory T cells. *J Immunol* **183:** 332–339.

Ochando JC, Homma C, Yang Y, Hidalgo A, Garin A, Tacke F, Angeli V, Li Y, Boros P, Ding Y, et al. 2006. Alloantigen-presenting plasmacytoid dendritic cells mediate tolerance to vascularized grafts. *Nat Immunol* **7:** 652–662.

Ohkura N, Hamaguchi M, Morikawa H, Sugimura K, Tanaka A, Ito Y, Osaki M, Tanaka Y, Yamashita R, Nakano N, et al. 2012. T cell receptor stimulation-induced epigenetic changes and Foxp3 expression are independent and complementary events required for Treg cell development. *Immunity* **37:** 785–799.

Ojo AO, Held PJ, Port FK, Wolfe RA, Leichtman AB, Young EW, Arndorfer J, Christensen L, Merion RM. 2003. Chronic renal failure after transplantation of a nonrenal organ. *N Engl J Med* **349:** 931–940.

Oliveira VG, Caridade M, Paiva RS, Demengeot J, Graca L. 2011. Sub-optimal $CD4^+$ T-cell activation triggers autonomous TGF-β-dependent conversion to $Foxp3^+$ regulatory T cells. *Eur J Immunol* **41:** 1249–1255.

Ouyang W, Liao W, Luo CT, Yin N, Huse M, Kim MV, Peng M, Chan P, Ma Q, Mo Y, et al. 2012. Novel Foxo1-dependent transcriptional programs control T_{reg} cell function. *Nature* **491:** 554–559.

Pandiyan P, Zheng L, Ishihara S, Reed J, Lenardo MJ. 2007. $CD4^+CD25^+Foxp3^+$ regulatory T cells induce cytokine deprivation-mediated apoptosis of effector $CD4^+$ T cells. *Nat Immunol* **8:** 1353–1362.

Pascual J, Bloom D, Torrealba J, Brahmbhatt R, Chang Z, Sollinger HW, Knechtle SJ. 2008. Calcineurin inhibitor withdrawal after renal transplantation with alemtuzumab: Clinical outcomes and effect on T-regulatory cells. *Am J Transplant* **8:** 1529–1536.

Penn I. 1990. Cancers complicating organ transplantation. *N Engl J Med* **323:** 1767–1769.

Perelson AS, Wiegel FW. 2009. Scaling aspects of lymphocyte trafficking. *J Theoret Biol* **257:** 9–16.

Peters JH, Hilbrands LB, Koenen HJ, Joosten I. 2008a. Ex vivo generation of human alloantigen-specific regulatory T cells from $CD4^{pos}CD25^{high}$ T cells for immunotherapy. *PLoS ONE* **3:** e2233.

Peters JH, Preijers FW, Woestenenk R, Hilbrands LB, Koenen HJ, Joosten I. 2008b. Clinical grade Treg: GMP isolation, improvement of purity by $CD127^{pos}$ depletion, Treg expansion, and Treg cryopreservation. *PLoS ONE* **3:** e3161.

Porter DL, Levine BL, Kalos M, Bagg A, June CH. 2011. Chimeric antigen receptor-modified T cells in chronic lymphoid leukemia. *N Engl J Med* **365:** 725–733.

Posselt AM, Szot GL, Frassetto LA, Masharani U, Tavakol M, Amin R, McElroy J, Ramos MD, Kerlan RK, Fong L, et al. 2010. Islet transplantation in type 1 diabetic patients using calcineurin inhibitor-free immunosuppressive protocols based on T-cell adhesion or costimulation blockade. *Transplantation* **90:** 1595–1601.

Putnam AL, Brusko TM, Lee MR, Liu W, Szot GL, Ghosh T, Atkinson MA, Bluestone JA. 2009. Expansion of human regulatory T-cells from patients with type 1 diabetes. *Diabetes* **58:** 652–662.

Putnam A, Safinia N, Medvec A, Laszkowska M, Wray M, Mintz M, Trotta E, Szot G, Liu W, Lares A, et al. 2013. Clinical grade manufacturing and therapeutic advantage of human alloantigen-reactive regulatory T cells in transplantation. *Am J Transplant* (to be published).

Restifo NP, Dudley ME, Rosenberg SA. 2012. Adoptive immunotherapy for cancer: Harnessing the T cell response. *Nat Rev Immunol* **12:** 269–281.

Rogers NJ, Lechler RI. 2001. Allorecognition. *Am J Transplant* **1:** 97–102.

Rosenblum MD, Gratz IK, Paw JS, Lee K, Marshak-Rothstein A, Abbas AK. 2011. Response to self antigen imprints regulatory memory in tissues. *Nature* **480:** 538–542.

Rudra D, deRoos P, Chaudhry A, Niec RE, Arvey A, Samstein RM, Leslie C, Shaffer SA, Goodlett DR, Rudensky AY. 2012. Transcription factor Foxp3 and its protein partners form a complex regulatory network. *Nat Immunol* **13:** 1010–1019.

Saadoun D, Rosenzwajg M, Joly F, Six A, Carrat F, Thibault V, Sene D, Cacoub P, Klatzmann D. 2011. Regulatory T-cell responses to low-dose interleukin-2 in HCV-induced vasculitis. *N Engl J Med* **365:** 2067–2077.

Sagoo P, Ali N, Garg G, Nestle FO, Lechler RI, Lombardi G. 2011. Human regulatory T cells with alloantigen specificity are more potent inhibitors of alloimmune skin graft damage than polyclonal regulatory T cells. *Sci Transl Med* **3:** 83ra42.

Sagoo P, Lombardi G, Lechler RI. 2012. Relevance of regulatory T cell promotion of donor-specific tolerance in solid organ transplantation. *Front Immunol* **3:** 184.

Samstein RM, Arvey A, Josefowicz SZ, Peng X, Reynolds A, Sandstrom R, Neph S, Sabo P, Kim JM, Liao W, et al. 2012. Foxp3 exploits a pre-existent enhancer landscape for regulatory T cell lineage specification. *Cell* **151:** 153–166.

Sanchez-Abarca LI, Gutierrez-Cosio S, Santamaria C, Caballero-Velazquez T, Blanco B, Herrero-Sanchez C, Garcia JL, Carrancio S, Hernandez-Campo P, Gonzalez FJ, et al. 2010. Immunomodulatory effect of 5-azacytidine (5-azaC): Potential role in the transplantation setting. *Blood* **115:** 107–121.

Sanchez-Fueyo A. 2011. Hot-topic debate on tolerance: Immunosuppression withdrawal. *Liver Transpl* **17:** S69–S73.

Sanchez-Fueyo A, Domenig CM, Mariat C, Alexopoulos S, Zheng XX, Strom TB. 2007. Influence of direct and indirect allorecognition pathways on $CD4^+CD25^+$ regulatory T-cell function in transplantation. *Transpl Int* **20:** 534–541.

Scholler J, Brady TL, Binder-Scholl G, Hwang WT, Plesa G, Hege KM, Vogel AN, Kalos M, Riley JL, Deeks SG, et al. 2012. Decade-long safety and function of retroviral-modified chimeric antigen receptor T cells. *Sci Transl Med* **4:** 132ra153.

Serra P, Amrani A, Yamanouchi J, Han B, Thiessen S, Utsugi T, Verdaguer J, Santamaria P. 2003. CD40 ligation releases immature dendritic cells from the control of regulatory $CD4^+CD25^+$ T cells. *Immunity* **19:** 877–889.

Soltys KA, Mazariegos GV, Squires RH, Sindhi RK, Anand R. 2007. Late graft loss or death in pediatric liver transplantation: An analysis of the SPLIT database. *Am J Transplant* **7:** 2165–2171.

Tang Q, Bluestone JA. 2008. The $Foxp3^+$ regulatory T cell: A jack of all trades, master of regulation. *Nat Immunol* **9:** 239–244.

Tang Q, Lee K. 2012. Regulatory T-cell therapy for transplantation: How many cells do we need? *Curr Opin Organ Transplant* **17:** 349–354.

Tang Q, Adams JY, Penaranda C, Melli K, Piaggio E, Sgouroudis E, Piccirillo CA, Salomon BL, Bluestone JA. 2008. Central role of defective interleukin-2 production in the triggering of islet autoimmune destruction. *Immunity* **28:** 687–697.

Tao R, Hancock WW. 2007. Regulating regulatory T cells to achieve transplant tolerance. *Hepatobiliary Pancreat Dis Int* **6:** 348–357.

Tao R, de Zoeten EF, Ozkaynak E, Chen C, Wang L, Porrett PM, Li B, Turka LA, Olson EN, Greene MI, et al. 2007. Deacetylase inhibition promotes the generation and function of regulatory T cells. *Nat Med* **13:** 1299–1307.

Textor SC, Taler SJ, Canzanello VJ, Schwartz L, Augustine JE. 2000. Posttransplantation hypertension related to calcineurin inhibitors. *Liver Transpl* **6:** 521–530.

Toso C, Edgar R, Pawlick R, Emamaullee J, Merani S, Dinyari P, Mueller TF, Shapiro AM, Anderson CC. 2009. Effect of different induction strategies on effector, regulatory and memory lymphocyte sub-populations in clinical islet transplantation. *Transpl Int* **22:** 182–191.

Tran GT, Hodgkinson SJ, Carter NM, Verma ND, Plain KM, Boyd R, Robinson CM, Nomura M, Killingsworth M, Hall BM. 2012. Interleukin-5 (IL-5) promotes induction of antigen specific $CD4^+CD25^+$ T regulatory cells that suppress autoimmunity. *Blood* **119:** 4441–4450.

Trzonkowski P, Bieniaszewska M, Juscinska J, Dobyszuk A, Krzystyniak A, Marek N, Mysliwska J, Hellmann A. 2009. First-in-man clinical results of the treatment of patients with graft versus host disease with human ex vivo expanded $CD4^+CD25^+CD127^-$ T regulatory cells. *Clin Immunol* **133:** 22–26.

Tsang JY, Tanriver Y, Jiang S, Xue SA, Ratnasothy K, Chen D, Stauss HJ, Bucy RP, Lombardi G, Lechler R. 2008. Conferring indirect allospecificity on $CD4^+CD25^+$ Tregs by TCR gene transfer favors transplantation tolerance in mice. *J Clin Invest* **118:** 3619–3628.

Tsang JY, Tanriver Y, Jiang S, Leung E, Ratnasothy K, Lombardi G, Lechler R. 2009. Indefinite mouse heart allograft survival in recipient treated with $CD4^+CD25^+$ regulatory T cells with indirect allospecificity and short term immunosuppression. *Transpl Immunol* **21:** 203–209.

Veerapathran A, Pidala J, Beato F, Yu XZ, Anasetti C. 2011. Ex vivo expansion of human Tregs specific for alloantigens presented directly or indirectly. *Blood* **118:** 5671–5680.

Verginis P, McLaughlin KA, Wucherpfennig KW, von Boehmer H, Apostolou I. 2008. Induction of antigen-specific regulatory T cells in wild-type mice: Visualization and targets of suppression. *Proc Natl Acad Sci* **105:** 3479–3484.

Waldmann H. 2008. Tolerance can be infectious. *Nat Immunol* **9:** 1001–1003.

Walker MR, Carson BD, Nepom GT, Ziegler SF, Buckner JH. 2005. De novo generation of antigen-specific $CD4^+$ $CD25^+$ regulatory T cells from human $CD4^+CD25^-$ cells. *Proc Natl Acad Sci* **102:** 4103–4108.

Walker MR, Kasprowicz DJ, Gersuk VH, Benard A, Van Landeghen M, Buckner JH, Ziegler SF. 2003. Induction of FoxP3 and acquisition of T regulatory activity by stimulated human $CD4^+CD25^-$ T cells. *J Clin Invest* **112:** 1437–1443.

Westermann J, Pabst R. 1992. Distribution of lymphocyte subsets and natural killer cells in the human body. *Clin Investig* **70:** 539–544.

Wieczorek G, Asemissen A, Model F, Turbachova I, Floess S, Liebenberg V, Baron U, Stauch D, Kotsch K, Pratschke J, et al. 2009. Quantitative DNA methylation analysis of FOXP3 as a new method for counting regulatory T cells in peripheral blood and solid tissue. *Cancer Res* **69:** 599–608.

Wiegel FW, Perelson AS. 2004. Some scaling principles for the immune system. *Immunol Cell Biol* **82:** 127–131.

Wise MP, Bemelman F, Cobbold SP, Waldmann H. 1998. Linked suppression of skin graft rejection can operate through indirect recognition. *J Immunol* **161:** 5813–5816.

Wohlfert EA, Grainger JR, Bouladoux N, Konkel JE, Oldenhove G, Ribeiro CH, Hall JA, Yagi R, Naik S, Bhairavabhotla R, et al. 2011. GATA3 controls $Foxp3^+$ regulatory T cell fate during inflammation in mice. *J Clin Invest* **121:** 4503–4515.

Xu L, Xu Z, Xu M. 2009. Glucocorticoid treatment restores the impaired suppressive function of regulatory T cells in patients with relapsing-remitting multiple sclerosis. *Clin Exp Immunol* **158:** 26–30.

Yamada A, Chandraker A, Laufer TM, Gerth AJ, Sayegh MH, Auchincloss H Jr, 2001. Recipient MHC class II expression is required to achieve long-term survival of murine cardiac allografts after costimulatory blockade. *J Immunol* **167:** 5522–5526.

Yamaguchi T, Wing JB, Sakaguchi S. 2011. Two modes of immune suppression by $Foxp3^+$ regulatory T cells under inflammatory or non-inflammatory conditions. *Semin Immunol* **23:** 424–430.

Yates SF, Paterson AM, Nolan KF, Cobbold SP, Saunders NJ, Waldmann H, Fairchild PJ. 2007. Induction of regulatory T cells and dominant tolerance by dendritic cells incapable of full activation. *J Immunol* **179:** 967–976.

Zeiser R, Leveson-Gower DB, Zambricki EA, Kambham N, Beilhack A, Loh J, Hou JZ, Negrin RS. 2008. Differential impact of mammalian target of rapamycin inhibition on $CD4^+CD25^+Foxp3^+$ regulatory T cells compared with conventional $CD4^+$ T cells. *Blood* **111:** 453–462.

Zhang N, Schroppel B, Lal G, Jakubzick C, Mao X, Chen D, Yin N, Jessberger R, Ochando JC, Ding Y, et al. 2009. Regulatory T cells sequentially migrate from inflamed tissues to draining lymph nodes to suppress the alloimmune response. *Immunity* **30:** 458–469.

Zheng Y, Josefowicz S, Chaudhry A, Peng XP, Forbush K, Rudensky AY. 2010. Role of conserved non-coding DNA elements in the *Foxp3* gene in regulatory T-cell fate. *Nature* **463:** 808–812.

Zhou X, Jeker LT, Fife BT, Zhu S, Anderson MS, McManus MT, Bluestone JA. 2008. Selective miRNA disruption in T reg cells leads to uncontrolled autoimmunity. *J Exp Med* **205:** 1983–1991.

Mesenchymal Stromal Cells in Transplantation Rejection and Tolerance

Karen English[1] and Kathryn J. Wood[2]

[1]Cellular Immunology Group, Institute of Immunology, National University of Ireland Maynooth, County Kildare, Ireland

[2]Transplantation Research Immunology Group, Nuffield Department of Surgical Sciences, John Radcliffe Hospital, University of Oxford, Oxford 0X3 9DU, United Kingdom

Correspondence: karen.english@nuim.ie

Mesenchymal stromal cells (MSCs) have recently emerged as promising candidates for cell-based immunotherapy in solid organ transplantation (SOT). In addition to immune modulation, MSCs possess proreparative properties and preclinical studies indicate that MSCs have the capacity to prolong graft survival and in some cases induce tolerance. Currently, the application of MSCs in SOT is being evaluated in phase I/II clinical trials. Whereas the mechanisms of action used by MSC immunomodulation have been somewhat elucidated in vitro, the data from preclinical transplant models have been unclear. Furthermore, the optimal timing, dose, and route of administration remain to be elucidated. Importantly, MSCs have the ability to sense their environment, which may influence their function. In this article, we discuss the impact of the local microenvironment on MSCs and the mechanisms of MSC immunomodulation in the setting of SOT.

Mesenchymal stromal cells (MSCs) are a subpopulation of multipotent cells originally identified in the bone marrow (Friedenstein et al. 1976). MSCs are characterized by their fibroblast-like appearance, colony forming unit capacity, and their rapid adherence to tissue culture plastic. Although MSCs are relatively easy to isolate, culture, and expand (from a number of tissues), the lack of a unique marker to identify MSCs has impacted the advancement of this research field as difficulties arise in comparing data using different MSC populations. In 2006, the International Society for Cellular Therapy proposed a set of phenotypic and functional criteria to define MSCs (Dominici et al. 2006), however, the discovery of new markers that specifically identify MSCs are eagerly awaited. MSCs have the capacity to differentiate into adipocytes, chondrocytes, and osteoblasts in vitro and in vivo (Pittenger et al. 1999). Based on the differentiation potential of MSCs, initially studies focused on the regenerative capacity of these cells (Mahmood et al. 2003; Murphy et al. 2003); however, over time, it became clear that MSCs mediated their effects predominantly through the production of trophic factors (Caplan and Dennis 2006; Prockop 2009). Indeed, some of these trophic

Cite this article as *Cold Spring Harb Perspect Med* doi: 10.1101/cshperspect.a015560

factors facilitate MSC modulation of immune responses.

One of the first reports describing MSC immunosuppressive capacity was in fact a transplant model that showed that allogeneic (donor derived) MSCs prolonged allogeneic (donor and third-party-derived) skin graft survival (Bartholomew et al. 2002). Around the same time, Di Nicola et al. (2002) showed that MSCs mediated their suppressive effect through secretion of soluble factors. A significant body of data now supports an immunosuppressive capacity for MSCs both in vitro and in vivo. At the outset, studies focused primarily on MSC suppression of the adaptive immune response showing that MSCs can directly inhibit T-cell function, shift the T-helper lymphocyte balance, induce T-cell apoptosis, and induce functional regulatory T cells (Treg) (Kong et al. 2009; Ge et al. 2010; Akiyama et al. 2012). With respect to B cells, the available data are sparse and in some cases contradictory, but some studies suggest that MSCs can also suppress B-cell proliferation and function (Comoli et al. 2008). Recent findings convincingly show that MSCs modulate multiple components of the innate immune system including complement, toll-like receptor (TLR) signaling, macrophages, dendritic cells neutrophils, mast cells, and natural killer cells (Spaggiari et al. 2006; English et al. 2008; Kim and Hematti 2009; Nemeth et al. 2009; Cutler et al. 2010; Choi et al. 2011). Therapeutic efficacy of MSC anti-inflammatory effects has been established in a number of preclinical models including graft versus host disease, sepsis, inflammatory bowel disease, and allergic airway disease (Polchert et al. 2008; Ren et al. 2008; Nemeth et al. 2009; Kavanagh and Mahon 2011; Akiyama et al. 2012). In the case of solid organ transplantation (SOT), MSCs exert their effects on two fronts through attenuation of ischemia reperfusion injury (Liu et al. 2012a) and through the prevention of allograft rejection (Casiraghi et al. 2008; Ding et al. 2009; Ge et al. 2010). Moreover, in some cases, MSC induce a state of tolerance (Ge et al. 2010; Casiraghi et al. 2012). The in vitro immunosuppressive capacity, combined with the proven therapeutic efficacy of MSCs in preclinical models, has paved the way for MSCs in clinical application. Further evidence of a protective role for MSCs in preclinical models of organ transplantation in combination with the reported safety of MSCs in clinical trials has prompted the evaluation of safety and efficacy of MSCs in SOT (Tan et al. 2012). Herein, we will discuss the underlying mechanisms of MSC immunomodulation in the context of ischemia reperfusion injury, prevention of allogeneic graft rejection, and induction of tolerance.

REJECTION

Mechanisms of Transplantation Rejection

Despite the significant achievements accomplished during the past 60 years in SOT, rejection remains the greatest barrier (Wood and Goto 2012; Wood et al. 2012). Whereas, the advent of immunosuppressive drugs has facilitated improved outcomes in graft survival and long term function, the toxicity and associated complications of nonspecific immunosuppression are substantial limiting factors (Halloran 2004). Thus, there is a significant unmet need for nontoxic immunosuppressive therapies.

The immune response to an allograft is an ongoing process involving both innate and adaptive components starting from the moment of reperfusion. In fact, the tissue injury associated with organ retrieval (ischemic injury) initiates the production of damage-associated molecular patterns (DAMPs) and subsequent activation of the complement system and innate immune cells (macrophages and neutrophils) through pathogen recognition receptors (PRRs) (Eltzschig and Eckle 2011) after reperfusion and initiates a local inflammatory environment. Activation of the innate immune response orchestrates the adaptive immune response. Although, graft rejection is multifactorial, alloantigen specific induction of T-cell proliferation and activation of T-cell effector functions is the major player in graft destruction. Antibody mediated rejection is triggered by alloantibody binding and complement activation and also significantly contributes to graft loss. Here, we discuss the effect of MSC on ischemia reperfusion injury and innate and

 Cite this article as *Cold Spring Harb Perspect Med* doi: 10.1101/cshperspect.a015560

adaptive components in the context of solid organ transplantation.

ISCHEMIA REPERFUSION INJURY

Effect of Mesenchymal Stromal Cells on Ischemia Reperfusion Injury

Ischemia reperfusion injury causes sterile inflammation and results in the production of a number of DAMPs including necrotic cells, cellular debris, heat shock proteins (HSPs), and high mobility group protein box-1 (HMGB-1) (Eltzschig and Eckle 2011). DAMPs activate PPRs such as TLRs, C-type lectin receptors, nucleotide-binding oligomerization domain (NOD) and NOD-like receptors, receptor for advanced glycation endproducts (RAGE), and retinoic acid inducible gene-I receptors. Signaling through these receptors results in activation of the inflammasome (Ogura et al. 2006) and the complement system, upregulating gene transcription and production of micro-RNAs (Eltzschig and Eckle 2011) involved in the inflammatory response. Together, these factors lead to the production of proinflammatory cytokines, the activation of platelets and endothelial cells, tissue hypoxia, and the recruitment of innate and eventually adaptive immune cells (Eltzschig and Eckle 2011).

MSCs are known to express a number of PRRs, including TLR1-9 (Pevsner-Fischer et al. 2007; Tomchuck et al. 2008; Opitz et al. 2009; Romieu-Mourez et al. 2009), NOD receptors (Kim et al. 2010; Sioud et al. 2010), and RAGE (Kume et al. 2005). These receptors are functionally active on MSCs, and binding to their respective ligands leads to alterations in MSC functions. For example, stimulation of NOD-like receptors on MSC leads to the production of interleukin (IL)-8 and vascular endothelial growth factor (Kim et al. 2010; Sioud et al. 2010). HMGB-1 signaling through RAGE induced MSC migration and inhibited MSC production of indoleamine 2,3-dioxygenase (IDO) (Lotfi et al. 2011). TLR3 and TLR4 activation of MSCs resulted in differential effects with TLR4 priming inducing a proinflammatory phenotype and secretion of IL-6, IL-8, and transforming growth factor β (TGF-β), whereas TLR3 priming induced anti-inflammatory MSCs producing IDO, prostaglandin E-2 (PGE-2), IL-4, and IL-1RA (Waterman et al. 2010). In regard to MSC immunosuppressive function, TLR3 and TLR4 enhanced MSC immunosuppression in vitro through IDO induction via IFN-β and protein kinase R signaling (Opitz et al. 2009). In addition, TLR2 but not NOD-1 activation of human MSCs resulted in the upregulation of the immune suppressive protein galectin-3 (Sioud et al. 2010). In contrast, Liotta et al. (2008) showed that TLR3 and TLR4 ligand binding attenuated MSC immunosuppressive effects. It is clear that MSCs are receptive to environmental cues that may be present during ischemia reperfusion injury, however, further research is required to understand how these DAMPs effect MSCs and whether or not they play a role in determining the function of MSCs.

MSCs are also responsive to complement and migrate in response to C1q, C3a, and C5a (Schraufstatter et al. 2009; Qiu et al. 2012), and high levels of C3 activation correlate with enhanced immunosuppressive capacity of MSCs (Moll et al. 2011). Importantly, MSCs express CD59, a complement regulatory protein, and also release complement factor H that protects them from complement lysis (Tu et al. 2010; Moll et al. 2011). In addition, stimulation through C3aR and C5aR protect MSCs from oxidative damage (Schraufstatter et al. 2009) and MSCs produce a number of antioxidants including hemeoxygenase-1 and superoxide dismutase (Kemp et al. 2010; Mougiakakos et al. 2011), and have been shown to suppress oxidative stress and inflammation in ischemia reperfusion injury models in vivo (Chen et al. 2011, 2012; Sun et al. 2011; Du et al. 2012). The exact mechanisms of action are unclear; however, MSC protection in these models was associated with increased expression of IL-10, heme oxygenase-1 (HO-1), and hepatocyte growth factor, decreased expression of the proinflammatory cytokines IL-1β, TNF-α, and interferon γ (IFN-γ), reduced reactive oxygen species, reduced apoptosis and decreased numbers of activated T cells, and infiltrating immune cells (Hara et al. 2011; Sun et al. 2011; Chen et al. 2012; Du et al. 2012).

MSC derived microvesicles combined with soluble factors have been shown to protect against ischemia reperfusion induced acute and chronic kidney injury through inhibition of apoptosis and stimulation of tubular epithelial-cell proliferation (Gatti et al. 2011). Importantly, a protective effect of MSCs in ischemia reperfusion transplant models has been reported. Administration of MSCs reduced intragraft inflammatory gene expression and recruitment of antigen presenting cells into the allograft in a prolonged cold ischemic kidney transplant model (Hara et al. 2011) and provided long-term protection from chronic allograft nephropathy (Franquesa et al. 2012a). Ischemia reperfusion also plays a key role in the recruitment of MSCs to transplanted organs (Casiraghi et al. 2012). MSC recruitment mediated by ischemia reperfusion injury in combination with an ongoing alloreactive response leads to premature graft dysfunction and fails to prolong graft survival (Casiraghi et al. 2012). This study showed that although MSCs administered posttransplant promoted neutrophil infiltration and complement deposition, infusion of MSCs pretransplant induced significant allograft survival through a Treg dependent mechanism (Fig. 1) (Casiraghi et al. 2012). The key observation of this study is that MSCs infused pretransplant localize in the lymphoid organs whereas MSC administered posttransplant are recruited to the graft (syngeneic or allogeneic). Overall, it seems that MSCs can exert protective effects in ischemic reperfusion injuries through anti-inflammatory and paracrine factors and this likely plays an important part in MSC enhancement of allograft survival. However, there is still a lot to learn with regard to the effect of the ischemic environment on MSC functions and how this might impact MSC therapy in solid organ transplantation.

TOLERANCE

Mesenchymal Stromal Cell Modulation of Macrophages

Macrophages and neutrophils are generally the first innate immune cells to infiltrate the graft postischemia reperfusion injury. Whereas neutrophils are present only in the graft during inflammatory episodes, macrophages are present throughout the life of the graft, infiltrating in response to ischemia reperfusion injury, and maintained in reduced numbers after resolution of tissue injury in the absence of rejection. Neutrophils and macrophages have been shown to play a role in graft rejection through tissue damage induced by effector function, production of proinflammatory cytokines, and activation of antigen specific T cells (Wyburn et al. 2005). On the other hand, macrophages are also known to play a role in tissue repair and can be repolarized from pro to anti-inflammatory (Sica and Mantovani 2012). Although it has not been shown in a SOT setting, MSCs have the capacity to re-educate monocytes/macrophages. MSCs induce alternatively activated macrophages down-regulating the production of TNF-α, IL-1α, IL-6, and IL-12p70, and increasing the production of IL-10 and enhancing phagocytic activity (Kim and Hematti 2009; Nemeth et al. 2009; Cutler et al. 2010; Choi et al. 2011) through production of IDO and PGE-2 (Maggini et al. 2010; Francois et al. 2012).

Two important studies build a picture of how MSCs orchestrate macrophage polarisation and the influence the local microenvironment has on that process. Nemeth and colleagues show that MSCs ameliorate sepsis through alternative activation of macrophages, showing that lipopolysaccharide (LPS) and TNF-α activate TLR4 and tumor necrosis factor receptor 1 on MSCs to activate NF-κB signaling. This, in turn, leads to the expression of cyclooxygenase (COX)-2 and synthesis of PGE-2 by MSCs that bind E-prostanoid 2/4 (EP2/EP4) receptors on macrophages resulting in increased production of IL-10 and facilitating the resolution of inflammation (Nemeth et al. 2009). In the second study, Prockop's group used a zymosan induced peritonitis model to show that MSCs exerted anti-inflammatory effects through the production of tumor necrosis factor α (TNF-α)-induced protein 6 (TSG-6), which subsequently limits TLR2/NF-κB signaling through direct interaction with CD44 expressed on the macrophage to initiate a negative feedback loop

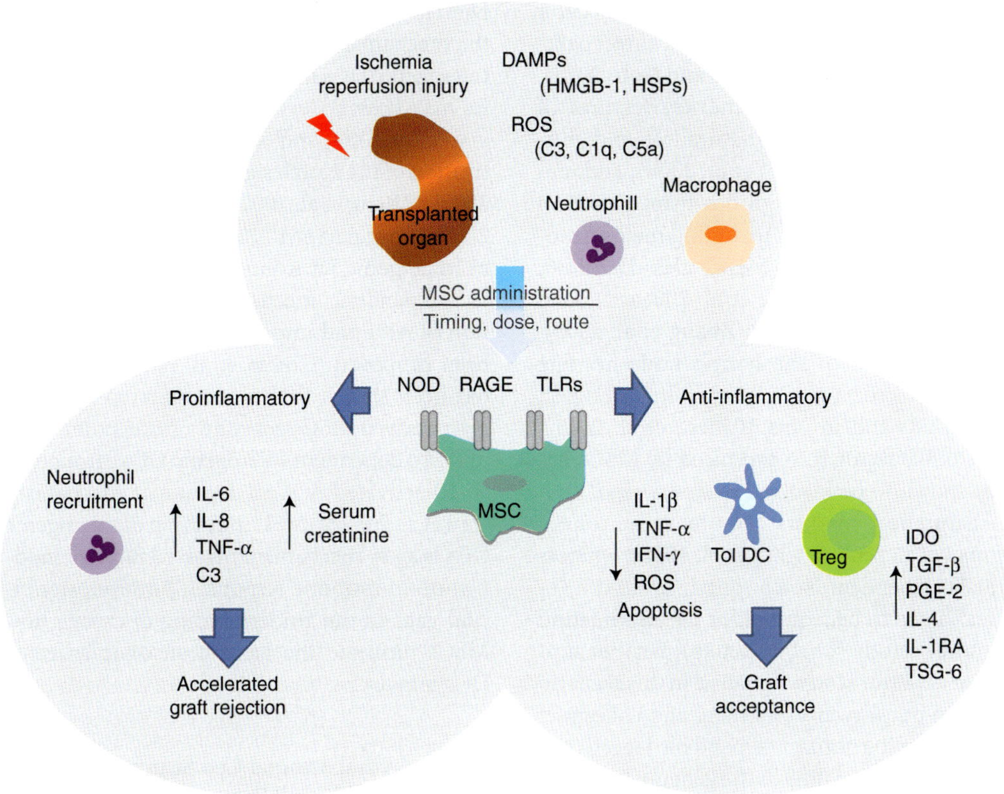

Figure 1. Influence of ischemia reperfusion injury on MSC function. Ischemia reperfusion injury (associated with organ retrieval) leads to the production of DAMPs including HMGB-1 and HSPs among others. Signaling through PRRs including TLRs, NOD receptors, and RAGE results in the activation of the complement system, recruitment of innate immune cells (neutrophils and macrophages), and the production of reactive oxygen species (ROS). Depending on the timing, dose, and route of administration, MSC given to patients in the context of SOT may encounter an ischemia reperfusion microenvironment. MSCs express a number of PRRs including NOD, RAGE, and TLRs, and activation of these receptors through DAMPs like HMGB-1 and HSPs may determine MSC function. Signaling through RAGE, NOD-like receptors, or TLR4 on MSC may lead to the promotion of a proinflammatory environment leading to the production of the proinflammatory cytokines IL-6, IL-8, and TNF-α, neutrophil recruitment, and complement deposition resulting in elevated serum creatinine and accelerated graft rejection. Alternatively, activation of TLR3 on MSC may induce an anti-inflammatory milieu, reducing the production of proinflammatory cytokines (IL-1β, TNF-α, and IFN-γ), decreasing ROS and apoptosis, and generating tolerogenic dendritic cells and Treg facilitating graft acceptance.

inhibiting the inflammatory response (Choi et al. 2011). Although MSC secretion of TSG-6 has been implicated in corneal allograft survival (Oh et al. 2012), it remains to be determined whether or not MSC derived TSG-6 can modulate macrophages in this setting. On the whole, microenvironmental cues present at the site of MSC activation seem to determine the particular mechanism of actions deployed by MSCs in modulating the immune response and resolving inflammation.

Mesenchymal Stromal Cell Modulation of Dendritic Cells

Dendritic cells (DCs) are sentinel cells and as such, present alloantigens activating antigen specific T cells. Both donor and recipient

DCs play a critical role triggering graft rejection through the direct, indirect, or semi-direct pathways of allorecognition (Wood and Goto 2012). MSCs can interfere with the key features of DC function: migration, maturation, and antigen presentation (English et al. 2008), and mediate these effects through down-regulation of DC maturation markers including major histocompatibility complex (MHC) class II, CD40, CD80, and CD86 (Nauta et al. 2006a; Djouad et al. 2007; Li et al. 2008b; Zhang et al. 2009), and modulation of the lymph node homing chemokine receptor CCR7 in vitro (English et al. 2008) and in vivo (Chiesa et al. 2011). The soluble factor IL-6 produced by MSCs has been shown to be involved in down-regulation of maturation markers (Nauta et al. 2006a; Djouad et al. 2007; English et al. 2008), whereas contact-dependent Notch signaling but not IL-6 was shown to be required for DC modulation in another study (Li et al. 2008b). Further support of a role for Notch signaling in this scenario has been put forward by Zhang and colleagues (2009) showing contact-dependent Jagged-2 (a Notch ligand) signaling in the generation of regulatory DC. In the context of SOT, MSCs have been shown to block DC maturation and function in a kidney allograft model (Ge et al. 2010); however, the exact mechanisms used by MSCs to achieve this effect remains to be elucidated.

Analogous to the effects of MSCs on macrophage polarization, MSCs can also promote the generation of tolerogenic DCs (Li et al. 2008a; Ge et al. 2009; Spaggiari et al. 2009; Zhang et al. 2009; Liu et al. 2012b) producing anti-inflammatory cytokines and displaying enhanced phagocytic activity (Zhang et al. 2009; Liu et al. 2012b) typical of tolerogenic DCs. MSC educated DCs have the capacity to suppress alloreactive responses and prolong islet allograft survival (Huang et al. 2010) and to induce a state of tolerance in the context of SOT (cardiac allograft) in the presence of low dose immunosuppression (Ge et al. 2009).

The mechanisms of action mediated by MSCs in the generation of tolerogenic DCs are likely influenced by the context in which MSCs see DCs. The key mediator in MSC modulation of DC maturation is IL-6 (Nauta et al. 2006a; Djouad et al. 2007; English et al. 2008), however, the mechanism involved in MSC promotion of tolerogenic DCs has been less clear. Other studies have shown a central role for PGE-2 (Spaggiari et al. 2009) and cell contact-dependent activation of the Notch signaling pathway (Li et al. 2008b; Zhang et al. 2009), but not IL-6 (Li et al. 2008b; Spaggiari et al. 2009) in MSC generation of tolerogenic DCs. Further evidence of a contact-dependent mechanism involving activation of AKT and impaired NF-κB signaling has been proposed (Chiesa et al. 2011). Finally, it has been shown that mouse embryonic fibroblast derived MSCs generate a novel population of IL-10 dependent tolerogenic DCs through an IL-10 activated SOCS3 dependent mechanism (Liu et al. 2012b). MSC induction of tolerogenic DCs is a key mechanism involved in MSC modulation of immune responses; however, significant gaps in our understanding of exactly how MSCs promote the generation of tolerogenic DCs remain.

Mesenchymal Stromal Cell Suppression of Allogeneic T-Cell Responses in Transplantation Rejection

Allogeneic T-cell proliferation and activation are prerequisites for allograft rejection and induction of tolerance in allogeneic organ transplantation is usually associated with Treg (Issa et al. 2011; Wood et al. 2012). A large body of data convincingly show that autologous and allogeneic MSCs modulate T-cell proliferation, activation, and function both in vitro and in vivo (Di Nicola et al. 2002; Glennie et al. 2005; English et al. 2007; Asari et al. 2009; Ding et al. 2009; English and Mahon 2011). Moreover, in vitro assays have also confirmed the capacity for MSCs to inhibit Th17 cell differentiation (Duffy et al. 2011; Tatara et al. 2011) or to shift the T helper cell balance in favor of a more anti-inflammatory phenotype (Batten et al. 2006; Bai et al. 2009; English et al. 2009; Fiorina et al. 2009; Ghannam et al. 2010). The mechanisms used by MSCs in mediating these effects vary between in vitro and in vivo models. However, many of these effects are mediated through soluble factors secreted by MSCs (English 2013).

IDO and PGE-2 have been implicated in MSC inhibition of Th17 differentiation (Duffy et al. 2011; Tatara et al. 2011). In the case of PGE-2, the steps involved in the process require contact-dependent COX-2 induction of PGE-2 and direct inhibition through EP4 (Duffy et al. 2011). MSCs can also mediate this effect through suppressing the Th17 transcription factor RORγt and upregulating Foxp3 to induce a Treg phenotype producing IL-10 (Ghannam et al. 2010). MSC derived TGF-β has been shown to play a partial role in shifting the balance of Th1/Th2/Th17 and Treg in an autoimmune disease model (Kong et al. 2009).

With regard to SOT, MSCs impaired alloreactive T-cell responses (Casiraghi et al. 2008, 2012; Ge et al. 2010) and inhibited the migration of activated T cells into the allograft (Eggenhofer et al. 2011b; Hara et al. 2011; Franquesa et al. 2012a). In particular, matrix metalloproteinase (MMP)2 and MMP9 secreted by MSCs facilitated cleavage of CD25 expressed on $CD4^+$ T cells inhibiting alloantigen driven proliferation and preventing islet allograft rejection (Ding et al. 2009). Other evidence suggests that MSC derived MMPs also cleave CC chemokine ligand (CCL2), which subsequently inhibits Th17 activation via a STAT3 dependent pathway (Rafei et al. 2009). In addition, MSCs mediated their effects in part through shifting the balance of T helper 1 cell phenotype to a more anti-inflammatory Th2 phenotpye producing IL-4 and IL-10 (Ge et al. 2010; Jia et al. 2012). Importantly, MSCs also have the capacity to expand or induce Treg in the setting of SOT (Casiraghi et al. 2008, 2012; Wang et al. 2009; Jia et al. 2012) and in some cases to generate a state of Treg-dependent tolerance (Ge et al. 2010; Casiraghi et al. 2012). Both of these studies elegantly show the importance of Treg in MSC induced tolerance using Treg depletion studies. Apart from the study by Ge et al. (2010), which identifies an important role for MSC derived IDO in the generation of Treg, the mechanisms of action mediated by MSCs remain to be elucidated. In vitro, the factors required for MSC induction of Treg are thought to involve cell contact, PGE-2, and TGF-β (English et al. 2009). Further evidence for a contact-dependent role was provided by Selmani and colleagues, showing that cell contact-dependent production of HLA-G5 was required for the expansion of Treg (Selmani et al. 2008). In vivo, MSC derived TGF-β was required for the generation of antigen specific Treg and overall, TGF-β seems to be the major soluble factor involved in MSC promotion of Treg in vivo (Zhao et al. 2008; Kong et al. 2009; Nemeth et al. 2010; Akiyama et al. 2012). MSC generation of Treg involves a number of different steps that are dependent on the specific environment or disease model. For example, Nemeth and colleagues (2010) showed that MSCs exposed to IL-4 and IL-13 (typically produced in allergic environment) produced TGF-β through an IL-4R/STAT6 dependent pathway. Alternatively, MSCs may induce Treg indirectly through modulation of innate immune cells. A study performed by Akiyama et al. examining the effect of MSCs in a mouse model of dextran sodium sulfate-induced colitis, unravelled a complex course of events that ultimately led to the expansion of Treg. Specifically, MSCs induced T effector cell apoptosis through FAS/FASL facilitated by MSC MCP-1 chemoattraction of T cells. Subsequently, macrophages produce TGF-β following phagocytosis of the apoptotic cell debris resulting in the expansion of Treg (Akiyama et al. 2012).

INTERACTION OF MESENCHYMAL STROMAL CELLS WITH ALLOREACTIVE B CELL RESPONSES

The triggering of antibody mediated rejection by alloantibody binding and complement activation is increasingly associated with graft loss (Wood and Goto 2012). The reported effects of MSCs on B-cell activation, proliferation, and function have been variable and in some cases contradictory. MSCs have been shown to inhibit B-cell proliferation (Augello et al. 2005; Corcione et al. 2006; Asari et al. 2009; Schena et al. 2010) and immunoglobulin (Ig) production (Corcione et al. 2006; Comoli et al. 2008; Rafei et al. 2008) in vitro. The mechanisms of action involve contact-dependent factors (Schena et al. 2010) including programmed death-1 [PD-1]/programmed death-ligand [PD-

L1]/PD-L2 (Augello et al. 2005) as well as soluble factors like MMP cleaved CCL2 (Rafei et al. 2008). MSC inhibition of plasma cells induced by LPS or plasmacytoid DCs was shown to be mediated through extracellular signal-related kinases (ERK) 1/2 and phosphorylation of p38 (Tabera et al. 2008; Asari et al. 2009). Moreover, MSCs induced cell cycle arrest in B cells (Corcione et al. 2006), and particularly under highly proliferative conditions (Traggiai et al. 2008). In contrast, MSCs also have the capacity to promote B-cell proliferation and survival (Tabera et al. 2008; Traggiai et al. 2008; Youd et al. 2010) and enhance Ig production (Rasmusson et al. 2007; Youd et al. 2010). Considerable variations are reported in these findings with MSCs increasing IgM but not IgA or IgG in another study (Traggiai et al. 2008) with differences in the mechanisms involved (contact-dependent versus soluble factors) depending on the source of B cells (purified versus peripheral blood mononuclear cells or mononuclear cells) (Rasmusson et al. 2007) and the ratio of MSCs to B cells in co-cultures (Franquesa et al. 2012b). These differences may well be resolved by ensuring purity of B cells (excluding T-cell help) and by further understanding the effect that TLR ligands (LPS, cytosine phosphodiester guanine, polyinosinic:polycytidylic acid) have on MSC activation and function. The effect of MSCs has also been examined in the B cell driven pathology systemic lupus erythematosus (SLE) both in mouse models and in patient samples. MSC enhanced survival and reduced serum creatinine, blood urea nitrogen, proteinuria, C3 deposition, and decreased circulating double-stranded deoxyribonucleic acid (dsDNA) antibodies as well as antigen specific IgM and IgG secretion (Zhou et al. 2008; Asari et al. 2009; Choi et al. 2011). In contrast, Schena et al. (2010) reported no effect of MSCs on survival, proteinuria, or dsDNA antibodies. Furthermore, MSCs were shown to negatively impact SLE through enhancing pathology, autoantibody production, and proteinuria (Youd et al. 2010). In the setting of SOT, much of the focus has been on the effect of MSCs on T cells (discussed above), and to our knowledge, no study has rigorously examined the impact of MSC on alloreactive B cells. Nonetheless, there are reports that suggest that MSCs reduce intragraft IgG deposits as well as circulating donor specific antibodies (Ge et al. 2009; Franquesa et al. 2012a) providing protection from injury (Franquesa et al. 2012a) and inducing allograft tolerance in the presence of immunosuppression (Ge et al. 2009). Notably, failure of MSCs to prolong allograft survival has been associated with MSC promotion of intragraft B cell infiltration (Seifert et al. 2012), and although administration of donor MSCs posttransplant lead to sensitization and premature graft dysfunction, this was not thought to be associated with antibody mediated humoral rejection (Casiraghi et al. 2012). Rather, syngeneic MSCs administered posttransplant (but not pretransplant) localized in the transplanted kidney in response to ischemia reperfusion injury and subsequently produced IL-6 and TNF-α promoting a proinflammatory environment facilitating neutrophil infiltration and C3 deposition (Casiraghi et al. 2012). These studies highlight the significant gap in our understanding of the effect of the microenvironment on MSC activation and function and particularly the effect this has on how MSCs see B cells and vice versa.

Immunogenicity of Allogeneic Mesenchymal Stromal Cells

Although, allogeneic MSCs were thought to be immune privileged, evidence now suggests that allogeneic MSCs are recognized by the innate and adaptive immune system (Griffin et al. 2013). A clear understanding of how recognition of allogeneic MSCs impacts their capacity for modulating immune responses in vivo is hampered by the lack of appropriate experimental data measuring antidonor T-cell and antibody responses to allogeneic MSC. Nevertheless, a small number of studies show that allogeneic MSC evoke antidonor T-cell and antibody responses in vivo in healthy animals as well as models of myocardial infarction and bone marrow transplantation (Eliopoulos et al. 2005; Beggs et al. 2006; Nauta et al. 2006b; Badillo et al. 2007; Poncelet et al. 2008; Zangi et al. 2009; Isakova et al. 2010; Schu et al. 2012).

 Cite this article as *Cold Spring Harb Perspect Med* doi: 10.1101/cshperspect.a015560

In the SOT setting, allogeneic MSC have been shown to accelerate graft rejection (Inoue et al. 2006; Popp et al. 2008; Renner et al. 2009; Eggenhofer et al. 2011a,b; Seifert et al. 2012). In some cases, accelerated graft rejection was attributed to administration of allogeneic MSC pretransplant (Eggenhofer et al. 2011b; Seifert et al. 2012) or syngeneic MSC posttransplant (Casiraghi et al. 2012). In other cases, allogeneic or syngeneic MSC were shown to attenuate immunosuppressive drugs like cyclosporine A (Inoue et al. 2006). Importantly, allogeneic MSCs have the capacity to work in synergy with immunosuppressive drugs including mycophenolate mofetil (Eggenhofer et al. 2011a,b) and rapamycin (Ge et al. 2009), and in these studies promote graft survival. Perhaps these studies emphasize the impact that different microenvironments may have on allogeneic MSC immunogenicity as well as the immunosuppressive function of both syngeneic and allogeneic MSCs in vivo. These conflicting reports highlight the significant need for further research in this area.

CLINICAL APPLICATION OF MESENCHYMAL STROMAL CELLS IN SOLID ORGAN TRANSPLANTATION

The application of MSC therapy in conjunction with a reduced immunosuppressive regimen is theoretically very appealing as MSCs not only promote the resolution of inflammation and enhance graft repair, but also may facilitate the induction of tolerance. Based on the safety and efficacy data generated in preclinical models and clinical trials utilizing MSC therapy for acute graft versus host disease (Casiraghi et al. 2008; LeBlanc et al. 2008; Ge et al. 2010), MSC therapy is currently being evaluated in SOT (Hoogduijn et al. 2010). Perico and colleagues (2011) provided the first report on MSCs in kidney transplant patients in a pilot study examining safety and feasibility. Culture expanded autologous bone marrow derived MSCs ($1.7 \times 10^6 - 2.0 \times 10^6$/kg body weight) were administered intravenously on day 7 postkidney (living donor) transplant in addition to T-cell depletion induction therapy in two patients. This study reported an increase in serum creatinine in both patients 7–14 days after MSC infusion. In addition, a focal inflammatory (granulocyte) infiltrate was observed in a graft biopsy taken from one of the patients, but acute graft rejection was ruled out. Importantly, both patients maintained stable graft function, which was associated with decreased memory $CD8^+$ T cells and increased Treg (Perico et al. 2011). Notably, similar effects were observed in a mouse model of kidney allograft transplantation, and this study elegantly showed the correlation of premature graft injury with posttransplant (but not pretransplant) syngeneic MSC infusion and localization of the MSCs primarily in the injured graft (Casiraghi et al. 2012). Indeed these studies may help to resolve the disparate findings with regard to MSC efficacy in prolonging graft survival (Inoue et al. 2006; Popp et al. 2008; Renner et al. 2009; Eggenhofer et al. 2011a,b; Seifert et al. 2012). This highlights the importance of timing of MSC administration and the requirement for a better understanding of the influence of the transplanted graft microenvironment (ischemia reperfusion, for example) on MSC function (Fig. 1). A large randomized controlled trial (106 patients over three arms) investigating the safety and efficacy of autologous bone marrow derived MSCs (in kidney transplantation) in combination with standard dose calcineurin inhibitors (CNI), or low dose CNI were compared with control groups receiving anti-IL-2 receptor antibody therapy in combination with standard dose CNI (Tan et al. 2012). This trial showed safety and efficacy with decreased incidence of acute rejection and glucocorticoid-resistant rejection, and increased estimated glomerular filtration rate levels and faster recovery of renal function in the first month post transplant as well as better estimated renal function at year 1. In addition, MSC treated groups revealed significantly reduced risk of opportunistic infections than the control group (Tan et al. 2012). Importantly, this large randomized controlled trial did not observe increased creatinine levels in MSC treated patients and this may be associated with the dose and/or timing ($1-2 \times 10^6$ on day 0 and day 14 versus $1.7-2 \times 10^6$ on day 7 posttransplant) of

MSC administration or with differences in immunosuppressive regimen (Perico et al. 2011; Tan et al. 2012). These are important factors that may significantly impact MSC efficacy in SOT and require careful consideration.

CONCLUDING REMARKS

Although the capacity for MSCs to modulate immune responses in vitro has been undeniable, the in vivo immunosuppressive function of MSCs has, at times, been ambiguous. Over time it has become clear that MSCs are receptive to microenvironmental cues, and indeed evidence has emerged to suggest that the interactions between MSCs and the local environment are fundamental in determining MSC activation and function. In vivo models have shown that MSCs are responsive to ischemia reperfusion injury and can provide protection in this setting. Moreover, MSCs have the capacity to prolong graft survival and, indeed, may induce tolerance. Together, these characteristics make MSCs an ideal candidate for application in SOT, and early data from clinical trials suggest that MSCs are safe and efficacious, but highlight the gap in our understanding of how exactly MSCs mediate their protective effects in vivo. To this end, we must endeavor to fill these gaps and enhance MSC therapeutic efficacy.

ACKNOWLEDGMENTS

Karen English is supported by an HRB Translational Medicine Postdoctoral Fellowship and a Marie Curie Career Integration Grant. Work from the authors' own laboratories is supported by grants from The Wellcome Trust, European Union FP7 Programme (OPTISTEM), and the Medical Research Council, United Kingdom.

REFERENCES

Akiyama K, Chen C, Wang D, Xu X, Qu C, Yamaza T, Cai T, Chen W, Sun L, Shi S. 2012. Mesenchymal-stem-cell-induced immunoregulation involves FAS-ligand-/FAS-mediated T cell apoptosis. *Cell Stem Cell* **10:** 544–555.

Asari S, Itakura S, Ferreri K, Liu CP, Kuroda Y, Kandeel F, Mullen Y. 2009. Mesenchymal stem cells suppress B-cell terminal differentiation. *Exp Hematol* **37:** 604–615.

Augello A, Tasso R, Negrini SM, Amateis A, Indiveri F, Cancedda R, Pennesi G. 2005. Bone marrow mesenchymal progenitor cells inhibit lymphocyte proliferation by activation of the programmed death 1 pathway. *Eur J Immunol* **35:** 1482–1490.

Badillo AT, Beggs KJ, Javazon EH, Tebbets JC, Flake AW. 2007. Murine bone marrow stromal progenitor cells elicit an in vivo cellular and humoral alloimmune response. *Biol Blood Marrow Transplant* **13:** 412–422.

Bai L, Lennon DP, Eaton V, Maier K, Caplan AI, Miller SD, Miller RH. 2009. Human bone marrow-derived mesenchymal stem cells induce Th2-polarized immune response and promote endogenous repair in animal models of multiple sclerosis. *Glia* **57:** 1192–1203.

Bartholomew A, Sturgeon C, Siatskas M, Ferrer K, McIntosh K, Patil S, Hardy W, Devine S, Ucker D, Deans R, et al. 2002. Mesenchymal stem cells suppress lymphocyte proliferation in vitro and prolong skin graft survival in vivo. *Exp Hematol* **30:** 42–48.

Batten P, Sarathchandra P, Antoniw JW, Tay SS, Lowdell MW, Taylor PM, Yacoub MH. 2006. Human mesenchymal stem cells induce T cell anergy and downregulate T cell allo-responses via the TH2 pathway: Relevance to tissue engineering human heart valves. *Tissue Eng* **12:** 2263–2273.

Beggs KJ, Lyubimov A, Borneman JN, Bartholomew A, Moseley A, Dodds R, Archambault MP, Smith AK, McIntosh KR. 2006. Immunologic consequences of multiple, high-dose administration of allogeneic mesenchymal stem cells to baboons. *Cell Transplant* **15:** 711–721.

Caplan AI, Dennis JE. 2006. Mesenchymal stem cells as trophic mediators. *J Cell Biochem* **98:** 1076–1084.

Casiraghi F, Azzollini N, Cassis P, Imberti B, Morigi M, Cugini D, Cavinato RA, Todeschini M, Solini S, Sonzogni A, et al. 2008. Pretransplant infusion of mesenchymal stem cells prolongs the survival of a semiallogeneic heart transplant through the generation of regulatory T cells. *J Immunol* **181:** 3933–3946.

Casiraghi F, Azzollini N, Todeschini M, Cavinato RA, Cassis P, Solini S, Rota C, Morigi M, Introna M, Maranta R, et al. 2012. Localization of mesenchymal stromal cells dictates their immune or proinflammatory effects in kidney transplantation. *Am J Transplant* **12:** 2373–2383.

Chen YT, Sun CK, Lin YC, Chang LT, Chen YL, Tsai TH, Chung SY, Chua S, Kao YH, Yen CH, et al. 2011. Adipose-derived mesenchymal stem cell protects kidneys against ischemia-reperfusion injury through suppressing oxidative stress and inflammatory reaction. *J Transl Med* **9:** 51.

Chen S, Chen L, Wu X, Lin J, Fang J, Chen X, Wei S, Xu J, Gao Q, Kang M. 2012. Ischemia postconditioning and mesenchymal stem cells engraftment synergistically attenuate ischemia reperfusion-induced lung injury in rats. *J Surg Res* **178:** 81–91.

Chiesa S, Morbelli S, Morando S, Massollo M, Marini C, Bertoni A, Frassoni F, Bartolome ST, Sambuceti G, Traggiai E, et al. 2011. Mesenchymal stem cells impair in vivo T-cell priming by dendritic cells. *Proc Natl Acad Sci* **108:** 17384–17389.

Choi H, Lee RH, Bazhanov N, Oh JY, Prockop DJ. 2011. Anti-inflammatory protein TSG-6 secreted by activated MSCs attenuates zymosan-induced mouse peritonitis by decreasing TLR2/NF-κB signaling in resident macrophages. *Blood* **118:** 330–338.

Comoli P, Ginevri F, Maccario R, Avanzini MA, Marconi M, Groff A, Cometa A, Cioni M, Porretti L, Barberi W, et al. 2008. Human mesenchymal stem cells inhibit antibody production induced in vitro by allostimulation. *Nephrol Dial Transplant* **23:** 1196–1202.

Corcione A, Benvenuto F, Ferretti E, Giunti D, Cappiello V, Cazzanti F, Risso M, Gualandi F, Mancardi GL, Pistoia V, et al. 2006. Human mesenchymal stem cells modulate B-cell functions. *Blood* **107:** 367–372.

Cutler AJ, Limbani V, Girdlestone J, Navarrete CV. 2010. Umbilical cord-derived mesenchymal stromal cells modulate monocyte function to suppress T cell proliferation. *J Immunol* **185:** 6617–6623.

Ding YC, Xu DM, Feng G, Bushell A, Muschel RJ, Wood KJ. 2009. Mesenchymal stem cells prevent the rejection of fully allogenic islet grafts by the immunosuppressive activity of matrix metalloproteinase-2 and -9. *Diabetes* **58:** 1797–1806.

Di Nicola M, Carlo-Stella C, Magni M, Milanesi M, Longoni PD, Matteucci P, Grisanti S, Gianni AM. 2002. Human bone marrow stromal cells suppress T-lymphocyte proliferation induced by cellular or nonspecific mitogenic stimuli. *Blood* **99:** 3838–3843.

Djouad F, Charbonnier LM, Bouffi C, Louis-Plence P, Bony C, Apparailly F, Cantos C, Jorgensen C, Noel D. 2007. Mesenchymal stem cells inhibit the differentiation of dendritic cells through an interleukin-6-dependent mechanism. *Stem Cells* **25:** 2025–2032.

Dominici M, Le Blanc K, Mueller I, Slaper-Cortenbach I, Marini FC, Krause DS, Deans RJ, Keating A, Prockop DJ, Horwitz EM. 2006. Minimal criteria for defining multipotent mesenchymal stromal cells. The International Society for Cellular Therapy position statement. *Cytotherapy* **8:** 315–317.

Du T, Cheng J, Zhong L, Zhao XF, Zhu J, Zhu YJ, Liu GH. 2012. The alleviation of acute and chronic kidney injury by human Wharton's jelly-derived mesenchymal stromal cells triggered by ischemia-reperfusion injury via an endocrine mechanism. *Cytotherapy* **14:** 1215–1227.

Duffy MM, Pindjakova J, Hanley SA, McCarthy C, Weidhofer GA, Sweeney EM, English K, Shaw G, Murphy JM, Barry FP, et al. 2011. Mesenchymal stem cell inhibition of T-helper 17 cell- differentiation is triggered by cell–cell contact and mediated by prostaglandin E_2 via the EP4 receptor. *Eur J Immunol* **41:** 2840–2851.

Eggenhofer E, Renner P, Soeder Y, Popp FC, Hoogduijn MJ, Geissler EK, Schlitt HJ, Dahlke MH. 2011a. Features of synergism between mesenchymal stem cells and immunosuppressive drugs in a murine heart transplantation model. *Transpl Immunol* **25:** 141–147.

Eggenhofer E, Steinmann JF, Renner P, Slowik P, Piso P, Geissler EK, Schlitt HJ, Dahlke MH, Popp FC. 2011b. Mesenchymal stem cells together with mycophenolate mofetil inhibit antigen presenting cell and T cell infiltration into allogeneic heart grafts. *Transpl Immunol* **24:** 157–163.

Eliopoulos N, Stagg J, Lejeune L, Pommey S, Galipeau J. 2005. Allogeneic marrow stromal cells are immune rejected by MHC class I- and class II-mismatched recipient mice. *Blood* **106:** 4057–4065.

Eltzschig HK, Eckle T. 2011. Ischemia and reperfusion—from mechanism to translation. *Nat Med* **17:** 1391–1401.

English K. 2013. Mechanisms of mesenchymal stromal cell immunomodulation. *Immunol Cell Biol* **91:** 19–26.

English K, Mahon BP. 2011. Allogeneic mesenchymal stem cells: Agents of immune modulation. *J Cell Biochem* **112:** 1963–1968.

English K, Barry FP, Field-Corbett CP, Mahon BP. 2007. IFN-γ and TNF-α differentially regulate immunomodulation by murine mesenchymal stem cells. *Immunol Lett* **110:** 91–100.

English K, Barry FP, Mahon BP. 2008. Murine mesenchymal stem cells suppress dendritic cell migration, maturation and antigen presentation. *Immunol Lett* **115:** 50–58.

English K, Ryan JM, Tobin L, Murphy MJ, Barry FP, Mahon BP. 2009. Cell contact, prostaglandin E_2 and transforming growth factor β1 play non-redundant roles in human mesenchymal stem cell induction of $CD4^+$ $CD25^{high}$ forkhead box $P3^+$ regulatory T cells. *Clin Exp Immunol* **156:** 149–160.

Fiorina P, Jurewicz M, Augello A, Vergani A, Dada S, La Rosa S, Selig M, Godwin J, Law K, Placidi C, et al. 2009. Immunomodulatory function of bone marrow-derived mesenchymal stem cells in experimental autoimmune type 1 diabetes. *J Immunol* **183:** 993–1004.

Francois M, Romieu-Mourez R, Li M, Galipeau J. 2012. Human MSC suppression correlates with cytokine induction of indoleamine 2,3-dioxygenase and bystander M2 macrophage differentiation. *Mol Ther* **20:** 187–195.

Franquesa M, Herrero E, Torras J, Ripoll E, Flaquer M, Goma M, Lloberas N, Anegon I, Cruzado JM, Grinyo JM, et al. 2012a. Mesenchymal stem cell therapy prevents interstitial fibrosis and tubular atrophy in a rat kidney allograft model. *Stem Cells Dev* **21:** 3125–3135.

Franquesa M, Hoogduijn MJ, Bestard O, Grinyo JM. 2012b. Immunomodulatory effect of mesenchymal stem cells on B cells. *Front Immunol* **3:** 212.

Friedenstein AJ, Gorskaja JF, Kulagina NN. 1976. Fibroblast precursors in normal and irradiated mouse hematopoietic organs. *Exp Hematol* **4:** 267–274.

Gatti S, Bruno S, Deregibus MC, Sordi A, Cantaluppi V, Tetta C, Camussi G. 2011. Microvesicles derived from human adult mesenchymal stem cells protect against ischaemia-reperfusion-induced acute and chronic kidney injury. *Nephrol Dial Transplant* **26:** 1474–1483.

Ge W, Jiang J, Baroja ML, Arp J, Zassoko R, Liu W, Bartholomew A, Garcia B, Wang H. 2009. Infusion of mesenchymal stem cells and rapamycin synergize to attenuate alloimmune responses and promote cardiac allograft tolerance. *Am J Transplant* **9:** 1760–1772.

Ge W, Jiang J, Arp J, Liu W, Garcia B, Wang H. 2010. Regulatory T-cell generation and kidney allograft tolerance induced by mesenchymal stem cells associated with indoleamine 2,3-dioxygenase expression. *Transplantation* **90:** 1312–1320.

Ghannam S, Pene J, Torcy-Moquet G, Jorgensen C, Yssel H. 2010. Mesenchymal stem cells inhibit human Th17 cell differentiation and function and induce a T regulatory cell phenotype. *J Immunol* **185:** 302–312.

Glennie S, Soeiro I, Dyson PJ, Lam EW, Dazzi F. 2005. Bone marrow mesenchymal stem cells induce division arrest anergy of activated T cells. *Blood* **105:** 2821–2827.

Griffin MD, Ryan AE, Alagesan S, Lohan P, Treacy O, Ritter T. 2013. Anti-donor immune responses elicited by allogeneic mesenchymal stem cells: What have we learned so far? *Immunol Cell Biol* **91:** 40–51.

Halloran PF. 2004. Immunosuppressive drugs for kidney transplantation. *N Engl J Med* **351:** 2715–2729.

Hara Y, Stolk M, Ringe J, Dehne T, Ladhoff J, Kotsch K, Reutzel-Selke A, Reinke P, Volk HD, Seifert M. 2011. In vivo effect of bone marrow-derived mesenchymal stem cells in a rat kidney transplantation model with prolonged cold ischemia. *Transpl Int* **24:** 1112–1123.

Hoogduijn MJ, Popp FC, Grohnert A, Crop MJ, van Rhijn M, Rowshani AT, Eggenhofer E, Renner P, Reinders ME, Rabelink TJ, et al. 2010. Advancement of mesenchymal stem cell therapy in solid organ transplantation (MISOT). *Transplantation* **90:** 124–126.

Huang Y, Chen P, Zhang CB, Ko GJ, Ruiz M, Fiorina P, Hussain MA, Wasowska BA, Rabb H, Womer KL. 2010. Kidney-derived mesenchymal stromal cells modulate dendritic cell function to suppress alloimmune responses and delay allograft rejection. *Transplantation* **90:** 1307–1311.

Inoue S, Popp FC, Koehl GE, Piso P, Schlitt HJ, Geissler EK, Dahlke MH. 2006. Immunomodulatory effects of mesenchymal stem cells in a rat organ transplant model. *Transplantation* **81:** 1589–1595.

Isakova IA, Dufour J, Lanclos C, Bruhn J, Phinney DG. 2010. Cell-dose-dependent increases in circulating levels of immune effector cells in rhesus macaques following intracranial injection of allogeneic MSCs. *Exp Hematol* **38:** 957–967.

Issa F, Chandrasekharan D, Wood KJ. 2011. Regulatory T cells as modulators of chronic allograft dysfunction. *Curr Opin Immunol* **23:** 648–654.

Jia Z, Jiao C, Zhao S, Li X, Ren X, Zhang L, Han ZC, Zhang X. 2012. Immunomodulatory effects of mesenchymal stem cells in a rat corneal allograft rejection model. *Exp Eye Res* **102:** 44–49.

Kavanagh H, Mahon BP. 2011. Allogeneic mesenchymal stem cells prevent allergic airway inflammation by inducing murine regulatory T cells. *Allergy* **66:** 523–531.

Kemp K, Hares K, Mallam E, Heesom KJ, Scolding N, Wilkins A. 2010. Mesenchymal stem cell-secreted superoxide dismutase promotes cerebellar neuronal survival. *J Neurochem* **114:** 1569–1580.

Kim J, Hematti P. 2009. Mesenchymal stem cell-educated macrophages: A novel type of alternatively activated macrophages. *Exp Hematol* **37:** 1445–1453.

Kim HS, Shin TH, Yang SR, Seo MS, Kim DJ, Kang SK, Park JH, Kang KS. 2010. Implication of NOD1 and NOD2 for the differentiation of multipotent mesenchymal stem cells derived from human umbilical cord blood. *PLoS ONE* **5:** e15369.

Kong QF, Sun B, Bai SS, Zhai DX, Wang GY, Liu YM, Zhang SJ, Li R, Zhao W, Sun YY, et al. 2009. Administration of bone marrow stromal cells ameliorates experimental autoimmune myasthenia gravis by altering the balance of Th1/Th2/Th17/Treg cell subsets through the secretion of TGF-β. *J Neuroimmunol* **207:** 83–91.

Kume S, Kato S, Yamagishi S, Inagaki Y, Ueda S, Arima N, Okawa T, Kojiro M, Nagata K. 2005. Advanced glycation end-products attenuate human mesenchymal stem cells and prevent cognate differentiation into adipose tissue, cartilage, and bone. *J Bone Miner Res* **20:** 1647–1658.

LeBlanc K, Frassoni F, Ball L, Locatelli F, Roelofs H, Lewis I, Lanino E, Sundberg B, Bernardo ME, Remberger M, et al. 2008. Mesenchymal stem cells for treatment of steroid-resistant, severe, acute graft-versus-host disease: A phase II study. *Lancet* **371:** 1579–1586.

Li H, Guo ZK, Jiang XX, Zhu H, Li XS, Mao N. 2008a. Mesenchymal stem cells alter migratory property of T and dendritic cells to delay the development of murine lethal acute graft-versus-host disease. *Stem Cells* **26:** 2531–2541.

Li YP, Paczesny S, Lauret E, Poirault S, Bordigoni P, Mekhloufi F, Hequet O, Bertrand Y, Ou-Yang JP, Stoltz JF, et al. 2008b. Human mesenchymal stem cells license adult $CD34^+$ hemopoietic progenitor cells to differentiate into regulatory dendritic cells through activation of the notch pathway. *J Immunol* **180:** 1598–1608.

Liotta F, Angeli R, Cosmi L, Fili L, Manuelli C, Frosali F, Mazzinghi B, Maggi L, Pasini A, Lisi V, et al. 2008. Toll-like receptors 3 and 4 are expressed by human bone marrow-derived mesenchymal stem cells and can inhibit their T-cell modulatory activity by impairing Notch signaling. *Stem Cells* **26:** 279–289.

Liu H, Liu S, Li Y, Wang X, Xue W, Ge G, Luo X. 2012a. The role of SDF-1-CXCR4/CXCR7 axis in the therapeutic effects of hypoxia-preconditioned mesenchymal stem cells for renal ischemia/reperfusion injury. *PLoS ONE* **7:** e34608.

Liu X, Qu X, Chen Y, Liao L, Cheng K, Shao C, Zenke M, Keating A, Zhao RC. 2012b. Mesenchymal stem/stromal cells induce the generation of novel IL-10-dependent regulatory dendritic cells by SOCS3 activation. *J Immunol* **189:** 1182–1192.

Lotfi R, Eisenbacher J, Solgi G, Fuchs K, Yildiz T, Nienhaus C, Rojewski MT, Schrezenmeier H. 2011. Human mesenchymal stem cells respond to native but not oxidized damage associated molecular pattern molecules from necrotic (tumor) material. *Eur J Immunol* **41:** 2021–2028.

Maggini J, Mirkin G, Bognanni I, Holmberg J, Piazzon IM, Nepomnaschy I, Costa H, Canones C, Raiden S, Vermeulen M, et al. 2010. Mouse bone marrow-derived mesenchymal stromal cells turn activated macrophages into a regulatory-like profile. *PLoS ONE* **5:** e9252.

Mahmood A, Lu D, Lu M, Chopp M. 2003. Treatment of traumatic brain injury in adult rats with intravenous administration of human bone marrow stromal cells. *Neurosurgery* **53:** 697–702; discussion 702–703.

Moll G, Jitschin R, von Bahr L, Rasmusson-Duprez I, Sundberg B, Lonnies L, Elgue G, Nilsson-Ekdahl K, Mougiakakos D, Lambris JD, et al. 2011. Mesenchymal stromal cells engage complement and complement receptor bearing innate effector cells to modulate immune responses. *PLoS ONE* **6:** e21703.

Mougiakakos D, Jitschin R, Johansson CC, Okita R, Kiessling R, Le Blanc K. 2011. The impact of inflammatory licensing on heme oxygenase-1-mediated induction of regulatory T cells by human mesenchymal stem cells. *Blood* **117:** 4826–4835.

Murphy JM, Fink DJ, Hunziker EB, Barry FP. 2003. Stem cell therapy in a caprine model of osteoarthritis. *Arthritis Rheum* **48:** 3464–3474.

Nauta AJ, Kruisselbrink AB, Lurvink E, Willemze R, Fibbe WE. 2006a. Mesenchymal stem cells inhibit generation and function of both $CD34^+$-derived and monocyte-derived dendritic cells. *J Immunol* **177:** 2080–2087.

Nauta AJ, Westerhuis G, Kruisselbrink AB, Lurvink EG, Willemze R, Fibbe WE. 2006b. Donor-derived mesenchymal stem cells are immunogenic in an allogeneic host and stimulate donor graft rejection in a nonmyeloablative setting. *Blood* **108:** 2114–2120.

Nemeth K, Leelahavanichkul A, Yuen PST, Mayer B, Parmelee A, Doi K, Robey PG, Leelahavanichkul K, Koller BH, Brown JM, et al. 2009. Bone marrow stromal cells attenuate sepsis via prostaglandin E_2-dependent reprogramming of host macrophages to increase their interleukin-10 production. *Nat Med* **15:** 42–49.

Nemeth K, Keane-Myers A, Brown JM, Metcalfe DD, Gorham JD, Bundoc VG, Hodges MG, Jelinek I, Madala S, Karpati S, et al. 2010. Bone marrow stromal cells use TGF-β to suppress allergic responses in a mouse model of ragweed-induced asthma. *Proc Natl Acad Sci* **107:** 5652–5657.

Ogura Y, Sutterwala FS, Flavell RA. 2006. The inflammasome: First line of the immune response to cell stress. *Cell* **126:** 659–662.

Oh JY, Lee RH, Yu JM, Ko JH, Lee HJ, Ko AY, Roddy GW, Prockop DJ. 2012. Intravenous mesenchymal stem cells prevented rejection of allogeneic corneal transplants by aborting the early inflammatory response. *Mol Ther* **20:** 2143–2152.

Opitz CA, Litzenburger UM, Lutz C, Lanz TV, Tritschler I, Koppel A, Tolosa E, Hoberg M, Anderl J, Aicher WK, et al. 2009. Toll-like receptor engagement enhances the immunosuppressive properties of human bone marrow-derived mesenchymal stem cells by inducing indoleamine-2,3-dioxygenase-1 via interferon-β and protein kinase R. *Stem Cells* **27:** 909–919.

Perico N, Casiraghi F, Introna M, Gotti E, Todeschini M, Cavinato RA, Capelli C, Rambaldi A, Cassis P, Rizzo P, et al. 2011. Autologous mesenchymal stromal cells and kidney transplantation: A pilot study of safety and clinical feasibility. *Clin J Am Soc Nephrol* **6:** 412–422.

Pevsner-Fischer M, Morad V, Cohen-Sfady M, Rousso-Noori L, Zanin-Zhorov A, Cohen S, Cohen IR, Zipori D. 2007. Toll-like receptors and their ligands control mesenchymal stem cell functions. *Blood* **109:** 1422–1432.

Pittenger MF, Mackay AM, Beck SC, Jaiswal RK, Douglas R, Mosca JD, Moorman MA, Simonetti DW, Craig S, Marshak DR. 1999. Multilineage potential of adult human mesenchymal stem cells. *Science* **284:** 143–147.

Polchert D, Sobinsky J, Douglas G, Kidd M, Moadsiri A, Reina E, Genrich K, Mehrotra S, Setty S, Smith B, et al. 2008. IFN-γ activation of mesenchymal stem cells for treatment and prevention of graft versus host disease. *Eur J Immunol* **38:** 1745–1755.

Poncelet AJ, Nizet Y, Vercruysse J, Hiel AL, Saliez A, Gianello P. 2008. Inhibition of humoral response to allogeneic porcine mesenchymal stem cell with 12 days of tacrolimus. *Transplantation* **86:** 1586–1595.

Popp FC, Eggenhofer E, Renner P, Slowik P, Lang SA, Kaspar H, Geissler EK, Piso P, Schlitt HJ, Dahlke MH. 2008. Mesenchymal stem cells can induce long-term acceptance of solid organ allografts in synergy with low-dose mycophenolate. *Transpl Immunol* **20:** 55–60.

Prockop DJ. 2009. Repair of tissues by adult stem/progenitor cells (MSCs): Controversies, myths, and changing paradigms. *Mol Ther* **17:** 939–946.

Qiu Y, Marquez-Curtis LA, Janowska-Wieczorek A. 2012. Mesenchymal stromal cells derived from umbilical cord blood migrate in response to complement C1q. *Cytotherapy* **14:** 285–295.

Rafei M, Hsieh J, Fortier S, Li MY, Yuan S, Birman E, Forner K, Boivin MN, Doody K, Tremblay M, et al. 2008. Mesenchymal stromal cell-derived CCL2 suppresses plasma cell immunoglobulin production via STAT3 inactivation and PAX5 induction. *Blood* **112:** 4991–4998.

Rafei M, Campeau PA, Aguilar-Mahecha A, Buchanan M, Williams P, Birman E, Yuan S, Young YK, Boivin MN, Forner K, et al. 2009. Mesenchymal stromal cells ameliorate experimental autoimmune encephalomyelitis by inhibiting CD4 Th17 T cells in a CC chemokine ligand 2-dependent manner. *J Immunol* **182:** 5994–6002.

Rasmusson I, Le Blanc K, Sundberg B, Ringden O. 2007. Mesenchymal stem cells stimulate antibody secretion in human B cells. *Scand J Immunol* **65:** 336–343.

Ren GW, Zhang LY, Zhao X, Xu GW, Zhang YY, Roberts AI, Zhao RC, Shi YF. 2008. Mesenchymal stem cell-mediated immunosuppression occurs via concerted action of chemokines and nitric oxide. *Cell Stem Cell* **2:** 141–150.

Renner P, Eggenhofer E, Rosenauer A, Popp FC, Steinmann JF, Slowik P, Geissler EK, Piso P, Schlitt HJ, Dahlke MH. 2009. Mesenchymal stem cells require a sufficient, ongoing immune response to exert their immunosuppressive function. *Transplant Proc* **41:** 2607–2611.

Romieu-Mourez R, Francois M, Boivin MN, Bouchentouf M, Spaner DE, Galipeau J. 2009. Cytokine modulation of TLR expression and activation in mesenchymal stromal cells leads to a proinflammatory phenotype. *J Immunol* **182:** 7963–7973.

Schena F, Gambini C, Gregorio A, Mosconi M, Reverberi D, Gattorno M, Casazza S, Uccelli A, Moretta L, Martini A, et al. 2010. Interferon-γ-dependent inhibition of B cell activation by bone marrow-derived mesenchymal stem cells in a murine model of systemic lupus erythematosus. *Arthritis Rheum* **62:** 2776–2786.

Schraufstatter IU, Discipio RG, Zhao M, Khaldoyanidi SK. 2009. C3a and C5a are chemotactic factors for human mesenchymal stem cells, which cause prolonged ERK1/2 phosphorylation. *J Immunol* **182:** 3827–3836.

Schu S, Nosov M, O'Flynn L, Shaw G, Treacy O, Barry F, Murphy M, O'Brien T, Ritter T. 2012. Immunogenicity of allogeneic mesenchymal stem cells. *J Cell Mol Med* **16:** 2094–2103.

Seifert M, Stolk M, Polenz D, Volk HD. 2012. Detrimental effects of rat mesenchymal stromal cell pre-treatment in a model of acute kidney rejection. *Front Immunol* **3:** 202.

Selmani Z, Naji A, Zidi I, Favier B, Gaiffe E, Obert L, Borg C, Saas P, Tiberghien P, Rouas-Freiss N, et al. 2008. Human leukocyte antigen-G5 secretion by human mesenchymal stem cells is required to suppress T lymphocyte and natural killer function and to induce CD4$^+$ CD25^{high} FOXP3$^+$ regulatory T cells. *Stem Cells* **26:** 212–222.

Sica A, Mantovani A. 2012. Macrophage plasticity and polarization: In vivo veritas. *J Clin Invest* **122:** 787–795.

Sioud M, Mobergslien A, Boudabous A, Floisand Y. 2010. Evidence for the involvement of galectin-3 in mesenchymal stem cell suppression of allogeneic T-cell proliferation. *Scand J Immunol* **71:** 267–274.

Spaggiari GM, Capobianco A, Becchetti S, Mingari MC, Moretta L. 2006. Mesenchymal stem cell-natural killer cell interactions: Evidence that activated NK cells are capable of killing MSCs, whereas MSCs can inhibit IL-2-induced NK-cell proliferation. *Blood* **107:** 1484–1490.

Spaggiari GM, Abdelrazik H, Becchetti F, Moretta L. 2009. MSCs inhibit monocyte-derived DC maturation and function by selectively interfering with the generation of immature DCs: Central role of MSC-derived prostaglandin E_2. *Blood* **113:** 6576–6583.

Sun CK, Yen CH, Lin YC, Tsai TH, Chang LT, Kao YH, Chua S, Fu M, Ko SF, Leu S, et al. 2011. Autologous transplantation of adipose-derived mesenchymal stem cells markedly reduced acute ischemia-reperfusion lung injury in a rodent model. *J Transl Med* **9:** 118.

Tabera S, Perez-Simon JA, Diez-Campelo M, Sanchez-Abarca LI, Blanco B, Lopez A, Benito A, Ocio E, Sanchez-Guijo FM, Canizo C, et al. 2008. The effect of mesenchymal stem cells on the viability, proliferation and differentiation of B-lymphocytes. *Haematologica* **93:** 1301–1309.

Tan J, Wu W, Xu X, Liao L, Zheng F, Messinger S, Sun X, Chen J, Yang S, Cai J, et al. 2012. Induction therapy with autologous mesenchymal stem cells in living-related kidney transplants: A randomized controlled trial. *JAMA* **307:** 1169–1177.

Tatara R, Ozaki K, Kikuchi Y, Hatanaka K, Oh I, Meguro A, Matsu H, Sato K, Ozawa K. 2011. Mesenchymal stromal cells inhibit Th17 but not regulatory T-cell differentiation. *Cytotherapy* **13:** 686–694.

Tomchuck SL, Zwezdaryk KJ, Coffelt SB, Waterman RS, Danka ES, Scandurro AB. 2008. Toll-like receptors on human mesenchymal stem cells drive their migration and immunomodulating responses. *Stem Cells* **26:** 99–107.

Traggiai E, Volpi S, Schena F, Gattorno M, Ferlito F, Moretta L, Martini A. 2008. Bone marrow-derived mesenchymal stem cells induce both polyclonal expansion and differentiation of B cells isolated from healthy donors and systemic lupus erythematosus patients. *Stem Cells* **26:** 562–569.

Tu Z, Li Q, Bu H, Lin F. 2010. Mesenchymal stem cells inhibit complement activation by secreting factor H. *Stem Cells Dev* **19:** 1803–1809.

Wang Y, Zhang A, Ye Z, Xie H, Zheng S. 2009. Bone marrow-derived mesenchymal stem cells inhibit acute rejection of rat liver allografts in association with regulatory T-cell expansion. *Transplant Proc* **41:** 4352–4356.

Waterman RS, Tomchuck SL, Henkle SL, Betancourt AM. 2010. A new mesenchymal stem cell (MSC) paradigm: Polarization into a pro-inflammatory MSC1 or an immunosuppressive MSC2 phenotype. *PLoS ONE* **5:** e10088.

Wood KJ, Goto R. 2012. Mechanisms of rejection: Current perspectives. *Transplantation* **93:** 1–10.

Wood KJ, Bushell A, Hester J. 2012. Regulatory immune cells in transplantation. *Nat Rev Immunol* **12:** 417–430.

Wyburn KR, Jose MD, Wu H, Atkins RC, Chadban SJ. 2005. The role of macrophages in allograft rejection. *Transplantation* **80:** 1641–1647.

Youd M, Blickarz C, Woodworth L, Touzjian T, Edling A, Tedstone J, Ruzek M, Tubo R, Kaplan J, Lodie T. 2010. Allogeneic mesenchymal stem cells do not protect NZBxNZW F1 mice from developing lupus disease. *Clin Exp Immunol* **161:** 176–186.

Zangi L, Margalit R, Reich-Zeliger S, Bachar-Lustig E, Beilhack A, Negrin R, Reisner Y. 2009. Direct imaging of immune rejection and memory induction by allogeneic mesenchymal stromal cells. *Stem Cells* **27:** 2865–2874.

Zhang B, Liu R, Shi D, Liu XX, Chen Y, Dou XW, Zhu XS, Lu CH, Liang W, Liao LM, et al. 2009. Mesenchymal stem cells induce mature dendritic cells into a novel Jagged-2-dependent regulatory dendritic cell population. *Blood* **113:** 46–57.

Zhao W, Wang Y, Wang D, Sun B, Wang G, Wang J, Kong Q, Wang Q, Peng H, Jin L, et al. 2008. TGF-β expression by allogeneic bone marrow stromal cells ameliorates diabetes in NOD mice through modulating the distribution of CD4$^+$ T cell subsets. *Cell Immunol* **253:** 23–30.

Zhou K, Zhang H, Jin O, Feng X, Yao G, Hou Y, Sun L. 2008. Transplantation of human bone marrow mesenchymal stem cell ameliorates the autoimmune pathogenesis in MRL/lpr mice. *Cell Mol Immunol* **5:** 417–424.

Immunological Challenges and Therapies in Xenotransplantation

Marta Vadori[1] and Emanuele Cozzi[1,2,3]

[1]CORIT (Consortium for Research in Organ Transplantation), Legnaro, 35020 Padua, Italy
[2]Transplant Immunology Unit, Padua University Hospital, 35128 Padua, Italy
[3]Department of Surgical, Oncological and Gastroenterological Sciences, University of Padua, Ospedale Giustinianeo, 35128 Padua, Italy

Correspondence: emanuele.cozzi@unipd.it

Xenotransplantation, or the transplantation of cells, tissues, or organs between different species, was proposed a long time ago as a possible solution to the worldwide shortage of human organs and tissues for transplantation. In this setting, the pig is currently seen as the most likely candidate species. In the last decade, progress in this field has been remarkable and includes a better insight into the immunological mechanisms underlying the rejection process. Several immunological hurdles nonetheless remain, such as the strong antibody-mediated and innate or adaptive cellular immune responses linked to coagulation derangements, precluding indefinite xenograft survival. This article reviews our current understanding of the immunological mechanisms involved in xenograft rejection and the potential strategies that may enable xenotransplantation to become a clinical reality in the not-too-distant future.

By "xenotransplantation," we conventionally refer to the transplantation of cells, tissues, or organs from one species to another. Current interest in xenotransplantation stems from the worldwide shortage of human organs, tissues, and cells for use in clinical transplantation. At least in theory, the imbalance between supply and demand could be wholly addressed if organs, tissues, or cells from other species could be transplanted into humans. The pig is currently considered the most appropriate candidate species because of its anatomical similarity, physiological compatibility, breeding characteristics, and for ethical reasons. Ongoing preclinical research in this field is consequently based on the use of pigs as donors and nonhuman primates as recipient species.

By now, we have gained a significant insight into the immunological processes underlying the rejection of porcine xenografts transplanted into primates. Considerable advances have also been made to elucidate the dysregulated coagulation occurring after porcine xenografts have been transplanted into primates. Despite the encouraging results obtained to date, especially in the field of cell xenotransplantation, several issues nonetheless remain to be addressed before any clinical application of xenotransplantation can proceed.

Cite this article as *Cold Spring Harb Perspect Med* doi: 10.1101/cshperspect.a015578

This review summarizes current knowledge in this field, focusing exclusively on the immune mechanisms underlying the rejection of cardiac, renal, and islet xenografts and on possible strategies to overcome these obstacles. The main emphasis is placed on the most clinically relevant pig-to-primate models. A comprehensive, accurate analysis of the coagulation derangements associated with xenotransplantation would be too lengthy for the format of this monograph; thus, for the sake of brevity, the reader is referred to other, excellent reviews recently published on the subject (Lin et al. 2009; Schmelzle et al. 2010; Cowan et al. 2011; Bulato et al. 2012).

MECHANISMS UNDERLYING XENOGRAFT REJECTION

Antibody-Mediated Xenograft Rejection

The rejection of a xenografted solid organ is characterized primarily by a picture compatible with a humorally driven immunological process. The humoral component of the immune response is a formidable barrier to short- and long-term organ survival. Hyperacute rejection (HAR) and acute humoral xenograft rejection (AHXR), also termed delayed xenograft rejection, are the main features of the humorally mediated xenograft rejection occurring when pig organs are transplanted into untreated primates.

HAR is a rapid, powerful process involving diffuse interstitial hemorrhage, edema, and thrombosis of the small vessels (Stevens and Platt 1992). This process is triggered by preexisting antibodies binding to xenograft antigens and prompting complement activation, graft endothelial cell activation and destruction, activation of the coagulation cascade, and graft rejection within minutes or hours. Preformed antipig antibodies are believed to be directed primarily against the terminal α3-galactose of the *N*-acetyllactosamine in glycoprotein and glycolipid carbohydrate chains (Galα3Galβ4Glc-Nac-R or αGal epitope) (Galili et al. 1988; Macher and Galili 2008). The synthesis of αGal is catalyzed by the α1-3 galactosyltransferase (α1-3 GalT), an enzyme expressed in nonprimate mammals, including the pig, and in New World monkeys. The density of αGal epitopes in pig organs goes from 1 to 3 $\times 10^7$ epitopes per cell, in endothelial and epithelial cells, respectively (Galili et al. 1988). These αGal epitopes have been identified in decellularized xenogeneic biological scaffolds, like the mammalian extracellular matrix used in surgical reconstructions, even after treatments to remove or mask antigenic epitopes (McPherson et al. 2000; Konakci et al. 2005; Stone et al. 2007). In human beings, and in Old World monkeys and apes, α1-3 GalT is inactive, the αGal epitope is not expressed, and there are high titers of anti-αGal antibodies due to exposure to similar epitopes expressed by bacteria hosted in the gut flora. A significant percentage of preformed total antibodies in human and nonhuman primates reacts with the αGal epitope (1%–8% of total IgM and 1%–2.4% of total IgG) (Parker et al. 1994; McMorrow et al. 1997).

A detailed study to ascertain the nature and residence of cells producing anti-αGal antibodies found them located primarily in the spleen, in both naïve and immunized nonhuman primates, and to a lesser extent in the lymph nodes (LN) and bone morrow (BM) (Xu et al. 2006). Splenectomy does not seem to affect the number of cells secreting anti-αGal antibodies in the LN and BM, except in the case of sensitized animals, when exposure to αGal considerably increases their number, but only in the BM. The cells that produce anti-αGal IgM and IgG are the surface Ig-positive B cells and mature plasma cells, respectively. Six months after exposure to porcine tissues or hematopoietic cells, cells secreting anti-αGal antibodies mainly produce IgG and are found in the BM as long-lived mature plasma cells.

The humoral response to sugars is primarily T-cell independent, although evidence from several lines of research suggests that natural and elicited anti-αGal humoral response is primarily T-cell dependent. First, a significantly reduced natural and elicited anti-αGal IgM were observed in an α1-3 GalT- and TCR-β-deficient mouse model lacking any αGal epitope expression and functional αβ T cells, by comparison with mice with normal αβ T cells (TCR-β^+) (Cretin et al. 2002). Second, blocking the CD40/CD154 pathway's interaction between B and T cells with an anti-CD154 agent inhibits the elicited anti-αGal IgM (Cretin et al. 2002) and IgG (Tanemura et al. 2000) response. Third, immunization with T-cell-independent agents expressing αGal epitopes, such as an αGal multivalent polyacrylamide polymer (Cretin et al. 2002) or glycolipids (Tanemura et al. 2000), induces a mild increase in IgM antibodies and no IgG. There is consequently evidence of a limited amount of anti-αGal IgM production occurring via T-cell-independent mechanisms (Tanemura et al. 2000; Cretin et al. 2002). The demonstration that immunizing αGalT-KO mice with pig cell membranes elicits anti-αGal IgM and IgG, induces an expansion of the anti-αGal B cell clones, and causes a strong in vitro T-cell stimulation even after αGal expression has been suppressed leads to the hypothesis that xenopeptides bearing αGal epitopes are internalized, processed by B cells, and presented to helper T cells. The resulting activation of helper T lymphocytes enables the B cells to complete their activation process, ultimately resulting in proliferation, isotype switching, and the generation of plasma cells and high-affinity anti-αGal antibodies (Tanemura et al. 2000). Finally, there are reports of follicular dendritic cells expressing complement receptors 1 and 2 being involved in the presentation of immune complexes to αGal-reactive B cells and being needed for antigen-specific anti-αGal response (Shimizu et al. 2007).

At this stage, thanks to the many approaches pursued with a view to removing preexisting anti-αGal antibodies and controlling their effector functions, HAR of a solid organ xenograft has become a rare event. Solid organ xenografts are nonetheless still rejected within days or months as a result of AHXR, even when the donor has been genetically engineered and expresses no αGal epitope (GalT-KO pigs). After the transplantation of GalT-KO kidneys, AHXR is characterized histologically by a thrombotic microangiopathic glomerulopathy with increasing IgM, IgG, C4d, and C5b-9 deposition in the glomeruli, with thrombi forming inside the injured glomeruli, loss of capillaries, and endothelial cell death (Shimizu et al. 2012). A similar picture can be seen after heart xenotransplantation (Shimizu et al. 2008), in which case, AHXR is characterized by antibody and complement deposition on the capillary walls, multiple microthrombi in the capillaries, myocardial ischemia, and necrosis. The pathogenesis of AHXR is assumed to be multifactorial, but preformed and induced antibodies directed against the endothelium are believed to be the primary factors triggering AHXR, resulting in endothelial activation and orienting the anticoagulative properties of the endothelium toward a procoagulative phenotype favoring thrombosis (Crikis et al. 2006).

The damage seen in the AHXR process, and in the thrombotic microangiopathy (TM) occurring even when porcine GalT-KO organs are used, gives the impression that αGal epitopes may still be expressed by GalT-KO organs, or else that other antibodies not directed against αGal are involved. It has now been shown, in fact, that another enzyme called iGb_3 synthase leads to αGal epitope production in GalT-KO animals (Sharma et al. 2003), although this has not been confirmed by other investigators (Diswall et al. 2010, 2011; Fang et al. 2012; Puga Yung et al. 2012; Tahiri et al. 2013). In any case, the levels of both IgM and IgG anti-αGal antibodies remain stable after GalT-KO xenografts have been transplanted into nonhuman primates (Chen et al. 2005; Kuwaki et al. 2005), suggesting that—with the coverage afforded by current immunosuppression—any remaining αGal epitopes are poorly immunogenic and may not be responsible for any graft damage. In this light, much attention has been paid to the influence of antibodies against non-αGal epitopes on humoral rejection. High levels of natural and elicited IgM and IgG against non-αGal are detected in xenografted primates and associated with the onset of AHXR, even when the potentially detrimental role of αGal antibodies is averted by absorption (Lam et al. 2004; Chen et al. 2006).

In healthy humans, albeit with some interindividual variability, it has been reported that 13% of IgM and 36% of IgG binding to pig endothelial cells are directed against non-αGal epitopes, and these antibodies can cause cell damage via complement fixation and antibody-dependent cell-mediated cytotoxicity (ADCC) (Baumann et al. 2007). Little is known about the kinetics of the anti-αGal and anti-non-αGal antibody response elicited in humans. When mouse fibroblasts were injected into humans as part of a gene therapy approach, there was a very rapid and sustained anti-αGal antibody response, with a 100-fold increase in antibody titers that dropped back within the second month, possibly as a consequence of the immunizing fibroblasts being eliminated. As for non-αGal antibodies, the immune response elicited took longer and never reached the same levels as the anti-αGal antibodies. In another study, pig ligaments lacking any αGal epitopes implanted into humans elicited an anti-non-αGal antibody response directed against both carbohydrate and protein structures (Breimer 2011; Galili 2012). Two months after the ligaments were implanted, there was a rise in non-αGal IgG that peaked at 6 mo and decreased thereafter (coinciding with the pig tissue being gradually replaced by recipient collagen), returning to pretransplant levels within 2 yr. Like the anti-αGal antibody response, the natural and elicited response against anti-non-αGal also appears to be T-cell dependent. This hypothesis is supported by the fact that there was no evidence of any elicited anti-non-αGal antibody response after blockade of the CD40–CD154 activation pathway and no subsequent T-cell help to B lymphocytes (Buhler et al. 2000). Similarly to what happens following the xenografting of wild-type organs, humoral response to GalT-KO pigs is not polyclonal—it is restricted. It is encoded, in this case, by V3-21 germline progenitors in the V_{H^3} family (Kiernan et al. 2008), an observation that may have therapeutic implications.

With regard to the specificity of anti-non-αGal antibodies, recent studies have reported that they are directed against carbohydrates (that are distinct from αGal, although they are structurally related) or proteins. When Yeh et al. (2010) analyzed naïve human or immunized baboon sera, reactivity against a panel of selected synthetic non-αGal saccharides showed that human and baboon sera contain antisaccharide antibodies directed against several specific entities including αGal-penta, α-LacNAc, Forssman antigen, P1, Pk, and Neu5Gc. There was only a minimal presence of anti-β-LacNAc antibodies, and this is a very important finding because GalT-KO animals express large amounts of this carbohydrate, appearing de novo as a consequence of αGal epitope removal. Such studies were also able to show that pigs do not express the Forssman antigen (a glycolipid against which humans have several antibodies), as shown by the presence of the high levels of anti-Forssman antibodies and the absence of Forssman antigens in pig tissues. Unexpectedly, the same study found that baboons immunized with GalT-KO pig cells had no significantly increased IgM and IgG binding to αGal-tri, or any of the saccharides in the panel considered, a finding that contrasts with the report from Diswall et al. (2010), who identified a different pattern of glycolipid expression in organs from GalT-KO pigs. The P_1 antigen (not seen in wild-type kidneys) was detected in the GalT-KO line, for instance. Similarly, the X_2 antigen was found expressed in the heart from the GalT-KO line, but not in the wild-type line analyzed. These antigens should not be seen as novel antigens appearing ex novo in the GalT-KO line, however, because the glycosyltransferases involved (namely, α1-4 galactosyltransferase and β_1, 3GalNacT) occur naturally in the wild-type line. Deletion of the α1-3 GalT enzyme may, however, have diverted the sugar metabolism toward an increased or de novo expression of these glycolipids in some tissues.

Another set of particularly important target epitopes includes the glycans carrying *N*-glycolylneuraminic acid (Neu5GC), or the so-called Hanganutziu–Deicher antigen, which is abundantly present in many mammals, including pigs and monkeys, but not in humans (Varki 2010). Healthy humans consequently have a highly variable polyclonal antibody profile against Neu5GC. Natural anti-Neu5GC antibodies are predominantly IgG (Padler-Karavani et al. 2008), but may be IgM and IgA too, albeit to a lesser extent. Anti-Neu5GC can be induced in humans after exposure to porcine tissues (Blixt et al. 2009; Scobie et al. 2013). Nonhuman primates express Neu5GC antigens, however, and are unable to elicit a humoral response against this epitope, making their use in relevant preclinical studies unfeasible; hence the development of surrogate rodent models, such as double-KO mice that lack both αGal and Neu5GC epitopes as a consequence of GalT and cytidine monophospho-*N*-acetylneuraminic acid hydrolase (CMAH) inactivation (Basnet et al. 2010). Extensive investigations have shown that anti-Neu5GC antibodies can induce complement-mediated cytotoxicity, albeit to a lesser extent than anti-αGal antibodies (Basnet et al. 2010), suggesting that anti-Neu5GC antibodies may be important in eliciting AHXR of GalT-KO organs. This hypothesis is further supported by the recent finding that GalT-KO pigs have an increased Neu5GC production (Park et al. 2011, 2012).

On the matter of xenogeneic proteins, two recent reports have identified several membrane proteins that are recognized by primates rejecting porcine cardiac xenografts. In particular, most recipients' elicited antixenograft repertoire included antibodies directed against fibronectin, several stress response and inflammation proteins, and also proteins involved in key endothelial cell functions (Byrne et al. 2008, 2011). These included proteins involved in regulating inflammation (e.g., annexin A2), hemostasis (e.g., CD9 and endothelial cell protein C receptor), or the complement cascade (e.g., CD46 and CD59), all potentially important functions in the context of AHXR. In theory at least, such elicited antibodies could block important cell functions, and the researchers speculated that it may be necessary to substitute these key porcine proteins with their human counterpart, instead of eliminating them, if porcine endothelial function needs to remain intact. A recent proteomic analysis identified several immunoreactive membrane proteins on GalT-KO pig liver endothelial cells recognized by natural human IgM and IgG (Burlak et al. 2012), and the investigators suggested that more than 800 different proteins may be recognized by preexisting IgG and/or IgM in human sera. However, they were unable to establish unequivocally whether antibodies bind to different proteins or to a limited number of carbohydrate epitopes expressed by a large number of different proteins.

Many groups have shown that the swine major histocompatibility complex (SLA) is recognized by anti-HLA antibodies in the sera of sensitized patients (Naziruddin et al. 1998; Diaz Varela et al. 2003), and that these antibodies may be cytotoxic to pig cells (Taylor et al. 1998; Mulder et al. 2010). The cross-reactivity between human and pig anti-MHC antibodies may be due to MHC epitopes conserved between the two species (Mulder et al. 2010). For a future clinical application of xenotransplantation, it may therefore be essential to select the most appropriate, matched donor pig haplotype.

Cell-Mediated Xenograft Rejection

The Role of T Lymphocytes

Although natural and elicited antibody responses have so far been considered the main barrier to successful xenotransplantation, the involvement of T cells in xenograft rejection has yet to be fully clarified. T cells contribute to the induction of antidonor antibodies, but it has not yet been unequivocally shown whether T cells alone are capable of mediating xenograft rejection.

T-cell antixenograft immune response was studied first in vitro (Yamada et al. 1995; Dorling et al. 1996), and it was found that T-cell responses against pig xenografts could be at least as strong as in the allotransplantation setting (Yamada et al. 1995; Lin et al. 2008). These studies showed that, as in the allogeneic response, human T cells are able to recognize porcine MHC molecules via direct and indirect recognition pathways (Yamada et al. 1995; Dorling et al. 1996). Direct T-cell xenoresponses accounted for >75% of the xenodirected lymphocyte response observed in mixed lymphocyte reaction (MLR) studies (Yamada et al. 1995; Tahara et al. 2010) and were mediated mainly by $CD4^+$ cells directed primarily against SLA-DR, but also against SLA-DQ molecules (Yamada et al. 1995; Dorling et al. 1996). In contrast, the same researchers reported little to no direct recognition of SLA class I antigens by $CD8^+$ human T cells. The strong direct T-cell reaction is not due to a higher frequency of human xenoreactive $CD4^+$ T cells than in the allogeneic counterpart, and there was no evidence of any differences in $CD8^+$ T-cell content between allo- and xenorelated direct reactions (Tahara et al. 2010). These studies also suggest that the $CD4^+$ helper function is needed for xenoreactive $CD8^+$ T cells to proliferate (Tahara et al. 2010). Indirect antipig T-cell responses appear to be mediated primarily by $CD4^+$ T cells (Dorling et al. 1996). The indirect antixenograft response is stronger than in the allogeneic counterpart, possibly because of the larger number of xenogeneic peptides presented by human antigen-presenting cells (Dorling et al. 1996).

As in allotransplantation, costimulatory signals are needed to fully activate the antixenograft T-cell responses via both direct and indirect pathways. In particular, CD40 and CD80/CD86 on antigen-presenting cells must interact with CD154 (CD40 L) and CD28 or cytotoxic T-lymphocyte antigen-4 (CTLA-4) on T cells. Extensive studies conducted by Rogers et al. (2003) have shown that porcine CD40, CD80, and CD86 are independently capable of costimulating human $CD4^+$ cells efficiently, albeit via different kinetics. In particular, porcine endothelial cells constitutively express CD80/CD86 (Koshika et al. 2011) and SLA class I, and possibly class II molecules as well (Choo et al. 1997). Porcine endothelial cells also trigger a direct, MHC-restricted $CD8^+$ T- and $CD4^+$ T-cell activation (Yamada et al. 1995; Dorling and Lechler 1998; Kim et al. 2010; Koshika et al. 2011; Wilhite et al. 2012). High levels of cytotoxicity have also been detected, mediated mainly by $CD4^+$ T cells, but also by $CD8^+$ T cells.

Analyzing in vitro lymphocyte proliferation induced by wild-type or GalT-KO endothelial cell shows that αGal epitope expression on the endothelium is associated with a greater proliferation of $CD4^+$ and $CD8^+$ T cells (Lin et al. 2008; Wilhite et al. 2012), suggesting an as-yet-unidentified role of αGal epitopes in sustaining T cell response. The absence of αGal is also associated with a significantly reduced secretion of INF-γ, TNF-α, IL-17A by $CD4^+$ T cells, and of INF-γ, granzyme-B, and the chemokine IP-10 by $CD8^+$ T cells (Wilhite et al. 2012), partially confirming earlier findings reported by Saethre et al. (2008), who showed that exposure to $Gal^{+/+}$ endothelial cells was associated with a significant release of human INF-γ, human and porcine proinflammatory IL-6 and IL-8, and several human β chemokines, whereas this picture was not seen after exposure to $Gal^{-/-}$ cells (Saethre et al. 2008). Unlike the report from Wilhite et al. (2012), however, the T-cell-recruiting α-chemokine IP-10 in the Saethre study was induced in cells lacking the αGal epitopes. Complement inhibition with Compstatin (a C3 inhibitor) or C5aR (a C5a antagonist) was able to abolish the production of human cytokines and chemokines, with the exception of the α-chemokine IP-10. Following exposure to human whole blood, $Gal^{+/+}$ (but not $Gal^{-/-}$) endothelial cells release porcine IL-6 and IL-8, a phenomenon that could be abolished by Compstatin. Taken together, these findings show the important role of the αGal epitope in the production of proinflammatory cytokines by human blood cells, as well as the central role of complement in mediating human immune effector functions and porcine cell activation. Recent evidence of a complement-induced T-cell and APC activation should also be borne in mind and warrants further analysis (Kwan et al. 2012).

Different cell infiltration patterns have been described after pig-to-primate solid organ xenotransplantation. In most cases, $CD8^+$ T cells with monocytes/macrophages, B cells, and some NK cells were the predominant cells detected in the graft at euthanasia (Ashton-Chess et al. 2003; Cozzi et al. 2003; Shimizu et al. 2012). $CD4^+$ T cells were only reported in a limited number of cases (Davila et al. 2006; Hisashi et al. 2008).

Taken together, in vitro and in vivo data convincingly point to the existence of a vigorous antixenograft T-cell immune response. Still, many investigators in the field share the impression that, with current immunosuppression, this cell-mediated immune response is less destructive than the response induced by activation of the humoral immune system, although the contribution of the T cells in the elicited antibody production should not be underestimated.

It is in the setting of nonvascularized xenografts such as porcine pancreatic islet xenotransplantation that T cells have been found to play the most significant part. Early studies on mice showed that fetal pig islets transplanted under the kidney capsule were rapidly rejected in normal but not in athymic (Karlsson-Parra et al. 1996) or TCR-deficient mice (Benda et al. 1998). The T cells infiltrating the graft were mainly $CD4^+$ cells, and no $CD8^+$ cells were detected. These studies also suggested that the xenogeneic islet rejection was a T-cell-dependent process, possibly via the stimulation of the effector activity of macrophages (Karlsson-Parra et al. 1996).

Returning to the clinically most relevant pig-to-primate models, porcine islets infused through the portal vein were lost *en masse* immediately after coming into contact with primate blood (Bennet et al. 2000) owing to the so-called instant blood-mediated inflammatory reaction (IBMIR), which is characterized by macroscopic coagulation, rapid platelet consumption, leukocyte infiltration, and the deposition of complement components (Goto et al. 2008; van der Windt et al. 2009). It has also been shown that islets from neonatal (Cardona et al. 2006) or adult pigs (Kirchhof et al. 2004) infused intraportally in nonimmunosuppressed primates and engrafted in the liver are destroyed within 3–5 d with a marked infiltration of $CD4^+$ and $CD8^+$ cells and macrophages (Kirchhof et al. 2004; Cardona et al. 2006).

When immunosuppressive therapy comprising basiliximab, FTY720, everolimus, and anti-CD154 was administered, adult islet survival was prolonged to up to 187 d (Hering et al. 2006). Despite the absence of any IgM or IgG and complement deposition, peri-/intragraft T-cell infiltration, both $CD4^+$ and $CD8^+$, and macrophages were apparent in the rejected grafts. The high levels of circulating indirectly activated donor-reactive T cells in rejecting recipients suggest a crucial role of such infiltrating cells in islet rejection, and their incomplete inhibition may have been the primary cause of graft rejection (Hering et al. 2006; Hering and Walawalkar 2009).

The Role of the Cells Involved in the Innate Immune Response

At least three types of cell—namely, neutrophils, natural killer (NK) cells, and macrophages—should be critically considered when analyzing the potential contribution to xenograft rejection of the cellular component of the innate immune system. For each of these, recruitment, adhesion, and *trans*-endothelial migration are mediated by different receptor–ligand interactions and finely regulated by chemotactic cytokines and chemokines released by host cells and the activated endothelium following xenotransplantation (Holgersson et al. 2002). Compatibility between human receptors and their porcine counterparts is therefore indispensable for the optimal migration and function of each of these cells.

Human naïve neutrophils have been found capable of directly recognizing and binding naïve xenogeneic endothelial cells more avidly than their allogeneic counterpart, irrespective of any presence of αGal, ICAM-1 (Sheikh et al. 2002), xenogeneic natural antibodies, or complement (Al-Mohanna et al. 1997; Cardozo et al. 2004). On the other hand, complement has been seen to increase neutrophil adhesion (Vercellotti et al. 1991). Human neutrophils activate porcine but not allogeneic endothelial cells (as shown by an increased P-selectin and VCAM-1 expression), making porcine endothelium more prone to human NK-driven cytotoxicity. In turn, porcine endothelium secretes an as-yet-unidentified soluble factor that is chemotactic for human neutrophils (Cardozo et al. 2004). The xenogeneic cell contact with porcine endothelial cell induces human neutrophil activation with an increased output of toxic reactive oxygen metabolites (ROM), irrespective of any presence of αGal epitopes, natural xenoantibodies, or complement (Al-Mohanna et al. 2005). Xenogeneic activation also prompts neutrophils to secrete proinflammatory cytokines, such as IL-1α/β, IL-6, and IL-8, involved in cellular and humoral adaptive immune responses and platelet activation, crucial players in the rejection process. As shown in some studies, polymorphonuclear neutrophils appear to be the first line of cells infiltrating and possibly damaging (van der Windt et al. 2007) cell and solid-organ xenografts soon after transplantation (Kirchhof et al. 2004; Hisashi et al. 2008; Ezzelarab et al. 2009; Shimizu et al. 2012). As shown in the majority of reports, however, this infiltration appears to be replaced by T cells and macrophages (Kirchhof et al. 2004; Hisashi et al. 2008; Ezzelarab et al. 2009; Shimizu et al. 2012). Although in vitro studies suggest that neutrophils have the potential to damage porcine cells and tissues, there is only limited evidence of their detrimental role in vivo beyond HAR.

Other important players in the antixenograft innate immune response are the NK cells, a subset of lymphocytes capable of killing cells recognized as non-self, including tumor cells and virus-infected cells. Several lines of research have suggested that, following adhesion to porcine grafts, NK cells may have a detrimental role in pig-to-primate xenotransplantation. First, it has been shown in vitro that porcine cells are susceptible to direct cell-mediated damage and to ADCC mediated by naïve or activated human NK cells (Horvath-Arcidiacono et al. 2006; Schneider and Seebach 2008; Kennett et al. 2010; Sommaggio et al. 2012). The direct cytotoxic activity of NK cells, which is inhibited primarily by human MHC class I molecules through inhibitory NK receptors, is not blocked by SLA class I products. Ligands on porcine cells, one of which has recently been identified (Lilienfeld et al. 2006), can interact efficiently with the activating receptors NKp44 or NKG2D, resulting in direct human NK cell cytotoxicity (Forte et al. 2005). Interaction between the porcine costimulatory molecule CD86 and a variant form of CD28 on hNK cells may also be involved in direct NK cytotoxicity (Costa et al. 2002), whereas the expression of αGal residues on pig endothelium does not appear to have a major role, because the use of cells lacking αGal expression did not prevent direct NK-mediated cell lysis (Baumann et al. 2004; Horvath-Arcidiacono et al. 2006). Second, it has also been shown that, on activation by porcine endothelial cells, NK cells produce INF-γ, ultimately potentiating T-cell activity (Xu et al. 2002). Third, NK cells rapidly infiltrate porcine xenografts perfused ex vivo with human blood (Kirk et al. 1993; Inverardi and Pardi 1994). In explanted xenografts, on the other hand, NK cells were found in small numbers (Davila et al. 2006; Hisashi et al. 2008; Shimizu et al. 2012) or were not reported (Ashton-Chess et al. 2003; Cozzi et al. 2003; McGregor et al. 2005; Ezzelarab et al. 2009), except for one case in which the particular features of the model may have heavily influenced the outcome (Itescu et al. 1998). It is worth noting, however, that the in vivo studies conducted to date cannot definitely rule out the possibility of NK cells being directly implicated in the untimely failure of xenografts.

Monocytes and differentiated macrophages are the third component of the innate immune system considered here. These phagocytic cells' involvement in xenograft rejection has been amply discussed, especially in islet xenotransplantation, although the reported degree and timing of infiltration have varied in the different models. After xenotransplantation, monocytes attracted by graft proinflammatory mediators migrate into the xenograft and infiltrate it by virtue of the cross-species interaction of adhesion molecules, including CD49d/CD29 and β_2-integrins (Hauzenberger et al. 2000; Schneider et al. 2009). Macrophage recruitment appears to be more intensive when xenogeneic rather than allogeneic cells are transplanted (Fox et al. 2001), and it seems to be T-cell independent. The role of αGal in increasing monocyte adhesion is still being debated because macrophage infiltration has been identified in adult porcine islet grafts known to express low levels of αGal epitopes, and in islets from GalT-KO donors (Thompson et al. 2011a). In xenotransplantation, macrophages exert their phagocytic action and modulate adaptive immunity by contributing to cell recruiting and antigen presentation. Upon contact with xenogeneic cells, macrophages have been seen to allow the recruitment of both $CD4^+$ and $CD8^+$ T cells. In turn, optimal macrophage activation requires a contribution from the $CD4^+$ T cells, possibly via the INF-γ pathway (Yi et al. 2002), ultimately resulting in graft rejection (Fox et al. 2001; Yi et al. 2003).

The macrophages' phagocytic activity is mediated by an ADCC mechanism or direct cell-to-cell contact (Jackson and Evans 2000). Like the case of NK cells, the "missing self" (Ljunggren and Karre 1990) has been suggested as the regulatory mechanism behind their phagocytic activity against xenografts. It has been shown, in fact, that the lack of any functional interaction between porcine CD47 (a membrane glycoprotein expressed ubiquitously on the cell surface of xenografts) with the human species-specific macrophage inhibitory receptor SIRPα makes porcine cells more susceptible to macrophage-mediated damage (Ide et al. 2007).

Macrophages activated by $CD4^+$ T cells have an important role in the destruction of pancreatic islet xenografts. In rodents, macrophages are reportedly the earliest infiltrating cell population after islet transplantation under the kidney capsule (Wallgren et al. 1995). Macrophages activated by $CD4^+$ T cells in islet-transplanted NOD-SCID have also been found to cause graft destruction within 8 d (Yi et al. 2003). After intraportal islet injection in nonhuman primates, T cells reportedly preceded the influx of macrophages (Kirchhof et al. 2004) into the graft. Together with T cells, macrophages might contribute to islet damage (Kirchhof et al. 2004; Cardona et al. 2006; Hering et al. 2006). They appear to have a pivotal role in cellular graft rejection, and they have also been identified soon after solid organ xenografting. They continue to be detectable up until the graft is rejected (Ashton-Chess et al. 2003; Cozzi et al. 2003; Hisashi et al. 2008; Ezzelarab et al. 2009; Shimizu et al. 2012), and they have been accused of contributing to the onset of graft failure and TM (Ezzelarab et al. 2009).

IMMUNE STRATEGIES TO EXTEND XENOGRAFT SURVIVAL

Preventing Humoral Rejection

Several strategies may be envisaged to deal with the detrimental role of the humoral antixenograft immune response, as briefly reviewed below.

Removing the Key Target Antigens

Removing the immunogenic epitopes recognized by preexisting or elicited antixenograft antibodies is an obvious approach that has been pursued successfully in the case of the αGal epitope (Dai et al. 2002; Lai et al. 2002; Phelps et al. 2003). The technology is available for preventing the expression of undesirable target antigens on pig cells (Hauschild et al. 2011).

At least two caveats should be borne in mind, however, when using such an approach. First, the number of epitopes potentially eliciting a humoral response to a xenograft is considerable, in theory at least (Burlak et al. 2012), making it unrealistic to try and delete them all. Second, deleting epitopes may sometimes result in an unpredictable de novo appearance of new target molecules that are recognized as foreign (Diswall et al. 2010). Therefore, apart from a handful of antigens (e.g., the αGal and possibly also the Neu5GC epitopes [Lutz et al. 2013], and a few others) whose deletion is considered crucial to help establish an advantageous immunological context, such as accommodation or tolerance, the application of this approach in xenotransplantation is probably of limited value.

Preventing the Activation of the Complement Cascade Triggered by Xenoreactive Antibodies

Several efforts have been made to eliminate the detrimental effect of the complement cascade's activation on the xenograft, including the use of specific soluble complement inhibitors, such as soluble complement receptor type 1 (Lam et al. 2005), or the cobra venom factor (Kobayashi et al. 1997), and the production of animals transgenic for human complement regulatory proteins (CRP), such as hCD55, hCD46, or hCD59 (Fodor et al. 1994; Cozzi and White 1995; McCurry et al. 1995; Cowan et al. 2000; Loveland et al. 2004; Menoret et al. 2004). This approach represents an elegant strategy for down-regulating local complement activation on the xenograft while preserving systemic complement activity as a first line of defense. The expression of hCRP by the xenograft cell surface may have a beneficial role on T-cell immunity too because molecules such as hCD55 can negatively modulate T-cell expansion and function (Heeger et al. 2005; Kwan et al. 2012). Judging from the data obtained so far, however, pig xenografts expressing CRPs are not protected from the overwhelming complement activation occurring in AHXR, even if CRPs are expressed on a GalT-KO background. This underlines the existing limitation of the use of CRP transgenic organs (Le Bas-Bernardet et al. 2011; Tazelaar et al. 2011). Indeed, it may be necessary to implement a multistep inhibition of the complement cascade, associated with a tight regulation of the coagulation system involved in complement activation (Huber-Lang et al. 2006).

Removing Xenoreactive Antibodies or Cells Producing Xenoreactive Antibodies

Removing or neutralizing xenoreactive antibodies with the aid of antigen-specific immunoadsorbants (Katopodis et al. 2002; Brandl et al. 2007), ex vivo organ perfusion, or nonspecific tools such as Sepharose beads conjugated with polyclonal antibodies against human Ig (Brenner et al. 2005) may help to protect grafts, at least temporarily, from offending antibodies. Strategies have also been attempted to delete antibody-producing cells using aspecific pharmacological immunosuppression (e.g., with cyclophosphamide) (Cozzi et al. 2000, 2003) or specifically targeting the B-cell lineage with anti-CD20 monoclonal antibodies (Vugmeyster et al. 2005; Mohiuddin et al. 2012), with ricin A–labeled target molecules (Tanemura et al. 2002), or even using the proteasome inhibitor bortezomib (Bauer et al. 2010). To date, however, findings indicate that none of these strategies is yet able to prolong xenograft survival indefinitely, suggesting that these approaches need to be further refined.

Inducing Accommodation and B-Cell Tolerance

Inducing accommodation—defined as long-term graft survival notwithstanding the continuing presence of xenoreactive antibodies and complement (Bach et al. 1991)—is a very appealing option for preventing antibody-mediated damage. Certain changes in the graft, such as the up-regulation of protective genes or changes in the antigens targeted by the humoral response after transplantation, may at least partially explain the mechanisms behind accommodation (Koch et al. 2004). Alternatively, changes in the recipient's antibody repertoire may underlie this fortunate situation. Be that as it may, accommodation is currently seen as a phenomenon requiring further investigation to better identify its potential in the primate xenotransplantation setting.

The induction of B-cell tolerance against xenogeneic antigens is another very attractive idea for preventing antibody-mediated damage. Inducing chimerism in the recipient has been considered indispensable to achieving tolerance. Elegant preliminary studies conducted in mice to induce a tolerance of B-cell clones producing anti-αGal antibodies have convincingly shown that inducing a mixed hematopoietic chimerism may lead to B-cell tolerance. This mixed hematopoietic chimerism was obtained by infusing allogeneic (Ohdan et al. 1999) αGal^+ T-cell-depleted bone marrow into αGal^- mice submitted to lethal whole-body irradiation (Yang et al. 1998) or to a nonmyeloablative regimen (Ohdan et al. 1999). The resulting chimerism was associated with an indefinite acceptance of αGal^+ grafts with no need for further immunosuppression (Ohdan et al. 1999). Tolerance early after transplantation was induced by anergy, requiring the persistence of αGal^+ cells, whereas clonal deletion or receptor editing was believed to be the mechanism involved in B-cell tolerization at subsequent stages (Kawahara et al. 2005). B-cell tolerance has also been achieved in relation to non-αGal epitopes (Aksentijevich et al. 1992). Using a microchimerism-based approach, Griesemer and colleagues recently succeeded in obtaining a specific humoral unresponsiveness to swine antigens in baboons. This is a promising strategy and its refinement may, at some stage, enable long-term xenograft acceptance in primates (Tseng et al. 2004; Griesemer et al. 2010).

 Cite this article as *Cold Spring Harb Perspect Med* doi: 10.1101/cshperspect.a015578

Preventing Cell-Mediated Rejection

Several strategies have also been developed to block innate and adaptive cell-mediated xenograft rejection.

Systemic Immunosuppression

Many immunosuppressive approaches have been attempted to improve xenograft survival, including conventional and also more recently developed molecules or biologics. The introduction of anti-CD154 monoclonal antibodies has contributed to a significantly longer solid-organ and islet xenograft survival (Kuwaki et al. 2005; Yamada et al. 2005; Cardona et al. 2006; Hering et al. 2006), while preventing elicited antibody responses (Ezzelarab et al. 2012). These results have been severely penalized by thromboembolic events that have precluded further use of anti-CD154 antibodies and prompted researchers to explore other therapeutic options, such as anti-CD40 agents or CTLA-4Ig-based strategies in combination with immunosuppressive drugs (Ezzelarab et al. 2009; Thompson et al. 2011b). Using an anti-CD40 antibody associated with belatacept extended neonatal porcine islet survival (Thompson et al. 2011b), although further studies are needed to confirm this finding. On the other hand, Ezzelarab et al. (2012) found in an artery-patch xenotransplantation model that CTLA-4Ig therapy was probably not as efficacious as CD154/CD40 inhibitors in preventing cellular and humoral response (Dons et al. 2012).

Local Immunosuppression Using Specifically Engineered Source Animals

Genetic engineering of the xenograft has also been proposed as a means to inhibit rejection locally, while containing systemic immunosuppression and its side effects. Several approaches have been considered, and novel pig lines expressing molecules that may block the recipient's immune cells have been produced.

When a mutant form of the CIITA gene (an essential coactivator for the transcription of MHC class II genes) was introduced on a GalT-KO/CD46/CD55 background, SLA II molecules were down-regulated and $CD4^+$ T-cell proliferation was inhibited in MLR studies (Ayares et al. 2011). Several porcine and human CTLA-4Ig transgenic animals were obtained (Martin et al. 2005; Phelps et al. 2009; Koshika et al. 2011), and hCTLA-4Ig transgenic pigs expressing the transgene selectively in neurons enabled a long-term survival of neural precursors in mice (Martin et al. 2005) and primates (Aron Badin et al. 2009). Because the constitutive expression of pCTLA-4Ig resulted in severely immunosuppressed animals, pigs with an inducible pCTLA-4Ig expression have recently been obtained (Klymiuk et al. 2012). As for the NK cells, transgenic pigs expressing HLA-E (a molecule inhibiting NK cell adhesion and cytotoxicity) have also been produced (Weiss et al. 2009). Transgenic pigs expressing human CD47 have likewise recently been generated (Tena et al. 2011). Although the potential of these genetic modifications still remains to be tested in vivo, their association with other strategies may lead to a better graft survival.

Using Mechanical Barriers

Encapsulating xenogeneic islets has been suggested as a strategy to prevent cell-mediated rejection and enable graft survival in the absence of immunosuppression. The crucial issues to tackle in this area include the biocompatibility of the encapsulating material, the mechanical stability of the capsules, and the selection of the implantation site (Dufrane et al. 2006b). Microencapsulation of pig islets in water-soluble polymers, such as alginate (Dufrane et al. 2006a), or macroencapsulation in subcutaneous devices permeable to glucose, insulin, and nutrients (but not to antibodies or the immune system's cellular components) have enabled islets to survive in primates for up to 6 mo without any immunosuppression (Dufrane et al. 2010). These findings suggest that, notwithstanding the currently limited life span of such devices, refining this approach may lead in the future to the successful clinical application of islet xenotransplantation.

T-Cell Tolerance and Tregs

The feasibility of inducing T-cell tolerance has also been explored. Early studies on T-cell tolerance in mice showed that transplanting fetal or neonatal pig thymic tissue into thymectomized mice could induce a central tolerance of pig skin grafts (Zhao et al. 1996). This strategy was successfully applied to the transplantation of GalT-KO pig kidneys into baboons, resulting in a normal renal function for up to 83 d (Yamada et al. 2005; Griesemer et al. 2009). This result was achieved by transplanting thymic tissue under the donor renal capsule (thymokidney) in recipients exposed to a protocol that included thymectomy, splenectomy, T-cell depletion, and whole-body irradiation. Doing without whole-body irradiation enabled a mean survival of more than 50 d. Thymic grafts supported thymopoiesis, and there was evidence of $CD4^+CD3^-$ baboon cells adjacent to porcine thymic epithelial cells (Griesemer et al. 2009). Donor-specific unresponsiveness with respect to the normal responses to allogeneic third parties was also detected in CTL assays. As an alternative approach to enabling T-cell tolerance in xenotransplantation, thymic lobe (vascularized thymic lobe, VTL) transplantation from either αGal^+ (Yamamoto et al. 2005) or GalT-KO donors (Yamada et al. 2005) has also been proposed. This procedure enabled early thymopoiesis, reconstitution of the recipient's naïve T-cell population, and donor-specific unresponsiveness, albeit for a limited time (Yamamoto et al. 2005). These regimens to induce T-cell tolerance clearly require further fine-tuning, but the results achieved to date suggest that this approach may soon become a viable option for primates.

Inducing tolerance through regulatory T-cell-based therapies also has its appeal for both allo- and xenotransplantation. Regulatory T cells have been found to suppress T-cell-direct (Wu et al. 2008) and -indirect xenogeneic responses to T cells (Nishimura et al. 2010), B cells (Singh et al. 2012), macrophage activation (Fu et al. 2008), and-antigen-presenting cell functions (Cederbom et al. 2000; Fu et al. 2008). Several types of regulatory T cells (Tregs) have been described, and those naturally occurring in the immune system (nTregs) account for nearly 3%–10% of the peripheral $CD4^+$ T cells (Dons et al. 2010; Muller et al. 2011). Phenotypically, nTregs are characterized by the constitutive presence of CD4, CD25, the IL-2 receptor α-chain, and the intracellular expression of transcription factor Forkhead box P3 (Foxp3), and by little or no expression of CD127, the IL-2 receptor α-chain. The suppressor activity of nTregs depends on cell-to-cell contact via important costimulatory molecules, which include CTLA-4, membrane-bound TGF-β latency-associated peptide (LAP), soluble inducible costimulatory molecule (ICOS), galectin-1, CD39, CD73, and PD-1, or it is mediated by regulatory cytokines such as IL-10, TGF-β, or IL-35 (Muller et al. 2011, 2012). Notably, the expression of hPD-L1 (the ligand of activated T cells PD-1) by porcine endothelial cells stimulates the proliferation of human $Foxp3^+CD4^+$ T cells both in vitro and in vivo, inducing the expression of CD73 by Tregs and promoting the production of IL-10, and reducing the proinflammatory Th1 and Th17 cytokines. The in vivo infusion of $hPD\text{-}L1^+$ endothelial cells prolongs the survival of porcine skin grafts, favoring the expansion of $Foxp3^+CD4^+$ T cells (Ding et al. 2011). Much the same results have been obtained in mice by grafting porcine endothelial cells transfected with ICOS-Ig, a soluble form of ICOS (Hodgson et al. 2011), suggesting that Treg-based strategies may represent a novel path toward achieving tolerance in xenotransplantation.

CONCLUSIONS

To sum up, despite the challenges that remain, the potential health benefits afforded by xenotransplantation make it a fascinating area of research that deserves to be pursued. Important preclinical data have been generated in recent years, making the initiation of clinical trials a realistic option for the not-too-distant future. At this point, many researchers in this field would agree that genetic engineering of the donor species may be needed to improve current results. Such genetic engineering should selec-

 Cite this article as *Cold Spring Harb Perspect Med* doi: 10.1101/cshperspect.a015578

tively address only major interspecies molecular discrepancies, like those involved in regulating the complement or coagulation cascades, eliminate highly immunogenic epitopes, or block major immune cell activation pathways. This should ultimately contribute to the development of the ideal milieu in which novel immunosuppressive drugs or innovative strategies designed to enable accommodation or tolerance will have the greatest chances of success in terms of achieving long-term xenograft survival. It has also become clear that, before xenotransplantation can enter the clinical arena, a multidisciplinary approach will be needed to comprehensively tackle the various issues relating to the use of xenografts to cure human diseases. The safety-related, ethical, and regulatory issues of xenotransplantation (which were not the object of this review) are already being addressed by highly specialized and dedicated investigators, and the multidisciplinary effort currently under way is expected to enable xenotransplantation to happen with a favorable risk/benefit ratio in the foreseeable future.

REFERENCES

Aksentijevich I, Sachs DH, Sykes M. 1992. Humoral tolerance in xenogeneic BMT recipients conditioned by a nonmyeloablative regimen. *Transplantation* **53:** 1108–1114.

Al-Mohanna F, Collison K, Parhar R, Kwaasi A, Meyer B, Saleh S, Allen S, al-Sedairy S, Stern D, Yacoub M. 1997. Activation of naive xenogeneic but not allogeneic endothelial cells by human naive neutrophils: A potential occult barrier to xenotransplantation. *Am J Pathol* **151:** 111–120.

Al-Mohanna F, Saleh S, Parhar RS, Khabar K, Collison K. 2005. Human neutrophil gene expression profiling following xenogeneic encounter with porcine aortic endothelial cells: The occult role of neutrophils in xenograft rejection revealed. *J Leukoc Biol* **78:** 51–61.

Aron Badin R, Padoan A, Vadori M, Boldrin M, Cavicchioli L, De Benedictis G, Fante F, Seveso M, Sgarabotto D, Jan C, et al. 2009. Transgenic porcine embryonic xenografts as a treatment for Parkinson's disease in non-human primates. *Xenotrasplantation* **16:** 543.

Ashton-Chess J, Roussel JC, Manez R, Ruiz C, Moreau A, Cozzi E, Minault D, Soulillou JP, Blancho G. 2003. Cellular participation in delayed xenograft rejection of hCD55 transgenic pig hearts by baboons. *Xenotransplantation* **10:** 446–453.

Ayares D, Phelps C, Vaught T, Ball S, Mendicino M, Ramsoondar J, Monahan J, Delong A, Walters A, Starzl T, et al. 2011. Multitrasgenic pigs for vascularized organ xenografts. *Xenotransplantation* **18:** 269.

Bach FH, Turman MA, Vercellotti GM, Platt JL, Dalmasso AP. 1991. Accommodation: A working paradigm for progressing toward clinical discordant xenografting. *Transplant Proc* **23:** 205–207.

Basnet NB, Ide K, Tahara H, Tanaka Y, Ohdan H. 2010. Deficiency of *N*-glycolylneuraminic acid and Galα1-3Galβ1-4GlcNAc epitopes in xenogeneic cells attenuates cytotoxicity of human natural antibodies. *Xenotransplantation* **17:** 440–448.

Bauer A, Postrach J, Thormann M, Blanck S, Faber C, Wintersperger B, Michel S, Abicht JM, Christ F, Schmitz C, et al. 2010. First experience with heterotopic thoracic pig-to-baboon cardiac xenotransplantation. *Xenotransplantation* **17:** 243–249.

Baumann BC, Forte P, Hawley RJ, Rieben R, Schneider MK, Seebach JD. 2004. Lack of galactose-α-1,3-galactose expression on porcine endothelial cells prevents complement-induced lysis but not direct xenogeneic NK cytotoxicity. *J Immunol* **172:** 6460–6467.

Baumann BC, Stussi G, Huggel K, Rieben R, Seebach JD. 2007. Reactivity of human natural antibodies to endothelial cells from Galα(1,3)Gal-deficient pigs. *Transplantation* **83:** 193–201.

Benda B, Sandberg JO, Holstad M, Korsgren O. 1998. T cells in islet-like cell cluster xenograft rejection: A study in the pig-to-mouse model. *Transplantation* **66:** 435–440.

Bennet W, Groth CG, Larsson R, Nilsson B, Korsgren O. 2000. Isolated human islets trigger an instant blood mediated inflammatory reaction: Implications for intraportal islet transplantation as a treatment for patients with type 1 diabetes. *Ups J Med Sci* **105:** 125–133.

Blixt O, Kumagai-Braesch M, Tibellb A, Groth C, Holgerssonc J. 2009. Anticarbohydrate antibody repertoires in patients transplanted with fetal pig islets revealed by glycan arrays. *Am J Transplant* **9:** 83–90.

Brandl U, Erhardt M, Michel S, Jockle H, Burdorf L, Bittmann I, Rossle M, Mordstein V, Brenner P, Hammer C, et al. 2007. Soluble Galα(1,3)Gal conjugate combined with hDAF preserves morphology and improves function of cardiac xenografts. *Xenotransplantation* **14:** 323–332.

Breimer ME. 2011. Gal/non Gal antigens in pig tissues and human non-Gal antibodies in the GalT-KO era. *Xenotransplantation* **18:** 215–228.

Brenner P, Schmoeckel M, Wimmer C, Rucker A, Eder V, Uchita S, Brandl U, Hinz M, Felbinger T, Meiser B, et al. 2005. Combination of hDAF-transgenic pig hearts and immunoadsorption in heterotopic xenotransplantation of immunosuppressed baboons. *Transplant Proc* **37:** 483–486.

Buhler L, Awwad M, Basker M, Gojo S, Watts A, Treter S, Nash K, Oravec G, Chang Q, Thall A, et al. 2000. High-dose porcine hematopoietic cell transplantation combined with CD40 ligand blockade in baboons prevents an induced anti-pig humoral response. *Transplantation* **69:** 2296–2304.

Bulato C, Radu C, Simioni P. 2012. Studies on coagulation incompatibilities for xenotransplantation. *Methods Mol Biol* **885:** 71–89.

Burlak C, Wang ZY, Chihara RK, Lutz AJ, Wang Y, Estrada JL, Tector AJ. 2012. Identification of human preformed antibody targets in GTKO pigs. *Xenotransplantation* **19:** 92–101.

Byrne GW, Stalboerger PG, Davila E, Heppelmann CJ, Gazi MH, McGregor HC, LaBreche PT, Davies WR, Rao VP, Oi K, et al. 2008. Proteomic identification of non-Gal antibody targets after pig-to-primate cardiac xenotransplantation. *Xenotransplantation* **15:** 268–276.

Byrne GW, Stalboerger PG, Du Z, Davis TR, McGregor CG. 2011. Identification of new carbohydrate and membrane protein antigens in cardiac xenotransplantation. *Transplantation* **91:** 287–292.

Cardona K, Korbutt GS, Milas Z, Lyon J, Cano J, Jiang W, Bello-Laborn H, Hacquoil B, Strobert E, Gangappa S, et al. 2006. Long-term survival of neonatal porcine islets in nonhuman primates by targeting costimulation pathways. *Nat Med* **12:** 304–306.

Cardozo LA, Rouw DB, Ambrose LR, Midulla M, Florey O, Haskard DO, Warrens AN. 2004. The neutrophil: The unnoticed threat in xenotransplantation? *Transplantation* **78:** 1721–1728.

Cederbom L, Hall H, Ivars F. 2000. $CD4^+CD25^+$ regulatory T cells down-regulate co-stimulatory molecules on antigen-presenting cells. *Eur J Immunol* **30:** 1538–1543.

Chen G, Qian H, Starzl T, Sun H, Garcia B, Wang X, Wise Y, Liu Y, Xiang Y, Copeman L, et al. 2005. Acute rejection is associated with antibodies to non-Gal antigens in baboons using Gal-knockout pig kidneys. *Nat Med* **11:** 1295–1298.

Chen G, Sun H, Yang H, Kubelik D, Garcia B, Luo Y, Xiang Y, Qian A, Copeman L, Liu W, et al. 2006. The role of anti-non-Gal antibodies in the development of acute humoral xenograft rejection of hDAF transgenic porcine kidneys in baboons receiving anti-Gal antibody neutralization therapy. *Transplantation* **81:** 273–283.

Choo JK, Seebach JD, Nickeleit V, Shimizu A, Lei H, Sachs DH, Madsen JC. 1997. Species differences in the expression of major histocompatibility complex class II antigens on coronary artery endothelium: Implications for cell-mediated xenoreactivity. *Transplantation* **64:** 1315–1322.

Costa C, Barber DF, Fodor WL. 2002. Human NK cell-mediated cytotoxicity triggered by CD86 and Galα 1,3-Gal is inhibited in genetically modified porcine cells. *J Immunol* **168:** 3808–3816.

Cowan PJ, Aminian A, Barlow H, Brown AA, Chen CG, Fisicaro N, Francis DM, Goodman DJ, Han W, Kurek M, et al. 2000. Renal xenografts from triple-transgenic pigs are not hyperacutely rejected but cause coagulopathy in non-immunosuppressed baboons. *Transplantation* **69:** 2504–2515.

Cowan PJ, Robson SC, d'Apice AJ. 2011. Controlling coagulation dysregulation in xenotransplantation. *Curr Opin Organ Transplant* **16:** 214–221.

Cozzi E, White DJ. 1995. The generation of transgenic pigs as potential organ donors for humans. *Nat Med* **1:** 964–966.

Cozzi E, Bhatti F, Schmoeckel M, Chavez G, Smith KG, Zaidi A, Bradley JR, Thiru S, Goddard M, Vial C, et al. 2000. Long-term survival of nonhuman primates receiving life-supporting transgenic porcine kidney xenografts. *Transplantation* **70:** 15–21.

Cozzi E, Vial C, Ostlie D, Farah B, Chavez G, Smith KG, Bradley JR, Thiru S, Davies HF, Wallwork J, et al. 2003. Maintenance triple immunosuppression with cyclosporin A, mycophenolate sodium and steroids allows prolonged survival of primate recipients of hDAF porcine renal xenografts. *Xenotransplantation* **10:** 300–310.

Cretin N, Bracy J, Hanson K, Iacomini J. 2002. The role of T cell help in the production of antibodies specific for Galα1–3Gal. *J Immunol* **168:** 1479–1483.

Crikis S, Cowan PJ, d'Apice AJ. 2006. Intravascular thrombosis in discordant xenotransplantation. *Transplantation* **82:** 1119–1123.

Dai Y, Vaught TD, Boone J, Chen SH, Phelps CJ, Ball S, Monahan JA, Jobst PM, McCreath KJ, Lamborn AE, et al. 2002. Targeted disruption of the α1,3-galactosyltransferase gene in cloned pigs. *Nat Biotechnol* **20:** 251–255.

Davila E, Byrne GW, LaBreche PT, McGregor HC, Schwab AK, Davies WR, Rao VP, Oi K, Tazelaar HD, Logan JS, et al. 2006. T-cell responses during pig-to-primate xenotransplantation. *Xenotransplantation* **13:** 31–40.

Diaz Varela I, Sanchez Mozo P, Centeno Cortes A, Alonso Blanco C, Valdes Canedo F. 2003. Cross-reactivity between swine leukocyte antigen and human anti-HLA-specific antibodies in sensitized patients awaiting renal transplantation. *J Am Soc Nephrol* **14:** 2677–2683.

Ding Q, Lu L, Zhou X, Zhou Y, Chou KY. 2011. Human PD-L1-overexpressing porcine vascular endothelial cells induce functionally suppressive human $CD4^+$ $CD25hiFoxp3^+$ Treg cells. *J Leukoc Biol* **90:** 77–86.

Diswall M, Angstrom J, Karlsson H, Phelps CJ, Ayares D, Teneberg S, Breimer ME. 2010. Structural characterization of α1,3-galactosyltransferase knockout pig heart and kidney glycolipids and their reactivity with human and baboon antibodies. *Xenotransplantation* **17:** 48–60.

Diswall M, Gustafsson A, Holgersson J, Sandrin MA, Breimer ME. 2011. Antigen-binding specificity of anti-αGal reagents determined by solid-phase glycolipid-binding assays. A complete lack of αGal glycolipid reactivity in α1,3GalT-KO pig small intestine. *Xenotransplantation* **18:** 28–39.

Dons EM, Raimondi G, Cooper DK, Thomson AW. 2010. Non-human primate regulatory T cells: Current biology and implications for transplantation. *Transplantation* **90:** 811–816.

Dons EM, Montoya C, Long CE, Hara H, Echeverri GJ, Ekser B, Ezzelarab C, Medellin DR, van der Windt DJ, Murase N, et al. 2012. T-cell-based immunosuppressive therapy inhibits the development of natural antibodies in infant baboons. *Transplantation* **93:** 769–776.

Dorling A, Lechler RI. 1998. T cell-mediated xenograft rejection: Specific tolerance is probably required for long term xenograft survival. *Xenotransplantation* **5:** 234–245.

Dorling A, Lombardi G, Binns R, Lechler RI. 1996. Detection of primary direct and indirect human anti-porcine T cell responses using a porcine dendritic cell population. *Eur J Immunol* **26:** 1378–1387.

Dufrane D, Goebbels RM, Saliez A, Guiot Y, Gianello P. 2006a. Six-month survival of microencapsulated pig islets and alginate biocompatibility in primates: Proof of concept. *Transplantation* **81:** 1345–1353.

Dufrane D, Steenberghe M, Goebbels RM, Saliez A, Guiot Y, Gianello P. 2006b. The influence of implantation site on the biocompatibility and survival of alginate encapsulated pig islets in rats. *Biomaterials* **27:** 3201–3208.

Dufrane D, Goebbels RM, Gianello P. 2010. Alginate macroencapsulation of pig islets allows correction of streptozotocin-induced diabetes in primates up to 6 months without immunosuppression. *Transplantation* **90:** 1054–1062.

Ezzelarab M, Garcia B, Azimzadeh A, Sun H, Lin CC, Hara H, Kelishadi S, Zhang T, Lin YJ, Tai HC, et al. 2009. The innate immune response and activation of coagulation in α1,3-galactosyltransferase gene-knockout xenograft recipients. *Transplantation* **87:** 805–812.

Ezzelarab MB, Ekser B, Echeverri G, Hara H, Ezzelarab C, Long C, Bajona P, Garcia B, Murase N, Ayares D, et al. 2012. Costimulation blockade in pig artery patch xenotransplantation—A simple model to monitor the adaptive immune response in nonhuman primates. *Xenotransplantation* **19:** 221–232.

Fang J, Walters A, Hara H, Long C, Yeh P, Ayares D, Cooper DK, Bianchi J. 2012. Anti-gal antibodies in α1,3-galactosyltransferase gene-knockout pigs. *Xenotransplantation* **19:** 305–310.

Fodor WL, Williams BL, Matis LA, Madri JA, Rollins SA, Knight JW, Velander W, Squinto SP. 1994. Expression of a functional human complement inhibitor in a transgenic pig as a model for the prevention of xenogeneic hyperacute organ rejection. *Proc Natl Acad Sci* **91:** 11153–11157.

Forte P, Lilienfeld BG, Baumann BC, Seebach JD. 2005. Human NK cytotoxicity against porcine cells is triggered by NKp44 and NKG2D. *J Immunol* **175:** 5463–5470.

Fox A, Mountford J, Braakhuis A, Harrison LC. 2001. Innate and adaptive immune responses to nonvascular xenografts: Evidence that macrophages are direct effectors of xenograft rejection. *J Immunol* **166:** 2133–2140.

Fu Y, Yi S, Wu J, Jimenez E, Simond D, Hawthorne WJ, O'Connell PJ. 2008. In vitro suppression of xenoimmune-mediated macrophage activation by human $CD4^+CD25^+$ regulatory T cells. *Transplantation* **86:** 865–874.

Galili U. 2012. Induced anti-non-gal antibodies in human xenograft recipients. *Transplantation* **93:** 11–16.

Galili U, Shohet SB, Kobrin E, Stults CL, Macher BA. 1988. Man, apes, and Old World monkeys differ from other mammals in the expression of α-galactosyl epitopes on nucleated cells. *J Biol Chem* **263:** 17755–17762.

Goto M, Tjernberg J, Dufrane D, Elgue G, Brandhorst D, Ekdahl KN, Brandhorst H, Wennberg L, Kurokawa Y, Satomi S, et al. 2008. Dissecting the instant blood-mediated inflammatory reaction in islet xenotransplantation. *Xenotransplantation* **15:** 225–234.

Griesemer AD, Hirakata A, Shimizu A, Moran S, Tena A, Iwaki H, Ishikawa Y, Schule P, Arn JS, Robson SC, et al. 2009. Results of Gal-knockout porcine thymokidney xenografts. *Am J Transplant* **9:** 2669–2678.

Griesemer A, Liang F, Hirakata A, Hirsh E, Lo D, Okumi M, Sykes M, Yamada K, Huang CA, Sachs DH. 2010. Occurrence of specific humoral non-responsiveness to swine antigens following administration of GalT-KO bone marrow to baboons. *Xenotransplantation* **17:** 300–312.

Hauschild J, Petersen B, Santiago Y, Queisser AL, Carnwath JW, Lucas-Hahn A, Zhang L, Meng X, Gregory PD, Schwinzer R, et al. 2011. Efficient generation of a biallelic knockout in pigs using zinc-finger nucleases. *Proc Natl Acad Sci* **108:** 12013–12017.

Hauzenberger E, Hauzenberger D, Hultenby K, Holgersson J. 2000. Porcine endothelium supports transendothelial migration of human leukocyte subpopulations: Anti-porcine vascular cell adhesion molecule antibodies as species-specific blockers of transendothelial monocyte and natural killer cell migration. *Transplantation* **69:** 1837–1849.

Heeger PS, Lalli PN, Lin F, Valujskikh A, Liu J, Muqim N, Xu Y, Medof ME. 2005. Decay-accelerating factor modulates induction of T cell immunity. *J Exp Med* **201:** 1523–1530.

Hering BJ, Walawalkar N. 2009. Pig-to-nonhuman primate islet xenotransplantation. *Transpl Immunol* **21:** 81–86.

Hering BJ, Wijkstrom M, Graham ML, Hardstedt M, Aasheim TC, Jie T, Ansite JD, Nakano M, Cheng J, Li W, et al. 2006. Prolonged diabetes reversal after intraportal xenotransplantation of wild-type porcine islets in immunosuppressed nonhuman primates. *Nat Med* **12:** 301–303.

Hisashi Y, Yamada K, Kuwaki K, Tseng YL, Dor FJ, Houser SL, Robson SC, Schuurman HJ, Cooper DK, Sachs DH, et al. 2008. Rejection of cardiac xenografts transplanted from α1,3-galactosyltransferase gene-knockout (GalT-KO) pigs to baboons. *Am J Transplant* **8:** 2516–2526.

Hodgson R, Christiansen D, Ziolkowski A, Mouhtouris E, Simeonovic CJ, Ierino FL, Sandrin MS. 2011. Prolonged xenograft survival induced by inducible costimulator-Ig is associated with increased Forkhead box $P3^+$ cells. *Transplantation* **91:** 1090–1097.

Holgersson J, Ehrnfelt C, Hauzenberger E, Serrander L. 2002. Leukocyte endothelial cell interactions in pig to human organ xenograft rejection. *Vet Immunol Immunopathol* **87:** 407–415.

Horvath-Arcidiacono JA, Porter CM, Bloom ET. 2006. Human NK cells can lyse porcine endothelial cells independent of their expression of Galα(1,3)-Gal and killing is enhanced by activation of either effector or target cells. *Xenotransplantation* **13:** 318–327.

Huber-Lang M, Sarma JV, Zetoune FS, Rittirsch D, Neff TA, McGuire SR, Lambris JD, Warner RL, Flierl MA, Hoesel LM, et al. 2006. Generation of C5a in the absence of C3: A new complement activation pathway. *Nat Med* **12:** 682–687.

Ide K, Wang H, Tahara H, Liu J, Wang X, Asahara T, Sykes M, Yang YG, Ohdan H. 2007. Role for CD47-SIRPα signaling in xenograft rejection by macrophages. *Proc Natl Acad Sci* **104:** 5062–5066.

Inverardi L, Pardi R. 1994. Early events in cell-mediated recognition of vascularized xenografts: Cooperative interactions between selected lymphocyte subsets and natural antibodies. *Immunol Rev* **141:** 71–93.

Itescu S, Kwiatkowski P, Artrip JH, Wang SF, Ankersmit J, Minanov OP, Michler RE. 1998. Role of natural killer cells, macrophages, and accessory molecule interactions in the rejection of pig-to-primate xenografts beyond the hyperacute period. *Hum Immunol* **59:** 275–286.

Jackson PG, Evans SR. 2000. Intraperitoneal macrophages and tumor immunity: A review. *J Surg Oncol* **75:** 146–154.

Karlsson-Parra A, Ridderstad A, Wallgren AC, Moller E, Ljunggren HG, Korsgren O. 1996. Xenograft rejection of porcine islet-like cell clusters in normal and natural killer cell-depleted mice. *Transplantation* **61:** 1313–1320.

Katopodis AG, Warner RG, Duthaler RO, Streiff MB, Bruelisauer A, Kretz O, Dorobek B, Persohn E, Andres H, Schweitzer A, et al. 2002. Removal of anti-Galα1,3Gal xenoantibodies with an injectable polymer. *J Clin Invest* **110:** 1869–1877.

Kawahara T, Shimizu I, Ohdan H, Zhao G, Sykes M. 2005. Differing mechanisms of early and late B cell hyporesponsiveness induced by mixed chimerism. *Am J Transplant* **5:** 2821–2829.

Kennett SB, Porter CM, Horvath-Arcidiacono JA, Bloom ET. 2010. Characterization of baboon NK cells and their xenogeneic activity. *Xenotransplantation* **17:** 288–299.

Kiernan K, Harnden I, Gunthart M, Gregory C, Meisner J, Kearns-Jonker M. 2008. The anti-non-gal xenoantibody response to xenoantigens on Gal knockout pig cells is encoded by a restricted number of germline progenitors. *Am J Transplant* **8:** 1829–1839.

Kim CH, Oh K, Kim DE, Lee SB, Yang JH, Lee G, Cho J, Lee DS. 2010. Both $CD45RA^+$ and $CD45RO^+$ human $CD4^+$ T cells drive direct xenogeneic T-cell responses against porcine aortic endothelial cells. *Xenotransplantation* **17:** 224–232.

Kirchhof N, Shibata S, Wijkstrom M, Kulick DM, Salerno CT, Clemmings SM, Heremans Y, Galili U, Sutherland DE, Dalmasso AP, et al. 2004. Reversal of diabetes in non-immunosuppressed rhesus macaques by intraportal porcine islet xenografts precedes acute cellular rejection. *Xenotransplantation* **11:** 396–407.

Kirk AD, Heinle JS, Mault JR, Sanfilippo F. 1993. Ex vivo characterization of human anti-porcine hyperacute cardiac rejection. *Transplantation* **56:** 785–793.

Klymiuk N, Bocker W, Schonitzer V, Bahr A, Radic T, Frohlich T, Wunsch A, Kessler B, Kurome M, Schilling E, et al. 2012. First inducible transgene expression in porcine large animal models. *FASEB J* **26:** 1086–1099.

Kobayashi T, Taniguchi S, Neethling FA, Rose AG, Hancock WW, Ye Y, Niekrasz M, Kosanke S, Wright LJ, White DJ, et al. 1997. Delayed xenograft rejection of pig-to-baboon cardiac transplants after cobra venom factor therapy. *Transplantation* **64:** 1255–1261.

Koch CA, Khalpey ZI, Platt JL. 2004. Accommodation: Preventing injury in transplantation and disease. *J Immunol* **172:** 5143–5148.

Konakci KZ, Bohle B, Blumer R, Hoetzenecker W, Roth G, Moser B, Boltz-Nitulescu G, Gorlitzer M, Klepetko W, Wolner E, et al. 2005. α-Gal on bioprostheses: Xenograft immune response in cardiac surgery. *Eur J Clin Invest* **35:** 17–23.

Koshika T, Phelps C, Fang J, Lee SE, Fujita M, Ayares D, Cooper DK, Hara H. 2011. Relative efficiency of porcine and human cytotoxic T-lymphocyte antigen 4 immunoglobulin in inhibiting human $CD4^+$ T-cell responses co-stimulated by porcine and human B7 molecules. *Immunology* **134:** 386–397.

Kuwaki K, Tseng YL, Dor FJ, Shimizu A, Houser SL, Sanderson TM, Lancos CJ, Prabharasuth DD, Cheng J, Moran K, et al. 2005. Heart transplantation in baboons using α1,3-galactosyltransferase gene-knockout pigs as donors: Initial experience. *Nat Med* **11:** 29–31.

Kwan WH, van der Touw W, Heeger PS. 2012. Complement regulation of T cell immunity. *Immunol Res* **54:** 247–253.

Lai L, Kolber-Simonds D, Park KW, Cheong HT, Greenstein JL, Im GS, Samuel M, Bonk A, Rieke A, Day BN, et al. 2002. Production of α-1,3-galactosyltransferase knockout pigs by nuclear transfer cloning. *Science* **295:** 1089–1092.

Lam TT, Paniagua R, Shivaram G, Schuurman HJ, Borie DC, Morris RE. 2004. Anti-non-Gal porcine endothelial cell antibodies in acute humoral xenograft rejection of hDAF-transgenic porcine hearts in cynomolgus monkeys. *Xenotransplantation* **11:** 531–535.

Lam TT, Hausen B, Hook L, Lau M, Higgins J, Christians U, Jacobsen W, Baluom M, Duthaler R, Katopodis A, et al. 2005. The effect of soluble complement receptor type 1 on acute humoral xenograft rejection in hDAF-transgenic pig-to-primate life-supporting kidney xenografts. *Xenotransplantation* **12:** 20–29.

Le Bas-Bernardet S, Tillou X, Poirier N, Dilek N, Chatelais M, Devalliere J, Charreau B, Minault D, Hervouet J, Renaudin K, et al. 2011. Xenotransplantation of galactosyltransferase knockout, CD55, CD59, CD39, and fucosyltransferase transgenic pig kidneys into baboons. *Transplant Proc* **43:** 3426–3430.

Lilienfeld BG, Garcia-Borges C, Crew MD, Seebach JD. 2006. Porcine UL16-binding protein 1 expressed on the surface of endothelial cells triggers human NK cytotoxicity through NKG2D. *J Immunol* **177:** 2146–2152.

Lin YJ, Hara H, Tai HC, Long C, Tokita D, Yeh P, Ayares D, Morelli AE, Cooper DK. 2008. Suppressive efficacy and proliferative capacity of human regulatory T cells in allogeneic and xenogeneic responses. *Transplantation* **86:** 1452–1462.

Lin CC, Cooper DK, Dorling A. 2009. Coagulation dysregulation as a barrier to xenotransplantation in the primate. *Transpl Immunol* **21:** 75–80.

Ljunggren HG, Karre K. 1990. In search of the "missing self": MHC molecules and NK cell recognition. *Immunol Today* **11:** 237–244.

Loveland BE, Milland J, Kyriakou P, Thorley BR, Christiansen D, Lanteri MB, Regensburg M, Duffield M, French AJ, Williams L, et al. 2004. Characterization of a CD46 transgenic pig and protection of transgenic kidneys against hyperacute rejection in non-immunosuppressed baboons. *Xenotransplantation* **11:** 171–183.

Lutz AJ, Li P, Estrada JL, Sidner RA, Chihara RK, Downey SM, Burlak C, Wang ZY, Reyes LM, Ivary B, et al. 2013. Double knockout pigs deficient in *N*-glycolylneuraminic acid and galactose α-1,3-galactose reduce the humoral barrier to xenotransplantation. *Xenotransplantation* **20:** 27–35.

Macher BA, Galili U. 2008. The Galα1,3Galβ1,4GlcNAc-R (α-Gal) epitope: A carbohydrate of unique evolution and clinical relevance. *Biochim Biophys Acta* **1780:** 75–88.

Martin C, Plat M, Nerriere-Daguin V, Coulon F, Uzbekova S, Venturi E, Conde F, Hermel JM, Hantraye P, Tesson L, et al. 2005. Transgenic expression of CTLA4-Ig by fetal pig neurons for xenotransplantation. *Transgenic Res* **14:** 373–384.

McCurry KR, Kooyman DL, Alvarado CG, Cotterell AH, Martin MJ, Logan JS, Platt JL. 1995. Human complement regulatory proteins protect swine-to-primate cardiac xenografts from humoral injury. *Nat Med* **1:** 423–427.

McGregor CG, Davies WR, Oi K, Teotia SS, Schirmer JM, Risdahl JM, Tazelaar HD, Kremers WK, Walker RC, Byrne RW, et al. 2005. Cardiac xenotransplantation: Recent preclinical progress with 3-month median survival. *J Thorac Cardiovasc Surg* **130:** 844–851.

McMorrow IM, Comrack CA, Sachs DH, DerSimonian H. 1997. Heterogeneity of human anti-pig natural antibodies cross-reactive with the Gal(α1,3)Galactose epitope. *Transplantation* **64:** 501–510.

McPherson TB, Liang H, Record RD, Badylak SF. 2000. Galα(1,3)Gal epitope in porcine small intestinal submucosa. *Tissue Eng* **6:** 233–239.

Menoret S, Plat M, Blancho G, Martinat-Botte F, Bernard P, Karam G, Tesson L, Renaudin K, Guillouet P, Weill B, et al. 2004. Characterization of human CD55 and CD59 transgenic pigs and kidney xenotransplantation in the pig-to-baboon combination. *Transplantation* **77:** 1468–14671.

Mohiuddin MM, Corcoran PC, Singh AK, Azimzadeh A, Hoyt RF Jr, Thomas ML, Eckhaus MA, Seavey C, Ayares D, Pierson RN III, et al. 2012. B-cell depletion extends the survival of GTKO.hCD46Tg pig heart xenografts in baboons for up to 8 months. *Am J Transplant* **12:** 763–771.

Mulder A, Kardol MJ, Arn JS, Eijsink C, Franke ME, Schreuder GM, Haasnoot GW, Doxiadis II, Sachs DH, Smith DM, et al. 2010. Human monoclonal HLA antibodies reveal interspecies crossreactive swine MHC class I epitopes relevant for xenotransplantation. *Mol Immunol* **47:** 809–815.

Muller YD, Seebach JD, Buhler LH, Pascual M, Golshayan D. 2011. Transplantation tolerance: Clinical potential of regulatory T cells. *Self Nonself* **2:** 26–34.

Muller YD, Ehirchiou D, Golshayan D, Buhler LH, Seebach JD. 2012. Potential of T-regulatory cells to protect xenografts. *Curr Opin Organ Transplant* **17:** 155–161.

Naziruddin B, Durriya S, Phelan D, Duffy BF, Olack B, Smith D, Howard T, Mohanakumar T. 1998. HLA antibodies present in the sera of sensitized patients awaiting renal transplant are also reactive to swine leukocyte antigens. *Transplantation* **66:** 1074–1080.

Nishimura T, Onda M, Takao S. 2010. CD4$^+$ CD25$^+$ regulatory T cells suppressed the indirect xenogeneic immune response mediated by porcine epithelial cell pulsed dendritic cells. *Xenotransplantation* **17:** 313–323.

Ohdan H, Yang YG, Shimizu A, Swenson KG, Sykes M. 1999. Mixed chimerism induced without lethal conditioning prevents T cell- and anti-Galα1,3Gal-mediated graft rejection. *J Clin Invest* **104:** 281–290.

Padler-Karavani V, Yu H, Cao H, Chokhawala H, Karp F, Varki N, Chen X, Varki A. 2008. Diversity in specificity, abundance, and composition of anti-Neu5Gc antibodies in normal humans: Potential implications for disease. *Glycobiology* **18:** 818–830.

Park JY, Park MR, Kwon DN, Kang MH, Oh M, Han JW, Cho SG, Park C, Kim DK, Song H, et al. 2011. α1,3-Galactosyltransferase deficiency in pigs increases sialyltransferase activities that potentially raise non-gal xenoantigenicity. *J Biomed Biotechnol* **2011:** 560850.

Park JY, Park MR, Bui HT, Kwon DN, Kang MH, Oh M, Han JW, Cho SG, Park C, Shim H, et al. 2012. α1,3-Galactosyltransferase deficiency in germ-free miniature pigs increases *N*-glycolylneuraminic acids as the xenoantigenic determinant in pig–human xenotransplantation. *Cell Reprogram* **14:** 353–363.

Parker W, Bruno D, Holzknecht ZE, Platt JL. 1994. Characterization and affinity isolation of xenoreactive human natural antibodies. *J Immunol* **153:** 3791–3803.

Phelps CJ, Koike C, Vaught TD, Boone J, Wells KD, Chen SH, Ball S, Specht SM, Polejaeva IA, Monahan JA, et al. 2003. Production of α1,3-galactosyltransferase-deficient pigs. *Science* **299:** 411–414.

Phelps CJ, Ball SF, Vaught TD, Vance AM, Mendicino M, Monahan JA, Walters AH, Wells KD, Dandro AS, Ramsoondar JJ, et al. 2009. Production and characterization of transgenic pigs expressing porcine CTLA4-Ig. *Xenotransplantation* **16:** 477–485.

Puga Yung GL, Li Y, Borsig L, Millard AL, Karpova MB, Zhou D, Seebach JD. 2012. Complete absence of the αGal xenoantigen and isoglobotrihexosylceramide in α1,3galactosyltransferase knock-out pigs. *Xenotransplantation* **19:** 196–206.

Rogers NJ, Jackson IM, Jordan WJ, Hawadle MA, Dorling A, Lechler RI. 2003. Cross-species costimulation: Relative contributions of CD80, CD86, and CD40. *Transplantation* **75:** 2068–2076.

Saethre M, Schneider MK, Lambris JD, Magotti P, Haraldsen G, Seebach JD, Mollnes TE. 2008. Cytokine secretion depends on Galα(1,3)Gal expression in a pig-to-human whole blood model. *J Immunol* **180:** 6346–6353.

Schmelzle M, Schulte Esch J II, Robson SC. 2010. Coagulation, platelet activation and thrombosis in xenotransplantation. *Curr Opin Organ Transplant* **15:** 212–218.

Schneider MK, Seebach JD. 2008. Current cellular innate immune hurdles in pig-to-primate xenotransplantation. *Curr Opin Organ Transplant* **13:** 171–177.

Schneider MK, Ghielmetti M, Rhyner DM, Antsiferova MA, Seebach JD. 2009. Human leukocyte transmigration across Galα(1,3)Gal-negative porcine endothelium is regulated by human CD18 and CD99. *Transplantation* **87:** 491–499.

Scobie L, Padler-Karavani V, Le Bas-Bernardet S, Crossan C, Blaha J, Matouskova M, Hector RD, Cozzi E, Vanhove B, Charreau B, et al. 2013. Long-term IgG response to porcine Neu5Gc antigens without transmission of PERV in burn patients treated with porcine skin xenografts. *J Immunol* **191:** 2907–2915.

Sharma A, Naziruddin B, Cui C, Martin MJ, Xu H, Wan H, Lei Y, Harrison C, Yin J, Okabe J, et al. 2003. Pig cells that lack the gene for α1-3 galactosyltransferase express low levels of the Gal antigen. *Transplantation* **75:** 430–436.

Sheikh S, Parhar R, Al-Mohanna F. 2002. Rapid static adhesion of human naive neutrophil to naive xenoendothelium under physiologic flow is independent of Galα1,3-gal structures. *J Leukoc Biol* **71:** 932–940.

Shimizu I, Kawahara T, Haspot F, Bardwell PD, Carroll MC, Sykes M. 2007. B-cell extrinsic CR1/CR2 promotes natural antibody production and tolerance induction of anti-αGAL-producing B-1 cells. *Blood* **109:** 1773–1781.

Shimizu A, Hisashi Y, Kuwaki K, Tseng YL, Dor FJ, Houser SL, Robson SC, Schuurman HJ, Cooper DK, Sachs DH, et al. 2008. Thrombotic microangiopathy associated with humoral rejection of cardiac xenografts from α1,3-galactosyltransferase gene-knockout pigs in baboons. *Am J Pathol* **172:** 1471–1481.

Shimizu A, Yamada K, Robson SC, Sachs DH, Colvin RB. 2012. Pathologic characteristics of transplanted kidney xenografts. *J Am Soc Nephrol* **23:** 225–235.

Singh AK, Seavey CN, Horvath KA, Mohiuddin MM. 2012. Ex-vivo expanded baboon $CD4^+$ CD25 Hi Treg cells suppress baboon anti-pig T and B cell immune response. *Xenotransplantation* **19:** 102–111.

Sommaggio R, Cohnen A, Watzl C, Costa C. 2012. Multiple receptors trigger human NK cell-mediated cytotoxicity against porcine chondrocytes. *J Immunol* **188:** 2075–2083.

Stevens RB, Platt JL. 1992. The pathogenesis of hyperacute xenograft rejection. *Am J Kidney Dis* **20:** 414–421.

Stone KR, Walgenbach AW, Turek TJ, Somers DL, Wicomb W, Galili U. 2007. Anterior cruciate ligament reconstruction with a porcine xenograft: A serologic, histologic, and biomechanical study in primates. *Arthroscopy* **23:** 411–419.

Tahara H, Ide K, Basnet N, Tanaka Y, Ohdan H. 2010. Determination of the precursor frequency and the reaction intensity of xenoreactive human T lymphocytes. *Xenotransplantation* **17:** 188–196.

Tahiri F, Li Y, Hawke D, Ganiko L, Almeida I, Levery S, Zhou D. 2013. Lack of iGb3 and Isoglobo-series glycosphingolipids in pig organs used for xenotransplantation: Implications for natural killer T-cell biology. *J Carbohydr Chem* **32:** 44–67.

Tanemura M, Yin D, Chong AS, Galili U. 2000. Differential immune responses to α-Gal epitopes on xenografts and allografts: Implications for accommodation in xenotransplantation. *J Clin Invest* **105:** 301–310.

Tanemura M, Ogawa H, Yin DP, Chen ZC, DiSesa VJ, Galili U. 2002. Elimination of anti-Gal B cells by α-Gal ricin1. *Transplantation* **73:** 1859–1868.

Taylor CJ, Tang KG, Smith SI, White DJ, Davies HF. 1998. HLA-specific antibodies in highly sensitized patients can cause a positive crossmatch against pig lymphocytes. *Transplantation* **65:** 1634–1641.

Tazelaar HD, Byrne GW, McGregor CG. 2011. Comparison of Gal and non-Gal-mediated cardiac xenograft rejection. *Transplantation* **91:** 968–975.

Tena A, Turcotte N, Leto Barone A, Arn S, Terlouw S, Dobrinsky J, Sachs D, Hawley R. 2011. Miniature swine expressing human CD47 to enhance bone marrow engraftment in non-human primates. *Xenotransplantation* **18:** 271.

Thompson P, Badell IR, Lowe M, Cano J, Song M, Leopardi F, Avila J, Ruhil R, Strobert E, Korbutt G, et al. 2011a. Islet xenotransplantation using Gal-deficient neonatal donors improves engraftment and function. *Am J Transplant* **11:** 2593–2602.

Thompson P, Cardona K, Russell M, Badell IR, Shaffer V, Korbutt G, Rayat GR, Cano J, Song M, Jiang W, et al. 2011b. CD40-specific costimulation blockade enhances neonatal porcine islet survival in nonhuman primates. *Am J Transplant* **11:** 947–957.

Tseng YL, Dor FJ, Kuwaki K, Ryan D, Wood J, Denaro M, Giovino M, Yamada K, Hawley R, Patience C, et al. 2004. Bone marrow transplantation from α1,3-galactosyltransferase gene-knockout pigs in baboons. *Xenotransplantation* **11:** 361–370.

van der Windt DJ, Bottino R, Casu A, Campanile N, Cooper DK. 2007. Rapid loss of intraportally transplanted islets: An overview of pathophysiology and preventive strategies. *Xenotransplantation* **14:** 288–297.

van der Windt DJ, Bottino R, Casu A, Campanile N, Smetanka C, He J, Murase N, Hara H, Ball S, Loveland BE, et al. 2009. Long-term controlled normoglycemia in diabetic non-human primates after transplantation with hCD46 transgenic porcine islets. *Am J Transplant* **9:** 2716–2726.

Varki A. 2010. Colloquium paper: Uniquely human evolution of sialic acid genetics and biology. *Proc Natl Acad Sci* **107:** 8939–8946.

Vercellotti GM, Platt JL, Bach FH, Dalmasso AP. 1991. Neutrophil adhesion to xenogeneic endothelium via iC3b. *J Immunol* **146:** 730–734.

Vugmeyster Y, Beyer J, Howell K, Combs D, Fielder P, Yang J, Qureshi F, Sandlund B, Kawaguchi L, Dummer W, et al. 2005. Depletion of B cells by a humanized anti-CD20 antibody PRO70769 in *Macaca fascicularis*. *J Immunother* **28:** 212–219.

Wallgren AC, Karlsson-Parra A, Korsgren O. 1995. The main infiltrating cell in xenograft rejection is a $CD4^+$ macrophage and not a T lymphocyte. *Transplantation* **60:** 594–601.

Weiss EH, Lilienfeld BG, Muller S, Muller E, Herbach N, Kessler B, Wanke R, Schwinzer R, Seebach JD, Wolf E, et al. 2009. HLA-E/human β_2-microglobulin transgenic pigs: Protection against xenogeneic human anti-pig natural killer cell cytotoxicity. *Transplantation* **87:** 35–43.

Wilhite T, Ezzelarab C, Hara H, Long C, Ayares D, Cooper DK, Ezzelarab M. 2012. The effect of Gal expression on pig cells on the human T-cell xenoresponse. *Xenotransplantation* **19:** 56–63.

Wu J, Yi S, Ouyang L, Jimenez E, Simond D, Wang W, Wang Y, Hawthorne WJ, O'Connell PJ. 2008. In vitro expanded human $CD4^+CD25^+$ regulatory T cells are potent suppressors of T-cell-mediated xenogeneic responses. *Transplantation* **85:** 1841–1848.

Xu XC, Goodman J, Sasaki H, Lowell J, Mohanakumar T. 2002. Activation of natural killer cells and macrophages by porcine endothelial cells augments specific T-cell xenoresponse. *Am J Transplant* **2:** 314–322.

Xu Y, Yang YG, Ohdan H, Ryan D, Harper D, Wu C, Kruger-Grey HS, Thall AD, Awwad M, Sykes M. 2006. Characterization of anti-Gal antibody-producing cells of baboons and humans. *Transplantation* **81:** 940–948.

Yamada K, Sachs DH, DerSimonian H. 1995. Human anti-porcine xenogeneic T cell response. Evidence for allelic specificity of mixed leukocyte reaction and for both direct and indirect pathways of recognition. *J Immunol* **155:** 5249–5256.

Yamada K, Yazawa K, Shimizu A, Iwanaga T, Hisashi Y, Nuhn M, O'Malley P, Nobori S, Vagefi PA, Patience C, et al. 2005. Marked prolongation of porcine renal xenograft survival in baboons through the use of α1,3-galactosyltransferase gene-knockout donors and the cotransplantation of vascularized thymic tissue. *Nat Med* **11:** 32–34.

Yamamoto S, Lavelle JM, Vagefi PA, Arakawa H, Samelson-Jones E, Moran S, Teranishi K, Kamano C, Fishman J,

Awwad M, et al. 2005. Vascularized thymic lobe transplantation in a pig-to-baboon model: A novel strategy for xenogeneic tolerance induction and T-cell reconstitution. *Transplantation* **80:** 1783–1790.

Yang YG, deGoma E, Ohdan H, Bracy JL, Xu Y, Iacomini J, Thall AD, Sykes M. 1998. Tolerization of anti-Galα1-3Gal natural antibody-forming B cells by induction of mixed chimerism. *J Exp Med* **187:** 1335–1342.

Yeh P, Ezzelarab M, Bovin N, Hara H, Long C, Tomiyama K, Sun F, Ayares D, Awwad M, Cooper DK. 2010. Investigation of potential carbohydrate antigen targets for human and baboon antibodies. *Xenotransplantation* **17:** 197–206.

Yi S, Feng X, Hawthorne WJ, Patel AT, Walters AN, O'Connell PJ. 2002. $CD4^+$ T cells initiate pancreatic islet xenograft rejection via an interferon-γ-dependent recruitment of macrophages and natural killer cells. *Transplantation* **73:** 437–446.

Yi S, Hawthorne WJ, Lehnert AM, Ha H, Wong JK, van Rooijen N, Davey K, Patel AT, Walters SN, Chandra A, et al. 2003. T cell-activated macrophages are capable of both recognition and rejection of pancreatic islet xenografts. *J Immunol* **170:** 2750–2758.

Zhao Y, Swenson K, Sergio JJ, Arn JS, Sachs DH, Sykes M. 1996. Skin graft tolerance across a discordant xenogeneic barrier. *Nat Med* **2:** 1211–1216.

Tolerance—Is It Worth It?

Erik B. Finger[1], Terry B. Strom[2], and Arthur J. Matas[1]

[1]Department of Surgery, University of Minnesota, Minneapolis, Minnesota 55455

[2]Harvard Medical School, Department of Medicine, The Transplant Institute, Beth Israel Deaconess Medical Center, Boston, Massachusetts 02215

Correspondence: matas001@umn.edu

We are entering an exciting time in the study of immunologic tolerance. Several cellular and molecular strategies have been developed that show promise in nonhuman transplant models and these approaches are just now appearing in clinical trials. Tolerance strategies that prevent immune rejection and obviate the need for immunosuppressive medications (with inherent risk of cancer, infection, and organ toxicity) would improve both graft and patient survival. Each tolerance protocol brings its own set of associated risks. As the results of these trials become available, we must continue to evaluate their successes and failures. The balance of these outcomes will help us answer the question: "Tolerance—Is it worth it?"

WHAT IS TOLERANCE?

In the words of Nobel Prize winner Peter Medawar, "'Immunological tolerance' may be described as a state of indifference or non-reactivity towards a substance that would normally be expected to excite an immunological response" (Medawar 1964). The concept of tolerance is further clarified by the editors of the *American Journal of Transplantation* as "a durable state of antigen-specific unresponsiveness, induced by exposure to antigen, in a patient who is otherwise fully immunologically competent" (Halloran et al. 2008).

The experimental evidence for tolerance began with the 1945 studies of Owen (1945) who showed persistence of postnatal red blood cell chimerism in freemartin cattle that share a common placenta and develop cross-placental common circulation. The significance of these observations to transplantation was made apparent by Billingham, Medawar, and colleagues, who showed tolerance to skin grafts from nonidentical dizygotic cattle twins (Anderson et al. 1951; Billingham et al. 1952). Interestingly, these observations occurred in a failed attempt to develop a technique to distinguish monozygotic versus dizygotic twins. The original hypothesis was that skin grafts between dizygotic twins would reject, whereas grafts between single-cell twins would not. To the contrary, skin grafts from both identical and dizygotic twins were accepted without rejection. Similar findings were shown by parabiosis in birds, in which embryonic cross circulation made siblings incapable of making antibodies against each other, or of rejecting skin grafts (Hasek and Hraba 1955).

Billingham et al. (1953) reported the first formal documentation of transplant tolerance

Cite this article as *Cold Spring Harb Perspect Med* doi: 10.1101/cshperspect.a015594

by demonstrating that inoculation of fetal mice or chickens with donor cells of a different strain of that species made these animals tolerant to skin grafts of the donor strain. This study is the first report of tolerance induction by mixed chimerism. However, there has been some debate as to whether these experiments show true tolerance (Halloran et al. 2008). Subsequent studies suggest that mixed chimerism may, in itself, be a form of immunosuppression, as this antigen's nonresponsiveness was not donor-specific (Hofmann et al. 2004). There is disagreement over the interpretation of these findings (Sachs et al. 2008). Additional evidence shows persistence of intact immune function remains with mixed chimerism (Ildstad et al. 1985; Rüedi et al. 1989), and that although nonresponsiveness in these studies was not strain specific, they were at least organ specific (Hofmann et al. 2004).

Despite conflicts in the formal definition of tolerance, a more practical definition may be that of operational tolerance. This may be defined as "a well-functioning graft lacking histological signs of rejection, in the absence of any immunosuppressive drugs (for at least 1 year), in an immunocompetent host" (Ashton-Chess et al. 2007; Orlando et al. 2010). Operational tolerance may further be clarified to include the absence of antibody-mediated rejection.

WHY IS TOLERANCE IMPORTANT?

Transplantation is the best, and in some cases only, treatment for many types of end-stage organ disease (Grinyó 2013). It has been shown to increase both quality and length of life for those affected by organ failure. Unfortunately, for the vast majority of transplant recipients there is an obligate requirement for lifelong immunosuppression. These immunosuppressive medications increase the risk of cancer, infection, and direct drug-associated organ toxicity, all of which contribute to reduced patient survival.

The most common cause of death in patients with functioning renal transplants is cardiovascular disease (24%), followed by infection (16%), and malignancy (12%) (Kahwaji et al. 2011). Each organ has a different distribution of causes of death after transplant, but infection and cancer are frequent etiologies in all (Chapman et al. 2013; Fishman 2013a). There is an ~50% overall risk of developing some form of cancer (including nonmelanoma skin cancer) attributed to immunosuppressive medications (Scherer et al. 2007). The association of individual cancers with immunosuppression varies, but relative risks of two- to 20-fold have been reported (Kasiske et al. 2004) with the greatest risk being non-Hodgkin's lymphomas and nonmelanoma skin cancers. The risks of Epstein–Barr virus (EBV)-associated posttransplant lymphoproliferative disease (PTLD) vary based on organ type, age at transplant, and EBV serostatus. A recent report from a large kidney transplant series measured annual incidence of 2.6 cases of PTLD per 1000 patient years and an increase in Hodgkins and non-Hodgkins lymphoma of 5.9 and 7.6 times, respectively, the rate in the general population (Morton et al. 2013).

Transplant patients are also at increased risk for both routine and opportunistic infections (Fishman 2007, 2013a). Of particular concern is the increased susceptibility to viral and fungal infections (Fishman 2002, 2013b; Winston et al. 1995; Winston and Busuttil 2002). Opportunistic infections are even more prevalent in patients undergoing antilymphocyte induction therapy (Issa and Fishman 2009). Infection is second only to cardiovascular disease, which is one of the most common causes of death in kidney transplant recipients (Djamali et al. 2006).

Immunosuppressive agents have a range of associated direct toxicities. Of particular concern is chronic nephropathy of the transplanted allograft in renal transplant recipients or the native kidneys in extrarenal transplant recipients (Ojo et al. 2003). New onset diabetes associated with calcineurin inhibitors is a well-described complication (Kasiske et al. 2003). With strict definition for new onset diabetes, the tacrolimus-associated incidence is in excess of 30% in kidney recipients and even higher in liver recipients (First et al. 2013). The degree of posttransplant hyperglycemia is predictive of cardiovascular and all-cause mortality (Valderhaug et al. 2011).

In addition to the adverse health effects of these medications, there is a continued problem of graft loss due to noncompliance. A review of multiple reports of compliance in kidney transplantation reported that approximately 22% of patients are noncompliant with their medications (Butler et al. 2004). Furthermore, 36% of graft losses are associated with prior noncompliance, and the odds ratio of graft failure is sevenfold higher for nonadherent patients (Butler et al. 2004). A report from Sweden documented that 16% of renal transplant recipients report having missed doses of immunosuppression in the preceding month (Lennerling and Forsberg 2012). The problem is even more pronounced in pediatric recipients, in which there is as much as a 75% nonadherence rate (Fredericks and Dore-Stites 2010). In kidney transplantation, long-term mortality rate was 48%–82% lower among transplant recipients than patients on the waiting list (Wolfe et al. 1999). Graft losses due to noncompliance or direct toxicity of these medications likely contribute to overall shortened patient lifespan.

Successful induction of transplant tolerance would eliminate immune rejection. Even with continued use of therapeutic immunosuppression, graft loss due to immune-mediated acute and chronic rejection shortens both graft and patient survival. Since the first successful kidney transplant in 1954 (Harrison et al. 1956), there has been a continuous improvement in transplant outcomes for all organ types (Organ Procurement 2012). Despite improvements in predominantly short-term outcomes over the last 50 years, improvements in chronic graft loss have not kept pace (Meier-Kriesche et al. 2004; Sykes 2007). Even in the absence of organ dysfunction, there can be subclinical rejection present in nearly half of biopsy specimens. Over time, severe chronic allograft nephropathy develops in 58.4% of patients (Nankivell et al. 2003). This damage, once established, is irreversible, resulting in declining renal function and graft failure. This resulting decline in glomerular filtration rate (GFR) is predictive of mortality following renal transplant (Moranne et al. 2013). In liver, lung, and heart transplants, immune-mediated graft loss leads to patient death unless retransplant is possible.

Tolerogenic strategies that lessen infectious complications, reduce cancer risk, avoid direct pharmacologic toxicity, and prevent graft loss due to acute and chronic immune destruction would improve graft and subsequent patient survival.

HOW CAN TOLERANCE BE ACHIEVED?

Many approaches have been pursued with the objective of tolerance in solid organ transplantation. These approaches can be divided into three main groups: (1) study of noncompliant patients that show spontaneous tolerance following discontinuation of immunosuppression, (2) patients with stable long-term graft function that undergo planned immunosuppression minimization and cessation, and (3) pretransplant-planned protocols for tolerance induction via pharmacologic or cellular therapy (Orlando et al. 2010). There are examples of successful withdrawal of immunosuppression with each of these approaches. Although they may not meet the strict definition of "tolerance," they fit within a more clinically relevant description of "clinical operational tolerance." Some of the most promising approaches are summarized below.

Study of Spontaneous Tolerance

There have been many reports of sporadic tolerance identified in noncompliant patients (summarized by Orlando et al. 2010). It is clear that organs differ in their response to cessation of immunosuppression out of noncompliance. There have been many reports of tolerance in kidney transplantation (Orlando et al. 2010), a single report in lung transplant (Di Cocco et al. 2009), and no reports in pancreas transplant. Liver transplant recipients are most likely to have continued graft function after cessation of immunosuppression. Multiple reports document durable liver allograft function after discontinuation by nonadherence (Orlando et al. 2009). There are other case reports or small series of tolerance following discontinuation under circumstances other than noncompliance. One report documented operational tolerance in heart transplant after total lymphoid irradiation for resistant rejection and subsequent withdrawal of all medications, including immunosuppression, because of drug intolerance (Comerci et al. 2009). Similar findings have been found in renal transplants with operational tolerance after total lymphoid irradiation (Strober et al. 1989, 2000). A number of case reports have documented spontaneous tolerance in a subset of patients undergoing immunosuppression withdrawal for treatment of PTLD (Orlando et al. 2010).

Planned Immunosuppression Weaning

Based on the findings of spontaneous tolerance in nonadherent patients, several studies have been performed that sought to wean patients off of immunosuppression as a planned treatment protocol. Orlando reviewed reports of operational tolerance in kidney (Orlando et al. 2010) and liver transplantation (Orlando et al. 2009) following planned immunosuppression withdrawal. In liver transplant, there is an approximate 22% success rate for protocols of withdrawal. In a particularly informative study in liver transplantation, it was found that pediatric patients may be uniquely amenable to immunosuppression weaning (Feng et al. 2012). In this study of planned immunosuppression weaning in patients with stable graft function on single agents after liver transplant for nonviral and nonautoimmune etiologies of ESLD, there was a 60% success rate in immunosuppression withdrawal.

The prevalence of potentially tolerant renal transplant recipients is clearly less than in liver transplant recipients, but the actual percentage of patients is harder to define. Most studies rely on a series of case reports (Roussey-Kesler et al. 2006) or assembled groups of patients already with stable graft function after nonadherence. Approximately 100 cases of operational tolerance in renal transplant have been reported (Orlando et al. 2010) with a denominator of many tens of thousands of total kidney transplants. There are a steadily increasing number of reports of different immune monitoring phenotypes or "signatures" that characterize tolerant recipients (Brouard et al. 2005, 2007; Newell and Larsen 2006; Einecke et al. 2010; Newell et al. 2010; Sagoo et al. 2010). It may be possible to use these combined signatures to identify candidates for immunosuppression weaning.

Pretransplant Tolerance Protocols

Mixed Chimerism

A promising approach developed by several groups makes use of strategies to induce mixed chimerism (Sachs et al. 2014). This is accomplished by a stem-cell infusion in combination with a transplant from the same donor. There are several variations to this approach. The earliest reports were in patients that first had bone marrow/stem cell transplantation and then later had a kidney transplant from the same donor (Sayegh et al. 1991; Helg et al. 1994; Jacobsen et al. 1994; Sorof et al. 1995; Butcher et al. 1999; Sellers et al. 2001; Light et al. 2002). Operational tolerance was reported in these recipients. Subsequent studies have documented some degree of success with a combined or simultaneous stem cell and kidney transplant from the same donor rather than the earlier sequential approach.

The group from Massachusetts General Hospital (MGH) showed tolerance in six patients undergoing simultaneous bone marrow and kidney transplants from human leukocyte antigen (HLA)-identical donors for the treatment of multiple myeloma in patients with renal failure (Fudaba et al. 2006). Three patients showed transient donor chimerism, but were successfully weaned off immunosuppression. Two patients had full chimerism, but immunosuppression was resumed for treatment of graft-versus-host disease (GVHD). One patient developed acute rejection and was treated with resumption of cyclosporine, but was later able to be weaned off immunosuppression once again.

This report was followed by a more rigorous non-HLA-identical series from the same group wherein they transplanted five patients with single haplotype mismatched sibling donors (Kawai et al. 2008). They performed a pretransplant

 Cite this article as *Cold Spring Harb Perspect Med* doi: 10.1101/cshperspect.a015594

nonmyeloablative conditioning protocol with cyclophosphamide, thymic irradiation, and humanized anti-CD2 monoclonal antibody. Induction therapy also consisted of short-course cyclosporine or tacrolimus (plus rituximab and steroids in a subset of patients). Four out of five patients showed operational tolerance and were weaned off of immunosuppression. The fifth developed early refractory antibody-mediated rejection and had graft failure. The tolerant patients showed transient microchimersim and persisting donor-specific hyporesponsiveness in in vitro assays (Kawai et al. 2008; LoCascio et al. 2010). In subsequent follow-up, there were three cases of acute or chronic GVHD, for which two recipients were returned to immunosuppression (Spitzer et al. 2011). Three patients had recurrence of multiple myeloma. This trial has since been extended to 10 patients; seven have shown long-term graft survival without maintenance immunosuppression (Markmann and Kawai 2012). Three failures occurred after withdrawal of immunosuppression from early humoral rejection, thrombotic microangiopathy, and acute rejection.

A group from Stanford reported on their series of 16 patients undergoing simultaneous HLA-identical kidney transplant and same-donor granulocyte-colony stimulating factor (GCSF)-mobilized CD34$^+$ stem-cell infusion (Scandling et al. 2012). They used a nonmyeloablative conditioning regimen of total lymphoid irradiation and antithymocyte globulin. Fifteen out of 16 developed mixed donor/recipient chimerism. Eight of these showed chimerism for more than 6 months and were weaned off immunosuppression. Four patients were undergoing immunosuppression wean. Four patients developed recurrent disease or rejection and were not weaned. All patients had functioning grafts at the end of the reporting period and there was no GVHD.

The groups from Northwestern and University of Louisville reported a novel approach to augment donor stem cell engraftment in conjunction with renal transplant (Leventhal et al. 2012). They make use of a bone marrow-derived CD8$^+$, TCR$^-$ "facilitator cell" population that promotes the engraftment of allogeneic hematopoietic stem cells in animal models without observed GVHD (Kaufman et al. 1994). To date, data on 15 patients undergoing their protocol have been reported (Leventhal et al. 2012, 2013). Recipients undergo a "low-intensity" conditioning regimen consisting of fludarabine, cyclophosphamide, and 200 cGy total-body irradiation. Subsequently, a combined HLA-mismatched renal transplant with same-donor hematopoietic stem-cell and facilitator-cell infusion is performed. Fourteen out of 15 showed initial macrochimerism that was eventually lost in three patients. There was one engraftment failure in a patient with panel reactive antibody (PRA) of 52%. Initial immunosuppression was tacrolimus and mycophenolate mofetil, which was successfully weaned off at 1 year in all patients with durable macrochimerism (12 patients). One patient lost their graft because of renal artery thrombosis after viral infection and sepsis. There were no cases of GVHD or engraftment syndrome. Of note, all patients that were successfully weaned off all immunosuppression had extremely high levels of donor-cell engraftment (94%–100% donor type), suggesting donor-type replacement rather than mixed chimerism. Two concerns with these reports were the lack of details on the engineered facilitator population and a lack of a control group without facilitator cells (Markmann and Kawai 2012). However, in a similar induction therapy for stem cell transplant, there was an engraftment failure rate of 13% and severe GVHD in 6% (Luznik et al. 2008; Markmann and Kawai 2012).

Regulatory T Cells

A different cellular therapy approach uses regulatory T cells (Treg) (Cobbold and Waldmann 2013). These lymphocytes develop in the thymus (natural Treg) or are induced in the periphery in response to antigen exposure in the presence of TGF-β (adaptive Treg) (reviewed in Josefowicz et al. 2012). Treg are phenotypically described by their expression of the transcription factor FoxP3. When these cells encounter their cognate antigens properly presented by antigen-presenting cells, they suppress the activation and function of nearby effector T cells

(Schmidt et al. 2012). Treg are important in a wide variety of counterinflammatory conditions and are a promising target for tolerance-induction strategies.

The first clinical use of Tregs has been in the prevention or treatment of GVHD. Trzonkowski et al. (2009) performed a trial in two patients who were failing standard treatment of GVHD following allogeneic stem cell transplant. Treg were isolated from the stem cell donor and infused back into the recipient. In one patient with chronic GVHD, they were able to stabilize symptoms and wean immunosuppression medications. The long-term durability was not reported. In a second patient with acute GVHD, they were able to temporize symptoms with repeated dosing of Treg. However, they ran out of donor material, were unable to infuse further cells, and the patient relapsed and succumbed to the disease.

A second trial reported by Di Ianni et al. (2011), attempted to use Treg paired with conventional T-cell infusion to prevent GVHD in patients at high risk for GVHD, and to more quickly restore cellular immunity following stem cell transplant. This study consisted of 28 patients with high-risk hematologic malignancies that underwent haploidentical hematopoietic stem cell transplant. The recipients underwent induction conditioning with total-body irradiation and treatment with thiopeta, fludarabine, and cyclophosphamide. GCSF-mobilized, CD34-enriched hematopoietic stem cells from the donors were infused along with ex vivo expanded Treg and conventional T cells. Twenty-six out of 28 had sustained full donor-type engraftment. Two out of 26 developed grade 2, or better, acute GVHD. No chronic GVHD was observed and there was improved T-cell reconstitution and decreased cytomegalovirus compared with standard hematopoietic stem cell transplantation (HST).

A third trial was performed at the University of Minnesota (Brunstein et al. 2011). This study was a phase I dose escalation trial of ex vivo expanded umbilical cord Treg. Patients underwent a conditioning regimen consisting of total-body irradiation and treatment with fludarabine and cyclophosphamide. Patients received two partially HLA-matched umbilical cord blood infusions followed by infusion of Treg isolated and expanded from a third umbilical cord blood sample that was 4-6/6 HLA matched to the patient. Induction immunosuppression was cyclosporine and mycophenolate mofetil (MMF) in the first group of patients and then sirolimus and MMF in the second group. There was no increased risk of infection, relapse of disease, or mortality. There was a decrease in severe GVHD from 61% to 43%.

In an interesting but theoretically contradictory approach, trials are beginning that combine rapamycin and IL-2 administration in diabetes, GVHD, and composite tissue transplants. This combination has a number of Treg-friendly benefits. Treg are IL-2 dependent and IL-2 rather surprisingly serves as a death factor for $CD8^+$ memory T cells (Ku et al. 2000) or newly activated effector T cells (Li et al. 2001) by opposing cytokines. Low-dose IL-2 administration in diabetes-prone nonobese diabetic (NOD) mice promotes Treg survival and protects against the development of diabetes (Tang et al. 2008). The combination of IL-2 and rapamycin favor Treg survival and Teffector apoptosis by activating T cells toward an apoptotic pathway rather than clonal expansion (Rabinovitch et al. 2002).

Triple therapy with doses of IL-2Ig that saturate the high affinity, but not intermediate affinity IL-2 receptor, mutant antagonist type IL-15Ig, and rapamycin tilts the balance of immunity from rejection to or toward tolerance in a variety of daunting murine transplant (Zheng et al. 2003) and new onset type 1 diabetes models (Koulmanda et al. 2007). Triple therapy also results in long-term islet allograft survival in nonhuman primates (Koulmanda et al. 2012). The mechanism is hypothesized to be selective depletion of alloreactive T cells and stabilization/expansion of Tregs. Recent clinical trials of low-dose IL-2 therapy in corticosteroid-resistant GVHD (Koreth et al. 2011) and virus-induced vasculitis (Saadoun et al. 2011) have produced exciting results, whereas higher doses produce toxicity by activating innate immune cells bearing the intermediate affinity IL-2 receptor despite expansion of Tregs.

 Cite this article as *Cold Spring Harb Perspect Med* doi: 10.1101/cshperspect.a015594

Currently, there are several trials of Treg in solid organ transplantation under way, including a large multinational trial (www.onestudy.org) and other center-based trials. Although there is considerable hope that Treg may be important in the induction of tolerance, many questions persist. How many Treg are needed to prevent rejection? What is the source of Treg? Donor cells? Recipient cells? Do Treg need to be antigen-specific, or would polyclonal expanded cells be sufficient? How long do these cells last once infused? Based on experimental studies in animal models, there may be some benefit to combining Treg with stem-cell infusion. As has been seen with facilitator cells and hematopoetic stem cells, Treg may help with establishment of mixed chimerism and prevent induction of GVHD. Thus, the Treg themselves may not be tolerogenic, but they may support persistence of mixed chimerism that induces tolerant phenotype.

Costimulatory Blockade

Costimulatory molecules provide another attractive target for tolerance-induction protocols (Maltzman and Turka 2013). There are abundant examples in mice in which a blockade of costimulatory molecules prolongs graft survival (reviewed in Kinnear et al. 2013). The three main pathways that have approached as clinical targets have been the CD28:CD80/CD86, CD40:CD154, and LFA-1:ICAM receptor:ligand pairings.

The CD28 costimulatory pathway was the first of these pathways to have a biologic agent approved for clinical use in transplant. In murine experimental models, CTLA-4Ig administration leads to prolonged or indefinite graft survival (Lenschow et al. 1992; Baliga et al. 1994; Pearson et al. 1994, 1997; Judge et al. 1996). CTLA-4Ig works via two pathways to inhibit alloreactive T-cell function. First, CTLA-4Ig binds to CD80 and CD86 on antigen-presenting cells and inhibits binding of these ligands to CD28 molecules on T cells. This inhibition reduces CD28-mediated costimulation during antigen recognition. Second, CTLA-4Ig binding to CD80/86 molecules also induces indoleamine 2,3-dioxygenase (IDO) expression in dendritic cells via an IFN-γ-dependent pathway. Up-regulation of IDO leads to changes in tryptophan metabolism that suppresses T-cell function (Grohmann et al. 2002).

In 2011, belatacept (human CTLA-4Ig) was approved by the Food and Drug Administration for use in kidney transplantation for the prevention of acute rejection in combination with basiliximab, mycophenolate mofetil, and corticosteroids. A number of trials have reported the outcomes of belatacept-based immunosuppression protocols in comparison to cyclosporine-based regimens. One- and 3-year outcomes have shown similar graft and patient survival between cyclosporine and belatacept protocols, but slightly more rejection with belatacept (Durrbach et al. 2010; Vincenti et al. 2010; Rostaing et al. 2011; Pestana et al. 2012). However, belatacept-treated patients had better preservation of kidney function with a higher GFR than the cyclosporine-treated patients. In diabetic patients, there was an increase in patient and graft survival with belatacept, but this did not achieve statistical significance (Rostaing et al. 2011). In each of the studies, there was an increased risk of PTLD with belatacept. However, this was seen predominantly in EBV-naïve patients. Currently, belatacept is not recommended for use in transplant recipients who are EBV negative. In the 3-year follow-up, there was an increase in the incidence of tuberculosis infection (Pestana et al. 2012).

A second potential target for costimulatory blockade is the CD40:CD154 pathway. Blockade of this pathway results in prolongation of transplant survival in mouse (Parker et al. 1995; Hancock et al 1996; Larsen et al. 1996; Larsen and Pearson 1997; Shimizu et al. 2000) and nonhuman primate models (Kenyon et al. 1999; Kirk et al. 1999; Koulmanda et al. 2006). However, clinical trials in humans were halted because of safety concerns over thromboembolic events (Sidiropoulos and Boumpas 2004; Couzin 2005). These thromboembolic events were validated in nonhuman primate models and the mechanism was postulated to be caused by unanticipated binding to CD154 on vascular endothelium and platelets (Kawai et al. 2000). Although most preclinical studies reveal anti-CD154 to be more potent than most anti-CD40 mAbs as immune therapy, Reimann has developed an anti-CD40 mAb with immune-inhibitory functions as evidenced by the ability to block IgG production by lipopolysaccharide-activated B cells. This unique antibody has enabled drug-free nonhuman primate islet allograft survival, suggesting that optimization of this approach may have great promise (Lowe et al. 2012).

The third target, efalizumab, is a humanized anti-LFA-1 (CD11a/CD18) monoclonal antibody that was approved for the treatment of psoriasis (Dedrick et al. 2002; Stern 2003), and has been effective at reducing the severity of the disease in moderate to severe psoriasis (Gordon et al. 2003; Lebwohl et al. 2003). The antibody blocks the cell-adhesion molecule LFA-1 from binding to its ligand ICAM-1. LFA-1-mediated cell adhesion is important in a number of steps in the immune process including leukocyte adhesion to endothelium and the strengthening of T-cell/antigen-presenting cell interactions (Dedrick et al. 2002). Anti-LFA-1 was tested in a clinical trial in kidney transplant recipients (Vincenti et al. 2007). It was used in combination with cyclosporine, MMF, and prednisone. Although graft and patient survival was excellent (95% and 97% at 6 months), there was an 11% rejection risk and an 8% PTLD incidence. Other groups have used efalizumab as adjunct therapy in islet transplant trials (Posselt et al. 2010; Turgeon et al. 2010) as part of a maintenance immunosuppression regimen. The use of this agent was halted because of an association with progressive multifocal leukoencephalopathy (PML), a serious and usually fatal central nervous system infection caused by JC polyoma virus (Carson et al. 2009). However, PML has also been associated with other immunosuppressive mediations that remain in common use (Neff et al. 2008; Carson et al. 2009; Weber et al. 2011).

Other Possibilities

A number of additional cellular therapy approaches with promising results in preclinical studies are entering into clinical trial (reviewed in Page et al. 2012; Tang and Bluestone 2013). Tolerogenic macrophage, dendritic cells, and mesenchymal stromal cells have all shown promise in transplant models. Transplant acceptance-inducing cells are a form of regulatory macrophage. Two trials using infusion of donor-derived splenic monocytes in kidney transplant were performed (Hutchinson et al. 2008a,b). Standard immunosuppression induction was initiated and patients were weaned to low-dose monotherapy (or completely off of immunosuppression in one patient). There was an elevated incidence of early acute rejection, but several patients were able to wean down to low-dose single agent. Additional cell types with tolerogenic properties include mesenchymal stromal cells (English and Wood 2013) and tolerogenic dendritic cells (Page et al. 2012). These cell types have been shown effective in animal models and are also entering into clinical testing.

AT WHAT COST?

Highlighted above are a number of approaches that have been, or are soon to be, put into clinical trials as potentially tolerogenic therapies in transplantation. Despite some promising outcomes with several of these approaches, they each pose risk of adverse events and outcomes. For each approach there are specific complications, but these may represent more generalized concerns.

Risks of Stem Cell Infusion

A particular concern of the mixed chimerism approaches is the toxicity of the induction protocols. These protocols make use of chemotherapeutic medications and systemic irradiation as part as their engraftment protocol. These treatments come with both short- and long-term sequella. Although the immediate health outcomes reported seem to decrease with expected outcomes after transplant, we do not know the long-term health impact. An additional risk is the development of GVHD, a potentially life-threatening outcome, which was seen in several of the combined hematopoietic stem cell transplantation (HSCT)/kidney recipients.

Although we do not have long-term outcomes on all patients in the mixed chimerism trials, we can look at similar protocols for HSCT treatment of hematologic malignancies. Apart from recurrent disease, toxicity of the induction protocols and GVHD are the biggest long-term risk factors for these patients (Tichelli et al. 2009, 2012). HSCT recipients have increased risk of endocrine, musculoskeletal, and cardiovascular disorders in addition to increased risk of malignancy. These factors have an impact on morbidity and mortality following treatment. Patients alive at 5 years posttransplant still have a 30% reduction in life expectancy compared with the general population (Bhatia 2011).

Risks of PTLD

The risk of PTLD is increased in several of the trials described, particularly with the use of belatacept and efalizumab. This risk may be inherent to the tolerance approach, as the recipient's immune system is less able to fight off the subset of PTLDs that are donor cell in origin (Olagne et al. 2011). The concern of PTLD is greatest in patients who are EBV-naïve. Avoiding this population of patients may mitigate some of these concerns.

Off-Target Effects

A particular concern in trials of Treg is whether the suppression of allogeneic immune responses will lead to suppression of other off-target immune functions. That is, will Tregs that suppress rejection of liver transplant also suppress antiviral responses to the hepatitis virus that caused the disease (Dolganiuc and Szabo 2008)? Or will they suppress antitumor responses to hepatocellular carcinoma present in the liver explant (Lin et al. 2013)? We know that Treg are capable of bystander suppression (Tang and Bluestone 2008). We do not know if this suppression would be of harm to the transplant recipient (or maybe of benefit if the disease was an autoimmune process).

Other unanticipated off-target effects were seen with costimulation blockade trials. Antibodies against CD154 resulted in increased thromboembolic events. The unexpected complication was likely due to expression of this molecule on unanticipated cell types (platelets and endothelial cells). Other agents that block CD40 may have similar efficacy, but avoid these complications. Use of efalizumab was halted when there was a concern for increased risk of PML. This risk may be common to many immunosuppressant protocols, but serves as an example of another unexpected off-target effect.

Economic Considerations

A final consideration applicable to all of these approaches is an economic one. These protocols are expensive. Each protocol costs several million dollars to enroll a limited number of patients. In our own center, the cost of Treg isolation and expansion alone is ~$45 K per transplant. Counterbalancing this cost are the expenses of traditional transplants. Financial analysis is most complete for end-stage renal disease (ESRD) and renal transplantation. Data from the USRDS (2012) report shows that the average Medicare cost of ESRD in 2010 was $87,561 per person per year. Despite the cost of the transplant procedure, it is estimated that the transplant-associated savings are approximately $200,000 over 5 years (reviewed in Axelrod 2013). The cost of the transplant procedures themselves would likely be similar for standard therapy and tolerance protocols. The economic benefits of tolerance protocols would be generated by reduction in immunosuppression costs. Annual Medicare costs for injectable medications and immunosuppression following transplant was $10,000 in 2010 (USRDS 2012). Additionally, there would be significant cost savings by potentially avoiding costs associated with long-term viral and bacterial prophylaxis and treatment of opportunistic infections. Tolerance approaches could be financially advantageous if the cost savings associated with avoiding immunosuppression and chemoprophylaxis increased graft survival, and reduced need for retransplants or dialysis exceeds the additional costs of these procedures.

CONCLUSIONS

It is clear that transplantation improves quality and length of life for patients with end-stage organ diseases. Unfortunately, the vast majority of transplant recipients require life-long immunosuppression. This immunosuppression increases risk for cancer, infection, and other direct organ toxicity. Despite continued use of these medications, functional organ lifespans are finite. There is a persisting risk of immune and nonimmune-mediated graft loss that results in decreased patient survival. Strategies that enable long-lasting graft function obviate need for immunosuppression medication, and maintained immune function could increase patient longevity. The question remains, are the costs of these tolerance-induction protocols worth the benefits of immunosuppression avoidance?

There are many considerations when attempting to answer "Is it worth it?" These considerations may not be the same for all organ systems or even for different recipients. For example, it may be worth the risks of total body irradiation and GVHD present in mixed chimerism protocols for a 50-year-old who gets a heart transplant and may not be a candidate for retransplant should their heart fail at the age of 60 from chronic rejection. On the other hand, a 25-year-old living donor kidney recipient might not want to take those risks if they know that if their graft fails they have the option of dialysis, and perhaps have another potential live donor for that eventuality. The consequences of transplant failure for lifesaving organs (heart, lung, and liver) may make these recipients more suitable for higher-risk tolerance protocols than for organs with alternative therapies (insulin for pancreas transplants and dialysis for kidney transplants). Additional factors may influence the risk/benefit balance. Highly sensitized patients with donor-specific anti-HLA antibodies may take the risk of a tolerance protocol given their likely reduced graft survival compared with an unsensitized recipient receiving a 0-HLA mismatch graft. It may be one thing to ask a patient with hematologic malignancy to undergo a risky combined stem cell and kidney transplant with irradiation and chemotherapy, but a completely different consideration for a relatively healthy person needing a preemptive kidney transplant. Clearly, each tolerance protocol has a set of risks that must be balanced with the long-term benefits. This risk/benefit analysis would need to be individualized to the organ type, disease process, and even individual patient circumstances, and would also depend on the efficacy and costs of the protocol developed.

The last consideration is: What are the consequences of failure? Under current practice, transplant outcomes are pretty good (at least for short- and medium-term outcomes). Each of the tolerance protocols will have some failure rate. Is the long-term benefit in the successes worth the risks of graft loss, patient death, or other complications in the failures?

The evaluation of each tolerance protocol will be influenced by the success rates and the implications of failure. If protocol failure means loss of graft and need for retransplant in a kidney recipient, the risk may be warranted. If failure means loss of graft and death in a heart transplant recipient, the considerations will be different. If failure means resumption of immunosuppression, the down side is low. If protocol failure means graft loss or death, the implications are more considerable. For example, in the MGH series of patients undergoing combined bone marrow and kidney transplant for multiple myeloma, seven out of 10 patients were successfully weaned off of immunosuppression with stable long-term graft function (Markmann and Kawai 2012). However, three out of 10 failed their protocol and lost their grafts. Is a 70% long-term survival rate free of immunosuppression worth the 30% risk of graft loss? Also, what do we mean by "long-term graft survival?" The MGH experience has shown us that success measured at 1 year may lead to failures after five. Each organ, disease, and patient will have differing answers to these questions.

The answer to "Tolerance—Is it worth it?" is clear—"It depends." The up side of a nontoxic protocol that enables long-term graft and patient survival without immunosuppression would be tremendous. But currently the risks are significant. As more results of ongoing and future trials become available, we must continually evaluate what are our successes and what are our failures? The decision of "Is it worth it" depends on the balance of the two.

REFERENCES

*Reference is also in this collection.

Anderson D, Billingham RE, Lampkin GH, Medawar PB. 1951. The use of skin grafting to distinguish between monozygotic and dizygotic twins in cattle. *Heredity* **5:** 379–397.

Ashton-Chess J, Giral M, Brouard S, Soulillou JP. 2007. Spontaneous operational tolerance after immunosuppressive drug withdrawal in clinical renal allotransplantation. *Transplantation* **84:** 1215–1219.

Axelrod DA. 2013. Economic and financial outcomes in transplantation: Whose dime is it anyway? *Curr Opin Organ Transplant* **18:** 222–228.

Baliga P, Chavin KD, Qin L, Woodward J, Lin J, Linsley PS, Bromberg JS. 1994. CTLA4Ig prolongs allograft survival while suppressing cell-mediated immunity. *Transplantation* **58:** 1082–1090.

Bhatia S. 2011. Long-term health impacts of hematopoietic stem cell transplantation inform recommendations for follow-up. *Exp Rev Hematol* **4:** 437–452; quiz 453–454.

Billingham RE, Lampkin GH, Medawar PB, Williams HLL. 1952. Tolerance to homografts, twin diagnosis, and the freemartin condition in cattle. *Heredity* **6:** 201–212.

Billingham RE, Brent L, Medawar PB. 1953. Actively acquired tolerance of foreign cells. *Nature* **172:** 603–606.

Brouard S, Dupont A, Giral M, Louis S, Lair D, Braudeau C, Degauque N, Moizant F, Pallier A, Ruiz C, et al. 2005. Operationally tolerant and minimally immunosuppressed kidney recipients display strongly altered blood T-cell clonal regulation. *Am J Transplant* **5:** 330–340.

Brouard S, Mansfield E, Braud C, Li L, Giral M, Hsieh SC, Baeten D, Zhang M, Ashton-Chess J, Braudeau C, et al. 2007. Identification of a peripheral blood transcriptional biomarker panel associated with operational renal allograft tolerance. *Proc Natl Acad Sci* **104:** 15448–15453.

Brunstein CG, Miller JS, Cao Q, McKenna DH, Hippen KL, Curtsinger J, Defor T, Levine BL, June CH, Rubinstein P, et al. 2011. Infusion of ex vivo expanded T regulatory cells in adults transplanted with umbilical cord blood: Safety profile and detection kinetics. *Blood* **117:** 1061–1070.

Butcher JA, Hariharan S, Adams MB, Johnson CP, Roza AM, Cohen EP. 1999. Renal transplantation for end-stage renal disease following bone marrow transplantation: A report of six cases, with and without immunosuppression. *Clin Transplant* **13:** 330–335.

Butler JA, Roderick P, Mullee M, Mason JC, Peveler RC. 2004. Frequency and impact of nonadherence to immunosuppressants after renal transplantation: A systematic review. *Transplantation* **77:** 769–776.

Carson KR, Focosi D, Major EO, Petrini M, Richey EA, West DP, Bennett CL. 2009. Monoclonal antibody-associated progressive multifocal leucoencephalopathy in patients treated with rituximab, natalizumab, and efalizumab: A review from the Research on Adverse Drug Events and Reports (RADAR) Project. *Lancet Oncol* **10:** 816–824.

* Chapman JR, Webster AC, Wong G. 2013. Cancer in the transplant recipient. *Cold Spring Harb Perspect Med* **3:** a015677.

* Cobbold SP, Waldmann H. 2013. Regulatory cells and transplantation tolerance. *Cold Spring Harb Perspect Med* **3:** a015545.

Comerci GD Jr, Williams TM, Kellie S. 2009. Immune tolerance after total lymphoid irradiation for heart transplantation: Immunosuppressant-free survival for 8 years. *J Heart Lung Transplant* **28:** 743–745.

Couzin J. 2005. Drug discovery. Magnificent obsession. *Science* **307:** 1712–1715.

Dedrick RL, Walicke P, Garovoy M. 2002. Anti-adhesion antibodies efalizumab, a humanized anti-CD11a monoclonal antibody. *Transpl Immunol* **9:** 181–186.

Di Cocco P, Bonanni L, D'Angelo M, Clemente K, Greco S, Rizza V, Mazzotta C, Scelzo C, Famulari A, Pisani F, et al. 2009. Clinical operational tolerance after solid organ transplantation. *Transplant Proc* **41:** 1278–1282.

Di Ianni M, Falzetti F, Carotti A, Terenzi A, Castellino F, Bonifacio E, Del Papa B, Zei T, Ostini RI, Cecchini D, et al. 2011. Tregs prevent GVHD and promote immune reconstitution in HLA-haploidentical transplantation. *Blood* **117:** 3921–3928.

Djamali A, Samaniego M, Muth B, Muehrer R, Hofmann RM, Pirsch J, Howard A, Mourad G, Becker BN. 2006. Medical care of kidney transplant recipients after the first posttransplant year. *Clin J Am Soc Nephrol* **1:** 623–640.

Dolganiuc A, Szabo G. 2008. T cells with regulatory activity in hepatitis C virus infection: What we know and what we don't. *J Leukocyte Biol* **84:** 614–622.

Durrbach A, Pestana JM, Pearson T, Vincenti F, Garcia VD, Campistol J, Rial Mdel C, Florman S, Block A, Di Russo G, et al. 2010. A phase III study of belatacept versus cyclosporine in kidney transplants from extended criteria donors (BENEFIT-EXT study). *Am J Transplant* **10:** 547–557.

Einecke G, Reeve J, Sis B, Mengel M, Hidalgo L, Famulski KS, Matas A, Kasiske B, Kaplan B, Halloran PF. 2010. A molecular classifier for predicting future graft loss in late kidney transplant biopsies. *J Clin Invest* **120:** 1862–1872.

* English K, Wood KJ. 2013. Mesenchymal stromal cells in transplantation rejection and tolerance. *Cold Spring Harb Perspect Med* **3:** a015560.

Feng S, Ekong UD, Lobritto SJ, Demetris AJ, Roberts JP, Rosenthal P, Alonso EM, Philogene MC, Ikle D, Poole KM, et al. 2012. Complete immunosuppression withdrawal and subsequent allograft function among pediatric recipients of parental living donor liver transplants. *JAMA* **307:** 283–293.

First MR, Dhadda S, Croy R, Holman J, Fitzsimmons WE. 2013. New-onset diabetes after transplantation (NODAT): An evaluation of definitions in clinical trials. *Transplantation* **96:** 58–64.

Fishman JA. 2002. Overview: Fungal infections in the transplant patient. *Transpl Infect Dis* **4:** 3–11.

Fishman JA. 2007. Infection in solid-organ transplant recipients. *N Engl J Med* **357:** 2601–2614.

* Fishman JA. 2013a. Opportunistic infections—Coming to the limits of immunosuppression? *Cold Spring Harb Perspect Med* doi: 10.1101/cshperspect.a015669.

Fishman JA. 2013b. Overview: Cytomegalovirus and the herpesviruses in transplantation. *Am J Transplant* **13:** 1–8; quiz 8.

Fredericks EM, Dore-Stites D. 2010. Adherence to immunosuppressants: How can it be improved in adolescent organ transplant recipients? *Curr Opin Organ Transplant* **15:** 614–620.

Fudaba Y, Spitzer TR, Shaffer J, Kawai T, Fehr T, Delmonico F, Preffer F, Tolkoff-Rubin N, Dey BR, Saidman SL, et al. 2006. Myeloma responses and tolerance following combined kidney and nonmyeloablative marrow transplantation: In vivo and in vitro analyses. *Am J Transplant* **6:** 2121–2133.

Gordon KB, Papp KA, Hamilton TK, Walicke PA, Dummer W, Li N, Bresnahan BW, Menter A, Efalizumab Study Group. 2003. Efalizumab for patients with moderate to severe plaque psoriasis: A randomized controlled trial. *JAMA* **290:** 3073–3080.

* Grinyó JM. 2013. Why is organ transplantation clinically important? *Cold Spring Harb Perspect Med* **3:** a014985.

Grohmann U, Orabona C, Fallarino F, Vacca C, Calcinaro F, Falorni A, Candeloro P, Belladonna ML, Bianchi R, Fioretti MC, et al. 2002. CTLA-4Ig regulates tryptophan catabolism in vivo. *Nat Immunol* **3:** 1097–1101.

Halloran PF, Bromberg J, Kaplan B, Vincenti F. 2008. Tolerance versus immunosuppression: A perspective. *Am J Transplant* **8:** 1365–1366.

Hancock WW, Sayegh MH, Zheng XG, Peach R, Linsley PS, Turka LA. 1996. Costimulatory function and expression of CD40 ligand, CD80, and CD86 in vascularized murine cardiac allograft rejection. *Proc Natl Acad Sci* **93:** 13967–13972.

Harrison JH, Merrill JP, Murray JE. 1956. Renal homotransplantation in identical twins. *Surgical Forum* **6:** 432–436.

Hasek M, Hraba T. 1955. Artificial production of immunological tolerance: Immunological effects of experimental embryonal parabiosis. *Nature* **175:** 764–765.

Helg C, Chapuis B, Bolle JF, Morel P, Salomon D, Roux E, Antonioli V, Jeannet M, Leski M. 1994. Renal transplantation without immunosuppression in a host with tolerance induced by allogeneic bone marrow transplantation. *Transplantation* **58:** 1420–1422.

Hofmann B, Tao K, Mai L, West LJ. 2004. Acceptance of related and unrelated cardiac allografts in neonatally tolerized mice is cardio-specific and transferable by regulatory CD4$^+$ T cells. *J Heart Lung Transplant* **23:** 1069–1076.

Hutchinson JA, Brem-Exner BG, Riquelme P, Roelen D, Schulze M, Ivens K, Grabensee B, Witzke O, Philipp T, Renders L, et al. 2008a. A cell-based approach to the minimization of immunosuppression in renal transplantation. *Transpl Int* **21:** 742–754.

Hutchinson JA, Riquelme P, Brem-Exner BG, Schulze M, Matthäi M, Renders L, Kunzendorf U, Geissler EK, Fändrich F. 2008b. Transplant acceptance-inducing cells as an immune-conditioning therapy in renal transplantation. *Transpl Int* **21:** 728–741.

Ildstad ST, Wren SM, Bluestone JA, Barbieri SA, Sachs DH. 1985. Characterization of mixed allogeneic chimeras. Immunocompetence, in vitro reactivity, and genetic specificity of tolerance. *J Exp Med* **162:** 231–244.

Issa NC, Fishman JA. 2009. Infectious complications of antilymphocyte therapies in solid organ transplantation. *Clin Infect Dis* **48:** 772–786.

Jacobsen N, Taaning E, Ladefoged J, Kristensen JK, Pedersen FK. 1994. Tolerance to an HLA-B,DR disparate kidney allograft after bone-marrow transplantation from same donor. *Lancet* **343:** 800.

Josefowicz SZ, Lu LF, Rudensky AY. 2012. Regulatory T cells: Mechanisms of differentiation and function. *Ann Rev Immunol* **30:** 531–564.

Judge TA, Tang A, Turka LA. 1996. Immunosuppression through blockade of CD28:B7-mediated costimulatory signals. *Immunol Res* **15:** 38–49.

Kahwaji J, Bunnapradist S, Hsu JW, Idroos ML, Dudek R. 2011. Cause of death with graft function among renal transplant recipients in an integrated healthcare system. *Transplantation* **91:** 225–230.

Kasiske BL, Snyder JJ, Gilbertson D, Matas AJ. 2003. Diabetes mellitus after kidney transplantation in the United States. *Am J Transplant* **3:** 178–185.

Kasiske BL, Snyder JJ, Gilbertson DT, Wang C. 2004. Cancer after kidney transplantation in the United States. *Am J Transplant* **4:** 905–913.

Kaufman CL, Colson YL, Wren SM, Watkins S, Simmons RL, Ildstad ST. 1994. Phenotypic characterization of a novel bone marrow-derived cell that facilitates engraftment of allogeneic bone marrow stem cells. *Blood* **84:** 2436–2446.

Kawai T, Andrews D, Colvin RB, Sachs DH, Cosimi AB. 2000. Thromboembolic complications after treatment with monoclonal antibody against CD40 ligand. *Nat Med* **6:** 114.

Kawai T, Cosimi AB, Spitzer TR, Tolkoff-Rubin N, Suthanthiran M, Saidman SL, Shaffer J, Preffer FI, Ding R, Sharma V, et al. 2008. HLA-mismatched renal transplantation without maintenance immunosuppression. *N Engl J Med* **358:** 353–361.

Kenyon NS, Chatzipetrou M, Masetti M, Ranuncoli A, Oliveira M, Wagner JL, Kirk AD, Harlan DM, Burkly LC, Ricordi C. 1999. Long-term survival and function of intrahepatic islet allografts in rhesus monkeys treated with humanized anti-CD154. *Proc Natl Acad Sci* **96:** 8132–8137.

Kinnear G, Jones ND, Wood KJ. 2013. Costimulation blockade: Current perspectives and implications for therapy. *Transplantation* **95:** 527–535.

Kirk AD, Burkly LC, Batty DS, Baumgartner RE, Berning JD, Buchanan K, Fechner JH Jr, Germond RL, Kampen RL, Patterson NB, et al. 1999. Treatment with humanized monoclonal antibody against CD154 prevents acute renal allograft rejection in nonhuman primates. *Nature Med* **5:** 686–693.

Koreth J, Matsuoka K, Kim HT, McDonough SM, Bindra B, Alyea EP III, Armand P, Cutler C, Ho VT, Treister NS, et al. 2011. Interleukin-2 and regulatory T cells in graft-versus-host disease. *N Engl J Med* **365:** 2055–2066.

Koulmanda M, Smith RN, Qipo A, Weir G, Auchincloss H, Strom TB. 2006. Prolonged survival of allogeneic islets in cynomolgus monkeys after short-term anti-CD154-based therapy: Nonimmunologic graft failure? *Am J Transplant* **6:** 687–696.

Koulmanda M, Budo E, Bonner-Weir S, Qipo A, Putheti P, Degauque N, Shi H, Fan Z, Flier JS, Auchincloss H Jr, et al. 2007. Modification of adverse inflammation is required to cure new-onset type 1 diabetic hosts. *Proc Natl Acad Sci* **104:** 13074–13079.

Koulmanda M, Qipo A, Fan Z, Smith N, Auchincloss H, Zheng XX, Strom TB . 2012. Prolonged survival of allogeneic islets in cynomolgus monkeys after short-term triple therapy. *Am J Transplant* **12:** 1296–1302.

Ku CC, Murakami M, Sakamoto A, Kappler J, Marrack P. 2000. Control of homeostasis of $CD8^+$ memory T cells by opposing cytokines. *Science* **288:** 675–678.

Larsen CP, Pearson TC. 1997. The CD40 pathway in allograft rejection, acceptance, and tolerance. *Curr Opin Immunol* **9:** 641–647.

Larsen CP, Elwood ET, Alexander DZ, Ritchie SC, Hendrix R, Tucker-Burden C, Cho HR, Aruffo A, Hollenbaugh D, Linsley PS, et al. 1996. Long-term acceptance of skin and cardiac allografts after blocking CD40 and CD28 pathways. *Nature* **381:** 434–438.

Lebwohl M, Tyring SK, Hamilton TK, Toth D, Glazer S, Tawfik NH, Walicke P, Dummer W, Wang X, Garovoy MR, et al. 2003. A novel targeted T-cell modulator, efalizumab, for plaque psoriasis. *N Engl J Med* **349:** 2004–2013.

Lennerling A, Forsberg A. 2012. Self-reported non-adherence and beliefs about medication in a Swedish kidney transplant population. *Open Nurs J* **6:** 41–46.

Lenschow DJ, Zeng Y, Thistlethwaite JR, Montag A, Brady W, Gibson MG, Linsley PS, Bluestone JA. 1992. Long-term survival of xenogeneic pancreatic islet grafts induced by CTLA-4Ig. *Science* **257:** 789–792.

Leventhal J, Abecassis M, Miller J, Gallon L, Ravindra K, Tollerud DJ, King B, Elliott MJ, Herzig G, Herzig R, et al. 2012. Chimerism and tolerance without GVHD or engraftment syndrome in HLA-mismatched combined kidney and hematopoietic stem cell transplantation. *Sci Transl Med* **4:** 124ra28.

Leventhal J, Abecassis M, Miller J, Gallon L, Tollerud D, Elliott MJ, Bozulic LD, Houston C, Sustento-Reodica N, Ildstad ST. 2013. Tolerance induction in HLA disparate living donor kidney transplantation by donor stem cell infusion: Durable chimerism predicts outcome. *Transplantation* **95:** 169–176.

Li XC, Demirci G, Ferrari-Lacraz S, Groves C, Coyle A, Malek TR, Strom TB. 2001. IL-15 and IL-2: A matter of life and death for T cells in vivo. *Nat Med* **7:** 114–118.

Light J, Salomon DR, Diethelm AG, Alexander JW, Hunsicker L, Thistlethwaite R, Reinsmoen N, Stablein DM. 2002. Bone marrow transfusions in cadaver renal allografts: Pilot trials with concurrent controls. *Clin Transplant* **16:** 317–324.

Lin SZ, Chen KJ, Xu ZY, Chen H, Zhou L, Xie HY, Zheng SS. 2013. Prediction of recurrence and survival in hepatocellular carcinoma based on two Cox models mainly determined by $FoxP3^+$ regulatory T cells. *Clin Transplant* 16: 317–324.

LoCascio SA, Morokata T, Chittenden M, Preffer FI, Dombkowski DM, Andreola G, Crisalli K, Kawai T, Saidman SL, Spitzer TR, et al. 2010. Mixed chimerism, lymphocyte recovery, and evidence for early donor-specific unresponsiveness in patients receiving combined kidney and bone marrow transplantation to induce tolerance. *Transplantation* **90:** 1607–1615.

Lowe M, Badell IR, Thompson P, Martin B, Leopardi F, Strobert E, Price AA, Abdulkerim HS, Wang R, Iwakoshi NN, et al. 2012. A novel monoclonal antibody to CD40 prolongs islet allograft survival. *Am J Transplant* **12:** 2079–2087.

Luznik L, O'Donnell PV, Symons HJ, Chen AR, Leffell MS, Zahurak M, Gooley TA, Piantadosi S, Kaup M, Ambinder RF, et al. 2008. HLA-haploidentical bone marrow transplantation for hematologic malignancies using nonmyeloablative conditioning and high-dose, posttransplantation cyclophosphamide. *Biol Blood Marrow Transplant* **14:** 641–650.

* Maltzman JS, Turka LA. 2013. T-cell costimulatory blockade in organ transplantation. *Cold Spring Harb Perspect Med* doi: 10.1101/cshperspect.a015529.

Markmann JF, Kawai T. 2012. The quest for transplantation tolerance: Have we finally sipped from the cup? *Sci Transl Med* **4:** 124fs5.

Medawar PB. 1964. *The Nobel Prize in physiology or medicine 1960.* Elsevier, Amsterdam.

Meier-Kriesche HU, Schold JD, Srinivas TR, Kaplan B. 2004. Lack of improvement in renal allograft survival despite a marked decrease in acute rejection rates over the most recent era. *Am J Transplant* **4:** 378–383.

Moranne O, Maillard N, Fafin C, Thibaudin L, Alamartine E, Mariat C. 2013. Rate of renal graft function decline after one year is a strong predictor of all-cause mortality. *Am J Transplant* **13:** 695–706.

Morton M, Coupes B, Roberts SA, Klapper PE, Byers RJ, Vallely PJ, Ryan K, Picton ML. 2013. Epidemiology of posttransplantation lymphoproliferative disorder in adult renal transplant recipients. *Transplantation* **95:** 470–478.

Nankivell BJ, Borrows RJ, Fung CL, O'Connell PJ, Allen RD, Chapman JR. 2003. The natural history of chronic allograft nephropathy. *N Engl J Med* **349:** 2326–2333.

Neff RT, Hurst FP, Falta EM, Bohen EM, Lentine KL, Dharnidharka VR, Agodoa LY, Jindal RM, Yuan CM, Abbott KC. 2008. Progressive multifocal leukoencephalopathy and use of mycophenolate mofetil after kidney transplantation. *Transplantation* **86:** 1474–1478.

Newell KA, Larsen CP. 2006. Tolerance assays: Measuring the unknown. *Transplantation* **81:** 1503–1509.

Newell KA, Asare A, Kirk AD, Gisler TD, Bourcier K, Suthanthiran M, Burlingham WJ, Marks WH, Sanz I, Lechler RI, et al. 2010. Identification of a B cell signature associated with renal transplant tolerance in humans. *J Clin Invest* **120:** 1836–1847.

Ojo AO, Held PJ, Port FK, Wolfe RA, Leichtman AB, Young EW, Arndorfer J, Christensen L, Merion RM. 2003. Chronic renal failure after transplantation of a nonrenal organ. *N Engl J Med* **349:** 931–940.

Olagne J, Caillard S, Gaub MP, Chenard MP, Moulin B. 2011. Post-transplant lymphoproliferative disorders: Determination of donor/recipient origin in a large cohort of kidney recipients. *Am J Transplant* **11:** 1260–1269.

Organ Procurement Transplantation Network (SRTR) SRoTR. 2012. OPTN/SRTR 2011 Annual Data Report. Department of Health and Human Services, Health Resources and Services Administration, Healthcare Systems Bureau, Division of Transplantation, Washington, DC.

Orlando G, Soker S, Wood K. 2009. Operational tolerance after liver transplantation. *J Hepatol* **50:** 1247–1257.

Orlando G, Hematti P, Stratta RJ, Burke GW III, Di Cocco P, Pisani F, Soker S, Wood K. 2010. Clinical operational tolerance after renal transplantation: Current status and future challenges. *Ann Surg* **252:** 915–928.

Owen RD. 1945. Immunogenetic consequences of vascular anastomoses between bovine twins. *Science* **102:** 400–401.

Page EK, Dar WA, Knechtle SJ. 2012. Tolerogenic therapies in transplantation. *Front Immunol* **3:** 198.

Parker DC, Greiner DL, Phillips NE, Appel MC, Steele AW, Durie FH, Noelle RJ, Mordes JP, Rossini AA. 1995. Survival of mouse pancreatic islet allografts in recipients treated with allogeneic small lymphocytes and antibody to CD40 ligand. *Proc Natl Acad Sci* **92:** 9560–9564.

Pearson TC, Alexander DZ, Winn KJ, Linsley PS, Lowry RP, Larsen CP. 1994. Transplantation tolerance induced by CTLA4-Ig. *Transplantation* **57:** 1701–1706.

Pearson TC, Alexander DZ, Corbascio M, Hendrix R, Ritchie SC, Linsley PS, Faherty D, Larsen CP. 1997. Analysis of the B7 costimulatory pathway in allograft rejection. *Transplantation* **63:** 1463–1469.

Pestana JO, Grinyo JM, Vanrenterghem Y, Becker T, Campistol JM, Florman S, Garcia VD, Kamar N, Lang P, Manfro RC, et al. 2012. Three-year outcomes from BENEFIT-EXT: A phase III study of belatacept versus cyclosporine in recipients of extended criteria donor kidneys. *Am J Transplant* **12:** 630–639.

Posselt AM, Bellin MD, Tavakol M, Szot GL, Frassetto LA, Masharani U, Kerlan RK, Fong L, Vincenti FG, Hering BJ, et al. 2010. Islet transplantation in type 1 diabetics using an immunosuppressive protocol based on the anti-LFA-1 antibody efalizumab. *Am J Transplant* **10:** 1870–1880.

Rabinovitch A, Suarez-Pinzon WL, Shapiro AM, Rajotte RV, Power R. 2002. Combination therapy with sirolimus and interleukin-2 prevents spontaneous and recurrent autoimmune diabetes in NOD mice. *Diabetes* **51:** 638–645.

Rostaing L, Neumayer HH, Reyes-Acevedo R, Bresnahan B, Florman S, Vitko S, Heifets M, Xing J, Thomas D, Vincenti F. 2011. Belatacept-versus cyclosporine-based immunosuppression in renal transplant recipients with preexisting diabetes. *Clin J Am Soc Nephrol* **6:** 2696–2704.

Roussey-Kesler G, Giral M, Moreau A, Subra JF, Legendre C, Noel C, Pillebout E, Brouard S, Soulillou JP. 2006. Clinical operational tolerance after kidney transplantation. *Am J Transplant* **6:** 736–746.

Rüedi E, Sykes M, Ildstad ST, Chester CH, Althage A, Hengartner H, Sachs DH, Zinkernagel RM. 1989. Antiviral T cell competence and restriction specificity of mixed allogeneic (P1 + P2–P1) irradiation chimeras. *Cell Immunol* **121:** 185–195.

Saadoun D, Rosenzwajg M, Joly F, Six A, Carrat F, Thibault V, Sene D, Cacoub P, Klatzmann D. 2011. Regulatory T-cell responses to low-dose interleukin-2 in HCV-induced vasculitis. *N Engl J Med* **365:** 2067–2077.

Sachs DH, Kawai T, Colvin RB, Fishman JA, Sykes M, Cosimi AB. 2008. Response to "Tolerance versus immunosuppression: A perspective." *Am J Transplant* **8:** 1573–1574.

* Sachs DH, Kawai T, Sykes M. 2014. Induction of tolerance through mixed chimerism. *Cold Spring Harb Perspect Med* doi: 10.1101/cshperspect.a015529.

Sagoo P, Perucha E, Sawitzki B, Tomiuk S, Stephens DA, Miqueu P, Chapman S, Craciun L, Sergeant R, Brouard S, et al. 2010. Development of a cross-platform biomarker signature to detect renal transplant tolerance in humans. *J Clin Invest* **120:** 1848–1861.

Sayegh MH, Fine NA, Smith JL, Rennke HG, Milford EL, Tilney NL. 1991. Immunologic tolerance to renal allografts after bone marrow transplants from the same donors. *Ann Intern Med* **114:** 954–955.

Scandling JD, Busque S, Dejbakhsh-Jones S, Benike C, Sarwal M, Millan MT, Shizuru JA, Lowsky R, Engleman EG, Strober S. 2012. Tolerance and withdrawal of immunosuppressive drugs in patients given kidney and hematopoietic cell transplants. *Am J Transplant* **12:** 1133–1145.

Scherer MN, Banas B, Mantouvalou K, Schnitzbauer A, Obed A, Krämer BK, Schlitt HJ. 2007. Current concepts and perspectives of immunosuppression in organ transplantation. *Langenbecks Arch Surg* **392:** 511–523.

Schmidt A, Oberle N, Krammer PH. 2012. Molecular mechanisms of treg-mediated T cell suppression. *Front Immunol* **3:** 51.

Sellers MT, Deierhoi MH, Curtis JJ, Gaston RS, Julian BA, Lanier DC Jr, Diethelm AG. 2001. Tolerance in renal transplantation after allogeneic bone marrow transplantation-6-year follow-up. *Transplantation* **71:** 1681–1683.

Shimizu K, Schonbeck U, Mach F, Libby P, Mitchell RN. 2000. Host CD40 ligand deficiency long-term allograft survival and donor-specific tolerance in mouse cardiac transplantation but does not prevent graft arteriosclerosis. *J Immunol* **165:** 3506–3518.

Sidiropoulos PI, Boumpas DT. 2004. Lessons learned from anti-CD40L treatment in systemic lupus erythematosus patients. *Lupus* **13:** 391–397.

Sorof JM, Koerper MA, Portale AA, Potter D, DeSantes K, Cowan M. 1995. Renal transplantation without chronic immunosuppression after T cell-depleted, HLA-mismatched bone marrow transplantation. *Transplantation* **59:** 1633–1635.

Spitzer TR, Sykes M, Tolkoff-Rubin N, Kawai T, McAfee SL, Dey BR, Ballen K, Delmonico F, Saidman S, Sachs DH, et al. 2011. Long-term follow-up of recipients of combined human leukocyte antigen-matched bone marrow and kidney transplantation for multiple myeloma with end-stage renal disease. *Transplantation* **91:** 672–676.

Stern RS. 2003. A promising step forward in psoriasis therapy. *JAMA* **290:** 3133–3135.

Strober S, Dhillon M, Schubert M, Holm B, Engleman E, Benike C, Hoppe R, Sibley R, Myburgh JA, Collins G, et al. 1989. Acquired immune tolerance to cadaveric renal allografts. A study of three patients treated with total lymphoid irradiation. *N Engl J Med* **321:** 28–33.

Strober S, Benike C, Krishnaswamy S, Engleman EG, Grumet FC. 2000. Clinical transplantation tolerance twelve years after prospective withdrawal of immunosuppressive drugs: Studies of chimerism and anti-donor reactivity. *Transplantation* **69:** 1549–1554.

Sykes M. 2007. Immune tolerance: Mechanisms and application in clinical transplantation. *J Intern Med* **262:** 288–310.

Tang Q, Bluestone JA. 2008. The Foxp3$^+$ regulatory T cell: A jack of all trades, master of regulation. *Nat Immunol* **9:** 239–244.

* Tang Q, Bluestone JA. 2013. Regulatory T cell therapy in transplantation: Moving to the clinic. *Cold Spring Harb Perspect Med* **3:** a015552.

Tang Q, Adams JY, Penaranda C, Melli K, Piaggio E, Sgouroudis E, Piccirillo CA, Salomon BL, Bluestone JA. 2008. Central role of defective interleukin-2 production in the triggering of islet autoimmune destruction. *Immunity* **28:** 687–697.

Tichelli A, Rovó A, Socié G. 2012. Late effects after hematopoietic stem cell transplantation—Critical issues. *Curr Probl Dermatol* **43:** 132–149.

Tichelli A, Rovó A, Passweg J, Schwarze CP, Van Lint MT, Arat M, Socié G, Late Effects Working Party of the European Group for Blood Marrow Transplantation. 2009. Late complications after hematopoietic stem cell transplantation. *Expert Rev Hematol* **2:** 583–601.

Trzonkowski P, Bieniaszewska M, Juscinska J, Dobyszuk A, Krzystyniak A, Marek N, Myśliwska J, Hellmann A. 2009. First-in-man clinical results of the treatment of patients with graft versus host disease with human ex vivo expanded $CD4^+$ $CD25^+$ $CD127^-$ T regulatory cells. *Clin Immunol* **133:** 22–26.

Turgeon NA, Avila JG, Cano JA, Hutchinson JJ, Badell IR, Page AJ, Adams AB, Sears MH, Bowen PH, Kirk AD, et al. 2010. Experience with a novel efalizumab-based immunosuppressive regimen to facilitate single donor islet cell transplantation. *Am J Transplant* **10:** 2082–2091.

USRDS. 2012. USRDS 2012 Annual Data Report: Atlas of chronic kidney disease and end-stage renal disease in the United States. National Institutes of Health, National Institute of Diabetes and Digestive and Kidney Diseases, Bethesda, MD.

Valderhaug TG, Hjelmesaeth J, Hartmann A, Roislien J, Bergrem HA, Leivestad T, Line PD, Jenssen T. 2011. The association of early post-transplant glucose levels with long-term mortality. *Diabetologia* **54:** 1341–1349.

Vincenti F, Mendez R, Pescovitz M, Rajagopalan PR, Wilkinson AH, Butt K, Laskow D, Slakey DP, Lorber MI, Garg JP, et al. 2007. A phase I/II randomized open-label multicenter trial of efalizumab, a humanized anti-CD11a, anti-LFA-1 in renal transplantation. *Am J Transplant* **7:** 1770–1777.

Vincenti F, Charpentier B, Vanrenterghem Y, Rostaing L, Bresnahan B, Darji P, Massari P, Mondragon-Ramirez GA, Agarwal M, Di Russo G, et al. 2010. A phase III study of belatacept-based immunosuppression regimens versus cyclosporine in renal transplant recipients (BENEFIT study). *Am J Transplant* **10:** 535–546.

Weber SC, Uhlenberg B, Raile K, Querfeld U, Muller D. 2011. Polyoma virus-associated progressive multifocal leukoencephalopathy after renal transplantation: Regression following withdrawal of mycophenolate mofetil. *Ped Transplant* **15:** E19–E24.

Winston DJ, Busuttil RW. 2002. Randomized controlled trial of oral itraconazole solution versus intravenous/oral fluconazole for prevention of fungal infections in liver transplant recipients. *Transplantation* **74:** 688–695.

Winston DJ, Emmanouilides C, Busuttil RW. 1995. Infections in liver transplant recipients. *Clin Infect Dis* **21:** 1077–1089; quiz 90–91.

Wolfe RA, Ashby VB, Milford EL, Ojo AO, Ettenger RE, Agodoa LY, Held PJ, Port FK. 1999. Comparison of mortality in all patients on dialysis, patients on dialysis awaiting transplantation, and recipients of a first cadaveric transplant. *N Engl J Med* **341:** 1725–1730.

Zheng XX, Sanchez-Fueyo A, Sho M, Domenig C, Sayegh MH, Strom TB. 2003. Favorably tipping the balance between cytopathic and regulatory T cells to create transplantation tolerance. *Immunity* **19:** 503–514.

Overview of the Indications and Contraindications for Liver Transplantation

Stefan Farkas, Christina Hackl, and Hans Jürgen Schlitt

Department of Surgery, University Medical Center Regensburg, D-93053 Regensburg, Germany

Correspondence: hans.schlitt@klinik.uni-regensburg.de

Liver transplantation is the only definitive treatment option for patients with irrevocable acute or chronic liver failure. In the last four decades, liver transplantation has developed from an experimental approach with a very high mortality to an almost routine procedure with good short- and long-term survival rates. Here, we present an up-to-date overview of the indications and contraindications for liver transplantation. It is shown how the evaluation of a candidate and finally listing for transplantation has to be performed in a multidisciplinary setting. Meticulous listing, timing, and organ allocation are the crucial factors to achieve an optimal outcome for the individual patient on the one hand, and reasonably using the limited deceased donor pool on the other hand. Living-donor liver transplantation is demanding but necessarily increasing. Because patients after liver transplantation need lifelong aftercare, it is important for primary care clinicians to understand the basic medical problems and risks.

Despite significant improvements in the medical management of the complications of liver cirrhosis including hepatocellular carcinoma, liver transplantation (LTx) remains the only definitive treatment option for patients with end-stage liver disease. Significantly improved graft and patient survival rates have been observed over time and, in the last 15 years, are relatively stable, with an overall survival rate of 85% in the first year and ~75% at 5 yr (Kim et al. 2013). However, ~10% of patients listed for LTx die on the waiting list (Kim et al. 2006), and many potential candidates are not listed because of the shortage of deceased donor organs (Fig. 1). Other patients are not candidates because of comorbidities, psychosocial issues, and medical issues like hepatocellular carcinoma exceeding a designated size. Acute liver failure is less common but has an excellent outcome if the patient is transplanted promptly.

HISTORY

The pioneer of human orthotopic liver transplantation, Thomas E. Starzl, learned about experimental auxiliary liver transplant models in dogs while attending a lecture by C. Stuart Welch in 1957 (Starzl 2012). After discussing and refining these canine models, Starzl was the first to attempt an orthotopic liver transplant into a 3-yr-old human recipient suffering from biliary atresia in 1963 (Starzl et al. 1963). The patient did not survive the operation. After several equally unsuccessful attempts, Starzl

Cite this article as *Cold Spring Harb Perspect Med* doi: 10.1101/cshperspect.a015602

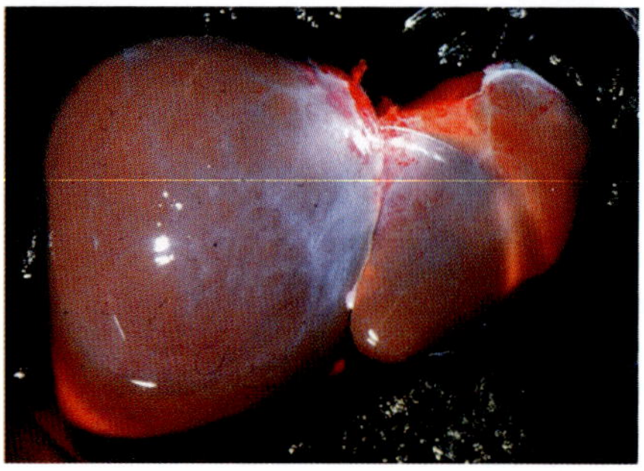

Figure 1. Liver of a deceased organ donor.

succeeded in performing an orthotopic liver transplant into a patient diagnosed with hepatoblastoma in 1967 (Starzl et al. 1968). This patient survived for 18 mo before dying from metastatic disease. During the following years, major breakthroughs such as the expansion of the organ donor pool by introduction of the brain-dead criteria in 1968 (Shapiro 1968), refined surgical technique, and, last but not least, introduction of new immunosuppressive medication such as cyclosporine in 1979 (Starzl et al. 1981a,b) led to a significant increase in liver transplantation. In 1983, the NIH declared that liver transplantation was a valid therapy for end-stage liver disease (National Institutes of Health 1984), and, a few years subsequently, the United Network for Organ Sharing (UNOS) was founded (United States Congress 1984). Already in 1967, the Eurotransplant (ET) International Foundation was founded in Leiden, The Netherlands. In 1988, Rudolf Pichlmayr was the first to perform a split-liver transplantation, offering one liver to two recipients (Pichlmayr et al. 1988). In 1988 and 1989, living-donor liver donation was successfully introduced in the adult-to-pediatric and adult-to-adult settings (Fig. 2) (Broelsch et al. 1990). As of today, approximately 6200 and 1700 liver transplants are performed each year within the UNOS and ET networks, respectively.

INDICATIONS FOR LIVER TRANSPLANTATION

Indications for liver transplantation are manifold and can be classified into end-stage liver disease, acute liver failure, and certain benign and malignant liver tumors. An overview is given in Table 1. Liver transplantation should be considered for any patient in whom anticipated overall survival exceeds life expectancy of the underlying disease or where a significant increase in quality of life can be achieved. Within the UNOS and ET networks, liver cirrhosis caused by chronic viral hepatitis and alcohol abuse is the major cause for end-stage liver disease, accounting for ~70% of liver transplantations (Zakhari 2013).

Liver transplantation for malignant disease is a medical and ethical challenge with regard to long-term oncological outcome under immunosuppressive therapy and allocation justice because of organ shortage. Childhood hepatoblastoma (Meyers et al. 2012), epitheloid hemangioendothelioma (Grossman and Millis 2010), and limited hepatocellular carcinoma (HCC) (Dhir et al. 2012; Lim et al. 2012), within the Milan criteria (one lesion ≤5 cm, or two to three lesions each ≤3 cm, no extrahepatic lesion and no vascular invasion), are standard indications for liver transplantation. Patients diagnosed with HCC—often presenting in good clinical condition—therefore receive additional MELD (model of end-stage liver disease) points according to HCC tumor size and waiting time to enable transplantation before the tumor exceeds the Milan criteria (Earl and Chapman 2013). Surgical resection and therapeutic interventions to control HCC progress during the waiting period as well as liver transplantation for patients exceeding the Milan criteria are a

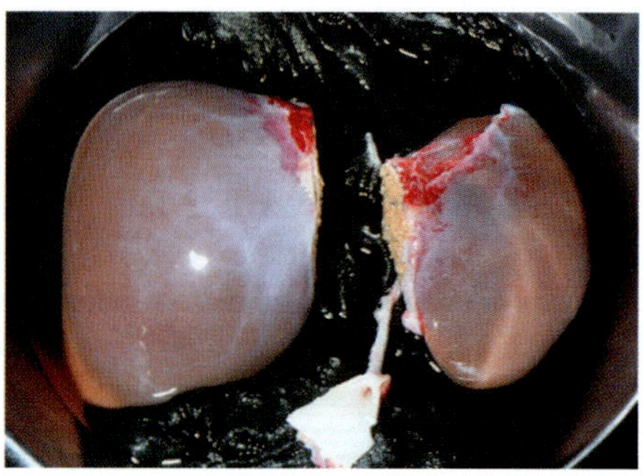

Figure 2. Splitting of liver for two recipients: one adult and one pediatric.

Table 1. Indications for liver transplantation

Acute liver failure
Hepatitis A/B
Intoxication (e.g., acetaminophen, death cap)
Wilson's disease
Budd–Chiari syndrome
Chronic liver failure: Noncholestatic cirrhosis
Hepatitis B/C
Autoimmune hepatitis
Alcohol-induced cirrhosis
Chronic liver failure: Cholestatic cirrhosis
Primary biliary cirrhosis (PBC)
Primary sclerosing cholangitis (PSC)
Secondary biliary cirrhosis
Chronic liver failure: Metabolic
Wilson's disease
Hemochromatosis
α-1 Antitrypsin deficiency
Amyloidosis
Cystic fibrosis
Tyrosinemia
Chronic liver failure: Vascular
Budd–Chiari syndrome
Other indications
Primary oxalosis
Gycogen storage diseases
Hyperlipidemia
Polycystic liver disease
Malignant disease
Hepatocellular carcinoma (HCC, within Milan criteria)
Fibrolamellar carcinoma (FLC)
Hepatoblastoma
Epitheloid hemangioendothelioma
Cholangiocellular adenocarcinoma
Neuroendocrine liver metastases
Benign liver tumors
Adenomatosis
Liver transplantation in pediatric patients
Biliary atresia
Byler's disease
Alagille's syndrome
Neonatal hepatitis/neonatal viral hepatitis
Autoimmune hepatitis
Hepatoblastoma

focus of ongoing clinical research (Yao 2008). Living-donor liver transplantation may here offer a treatment option for selected HCC patients to minimize waiting time or enable liver transplantation in tumors exceeding the Milan criteria (Grant et al. 2013). In contrast, patients diagnosed with intrahepatic cholangiocellular adenocarcinoma (CCA) have shown poor long-term overall survival after liver transplantation and are not transplant candidates (Ali et al. 2011). However, recent clinical trials by the Mayo Clinic evaluating a multimodality treatment concept for CCA combining neoadjuvant radiochemotherapy and liver transplantation have established CCA as an indication for liver transplantation in selected patients with unresectable hilar CCA or CCA arising in patients with primary sclerosing cholangitis (Rosen et al. 2010; Darwish Murad et al. 2012). Clinical trials evaluating liver transplantation for selected patients with neuroendocrine hepatic metastases have shown long-term graft and patient survival comparable with patients transplanted for HCC (Gedaly et al. 2011). In individual cases, patients with neuroendocrine liver metastases thus are eligible for liver transplantation. Extrahepatic malignancies as well as hepatic metastases from non-neuroendocrine tumors so far remain absolute contraindications for liver transplantation. However, a recent Norwegian pilot study evaluating liver transplantation for unresectable colorectal liver metastases showed that overall survival exceeds by far the reported outcome for chemotherapy, is comparable with overall survival after liver resection for resectable colorectal liver metastases, and is comparable with overall survival after repeat liver transplantation for nonmalignant diseases (Hagness et al. 2013). Thus, ongoing improvements in multimodality cancer therapy may in the future widen indications for liver transplantation in malignant disease.

CONTRAINDICATIONS FOR LIVER TRANSPLANTATION

Absolute and relative contraindications for liver transplantation are shown in Table 2. Abstinence from alcohol and drug abuse for a minimum of

Table 2. Contraindications for liver transplantation

Absolute contraindication	Relative contraindications
Active alcohol abuse	Psychosocial conditions
Uncontrolled systemic infections	Advanced age
Uncontrolled extrahepatic malignancy	Severe hepatopulmonary or severe hepatorenal syndrome
Uncontrolled/limiting medical conditions	Severe obesity/ malnutrition

6 mo is required in the UNOS, ET, and other transplant programs. This mandatory duration of abstinence is a matter of debate because the 6-mo threshold has shown to be insufficient for predicting long-term graft and patient survival (Rice and Lucey 2013). Although uncontrolled systemic infections, which exclude patient survival under immunosuppression, and AIDS-defining symptoms in HIV patients are absolute contraindications, the possibility of performing a liver transplantation in patients diagnosed with infections such as HBV and HIV and controllable local infections must be assessed for each individual patient (Grossi 2003; Tavio et al. 2011; Campsen et al. 2013). Further absolute contraindications for liver transplantation are life-limiting medical conditions such as advanced cadiovascular, pulmonary, or neurologic disorders; intrahepatic CCA; hepatic metastases other than neuroendocrine metastases in selected patients; and extrahepatic malignancy (see also Lakkis and Lechler 2013).

Patients diagnosed with HCC exceeding the Milan criteria can still be candidates for liver transplantation, depending on local or national allocation guidelines (Yao 2008; Grant et al. 2013). In cancer survivors with complete remission, a tumor-free survival of 2–5 yr, depending on the type of malignancy, is required before listing for liver transplantation (Bachir and Larson 2012).

Relative contraindications may be psychosocial conditions resulting in poor compliance, advanced age, and severe hepatopulmonary or hepatorenal syndrome that may not be cured or improved after liver transplantation, as well as severe obesity or severe malnutrition. Here, the indication must be assessed individually for each patient.

EVALUATION

From a transplant-physician viewpoint, patients should be referred for consideration for liver transplantation as early as possible. Referral does not automatically mean listing of the patient, but allows thorough assessment of candidacy for transplant before listing. If the interdisciplinary decision is made that the patient is eventually a candidate for listing, an extensive patient evaluation is performed (Table 3).

After exclusion of absolute contraindications and discussion of relative contraindications, several medical, psychosocial, and ethical questions have to be answered:

- Can the patient survive the operation and postoperative period?
- What gain of lifetime and what quality of life will the patient have after transplantation?
- Will the patient be compliant regarding the medical regimen?
- In patients with alcoholic liver disease and/or a history of drug abuse, besides being abstinent for at least 6 mo, what is the chance of the patient staying abstinent lifelong?
- Psychosocial issues: Do psychological disorders or lack of social support compromise long-term outcome?
- Is living donation an option?

Some comorbidities are discovered only at the evaluation process but can also be corrected before transplantation. As the transplant patient age has increased throughout the years, cardiovascular disease is common and has to be treated before listing. An increasing number of patients in Western countries suffer from morbid obesity and have to be critically evaluated concerning perioperative risk and probable need of weight loss with or without bariatric surgery. Up to one-third of patients with cirrhosis develop diabetes. These patients require good metabolic control before listing because the re-

Table 3. Evaluation for liver transplantation

General assessments
Previous medical/surgical history
Physical examination
Vaccinations: Hepatitis A/B, tetanus, diphtheria, polio
Psychosocial assessment by psychiatrist
Assessment by hepatologist
Consult with transplant surgeon
Blood tests
ABO and Rhesus blood group, full blood count, electrolytes, BUN, creatinine, creatinine-clearance, coagulation (INR, APTT, factor II, factor V), liver function tests, glucose, ferritine, transferrine, protein electrophoresis, thyroid hormones, tumor marker (AFP, CA19-9, CEA, CA125 in females; PSA in males), isotypes of antibodies (IgA, IgG, IgM, ANA, AMA)
Urinalysis and microscopy
Microbiology/virology
Hepatitis serology (Hepatitis A/B/C/D), CMV IgG/IgM, EBV IgG/IgM, VZV IgG/IgM, HIV, tuberculosis
Screening for infections/tumors
Maxillofascial/dental examination
ENT examination
Dermatological examination
Gynecology/urology examination
Gastroscopy
Coloscopy in patients at risk or >40 yr
Preparation for surgery
Pulmonary function test and blood gas analysis
EKG and echocardiography
Cardiac catheterization in patients at risk or >65 yr
Assessment by transplant anesthetist
Radiologic assessment
X ray of paranasal sinuses
Thoracic and abdominal three-phase angio-CT scan
Mammography in female patients at risk or >40 yr
Duplex-sonography of the liver
Final assessment
Interdisciplinary review by transplant board (hepatologist, transplant surgeon, transplant anesthetist, psychiatrist)

quired immunosuppression itself increases the risk of and probably worsens preexisting diabetes mellitus (Prokai et al. 2012).

Portopulmonary hypertension and hepatopulmonary syndrome may be reversed by liver transplantation. However, uncontrolled portopulmonary hypertension during the transplantation procedure may cause severe anesthesia problems and severe damage to the graft due to elevated central venous pressure.

Hepatorenal syndrome may also be reversed after transplantation, but a simultaneous listing for liver and kidney may be considered if persistence of chronic kidney disease is suspected (Angeli and Gines 2012). Because the results of a simultaneous liver and kidney transplantation are critical (Papafragkakis et al. 2010), with an increased loss of kidneys in the Eurotransplant area, the following solution was found. Patients listed simultaneously for liver and kidney can be first transplanted with a liver alone and receive extra points on the kidney waiting list. If they are still in need of a kidney transplant (clearance is <15 mL/min) 90–360 d after liver transplantation, they have the chance to receive a kidney transplant within 3 mo. Thus, loss of kidneys during simultaneous transplantation, early death with a functioning kidney graft, and unnecessary kidney transplantations should be prevented. This approach has to be evaluated during the next years.

The listing criteria by the UNOS state that LTx candidates can be listed for kidney transplantation too if they have a documented stage 4–5 chronic kidney disease, acute kidney injury with a glomerular filtration rate lower than 25 mL/min continuously for 6 wk, or a metabolic disease such as hyperoxaluria.

Patients with HCC who are eligible for an exceptional MELD (eMELD) have to be evaluated concerning the specificity of the lesion and the size and number of tumors and vascular infiltration in the liver. Furthermore, extrahepatic spread has to be excluded. For that purpose, imaging with MRI and CT and—for specificity of the typical arterial hyperperfusion in the HCC lesion—contrast enhanced ultrasound (CEUS) is recommended. Thus, it can be evaluated if the patient fits, for example, in the Milan criteria or the UNOS criteria for listing with exceptional MELD. These patients should be also evaluated concerning bridging strategies until the expected time of transplantation. Here interventional radiologists and oncological liver surgeons should be included.

LISTING AND TIMING AND ALLOCATION OF LTX

Listing

After extensive evaluation, the final listing of the patient is an interdisciplinary process. In the liver transplantation board, transplant surgeons, transplant hepatologists, anesthesiologists, radiologists, transplant coordinators, and psychiatrists should be included. All participate in the decision of listing the patient formally for transplantation (Dawwas and Gimson 2009).

Timing and Allocation

Timing is crucial for the success of liver transplantation. On the one hand, best results are achieved if the patient is not decompensated and still in a good general condition. On the other hand, the decompensated and sickest patients are the ones who most urgently need transplantation—but have the worst outcome. Owing to organ shortage, different allocation solutions are in use but are currently intensively discussed. A model for the sickest-first policy, MELD, was implemented in the allocation in the UNOS area in 2002 and in the Eurotransplant area in 2007. It consists of serum creatinine, international normalized ratio (INR), and bilirubin (patient-based allocation). The MELD calculator can be found, for example, on the following website (www.mayoclinic.org/meld/mayomodel6.htm). The MELD was originally developed to predict 3-mo survival after transjugular intrahepatic portosystemic shunt replacement (Malinchoc et al. 2000; Said et al. 2004). Since implementation of the MELD system, waiting list mortality has declined. However, patients with very high laboratory MELD scores (>35) are normally ICU-bound, on dialysis, and often require vasopressor support and artificial ventilation. If these patients are transplanted and not temporarily delisted until recovery, the 1-yr mortality is very high. Studies are ongoing to assess whether the addition of parameters that are associated with poor patient outcome (serum sodium, renal failure, and serum albumin) will improve the predictive ability of MELD (Biggins et al. 2005; Kim et al. 2008).

"Center-based allocation" is in use especially in countries with few transplant programs, for example, in Australia, the United Kingdom, and Austria. Moreover, it is used parallel to the MELD system for extended criteria donors and for recipients who are stable but not well represented in the MELD system. The advantage of center-based allocation is that the physicians can match the organ to the patient, which results in a relatively good outcome although extended criteria organs are used.

"High-urgency allocation" is generally only possible in case of acute liver failure (see above) or for retransplantation.

DECEASED-DONOR LIVER TRANSPLANTATION (DDLT)

The deceased-donor pool is limited. Thus, in the last decade, different strategies have been implemented to increase the pool of deceased liver donation and transplantation. Donation after brain death is widely accepted and well established. Donation after cardiac death (DCD) is well established, for example, in the UNOS area, but still controversial in Europe. The results after DCD transplantation are not as good compared with donation after brain death; however, it seems to be a feasible source to increase the donor pool. In addition, older donors are increasingly accepted, because in some countries the majority of donors are >50 yr. Absolute contraindications for deceased organ donation are cancer and uncontrolled infections (e.g., hepatitis). Cardiac arrest, hypotension, high sodium, and liver steatosis are not absolute contraindications. Macroscopic evaluation of the liver at time of retrieval (before and after perfusion with cold storage solution) is very important to assess organ quality. Frozen sections may not always be available at the donation site and are not optimal to evaluate the degree of steatosis. Extension of the donor pool by use of marginal grafts may result in higher complication rates such as early graft failure, biliary complications (ischemic type biliary lesions), and need for retransplantation (Busuttil and Tanaka 2003).

Another technique to increase the donor pool is split-liver transplantation. Optimal grafts can be split into an extended right lobe for an adult recipient and a left lateral lobe for a pediatric recipient (Fig. 3). Also, so-called full splits (right and full left lobe for two adult recipients) are used in some cases (Gundlach et al. 2000). These techniques—if used in experienced centers—do have good long-term results but are often limited by logistic and thus ischemic time problems if the split is allocated to two different centers.

LIVING-DONOR LIVER TRANSPLANTATION (LDLT)

Living-donor liver transplantation (LDLT) is an established method with increasing numbers worldwide. In Asian countries, close to 90% of liver transplantations are from living donors because of social and religious factors. In western countries and especially in the UNOS area, some recent donor deaths led to a decline in LDLT numbers. The advantage of LDLT is the use of an optimal healthy donor, minimal ischemic time, elective surgery, and timing of transplantation owing to the recipient's need and medical stability and not to deceased organ availability. A further advantage of LDLT is the possibility of ABO-incompatible transplantation (Song et al. 2013). However, living donation is not without risk for the healthy donor, and LDLT is surgically more demanding than whole-organ transplantation. The remaining liver in the donor regenerates within 3 mo to 90% of its original volume. The donor has a risk of a 30% morbidity and a mortality risk of up to 0.8% (Ghobrial et al. 2008). Especially, the bile duct system has to be evaluated intraoperatively by cholangiogram, and aberrant bile ducts have to be taken care of meticulously (Jeon et al. 2013). In addition, the venous outflow in the recipient is crucial, and

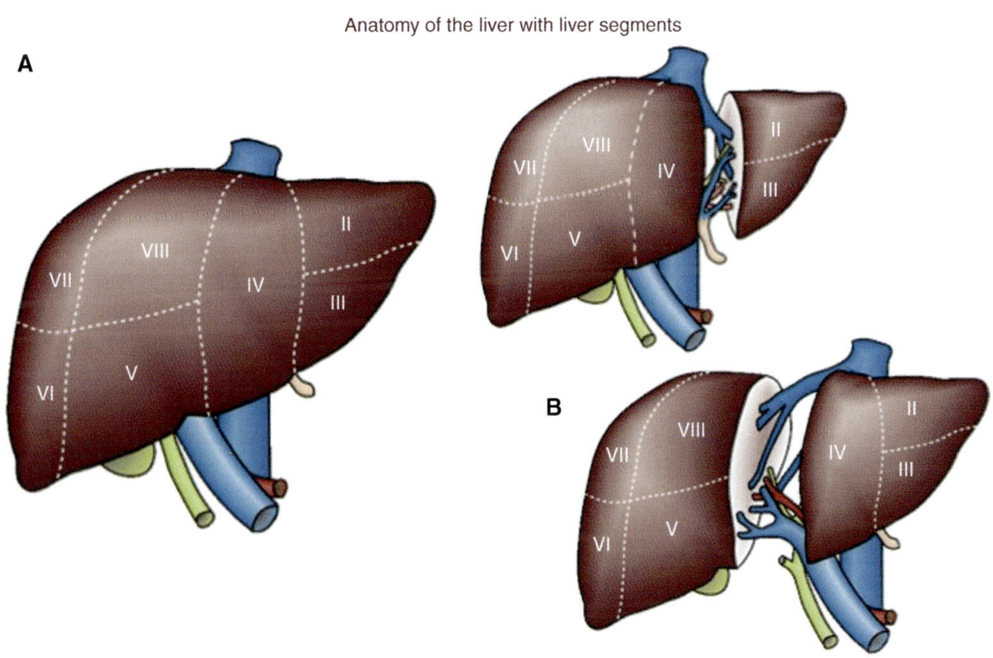

Figure 3. Anatomy of the liver with liver segments (seg). (*A*) Left lateral splitting or living donation of seg II/III for a pediatric recipient. (*B*) Living donation of the right lobe (seg V, VI, VII, VIII) for an adult recipient.

Table 4. Supplemental evaluation of the donor for living donor liver transplantation

Blood tests	
Fibrinogen, AT III, protein C, protein S, factors VII and VIII, CRP, triglycerides, cholesterol, LDL, HDL	
Virology	
Anti-HAV, HbsAg, anti-HBc, HBV-DNA, anti-HBe, anti-HBs, anti-HCV, HCV-RNA	
Other	
MRSA-screening, stress-EKG, CT-volumetry of liver segments to be donated, and remaining liver	
Liver biopsy in patients with elevated GGT or BMI >30	
Final discussion in ethics board confirming informed consent and voluntariness of liver donation.	

branches of the middle vein may have to be reconstructed. Moreover, in adult LDLT, small-for-size syndrome is sometimes a problem. Thus, a precise imaging evaluation with three-phase CT angiography and volume measurement of the future liver remnant of the donor as well as of the graft size have to be performed and calculated with the patient body weight. For all LDLT, careful selection and extensive evaluation of the donor are very important (Table 4) (Berg et al. 2007). All ethical, legal, and insurance aspects have to be cleared with the authorities and well documented. Altogether, a risk adjustment for the donor and the recipient has to be performed multidisciplinarily. The risk of the recipient must equal or exceed the risk of the donor (Thuluvath and Yoo 2004).

LIVER TRANSPLANTATION—TECHNIQUES

The transplant operation itself has been significantly standardized and optimized during the last two decades. Nowadays liver transplantation in a stable donor with a standard donor organ is a routine operation and may be performed without any blood transfusion and only few plasma replacements. However, the standards developed for liver transplantation after deceased donation are different based on surgical training.

First, the native liver has to be explanted via right angular incision. Here, two different methods are used: replacement of the inferior vena cava (IVC) or so-called piggyback, in which the inferior vena cava is preserved. The clamping of the IVC needed for resection can decrease blood pressure and can cause cardiac problems difficult to manage for the anesthetist. Moreover, renal insufficiency may be a problem because of clamping of the IVC causing reduced perfusion of the kidneys. Recently, in a Cochrane database analysis, the piggyback technique versus standard technique was reviewed, and they found two randomized trials with a total of 106 patients (Gurusamy et al. 2011b). Warm ischemic time was shorter in the piggyback method because only one caval anastomosis had to be performed. However, the proportion of patients who developed chest complications was significantly higher with the piggyback method. There was no significant difference in postoperative death, primary graft nonfunction, complications related to the blood vessels, kidney failure, blood transfusion requirements, or ICU stay or hospital stay between the two groups. In our institution, we prefer replacement of the IVC because we see a substantial shortening of operation time.

To reduce the effects of clamping, the IVC and also the portal vein—which may cause mesenterial congestion—veno-venous bypass was once widely used and has now developed a revival in some centers. There are different techniques described for the bypass: open or percutaneous, heparin-coated or no heparin-coated. This was reviewed as well in a Cochrane database analysis, and no difference regarding renal failure or blood transfusions was found between the groups in two randomized trials (Gurusamy et al. 2011a). The operating time was significantly shorter in the percutaneous bypass group. To aim at a short operating time, we perform no veno-venous bypass at all at our center.

Another point of discussion is the technique of flushing and reperfusion for liver transplantation. Marginal organs are prone to ischemic time biliary lesions (ITBL) due to reduced flow in the pericanalicular arteries of the bile ducts. Here some studies hint that the way and sequence of reperfusion influence the rate of ITBL (Heidenhain et al. 2006; Farid et al. 2011). The Cochrane database analysis searched for comparisons that included initial hepatic artery flush versus initial portal vein flush or simultaneous flushing and different types of blood venting or use of perfusion fluid (Gurusamy et al. 2012). There was no significant difference in mortality, graft survival, or severe morbidity rates in any of the comparisons.

Bile duct reconstruction can be performed with a duct-to-duct or duct-to-small-bowel anastomosis. Duct-to-small-bowel is recommended for patients with PSC. Placing a T-tube or not in the bile duct remains controversial (Gastaca et al. 2013). The bile duct system and the common bile duct remain the Achilles heel of transplantation, because they are most sensitive to early ischemic injury already during cold storage. We developed a histological bile duct risk score that predicts biliary complications and may help decision making in LTx (Brunner et al. 2013).

POSTTRANSPLANT CARE

Lifelong aftercare is crucial for long-term graft and patient survival after liver transplantation. During the early postoperative phase, daily blood tests are necessary for surveillance of liver function, coagulation, electrolytes, and target blood levels of immunosuppressive drugs. Factor VII, being produced by the liver and having a half-life of 6 h, is an excellent parameter to assess liver function on a daily basis. Prophylactic antibiotic therapy is given for 3 d perioperatively. Prophylactic treatment for CMV and PCP infections should be given according to the donor/recipient risk profile (Lautenschlager 2009). Regular duplex-sonography is performed to assess liver perfusion.

Daily immunosuppressive therapy is mandatory to prevent organ rejection. Especially in the early posttransplant phase, immunosuppressive therapy consists of complex combinations of various drugs and needs to be adapted for each patient individually. Components are antilymphocyte antibodies, steroids, calcineurin inhibitors, and inhibitors of B- and T-cell proliferation (Scherer et al. 2007). Standard immunosuppressive combination regimens are basiliximab, MMF, cyclosporine, and prednisolone or basiliximab, MMF, sirolimus/everolimus, and prednisolone (Farkas et al. 2009). The chimeric monoclonal T-cell IL-2-receptor antibody basiliximab is given on d 0 and 4 after liver transplant for induction therapy. Mycophenolate mofetil (MMF), a reversible inhibitor of inosine monophosphate dehydrogenase in purine synthesis, reduces proliferation of B and T cells (Schlitt et al. 2013). Calcineurin inhibitors such as cyclosporine inhibit T-cell production and excretion of IL-2. mTor inhibitors such as sirolimus and everolimus also inhibit the proliferation of B and T cells. However, in contrast to calcineurin inhibitors, mTor inhibitors show no renal toxicity (Schnitzbauer et al. 2010). The main side effects are bone marrow toxicity, inhibition of wound healing, and reduction of liver function. On the other hand, mTor inhibitors lower cancer risk in selected patients and therefore are recommended especially for patients transplanted for malignant disease (Law 2005). Steroids inhibit T-cell activation and block IL-1 and IL-2 synthesis. Steroids are the backbone of all immunosuppressive regimens and are given already before reperfusion of the transplanted organ intraoperatively. In the early postoperative phase, steroids are given in high doses but should be gradually reduced. In many patients, steroids can be tapered 6 mo after transplantation (Shapiro 2004), and long-term immunosuppressive therapy often can consist of monotherapy. Because immunosuppressive drugs can interact with many other medications and dietary components, target levels must be checked lifelong on a regular basis, and interactions must be considered when new medications must be introduced.

Although patients must be closely monitored by their transplant center in the early postoperative phase, long-term lifelong aftercare can also, at least in part, be conducted by general practitioners and primary care hospitals. The main focuses of aftercare must be screening for complications and side effects of immunosuppressive therapy, recognition and treatment of acute or chronic graft rejection, recognition and treatment of biliary complications, malignant disease, and recurrence of the primary liver disease. Details are shown in Table 2. Especially, prevention, recognition, and treatment of infections including opportunistic infections are crucial for long-term patient survival because patients under immunosuppressive therapy often do not show typical clinical symptoms or leukocytosis. Infections can worsen rapidly to septic conditions, and infections present the leading cause of mortality after liver transplantation (Fishman and Rubin 1998; Torbenson et al. 1998).

For prevention of malignant disease, patients must be advised to use high sun-protection-factor sunscreen and to quit cigarette smoking. Furthermore, preventive screening examinations should be performed on a frequent basis (see Table 5) (Watt et al. 2009; Chak and Saab 2010). In patients transplanted for hepatitis B/C, PBC, PSC, autoimmune hepatitis, and hemochromatosis, screening for disease recurrence must be performed. Administration of hepatitis B immune globulin during transplant surgery and at regular intervals, in combination with antiviral therapy, can prevent HBC recurrence (Cholongitas et al. 2011). In contrast, treatment with pegylated interferone monotherapy or combined with ribavirin, owing to liver toxicity, is not started until HCV recurrence is proven (Gurusamy et al. 2010). Patients transplanted for alcoholic liver disease must stay sober after liver transplantation, and also patients transplanted for other indications should cease alcohol consumption.

Table 5. Overview of posttransplant aftercare

General assessments
Physical examination
Dental examination
Vaccinations: influenza, pneumonia, hepatitis A/B; live virus vaccinations should be avoided.
Screening for nonmalignant disease and immunosuppressant side effects
Hypertension
Renal dysfunction
Diabetes
Cardiovascular disease
Infection/opportunistic infection
Screening for malignant disease
Annual skin cancer screening
Annual gynecology and mammography screening in women
Annual urologic and PSA screening in men
Colonoscopy at regular intervals (depending on risk factor)
Annual abdominal and pelvic ultrasound
In some centers: Annual chest + abdominal CT/MRI
Screening for recurrence of primary liver disease
Hepatitis B/C, PBC, PSC, autoimmune hepatitis, hemochromatosis, malignancy

FUTURE PERSPECTIVES OF LIVER TRANSPLANTATION

The scarcity of deceased donor organs will remain one of the main problems for patients on the waiting list for liver transplantation. In addition, patients with end-stage liver disease are affected who cannot enter the waiting list at all because of strict allocation rules made to counterbalance the lack of organs. Extension of the deceased donor pool is essential but will reach a limit. Living-donation liver transplantation, which is standard procedure in Asian countries, will have to increase also in Western countries to cover the need for lifesaving organs.

Future immunosuppressive strategies in liver transplantation have to imply three main goals:

- Reduction of side effects like renal insufficiency,
- reduction of cancer recurrence of hepatocellular carcinoma and de novo cancer after transplantation, and
- finally, and optimally, induction of tolerance.

For that purpose, mTOR inhibitors are very interesting because they block the central pathway for vital aspects of tumor development, including angiogenesis and cell growth. mTOR inhibitors have anticancer activities, which may prove critical in the fight against high cancer recurrence and de novo cancer. They furthermore provide the capacity to interfere with fibrotic processes that often accompany transplant rejection and to influence the preferential development of immunological tolerance. Studies are ongoing that try to induce tolerance by either stem cell therapy (Dahlke et al. 2009; Dillmann et al. 2012) or by transfusion of regulating cells in the setting of living donation (www.onestudy.org).

REFERENCES

**Reference is also in this collection.*

Ali JM, Bonomo L, Brais R, Griffiths WJ, Lomas DJ, Huguet EL, Praseedom RK, Jamieson NV, Jah A. 2011. Outcomes and diagnostic challenges posed by incidental cholangiocarcinoma after liver transplantation. *Transplantation* **91:** 1392–1397.

Angeli P, Gines P. 2012. Hepatorenal syndrome, MELD score and liver transplantation: An evolving issue with relevant implications for clinical practice. *J Hepatol* **57:** 1135–1140.

Bachir NM, Larson AM. 2012. Adult liver transplantation in the United States. *Am J Med Sci* **343:** 462–469.

Berg CL, Gillespie BW, Merion RM, Brown RS Jr, Abecassis MM, Trotter JF, Fisher RA, Freise CE, Ghobrial RM, Shaked A, et al. 2007. Improvement in survival associated with adult-to-adult living donor liver transplantation. *Gastroenterology* **133:** 1806–1813.

Biggins SW, Rodriguez HJ, Bacchetti P, Bass NM, Roberts JP, Terrault NA. 2005. Serum sodium predicts mortality in patients listed for liver transplantation. *Hepatology* **41:** 32–39.

Broelsch CE, Emond JC, Whitington PF, Thistlethwaite JR, Baker AL, Lichtor JL. 1990. Application of reduced-size liver transplants as split grafts, auxiliary orthotopic grafts, and living related segmental transplants. *Ann Surg* **212:** 368–375.

Brunner SM, Junger H, Ruemmele P, Schnitzbauer AA, Doenecke A, Kirchner GI, Farkas SA, Loss M, Scherer MN, Schlitt HJ, et al. 2013. Bile duct damage after cold storage of deceased donor livers predicts biliary complications after liver transplantation. *J Hepatol* **58:** 1133–1139.

Busuttil RW, Tanaka K. 2003. The utility of marginal donors in liver transplantation. *Liver Transpl* **9:** 651–663.

Campsen J, Zimmerman M, Trotter J, Hong J, Freise C, Brown R, Cameron A, Ghobrial M, Kam I, Busuttil R, et al. 2013. Liver transplantation for hepatitis B liver disease and concomitant hepatocellular carcinoma in the United States with hepatitis B immunoglobulin and nucleoside/nucleotide analogues. *Liver Transpl* doi: 10.1002/lt.23703.

Chak E, Saab S. 2010. Risk factors and incidence of de novo malignancy in liver transplant recipients: A systematic review. *Liver Int* **30:** 1247–1258.

Cholongitas E, Goulis J, Akriviadis E, Papatheodoridis GV. 2011. Hepatitis B immunoglobulin and/or nucleos(t)ide analogues for prophylaxis against hepatitis B virus recurrence after liver transplantation: A systematic review. *Liver Transpl* **17:** 1176–1190.

Dahlke MH, Hoogduijn M, Eggenhofer E, Popp FC, Renner P, Slowik P, Rosenauer A, Piso P, Geissler EK, Lange C, et al. 2009. Toward MSC in solid organ transplantation: 2008 position paper of the MISOT study group. *Transplantation* **88:** 614–619.

Darwish Murad S, Kim WR, Therneau T, Gores GJ, Rosen CB, Martenson JA, Alberts SR, Heimbach JK. 2012. Predictors of pretransplant dropout and posttransplant recurrence in patients with perihilar cholangiocarcinoma. *Hepatology* **56:** 972–981.

Dawwas MF, Gimson AE. 2009. Candidate selection and organ allocation in liver transplantation. *Semin Liver Dis* **29:** 40–52.

Dhir M, Lyden ER, Smith LM, Are C. 2012. Comparison of outcomes of transplantation and resection in patients with early hepatocellular carcinoma: A meta-analysis. *HPB (Oxford)* **14:** 635–645.

Dillmann J, Popp FC, Fillenberg B, Zeman F, Eggenhofer E, Farkas S, Scherer MN, Koller M, Geissler EK, Deans R, et al. 2012. Treatment-emergent adverse events after infusion of adherent stem cells: The MiSOT-I score for solid organ transplantation. *Trials* **13:** 211.

Earl TM, Chapman WC. 2013. Transplantation for hepatocellular carcinoma: The North American experience. *Recent Results Cancer Res* **190:** 145–164.

Farid WR, de Jonge J, Slieker JC, Zondervan PE, Thomeer MG, Metselaar HJ, de Bruin RW, Kazemier G. 2011. The importance of portal venous blood flow in ischemic-type biliary lesions after liver transplantation. *Am J Transplant* **11:** 857–862.

Farkas SA, Schnitzbauer AA, Kirchner G, Obed A, Banas B, Schlitt HJ. 2009. Calcineurin inhibitor minimization protocols in liver transplantation. *Transpl Int* **22:** 49–60.

Fishman JA, Rubin RH. 1998. Infection in organ-transplant recipients. *N Engl J Med* **338:** 1741–1751.

Gastaca M, Andres V, Ruiz P, Ventoso A, de Urbina JO. 2013. T-tube or no T-tube in cadaveric orthotopic liver transplantation: The type of tube really matters. *Ann Surg* doi: 10.1097/SLA.0b013e31829d56c0.

Gedaly R, Daily MF, Davenport D, McHugh PP, Koch A, Angulo P, Hundley JC. 2011. Liver transplantation for the treatment of liver metastases from neuroendocrine tumors: An analysis of the UNOS database. *Arch Surg* **146:** 953–958.

Ghobrial RM, Freise CE, Trotter JF, Tong L, Ojo AO, Fair JH, Fisher RA, Emond JC, Koffron AJ, Pruett TL, et al. 2008. Donor morbidity after living donation for liver transplantation. *Gastroenterology* **135:** 468–476.

Grant RC, Sandhu L, Dixon PR, Greig PD, Grant DR, McGilvray ID. 2013. Living vs. deceased donor liver transplantation for hepatocellular carcinoma: A systematic review and meta-analysis. *Clin Transplant* **27:** 140–147.

Grossi P. 2003. Liver transplantation in HIV-positive individuals: A new paradigm. *Transplant Proc* **35:** 1005–1006.

Grossman EJ, Millis JM. 2010. Liver transplantation for non-hepatocellular carcinoma malignancy: Indications, limitations, and analysis of the current literature. *Liver Transpl* **16:** 930–942.

Gundlach M, Broering D, Topp S, Sterneck M, Rogiers X. 2000. Split-cava technique: Liver splitting for two adult recipients. *Liver Transpl* **6:** 703–706.

Gurusamy KS, Tsochatzis E, Xirouchakis E, Burroughs AK, Davidson BR. 2010. Antiviral therapy for recurrent liver graft infection with hepatitis C virus. *Cochrane Database Syst Rev* CD006803.

Gurusamy KS, Koti R, Pamecha V, Davidson BR. 2011a. Veno-venous bypass versus none for liver transplantation. *Cochrane Database Syst Rev* CD007712.

Gurusamy KS, Pamecha V, Davidson BR. 2011b. Piggy-back graft for liver transplantation. *Cochrane Database Syst Rev* CD008258.

Gurusamy KS, Naik P, Abu-Amara M, Fuller B, Davidson BR. 2012. Techniques of flushing and reperfusion for liver transplantation. *Cochrane Database Syst Rev* **3:** CD007512.

Hagness M, Foss A, Line PD, Scholz T, Jorgensen PF, Fosby B, Boberg KM, Mathisen O, Gladhaug IP, Egge TS, et al. 2013. Liver transplantation for nonresectable liver metastases from colorectal cancer. *Ann Surg* **257:** 800–806.

Heidenhain C, Heise M, Jonas S, Ben-Asseur M, Puhl G, Mittler J, Thelen A, Schmidt S, Langrehr J, Neuhaus P. 2006. Retrograde reperfusion via vena cava lowers the risk of initial nonfunction but increases the risk of ischemic-type biliary lesions in liver transplantation—A randomized clinical trial. *Transpl Int* **19:** 738–748.

Jeon YM, Lee KW, Yi NJ, Lee JM, Hong G, Choi Y, Park MS, Kim H, Suh KS. 2013. The right posterior bile duct anatomy of the donor is important in biliary complications of the recipients after living-donor liver transplantation. *Ann Surg* **257:** 702–707.

Kim WR, Therneau TM, Benson JT, Kremers WK, Rosen CB, Gores GJ, Dickson ER. 2006. Deaths on the liver transplant waiting list: An analysis of competing risks. *Hepatology* **43:** 345–351.

Kim WR, Biggins SW, Kremers WK, Wiesner RH, Kamath PS, Benson JT, Edwards E, Therneau TM. 2008. Hyponatremia and mortality among patients on the liver-transplant waiting list. *N Engl J Med* **359:** 1018–1026.

Kim WR, Stock PG, Smith JM, Heimbach JK, Skeans MA, Edwards EB, Harper AM, Snyder JJ, Israni AK, Kasiske BL. 2013. OPTN/SRTR 2011 Annual Data Report: Liver. *Am J Transplant* **13:** 73–102.

* Lakkis FG, Lechler RI. 2013. Origin and biology of the allogeneic response. *Cold Spring Harb Perspect Med* **3:** a014993.

Lautenschlager I. 2009. CMV infection, diagnosis and antiviral strategies after liver transplantation. *Transpl Int* **22:** 1031–1040.

Law BK. 2005. Rapamycin: An anti-cancer immunosuppressant? *Crit Rev Oncol Hematol* **56:** 47–60.

Lim KC, Chow PK, Allen JC, Siddiqui FJ, Chan ES, Tan SB. 2012. Systematic review of outcomes of liver resection for early hepatocellular carcinoma within the Milan criteria. *Br J Surg* **99:** 1622–1629.

Malinchoc M, Kamath PS, Gordon FD, Peine CJ, Rank J, ter Borg PC. 2000. A model to predict poor survival in patients undergoing transjugular intrahepatic portosystemic shunts. *Hepatology* **31:** 864–871.

Meyers RL, Tiao GM, Dunn SP, Langham MR Jr. 2012. Liver transplantation in the management of unresectable hepatoblastoma in children. *Front Biosci* **4:** 1293–1302.

National Institutes of Health. 1984. National Institutes of Health Consensus Development Conference Statement: Liver transplantation—June 20–23, 1983. *Hepatology* **4:** 107S–110S.

Papafragkakis H, Martin P, Akalin E. 2010. Combined liver and kidney transplantation. *Curr Opin Organ Transplant* **15:** 263–268.

Pichlmayr R, Ringe B, Gubernatis G, Hauss J, Bunzendahl H. 1988. Transplantation of a donor liver to 2 recipients (splitting transplantation)—A new method in the further development of segmental liver transplantation. *Langenbecks Archiv fur Chirurgie* **373:** 127–130.

Prokai A, Fekete A, Pasti K, Rusai K, Banki NF, Reusz G, Szabo AJ. 2012. The importance of different immunosuppressive regimens in the development of posttransplant diabetes mellitus. *Pediatr Diabetes* **13:** 81–91.

Rice JP, Lucey MR. 2013. Should length of sobriety be a major determinant in liver transplant selection? *Curr Opin Organ Transplant* **18:** 259–264.

Rosen CB, Heimbach JK, Gores GJ. 2010. Liver transplantation for cholangiocarcinoma. *Transplant Int* **23:** 692–697.

Said A, Williams J, Holden J, Remington P, Gangnon R, Musat A, Lucey MR. 2004. Model for end stage liver disease score predicts mortality across a broad spectrum of liver disease. *J Hepatol* **40:** 897–903.

Scherer MN, Banas B, Mantouvalou K, Schnitzbauer A, Obed A, Kramer BK, Schlitt HJ. 2007. Current concepts and perspectives of immunosuppression in organ transplantation. *Langenbecks Arch Surg* **392:** 511–523.

Schlitt HJ, Jonas S, Ganten TM, Grannas G, Moench C, Rauchfuss F, Obed A, Tisone G, Pinna AD, Gerunda GE, et al. 2013. Effects of mycophenolate mofetil introduction in liver transplant patients: Results from an observational, non-interventional, multicenter study (LOBSTER). *Clin Transplant* **27:** 368–378.

Schnitzbauer AA, Scherer MN, Rochon J, Sothmann J, Farkas SA, Loss M, Geissler EK, Obed A, Schlitt HJ. 2010. Study protocol: A pilot study to determine the safety and efficacy of induction-therapy, de novo MPA and delayed mTOR-inhibition in liver transplant recipients with impaired renal function. PATRON-study. *BMC Nephrol* **11:** 24.

Shapiro HA. 1968. Brain death and organ transplantation. *J Forensic Med* **15:** 89–90.

Shapiro R. 2004. Low toxicity immunosuppressive protocols in renal transplantation. *Keio J Med* **53:** 18–22.

Song GW, Lee SG, Hwang S, Ahn CS, Moon DB, Kim KH, Ha TY, Jung DH, Park GC, Namgung JM, et al. 2013. Successful experiences of ABO-incompatible adult living donor liver transplantation in a single institute: No immunological failure in 10 consecutive cases. *Transplant Proc* **45:** 272–275.

Starzl TE. 2012. The long reach of liver transplantation. *Nat Med* **18:** 1489–1492.

Starzl TE, Marchioro TL, Vonkaulla KN, Hermann G, Brittain RS, Waddell WR. 1963. Homotransplantation of the liver in humans. *Surg Gynecol Obstet* **117:** 659–676.

Starzl TE, Groth CG, Brettschneider L, Penn I, Fulginiti VA, Moon JB, Blanchard H, Martin AJ Jr, Porter KA. 1968. Orthotopic homotransplantation of the human liver. *Ann Surg* **168:** 392–415.

Starzl TE, Iwatsuki S, Klintmalm G, Schroter GP, Weil R III, Koep LJ, Porter KA. 1981a. Liver transplantation, 1980, with particular reference to cyclosporin-A. *Transplant Proc* **13:** 281–285.

Starzl TE, Klintmalm GB, Porter KA, Iwatsuki S, Schroter GP. 1981b. Liver transplantation with use of cyclosporin a and prednisone. *N Engl J Med* **305:** 266–269.

Tavio M, Grossi P, Baccarani U, Scudeller L, Pea F, Berretta M, Adani G, Vivarelli M, Riva A, Tirelli U, et al. 2011. HIV-infected patients and liver transplantation: Who, when and why. *Current HIV Res* **9:** 120–127.

Thuluvath PJ, Yoo HY. 2004. Graft and patient survival after adult live donor liver transplantation compared to a matched cohort who received a deceased donor transplantation. *Liver Transpl* **10:** 1263–1268.

Torbenson M, Wang J, Nichols L, Jain A, Fung J, Nalesnik MA. 1998. Causes of death in autopsied liver transplantation patients. *Mod Pathol* **11:** 37–46.

United States Congress. 1984. National Organ Transplant Act: Public Law 98-507. *US Statut Large* **98:** 2339–2348.

Watt KD, Pedersen RA, Kremers WK, Heimbach JK, Sanchez W, Gores GJ. 2009. Long-term probability of and mortality from de novo malignancy after liver transplantation. *Gastroenterology* **137:** 2010–2017.

Yao FY. 2008. Liver transplantation for hepatocellular carcinoma: Beyond the Milan criteria. *Am J Transplant* **8:** 1982–1989.

Zakhari S. 2013. Bermuda Triangle for the liver: Alcohol, obesity, and viral hepatitis. *J Gastroenterol Hepatol* **28:** 18–25.

Pancreas Transplantation: Solid Organ and Islet

Shruti Mittal, Paul Johnson, and Peter Friend

Nuffield Department of Surgical Sciences, University of Oxford, Oxford OX3 9DU, United Kingdom

Correspondence: peter.friend@nds.ox.ac.uk

Transplantation of the pancreas, either as a solid organ or as isolated islets of Langerhans, is indicated in a small proportion of patients with insulin-dependent diabetes in whom severe complications develop, particularly severe glycemic instability and progressive secondary complications (usually renal failure). The potential to reverse diabetes has to be balanced against the morbidity of long-term immunosuppression. For a patient with renal failure, the treatment of choice is often a simultaneous transplant of the pancreas and kidney (SPK), whereas for a patient with glycemic instability, specifically hypoglycemic unawareness, the choice between a solid organ and an islet transplant has to be individual to the patient. Results of SPK transplantation are comparable to other solid-organ transplants (kidney, liver, heart) and there is evidence of improved quality of life and life expectancy, but the results of solitary pancreas transplantation and islets are inferior with respect to graft survival. There is some evidence of benefit with respect to the progression of secondary diabetic complications in patients with functioning transplants for several years.

Effective medical therapy with exogenous insulin has been available for patients with diabetes since 1922, and the majority of patients are well managed on this basis. However, a proportion of patients develop life-threatening complications of diabetes that fail to respond to the most careful insulin regimes. These patients, in whom there is a failure of effective medical therapy, fall broadly into two categories: progression of diabetes renal failure and complications and instability of glycemic control.

Progression of Diabetes-Related Renal Failure and Other Secondary Complications

An unequivocal body of evidence shows that close glycemic control is beneficial to the patient in relation to the onset and progression of secondary complications (nephropathy, neuropathy, and retinopathy) (The Diabetes Control and Complications Trial Research Group 1993). However, tight glycemic control using exogenous insulin is associated with a higher incidence of severe hypoglycemic episodes, making perfect glycemic control unachievable.

Life-Threatening Instability of Glycemic Control

Some patients with difficult-to-manage diabetes experience hypoglycemic unawareness; not only are these patients liable to hypoglycemic events, but they also have no warning of such an episode and are therefore unable to react in time to prevent hypoglycemic coma. Patients with frequent hypoglycemic unawareness re-

Cite this article as *Cold Spring Harb Perspect Med* doi: 10.1101/cshperspect.a015610

quire constant supervision, and many cannot live independently or maintain employment.

It is reasonable to hypothesize that β-cell transplantation should be the most effective way to maintain physiological glycemic control and also prevent hypoglycemia and hypoglycemic unawareness. However, if transplantation is to be considered the treatment of choice in such patients, then two essential prerequisites must be fulfilled. The first is evidence that transplantation is more effective than other means to maintain stable glycemic control in brittle diabetic patients. The second is evidence that β-cell transplantation is beneficial with respect to life expectancy, quality of life, or disease progression in patients with renal failure and other secondary complications.

The potential benefit of pancreas transplantation was recognized from the early days of solid organ allografting. Following failed initial attempts at xenotransplantation of fragmented pancreases (Pybus 1924) and the development of immunosuppressive therapies, the first transplanted pancreas was reported by Kelly et al. (1967). The morbidity and mortality associated with these early transplants were extremely high: In a publication in 1970, only two patients out of a series of 10 were alive (Lillehei et al. 1970). In the decades that followed, the morbidity and mortality of solid organ transplantation have decreased greatly, and the rate of success has improved to a level now comparable to kidney and liver transplantation. However, much of the complexity and morbidity of pancreas transplantation still relates to the exocrine component of the gland, and the concept of transplanting only the 2%–3% of cells within the pancreas that produce insulin is clearly attractive. In 1974, Paul Lacy performed the first clinical allotransplant using isolated islets of Langerhans; although initially insulin independent, the recipient subsequently died from sepsis (Karl et al. 1977). It was not until isolation and transplantation of viable islets were first shown to be consistently successful in a report from Edmonton that islet cell transplantation became widespread (Shapiro et al. 2000). In the current era, therefore, there are two therapeutic options for patients with diabetes who require transplantation.

CURRENT PRACTICE AND INDICATIONS FOR TRANSPLANTATION (TABLE 1)

Diabetes and Renal Failure

The large majority of patients undergoing solid-organ pancreas transplantation are those with chronic renal failure secondary to diabetes. These patients, on or approaching the need for dialysis, are candidates for kidney transplantation even in the absence of a pancreas transplant. Kidney transplantation alone for diabetic renal failure has a relatively poor prognosis compared with other indications for this procedure (Cosio et al. 2008; Kuo et al. 2010; Taber et al. 2013); although never subjected to a randomized study, there is increasing evidence that, in patients suitably assessed for the larger procedure, simultaneous pancreas and kidney (SPK) transplantation is associated with improved quality of life (Speight et al. 2010) and life expectancy (van Dellen et al. 2013). The decision as to whether such patients are better advised to undergo kidney transplantation alone, and possible subsequent pancreas after kidney (PAK) transplantation, instead of the combined operation depends on several factors. These include the availability of a potential living donor, the urgency to avoid/come off dialysis, and the likely waiting time for an SPK. There are marked national differences: for example, in the United Kingdom, the majority of patients are listed for an SPK—this reflects the relatively high priority given to such patients in the organ allocation system and the shorter waiting time that results. Currently, SPK and PAK are offered to people with insulin-dependent diabetes (type 1 or type 2) with chronic renal failure. Good results have also been achieved in selected patients following both simultaneous islet kidney (SIK) and islet after kidney (IAK) transplantation (Gerber et al. 2008).

Diabetes and Hypoglycemia

In patients in whom the primary indication for transplantation is unstable glycemic control, particularly hypoglycemic unawareness, the choice is between solid-organ pancreas transplantation alone (PTA) or islet transplantation.

 Cite this article as *Cold Spring Harb Perspect Med* doi: 10.1101/cshperspect.a015610

This is, essentially, a choice between a much higher-morbidity procedure with a higher likelihood of insulin independence (solid-organ transplantation) versus a much lower-morbidity procedure with a lower chance of long-term function or insulin independence. As is discussed below, insulin independence is less essential if resolution of life-threatening hypoglycemia is the primary indication for transplantation. Indeed, resolution of hypoglycemia unawareness is increasingly seen as the metric of success in islet transplantation rather than insulin independence. This interface between solid-organ and islet transplantation is dynamic, and, with recent improvements in the outcome of islet transplantation in the best centers, the 1-yr and 5-yr graft survival data are much closer between the two modalities. Patients being offered transplantation for hypoglycemic unawareness should be considered for both treatment options. These should be considered as complementary rather than competing treatments, and the decision as to which procedure to perform should be tailored to individual patient needs and preferences.

Diabetes and Other Complications

The role of PTA in the management of patients who have neither established renal failure (yet) nor hypoglycemic unawareness remains controversial. Offering pancreas transplantation to such a patient is at least partly based on the assumption that benefits will accrue with respect to the progression of secondary diabetic complications and that this benefit will outweigh the complications of long-term immunosuppression. There is evidence (see below) that a successful pancreas transplant may, in the medium term, lead to stabilization or improvement of retinopathy, neuropathy, and cardiovascular disease. However, in a patient who does not require a kidney transplant, this necessitates the patient exchanging the complications of diabetes (metabolic instability, quality of life, life expectancy, secondary complications) with those of the procedure and long-term immunosuppression (operative risk, graft failure, opportunistic infection, cardiovascular disease, cancer). In the case of a patient with renal failure who would otherwise be receiving a kidney transplant, this balance does not apply because the patient is already committed to the long-term risks of immunosuppression. This is the primary reason that solid-organ pancreas transplantation in the absence of a kidney transplant is much less commonly performed than the combined operation.

This is an area where better evidence is needed; registry analysis is limited by the range and completeness of the recorded data and is unlikely to contribute more than graft and patient survival. Any beneficial effects on secondary complications are likely to require several years to become manifest, necessitating close and repeated measurements of (for example) retinal disease/visual acuity, cardiovascular status, neurological function, and renal function (after PTA). The relatively small numbers of transplants performed by most centers would require this longitudinal study to be performed on a multicenter basis over 5–10 yr.

SECONDARY COMPLICATIONS OF DIABETES

A crucial question that has not yet been answered definitively is whether successful β-cell transplantation leads to stabilization or improvement of the secondary complications of diabetes. Many small studies have addressed this issue in both solid-organ and islet transplantation, but the general quality of evidence is relatively poor, and there is a need for larger clinical trials.

Diabetic neuropathy can have a significant impact on the quality of life of patients with diabetes. Autonomic neuropathy causes debilitating gastrointestinal problems that are very difficult to manage, and peripheral neuropathy may cause problems with mobility, pain, or unrecognized injuries that may lead to amputations. Improvements in diabetic neuropathy have been observed with good glycemic control (Callaghan et al. 2012); however, evidence in the context of solid-organ pancreas transplantation is limited. Kennedy et al. (1990) showed improvements in neurological function after

24 mo of stable transplant pancreatic function. These improvements were seen in motor, sensory, and autonomic neurology and persisted to 10 yr posttransplant, although improvements in autonomic neuropathy were comparatively modest (Navarro et al. 1990). The Edmonton group were unable to show a statistically significant difference in neuropathy between an islet transplant and the best medical therapy group in a crossover trial (Thompson et al. 2011). One obstacle to research in this area is the lack of a gold standard outcome measure. The recent establishment of confocal corneal microscopy as a reliable, noninvasive, readily repeatable investigation for detecting new nerve growth may address this, and early evidence after pancreas transplantation is promising (Shtein and Callaghan 2013; Tavakoli et al. 2013).

Diabetic retinopathy is the leading cause of blindness in young people, and impairment in vision can severely limit independence and opportunities for employment. Giannarelli et al. (2006) studied 33 Type 1 diabetic patients who underwent PTA, matched with 35 controls and followed for a mean of 30 mo, and found significant improvement in the incidence of proliferative retinopathy and the need for laser treatment, and also in nonproliferative retinopathy. However, a high proportion of patients have advanced retinopathy at the time of pancreas transplantation, making stabilization rather than improvement a more realistic goal. Stabilization of disease progression has been observed in several studies (Chow et al. 1999), although the absence of a large comparative study means that it remains unclear whether this is a deviation from the natural progression of advanced retinopathy (Sosna et al. 1998). Data from islet transplantation are similarly sparse; Lee et al. (2005) examined eight patients and found improvement in one and stabilization in seven patients with diabetic retinopathy. Data from the cross-over study performed by the Edmonton group showed significant differences in progression between an islet transplant and best medical therapy (Thompson et al. 2008). However, in order to establish the real effect of transplantation on vision, a large prospective study involving longitudinal retinal images and a relevant control group will be necessary. It has been suggested that rejection episodes may have a deleterious effect on retinopathy, causing sudden deterioration (Scheider et al. 1991). In addition, the increased risk of cataracts following transplantation must also be considered. However, the indications for transplantation would expand significantly if it were proven definitively that transplantation is effective in preventing the progression of diabetic retinopathy.

The effect of pancreas transplantation on renal function is highly relevant to clinical practice but is confounded by the nephrotoxic effect of calcineurin inhibitor drugs, with which most pancreas transplant patients are treated. In patients with already significant diabetic nephropathy (glomerular filtration rates <40 mL/min), the drug-induced detriment in renal function may advance the need for dialysis. PTA transplantation is therefore contraindicated, but such patients are some years away from dialysis and not yet eligible for a kidney under current allocation systems. There is currently no satisfactory route for the management of this group of patients, but if there were evidence that pancreas transplantation prevents the progression of diabetic renal disease, this may justify much earlier transplantation in such patients. In addition, these patients may benefit from the use of calcineurin-inhibitor-avoiding immunosuppression, although there is little published literature in this area. In a much-cited study, Fioretto et al. (1998) followed eight nonuremic patients who underwent PTA. Although the glomerular filtration rate declined by 5 yr (108 to 74 mL/min), it remained stable at 10 yr. Protocol biopsies at 5 and 10 yr showed a decrease in the glomerular and tubular basement membrane thickness at 10 yr, with the mesangial fractional volume increased at 5 yr but decreased by 10 yr. In islet recipients, the Edmonton group showed a slower decline in GFR after transplantation compared with best medical therapy (Thompson et al. 2011).

Cardiac disease is the leading cause of death in patients with diabetic renal failure. There is evidence that pancreas–kidney transplantation reduces the cardiovascular death rate (La Rocca et al. 2001), and functional studies have shown improvement in blood pressure and dyslipidemia compared with kidney transplantation alone (Luan et al. 2007). Benefits have also been shown with respect to systolic and diastolic ventricular function (Fiorina et al. 2000; La Rocca et al. 2001) and endothelial dysfunction (Fiorina et al. 2003; Stadler et al. 2009). However, low cardiac-event rates after transplantation may also reflect stringent cardiac screening and preoperative optimization. Long-term studies linking postoperative functional investigations and clinical end points are needed; current research is addressing cardiac screening techniques and markers of cardiac risk. Fiorina et al. (2003) compared 21 patients with successful islet–kidney transplantation with 13 patients in whom the islet transplant failed, showing benefits with respect to survival and cardiovascular and endothelial function.

BENEFITS OF PANCREAS AND ISLET TRANSPLANTATION

Although there have been no large-scale randomized controlled trials to establish either whether transplant therapies are more effective than other means of maintaining stable glycemic control, or whether replacement of β-cell mass is beneficial in patients with secondary complications, ultimately, the object of both solid pancreas and islet transplantation is to enhance quality of life and life expectancy.

For a patient with hypoglycemic unawareness, both quality of life and life expectancy are likely to be improved by resolution of hypoglycemic unawareness, even if they are not rendered insulin independent. Although life-expectancy data are not yet available, emerging quality-of-life data do, indeed, suggest a significant benefit from successful islet transplantation (Bassi and Fiorina 2011) and solid-organ transplantation (Speight et al. 2010).

For patients undergoing transplantation because of secondary diabetic complications (particularly kidney damage), any benefit is likely to require longer-term function (several years) and insulin independence (probably). There is considerable evidence (although not from randomized trials) that successful solid-organ transplantation increases life expectancy. Tyden et al. (1999) showed that, in a reasonably homogeneous group of patients with diabetic renal failure, those who achieve stable function of a combined pancreas and kidney transplant had substantially improved 10-yr survival compared with the group with stable kidney transplant function alone. Registry analyses confirm the benefit of SPK transplantation versus kidney transplantation alone (and of kidney transplantation versus dialysis). Ojo et al. (2001) showed an almost doubling of life expectancy in patients undergoing the combined operation compared with those who underwent deceased-donor kidney transplantation alone for diabetic renal failure. Clearly, however, these groups were not matched for age or morbidity, although all patients were selected as suitable for transplantation of one or both organs.

SUCCESS OF PANCREAS AND ISLET TRANSPLANTATION

The outcomes of both solid pancreas and islet transplantation have improved progressively. One- and 5-yr results of SPK transplantation are now close to those of kidney, liver, and heart transplantation (Kandaswamy et al. 2013). Success, however, requires definition. Conventionally, a successful solid-organ pancreas transplant is defined by insulin independence. This is a rational policy based on the assumption that only perfect glycemic control is likely to achieve the benefits in relation to life expectancy and secondary complications. However, as discussed above, in the case of patients with unstable glycemic control undergoing islet transplantation, the primary objective is the resolution of hypoglycemia awareness, and not the restoration of normoglycemia. There is now good evidence that this does not require insulin independence but rather the consistent production of C peptide (Ryan et al. 2004). For this reason, lack of hypoglycemia unawareness is the generally accepted definition of success in islet transplantation.

Current pancreas graft survival in patients undergoing SPK is 85% at 1 yr and 73% at 5 yr (Figs. 1–3) (Kandaswamy et al. 2013). Nearly all

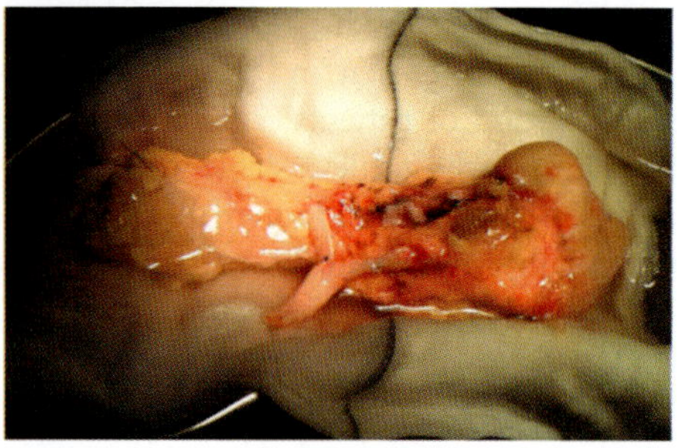

Figure 1. Back-table preparation of pancreas graft for implantation. The segment of donor duodenum is shortened, the donor spleen is removed, and retroperitoneal tissue is removed/ligated. The arterial supply via the superior mesenteric and splenic arteries is attached to a Y graft, usually from the bifurcation of the donor common iliac artery in order to provide a single arterial inflow. Venous outflow via portal vein (shown). (Photograph courtesy of James Gilbert, Oxford Transplant Centre.)

the improvement in graft outcome in recent decades has been achieved during the first 12 mo postoperatively with much less in subsequent years (in line with other solid-organ transplant outcomes). In solid-organ transplantation, SPK outcomes are consistently superior to those from PTA or PAK transplantation (Gruessner et al. 2012). Patients undergoing PTA can expect 1- and 5-yr pancreas graft survival of 80% and 58%, and those undergoing PAK can expect 77% and 56% (Kandaswamy et al. 2013). It is generally assumed that the main reason for the discrepancy between SPK and PTA/PAK outcomes is the presence of a kidney from the same donor in the former. This allows renal function to be used as a surrogate biomarker for rejection; not only is it possible to measure renal function in a more sensitive manner, but also the kidney is amenable to biopsy. The greatest risk of graft loss still occurs within the first year posttransplant, and, in particular, within the first 3 mo, when failure is often associated with reperfusion pancreatitis and sepsis due to ischemia-reperfusion injury, a potentially fatal complication. Pancreas transplantation, with or without kidney transplantation, has a patient survival rate of 95% at 1 yr and 85% at 5 yr.

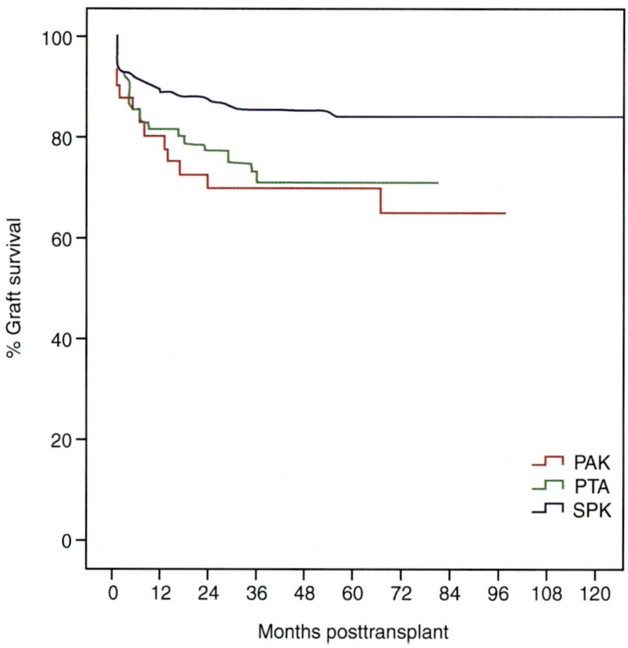

Figure 2. Pancreas graft survival (defined by insulin independence) and patient survival after transplantation at the Oxford Transplant Centre. SPK, Simultaneous pancreas kidney transplant; PTA, pancreas transplant alone; PAK, pancreas after kidney.

Cite this article as *Cold Spring Harb Perspect Med* doi: 10.1101/cshperspect.a015610

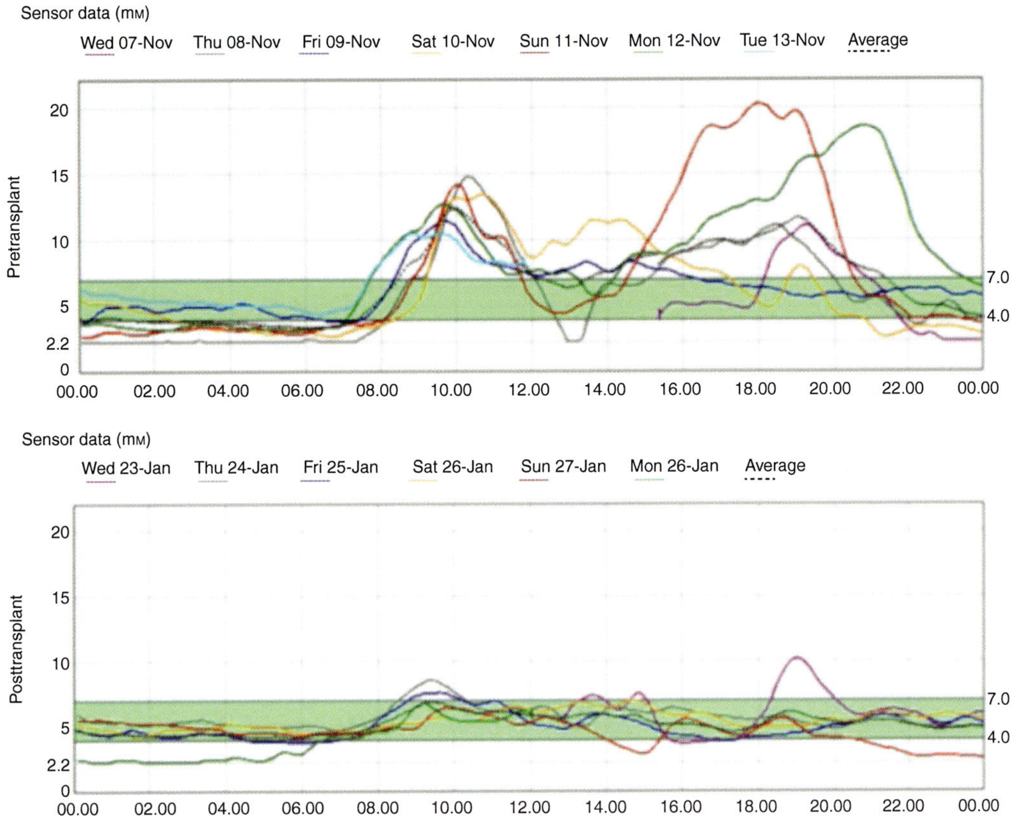

Figure 3. Continuous glucose monitoring before and after islet transplantation performed within the Oxford Islet Transplant Programme, showing improved glycemic control.

The results of islet transplantation have improved greatly over the past decade, in terms both of resolution of hypoglycemic unawareness and achievement of insulin independence. In the 1990s, the Islet Transplant Registry reported an insulin-independence rate of only 11% in almost 500 islet transplant recipients. However, the introduction of the steroid-free, repeat infusion Edmonton Protocol significantly improved outcomes (Shapiro et al. 2000), which were confirmed by the multicenter Immune Tolerance Network (ITN) trial results of 58% 1-yr insulin independence in 36 patients (Shapiro et al. 2006). Longer-term results of the original Edmonton Protocol showed a high rate of attrition, with only ~10% of 65 patients remaining insulin independent 5 yr later (Ryan et al. 2005a), although ~80% of patients remained C-peptide positive with well-controlled HbA1c. In recent years, the results of islet transplantation have improved. The Islet Transplant Registry reported a 3-yr graft function of only 19% in 2001, whereas in 2009, the Collaborative Islet Transplant Registry (CITR), with data from islet transplant centers of variable experience, reported that this had risen to 45%.

Islet transplantation has been seen to be even more successful in some centers of excellence. The Minneapolis team achieved 5-yr insulin-independence rates of 50% in 25 patients using single islet infusions (Ryan et al. 2005a), albeit in recipients with low body-mass indices. The Lille group reported 57% insulin independence at 3.3 yr posttransplantation in 14 patients (Campbell et al. 2007). The Edmonton team recently presented excellent 5-yr outcomes using alemtuzumab as the induction agent. Although the longer-term results of islet transplantation

Table 1. β-Cell transplantation indications, contraindications, risks, and benefits

Procedure	SPK	PTA	PAK	Islet
Indication	Insulin-treated diabetes Chronic renal failure with GFR < 20 mL/min or on dialysis	Insulin-treated diabetes Significant diabetic complications Severe diabetic complications but normal or near-normal renal function Frequent and severe episodes of hypoglycemia (more than two severe hypoglycemic episodes within last 24 mo); assessed by diabetologist to have disabling hypoglycemia *or* hypogylcemic unawareness *or* significant impairment of quality of life due to diabetes	Insulin-treated diabetes Stable function of previous renal allograft Meet criteria for PTA	Insulin-treated diabetes Frequent and severe episodes of hypoglycemia (more than two severe hypoglycemic episodes within last 24 mo); assessed by diabetologist to have disabling hypoglycemia *or* hypogylcemic unawareness *or* significant impairment of quality of life due to diabetes
Contraindications	**Relative:** Cerebrovascular accident with long-term impairment Active infection hepatitis B or C virus Body mass index >30 kg/m^2 Insulin requirements >1.5 units/kg per day Extensive aorta/iliac and/or peripheral vascular disease Continued abuse of alcohol or drugs		**Absolute:** Excessive cardiovascular risk (significant noncorrectable coronary artery disease; left ventricular ejection fraction $<50\%$; myocardial infarction within 6 mo) Uncurable malignancy (excluding localized skin cancer) Active sepsis or peptic ulcer Major psychiatric history likely to result in nonadherence Inability to withstand surgery and immunosuppression	
Benefits	Insulin independence Good pancreas and kidney graft outcomes	Insulin independence Cure of hypoglycemia	Insulin independence Early dialysis independence	Cure of hypoglycemia Less invasive
Risks	Operative morbidity and mortality	Risk of graft failure Higher morbidity procedure	Sensitization Poorer pancreas graft outcomes	Less likely to achieve insulin independence Often need more than one infusion

SPK, Simultaneous pancreas kidney transplant; PTA, pancreas transplant alone; PAK, pancreas after kidney.

Cite this article as *Cold Spring Harb Perspect Med* doi: 10.1101/cshperspect.a015610

are inferior to those of solid-organ transplantation, the much lower morbidity of islet transplantation does enable the possibility of repeat transplantation. However, this carries the risk of sensitizing the recipient to HLA antigens (Campbell et al. 2007; Naziruddin et al. 2012), which may disadvantage any patient who requires kidney transplantation at a subsequent stage.

β-CELL TRANSPLANTATION—THE PROCESS

Organ Donation

Donor selection for pancreas transplantation is more stringent than for other organs, thus limiting the number of potential donors available. Conservative acceptance practices mean that offered pancreases are often declined; indeed, organs may be declined on the basis of donor alcohol use, family history of diabetes, or serum amylase, none of which has been found to be associated with graft outcome (Axelrod et al. 2010). Many donor organs are found to be abnormal at the time of retrieval with fibrosis or fat deposition within the gland. Although steatotic organs often provide good immediate function if transplanted, these organs are liable to severe reperfusion pancreatitis with associated morbidity, and fatty organs are therefore usually discarded. Similarly, donors with a body mass index (BMI) >30 are usually declined for solid-organ donation although the evidence base for doing so is limited (Humar et al. 2004; Axelrod et al. 2010). Organ donors up to the age of 60 yr are considered for solid-organ transplantation, but, because donor age has frequently been identified as predictive of graft outcome, pancreases from younger donors are preferred (Proneth et al. 2013). The pancreas is very vulnerable to injury sustained during retrieval, which leads to a higher discard rate of retrieved organs than is seen in other transplant types. For all these reasons, the conversion rate from potential donor to transplant is very low; data from the Scientific Registry of Transplant Recipients show that as little as 20% of potential donor pancreases are actually retrieved, and of these a further 30% are discarded after retrieval (Kandaswamy et al. 2013).

Although historically older and higher-BMI donors were preferred for islet transplantation, this was based on isolation rates rather than islet physiology (Lakey et al. 1996; Nano et al. 2005). It is now clear that the optimal donor for islets is younger (Niclauss et al. 2011); this is supported by "in vitro" data of human islets (Lakey et al. 1996; Ihm et al. 2007). Fibrosis and fatty infiltration of the pancreatic parenchyma can prevent pancreas digestion during islet isolation, but fat overlying the pancreas is not in itself a contraindication because this can be removed before digestion.

Cold ischemia is deleterious in both solid organ and islet transplantation, although the recommended limit is shorter (8 h) in the case of islet transplantation (Lakey et al. 1995; Axelrod et al. 2010). Organs donated after circulatory death (DCD) are used in increasing numbers in solid organ transplantation and have been shown to produce results that are comparable in quality to organs removed from brain-dead donors (Muthusamy et al. 2012). However, this is achieved by a more selective approach, particularly in terms of donor age and BMI. An objective analysis of donor risk factors has been performed, leading to a donor risk index (Axelrod et al. 2010). The factors associated with the greatest risk (hazard ratio) are as follows: donor age (1.56), DCD status (1.39), black race (1.27), cerebrovascular cause of donor death (1.23), serum creatinine >2.5 mg/dL (1.22). Both donor and recipient risk factors need to be incorporated into new organ allocation schemes to optimize the usage of the limited supply of donor pancreases, as well as clinical outcomes for patients undergoing either form of transplantation (Berney and Johnson 2010).

Organ preservation for both solid-organ and islet transplants is performed by static cold storage, most commonly with University of Wisconsin solution, although some units use Celsior or HTK. Despite several trials, no consistent evidence of benefit has been shown, although some studies found a higher incidence of graft pancreatitis, a decreased rate of insulin independence at hospital discharge, and a higher rate of graft loss in HTK-stored pancreases (Stewart et al. 2009). The potential benefits of oxygen carriage using the two-layer technique for islet transplantation (Kawamura et al. 1992) did not to translate from the small animal models to clinical evidence (Kin et al. 2008).

Living Donation and Pancreas Transplantation

Sutherland et al. (2012) reported 163 living-donor pancreas transplants using segmental (body and tail of pancreas) organs; these were mainly solid organ, but a small number of islet transplants have been performed. The majority of these were performed at the University of Minnesota; others were performed in countries where deceased donors are rarely available, particularly Japan. There is little evidence of any long-term advantage over deceased-donor transplantation, and subsequent metabolic studies in donors detected significant and concerning metabolic abnormalities (Robertson et al. 2003). This procedure is very rarely performed now.

Solid-Organ Implantation

The surgical techniques of solid-organ transplantation have changed substantially over the years. The majority of transplant units around the world now transplant the solid pancreas together with a segment of duodenum. The arterial supply via the superior mesenteric and splenic arteries is attached to a Y graft usually from the bifurcation of the donor common iliac artery in order to provide a single arterial inflow. Most units transplant the pancreas with venous drainage to the common iliac vein or inferior vena cava and arterial inflow from the common iliac artery. A much smaller proportion of units advocate venous drainage into the portal venous system. Although associated with more physiological systemic levels of insulin (lower), there is no evidence of substantial benefit with respect to graft or patient survival or other parameters (Bazerbachi et al. 2012).

Drainage of the exocrine excretions of the pancreas has historically been the Achilles heel of pancreas transplantation. Numerous technical modifications have included (1) transplantation of the body and tail of the gland (segmental) with occlusion of the pancreatic duct, or drainage into the bladder, the jejunum or the stomach; and (2) transplantation of the whole organ with a "button" of duodenum or duodenal segment, drained into the bladder, jejunum, or duodenum. The expansion of clinical transplantation in the 1990s was based largely on anastomosis of the donor duodenum to the bladder (Sollinger et al. 2009), which had advantages for monitoring (as explained below). In recent years, bladder drainage has been largely supplanted by enteric drainage in which the donor duodenum is anastomosed to the proximal jejunum either directly or via a roux loop. This is clearly a more physiological technique but one that renders the pancreas less easily monitored.

Islet Cell Isolation and Implantation

The most challenging part of islet transplantation remains the islet isolation. This is a two-stage procedure involving pancreas digestion using a combination of enzymatic and mechanical dissociation, and islet purification using density-gradient separation (Ricordi et al. 1989). Although refined in recent decades, this remains a relatively nonspecific, unpredictable method with only ~30% of pancreases resulting in transplantable yields even in leading centers. Current research efforts are investigating the ultrastructure of the pancreatic exocrine–islet interface in different donor types, so that novel targeted enzyme blends can be developed (Hughes et al. 2006). Islet isolation requires specialized expertise, and, in order to focus skills, many programs are based on a hub-and-spoke model, in which each isolation laboratory provides islets for several implanting centers (Kempf et al. 2005).

Although the original Edmonton Protocol involved the transplantation of "fresh" islets immediately following islet isolation, most centers have now adopted a period of at least 24 h of islet culture before transplantation (Shapiro et al. 2000). This not only improves the logistics of the transplant, but also enables additional pretransplant islet assessment to be undertaken and enables the islets to "recover" following isolation. This is particularly useful for preparations involving larger islets, where central necrosis can be monitored during culture.

Although a few cases of intramuscular islet autotransplants have been reported (Dardenne et al. 2012), the majority of clinical islet allotransplants are performed via the portal vein into the liver sinusoids. This is performed either by a percutaneous, radiologically guided *trans*-hepatic approach, or via cannulation of a mesenteric vessel approached laparoscopically or via a mini-laparotomy. Recipients are anticoagulated, and heparin is added to the islet infusion to reduce the immediate inflammatory response, which is believed to be responsible for early islet loss (Cabric et al. 2007). If performed *trans*-hepatically, embolization of the puncture site in the liver is performed to prevent bleeding (Owen et al. 2003).

Immunosuppression

Immunosuppression in pancreas transplantation follows a similar pattern to other solid-organ transplants. The majority of units use biological induction (thymoglobulin, alemtuzumab, or basiliximab) followed by a combination of tacrolimus and mycophenolate, and the use of steroids is advocated by some but not all units. There is no conclusive evidence regarding optimal induction therapy; single center studies are small, and of the two multicenter trials, the largest examined an agent (daclizumab) that is no longer available, and the other used variable induction agents (Niederhaus et al. 2013). The use of non-nephrotoxic medication is attractive, but experience in the use of sirolimus following pancreas transplantation remains limited (Kandula et al. 2012).

Similarly, various immunosuppressive regimes have been used for islet transplantation. A feature of the successful Edmonton Protocol was a steroid-free regime, using tacrolimus and sirolimus as maintenance (Shapiro et al. 2000). Since then, many centers have replaced sirolimus with mycophenolate. The use of alemtuzumab-induction therapy is associated with encouraging longer-term function.

In the future, the application of currently experimental methods of immunomodulation are likely to change the practice of pancreas transplantation not only by reducing the morbidity of immunosuppression but also by reducing the adverse effects of unrecognized rejection. Treatment with T-regulatory cells and/or mesenchymal stem cells may substantially alter the practice of pancreas transplantation.

Graft Monitoring

Monitoring of both solid-organ and islet transplants is a major challenge. The lack of any simple marker of early graft injury is problematic, with no equivalent to the glomerular filtration rate in kidney transplantation. Hyperglycemia is a relatively late event and often signifies substantial and irreversible graft injury. Glycemic stability is closely monitored and graft function is measured by basal and stimulated C-peptide levels. In islet transplantation, several useful metabolic scores have been introduced to assess graft function objectively. These include the β-score, a composite of fasting glucose, HbA(1c), stimulated C-peptide, and insulin requirement (Ryan et al. 2005b); the HYPO score, which primarily assesses resolution of hypoglycemia; and the liability index (Ryan et al. 2004).

Surrogate measures, including those that relate to the transplanted exocrine tissue, are used in the monitoring of solid organ grafts. Rises in serum amylase and lipase are commonly associated with pancreatitis, but the relevance of changes in the context of graft rejection is uncertain. In combined pancreas–kidney transplantation, kidney graft function (and rejection) is used as a surrogate for changes presumed to be occurring within the pancreas also (although it is recognized that discordant rejection can occur) (Troxell et al. 2010). When the pancreas is transplanted alone, monitoring of pancreas function is problematic. This may explain the discrepancy in outcomes—unrecognized rejection may account for poorer graft survival in this group.

Bladder drainage of exocrine secretions enables amylase production to be used as a measure of transplant function and also allows biopsy of the donor duodenum via cystoscopy. Alterations in levels of urinary amylase are associated with impairment in pancreas graft function (Prieto et al. 1987), but this must be balanced against the substantial complications of bladder drainage, both urological (cystitis) and biochemical (bicarbonate loss and dehydration). Percutaneous, radiologically guided pancreatic transplant biopsy of enterically drained transplants is increasingly advocated and (despite earlier concerns) appears to be safe (Klassen et al. 2002). However, this provides the means of diagnosis of rejection but does not solve the problem of monitoring; the challenge of detecting the early manifestations of rejection remains (see review in Margreiter et al. 2013). The development of donor-specific antibodies posttransplant is associated with rejection and graft failure in the pancreas, as in other solid-organ transplants, and monitoring is advocated.

Radiological and radioisotopic methods to measure β-cell mass and monitor graft function are of experimental interest but not in clinical use. Metabolic tests indicate graft function and reserve but are unwieldy for the purpose of frequent monitoring. Immunological blood monitoring may give advance warning of an immune reaction, but such methods are not yet in routine clinical use.

Morbidity

Solid-organ and islet transplantation share similar long-term risks with respect to immunosuppression–opportunistic infections, specific drug side effects, and malignancy. Solid-organ pancreas transplantation is associated with much higher procedure-specific morbidity and mortality with many complications requiring reoperation, including hemorrhage, thrombosis, pancreatitis, and sepsis. Venous thrombosis, once a major cause of morbidity (Stockland et al. 2009), has been partially overcome by careful monitoring of coagulation using thromboelastography or other methods—many patients undergoing SPK are relatively hypercoagulable (Muthusamy et al. 2010). Ischemia-reperfusion injury is manifest by reperfusion pancreatitis leading to sepsis and, sometimes, exocrine leakage. Subsequent complications include mycotic aneurysm formation due to infection of vascular anastomoses—this appears to be a greater risk in pancreas transplantation, possibly because of contamination by enteric organisms at the time of organ retrieval or at the time of enteric anastomosis.

In contrast, islet transplantation is a safe procedure. The two most serious (but rare) procedure-related complications are portal vein thrombosis and bleeding. Although thrombosis of the main portal vein has been reported after islet transplantation using unpurified islets, the risk of this complication is now reduced substantially owing to more rigorous approaches to purification (reducing the volume of cells infused), anticoagulation, and portal pressure monitoring during infusion (Kawahara et al. 2011). However, segmental thrombosis has been reported in up to 5% of patients (Owen et al. 2003). This risk has been reduced by adding heparin to the islet preparation and by systemic anticoagulation posttransplantation. Hemorrhage from the hepatic puncture site was reported in up to 8% of cases (Owen et al. 2003), but this complication has been reduced with the introduction of techniques to embolize the catheter tract following islet infusion.

Challenges in β-Cell Transplantation

Currently, the morbidity of solid-organ pancreas transplantation restricts pancreas transplantation to relatively younger and fitter patients (very few patients older than 60 yr are accepted for solid-organ pancreas transplantation). Access to islet transplantation is restricted by different parameters, including body mass index and renal function. As the mortality and morbidity of solid pancreas transplantation diminish and the longer-term outcomes of both solid-organ and islet transplantation improve, the appropriate indications for both procedures will expand, particularly with the increasing incidence of diabetes as well as evidence that transplantation is suitable not only for type 1 diabetics but also for selected insulin-dependent patients with type 2 diabetes (nonobese, non-insulin resistant).

 Cite this article as *Cold Spring Harb Perspect Med* doi: 10.1101/cshperspect.a015610

There is a clear challenge to improve the utilization of potential donor organs to the level currently achieved in kidney and liver transplantation. This will require new approaches to organ retrieval, preservation, assessment, and the management of ischemia-reperfusion injury.

The process of islet isolation and culture still results in more than 30% of donor organs not producing a transplantable islet yield, and this is compounded by the fact that many patients still require two infusions to achieve optimal function. In the future, novel enzyme blends will specifically target the pancreatic matrix of the full range of different donor types.

Developments in immunosuppression will reduce the morbidity of transplantation and expand the criteria for transplantation. Tolerance strategies, including T-regulatory cells, will reduce the long-term risks and further shift the risk–benefit threshold of pancreatic transplantation. Novel methods are needed for graft monitoring by immunological or biochemical biomarkers.

It is widely assumed that the development of stem cell therapy will displace allotransplantation in the management of diabetes. Although this is a reasonable long-term vision, the development of insulin-producing cells is not yet close to the point at which clinical implementation is feasible.

REFERENCES

Axelrod DA, Sung RS, Meyer KH, Wolfe RA, Kaufman DB. 2010. Systematic evaluation of pancreas allograft quality, outcomes and geographic variation in utilization. *Am J Transplant* **10:** 837–845.

Bassi R, Fiorina P. 2011. Impact of islet transplantation on diabetes complications and quality of life. *Curr Diab Rep* **11:** 355–363.

Bazerbachi F, Selzner M, Marquez MA, Norgate A, Aslani N, McGilvray ID, Schiff J, Cattral MS. 2012. Portal venous versus systemic venous drainage of pancreas grafts: Impact on long-term results. *Am J Transplant* **12:** 226–232.

Berney T, Johnson PR. 2010. Donor pancreata: Evolving approaches to organ allocation for whole pancreas versus islet transplantation. *Transplantation* **90:** 238–243.

Cabric S, Sanchez J, Lundgren T, Foss A, Felldin M, Källen R, Salmela K, Tibell A, Tufveson G, Larsson R, et al. 2007. Islet surface heparinization prevents the instant blood-mediated inflammatory reaction in islet transplantation. *Diabetes* **56:** 2008–2015.

Callaghan BC, Little AA, Feldman EL, Hughes RA. 2012. Enhanced glucose control for preventing and treating diabetic neuropathy. *Cochrane Database Syst Rev* **6:** CD007543.

Campbell PM, Senior PA, Salam A, Labranche K, Bigam DL, Kneteman NM, Imes S, Halpin A, Ryan EA, Shapiro AM. 2007. High risk of sensitization after failed islet transplantation. *Am J Transplant* **7:** 2311–2317.

Chow V, Pai R, Chapman J, O'Connell P, Allen R, Mitchell P, Nankivell BJ. 1999. Diabetic retinopathy after combined kidney–pancreas transplantation. *Clin Transplant* **13:** 356–362.

Cosio FG, Hickson LJ, Griffin MD, Stegall MD, Kudva Y. 2008. Patient survival and cardiovascular risk after kidney transplantation: The challenge of diabetes. *Am J Transplant* **8:** 593–599.

Dardenne S, Sterkers A, Leroy C, Da Mata L, Zerbib P, Pruvot FR, Pattou F, Truant S. 2012. Laparoscopic spleen-preserving distal pancreatectomy followed by intramuscular autologous islet transplantation for traumatic pancreatic transection in a young adult. *JOP* **13:** 285–288.

Fioretto P, Steffes MW, Sutherland DE, Goetz FC, Mauer M. 1998. Reversal of lesions of diabetic nephropathy after pancreas transplantation. *N Engl J Med* **339:** 69–75.

Fiorina P, La Rocca E, Astorri E, Lucignani G, Rossetti C, Fazio F, Giudici D, di Carlo V, Cristallo M, Pozza G, et al. 2000. Reversal of left ventricular diastolic dysfunction after kidney–pancreas transplantation in type 1 diabetic uremic patients. *Diabetes Care* **23:** 1804–1810.

Fiorina P, Folli F, Bertuzzi F, Maffi P, Finzi G, Venturini M, Socci C, Davalli A, Orsenigo E, Monti L, et al. 2003. Long-term beneficial effect of islet transplantation on diabetic macro-/microangiopathy in type 1 diabetic kidney-transplanted patients. *Diabetes Care* **26:** 1129–1136.

Gerber PA, Pavlicek V, Demartines N, Zuellig R, Pfammatter T, Wuthrich R, et al. 2008. Simultaneous islet–kidney vs pancreas–kidney transplantation in type 1 diabetes mellitus: A 5 year single centre follow-up. *Diabetologia* **51:** 110–119.

Giannarelli R, Coppelli A, Sartini MS, Del Chiaro M, Vistoli F, Rizzo G, Barsotti M, Del Prato S, Mosca F, Boggi U, et al. 2006. Pancreas transplant alone has beneficial effects on retinopathy in type 1 diabetic patients. *Diabetologia* **49:** 2977–2982.

Gruessner AC, Sutherland DE, Gruessner RW. 2012. Long-term outcome after pancreas transplantation. *Curr Opin Organ Transplant* **17:** 100–105.

Hughes SJ, Clark A, McShane P, Contractor HH, Gray DW, Johnson PR. 2006. Characterisation of collagen VI within the islet–exocrine interface of the human pancreas: Implications for clinical islet isolation? *Transplantation* **81:** 423–426.

Humar A, Ramcharan T, Kandaswamy R, Gruessner RW, Gruessner AG, Sutherland DE. 2004. The impact of donor obesity on outcomes after cadaver pancreas transplants. *Am J Transplant* **4:** 605–610.

Ihm SH, Moon HJ, Kang JG, Park CY, Oh KW, Jeong IK, Oh YS, Park SW. 2007. Effect of aging on insulin secretory function and expression of β cell function–related genes of islets. *Diabetes Res Clin Pract* **77:** S150–S154.

Kandaswamy R, Stock PG, Skeans MA, Gustafson SK, Sleeman EF, Wainright JL, Carrico RJ, Ghimire V, Snyder JJ,

Israni AK, et al. 2013. OPTN/SRTR 2011 Annual Data Report: Pancreas. *Am J Transplant* **13:** 47–72.

Kandula P, Fridell J, Taber TE, Sharfuddin A, Yaqub MS, Phillips CL, Chen J, Mujtaba M. 2012. Impact of tacrolimus–sirolimus maintenance immunosuppression on proteinuria and kidney function in pancreas transplant alone recipients. *Transplantation* **94:** 940–946.

Karl RC, Scharp DW, Ballinger WF, Lacy PE. 1977. Transplantation of insulin-secreting tissues. *Gut* **18:** 1062–1072.

Kawahara T, Kin T, Kashkoush S, Gala-Lopez B, Bigam DL, Kneteman NM, Koh A, Senior PA, Shapiro AM. 2011. Portal vein thrombosis is a potentially preventable complication in clinical islet transplantation. *Am J Transplant* **11:** 2700–2707.

Kawamura T, Kuroda Y, Saitoh Y. 1992. Long-term preservation of canine pancreas by a new simple cold storage method using perfluorochemical—The two-layer cold storage method (Euro-Collins' solution/perfluorochemical). *Kobe J Med Sci* **38:** 135–145.

Kelly WD, Lillehei RC, Merkel FK, Idezuki Y, Goetz FC. 1967. Allotransplantation of the pancreas and duodenum along with the kidney in diabetic nephropathy. *Surgery* **61:** 827–837.

Kempf MC, Andres A, Morel P, Benhamou PY, Bayle F, Kessler L, Badet L, Thivolet C, Penfornis A, Renoult E, et al. 2005. Logistics and transplant coordination activity in the GRAGIL Swiss–French multicenter network of islet transplantation. *Transplantation* **79:** 1200–1205.

Kennedy WR, Navarro X, Goetz FC, Sutherland DER, Najarian JS. 1990. Effects of pancreatic transplantation on diabetic neuropathy. *N Engl J Med* **322:** 1031–1037.

Kin T, Senior P, O'Gorman D, Richer B, Salam A, Shapiro AM. 2008. Risk factors for islet loss during culture prior to transplantation. *Transpl Int* **21:** 1029–1035.

Klassen DK, Weir MR, Cangro CB, Bartlett ST, Papadimitriou JC, Drachenberg CB. 2002. Pancreas allograft biopsy: Safety of percutaneous biopsy—Results of a large experience. *Transplantation* **73:** 553–555.

Kuo HT, Sampaio MS, Vincenti F, Bunnapradist S. 2010. Associations of pretransplant diabetes mellitus, new-onset diabetes after transplant, and acute rejection with transplant outcomes: An analysis of the Organ Procurement and Transplant Network/United Network for Organ Sharing (OPTN/UNOS) database. *Am J Kidney Dis* **56:** 1127–1139.

Lakey JR, Rajotte RV, Warnock GL, Kneteman NM. 1995. Human pancreas preservation prior to islet isolation. Cold ischemic tolerance. *Transplantation* **59:** 689–694.

Lakey JR, Warnock GL, Rajotte RV, Suarez-Alamazor ME, Ao Z, Shapiro AM, Kneteman NM. 1996. Variables in organ donors that affect the recovery of human islets of Langerhans. *Transplantation* **61:** 1047–1053.

La Rocca E, Fiorina P, di Carlo V, Astorri E, Rossetti C, Lucignani G, Fazio F, Guidici D, Cristallo M, Biachi G, et al. 2001. Cardiovascular outcomes after kidney–pancreas and kidney alone transplantation. *Kidney Int* **60:** 1964–1971.

Lee TC, Barshes NR, O'Mahony CA, Nguyen L, Brunicardi FC, Ricordi C, Alejandro R, Schock AP, Mote A, Goss JA. 2005. The effect of pancreatic islet transplantation on progression of diabetic retinopathy and neuropathy. *Transplant Procs* **37:** 2263–2265.

Lillehei RC, Simmons RL, Najarian JS, Goetz FC. 1970. Pancreatico-duodenal and renal allotransplantation in juvenile onset, insulin dependent, diabetes mellitus with terminal nephropathy. *Langenbecks Arch Chir* **326:** 88–105.

Luan FL, Miles CD, Cibrik DM, Ojo AO. 2007. Impact of simultaneous pancreas and kidney transplantation on cardiovascular risk factors in patients with type 1 diabetes mellitus. *Transplantation* **84:** 541–544.

Margreiter C, Pratschke J, Margreiter R. 2013. Immunological monitoring after pancreas transplantation. *Curr Opin Organ Transplant* **18:** 71–75.

Muthusamy AS, Giangrande PL, Friend PJ. 2010. Pancreas allograft thrombosis. *Transplantation* **90:** 705–707.

Muthusamy AS, Mumford L, Hudson A, Fuggle SV, Friend PJ. 2012. Pancreas transplantation from donors after circulatory death from the United Kingdom. *Am J Transplant* **12:** 2150–2156.

Nano R, Clissi B, Melzi R, Calori G, Maffi P, Antonioli B, Marzorati S, Aldrighetti L, Freschi M, Grochowiecki T, et al. 2005. Islet isolation for allotransplantation: Variables associated with successful islet yield and graft function. *Diabetologia* **48:** 906–912.

Navarro X, Kennedy WR, Loewenson RB, Sutherland DE. 1990. Influence of pancreas transplantation on cardiorespiratory reflexes, nerve conduction, and mortality in diabetes mellitus. *Diabetes* **39:** 802–806.

Naziruddin B, Wease S, Stablein D, Barton FB, Berney T, Rickels MR, Alejandro R. 2012. HLA class I sensitization in islet transplant recipients: Report from the Collaborative Islet Transplant Registry. *Cell Transplant* **21:** 901–908.

Niclauss N, Bosco D, Morel P, Demuylder-Mischler S, Brault C, Milliat-Guittard L, Colin C, Parnaud G, Muller YD, Giovannoni L, et al. 2011. Influence of donor age on islet isolation and transplantation outcome. *Transplantation* **91:** 360–366.

Niederhaus SV, Kaufman DB, Odorico JS. 2013. Induction therapy in pancreas transplantation. *Transpl Int* **26:** 704–714.

Ojo AO, Meier-Kriesche HU, Hanson JA, Leichtman A, Magee JC, Cibrik D, Wolfe RA, Port FK, Agodoa L, Kaufman DB, et al. 2001. The impact of simultaneous pancreas–kidney transplantation on long-term patient survival. *Transplantation* **71:** 82–90.

Owen RJ, Ryan EA, O'Kelly K, Lakey JR, McCarthy MC, Paty BW, Bigan DL, Kneteman NM, Korbutt GS, Rajotte RV, et al. 2003. Percutaneous transhepatic pancreatic islet cell transplantation in type 1 diabetes mellitus: Radiologic aspects. *Radiology* **229:** 165–170.

Prieto M, Sutherland DE, Fernandez-Cruz L, Heil J, Najarian JS. 1987. Experimental and clinical experience with urine amylase monitoring for early diagnosis of rejection in pancreas transplantation. *Transplantation* **43:** 73–79.

Proneth A, Schmitzbauer AA, Zeman F, Foerster JR, Holub I, Arbogast H, Bechstein WO, Becker T, Dietz C, Guba M, et al. 2013. Extended pancreas donor program—The EXPAND study rationale and study protocol. *Transplant Res* **2:** 12.

Pybus FC. 1924. Notes on suprarenal and pancreatic grafting. *Lancet* **ii:** 550–551.

Ricordi C, Lacy PE, Scharp DW. 1989. Automated islet isolation from human pancreas. *Diabetes* **38:** 140–142.

Robertson RP, Sutherland DE, Seaquist ER, Lanz KJ. 2003. Glucagon, catecholamine, and symptom responses to hypoglycemia in living donors of pancreas segments. *Diabetes* **52:** 1689–1694.

Ryan EA, Shandro T, Green K, Paty BW, Senior PA, Bigam D, Shapiro AM, Vantyghem MC. 2004. Assessment of the severity of hypoglycemia and glycemic lability in type 1 diabetic subjects undergoing islet transplantation. *Diabetes* **53:** 955–962.

Ryan EA, Paty BW, Senior PA, Bigam D, Alfadhli E, Kneteman NM, Lakey JR, Shapiro AM. 2005a. Five-year follow-up after clinical islet transplantation. *Diabetes* **54:** 2060–2069.

Ryan EA, Paty BW, Senior PA, Lakey JR, Bigam D, Shapiro AM. 2005b. β-Score: An assessment of β-cell function after islet transplantation. *Diabetes Care* **28:** 343–347.

Scheider A, Meyer-Schwickerath E, Nusser J, Land W, Landgraf R. 1991. Diabetic retinopathy and pancreas transplantation: A 3-year follow-up. *Diabetologia* **34:** S95–S99.

Shapiro AM, Lakey JR, Ryan EA, Korbutt GS, Toth E, Warnock GL, Kneteman NM, Rajotte RV. 2000. Islet transplantation in seven patients with type 1 diabetes mellitus using a glucocorticoid-free immunosuppressive regimen. *N Engl J Med* **343:** 230–238.

Shapiro AM, Ricordi C, Hering BJ, Auchincloss H, Lindblad R, Robertson RP, Secchi A, Brendel MD, Berney T, Brennan DC, et al. 2006. International trial of the Edmonton Protocol for islet transplantation. *N Engl J Med* **355:** 1318–1330.

Shtein RM, Callaghan BC. 2013. Corneal confocal microscopy as a measure of diabetic neuropathy. *Diabetes* **62:** 25–26.

Sollinger HW, Odorico JS, Becker YT, D'Alessandro AM, Pirsch JD. 2009. One thousand simultaneous pancreas–kidney transplants at a single center with 22-year follow-up. *Ann Surg* **250:** 618–630.

Sosna T, Saudek F, Dominek Z. 1998. Effect of successful combined renal and pancreatic transplantation on diabetic retinopathy. *Acta Univ Palacki Olomuc Fac Med* **141:** 75–77.

Speight J, Reaney MD, Woodcock AJ, Smith RM, Shaw JA. 2010. Patient-reported outcomes following islet cell or pancreas transplantation (alone or after kidney) in Type 1 diabetes: A systematic review. *Diabetic Med* **27:** 812–822.

Stadler M, Theuer E, Anderwald C, Hanusch-Enserer U, Auinger M, Bieglmayer C, Quehenberger P, Bischof M, Kästenbauer T, Wolzt M, et al. 2009. Persistent arterial stiffness and endothelial dysfunction following successful pancreas–kidney transplantation in Type 1 diabetes. *Diabet Med* **26:** 1010–1018.

Stewart ZA, Cameron AM, Singer AL, Dagher NN, Montgomery RA, Segev DL. 2009. Histidine-tryptophan-ketoglutarate (HTK) is associated with reduced graft survival in pancreas transplantation. *Am J Transplant* **9:** 217–221.

Stockland AH, Willingham DL, Paz-Fumagalli R, Grewal HP, McKinney JM, Hughes CB, Walser EM. 2009. Pancreas transplant venous thrombosis: Role of endovascular interventions for graft salvage. *Cardiovasc Intervent Radiol* **32:** 279–283.

Sutherland DE, Radosevich D, Gruessner R, Gruessner A, Kandaswamy R. 2012. Pushing the envelope: Living donor pancreas transplantation. *Curr Opin Organ Transplant* **17:** 106–115.

Taber DJ, Meadows HB, Pilch NA, Chavin KD, Baliga PK, Egede LE. 2013. Pre-existing diabetes significantly increases the risk of graft failure and mortality following renal transplantation. *Clin Transplant* **27:** 274–282.

Tavakoli M, Mitu-Pretorian M, Petropoulos IN, Fadavi H, Asghar O, Alam U, Ponirakis G, Jeziorska M, Marshall A, Efron N, et al. 2013. Corneal confocal microscopy detects early nerve regeneration in diabetic neuropathy after simultaneous pancreas and kidney transplantation. *Diabetes* **62:** 254–260.

The Diabetes Control and Complications Trial Research Group. 1993. The effect of intensive treatment of diabetes on the development and progression of long-term complications in insulin-dependent diabetes mellitus. *N Engl J Med* **329:** 977–986.

Thompson DM, Begg IS, Harris C, Ao Z, Fung MA, Meloche RM, Keown P, Meneilly GS, Shapiro RJ, Ho S, et al. 2008. Reduced progression of diabetic retinopathy after islet cell transplantation compared with intensive medical therapy. *Transplantation* **85:** 1400–1405.

Thompson DM, Meloche M, Ao Z, Paty B, Keown P, Shapiro RJ, Ho S, Worsley D, Fung M, Meneilly G, et al. 2011. Reduced progression of diabetic microvascular complications with islet cell transplantation compared with intensive medical therapy. *Transplantation* **91:** 373–378.

Troxell ML, Koslin DB, Norman D, Rayhill S, Mittalhenkle A. 2010. Pancreas allograft rejection: Analysis of concurrent renal allograft biopsies and posttherapy follow-up biopsies. *Transplantation* **90:** 75–84.

Tyden G, Bolinder J, Solders G, Brattstrom C, Tibell A, Groth CG. 1999. Improved survival in patients with insulin-dependent diabetes mellitus and end-stage diabetic nephropathy 10 years after combined pancreas and kidney transplantation. *Transplantation* **67:** 645–648.

van Dellen D, Worthington J, Mitu-Pretorian OM, Ghazanfar A, Forgacs B, Pararajasingam R, Campbell B, Parrott NR, Augustine T, Tavakoli A. 2013. Mortality in diabetes: Pancreas transplantation is associated with significant survival benefit. *Nephrol Dial Transplant* **28:** 1315–1322.

Overview of Clinical Lung Transplantation

Jonathan C. Yeung and Shaf Keshavjee

Toronto Lung Transplant Program, University Health Network, University of Toronto, Toronto, Ontario M5G 2C4, Canada

Correspondence: shaf.keshavjee@uhn.ca

Since the first successful lung transplant 30 years ago, lung transplantation has rapidly become an established standard of care to treat end-stage lung disease in selected patients. Advances in lung preservation, surgical technique, and immunosuppression regimens have resulted in the routine performance of lung transplantation around the world for an increasing number of patients, with wider indications. Despite this, donor shortages and chronic lung allograft dysfunction continue to prevent lung transplantation from reaching its full potential. With research into the underlying mechanisms of acute and chronic lung graft dysfunction and advances in personalized diagnostic and therapeutic approaches to both the donor lung and the lung transplant recipient, there is increasing confidence that we will improve short- and long-term outcomes in the near future.

The worldwide burden of pulmonary diseases continues to increase and now ranks firmly among the leading causes of death (Lozano et al. 2013). Since its introduction in the early 1980s, lung transplantation has matured into a successful therapy for selected patients with end-stage lung disease, because of careful investigation into donor management, organ preservation, recipient management, and immunosuppression (Christie et al. 2012). This review provides an overview of modern clinical lung transplantation and anticipated future directions.

HISTORY

Attempts at lung transplantation have occurred as early as 1946 when Vladimir Demikhov, a Soviet scientist, attempted single-lung transplantation in a dog (Langer 2011). This transplant ultimately failed from bronchial anastomotic dehiscence, and difficulties with this anastomosis would plague clinical lung transplantation for the next 40 years. Henri Metras, in 1950, reported the first successful dog lung transplant and the first bronchial artery and left atrial anastomoses (Metras 1950). In a nonhuman primate model, Haglin et al. (1963) performed lung reimplantation and showed that these lungs were able to maintain function postoperatively, despite denervation. On June 11, 1963, Hardy et al. (1963) reported the first human lung transplant; however, the patient died from kidney failure after 18 d. The first real survivor during this early era of lung transplantation was a patient of Fritz Derom's in Belgium (Derom et al. 1971). This patient, however, survived only 10 mo. The failure of this early experience in clinical lung transplantation can be summarized by inadequate immunosuppression and difficulties with the bronchial anastomosis.

Cite this article as *Cold Spring Harb Perspect Med* doi: 10.1101/cshperspect.a015628

The advent of cyclosporine brought about significant improvements in patient survival following liver and kidney transplantation (Calne et al. 1978). This led to a resurgence of interest in heart/lung transplantation in Stanford and lung transplantation in Toronto (Starzl et al. 1981). The first successful combined heart–lung transplant was completed by Reitz and colleagues and showed that a grafted lung could survive and function in a recipient (Reitz et al. 1982). Research performed by Cooper's group in Toronto showed that corticosteroid use appeared to be a significant factor in the weakness of the bronchial anastomosis (Goldberg et al. 1983). With the use of cyclosporine, corticosteroid use could be reduced, leading to improved bronchial healing. In 1986, the Toronto Lung Transplant Program reported the first successful single-lung transplantations for two patients with pulmonary fibrosis (Cooper et al. 1987). This team went on to perform the first successful double-lung transplant, first with an *en bloc* technique that used a tracheal anastomosis, then evolving to the bilateral sequential transplantation technique that not only improved airway healing, but also had the additional benefit of avoiding cardiopulmonary bypass, if desired (Patterson et al. 1988). This technique remains the standard technique in use to this day.

INDICATIONS AND CONTRAINDICATIONS TO LUNG TRANSPLANTATION

The indications for lung transplantation can be broadly separated into the following main categories of end-stage lung diseases: obstructive lung disease, septic lung disease, fibrotic lung disease, and vascular lung disease. Of these categories, chronic obstructive pulmonary disease (COPD), cystic fibrosis (CF), interstitial pulmonary fibrosis (IPF), and primary pulmonary arterial hypertension make up the most common indication in each category, respectively (Christie et al. 2012). Lung transplantation for pulmonary malignancy has also been shown to be effective in highly selected patients (Machuca and Keshavjee 2012).

The current International Society for Heart and Lung Transplantation (ISHLT) and also the American Thoracic Society (ATS) selection criteria include appropriate age, clinically and physiologically severe disease, ineffective or unavailable medical therapy, substantial limitations in activities of daily living, limited life expectancy, adequate cardiac function without significant coronary disease, ambulatory with rehabilitation potential, acceptable nutritional status, and satisfactory psychosocial profile and emotional support system (Orens et al. 2006). Patients who undergo lung transplantation subject themselves to lifelong immunosuppression and surveillance. Thus, candidates for lung transplantation who have smoking or drug dependency, psychiatric issues affecting compliance with postoperative care, or absence of a reliable social support network will generally not be listed for lung transplantation. Previous malignancy, particularly in the 2 yr leading up to potential transplantation, is also a contraindication because of the need for lifelong immunosuppression. Indeed, although immunosuppression will potentiate infection, chronic infection is an intrinsic part of septic lung diseases and creates a challenge in these patients. In cystic fibrosis patients, nontuberculous mycobacterial, multi-drug-resistant bacteria, and *Aspergillus* are commonly isolated (Helmi et al. 2003; Gilljam et al. 2010). An effort is made to eradicate or at least minimize these organisms before and following transplantation. Moreover, in the presence of septic lung disease, a bilateral lung transplant is the only operation that can be performed because leaving a septic lung in a patient to be immunosuppressed is not advisable. The replacement of both organs removes the burden of infectious organisms and allows for successful transplantation in the presence of most septic lung conditions. One notable exception is the *Burkholderia cepacia* complex (Bcc), an organism with a significant negative impact on posttransplant survival (Murray et al. 2008). In multiple reports, patients infected with Bcc, particularly the subspecies *Burkholderia cenocepacia*, have a clearly worse survival than those without. Consequently, some transplant centers consider Bcc infection an absolute contraindication to transplantation, but this remains controversial (Chaparro et al. 2001).

Patients with significant dysfunction of other vital organs also represent potential contraindications to lung transplantation. Owing to age, general prevalence, and smoking, cardiac dysfunction is the most common concomitant organ dysfunction in the lung transplant population. Noninvasive and invasive cardiac testing is performed during workup to ensure adequate cardiac function, and concomitant coronary arterial bypass grafting can be performed at the time of lung transplantation if the left ventricular function is preserved (Patel et al. 2003).

Historically, age >65 has been considered a relative contraindication to lung transplantation given the propensity toward worse outcomes in that group. However, 7.7% of lung transplants performed in the recent era (2004–2011) were in patients >65 (Christie et al. 2012). Another 21.6% were performed in patients between 60 and 65. Careful assessment of comorbidities, rather than age alone, determines candidacy for transplantation.

DONOR LUNG ALLOCATION

In the United States, lung allocation was initially based on wait times alone with no consideration of disease severity or potential benefit. Many recipients were placed early on the waiting list in order to accrue time, but when a lung was ultimately made available, the patient may not have yet deteriorated to a point requiring transplantation. On the other hand, patients in more dire need of a transplant would often not survive long enough to accrue enough time to be offered a graft. In 2005, a new strategy for lung allocation was introduced (Egan et al. 2006) known as the lung allocation score (LAS). This score is conceptually based on the degree of need for a transplant and the probability of posttransplant survival. To calculate this, a measure of urgency (expected number of days lived without a transplant during an additional wait-list year) and a measure of survival following transplantation (expected number of days lived in the first year posttransplant) are calculated using statistical models based on the patient's clinical and physiological characteristics. The urgency measure is subtracted from the benefit measure and then normalized to give an LAS. Higher scores are allocated lungs sooner. Since the introduction of the LAS system, wait-list times have decreased, and the number of transplants has increased (Iribarne et al. 2009). Moreover, LAS scores have gradually increased, representing the increasing urgency of patients getting listed.

PEDIATRIC LUNG TRANSPLANTATION

Lung transplantation for children has also evolved into acceptable therapy (Wells and Faro 2006). In the United States, more than 100 children currently await lung transplantation. The indications for pediatric lung transplant differ from those for adults and stratify based on age. Children <1 yr of age suffer largely from congenital defects such as congenital heart disease, surfactant dysfunction, and pulmonary vascular disorders (Benden et al. 2013). Cystic fibrosis (CF) becomes the primary indication after 1 yr of age, with the majority of adolescent lung transplants being performed for CF. Identifying acceptable candidates for pediatric lung transplantation is difficult, particularly in the very young. The worldwide experience is limited in this population, and the natural history of many of these congenital diseases is not always clear. Moreover, readiness of the parents to embark on the demanding preoperative and postoperative care necessary of pediatric lung transplantation must be carefully assessed. Allocation of organs to pediatric recipients uses the LAS system in patients 12 yr of age or older (Sweet 2009). For patients under 12 yr of age, the United States uses a two-priority system in addition to wait times (Colvin-Adams et al. 2012). Patients meeting criteria (Table 1) are priority 1 and receive consideration for graft allocation first following geographic and blood-type criteria considerations. In the very young, where isohemagglutinins have not yet developed, ABO-incompatible lung transplantation has been shown to be a feasible option (Grasemann et al. 2012).

BRIDGING TO LUNG TRANSPLANTATION

Often, patients needing lung transplantation become critically ill before a set of donor organs

Table 1. Pediatric priority 1 criteria

Respiratory failure as defined by one or more of the following:
Continuous mechanical ventilation
Fraction of inspired oxygen (FiO_2) >50% to maintain saturation levels of >90%
An arterial or capillary PCO_2 of >50 mm Hg or a venous PCO_2 of >56 mm Hg
Pulmonary hypertension as defined by any of the following and on medical therapy:
Pulmonary vein stenosis involving three or more vessels
Suprasystemic PA pressures
Cardiac index 2 $L/min/m^2$

Data from Colvin-Adams et al. (2012).

is available. Traditionally, intubation and ECLS (extracorporeal lung support) were the sole modalities available for bridging to transplantation. In the past few years, novel artificial lung technologies have been developed that allow critically ill patients to be maintained temporarily over months (Fischer et al. 2006). One of these devices is the Novalung, a hollow fiber oxygenator with a uniquely low *trans*-device perfusion resistance. This allows for pumpless function from the arterial to venous circulation. Patients in respiratory failure can be divided into those with hypercapnic respiratory failure, those with hypoxemic respiratory failure, or those with both. Because removal of CO_2 requires only a low membrane flow on the order of 0.5–1 L/min, this can be achieved using the pumpless arteriovenous cannulation. An extension of this strategy can be applied to patients with pulmonary hypertension awaiting transplantation. By inserting the device between the pulmonary artery and the left atrium, the right heart can be unloaded into the Novalung device (Strueber et al. 2009). This provides a decompressing and oxygenating shunt. This treatment strategy unloads the right ventricle, allowing it to recover, and also leads to recovery of renal and hepatic dysfunction. All of this has dramatically improved the survival to and after lung transplantation for patients with primary pulmonary arterial hypertension (22% wait-list mortality pre-Novalung, 0% wait-list mortality post-Novalung utilization) (de Perrot et al. 2011).

For patients who suffer from more than pure hypercapnic respiratory failure, full venovenous or venoarterial ECLS is a better choice. Novel bicaval dual lumen cannulas, such as the Avalon cannula, allow venous–venous ECMO to be started via a single cannulation site in the neck (Wang et al. 2008). In case series, patients awaiting lung transplantation have avoided intubation and could continue an exercise regimen while on ECMO (Garcia et al. 2011; Cypel and Keshavjee 2012).

DONOR SELECTION

Once a donor has been identified and allocated to a potential recipient, the donor organs must be checked for quality and procured from the donor. Like all of solid organ transplantation, lung transplantation is even more limited by the number of available donor organs. Shortages in lung transplantation, however, are compounded by a low utilization rate of offered donor organs. The United Network for Organ Sharing data reports that lungs were used from only 1708 of 8143 deceased donors in 2012—a utilization of only 21% (National Data 2012). As a comparison, 7419 of those same donors were kidney donors. The current International Society for Heart and Lung Transplantation (ISHLT) criteria outlining an ideal donor are based on strict criteria established during the development of lung transplantation (Table 2). Although this has helped establish safe clinical

Table 2. ISHLT criteria for lung acceptance

Age <55 yr
ABO compatibility
Clear chest radiograph
PaO_2 >300 on FiO_2 = 1.0, PEEP 5 cm H_2O
Tobacco history <20 pack/yr
Absence of chest trauma
No evidence of aspiration/sepsis
No prior cardiopulmonary surgery
Sputum gram stain—absence of organisms
Absence of purulent secretions at bronchoscopy

From Orens et al. (2003); reproduced, with permission, from the authors.

lung transplantation, this combination of stringent donor criteria coupled with an increased susceptibility of donor lungs to injury leads to a conservative and low utilization rate (Orens et al. 2003).

DONOR LUNG INJURY

Currently, the largest pool of donor organs is those retrieved from brain-death donors. In these donors, cessation of neurologic function results in a legal definition of death (Wijdicks et al. 2010), but organs remain viable for a period of time owing to preserved cardiac function and ICU support. Although this situation is seemingly ideal for organ transplantation, many factors can contribute to donor lung injury during the process of donor death and subsequent donor maintenance in the intensive care unit (ICU). Direct trauma, aspiration, pneumonia, and complications of ICU care such as ventilator-induced lung injury, atelectasis, oxygen toxicity, and volume overload are all common causes of injury. More importantly, it is well recognized that the process of brain death itself can injure potential donor organs. Following brain death, a systemic inflammatory response known as a "cytokine storm" transpires. Increased proinflammatory cytokines have been found in organs following brain death in rodent models (Takada et al. 1998) and in brain-dead patients (Lopau et al. 2000), and lung injury similar to acute respiratory distress syndrome (ARDS) can occur as a result of this systemic inflammation following brain death. A second "storm" that occurs following brain death is the catecholamine storm. In an attempt to protect cerebral perfusion during brain death, the body releases a large amount of catecholamines. This surge of catecholamines causes significant systemic hypertension, which results in elevated left-sided heart pressures and consequent interstitial edema and sometimes even alveolar hemorrhage in the lung, resulting in a state known as neurogenic pulmonary edema. This generally precludes the use of the lungs for transplantation because of poor oxygenation. However, neurogenic pulmonary edema is a leaky capillary syndrome that is fully recoverable. This is one area where removal of the lung from the inflammatory milieu of the brain-dead donor and a period of time for recovery and removal of extravascular water with ex vivo lung perfusion has had a significant impact on donor lung utilization as described below.

DONATION AFTER CARDIAC DEATH

As an alternative source for lungs, some transplant programs have begun to reexplore the use of donation after cardiac death (DCD) (Kootstra et al. 2002; Steinbrook 2007). Because the majority of patients succumb as a result of cardiac arrest, the use of DCDs could potentially open a completely new pool of donor organs of such magnitude that ultimately the entire demand could be met. Renewed interest in the potential use of lungs from DCDs followed a series of experiments in dogs by Egan and coworkers in the early nineties (Egan et al. 1991; Ulicny et al. 1993). His group showed that lung cells remain viable for a certain period after circulatory arrest (Alessandrini et al. 1994; D'Armini et al. 1994). The lung is the sole solid organ that is not dependent on perfusion for aerobic metabolism but rather uses a mechanism of passive diffusion through the alveoli for substrate delivery (Keshavjee et al. 1992; Date et al. 1993; Egan 2004; Van Raemdonck et al. 2004).

At the First International Workshop on DCDs in Maastricht in the Netherlands in 1995, four types of donors were identified, so-called Maastricht categories (Table 3) (Kootstra et al. 1995). Categories I (dead on arrival) and II (unsuccessful resuscitation) comprise the *uncontrolled* donors. Categories III (awaiting car-

Table 3. Maastricht categories of donation after cardiac death

I	Dead on arrival to hospital	Uncontrolled
II	Unsuccessful resuscitation	Uncontrolled
III	Awaiting cardiac arrest	Controlled
IV	Cardiac arrest following brain death	Controlled
V	Cardiac arrest in a hospital inpatient	Uncontrolled

Data from Sánchez-Fructuoso et al. (2000).

diac arrest) and IV (cardiac arrest in brain-dead donor) include the *controlled* donors. A fifth category, cardiac arrest in a hospital inpatient, has recently been added (Sánchez-Fructuoso et al. 2000).

Several centers worldwide have now adopted DCD programs in their clinical routine of lung transplantation, and most reported series deal with controlled donation after withdrawal from life support (category III) (Erasmus et al. 2006; Butt et al. 2007; Oto et al. 2007; Van Raemdonck et al. 2008). The total reported experience with this category now amounts to more than 100 patients worldwide. The majority of reported series using category III DCDs have comparable results to a series of donation after brain death (DBD) patients. A recent review of the U.S. experience using data retrospectively collected from the UNOS database has shown an overall survival after lung transplantation of 94%, 94%, 94%, 94%, and 87% at 1, 3, 6, 12, and 24 mo, respectively, for recipients receiving lungs from DCD donors, compared with 92%, 88%, 84%, 78%, and 69%, respectively, from DBD (Mason et al. 2008).

DONOR LUNG EVALUATION

Evaluation of an offered donor lung block is largely a clinical process. Other than blood group and organ size matching, the majority of the evaluation is relatively subjective and occurs at the time of retrieval (Boasquevisque et al. 2009). Before opening the chest, a chest X-ray is taken and a bronchoscopy performed to exclude gross infection or anatomical abnormalities. The gas exchange capacity of the donor lungs is assessed with an oxygen challenge. Once the chest is open, the surgeon performs a final gross physical evaluation by macroscopic observation and palpation to assess lung compliance and edema and to exclude intrinsic lung disease, areas of contusion, pneumonic infiltrates, or nodules. Observation of the ventilated lungs during deflation is used to further assess pulmonary compliance. If the lungs are deemed usable, they are retrieved and sent to the transplant center for implantation. Evaluation of the DCD donor is similar, but functional testing is limited to before cardiac arrest; therefore, lung evaluation is less assured.

ORGAN PRESERVATION

In the initial early experience of lung transplantation, donors were brought into an adjacent operating room, and the donor lungs were deflated and topically cooled in situ for preservation (Cooper et al. 1987). The lungs were then stored deflated and submerged in hypothermic saline solution and placed on ice for hypothermic preservation. Despite short ischemic times, immediate postoperative care was fraught with significant postoperative primary graft dysfunction and poor preservation of the pulmonary and bronchial microcirculation. Only with subsequent advances in lung preservation has lung transplantation been able to be established as a clinical routine. Modern lung preservation is a continuum that begins at the time of declaration of death and ends after reperfusion in the recipient. In the preretrieval phase, care is taken to maintain euvolemia and to avoid ventilator barotrauma. A methylprednisolone bolus of 1 g is given to the donor as an anti-inflammatory agent to mitigate brain-death-induced inflammation and also in part to replace any steroid deficiencies (Follette et al. 1998; Venkateswaran et al. 2008). At the time of retrieval, the lungs are prepared for transport flushed with a preservation solution and inflated with oxygen. Use of an extracellular-type flush preservation solution with low potassium, coupled with glucose and dextran, was found to be best for prolonged cold preservation (Keshavjee et al. 1989, 1992). This has become the standard of lung preservation in the majority of lung transplant programs. PGE_1 (prostaglandin E_1) is a vasodilator that is given before flush to reduce the pulmonary vascular resistance in order to help produce a homogeneous flush. It also has anti-inflammatory properties useful for lung preservation and reperfusion injury (De Perrot et al. 2001). A retrograde in situ lung flush is also performed with the same solution, again to improve the homogeneity of the flush (Varela et al. 1997; Van De Wauwer et al. 2009). The lungs are inflated with 50% oxygen before their removal from the body in order to maintain the delicate alveolar structure and to provide oxygen for metabolism (Van Raemdonck et al. 1997). This is an important distinction of lung preservation as compared with other organ preservation in that the lung is stored in an *aerobic hypothermic ischemic* state so that aerobic metabolism can continue during the storage phase, albeit at a reduced rate.

Recently, a novel strategy of clinical lung preservation has been developed by our group. Known as ex vivo lung perfusion, lungs are continuously perfused anterograde with an acellular perfusate and ventilated with room air with an ICU ventilator at normothermia (Cypel et al. 2008). An oxygenator fed with a hypoxic gas is used to deoxygenate the perfusate and add CO_2 to the perfusate. We have shown that donor lungs can be preserved for at least 12 h safely using this method. Because the lungs will function using this strategy of preservation, they can be further evaluated as to their suitability for transplantation, but, more importantly, can be treated actively to improve their performance (Cypel et al. 2009). DCD lungs, given the less than optimal opportunity for lung evaluation after cardiac arrest, can greatly benefit from this technique. Marginal lungs can also be resuscitated during this time using existing and future therapies (Yeung et al. 2012). In a recent clinical trial, marginal lungs normally rejected for transplantation were evaluated and resuscitated using ex vivo lung perfusion and then transplanted with resulting outcomes equivalent to contemporaneous controls (Cypel et al. 2011, 2012). Because the large majority of offered donor lungs are rejected because of injury, this expansion of the donor pool is significant and could greatly help meet the need for donor lungs. In the future, we envision that retrieved lungs could be sent to organ evaluation and repair centers for final assessment and treatment for repair before distribution to recipients around the country. A case report describing the retrieval of injured lungs at one center, EVLP repair at a second center, and then transplantation at a third center across multiple political jurisdictions acts as a proof-of-concept for this approach (Wigfield et al. 2012).

LUNG TRANSPLANT PROCEDURES

Different lung transplant procedures have been developed to meet certain situations. Most familiar are: the single-lung transplant, where either a left or a right lung is transplanted into a recipient and the contralateral native lung is left in place; and the bilateral lung transplant, where both lungs are transplanted. The decision to transplant one or two lungs depends on the indication for transplantation and recipient factors, as well as donor availability. Emerging data suggest that bilateral lung transplantation results in prolonged survival because it provides more functional lung tissue as a buffer against lung dysfunction (Thabut et al. 2008; Weiss et al. 2009). For the same reason, marginal donor organs are often more likely to be used when implanted as a bilateral lung transplant. However, double-lung transplantation is a more complex operation, which necessitates a longer operative time that needs to be balanced against potential added morbidity. In addition, in jurisdictions where donor availability is severely lacking, transplanting two recipients with single lungs may reduce wait-list mortality. Specific indications also call for double-lung transplantation. Because of the presence of infection, septic lung disease patients require a double-lung transplant to prevent infection of the lung graft from the native lung. Although not an absolute indication, bilateral lung transplantation for vascular lung disease with elevated pulmonary arterial pressures is best because it avoids difficulties with hemodynamic instability and severe primary graft dysfunction postoperatively.

Although the above-mentioned lung transplant procedures are by far the most common, alternative procedures have been developed that can be used in unique situations. The donor lung can be nonanatomically volume-reduced to fit into a smaller recipient (Noirclerc et al. 1992). For larger size discrepancies, an adult donor lung can be downsized anatomically to implant a lower or upper lobe on either side in a smaller recipient (bilateral lobar transplant) (Starnes et al. 1994). Another innovative operation is the split left-lung bilateral lobar transplantation technique, which allows for a large donor to donate a single left lung to a smaller recipient or child, where the left upper lobe and left lower lobe are implanted to the right and left chest, respectively (Couetil et al. 1997). Living donor lobar transplantation is applied in jurisdictions where the legal and cultural climate is not conducive toward deceased donor transplantation, such as in Japan. In this situation, the right lower lobe is removed from one donor and transplanted into the right chest, and the left lower lobe is removed from another donor and transplanted into the left chest. Date and colleagues in Japan have a large experience with this technique and have shown excellent outcomes (Date et al. 2004). Experience in the United States with this technique has waned since the introduction of the LAS allocation system.

Combined heart/lung transplantation is reserved for cases in which irreversible heart disease occurs in conjunction with irreversible lung disease. In most cases, severe right heart dysfunction can be reversed after isolated lung transplantation, and thus heart/lung transplantation is generally reserved for patients with left heart dysfunction.

SURGICAL TECHNIQUE

In general, most lung transplant centers avoid the routine use of cardiopulmonary bypass (CPB) during lung transplantation. For single-lung transplantation, the diseased lung is excised and the new lung implanted while the patient is supported on the contralateral native lung. If the patient cannot be supported this way because of significant hypoxia, hypercarbia, or hemodynamic instability, then CPB is instituted.

Bilateral lung transplants are performed as sequential single-lung transplants via a clamshell incision or bilateral anterolateral thoracotomies if preferred and technically feasible (Taghavi et al. 1999). The sequential transplant strategy not only allows for the avoidance of CPB in most (60%–75%) cases, but also avoids a tracheal anastomosis, which was used in the original *en bloc* double-lung transplant operation (Griffith et al. 1994). The tracheal anastomosis was fraught with anastomotic complications related to the precarious blood supply of the graft airway at that level. Following chest opening, mobilization of each of the native lungs is performed. The lung with the poorest function is removed first, and implantation of the first graft begins. Once complete, this first allograft is perfused and ventilated. The recipient is then dependent on this lung's function as the contralateral side is removed and the second lung implanted. Hemodynamic instability, elevated pulmonary arterial pressure, or hypoxemia necessitates the use of cardiopulmonary bypass. Ultimately, only ~25% of recipients with no pulmonary vascular disease will require CPB (de Hoyos et al. 1993).

POSTOPERATIVE CARE

Care of the lung recipient in the immediate postoperative period focuses on ventilatory and hemodynamic support and weaning. In uncomplicated cases, ventilator weaning can proceed quickly over the first 6–12 h. Because of increased vascular permeability and severed lymphatics of the transplanted lung, development of some element of pulmonary edema is not uncommon. Therefore, attempts to minimize pulmonary capillary wedge pressure and central venous pressure are made while balancing for the need to maintain systemic perfusion (Pilcher et al. 2005). Often, vasopressors and inotropes are used rather than infusion of volume.

The major fear postoperatively is failure of the transplanted lung to function. Such a scenario can be mimicked by volume overload, cardiac dysfunction, pneumonia, aspiration, antibody-mediated rejection, pulmonary thromboembolism, or technical complications such as occlusion of venous outflow (de Perrot et al. 2003). In the absence of these conditions, primary graft dysfunction (PGD) becomes the diagnosis of exclusion. PGD is the major cause of morbidity and mortality in the first 72 h following transplantation. It can be considered the expression of all injury acquired by the lung while in the donor through to the time of reperfusion. Generally, this injury presents as pulmonary edema resulting in decreased compliance and ineffective oxygenation. This pulmonary edema results from increased permeability of pulmonary capillaries and alveolar damage and is reminiscent of ARDS. The ISHLT has published a grading system for PGD similar to that for ARDS (Christie et al. 2005). Patients are scored on this grading system at the time immediately posttransplant, and then at 24, 48, and 72 h posttransplant. Those who have PGD 3 persisting to the 72-h point have the worst short- and long-term outcomes (Whitson et al. 2006).

Treatment of PGD is supportive. Careful use of sedation and paralysis, close hemodynamic monitoring, and avoidance of nosocomial pneumonia are important. Nitric oxide has been shown to improve oxygenation, reduce pulmonary artery pressure without affecting systemic pressure, and shorten the duration of mechanical ventilation in case series (Date et al. 1996; Ardehali et al. 2001). Its primary effect is likely due to improvement in V/Q matching because NO is delivered to alveoli that are ventilated, although NO also has important anti-inflammatory properties related to neutrophil, macrophage, and platelet function (Coggins and Bloch 2007). PGE_1 infusion can also be helpful in treating severe PGD because it down-regulates inflammation in the acutely reperfused lung. When maximal treatment fails, ECLS should be started as a bridge to recovery (Fischer et al. 2007). ECLS can allow for gas exchange while avoiding harsh ventilation strategies that can further injure the graft. Although the experience is limited, the literature suggests that ECLS should be started early in the onset of severe graft dysfunction, usually within 24 h of lung transplant severe PGD, and be limited to a short trial of 4–7 d (Fiser et al. 2001).

POSTOPERATIVE IMMUNOSUPPRESSION

Lung transplant immunosuppression generally consists of a three-drug regimen with the exact composition being center dependent. A calcineurin inhibitor (cyclosporine or tacrolimus), a nucleotide blocking agent (azathioprine or MMF), and corticosteroids make up the three drugs. Induction therapy, the use of a potent immunosuppression agent to deplete T cells such as anti-IL2R antibodies, anti-CD52 antibodies, or antithymocyte globulin, is controversial in lung transplantation. Use of induction therapy is thought to promote strong early immunosuppression during the time when graft rejection is most likely to occur and may allow for a less toxic maintenance regimen going forward. Although prospective, randomized, placebo-controlled trials did not show significant benefit, and large RCTs do not currently exist, registry data show that ~50% of lung transplant programs use induction therapy based on logical extrapolation from the demonstrated benefit of induction therapy in other organ randomized control trials (Hachem et al. 2008; Hartwig et al. 2008).

OUTCOMES

Although lung transplantation has been shown to confer increased survival to patients with end-stage lung disease, survival following lung transplantation is still only ~50% at 5 yr (Christie et al. 2010). The major causes of death following lung transplantation vary with the time following transplantation. Thirty-day mortality is generally related primarily to surgical issues, donor lung preservation, and primary graft dysfunction (PGD) (Studer et al. 2004). Infectious causes, malignancy, and chronic lung allograft dysfunction (CLAD) predominate in the subsequent posttransplant period. Despite significant improvements in short-term outcomes owing to improved preservation and surgical technique, the long-term survival after lung transplantation remains at around a 50% 5-yr survival because of the development of CLAD (Christie et al. 2012). Next to the shortage of donor organ supply, CLAD remains the most significant challenge to the long-term success of lung transplantation and should be a focus of research in this area.

CHRONIC LUNG ALLOGRAFT DYSFUNCTION

CLAD is a recent term coined to describe the increasingly complex and heterogeneous conditions causing long-term lung dysfunction following transplantation (Sato 2013). Historically, chronic lung dysfunction was thought to present only as bronchiolitis obliterans syndrome (BOS), where the progressive development of obliterative bronchiolitis led to a fall in FEV_1 and ultimately to graft loss. A novel subtype of BOS (now CLAD) was first recognized as a distinct entity in 2005 (Pakhale et al. 2005) and then subsequently functionally and histologically characterized in further detail by Sato et al. (2011, 2013). Now known as restrictive allograft syndrome (RAS), this form of CLAD presents histologically as fibrosis in the peripheral lung tissue rather than in small airways, resulting in a decline in total lung capacity in addition to a decline in FEV_1. This solidified the understanding that all chronic lung dysfunction is not BOS and set the stage to focus the attention of the lung transplant community on exploring various different potential forms of CLAD and the potential multiple underlying contributing pathophysiologic mechanisms. Other phenotypes of CLAD have now been described, including neutrophilic allograft syndrome (NRAD), and fibrous BOS (FBOS) (Vanaudenaerde et al. 2008).

The pathogenesis of BOS was initially thought to represent a type of chronic rejection, that is, an alloimmune-dependent process, but alternative alloimmune-independent processes are now thought to contribute to the development of BOS and CLAD in general. Primary graft dysfunction, viral and bacterial infections of the graft, and gastroesophageal reflux and aspiration have been shown to predispose a graft toward CLAD (Davis et al. 2003; Botha et al. 2008; Snyder et al. 2010). It is thought that these processes, in addition to damaging the graft directly, activate innate immune processes, which then propagate the injury via the adaptive immune system (Palmer et al. 2005). Autoimmune processes have also been shown to play a role. Exposure of typically cryptic antigens such as type-V collagen and K-α1 tubulin during the lung transplant process may help drive the CLAD process (Burlingham et al. 2007; Goers et al. 2008).

Treatment options for CLAD are currently limited. The use of azithromycin in a randomized controlled trial has been shown to slow the progression of CLAD, potentially through its anti-inflammatory, antibiotic, or promotility effects, but the exact mechanism is unclear (Yates et al. 2005; Murphy et al. 2008; Vos et al. 2011). Switching classes of immunosuppression may be of some benefit in slowing progression of the disease, but no therapy has yet been shown to reverse the disease process. The only definitive treatment for end-stage CLAD is retransplantation, but limited organ availability and generally poorer outcomes render this approach controversial.

Although CLAD limits long-term survival, survival alone is an incomplete measurement of transplant benefit. To patients, quality of life (QOL) can be as important as quantity of life, and lung transplantation can almost be considered a type of treatment in which, in some instances, quality of life is improved even if little or no gain in quantity of life occurs. Multiple longitudinal studies have recently shown improvements in QOL following lung transplantation, some as early as 3 mo posttransplant (Gross et al. 1995; Cohen et al. 1998; Lanuza et al. 2000). More than 90% of lung transplant survivors are satisfied with their decision to undergo transplantation.

CONCLUDING REMARKS

Lung transplantation has become firmly established as a treatment of end-stage lung disease. Increased adoption of lung transplantation is currently limited by the availability and suitability of donor organs. Unique to lung transplantation is the low utilization rate of the existing donor lungs because of the frequent findings of lung injury. The recently developed ex vivo lung perfusion preservation strategy is now established in Toronto and in selected centers in Europe, with trials in the United States underway. This technique improves evaluation and also allows for the delivery of therapy to treat injured donor lungs in a targeted fashion allowing for their use. Exploration and development of ex vivo repair strategies tailored to each donor lung will be important toward a "personalized medicine" approach to the diagnosis and treatment of the donor lung.

The major factor limiting positive long-term outcomes remains the development of

CLAD. Although the pathogenesis is still not completely elucidated, it is well accepted that this represents graft dysfunction resulting from multiple etiologies. Many novel mechanisms have recently been recognized to contribute toward CLAD, and subtypes with different clinical courses have been identified. Further research into the underlying mechanisms and development of clinical strategies to mitigate the development of CLAD will transform lung transplantation into a sustained and long-term treatment of lung disease.

REFERENCES

Alessandrini F, D'Armini AM, Roberts CS, Reddick RL, Egan TM. 1994. When does the lung die? II. Ultrastructural evidence of pulmonary viability after "death." *J Heart Lung Transplant* **13:** 748–757.

Ardehali A, Laks H, Levine M, Shpiner R, Ross D, Watson LD, Shvartz O, Sangwan S, Waters PF. 2001. A prospective trial of inhaled nitric oxide in clinical lung transplantation. *Transplantation* **72:** 112–115.

Benden C, Edwards LB, Kucheryavaya AY, Christie JD, Dipchand AI, Dobbels F, Kirk R, Rahmel AO, Stehlik J, Hertz MI, et al. 2013. The registry of the International Society for Heart and Lung Transplantation: Fifteenth pediatric lung and heart-lung transplantation report—2012. *J Heart Lung Transplant* **31:** 1087–1095.

Boasquevisque CH, Yildirim E, Waddel TK, Keshavjee S. 2009. Surgical techniques: Lung transplant and lung volume reduction. *Proc Am Thorac Soc* **6:** 66–78.

Botha P, Archer L, Anderson RL, Lordan J, Dark JH, Corris PA, Gould K, Fisher AJ. 2008. *Pseudomonas aeruginosa* colonization of the allograft after lung transplantation and the risk of bronchiolitis obliterans syndrome. *Transplantation* **85:** 771–774.

Burlingham WJ, Love RB, Jankowska-Gan E, Haynes LD, Xu Q, Bobadilla JL, Meyer KC, Hayney MS, Braun RK, Greenspan DS, et al. 2007. IL-17-dependent cellular immunity to collagen type V predisposes to obliterative bronchiolitis in human lung transplants. *J Clin Invest* **117:** 3498–3506.

Butt TA, Aitchison JD, Corris PA, Wardle J, Clark S, Dark JH. 2007. 140: Lung transplantation from deceased donors without pretreatment. *J Heart Lung Transplant* **26:** S110.

Calne RY, White DJ, Thiru S, Evans DB, McMaster P, Dunn DC, Craddock GN, Pentlow BD, Rolles K. 1978. Cyclosporin A in patients receiving renal allografts from cadaver donors. *Lancet* **2:** 1323–1327.

Chaparro C, Maurer J, Gutierrez C, Krajden M, Chan C, Winton T, Keshavjee S, Scavuzzo M, Tullis E, Hutcheon M, et al. 2001. Infection with *Burkholderia cepacia* in cystic fibrosis: Outcome following lung transplantation. *Am J Respir Crit Care Med* **163:** 43–48.

Christie JD, Carby M, Bag R, Corris P, Hertz M, Weill D. 2005. Report of the ISHLT Working Group on Primary Lung Graft Dysfunction part II: Definition. A consensus statement of the International Society for Heart and Lung Transplantation. *J Heart Lung Transplant* **24:** 1454–1459.

Christie JD, Edwards LB, Kucheryavaya AY, Aurora P, Dobbels F, Kirk R, Rahmel AO, Stehlik J, Hertz MI. 2010. The Registry of the International Society for Heart and Lung Transplantation: Twenty-seventh official adult lung and heart-lung transplant report—2010. *J Heart Lung Transplant* **29:** 1104–1118.

Christie JD, Edwards LB, Kucheryavaya AY, Benden C, Dipchand AI, Dobbels F, Kirk R, Rahmel AO, Stehlik J, Hertz MI, et al. 2012. The Registry of the International Society for Heart and Lung Transplantation: 29th Adult lung and heart-lung transplant report—2012. *J Heart Lung Transplant* **31:** 1073–1086.

Coggins MP, Bloch KD. 2007. Nitric oxide in the pulmonary vasculature. *Arterioscler Thromb Vasc Biol* **27:** 1877–1885.

Cohen L, Littlefield C, Kelly P, Maurer J, Abbey S. 1998. Predictors of quality of life and adjustment after lung transplantation. *Chest* **113:** 633–644.

Colvin-Adams M, Valapour M, Hertz M, Heubner B, Paulson K, Dhungel V, Skeans MA, Edwards L, Ghimire V, Waller C, et al. 2012. Lung and heart allocation in the United States. *Am J Transplant* **12:** 3213–3234.

Cooper JD, Pearson FG, Patterson GA, Todd TR, Ginsberg RJ, Goldberg M, DeMajo WA. 1987. Technique of successful lung transplantation in humans. *J Thorac Cardiovasc Surg* **93:** 173–181.

Couetil JP, Tolan MJ, Loulmet DF, Guinvarch A, Chevalier PG, Achkar A, Birmbaum P, Carpentier AF. 1997. Pulmonary bipartitioning and lobar transplantation: A new approach to donor organ shortage. *J Thorac Cardiovasc Surg* **113:** 529–537.

Cypel M, Keshavjee S. 2012. Extracorporeal membrane oxygenation as a bridge to lung transplantation. *ASAIO J* **58:** 441–442.

Cypel M, Yeung JC, Hirayama S, Rubacha M, Fischer S, Anraku M, Sato M, Harwood S, Pierre A, Waddell TK, et al. 2008. Technique for prolonged normothermic ex vivo lung perfusion. *J Heart Lung Transplant* **27:** 1319–1325.

Cypel M, Liu M, Rubacha M, Yeung JC, Hirayama S, Anraku M, Sato M, Medin J, Davidson BL, de Perrot M, et al. 2009. Functional repair of human donor lungs by IL-10 gene therapy. *Sci Transl Med* **1:** 4ra9.

Cypel M, Yeung JC, Liu M, Anraku M, Chen F, Karolak W, Sato M, Laratta J, Azad S, Madonik M, et al. 2011. Normothermic ex vivo lung perfusion in clinical lung transplantation. *N Engl J Med* **364:** 1431–1440.

Cypel M, Yeung JC, Machuca T, Chen M, Singer LG, Yasufuku K, de Perrot M, Pierre A, Waddell TK, Keshavjee S. 2012. Experience with the first 50 ex vivo lung perfusions in clinical transplantation. *J Thorac Cardiovasc Surg* **144:** 1200–1206.

D'Armini AM, Roberts CS, Griffith PK, Lemasters JJ, Egan TM. 1994. When does the lung die? I. Histochemical evidence of pulmonary viability after "death." *J Heart Lung Transplant* **13:** 741–747.

Date H, Matsumura A, Manchester JK, Cooper JM, Lowry OH, Cooper JD. 1993. Changes in alveolar oxygen and carbon dioxide concentration and oxygen consumption during lung preservation. The maintenance of aerobic metabolism during lung preservation. *J Thorac Cardiovasc Surg* **105:** 492–501.

Date H, Triantafillou AN, Trulock EP, Pohl MS, Cooper JD, Patterson GA. 1996. Inhaled nitric oxide reduces human lung allograft dysfunction. *J Thorac Cardiovasc Surg* **111:** 913–919.

Date H, Aoe M, Sano Y, Nagahiro I, Miyaji K, Goto K, Kawada M, Sano S, Shimizu N. 2004. Improved survival after living-donor lobar lung transplantation. *J Thorac Cardiovasc Surg* **128:** 933–940.

Davis RD Jr, Lau CL, Eubanks S, Messier RH, Jadjiliadis D, Steele MP, Palmer SM. 2003. Improved lung allograft function after fundoplication in patients with gastroesophageal reflux disease undergoing lung transplantation. *J Thorac Cardiovasc Surg* **125:** 533–542.

de Hoyos A, Demajo W, Snell G, Miller J, Winton T, Maurer JR, Patterson GA. 1993. Preoperative prediction for the use of cardiopulmonary bypass in lung transplantation. *J Thorac Cardiovasc Surg* **106:** 787–795.

de Perrot M, Fischer S, Liu M, Jin R, Bai XH, Waddell TK, Keshavjee S. 2001. Prostaglandin E1 protects lung transplants from ischemia-reperfusion injury: A shift from pro- to anti-inflammatory cytokines. *Transplantation* **72:** 1505–1512.

de Perrot M, Liu M, Waddell TK, Keshavjee S. 2003. Ischemia-reperfusion-induced lung injury. *Am J Respir Crit Care Med* **167:** 490–511.

de Perrot M, Granton JT, McRae K, Cypel M, Pierre A, Waddell TK, Yasufuku K, Hutcheon M, Chaparro C, Singer L, et al. 2011. Impact of extracorporeal life support on outcome in patients with idiopathic pulmonary arterial hypertension awaiting lung transplantation. *J Heart Lung Transplant* **30:** 997–1002.

Derom F, Barbier F, Ringoir S, Versieck J, Rolly G, Berzsenyi G, Vermeire P, Vrints L. 1971. Ten-month survival after lung homotransplantation in man. *J Thorac Cardiovasc Surg* **61:** 835–846.

Egan TM. 2004. Non-heart-beating donors in thoracic transplantation. *J Heart Lung Transplant* **23:** 3–10.

Egan TM, Lambert CJ Jr, Reddick R, Ulicny KS Jr, Keagy BA, Wilcox BR. 1991. A strategy to increase the donor pool: Use of cadaver lungs for transplantation. *Ann Thorac Surg* **52:** 1113–1120.

Egan TM, Murray S, Bustami RT, Shearon TH, McCullough KP, Edwards LB, Coke MA, Garrity ER, Sweet SC, Heiney DA, et al. 2006. Development of the new lung allocation system in the United States. *Am J Transplant* **6:** 1212–1227.

Erasmus ME, van der Bij W, Verschuuren EAM. 2006. Non-heart-beating lung donation in the Netherlands: The first experience. *J Heart Lung Transplant* **25:** S63.

Fischer S, Simon AR, Welte T, Hoeper MM, Meyer A, Tessmann R, Gohrbandt B, Gottlieb J, Haverich A, Strueber M. 2006. Bridge to lung transplantation with the novel pumpless interventional lung assist device NovaLung. *J Thorac Cardiovasc Surg* **131:** 719–723.

Fischer S, Bohn D, Rycus P, Pierre AF, de Perrot M, Waddell TK, Keshavjee S. 2007. Extracorporeal membrane oxygenation for primary graft dysfunction after lung transplantation: Analysis of the Extracorporeal Life Support Organization (ELSO) registry. *J Heart Lung Transplant* **26:** 472–477.

Fiser SM, Kron IL, McLendon Long S, Kaza AK, Kern JA, Tribble CG. 2001. Early intervention after severe oxygenation index elevation improves survival following lung transplantation. *J Heart Lung Transplant* **20:** 631–636.

Follette DM, Rudich SM, Babcock WD. 1998. Improved oxygenation and increased lung donor recovery with high-dose steroid administration after brain death. *J Heart Lung Transplant* **17:** 423–429.

Garcia JP, Kon ZN, Evans C, Wu Z, Iacono AT, McCormick B, Griffith BP. 2011. Ambulatory veno-venous extracorporeal membrane oxygenation: Innovation and pitfalls. *J Thorac Cardiovasc Surg* **142:** 755–761.

Gilljam M, Schersten H, Silverborn M, Jonsson B, Ericsson Hollsing A. 2010. Lung transplantation in patients with cystic fibrosis and *Mycobacterium abscessus* infection. *J Cyst Fibros* **9:** 272–276.

Goers TA, Ramachandran S, Aloush A, Trulock E, Patterson GA, Mohanakumar T. 2008. De novo production of K-α1 tubulin-specific antibodies: Role in chronic lung allograft rejection. *J Immunol* **180:** 4487–4494.

Goldberg M, Lima O, Morgan E, Ayabe HA, Luk S, Ferdman A, Peters WJ, Cooper JD. 1983. A comparison between cyclosporin A and methylprednisolone plus azathioprine on bronchial healing following canine lung autotransplantation. *J Thorac Cardiovasc Surg* **85:** 821–826.

Grasemann H, de Perrot M, Bendiak GN, Cox P, van Arsdell GS, Keshavjee S, Solomon M. 2012. ABO-incompatible lung transplantation in an infant. *Am J Transplant* **12:** 779–781.

Griffith BP, Magee MJ, Gonzalez IF, Houel R, Armitage JM, Hardesty RL, Hattler BG, Ferson PF, Landreneau RJ, Keenan RJ. 1994. Anastomotic pitfalls in lung transplantation. *J Thorac Cardiovasc Surg* **107:** 743–753.

Gross CR, Savik K, Bolman RM III, Hertz MI. 1995. Long-term health status and quality of life outcomes of lung transplant recipients. *Chest* **108:** 1587–1593.

Hachem RR, Edwards LB, Yusen RD, Chakinala MM, Alexander Patterson G, Trulock EP. 2008. The impact of induction on survival after lung transplantation: An analysis of the International Society for Heart and Lung Transplantation Registry. *Clin Transplant* **22:** 603–608.

Haglin J, Telander RL, Muzzall RE, Kiser JC, Strobel CJ. 1963. Comparison of lung autotransplantation in the primate and dog. *Surg Forum* **14:** 196–198.

Hardy JD, Webb WR, Dalton ML Jr, Walker GR Jr. 1963. Lung homotransplantation in man. *JAMA* **186:** 1065–1074.

Hartwig MG, Snyder LD, Appel JZ II, Cantu E III, Lin SS, Palmer SM, Davis RD. 2008. Rabbit anti-thymocyte globulin induction therapy does not prolong survival after lung transplantation. *J Heart Lung Transplant* **27:** 547–553.

Helmi M, Love RB, Welter D, Cornwell RD, Meyer KC. 2003. *Aspergillus* infection in lung transplant recipients with cystic fibrosis: Risk factors and outcomes comparison to other types of transplant recipients. *Chest* **123:** 800–808.

Iribarne A, Russo MJ, Davies RR, Hong KN, Gelijns AC, Bacchetta MD, D'Ovidio F, Arcasoy S, Sonett JR. 2009.

Despite decreased wait-list times for lung transplantation, lung allocation scores continue to increase. *Chest* **135:** 923–928.

Keshavjee SH, Yamazaki F, Cardoso PF, McRitchie DI, Patterson GA, Cooper JD. 1989. A method for safe twelve-hour pulmonary preservation. *J Thorac Cardiovasc Surg* **98:** 529–534.

Keshavjee SH, Yamazaki F, Yokomise H, Cardoso PF, Mullen JB, Slutsky AS, Patterson GA. 1992. The role of dextran 40 and potassium in extended hypothermic lung preservation for transplantation. *J Thorac Cardiovasc Surg* **103:** 314–325.

Kootstra G, Daemen JH, Oomen AP. 1995. Categories of non-heart-beating donors. *Transplant Proc* **27:** 2893–2894.

Kootstra G, Kievit J, Nederstigt A. 2002. Organ donors: Heartbeating and non-heartbeating. *World J Surg* **26:** 181–184.

Langer RM. 2011. Vladimir P. Demikhov, a pioneer of organ transplantation. *Transplant Proc* **43:** 1221–1222.

Lanuza DM, Lefaiver C, McCabe M, Farcas GA, Garrity E Jr, 2000. Prospective study of functional status and quality of life before and after lung transplantation. *Chest* **118:** 115–122.

Lopau K, Mark J, Schramm L, Heidbreder E, Wanner C. 2000. Hormonal changes in brain death and immune activation in the donor. *Transpl Int* **13:** S282–S285.

Lozano R, Naghavi M, Foreman K, Lim S, Shibuya K, Aboyans V, Abraham J, Adair T, Aggarwal R, Ahn SY, et al. 2013. Global and regional mortality from 235 causes of death for 20 age groups in 1990 and 2010: A systematic analysis for the Global Burden of Disease Study 2010. *Lancet* **380:** 2095–2128.

Machuca TN, Keshavjee S. 2012. Transplantation for lung cancer. *Curr Opin Organ Transplant* **17:** 479–484.

Mason DP, Murthy SC, Gonzalez-Stawinski GV, Budev MM, Mehta AC, McNeill AM, Pettersson GB. 2008. Early experience with lung transplantation using donors after cardiac death. *J Heart Lung Transplant* **27:** 561–563.

Metras H. 1950. Note preliminarie sur la greffe total du poumon chez le chien. *CR Acad Sci III* **231:** 1176–1177.

Murphy DM, Forrest IA, Corris PA, Johnson GE, Small T, Jones D, Fisher AJ, Egan JJ, Cawston TE, Lordan JL, et al. 2008. Azithromycin attenuates effects of lipopolysaccharide on lung allograft bronchial epithelial cells. *J Heart Lung Transplant* **27:** 1210–1216.

Murray S, Charbeneau J, Marshall BC, LiPuma JJ. 2008. Impact of *Burkholderia*infection on lung transplantation in cystic fibrosis. *Am J Respir Crit Care Med* **178:** 363–371.

National Data. 2012. http://www.unos.org/data/about/viewDataReports.asp (accessed January 2, 2013).

Noirclerc M, Shennib H, Giudicelli R, Latter D, Metras D, Colt HG, Mulder D. 1992. Size matching in lung transplantation. *J Heart Lung Transplant* **11:** S203–S208.

Orens JB, Boehler A, de Perrot M, Estenne M, Glanville AR, Keshavjee S, Kotloff R, Morton J, Studer SM, Van Raemdonck D, et al. 2003. A review of lung transplant donor acceptability criteria. *J Heart Lung Transplant* **22:** 1183–1200.

Orens JB, Estenne M, Arcasoy S, Cont JV, Corris P, Egan JJ, Egan T, Keshavjee S, Knoop C, Kotloff R, et al. 2006. International guidelines for the selection of lung transplant candidates: 2006 update—A consensus report from the Pulmonary Scientific Council of the International Society for Heart and Lung Transplantation. *J Heart Lung Transplant* **25:** 745–755.

Oto T, Levvey B, McEgan R, Davies A, Pilcher D, Williams T, Marasco S, Rosenfeldt F, Snell G. 2007. A practical approach to clinical lung transplantation from a Maastricht Category III donor with cardiac death. *J Heart Lung Transplant* **26:** 196–199.

Pakhale SS, Hadjiliadis D, Howell DN, Palmer SM, Gutierrez C, Waddell TK, Chaparro C, Davis RD, Keshavjee S, Hutcheon MA, et al. 2005. Upper lobe fibrosis: A novel manifestation of chronic allograft dysfunction in lung transplantation. *J Heart Lung Transplant* **24:** 1260–1368.

Palmer SM, Burch LH, Trindade AJ, Davis RD, Herczyk WF, Reinsmoen NL, Schwartz DA. 2005. Innate immunity influences long-term outcomes after human lung transplant. *Am J Respir Crit Care Med* **171:** 780–785.

Patel VS, Palmer SM, Messier RH, Davis RD. 2003. Clinical outcome after coronary artery revascularization and lung transplantation. *Ann Thorac Surg* **75:** 372–377.

Patterson GA, Cooper JD, Goldman B, Weisel RD, Pearson FG, Waters PF, Todd TR, Scully H, Goldberg M, Ginsberg RJ. 1988. Technique of successful clinical double-lung transplantation. *Ann Thorac Surg* **45:** 626–633.

Pilcher DV, Scheinkestel CD, Snell GI, Davey-Quinn A, Bailey MJ, Williams TJ. 2005. High central venous pressure is associated with prolonged mechanical ventilation and increased mortality after lung transplantation. *J Thorac Cardiovasc Surg* **129:** 912–918.

Reitz BA, Wallwork JL, Hunt SA, Pennock JL, Billingham ME, Oyer PE, Stinson EB, Shumway NE. 1982. Heart–lung transplantation: Successful therapy for patients with pulmonary vascular disease. *N Engl J Med* **306:** 557–564.

Sánchez-Fructuoso AI, Prats D, Torrente J, Pérez-Contín MJ, Fernández C, Alvarez J, Barrientos A. 2000. Renal transplantation from non-heart beating donors: A promising alternative to enlarge the donor pool. *J Am Soc Nephrol* **11:** 328–350.

Sato M. 2013. Chronic lung allograft dysfunction after lung transplantation: The moving target. *Gen Thorac Cardiovasc Surg* **61:** 67–78.

Sato M, Waddell TK, Wagnetz U, Roberts HC, Hwang DM, Haroon A, Wagnetz D, Chaparro C, Singer LG, Hutcheon MA, et al. 2011. Restrictive allograft syndrome (RAS): A novel form of chronic lung allograft dysfunction. *J Heart Lung Transplant* **30:** 735–742.

Sato M, Hwang DM, Waddell TK, Singer LG, Keshavjee S. 2013. Progression pattern of restrictive allograft syndrome after lung transplantation. *J Heart Lung Transplant* **32:** 23–30.

Snyder LD, Finlen-Copeland CA, Turbyfill WJ, Howell D, Willner DA, Palmer SM. 2010. Cytomegalovirus pneumonitis is a risk for bronchiolitis obliterans syndrome in lung transplantation. *Am J Respir Crit Care Med* **181:** 1391–1396.

Starnes VA, Barr ML, Cohen RG. 1994. Lobar transplantation. Indications, technique, and outcome. *J Thorac Cardiovasc Surg* **108:** 403–410.

Starzl TE, Klintmalm GB, Porter KA, Iwatsuki S, Schroter GP. 1981. Liver transplantation with use of cyclosporin a and prednisone. *N Engl J Med* **305:** 266–269.

Steinbrook R. 2007. Organ donation after cardiac death. *N Engl J Med* **357:** 209–213.

Strueber M, Hoeper MM, Fischer S, Cypel M, Warnecke G, Gottlieb J, Pierre A, Welte T, Haverich A, Simon AR, et al. 2009. Bridge to thoracic organ transplantation in patients with pulmonary arterial hypertension using a pumpless lung assist device. *Am J Transplant* **9:** 853–857.

Studer SM, Levy RD, McNeil K, Orens JB. 2004. Lung transplant outcomes: A review of survival, graft function, physiology, health-related quality of life and cost-effectiveness. *Eur Respir J* **24:** 674–685.

Sweet SC. 2009. Update on pediatric lung allocation in the United States. *Pediatr Transplant* **13:** 808–813.

Taghavi S, Bîrsan T, Pereszlenyi A, Kupilik N, Deviatko E, Wisser W, Steltzer H, Klepetko W. 1999. Bilateral lung transplantation via two sequential anterolateral thoracotomies. *Eur J Cardiothorac Surg* **15:** 658–662.

Takada M, Nadeau KC, Hancock WW, Mackenzie HS, Shaw GD, Waaga AM, Chandraker A, Sayegh MH, Tilney NL. 1998. Effects of explosive brain death on cytokine activation of peripheral organs in the rat. *Transplantation* **65:** 1533–1542.

Thabut G, Christie JD, Ravaud P, Castier Y, Brugière O, Fournier M, Mal H, Lesèche G, Porcher R. 2008. Survival after bilateral versus single lung transplantation for patients with chronic obstructive pulmonary disease: A retrospective analysis of registry data. *Lancet* **371:** 744–751.

Ulicny KS Jr, Egan TM, Lambert CJ Jr, Reddick RL, Wilcox BR. 1993. Cadaver lung donors: Effect of preharvest ventilation on graft function. *Ann Thorac Surg* **55:** 1185–1191.

Vanaudenaerde BM, Meyts I, Vos R, Geudens N, De Wever W, Verbeken EK, Van Raemdonck DE, Dupont LJ, Verleden GM. 2008. A dichotomy in bronchiolitis obliterans syndrome after lung transplantation revealed by azithromycin therapy. *Eur Respir J* **32:** 832–843.

Van De Wauwer C, Neyrinck AP, Geudens N, Rega FR, Verleden GM, Verbeken E, Lerut TE, Van Raemdonck DE. 2009. Retrograde flush following warm ischemia in the non-heart-beating donor results in superior graft performance at reperfusion. *J Surg Res* **154:** 118–125.

Van Raemdonck DE, Jannis NC, Rega FR, De Leyn PR, Flameng WJ, Lerut TE. 1997. Extended preservation of ischemic pulmonary graft by postmortem alveolar expansion. *Ann Thorac Surg* **64:** 801–808.

Van Raemdonck DE, Rega FR, Neyrinck AP, Jannis N, Verleden GM, Lerut TE. 2004. Non-heart-beating donors. *Semin Thorac Cardiovasc Surg* **16:** 309–321.

Van Raemdonck D, Verleden GM, Dupont L, Ysebaert D, Monbaliu D, Neyrinck A, Coosemans W, Decaluwe H, De Lyn P, Nafteux P, et al. 2008. Initial experience with lung transplantation from non-heart-beating donors. *J Heart Lung Transplant* **27:** S198–S199.

Varela A, Cordoba M, Serrano-Fiz S, Burgos R, Montero CG, Téllez G, Novoa N, Castedo E, Tebar E, Téllez J, et al. 1997. Early lung allograft function after retrograde and antegrade preservation. *J Thorac Cardiovasc Surg* **114:** 1119–1120.

Venkateswaran RV, Patchell VB, Wilson IC, Mascaro JG, Thompson RD, Quinn DW, Stockley RA, Coote JH, Bonser RS. 2008. Early donor management increases the retrieval rate of lungs for transplantation. *Ann Thorac Surg* **85:** 278–286.

Vos R, Vanaudenaerde BM, Verleden SE, De Vleeschauwer SI, Willems-Widyastuti A, Van Raemdonck DE, Schoonis A, Nawrot TS, Dupont LJ, Verleden GM. 2011. A randomised controlled trial of azithromycin to prevent chronic rejection after lung transplantation. *Eur Respir J* **37:** 164–172.

Wang D, Zhou X, Liu X, Sidor B, Lynch J, Zwischenberger JB. 2008. Wang-Zwische double lumen cannula—Toward a percutaneous and ambulatory paracorporeal artificial lung. *ASAIO J* **54:** 606–611.

Weiss ES, Allen JG, Merlo CA, Conte JV, Shah AS. 2009. Survival after single versus bilateral lung transplantation for high-risk patients with pulmonary fibrosis. *Ann Thorac Surg* **88:** 1616–1625.

Wells A, Faro A. 2006. Special considerations in pediatric lung transplantation. *Semin Respir Crit Care Med* **27:** 552–560.

Whitson BA, Nath DS, Johnson AC, Walker AR, Prekker ME, Radosevich DM, Herrington CS, Dahlberg PS. 2006. Risk factors for primary graft dysfunction after lung transplantation. *J Thorac Cardiovasc Surg* **131:** 73–80.

Wigfield CH, Cypel M, Yeung J, Waddell T, Alex C, Johnson C, Keshavjee S, Love RB. 2012. Successful emergent lung transplantation after remote ex vivo perfusion optimization and transportation of donor lungs. *Am J Transplant* **12:** 2838–2844.

Wijdicks EF, Varelas PN, Gronseth GS, Greer DM. 2010. Evidence-based guideline update: Determining brain death in adults: Report of the Quality Standards Subcommittee of the American Academy of Neurology. *Neurology* **74:** 1911–1918.

Yates B, Murphy DM, Forrest IA, Ward C, Rutherford Rm, Fisher AJ, Lordan JL, Dark JH, Corris PA. 2005. Azithromycin reverses airflow obstruction in established bronchiolitis obliterans syndrome. *Am J Respir Crit Care Med* **172:** 772–775.

Yeung JC, Wagnetz D, Cypel M, Rubacha M, Koike T, Chun YM, Hu J, Waddell TK, Hwang DM, Liu M, et al. 2012. Ex vivo adenoviral vector gene delivery results in decreased vector-associated inflammation pre- and post-lung transplantation in the pig. *Mol Ther* **20:** 1204–1211.

Heart Transplantation: Challenges Facing the Field

Makoto Tonsho[1], Sebastian Michel[2], Zain Ahmed[1], Alessandro Alessandrini[1], and Joren C. Madsen[1,2,3]

[1]MGH Transplantation Center, Massachusetts General Hospital, Boston, Massachusetts 02114

[2]Transplantation Biology Research Center, Department of Surgery, Massachusetts General Hospital, Boston, Massachusetts 02114

[3]Division of Cardiac Surgery, Department of Surgery, Massachusetts General Hospital, Boston, Massachusetts 02114

Correspondence: jcmadsen@partners.org

There has been significant progress in the field of heart transplantation over the last 45 years. The 1-yr survival rates following heart transplantation have improved from 30% in the 1970s to almost 90% in the 2000s. However, there has been little change in long-term outcomes. This is mainly due to chronic rejection, malignancy, and the detrimental side effects of chronic immunosuppression. In addition, over the last decade, new challenges have arisen such as increasingly complicated recipients and antibody-mediated rejection. Most, if not all, of these obstacles to long-term survival could be prevented or ameliorated by the induction of transplant tolerance wherein the recipient's immune system is persuaded not to mount a damaging immune response against donor antigens, thus eliminating the need for chronic immunosuppression. However, the heart, as opposed to other allografts like kidneys, appears to be a tolerance-resistant organ. Understanding why organs like kidneys and livers are prone to tolerance induction, whereas others like hearts and lungs are tolerance-resistant, could aid in our attempts to achieve long-term, immunosuppression-free survival in human heart transplant recipients. It could also advance the field of pig-to-human xenotransplantation, which, if successful, would eliminate the organ shortage problem. Of course, there are alternative futures to the field of heart transplantation that may include the application of total mechanical support, stem cells, or bioengineered whole organs. Which modality will be the first to reach the ultimate goal of achieving unlimited, long-term, circulatory support with minimal risk to longevity or lifestyle is unknown, but significant progress in being made in each of these areas.

Although the first human-to-human heart transplant was performed in 1967, heart transplantation did not become the treatment of choice for patients with end-stage heart failure until the 1980s, when the use of cyclosporine (CyA) was extended to heart transplant recipients, resulting in a dramatic improvement in patient survival. One-year survival following heart

Cite this article as *Cold Spring Harb Perspect Med* doi: 10.1101/cshperspect.a015636

transplantation in the era 1967–1973 was 30%, in the era 1974–1980, it was 60%, and in the current era, it approaches a remarkable 90% (Stehlik et al. 2012; Colvin-Adams et al. 2013). Although progress has clearly been made over the last 45 years, there are still serious challenges facing the field, which limit the application and the success of heart transplantation. Some barriers are well known, such as (1) the shortage of donor organs, which greatly limits the number of patients able to receive a heart transplant; (2) cardiac allograft vasculopathy (CAV) and malignancy, which compromise the long-term survival of heart transplant recipients; and (3) drug-induced complications from chronic immunosuppression including diabetes mellitus, kidney disease, hypertension, and obesity, which contribute to patient morbidity and mortality. Other challenges, such as increasingly complicated recipients and antibody-mediated rejection (AMR), have only become evident over the last decade as the recipient demographics have changed and the use of mechanical circulatory support (MSC) devices has increased (Hunt and Haddad 2008; Kobashigawa 2012). Together, these obstacles account for the fact that there has been no increase in the number of adult heart transplants performed over the last decade (~4000 documented worldwide transplants/year) despite almost a 20% increase in the number of new adults on the waiting list (Colvin-Adams et al. 2013) and the fact that the 5-yr survival of patients lucky enough to receive a heart is still only ~70%, with a disappointing median survival of 11 yr and an annual attrition rate of 3%–4%, which has not changed significantly in the last three decades (Stehlik et al. 2012; Colvin-Adams et al. 2013).

Strategies that have been and are being developed to overcome these challenges have focused on either controlling the human immune system more effectively and specifically with newer immunosuppressive agents such as rapamycin and rituximab or, alternatively, attempting to harness the immune system to achieve a state of transplant tolerance in which the recipient is induced not to mount a damaging immune response against the donor heart and remains free of chronic immunosuppression. In this article, we review how the field has changed over the last decade, focusing on the new and old barriers facing heart transplant recipients. We then discuss a particular avenue of research that exemplifies the potential for immune tolerance in overcoming these barriers and achieving long-term, immunosuppression-free heart allograft survival.

CHANGES AND CHALLENGES IN THE FIELD

Recipient Demographics

Over the last decade, the demographics of heart recipients have shifted in ways that have brought new challenges to transplant clinicians. A greater proportion of patients in their sixties and seventies along with their age-related comorbidities are being transplanted. These patients tend to have higher risks of infection and CAV (Kobashigawa 2012). At the other end of the spectrum, advances in congenital heart surgery have led to a greater proportion of younger patients with congenital heart disease (CHD) surviving past childhood and developing heart failure later in life. These patients can have complex cardiopulmonary anatomy and usually have undergone multiple previous median sternotomies, which increases the risk of postoperative bleeding and mortality. Indeed, CHD is one of the strongest risk factors for 1-yr mortality after heart transplantation in adults (Stehlik et al. 2012).

Immunosuppression

The past decade has seen changes in what is considered to be standard, triple-drug, maintenance immunosuppression for the conventional heart transplant recipient. Corticosteroids (usually prednisone) remain the backbone of most immunosuppressive regimens. However, mycophenolate mofetil (MMF) has replaced azathioprine as the most commonly used antiproliferative agent, and tacrolimus (TAC) has replaced CyA as the most commonly used calcineurin inhibitor (CNI). The MMF/TAC combination seems to possess the optimum risk–

benefit ratio in preventing acute rejection (AR) and perhaps CAV even though it does not appear to improve long-term survival (Kobashigawa et al. 2006; Guethoff et al. 2013).

There are several important unanswered questions concerning immunosuppression for heart transplant recipients that require further study. For example, which recipients should receive induction therapy and using what agent? Although a survival benefit has not been clearly documented (Hershberger et al. 2005), half of all transplant programs currently use induction therapy, most commonly a short course of antithymocyte globulin (ATG) or anti-CD25 monoclonal antibody (basiliximab) (Stehlik et al. 2012). The general consensus is that the selective use of an induction agent is appropriate in highly sensitized patients or in patients with perioperative renal failure where delaying CNI therapy is beneficial. However, clear supporting data are lacking (Aliabadi et al. 2013).

The role for some of the newer immunosuppressive agents in heart transplantation is also being investigated. Several clinical trials have shown that inhibitors of the mammalian target of rapamycin (mTOR), such as sirolimus and everolimus, have been effective in preventing acute rejection (AR) (Eisen et al. 2003), mitigating CAV (Mancini et al. 2003), and improving outcomes in recipients with malignancies (Valantine 2007). They may allow for CNI minimization or elimination, which could avoid the progressive nephropathy associated with chronic CNI use (Zuckermann et al. 2012). Rituximab, a chimeric anti-CD20 (anti-B-cell) monoclonal antibody, has recently been shown to attenuate CAV in CNI-treated nonhuman primates (Kelishadi et al. 2010). An NIAID-sponsored trial (U01AI063623) is currently under way to determine whether preemptive rituximab will ameliorate CAV in human recipients. Bortezomib, a proteasome inhibitor that depletes plasma cells, has shown efficacy in the treatment of AMR and desensitization in kidney recipients (Walsh et al. 2012). In a recent pilot study, bortezomib and plasmapheresis appeared to decrease circulating antibodies in sensitized patients awaiting heart transplantation (Patel et al. 2011).

Antibody-Mediated Rejection

Antibody-mediated rejection (AMR) is a particularly challenging form of rejection in heart transplant recipients. The absence of practice guidelines for surveillance and diagnosis has resulted in it only recently being recognized as an important clinical entity. AMR results from alloantibody targeting donor antigens on capillary endothelium. It is increasingly recognized as a major cause of allograft failure and is associated with a greater risk of CAV and death (Nair et al. 2011). Prevention of AMR is dependent on identifying the sensitized patient before transplantation. This process has been assisted in recent years by the use of solid phase assays, which more accurately detects anti-HLA antibodies. In turn, this information permits virtual cross-matching, which identifies and rules out those prospective donors with HLA types that correspond to the specificities of the recipient's high-level anti-HLA antibodies without the need for complement-dependent cytotoxicity assays (Stehlik et al. 2009). Advances in assessing anti-HLA antibodies in the recipient and the use of virtual cross-matching have allowed for better choices of suitable organ donors.

At present, the guidelines for the diagnosis of AMR rely solely on the presence of antibody-mediated injury on endomyocardial biopsy and not on the presence of circulating alloantibody, which may be bound to the donor tissue (Berry et al. 2011). The treatment of AMR depends on the patient's presentation, the degree of cardiac dysfunction, and the detection of alloantibody (Kittleson and Kobashigawa 2012). Protocols differ by center because there is still a lack of randomized trials for AMR therapy (Kobashigawa et al. 2011). However, in most centers, patients with AMR and a significantly reduced ejection fraction are treated with intravenous corticosteroids and ATG. Patients presenting in cardiogenic shock can require plasmapheresis, intravenous immune globulin (IVIg), heparin, and mechanical support (Kittleson and Kobashigawa 2012). The long-term management of AMR is also complicated because patients can be left with a low ejection fraction, restrictive physiology, and accelerated CAV. Some institutions are treating these patients with rituximab, bortezomib, and photopheresis, and if necessary, redo transplantation (Kobashigawa et al. 2011).

Surgical Technique and Organ Preservation

The most significant technical advance in the heart transplantation surgery over the last decade has been related to the method of reestablishing systemic venous return. The original orthotopic heart transplant operation introduced by Lower and Shumway (1960) incorporated a biatrial technique in which cuffs of the left and right atria were preserved in the recipient and anastomosed to the corresponding atria of the donor heart. However, over the last decade, a bicaval method of systemic venous return has gained favor. The recipient's right atrium is completely resected, and the remaining superior and inferior vena cavae are anastomosed directly to the corresponding donor structures. The reason for the switch is that the conventional biatrial technique puts the sinoatrial node at risk of injury, in addition to adversely impacting atrial hemodynamics and contributing to an increased risk of atrial arrhythmias in the postoperative period (Freimark et al. 1995; Leyh et al. 1995; Brandt et al. 1997). The bicaval technique eliminates the right atrial suture line, preserves right atrial morphology, and maintains the sinoatrial node and tricuspid valve function (Traversi et al. 1998; Aziz et al. 1999). A meta-analysis of 41 papers comparing bicaval to biatrial anastomoses found significant benefits for the bicaval technique in terms of early atrial pressure, tricuspid valve regurgitation, return to sinus rhythm, frequency of permanent pacemaker implantation, and even perioperative mortality. However, long-term outcomes were less disparate between the groups (Jacob and Sellke 2009).

In the area of donor heart preservation, a promising new technology is currently being evaluated in which normothermic perfusion provides continuous warm blood flow to the beating donor heart during transportation (Ghodsizad et al. 2012). This switch from conventional cold, static storage may not only decrease reperfusion injury and primary graft dysfunction but may also allow greater utilization of available organs.

TRANSPLANT TOLERANCE

As described above, despite improvements in early posttransplant survival over the last three decades, a relentless annual attrition rate continues to plague recipients of previously successful heart allografts, resulting in a median survival of only 11 yr (Stehlik et al. 2012). Although infection accounts for most recipient deaths 1-yr posttransplant, CAV and malignancy account for most cardiac recipient deaths after 5 yr (Stehlik et al. 2012). These sobering statistics emphasize the limitations of chronically administered immunosuppression and make clear the need for strategies that achieve long-term graft survival without the use of chronic immunosuppression. Inducing a state of tolerance has the potential to prevent or ameliorate the three greatest contributors to heart transplant recipient mortality, namely, infection, CAV, and cancer, while at the same time eliminating drug-specific morbidities.

Tolerance of kidney allografts has been achieved in nonhuman primates (NHPs) (Kawai et al. 1995, 1999, 2004) and in humans (Kawai et al. 2008) by using a combination of nonmyeloablative conditioning and donor bone marrow transplantation that results in transient mixed chimerism. However, mixed chimerism protocols that achieve long-term tolerance of kidney allografts in NHPs fail to induce tolerance in recipients of heart allografts (Kawai et al. 2002). The reasons for this organ-specific difference are not clear. However, it is clear that all transplanted organs are *not* created equal. Not only does the strength of the immune response to a particular organ vary with the organ transplanted, but also the nature of the response itself, rejection versus tolerance, varies from organ to organ. In most experimental models of transplantation, heart and lung allografts evoke a stronger rejection response than kidney and liver allografts. Moreover, under the right circumstances, kidney and liver allografts can promote a state of unresponsiveness instead of inciting an aggressive alloresponse and thus can be consid-

 Cite this article as *Cold Spring Harb Perspect Med* doi: 10.1101/cshperspect.a015636

ered "tolerance-prone" organs. The same cannot be said for heart and lung allografts, which are, for the most part, "tolerance resistant." Not only do tolerance-prone kidney and liver allografts appear to contribute to the actual process of tolerance induction, but also they possess the unique ability to confer unresponsiveness upon cotransplanted, tolerance-resistant organs like hearts. The mechanisms underlying this phenomenon are unclear, but understanding them could aid into our attempts to bring tolerance to the clinic. Below, we review organ-specific differences in allograft rejection and tolerance, focusing on ways we might harness the tolerogenicity of kidney allografts to achieve long-term, immunosuppression-free survival of more stringent heart allografts.

ORGAN-SPECIFIC DIFFERENCES IN REJECTION

The most extreme examples of organ-specific differences in transplantation are experimental models in which kidney and liver allografts are accepted spontaneously (without the use of immunosuppression), whereas other allografts such as heart, intestine, and skin transplanted across the same MHC barrier are rejected acutely (Russell et al. 1978; Dahmen et al. 1994; Qian et al. 1994; Zhang et al. 1996; Bickerstaff et al. 2001; Cook et al. 2008; Li et al. 2008; Miyajima et al. 2011; Wang et al. 2011). Zhang et al. (1996) compared liver, kidney, and heart transplantation in three different MHC disparate mouse strain combinations without treatment. The differences in the patterns of rejection between organs were remarkably consistent (Table 1). The majority of liver allografts in each strain combination were spontaneously accepted long term, whereas heart grafts transplanted across identical histocompatibility barriers were all rejected in <10 d. The pattern of kidney allograft rejection was mixed, with 20%–50% of organs surviving long term (Table 1) (Zhang et al. 1996). Our results (Madsen et al. 1997; Miyajima et al. 2011) and others (Bickerstaff et al. 2001; Cook et al. 2008; Wang et al. 2011) in mice support the fact that kidney allografts have a significantly prolonged survival compared with heart allografts transplanted across the same MHC barrier. Organ-specific differences in rejection responses extend to human transplantation. For example, the graft half-life for heart allografts is 11 yr (Stehlik et al. 2012), whereas the graft half-life for lung allografts is only 5 yr (Christie et al. 2012). Thus, the organ-specific differences in transplantation have clinical significance and deserve further study.

Table 1. Proportion of liver, kidney, and heart allografts surviving >100 d in fully MHC disparate murine recipients

Strain combination	Liver	Kidney	Heart
C57BL/6 into BALB/c ($H\text{-}2^b$) ($H\text{-}2^d$)	72%	20%	0%
BALB/c into CBA ($H\text{-}2^d$) ($H\text{-}2^k$)	57%	33%	0%
C57BL/6 into C3H/HeN ($H\text{-}2^b$) ($H\text{-}2^k$)	73%	50%	0%

Recipients received no treatment; $n \geq 6$ recipients/group (from Zhang et al. 1996).

ORGAN-SPECIFIC DIFFERENCES IN TOLERANCE INDUCTION

Our laboratory has compared the immunobiology of heart, kidney, and lung transplantation in MHC inbred miniature swine (Madsen 1998). These large animals provide the only preclinical model in which organ transplants can be performed across the same histocompatibility barrier reproducibly (Sachs 1992). In brief, when porcine recipients were transplanted with MHC class I disparate hearts and treated with 12 d of CyA, they all rejected within 55 d and showed the florid intimal proliferation of CAV on necropsy (Madsen et al. 1996). In contrast, when swine were transplanted with class I disparate kidney allografts and treated with the same course of CyA, they all became tolerant to donor antigen and maintained excellent renal function long term, in some instances for >2 yr (Fig. 1) (Rosengard et al. 1992). The survival of lungs transplanted across the same class I barrier with 12 d of CyA were in between that of hearts and kidneys, with graft survival ranging from 67 to >605 d and two-thirds developing obliterative bronchiolitis (Allan et al. 2002). A similar

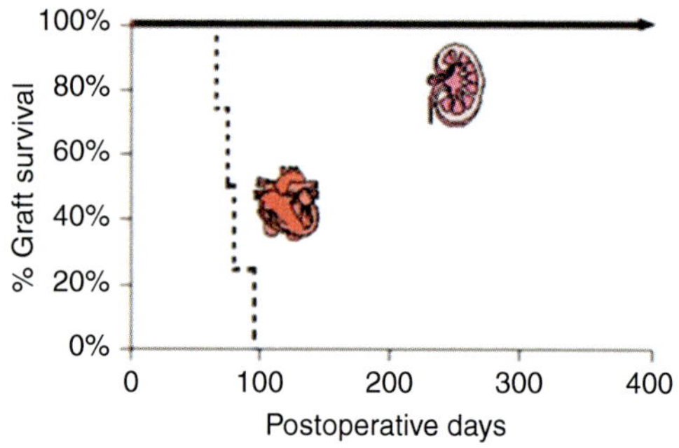

Figure 1. Heart versus kidney transplantation in MHC class I disparate swine treated with a 12-d course of CyA.

hierarchy was observed when organs were transplanted across a more rigorous two-haplotype full MHC barrier and FK506 was substituted for CyA. Long-term tolerance was induced in renal allografts (Utsugi et al. 2001), whereas lung survival ranged 100–500 d and hearts survived 55–134 d. Thus, there are also organ-specific differences that affect the process of tolerance induction. In swine, kidneys are easier to tolerize than lungs, which are easier to tolerize than hearts. This pattern seems to be true in most experimental transplant models. These observations emphasize that protocols designed to induce tolerance may not be directly transferable from one organ system to another and that the preclinical testing of tolerance protocols for human transplantation must proceed in an organ-specific manner (Massicot-Fisher et al. 2001).

MECHANISMS OF KIDNEY- AND LIVER-SPECIFIC TOLERANCE INDUCTION

Murine models of spontaneous kidney and liver allograft acceptance have been well studied and are arguably our best source of information regarding potential mechanisms of organ-specific tolerance induction.

Kidney Allografts

Spontaneous renal allograft survival was first reported 35 years ago by Russell et al. (1978). Since then, several studies have shown that MHC-disparate kidney allografts transplanted across certain strain combinations are accepted spontaneously (Russell et al. 1978; Bickerstaff et al. 2001; Cook et al. 2008; Miyajima et al. 2011; Wang et al. 2011). We have recently shown that spontaneously accepted kidney allografts showed prominent periarterial lymphoid sheaths containing nodules of $CD3^+Foxp3^+$ T cells, $CD4^+$ T cells, DCs, B cells, and indoleamine-pyrrole 2,3-dioxygenase $(IDO)^+$ cells (Miyajima et al. 2011). The majority of the cells were $CD3^+$, and ~20%–30% of them were $Foxp3^+$. These regulatory T-cell-rich organized lymphoid structures, which we term "TOLS," are distinct from tertiary lymphoid structures (TLOs) found in chronic inflammation in that they lack high endothelial venules ($MECA79^-$). Similar structures have been identified in tolerant pig and nonhuman primate kidney allografts (E Farkash, A Alessandrini, and RB Colvin, unpubl.). We have recently shown that $Foxp3^+$ regulatory T cells (Tregs) are necessary to maintain unresponsiveness in spontaneously accepted MHC mismatched mouse kidney allografts (Miyajima et al. 2011). Administering diphtheria toxin (DT) to kidney recipients expressing the human diphtheria toxin receptor (DTR) under the control of the *foxp3* gene ($B6.Foxp3^{DTR}$) allowed us to transiently deplete $Foxp3^+$ Tregs without morbidity (Kim et al. 2007). Treg depletion in long-term surviving kidney allograft recipients triggered acute cellular rejection, manifested by a sudden increase in BUN. The previously identified TOLS disintegrated after Treg depletion and was accompanied by widespread $CD8^+$ interstitial mononuclear inflammation, tubulitis, and endarteritis, indicating acute cellular rejection (Miyajima et al. 2011). Although there is convincing evidence that Tregs play a role in mediating the spontaneous acceptance of renal allografts (Bickerstaff et al. 2001; Cook et al. 2008; Miyajima et al. 2011; Wang et al. 2011), the mechanisms by which cells or cell products intrinsic to kidney but not heart allografts promote or expand Tregs is unclear.

In addressing this question, we know that there are two cell populations present in kidneys with the capacity to down-regulate alloimmune responses: (1) plasmacytoid dendritic cells (pDCs), which have been shown to be capable of promoting the generation of Tregs and inducing tolerance to heart allografts in mice (Abe et al. 2005; Ochando et al. 2006; Gehrie et al. 2011); and (2) renal tubular epithelial cells (RTECs), which have been shown to be capable of inducing T-cell unresponsiveness to self- and alloantigens in mice and humans (Hagerty and Allen 1992; Kirby et al. 1993; Singer et al. 1993; Hadley et al. 1996; Deckers et al. 1997; Frasca et al. 1998). These cell types form the basis for two nonmutually exclusive hypotheses to explain spontaneous kidney allograft acceptance.

The pDC Hypothesis

The "pDC hypothesis" predicts that donor pDCs transferred within the kidney allograft traffic to the host thymus and lymph nodes, where they facilitate the activation/expansion of donor-specific Tregs (Fig. 2A). It is known that abundant CD11c$^+$ DCs are present in normal mouse kidneys (Soos et al. 2006). We have recently shown that CD11c$^+$ cells isolated from naïve, untransplanted kidneys (from DBA/2 and BALB/c mice) contain a subpopulation of PDCA-1$^+$B220$^+$ pDCs not found in hearts from the same animals (A Alessandrini and RB Colvin, unpubl.). The presence of PDCA-1$^+$B220$^+$ pDCs in DBA/2 kidneys but not hearts may explain why DBA/2 kidneys are spontaneously accepted when transplanted into B6 wild-type recipients, whereas DBA/2 hearts are uniformly rejected (Bickerstaff et al. 2001; Cook et al. 2008).

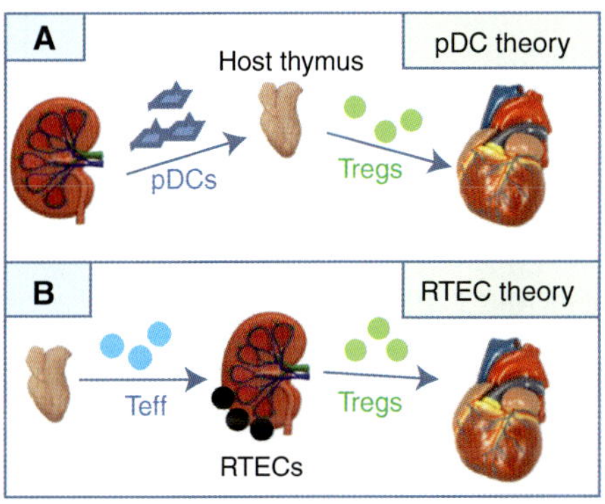

Figure 2. Hypothetical models explaining kidney-induced cardiac allograft tolerance.

The RTEC Hypothesis

The "RTEC hypothesis" predicts that RTECs, intrinsic to the donor kidney, down-regulate or inactivate effector T cells emigrating from the host thymus or convert them to Tregs, thus shifting the balance of the immune response away from rejection and toward tolerance (Fig. 2B). There is precedent for this theory in that it has been previously shown that Foxp3$^+$ cells are enriched in the tubules of mouse (Brown et al. 2007) and human (Veronese et al. 2007) renal allografts. Frasca and colleagues (1998) have shown that IFN-γ-treated human RTECs induce allospecific tolerance via a class II pathway. IFN-γ induces the T-cell inhibitor molecules PD-L1 and IDO on RTECs (Schoop et al. 2004; Mohib et al. 2007), and PD-L1 can induce the generation of allogeneic Foxp3$^+$ Tregs (Krupnick et al. 2005), in part by down-regulating PTEN (Francisco et al. 2009). Importantly, Amarnath et al. (2011) have recently shown that PD1 signaling results in the conversion of human TH1 cells into Treg cells. Finally, RTECs are known to produce and activate TGF-β (Robertson et al. 2001), which is a major inducer of Foxp3$^+$ Treg (Zheng 2008) and tolerogenic pDC (Pallotta et al. 2011) generation. High levels of TGF-β have been documented in accepted DBA/2 kidneys (Bickerstaff et al. 2001; Cook et al. 2008). Together, these findings support the possibility that RTECs contribute to spontaneous tolerance of kidney allografts.

Liver Allografts

Like kidney allografts, liver allografts are spontaneously accepted across certain mouse and rat strain combinations (Dahmen et al. 1994; Qian et al. 1994; Sriwatanawongsa et al. 1995; Zhang et al. 1996; Li et al. 2008). Like kidney allografts, depletion of recipient Tregs using anti-CD25 antibody induces acute liver allograft rejection (Li et al. 2008). And, like kidneys, the mechanism(s) by which Treg-mediated tolerance is induced by liver allografts is not known. However, the inherent tolerogenic properties of murine livers would suggest that, as in kidneys, cells or cell products intrinsic to the liver play a role. Three cell populations present in liver have been shown capable of suppressing an alloimmune response. They are (1) pDCs, located within the periportal areas and around the central veins (Sumpter et al. 2007); (2) liver sinusoidal endothelial cells (LSECs), which generate anti-inflammatory cytokines, such as IL-10 (Knolle et al. 1995), and diminish the survival $CD8^+$ T cells (Limmer et al. 2000; Bowen et al. 2004; von Oppen et al. 2009); and (3) hepatic stellate cells (HSCs), which can promote the induction of Tregs (Fig. 3) (Yang et al. 2009).

The pDC Hypothesis

The "pDC hypothesis" predicts that, like renal pDCs, liver pDCs migrate to the host thymus and lymph nodes, resulting in the activation/expansion of donor-specific Tregs. Support for this hypothesis comes from the observation that hepatic dendritic cells are less immunogenic than splenic dendritic cells and that only 5% of splenic DCs are made up of pDCs, whereas 19% of the liver DC population is made up of pDCs (Pillarisetty et al. 2004). Support also comes from that fact that circulating pDCs were increased relative to myeloid DCs (mDCs) in operationally tolerant pediatric liver allograft recipients as compared with patients maintained on chronic immunosuppression (Mazariegos et al. 2005).

The LSEC Hypothesis

The "LSEC hypothesis" predicts that these cells regulate the immune response by secreting IL-10 (Knolle et al. 1995). LSECs also constitutively express MHC class I molecules and, via cross-presentation of antigen to $CD8^+$ T cells, are able to induce $CD8^+$ T-cell tolerance rather than immunity (Limmer et al. 2000; von Oppen et al. 2009). LSEC-primed, naïve $CD8^+$ T cells are initially induced to proliferate, to release cytokines, such as IL-2 and IFN-γ, and to express CD69 and CD25, but eventually begin to secrete low levels of the cytokines and show low cytotoxicity activity (Limmer et al. 2000). The induction of $CD8^+$ T-cell tolerance has correlated with and has been shown to be dependent on the induction of the negative costimulatory molecule PD-L1 by LSECs (Diehl et al. 2008). Whether LSECs are able to induce Tregs or convert Teff to Tregs is not yet known, but their contribution to the down-regulation of an immune response begs further study in this area.

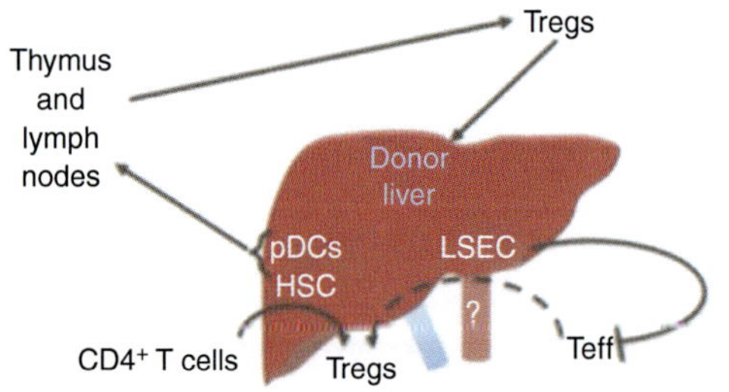

Figure 3. Hypothetical models explaining the spontaneous acceptance of liver allografts.

The HSC Hypothesis

Finally, the "HSC hypothesis" predicts that hepatic stellate cells (HSCs) are able to convert naïve or effector T cells to Tregs. HSCs store vitamin A and can produce TGF-β in response to inflammation and injury (Diehl et al. 2008; Tiegs and Lohse 2010). Vitamin A–derived retinoic acid and TGF-β have been shown to participate in the conversion of $CD4^+$ T cells to Tregs (Xiao et al. 2008; Liu et al. 2011). It has been further shown that activated HSCs express PD-L1 (Yu et al. 2004). There is precedent for the HSC theory because HSCs are able to confer unresponsiveness and long-term survival to islet allografts by inducing Tregs (Yang et al. 2009) and myeloid-derived suppressor cells (Chou et al. 2011a,b).

TOLERANCE-PRONE ORGANS CAN CONFER UNRESPONSIVENESS UPON TOLERANCE-RESISTANT ORGANS

Although, as mentioned above, some organs are known to be tolerance prone (liver and kidney), whereas others are tolerance resistant (heart and lung), less is known about why tolerance-prone organs are able to confer a survival advantage upon another organ allograft procured from the same donor and cotransplanted into the same recipient. This phenomenon was first described in pigs and termed the "liver effect" by Calne et al. (1969). It is now clear that a similar effect occurs in human recipients of liver transplants (Calne et al. 1969; Hart et al. 1971; Calne and Davies 1994; Rasmussen et al. 1995; Praseedom et al. 2001) and to a lesser extent in kidney allograft recipients (see below).

Kidney-Induced Cardiac Allograft Tolerance (KICAT)

We have studied the phenomenon of kidney-induced cardiac allograft tolerance (KICAT) in MHC inbred miniature swine by taking advantage of the ability of kidney allografts to induce long-term stable tolerance of cotransplanted heart allografts, which, if transplanted alone, would reject acutely (Table 2, No. 1). Recipients cotransplanted with the heart and kidney from the same class I disparate donor all developed rapid and stable tolerance to the donor, which led to long-term survival of both kidney and heart allografts with no evidence of rejection on serial biopsies. In addition, there was no alloantibody formation, loss of antidonor responsiveness in cell-mediated lympholysis (CML) assays, and prevention of CAV (Table 2, No. 2). Neither heart nor kidney allograft survival was affected by the placement of donor-specific and third-party skin grafts despite the return of antidonor CML reactivity, which suggested the ongoing effects of regulatory T cells (Tregs) (Table 2, No. 3).

Kidney-Specific Elements Responsible for KICAT

Donor Antigen Load

To test whether KICAT was simply due to the additional donor antigen presented by the cotransplanted kidney, we compared heart/kidney recipients to recipients grafted with two class I disparate hearts. Although the allografts in recipients of two hearts survived longer than the allograft in recipients of one heart, (1) the double-heart allografts showed high-grade rejection with early and severe CAV on serial biopsies; (2)

Table 2. Results in class I disparate swine treated with 12 d of CyA

No.	Experiment	Heart survival[a]	Heart rejection biopsies[b]	CAV[c]	CML/Ab response[d]	References
1	Heart alone	++	++++	++++	+/+	Madsen et al. 1996
2	Heart and kidney	++++	–	–	–/–	Madsen et al. 1998
3	Heart and kidney and MHC-disparate skin	++++	–	–	+/–	Madsen et al. 1998
4	Double hearts	++++	+++	++	+/+	Yamada et al. 2000
5	Heart and kidney nephrectomized early	+++	+++	+++	+/–	Mezrich et al. 2005
6	Heart and kidney nephrectomized late	++++	–	++	–/–	Mezrich et al. 2005
7	Irradiated heart and kidney	++++	+++	+++	+/–	Mezrich et al. 2003b
8	Irradiated heart and shielded kidney	++++	–	–	–/–	Mezrich et al. 2003b
9	Heart and kidney thymectomized early	+++	+++	+++	+/–	Yamada et al. 1999
10	Heart and kidney thymectomized late	++++	–	++	–/– –	Mezrich et al. 2005

[a]Graded on a relative scale from + (≤10 d) to ++++ (>100).
[b]Graded on a relative scale from + (ISHLT grade 2) to ++++ (ISHLT grade 4) on serial biopsies.
[c]Graded on a relative scale from + (rare, minimal luminal obstruction) to ++++ (frequent, significant luminal obstruction).
[d]Graded as + (response present) or – (response absent).

antidonor responsiveness in CML assays was maintained in the double-heart recipients; and (3) alloantibody was generated in the double-heart recipients (Table 2, No. 4). Thus, although augmentation of donor-antigen load could delay rejection, it could not induce tolerance.

Donor Nephrectomy

Donor kidney graftectomy on postoperative day (POD) 8 led to early cardiac allograft rejection. One recipient rejected its heart allograft on POD 29 with severe CAV. The other two showed prolonged heart allograft survival but showed high-grade interstitial rejection with severe CAV as early as POD 32 and never lost antidonor CML responsiveness (Table 2, No. 5). Furthermore, 2 wk after donor and third-party skin grafts were placed on the extended survivors, the heart allografts were acutely rejected. Thus, the donor kidney must remain in the recipient for >8 d to achieve KICAT. When donor nephrectomy was performed late (>100 d) rejection was not seen, although eventually mild CAV developed (Table 2, No. 6). However, within 2 wk of donor and third-party skin grafting, the heart allografts were rejected. These results suggest that the kidney allograft participates in the process of both tolerance induction and maintenance. More specifically, the skin grafting data suggest that elements associated with the donor kidney were able to actively suppress the antidonor response of circulating class I cytotoxic T lymphocyte precursors (CTLps), which were generated by the skin grafts (documented by CML) and which had the capacity to mediate acute rejection after the kidney allograft was removed (Madsen 1998; Christie et al. 2012).

Donor Kidney Irradiation

Cells of hematopoietic origin are extremely sensitive to radiation, with an LD_{50} estimated at 300–500 rads, but renal parenchymal cells are able to tolerate doses of 1000–2000 rads of unfractionated irradiation (Russell et al. 1978; Massicot-Fisher et al. 2001). Recipients of heart and kidney allografts from donors irradiated with 1000 rads before organ procurement failed to develop tolerance, maintaining strong antidonor CML responses despite having functioning, life-sustaining kidney allografts (Table 2, No. 7). Irradiating the donor heart while shielding the donor kidney before combined transplantation led to KICAT, indicating that it was unlikely that nonspecific irradiation-associated inflammation within the cardiac allograft was the reason for failure of KICAT. Instead, a radiosensitive, lymphohematopoietic cell population intrinsic to the donor kidney but not heart appeared necessary for the development of KICAT.

INITIAL ATTEMPTS TO DETERMINE THE ROLE FOR TREGS IN MEDIATING KICAT IN MINIATURE SWINE

Host Thymectomy

Regulatory T cells are primarily generated in the thymus (Wood and Sakaguchi 2003). To determine whether KICAT is dependent on an intact host thymus, total thymectomies were performed in recipients 21 d before heart and kidney transplants. Two of three thymectomized heart/kidney recipients rejected their heart grafts within 100 d, and the third animal had severe rejection when euthanized for seizures. Unlike the euthymic heart/kidney recipients, thymectomized heart/kidney recipients showed CAV, as well as persistent antidonor CML activity (Table 2, No. 9). The kidney grafts in these animals continued to function with stable creatinine values, but there was evidence of tubulitis and vasculopathy at necropsy. When the thymus was removed late (>100 d) no rejection was observed and CML assays remained unresponsive, although CAV eventually developed (Table 2, No. 10). These data suggest that, like the donor kidney, the host thymus plays a critical role during the induction phase of KICAT but has less influence late after transplantation.

Coculture Suppression Assays

We have found that primed naïve peripheral blood leukocytes (PBLs) cocultured with naïve cells augmented lysis of class I mismatched target cells. In contrast, primed PBLs from tolerant heart/kidney animals completely suppressed lysis of the same targets by naïve cells. Suppression was lost following removal of $CD25^+$ T cells from the tolerant heart/kidney PBL population but was reestablished by incubation of naïve cells with $CD25^+$ T cells from tolerant heart/kidney animals. In summary, these preliminary data suggest that $CD25^+$ T cells in PBLs from tolerant swine contain regulatory T cells but that they require priming to fully suppress the response of naïve-matched T cells in coculture CML (Mezrich et al. 2003a).

Putative Intragraft Tregs

We have examined the phenotypes of graft infiltrating lymphocytes (GILs) retrieved from isolated kidneys transplanted across a class I barrier that accepted versus rejected (Giangrande et al. 1997). The number of cells expressing the phenotypic markers of Tregs (which in swine are $CD4^+CD8^+CD25^{high}$) were substantially higher in acceptor than in rejector transplants at all time points.

KICAT in Fully MHC-Mismatched Swine

We have recently repeated these studies using donor/recipient pairs fully mismatched at the MHC and have again shown a dramatic difference in outcomes between recipients of isolated class I and II mismatched hearts, which reject their grafts by POD 35, and recipients of cotransplanted class I and II mismatched heart and kidneys, whose heart allografts continue to contract strongly, show no rejection on serial biopsies, and no evidence of alloantibody for >350 d (Madariaga et al. 2013).

KICAT in Nonhuman Primates

We have also extended these studies into cynomolgus monkeys using a mixed chimerism tolerance induction strategy and found that the majority of isolated heart recipients developed antidonor cellular and humoral immunity and lost their grafts to severe rejection with CAV by d 175. In contrast, recipients of heart and kidney allografts survived for >380–745 d with strongly contracting cardiac allografts (Tonsho et al. 2013). Impressively, every heart/kidney recipient that successfully completed its mixed chimerism conditioning regimen became tolerant. To our knowledge, these heart/kidney recipients represent the first nonhuman primates to become tolerant of cardiac allografts.

Summary

Together, these results underscore the robustness, consistency, and the clinical potential of kidney-induced cardiac allograft tolerance. Based on these findings and others from studies in isolated class I mismatched kidney transplant models (Ierino et al. 1999; Wu et al. 2003; Griesemer et al. 2008; Okumi et al. 2013), we hypothesize that cells or cell products intrinsic to kidney, but not heart allografts, promote the activation/expansion of host Tregs, which mediate tolerance of the heart grafts (for review, see Mezrich et al. 2004). The intrinsic mechanisms that allow a kidney but not a heart allograft to amplify regulatory mechanisms in the host following transplantation are unknown. Based on the findings in murine models described above, we are currently investigating whether pDCs or RTECs are the cells responsible for KICAT in large animals.

COMBINED ORGAN TRANSPLANTATION IN HUMAN RECIPIENTS

Review of Combined Heart and Kidney Transplantation in Patients

Given the remarkable effects a cotransplanted kidney allograft has on the survival of heart allografts in large animals undergoing tolerance-induction regimens, it is logical to expect to see a survival advantage in human recipients transplanted with combined heart and kidney allografts. However, the data are conflicting.

We have reviewed and summarized the world experience in combined heart and kidney transplantation in Table 3. The first clinical use of combined heart and kidney transplantation was reported in 1978 by Norman et al. (1978). Although combined heart and kidney trans-

Table 3. Outcomes of heart versus heart/kidney transplantation in humans

References	Year	Patients	Follow-up	Freedom from acute heart rejection	Overall survival
Czer et al. 2011	2011	30	5 yr	Heart/kidney: 96% at 1 mo 96% at 1 yr 91% at 5 yr Heart: 99% at 1 mo 95% at 1 yr 88% at 5 yr 81% at 10 yr Conclusion: No significant difference	Heart/kidney: 93% at 1 mo 87% at 1 yr 68% at 5 yr 51% at 10 yr Heart: 98% at 1 mo 93% at 1 yr 76% at 5 yr 53% at 10 yr Conclusion: No significant difference
Gill et al. 2009	2009	263	4 yr	Heart/kidney: 85% at 1 yr Heart: 69% at 1 yr Conclusion: Significantly less rejection in the combined group	Heart/kidney: 84% at 1 yr 77% at 4 yr Heart: 87% at 1 yr 77% at 4 yr Conclusion: Significantly lower risk for combined group on multivariate analysis
Bruschi et al. 2007	2007	9	10 yr	Heart/kidney: 56% at 3 mo	Heart/kidney: 89% at 1 yr 78% at 5 yr 65% at 10 yr Heart: 87% at 1 yr 78% at 5 yr 63% at 10 yr Conclusion: No significant difference
Vermes et al. 2009	2009	67	10 yr	Heart/kidney: 73% at 1 yr 73% at 3 yr 73% at 5 yr 73% at 10 yr	Heart/kidney: 62% at 1 yr 60% at 3 yr 53% at 5 yr 47% at 10 yr Heart: 71% at 1 yr 65% at 3 yr 60% at 5 yr 47% at 10 yr Conclusion: No significant difference

Continued

 Cite this article as *Cold Spring Harb Perspect Med* doi: 10.1101/cshperspect.a015636

Table 3. *Continued*

References	Year	Patients	Follow-up	Freedom from acute heart rejection	Overall survival
Hermsen et al. 2007	2007	19	5 yr	Heart/kidney: 57% at 1 yr 43% at 5 yr Heart: 28% at 1 yr 20% at 5 yr Conclusion: Significantly less rejection in the combined group	Heart/kidney: 90% at 1 yr 82% at 5 yr Heart: 88% at 1 yr 75% at 5 yr Conclusion: No significant difference
Wang et al. 2006	2006	16	10 yr	Heart/kidney: 100% at 10 yr	Heart/kidney: 83% at 1 yr 83% at 5 yr 55% at 10 yr Heart: 63% at 5 yr 46% at 10 yr
Groetzner et al. 2005	2005	13	5 yr	Incidence of acute heart rejection episode Heart/kidney: 0.02/100 patient-days Heart: 0.04/100 patient-days Conclusion: No significant difference	Heart/kidney: 92% at 1 yr 92% at 5 yr Heart: 88% at 1 yr 84% at 5 yr Conclusion: No significant difference
Trachiotis et al. 2003	2003	8	10 yr	Heart/kidney: 50% at 3 yr Heart: 50% at 4 yr	Heart/kidney: 88% at 1 yr 88% at 5 yr 88% at 10 yr Heart: 91% at 1 yr 74% at 5 yr 52% at 10 yr Conclusion: No significant difference
Luckraz et al. 2003	2003	13	10 yr	Incidence of acute heart rejection episodes per 100 patient-days was significantly lower in the combined group than that in the isolated heart transplant group	Heart/kidney: 77% at 1 yr 67% at 10 yr

Continued

Table 3. *Continued*

References	Year	Patients	Follow-up	Freedom from acute heart rejection	Overall survival
					Heart: 82% at 1 yr 58% at 10 yr Conclusion: No significant difference
Leeser et al. 2001	2001	13	5 yr	Heart/kidney: 85% Heart: 66% Conclusion: Significantly less rejection in the combined group	Heart/kidney: 92% at 1 mo 77% at 1 yr 60% at 5 yr Heart: 91% at 1 mo 80% at 1 yr 67% at 5 yr Conclusion: No significant difference
Vermes et al. 2001	2001	12	12 yr	Heart/kidney: 90% at 6 mo 70% at 1 yr Heart: 65% at 6 mo 49% at 1 yr Conclusion: Significantly less rejection in the combined group	Heart/kidney: 66% at 1 yr 55% at 5 yr 28% at 12 yr Heart: 66% at 1 yr 44% at 5 yr 32% at 12 yr Conclusion: No significant difference
Blanche et al. 2001	2001	10	5 yr	Heart/kidney: 90% at 30 d 80% at 1 yr 80% at 2 yr Conclusion: Suggestive of lower rates of acute rejection than in hearts alone	Heart/kidney: 100% at 1 yr 88% at 2 yr 55% at 5 yr Heart: 92% at 1 yr 84% at 2 yr 71% at 5 yr Conclusion: No significant difference
Col et al. 1998	1998	6	2 yr	Heart/kidney: 66% at 2 yr	Heart/kidney: 67% at 2 yr
Kocher et al. 1998	1998	9	4 yr	67% at 4 yr Conclusion: No significant difference	Heart/kidney: 89% at 4 yr Conclusion: No significant difference

Continued

Table 3. *Continued*

References	Year	Patients	Follow-up	Freedom from acute heart rejection	Overall survival
Narula et al. 1997	1997	82	2 yr	Heart/kidney: 88% at 1 mo 71% at 6 mo 66% at 12 mo Heart: 64% at 1 mo 44% at 6 mo 39% at 12 mo Conclusion: Significantly less rejection in the combined group	Heart/kidney: 92% at 1 mo 76% at 12 mo 67% at 24 mo Heart: 92% at 1 mo 86% at 6 mo 83% at 12 mo 79% at 24 mo Conclusion: No significant difference
Colucci et al. 1997	1997	6	3.5 yr	Heart/kidney: 100% at 3.5 yr	Heart/kidney: 100% at 3.5 yr
Gonwa et al. 1992	1992	3	6 mo	Heart/kidney: 100% at 3 yr	Heart/kidney: 100% at 6 mo

plantations account for only 1.5% of heart transplants performed annually, the last decade has seen a significant rise in the number of combined heart and kidney transplants conducted in humans, with excellent short-term successes (Cecka and Terasaki 2003). Indeed, since 1994, the number of patients in the United States requiring combined heart and kidney transplants has more than tripled to 78 recipients in 2012 (Stehlik et al. 2012).

Data regarding long-term outcomes in recipients of these transplants are now becoming available. The recent publication by Czer et al. (2011) is the latest in a series of papers examining the results of heart and kidney cotransplantation. The investigators compared the outcomes of 30 patients receiving heart and kidney transplants with those of 440 patients receiving isolated heart allografts and found that the heart/kidney recipients had the same excellent long-term survival rates and low incidence of cellular rejection (ISHLT grade >2) as recipients of isolated hearts. Although there were no statistical differences between the two groups in terms of overall outcomes, cellular and antibody-mediated rejection of the cardiac allograft was not observed more than 5 yr following heart/kidney transplantation, whereas it was observed 5 yr after heart-alone transplantation (Table 3) (Czer et al. 2011). Thus, combined heart/kidney transplantation may confer a late immunological advantage.

In contrast, the most recent cohort analysis using the OPTN/UNOS database showed that the risk of death was 44% lower in heart/kidney recipients compared with heart-alone recipients after adjusting for potential confounders on multivariate analysis. Furthermore, acute rejection rates at 1 yr posttransplant were lower in heart/kidney recipients (Table 3) (Gill et al. 2009).

Taken together, the world's literature suggests that recipients of combined heart and kidney allografts have less acute and possibly less chronic posttransplant rejection than recipients of isolated hearts (Rasmussen et al. 1995; Narula et al. 1997; Opelz et al. 2002; Goldfarb 2003; Luckraz et al. 2003; Hermsen et al. 2007). However, most studies do not show an increase in long-term survival following heart and kidney cotransplantation. The immunological advantage of heart and kidney recipients may not translate into a long-term survival benefit in human recipients because any tolerogenic effect engendered by the kidney allograft could be abrogated by the presence of continuous immunosuppression.

Review of Combined Heart and Liver Transplantation in Patients

Although many less patients have undergone heart and liver cotransplantation, the data that are available suggest that there is also an immunological benefit conferred upon the heart by the cotransplanted liver allograft. In a single-center study, Raichlin et al. (2009) reported that although 1-, 5-, and 10-yr survival rates in combined heart/liver recipients were similar to those in isolated heart recipients, the heart/kidney recipients showed significantly less acute cardiac allograft rejection. Moreover, heart/liver recipients showed no evidence of angiographic CAV, which was diagnosed in 38% of comparable isolated heart recipients (Raichlin et al. 2009). In a recent follow-up study using intravascular ultrasound (IVUS) to interrogate the coronary arteries, the same group showed that patients with heart/liver transplants had slower progression of CAV, which translated to better coronary-related clinical outcomes in unadjusted and adjusted analysis (Topilsky et al. 2013). These data suggest that the cotransplanted liver can modify the recipient's immunological risk factors and improve the natural history of CAV. Finally, a protective "liver effect" can be documented by the fact that heart–liver recipients require less immunosuppression than recipients of isolated heart and still maintain low rates of acute rejection (Te et al. 2008).

Summary

Kidney-induced cardiac allograft tolerance has proven effective across two large animal models and across two different tolerance-induction protocols. Understanding how tolerance-prone organs confer unresponsiveness upon tolerance-resistant organs could provide important mechanistic information that is relevant to our attempts at inducing tolerance in human transplant recipients. Data from studies in mice, swine, and nonhuman primates suggest that cells or cell products resident within kidneys and livers but not hearts actively participate in the induction of tolerance by amplifying the contributions of Tregs. Isolating and harnessing these cells or cell products could provide an avenue toward the induction of tolerance in humans. Of course, cotransplanting a kidney with a heart allograft to induce tolerance in a recipient that does not have irreversible end-stage renal failure is an untenable strategy. Instead, one might imagine achieving tolerance of isolated heart allografts without sacrificing a donor kidney, possibly by obtaining the to-be-identified, KICAT-mediating cells via a simple kidney biopsy at the time of organ procurement and expanding them in vitro for subsequent use in a delayed tolerance protocol that our group described previously (Yamada et al. 2012). Indeed, this strategy could be applicable to all deceased-donor transplantation and to all tolerance-resistant organs/tissues including islets and vascularized composite allografts.

CONCLUSIONS

The future of heart transplantation must focus on the disappointing late outcomes that have not changed in decades and the donor organ shortage. Achieving tolerance could address both challenges by eliminating the need for chronic immunosuppression (Kawai et al. 2008) and by providing a potential avenue to pig-to-human xenotransplantation (Kuwaki et al. 2005). However, there are other exciting possibilities on the horizon that could achieve the same end including stem cell therapy (Shiba et al. 2012), advanced mechanical assistance (Abraham and Smith 2013), and whole-organ bioengineering (Ott et al. 2008). Which modality will be the first to reach the ultimate goal of achieving unlimited, long-term, circulatory support with minimal risk to longevity or lifestyle is unknown. However, significant advances are being made and will continue to be made in each of these areas.

ACKNOWLEDGMENTS

This work is supported in part by grants from the National Heart, Lung, and Blood Institute (P01HL18646) and the National Institute of Allergy and Infectious Disease (U01AI94374, R01AI081734-01A1) of the National Institutes of Health. We acknowledge C06RR020135-01 for construction of the facility used for production and maintenance of miniature swine and are indebted to J. Scott Arn for herd management and quality-control typing.

REFERENCES

Abe M, Wang Z, de CA, Thomson AW. 2005. Plasmacytoid dendritic cell precursors induce allogeneic T-cell hyporesponsiveness and prolong heart graft survival. *Am J Transplant* **5:** 1808–1819.

Abraham WT, Smith SA. 2013. Devices in the management of advanced, chronic heart failure. *Nat Rev Cardiol* **10:** 98–110.

Aliabadi A, Grommer M, Cochrane A, Salameh O, Zuckermann A. 2013. Induction therapy in heart transplantation: Where are we now? *Transpl Int* **26:** 684–695.

Allan JS, Wain JC, Schwarze ML, Houser SL, Benjamin LC, Madsen JC, Sachs DH. 2002. Modeling chronic lung allograft rejection in miniature swine. *Transplantation* **73:** 447–453.

Amarnath S, Mangus CW, Wang JC, Wei F, He A, Kapoor V, Foley JE, Massey PR, Felizardo TC, Riley JL, et al. 2011. The PDL1–PD1 axis converts human TH1 cells into regulatory T cells. *Sci Transl Med* **3:** 111ra120.

Aziz T, Burgess M, Khafagy R, Wynn Hann A, Campbell C, Rahman A, Deiraniya A, Yonan N. 1999. Bicaval and standard techniques in orthotopic heart transplantation: Medium-term experience in cardiac performance and survival. *J Thorac Cardiovasc Surg* **118:** 115–122.

Berry GJ, Angelini A, Burke MM, Bruneval P, Fishbein MC, Hammond E, Miller D, Neil D, Revelo MP, Rodriguez ER, et al. 2011. The ISHLT working formulation for pathologic diagnosis of antibody-mediated rejection in heart transplantation: Evolution and current status (2005–2011). *J Heart Lung Transplant* **30:** 601–611.

Bickerstaff AA, Wang JJ, Pelletier RP, Orosz CG. 2001. Murine renal allografts: Spontaneous acceptance is associated with regulated T cell-mediated immunity. *J Immunol* **167:** 4821–4827.

Blanche C, Kamlot A, Blanche DA, Kearney B, Wong AV, Czer LS, Trento A. 2001. Combined heart–kidney transplantation with single-donor allografts. *J Thorac Cardiovasc Surg* **122:** 495–500.

Bowen DG, Zen M, Holz L, Davis T, McCaughan GW, Bertolino P. 2004. The site of primary T cell activation is a determinant of the balance between intrahepatic tolerance and immunity. *J Clin Invest* **114:** 701–712.

Brandt M, Harringer W, Hirt SW, Walluscheck KP, Cremer J, Sievers HH, Haverich A. 1997. Influence of bicaval anastomoses on late occurrence of atrial arrhythmia after heart transplantation. *Ann Thorac Surg* **64:** 70–72.

Brown K, Moxham V, Karegli J, Phillips R, Sacks SH, Wong W. 2007. Ultra-localization of Foxp3$^+$ T cells within renal allografts shows infiltration of tubules mimicking rejection. *Am J Pathol* **171:** 1915–1922.

Bruschi G, Busnach G, Colombo T, Radaelli L, Pedrazzini G, Garatti A, Sansalone CV, Frigerio M, Vitali E. 2007. Long-term follow-up of simultaneous heart and kidney transplantation with single donor allografts: Report of nine cases. *Ann Thorac Surg* **84:** 522–527.

Calne R, Davies H. 1994. Organ graft tolerance: The liver effect. *Lancet* **343:** 67–68.

Calne RY, Sells RA, Pena JR. 1969. Induction of immunological tolerance by porcine liver allografts. *Nature* **223:** 472–476.

Cecka JM, Terasaki PI. 2003. *Clinical transplants 2002*. University of California, Los Angeles.

Chou HS, Hsieh CC, Yang HR, Wang L, Arakawa Y, Brown K, Wu Q, Lin F, Peters M, Fung JJ, et al. 2011a. Hepatic stellate cells regulate immune response by way of induction of myeloid suppressor cells in mice. *Hepatology* **53:** 1007–1019.

Chou HS, Hsieh CC, Charles R, Wang L, Wagner T, Fung JJ, Qian S, Lu LL. 2011b. Myeloid-derived suppressor cells protect islet transplants by B7-H1 mediated enhancement of T regulatory cells. *Transplantation* **93:** 272–282.

Christie JD, Edwards LB, Kucheryavaya AY, Benden C, Dipchand AI, Dobbels F, Kirk R, Rahmel AO, Stehlik J, Hertz MI, et al. 2012. The registry of the international society for heart and lung transplantation: 29th adult lung and heart–lung transplant report—2012. *J Heart Lung Transplant* **31:** 1073–1086.

Col VJ, Jacquet L, Squifflet JP, Goenen M, Noirhomme P, Goffin E, Pirson Y. 1998. Combined heart–kidney transplantation: Report on six cases. *Nephrol Dial Transplant* **13:** 723–727.

Colucci V, Quaini E, Magnani P, Colombo T, De Carlis L, Grassi M, Merli M, Pellegrini A. 1997. Combined heart and kidney transplantation: An effective therapeutic option—Report of six cases. *Eur J Cardiothorac Surg* **12:** 654–658.

Colvin-Adams M, Smith JM, Heubner BM, Skeans MA, Edwards LB, Waller C, Snyder JJ, Israni AK, Kasiske BL. 2013. OPTN/SRTR 2011 Annual Data Report: Heart. *Am J Transplant* **13:** 119–148.

Cook CH, Bickerstaff AA, Wang JJ, Nadasdy T, Della Pelle P, Colvin RB, Orosz CG. 2008. Spontaneous renal allograft acceptance associated with "regulatory" dendritic cells and IDO. *J Immunol* **180:** 3103–3112.

Czer LS, Ruzza A, Vespignani R, Jordan S, De Robertis MA, Mirocha J, Gallagher SP, Patel K, Schwarz ER, Kass RM, et al. 2011. Survival and allograft rejection rates after combined heart and kidney transplantation in comparison with heart transplantation alone. *Transplant Proc* **43:** 3869–3876.

Dahmen U, Qian S, Rao AS, Demetris AJ, Fu F, Sun H, Gao L, Fung JJ, Starzl TE. 1994. Split tolerance induced by orthotopic liver transplantation in mice. *Transplantation* **58:** 1–8.

Deckers JGM, Boonstr JG, Van der Kooij SW, Daha MR, van der Woude FJ. 1997. Tissue-specific characteristics of cytotoxic graft-infiltrating T cells during renal allograft rejection. *Transplantation* **64:** 178–181.

Diehl L, Schurich A, Grochtmann R, Hegenbarth S, Chen L, Knolle PA. 2008. Tolerogenic maturation of liver sinusoidal endothelial cells promotes B7-homolog 1-dependent CD8$^+$ T cell tolerance. *Hepatology* **47:** 296–305.

Eisen HJ, Tuzcu EM, Dorent R, Kobashigawa J, Mancini D, Valantine-von Kaeppler HA, Starling RC, Sørensen K, Hummel M, Lind JM, et al. 2003. Everolimus for the prevention of allograft rejection and vasculopathy in cardiac-transplant recipients. *N Engl J Med* **349:** 847–858.

Francisco LM, Salinas VH, Brown KE, Vanguri VK, Freeman GJ, Kuchroo VK, Sharpe AH. 2009. PD-L1 regulates the development, maintenance, and function of induced regulatory T cells. *J Exp Med* **206:** 3015–3029.

Frasca L, Marelli-Berg F, Imami N, Potolicchio I, Carmichael P, Lombardi G, Lechler R. 1998. Interferon-γ-treated renal tubular epithelial cells induce allospecific tolerance. *Kidney Int* **53:** 679–689.

Freimark D, Silverman JM, Aleksic I, Crues JV, Blanche C, Trento A, Admon D, Queral CA, Harasty DA, Czer LS. 1995. Atrial emptying with orthotopic heart transplantation using bicaval and pulmonary venous anastomoses: A magnetic resonance imaging study. *J Am Coll Cardiol* **25:** 932–936.

Gehrie E, Van der Touw W, Bromberg JS, Ochando JC. 2011. Plasmacytoid dendritic cells in tolerance. *Methods Mol Biol* **677:** 127–147.

Ghodsizad A, Bordel V, Ungerer M, Karck M, Bekeredjian R, Ruhparwar A. 2012. Ex vivo coronary angiography of a donor heart in the organ care system. *Heart Surg Forum* **15:** E161–E163.

Giangrande I, Yamada K, Arn S, Lorf T, Sachs DH, LeGuern C. 1997. Selective increase in CD4-positive graft-infiltrating mononuclear cells among the infiltrates in class I disparate kidney grafts undergoing rejection. *Transplantation* **63:** 722–728.

Gill J, Shah T, Hristea I, Chavalitdhamrong D, Anastasi B, Takemoto SK, Bunnapradist S. 2009. Outcomes of simultaneous heart–kidney transplant in the US: A retrospective analysis using OPTN/UNOS data. *Am J Transplant* **9:** 844–852.

Goldfarb DA. 2003. Prolongation of long-term kidney graft survival by a simultaneous liver transplant: The liver does it, and the heart does it too. *J Urol* **169:** 2435–2436.

Gonwa TA, Husberg BS, Klintmalm GB, Mai ML, Goldstein RM, Capehart JE, Miller AH, Johnston SB, Alivizatos PA. 1992. Simultaneous heart and kidney transplantation: A report of three cases and review of the literature. *J Heart Lung Transplant* **11:** 152–155.

Griesemer AD, LaMattina JC, Okumi M, Etter JD, Shimizu A, Sachs DH, Yamada K. 2008. Linked suppression across an MHC-mismatched barrier in a miniature swine kidney transplantation model. *J Immunol* **181:** 4027–4036.

Groetzner J, Kaczmarek I, Mueller M, Huber S, Deutsch A, Daebritz S, Arbogast H, Meiser B, Reichart B. 2005. Freedom from graft vessel disease in heart and combined heart- and kidney-transplanted patients treated with tacrolimus-based immunosuppression. *J Heart Lung Transplant* **24:** 1787–1792.

Guethoff S, Meiser BM, Groetzner J, Eifert S, Grinninger C, Ueberfuhr P, Reichart B, Hagl C, Kaczmarek I. 2013. Ten-year results of a randomized trial comparing tacrolimus versus cyclosporine a in combination with mycophenolate mofetil after heart transplantation. *Transplantation* **95:** 629–634.

Hadley GA, Rostapshova EA, Bartlett ST. 1996. Dominance of tissue-restricted cytotoxic T lymphocytes in the response to allogeneic renal epithelial cell lines. *Transplantation* **62:** 75–83.

Hagerty DT, Allen PM. 1992. Processing and presentation of self and foreign antigens by the renal proximal tubule. *J Immunol* **148:** 2324–2330.

Hart AJ, Smellie WA, Calne RY. 1971. Fate of kidney allografts from cadavers whose livers were also transplanted. *Lancet* **1:** 103–105.

Hermsen JL, Nath DS, del Rio AM, Eickstaedt JB, Wigfield C, Lindsey JD, Edwards NM. 2007. Combined heart–kidney transplantation: The University of Wisconsin experience. *J Heart Lung Transplant* **26:** 1119–1126.

Hershberger RE, Starling RC, Eisen HJ, Bergh CH, Kormos RL, Love RB, Van Bakel A, Gordon RD, Popat R, Cockey L, et al. 2005. Daclizumab to prevent rejection after cardiac transplantation. *N Engl J Med* **352:** 2705–2713.

Hunt SA, Haddad F. 2008. The changing face of heart transplantation. *J Am Coll Cardiol* **52:** 587–598.

Ierino FL, Yamada K, Hatch T, Rembert J, Sachs DH. 1999. Peripheral tolerance to class I mismatched renal allografts in miniature swine: Donor antigen-activated peripheral blood lymphocytes from tolerant swine inhibit antidonor CTL reactivity. *J Immunol* **162:** 550–559.

Jacob S, Sellke F. 2009. Is bicaval orthotopic heart transplantation superior to the biatrial technique? *Interact Cardiovasc Thorac Surg* **9:** 333–342.

Kawai T, Cosimi AB, Colvin RB, Powelson J, Eason J, Kozlowski T, Sykes M, Monroy R, Tanaka M, Sachs DH. 1995. Mixed allogeneic chimerism and renal allograft tolerance in cynomolgus monkeys. *Transplantation* **59:** 256–262.

Kawai T, Poncelet A, Sachs DH, Mauiyyedi S, Boskovic S, Wee SL, Ko DS, Bartholomew A, Kimikawa M, Hong HZ, et al. 1999. Long-term outcome and alloantibody production in a non-myeloablative regimen for induction of renal allograft tolerance. *Transplantation* **68:** 1767–1775.

Kawai T, Cosimi AB, Wee SL, Houser S, Andrews D, Sogawa H, Phelan J, Boskovic S, Nadazdin O, Abrahamian G, et al. 2002. Effect of mixed hematopoietic chimerism on cardiac allograft survival in cynomolgus monkeys. *Transplantation* **73:** 1757–1764.

Kawai T, Sogawa H, Boskovic S, Abrahamian G, Smith RN, Wee SL, Andrews D, Nadazdin O, Koyama I, Sykes M, et al. 2004. CD154 blockade for induction of mixed chimerism and prolonged renal allograft survival in nonhuman primates. *Am J Transplant* **4:** 1391–1398.

Kawai T, Cosimi AB, Spitzer TR, Tolkoff-Rubin N, Suthanthiran M, Saidman SL, Shaffer J, Preffer FI, Ding R, Sharma V, et al. 2008. HLA-mismatched renal transplantation without maintenance immunosuppression. *N Engl J Med* **358:** 353–361.

Kelishadi SS, Azimzadeh AM, Zhang T, Stoddard T, Welty E, Avon C, Higuchi M, Laaris A, Cheng XF, McMahon C, et al. 2010. Preemptive $CD20^+$ B cell depletion attenuates cardiac allograft vasculopathy in cyclosporine-treated monkeys. *J Clin Invest* **120:** 1275–1284.

Kim JM, Rasmussen JP, Rudensky AY. 2007. Regulatory T cells prevent catastrophic autoimmunity throughout the lifespan of mice. *Nat Immunol* **8:** 191–197.

Kirby JA, Rajasekar MR, Lin Y, Proud G, Taylor RM. 1993. Interaction between T lymphocytes and kidney epithelial cells during renal allograft rejection. *Kidney Int Suppl* **39:** S124–S128.

Kittleson MM, Kobashigawa JA. 2012. Antibody-mediated rejection. *Curr Opin Organ Transplant* **17:** 551–557.

Knolle P, Schlaak J, Uhrig A, Kempf P, Meyer zum Buschenfelde KH, Gerken G. 1995. Human Kupffer cells secrete IL-10 in response to lipopolysaccharide (LPS) challenge. *J Hepatol* **22:** 226–229.

Kobashigawa JA. 2012. The future of heart transplantation. *Am J Transplant* **12:** 2875–2891.

Kobashigawa JA, Miller LW, Russell SD, Ewald GA, Zucker MJ, Goldberg LR, Eisen HJ, Salm K, Tolzman D, Gao J, et al. 2006. Tacrolimus with mycophenolate mofetil (MMF) or sirolimus vs. cyclosporine with MMF in cardiac transplant patients: 1-Year report. *Am J Transplant* **6:** 1377–1386.

Kobashigawa J, Crespo-Leiro MG, Ensminger SM, Reichenspurner H, Angelini A, Berry G, Burke M, Czer L, Hiemann N, Kfoury AG, et al. 2011. Report from a consensus conference on antibody-mediated rejection in heart transplantation. *J Heart Lung Transplant* **30:** 252–269.

Kocher AA, Schlechta B, Kopp CW, Ehrlich M, Ankersmit J, Ofner P, Langer F, Berlakovich GA, Grimm M, Wolner E, et al. 1998. Combined heart and kidney transplantation using a single donor: A single center's experience with nine cases. *Transplantation* **66:** 1760–1763.

Krupnick AS, Gelman AE, Barchet W, Richardson S, Kreisel FH, Turka LA, Colonna M, Patterson GA, Kreisel D. 2005. Murine vascular endothelium activates and induces the generation of allogeneic $CD4^+25^+Foxp3^+$ regulatory T cells. *J Immunol* **175:** 6265–6270.

Kuwaki K, Tseng YL, Dor FJ, Shimizu A, Houser SL, Sanderson TM, Lancos CJ, Prabharasuth DD, Cheng J, Moran K, et al. 2005. Heart transplantation in baboons using α-1,3-galactosyltransferase gene-knockout pigs as donors: Initial experience. *Nat Med* **11:** 29–31.

Leeser DB, Jeevanandam V, Furukawa S, Eisen H, Mather P, Silva P, Guy S, Forster CE III. 2001. Simultaneous heart and kidney transplantation in patients with end-stage heart and renal failure. *Am J Transplant* **1:** 89–92.

Leyh RG, Jahnke AW, Kraatz EG, Sievers HH. 1995. Cardiovascular dynamics and dimensions after bicaval and standard cardiac transplantation. *Ann Thorac Surg* **59:** 1495–1500.

Li W, Kuhr CS, Zheng XX, Carper K, Thomson AW, Reyes JD, Perkins JD. 2008. New insights into mechanisms of spontaneous liver transplant tolerance: The role of Foxp3-expressing $CD25^+CD4^+$ regulatory T cells. *Am J Transplant* **8:** 1639–1651.

Limmer A, Ohl J, Kurts C, Ljunggren HG, Reiss Y, Groettrup M, Momburg F, Arnold B, Knolle PA. 2000. Efficient presentation of exogenous antigen by liver endothelial cells to $CD8^+$ T cells results in antigen-specific T-cell tolerance. *Nat Med* **6:** 1348–1354.

Liu SM, Lee DH, Sullivan JM, Chung D, Jäger A, Shum BO, Sarvetnick NE, Anderson AC, Kuchroo VK. 2011. Differential IL-21 signaling in APCs leads to disparate Th17 differentiation in diabetes-susceptible NOD and diabetes-resistant NOD.Idd3 mice. *J Clin Invest* **121:** 4303–4310.

Lower RR, Shumway NE. 1960. Studies on orthotopic homotransplantation of the canine heart. *Surg Forum* **11:** 18–19.

Luckraz H, Parameshwar J, Charman SC, Firth J, Wallwork J, Large S. 2003. Short- and long-term outcomes of combined cardiac and renal transplantation with allografts from a single donor. *J Heart Lung Transplant* **22:** 1318–1322.

Madariaga ML, Michel SG, Tasaki M, Villani V, La Muraglia GM II, Sihag S, Gottschall J, Farkash EA, Shimizu A, Allan JS, et al. 2013. Induction of cardiac allograft tolerance across a full MHC barrier in miniature swine by donor kidney cotransplantation. *Am J Transplant* doi: 10.1111/ajt.12423.

Madsen JC. 1998. Cardiac allograft vasculopathy in miniature swine: Utility of a large animal model. *Graft* **1:** 41–44.

Madsen JC, Sachs DH, Fallon JT, Weissman NJ. 1996. Cardiac allograft vasculopathy in partially inbred miniature swine. I. Time course, pathology, and dependence on immune mechanisms. *J Thorac Cardiovasc Surg* **111:** 1230–1239.

Madsen JC, Morris PJ, Wood KJ. 1997. Immunogenetics of heart transplantation in rodents. *Transplant Rev* **11:** 141–150.

Madsen JC, Yamada K, Allan JS, Choo JK, Erhorn AE, Pins MR, Vesga L, Slisz JK, Sachs DH. 1998. Transplantation tolerance prevents cardiac allograft vasculopathy in major histocompatibility complex class I—Disparate miniature swine. *Transplantation* **65:** 304–313.

Mancini D, Pinney S, Burkhoff D, LaManca J, Itescu S, Burke E, Edwards N, Oz M, Marks AR. 2003. Use of rapamycin slows progression of cardiac transplantation vasculopathy. *Circulation* **108:** 48–53.

Massicot-Fisher J, Noel P, Madsen JC. 2001. Recommendations of the NHLBI heart and lung tolerance working group. *Transplantation* **72:** 1467–1470.

Mazariegos GV, Zahorchak AF, Reyes J, Chapman H, Zeevi A, Thomson AW. 2005. Dendritic cell subset ratio in tolerant, weaning and non-tolerant liver recipients is not affected by extent of immunosuppression. *Am J Transplant* **5:** 314–322.

Mezrich JD, Kesselheim JA, Johnston DR, Yamada K, Sachs DH, Madsen JC. 2003a. The role of regulatory cells in miniature swine rendered tolerant to cardiac allografts by donor kidney cotransplantation. *Am J Transplant* **3:** 1107–1115.

Mezrich JD, Yamada K, Lee RS, Mawulawde K, Benjamin LC, Schwarze ML, Maloney ME, Amoah HC, Houser SL, Sachs DH, et al. 2003b. Induction of tolerance to heart transplants by simultaneous cotransplantation of donor kidneys may depend on a radiation-sensitive renal-cell population. *Transplantation* **76:** 625–631.

Mezrich J, Yamada K, Sachs DH, Madsen JC. 2004. Regulatory T cells generated by the kidney may mediate the beneficial immune effects of combining kidney with heart transplantation. *Surgery* **135:** 473–478.

Mezrich JD, Benjamin LC, Sachs JA, Houser SL, Vagefi PA, Sachs DH, Madsen JC, Yamada K. 2005. Role of the thymus and kidney graft in the maintenance of tolerance to heart grafts in miniature swine. *Transplantation* **79:** 1663–1673.

Miyajima M, Chase CM, Alessandrini A, Farkash EA, Della Pelle P, Benichou G, Graham JA, Madsen JC, Russell PS, Colvin RB. 2011. Early acceptance of renal allografts in mice is dependent on Foxp3^{+} cells. *Am J Pathol* **178:** 1635–1645.

Mohib K, Guan Q, Diao H, Du C, Jevnikar AM. 2007. Proapoptotic activity of indoleamine 2,3-dioxygenase expressed in renal tubular epithelial cells. *Am J Physiol Renal Physiol* **293:** F801–F812.

Nair N, Ball T, Uber PA, Mehra MR. 2011. Current and future challenges in therapy for antibody-mediated rejection. *J Heart Lung Transplant* **30:** 612–617.

Narula J, Bennett LE, DiSalvo TG, Hosenpud JD, Semigran MJ, Dec GW. 1997. Outcomes in recipients of combined heart–kidney transplantation. Multi-organ, same-donor transplant study of the ISHLT/UNOS scientific registry. *Transplantation* **63:** 861–867.

Norman JC, Brook MI, Cooley DA, Klima T, Kahan BD, Frazier OH, Keats AS, Hacker J, Massin EK, Duncan JM, et al. 1978. Total support of the circulation of a patient with post-cardiotomy stone-heart syndrome by a partial artificial heart (ALVAD) for 5 days followed by heart and kidney transplantation. *Lancet* **1:** 1125–1127.

Ochando JC, Homma C, Yang Y, Hidalgo A, Garin A, Tacke F, Angeli V, Li Y, Boros P, Ding Y, et al. 2006. Alloantigen-presenting plasmacytoid dendritic cells mediate tolerance to vascularized grafts. *Nat Immunol* **7:** 652–662.

Okumi M, Scalea JR, Gillon BC, Tasaki M, Villani V, Cormack T, Hirakata A, Shimizu A, Sachs DH, Yamada K. 2013. The induction of tolerance of renal allografts by adoptive transfer in miniature swine. *Am J Transplant* **13:** 1193–1202.

Opelz G, Margreiter R, Dohler B. 2002. Prolongation of long-term kidney graft survival by a simultaneous liver transplant: The liver does it, and the heart does it too. *Transplantation* **74:** 1390–1394.

Ott HC, Matthiesen TS, Goh SK, Black LD, Kren SM, Netoff TI, Taylor DA. 2008. Perfusion-decellularized matrix: Using nature's platform to engineer a bioartificial heart. *Nat Med* **14:** 213–221.

Pallotta MT, Orabona C, Volpi C, Vacca C, Belladonna ML, Bianchi R, Servillo G, Brunacci C, Calvitti M, Bicciato S, et al. 2011. Indoleamine 2,3-dioxygenase is a signaling protein in long-term tolerance by dendritic cells. *Nat Immunol* **12:** 870–878.

Patel J, Everly M, Chang D, Kittleson M, Reed E, Kobashigawa J. 2011. Reduction of alloantibodies via proteasome inhibition in cardiac transplantation. *J Heart Lung Transplant* **30:** 1320–1326.

Pillarisetty VG, Shah AB, Miller G, Bleier JI, DeMatteo RP. 2004. Liver dendritic cells are less immunogenic than spleen dendritic cells because of differences in subtype composition. *J Immunol* **172:** 1009–1017.

Praseedom RK, McNeil KD, Watson CJ, Alexander GJ, Calne RY, Wallwork J, Friend PF. 2001. Combined transplantation of the heart, lung, and liver. *Lancet* **358:** 812–813.

Qian S, Demetris AJ, Murase N, Rao AS, Fung JJ, Starzl TE. 1994. Murine liver allograft transplantation: Tolerance and donor cell chimerism. *Hepatology* **19:** 916–924.

Raichlin E, Daly RC, Rosen CB, McGregor CG, Charlton MR, Frantz RP, Clavell AL, Rodeheffer RJ, Pereira NL, Kremers WK, et al. 2009. Combined heart and liver transplantation: A single-center experience. *Transplantation* **88:** 219–225.

Rasmussen A, Davies HF, Jamieson NV, Evans DB, Calne RY. 1995. Combined transplantation of liver and kidney from the same donor protects the kidney from rejection and improves kidney graft survival. *Transplantation* **59:** 919–921.

Robertson H, Wong WK, Burt AD, Mohamed MA, Talbot D, Kirby JA. 2001. Relationship between TGF-β1, intratubular CD103 positive T cells and acute renal allograft rejection. *Transplant Proc* **33:** 1159.

Rosengard BR, Ojikutu CA, Guzzetta PC, Smith CV, Sundt TM III, Nakajima K, Boorstein SM, Hill GS, Fishbein JM, Sachs DH. 1992. Induction of specific tolerance to class I disparate renal allografts in miniature swine with cyclosporine. *Transplantation* **54:** 490–497.

Russell PS, Chase CM, Colvin RB, Plate JM. 1978. Kidney transplants in mice. An analysis of the immune status of mice bearing long-term, H-2 incompatible transplants. *J Exp Med* **147:** 1449–1468.

Sachs DH. 1992. MHC homozygous miniature swine. In *Swine as models in biomedical research* (ed. Swindle MM, et al.), pp. 3–15. Iowa State University Press, Ames, IA.

Schoop R, Wahl P, Le Hir M, Heemann U, Wang M, Wuthrich RP. 2004. Suppressed T-cell activation by IFN-γ-induced expression of PD-L1 on renal tubular epithelial cells. *Nephrol Dial Transplant* **19:** 2713–2720.

Shiba Y, Fernandes S, Zhu WZ, Filice D, Muskheli V, Kim J, Palpant NJ, Gantz J, Moyes KW, Reinecke H, et al. 2012. Human ES-cell-derived cardiomyocytes electrically couple and suppress arrhythmias in injured hearts. *Nature* **489:** 322–325.

Singer GG, Yokoyama H, Bloom RD, Jevnikar AM, Nabavi N, Kelley VR. 1993. Stimulated renal tubular epithelial cells induce anergy in CD4^{+} T cells. *Kidney Int* **44:** 1030–1035.

Soos TJ, Sims TN, Barisoni L, Lin K, Littman DR, Dustin ML, Nelson PJ. 2006. CX3CR1^{+} interstitial dendritic cells form a contiguous network throughout the entire kidney. *Kidney Int* **70:** 591–596.

Sriwatanawongsa V, Davies HS, Calne RY. 1995. The essential roles of parenchymal tissues and passenger leukocytes in the tolerance induced by liver grafting in rats. *Nat Med* **1:** 428–432.

Stehlik J, Islam N, Hurst D, Kfoury AG, Movsesian MA, Fuller A, Delgado JC, Hammond ME, Gilbert EM, Renlund DG, et al. 2009. Utility of virtual crossmatch in sensitized patients awaiting heart transplantation. *J Heart Lung Transplant* **28:** 1129–1134.

Stehlik J, Edwards LB, Kucheryavaya AY, Benden C, Christie JD, Dipchand AI, Dobbels F, Kirk R, Rahmel AO, Hertz MI, et al. 2012. The Registry of the International Society for Heart and Lung Transplantation: 29th official adult heart transplant report—2012. *J Heart Lung Transplant* **31:** 1052–1064.

Sumpter TL, Abe M, Tokita D, Thomson AW. 2007. Dendritic cells, the liver, and transplantation. *Hepatology* **46:** 2021–2031.

Te HS, Anderson AS, Millis JM, Jeevanandam V, Jensen DM. 2008. Current state of combined heart–liver transplantation in the United States. *J Heart Lung Transplant* **27:** 753–759.

Tiegs G, Lohse AW. 2010. Immune tolerance: What is unique about the liver. *J Autoimmun* **34:** 1–6.

Tonsho M, Benichou G, Boskovic S, Nadazdin O, Smith N, Colvin R, Sachs D, Cosimi A, Kawai T, Madsen J. 2013. Successful tolerance induction of cardiac allografts in nonhuman primates through donor kidney co-transplantation. *2013 American Transplant Congress*, Seattle, WA, May 18–22.

Topilsky Y, Raichlin E, Hasin T, Boilson BA, Schirger JA, Pereira NL, Edwards BS, Clavell AL, Rodeheffer RJ, Frantz RP, et al. 2013. Combined heart and liver transplant attenuates cardiac allograft vasculopathy compared with isolated heart transplantation. *Transplantation* **95:** 859–865.

Trachiotis GD, Vega JD, Johnston TS, Berg A, Whelchel J, Smith AL, Lutz J, Kanter KR. 2003. Ten-year follow-up in patients with combined heart and kidney transplantation. *J Thorac Cardiovasc Surg* **126:** 2065–2071.

Traversi E, Pozzoli M, Grande A, Forni G, Assandri J, Viganò M, Tavazzi L. 1998. The bicaval anastomosis technique for orthotopic heart transplantation yields better atrial function than the standard technique: An echocardiographic automatic boundary detection study. *J Heart Lung Transplant* **17:** 1065–1074.

Utsugi R, Barth RN, Lee RS, Kitamura H, LaMattina JC, Ambroz J, Sachs DH, Yamada K. 2001. Induction of transplantation tolerance with a short course of tacrolimus (FK506). I: Rapid and stable tolerance to two-haplotype fully MHC-mismatched kidney allografts in miniature swine. *Transplantation* **71:** 1368–1379.

Valantine H. 2007. Is there a role for proliferation signal/mTOR inhibitors in the prevention and treatment of de novo malignancies after heart transplantation? Lessons learned from renal transplantation and oncology. *J Heart Lung Transplant* **26:** 557–564.

Vermes E, Kirsch M, Houël R, Legouvelo S, Benvenuti C, Aptecar E, Le Besnerais P, Lang P, Abbou C, Loisance D. 2001. Immunologic events and long-term survival after combined heart and kidney transplantation: A 12-year single-center experience. *J Heart Lung Transplant* **20:** 1084–1091.

Vermes E, Grimbert P, Sebbag L, Barrou B, Pouteil-Noble C, Pavie A, Obadia JF, Loisance D, Lang P, Kirsch M. 2009. Long-term results of combined heart and kidney transplantation: A French multicenter study. *J Heart Lung Transplant* **28:** 440–445.

Veronese F, Rotman S, Smith RN, Pelle TD, Farrell ML, Kawai T, Benedict Cosimi A, Colvin RB. 2007. Pathological and clinical correlates of $FOXP3^+$ cells in renal allografts during acute rejection. *Am J Transplant* **7:** 914–922.

von Oppen N, Schurich A, Hegenbarth S, Stabenow D, Tolba R, Weiskirchen R, Geerts A, Kolanus W, Knolle P, Diehl L. 2009. Systemic antigen cross-presented by liver sinusoidal endothelial cells induces liver-specific CD8 T-cell retention and tolerization. *Hepatology* **49:** 1664–1672.

Walsh RC, Alloway RR, Girnita AL, Woodle ES. 2012. Proteasome inhibitor-based therapy for antibody-mediated rejection. *Kidney Int* **81:** 1067–1074.

Wang SS, Chou NK, Chi NH, Hsu RB, Huang SC, Chen YS, Yu HY, Ko WJ, Chu SH, Tsai MK, et al. 2006. Simultaneous heart and kidney transplantation for combined cardiac and renal failure. *Transplant Proc* **38:** 2135–2137.

Wang C, Cordoba S, Hu M, Bertolino P, Bowen DG, Sharland AF, Allen RD, Alexander SI, McCaughan GW, Bishop GA. 2011. Spontaneous acceptance of mouse kidney allografts is associated with increased Foxp3 expression and differences in the B and T cell compartments. *Transpl Immunol* **24:** 149–156.

Wood KJ, Sakaguchi S. 2003. Regulatory T cells in transplantation tolerance. *Nat Rev* **3:** 199–210.

Wu A, Yamada K, Ierino FL, Vagefi PA, Sachs DH. 2003. Regulatory mechanism of peripheral tolerance: In vitro evidence for dominant suppression of host responses during the maintenance phase of tolerance to renal allografts in miniature swine. *Transpl Immunol* **11:** 367–374.

Xiao S, Jin H, Korn T, Liu SM, Oukka M, Lim B, Kuchroo VK. 2008. Retinoic acid increases $Foxp3^+$ regulatory T cells and inhibits development of Th17 cells by enhancing TGF-β-driven Smad3 signaling and inhibiting IL-6 and IL-23 receptor expression. *J Immunol* **181:** 2277–2284.

Yamada K, Choo JK, Allan JS, Erhorn AE, Menard MT, Mawulawde K, Slisz JK, Aretz HT, Shimizu A, Sachs DH, et al. 1999. The effect of thymectomy on tolerance induction and cardiac allograft vasculopathy in a miniature swine heart/kidney transplantation model. *Transplantation* **68:** 485–491.

Yamada K, Mawulawde K, Menard MT, Shimizu A, Aretz HT, Choo JK, Allison KS, Slisz JK, Sachs DH, Madsen JC. 2000. Mechanisms of tolerance induction and prevention of cardiac allograft vasculopathy in miniature swine: The effect of augmentation of donor antigen load. *J Thorac Cardiovasc Surg* **119:** 709–719.

Yamada Y, Boskovic S, Aoyama A, Murakami T, Putheti P, Smith RN, Ochiai T, Nadazdin O, Koyama I, Boenisch O, et al. 2012. Overcoming memory T-cell responses for induction of delayed tolerance in nonhuman primates. *Am J Transplant* **12:** 330–340.

Yang HR, Chou HS, Gu X, Wang L, Brown KE, Fung JJ, Lu L, Qian S. 2009. Mechanistic insights into immunomodulation by hepatic stellate cells in mice: A critical role of interferon-γ signaling. *Hepatology* **50:** 1981–1991.

Yu MC, Chen CH, Liang X, Wang L, Gandhi CR, Fung JJ, Lu L, Qian S. 2004. Inhibition of T-cell responses by hepatic stellate cells via B7-H1-mediated T-cell apoptosis in mice. *Hepatology* **40:** 1312–1321.

Zhang Z, Zhu L, Quan D, Garcia B, Ozcay N, Duff J, Stiller C, Lazarovits A, Grant D, Zhong R. 1996. Pattern of liver, kidney, heart, and intestine allograft rejection in different mouse strain combinations. *Transplantation* **62:** 1267–1272.

Zheng SG. 2008. The critical role of TGF-β1 in the development of induced $Foxp3^+$ regulatory T cells. *Int J Clin Exp Med* **1:** 192–202.

Zuckermann A, Keogh A, Crespo-Leiro MG, Mancini D, Vilchez FG, Almenar L, Brozena S, Eisen H, Tai SS, Kushwaha S. 2012. Randomized controlled trial of sirolimus conversion in cardiac transplant recipients with renal insufficiency. *Am J Transplant* **12:** 2487–2497.

Clinical Aspects: Focusing on Key Unique Organ-Specific Issues of Renal Transplantation

Sindhu Chandran[1] **and Flavio Vincenti**[2]

[1]Division of Nephrology, Department of Medicine, Connie Frank Transplant Center at UCSF, San Francisco, California 04143

[2]Department of Medicine and Department of Surgery, Deborah Faiman Endowed Chair in Kidney Transplantation, UCSF, San Francisco, California 04143

Correspondence: flavio.vincenti@ucsfmedctr.org

Kidney transplantation is the treatment of choice for patients with end-stage renal disease because of improved patient survival and quality of life as compared with dialysis. Successful transplantation requires the prompt recognition and appropriate management of both the immediate posttransplant surgical and medical complications as well as subsequent issues like recurrent disease and chronic rejection that affect long-term graft survival. Guidelines for understanding and managing some of the more important early and late kidney-specific transplant problems, including urologic complications, delayed graft function, acute and chronic rejection, BK polyoma virus infection, and recurrent glomerular disease, are reviewed.

Since the first successful kidney transplant in 1954, kidney transplantation has evolved to become the preferred therapeutic modality for patients with end-stage renal disease, conferring superior patient survival and quality of life as compared with dialysis. Over the past 15 years, 90-d, 6-mo, and 1-, 3-, and 5-yr results have shown ongoing improvement for both living and deceased donor kidney transplant recipients, and the 2011 Annual Data Report of the Scientific Registry of Transplant Recipients suggested for the first time an improvement in 10-yr results (Annual Data Report 2011). These improvements have been driven by a decrease in the rate of graft failure and return to dialysis. Three-month and 1-yr graft survival for patients receiving a kidney transplant from a deceased donor in 2010 were 96.2% and 92.9%, respectively.

As of June 30, 2011, 164,200 adults in the United States were surviving with a functioning kidney graft, about twice as many as a decade earlier. These patients present the physician with a complex set of medical issues that require intensive management to allow for longevity of both the patient and the allograft. The evaluation and management of the more commonly encountered post-kidney transplant problems are best organized by the time posttransplantation. Most acute events occur within the first 3 mo, and patients are usually clinically stable after the end of the first year. Early issues occurring within the first year are usually recognized and managed by the transplant physicians. Suc-

Cite this article as *Cold Spring Harb Perspect Med* doi: 10.1101/cshperspect.a015644

cessful management of subsequent issues requires heightened vigilance on the part of the community nephrologists who are often taking care of these patients, combined with prompt responses and appropriate interventions by the transplant center.

THE FIRST 3 MONTHS

Most kidney transplant recipients are discharged home before the end of the first week. The bladder catheter is usually removed before discharge and patients are voiding spontaneously. Graft function is closely monitored and usually still improving. Immunosuppressive therapy is adjusted frequently based on drug levels and the presence of side effects.

Urologic Issues

Urinary Retention

Urinary retention occurs commonly owing to prostatic enlargement or neurogenic bladder, because these diagnoses can be missed if recipients were anuric before the surgery. If the post-void residual urine volume is >100 mL, the catheter may need to be replaced and the patient may need training in self-catheterization. α-Blocking agents such as tamsulosin or terazosin are often started in older male recipients before catheter removal. Ideally, patients with suspected urologic issues should undergo a pretransplant urologic evaluation.

Urine Leak

Urinary complications are the most frequent technical adverse events following renal transplantation. In comparative analyses, the incidence of urine leak is 0%–9.3% (Pleass et al. 1995; Veale et al. 2007). Most urine leaks are ureteral rather than vesical in origin because most surgeons now use an extravesical ureteroneocystostomy for ureteral implantation, resulting in a shorter ureter, decreased likelihood of ischemia, and a limited cystotomy that rarely leads to leakage from the bladder. Early urine leaks can be of two types. Those occurring within the first 1–4 d are usually related to technical problems with the implantation such as excessive tension at the anastomosis. The second type, presenting 5–10 d posttransplantation, is due to distal ureteral ischemia. The sole arterial supply of the transplant ureter is the descending branch of the main renal artery or its lower polar branch. Damage to these vessels can be sustained during organ recovery, with any trivial mishandling of the ureter jeopardizing the tiny periureteral arterial branches and threatening the anastomosis. It is not always possible to discern intraoperatively whether a ureter will develop ischemic necrosis or stricture.

Clinically, urine leaks often present with excruciating abdominal pain, decreased urine output, and a rise in serum creatinine. Ultrasound may show a fluid collection and hydronephrosis or hydroureter, but not always. Aspiration of the fluid and measurement of fluid creatinine is one method of diagnosis. A Mag3 renogram (Fig. 1) is often helpful but can be falsely negative in the case of sluggish/delayed graft function or if the bladder is decompressed by a Foley catheter. Urine leaks can be treated conservatively, in which case a perinephric drain, ureteral stent, and bladder catheter are left in place until the

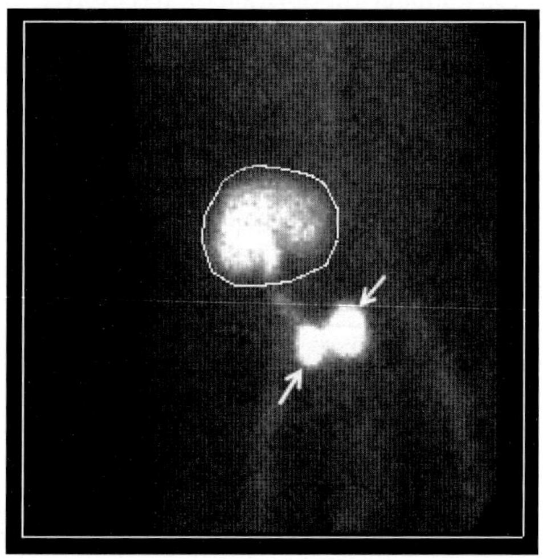

Figure 1. Diagnosis of a urine leak. Mag3 renogram shows a urine leak (*top* arrow) near the ureterovesical anastomosis. The isotope tracer has been excreted by the kidney (circle) and the bladder is contracted (*bottom* arrow).

leak appears to have healed. If the leak persists or is considered unlikely to heal, then reexploration is indicated to reimplant the ureter into the bladder or perform a uretero-ureterostomy to one of the native ureters.

Ureteral Obstruction

Ureteral obstruction can present early or subsequently after kidney transplantation, with reported incidence rates of 1%–8.3% (Veale et al. 2007). Although some cases of ureteral obstruction are caused by edema, fluid collections, adhesions, tumor, stones, blood clots, or compression by the round ligament or the spermatic cord, are due mostly to strictures, which are believed to be related to ischemia or to rejection events. Potential technical causes of obstruction are kinking within the submucosal tunnel or delayed necrosis of ureteral tips from an excessively tight closure at the seromuscular layer of the bladder. The diagnosis is usually suggested by an ultrasound showing hydronephrosis and/or hydroureter (Fig. 2A) and confirmed by an antegrade pyelogram (Fig. 2B). BK viral infections have also been implicated in the development of subsequent ureteral strictures. Early cases occurring within the first few weeks after transplantation are often managed with a ureteral stent and may not recur after the stent is removed. Recurrent or subsequent cases usually need ureteral revision.

Delayed Graft Function

There are more than 10 definitions of delayed renal allograft function or DGF recorded in the literature based on a range of clinical criteria. Most commonly, DGF has been defined as the use of dialysis within 7 d of the transplant. This definition has shortfalls because dialysis may be used in the first week after transplant without confirmation of kidney damage. Nevertheless, it offers a standard by which center-specific outcomes can be reported, and epidemiologic analyses and intercenter comparisons performed.

The reported incidence of DGF in deceased donors increased from 14.7% between 1985 and 1992 to 21.3% in 2008 despite the progress in acute rejection treatment, and occurred contemporaneously with the use of expanded criteria donors (ECDs) and donation after cardiac death (DCD) (Siedlecki et al. 2011). Most cases are due to ischemic acute tubular necrosis (ATN). Other causes include accelerated cellular or humoral rejection, vascular compromise, urinary obstruction or urine leak, nephrotoxicity from calcineurin inhibitors (CNI), and thrombotic microangiopathy. Donor-specific risk factors for DGF include both immunologic (e.g., absence of T-cell antibody induction therapy, female donor) and nonimmunologic (donor age, weight, prolonged cold/warm ischemia time) factors. Distinct recipient characteristics also confer risk of DGF. Maintenance dialysis

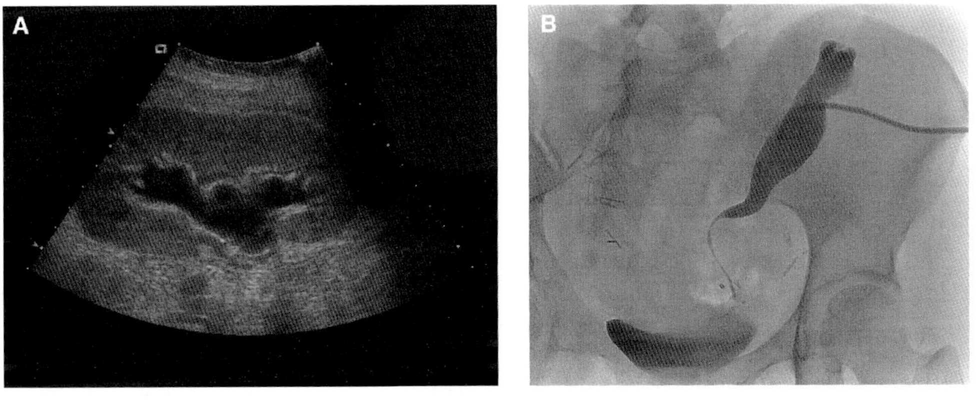

Figure 2. Diagnosis of ureteral obstruction. (*A*) Transplant renal ultrasound shows hydronephrosis. (*B*) Antegrade pyelogram in the same patient shows distal long segment ureteral stricture.

before transplant is perhaps the single greatest contributor. Other factors include obesity, diabetes, age >55 yr, male sex, African American race, prolonged wait period, presensitization, and small-for-size organs. Many of these donor and recipient characteristics are included in a nomogram used to obtain a composite risk assessment before the initiation of implantation surgery (Irish et al. 2010).

After competing prerenal, renal, and postrenal diagnoses are excluded, patients suspected of DGF should preferably be biopsied 5–14 d posttransplantation to rule out acute rejection as a cause of graft dysfunction. DGF associated with ischemic ATN is characterized by ischemia-reperfusion injury and immunologic dysregulation. Endothelial injury has been shown to up-regulate the expression of donor HLA antigens, adhesion molecules, and costimulatory molecules, thus increasing the risk for acute rejection.

Recent outcome studies have not shown a negative effect of DGF on long-term graft survival in the absence of acute rejection (Kayler et al. 2011). However, DGF consumes financial and programmatic resources in terms of prolonged hospitalization, dialysis treatments, and the higher utilization of radiologic studies and biopsies. The effect of supportive parameters on reducing the duration of DGF remains unknown. Posttransplant hemodialysis should be offered when clinically indicated. Hemodynamic instability and nephrotoxins should be avoided. It is important to achieve and maintain adequate immunosuppression in the presence of ATN. Although it has been theorized that a delay in CNI introduction might shorten the course of DGF by avoiding nephrotoxicity, this is unproven and the long-term benefits of such an approach remain debatable. Use of antilymphocyte therapy after DGF is established may not treat the DGF, but will reduce the rejection rate and minimize the negative impact of acute rejection in association with DGF.

The complexity of pathological mechanisms that cause DGF offers many potential targets for therapy to inhibit oxidant stress, encourage vasodilation, and blunt the immune response. There are several ongoing clinical trials currently exploring different targets with the primary end point of DGF.

Acute Rejection

With the use of potent induction and maintenance immunosuppressive agents, the incidence of acute rejection within the first year following transplantation has fallen dramatically, with current rates <15% in most centers. When acute rejection occurs, it remains an important clinical problem, increasing the relative risk for early transplant failure as well as contributing to the development of chronic rejection and late graft loss.

Acute rejection in kidney transplants is suspected whenever there is a rise in serum creatinine and is diagnosed by a kidney biopsy. The Banff Classification of Renal Allograft Pathology was originally developed in 1991 and was merged together with the most useful parts of the CCTT classification in 1997, to provide the framework on which today's schema is built. Criteria for antibody-mediated rejection were added in 2003 and updated in 2007 (Solez et al. 2008).

Current markers for acute rejection are imperfect. A rise in serum creatinine lacks sensitivity and specificity for the diagnosis of acute rejection. A kidney biopsy, on the other hand, is invasive and its interpretation subject to reader bias as well as sampling error. Newer assays that look at biomarkers in blood and urine are less invasive and may yield earlier and more precise measurements. Gene expression patterns in plasma, peripheral blood lymphocytes, and biopsy tissue using techniques such as PCR and DNA microarray are currently being evaluated to detect and distinguish different types of rejection. Lack of regulatory T cells (or Tregs) during acute rejection results in unimpeded effector T-cell activity and impaired graft function. Urine levels of the mRNA for the *FOXP3* gene, a transcription factor for Tregs, have been found to be positively correlated with graft survival (Muthukumar et al. 2005). A urine test for FOXP3 has been commercially developed and was released in 2012 for limited clinical use. In another landmark study (Li et al. 2012), a five-gene signature in peripheral blood lymphocytes composed of DUSP1, PBEF1, PSEN1, MAPK9, and NKTR was found to diagnose acute rejection with 91% sensitivity and 94% specificity in a largely pediatric population. Although this signature needs validation in larger and more diverse cohorts as well as correlation with graft outcome, it holds great potential as a diagnostic tool that can add to or replace the kidney biopsy as well as monitor the response to therapy.

Accelerated Acute Rejection

Accelerated acute rejection occurs because of preformed cytotoxic anti-HLA antibodies. It was originally described before adequate cross-match techniques and assays for antibodies were available. Humoral vascular rejection occurring immediately after reperfusion is called hyperacute rejection. Hyperacute rejection is an exceedingly rare event in the modern era but must still be considered when evaluating early graft dysfunction in patients who were highly sensitized before the transplant. Patients with hyperacute or delayed hyperacute rejection (occurring days after the transplant due to an an amnestic antibody response) may be febrile, and the kidney is usually swollen and tender upon palpation. Ultrasound may show impaired flow. Surgical exploration is indicated to exclude vascular compromise and obtain a biopsy to confirm the diagnosis.

Acute T-Cell-Mediated Rejection

Acute T-cell-mediated rejection before the end of the first week is uncommon, especially in the current era when most patients get antibody induction. It is usually asymptomatic and manifested by either a rise in serum creatinine or by a failure of renal function to improve in a patient with DGF. Ultrasound may show elevated resistive indices, but this finding is not specific. The diagnosis is made by a kidney biopsy showing interstitial infiltration with mononuclear cells and occasionally eosinophils (measured by the i-score), disruption of the tubular basement membranes by the infiltrating cells or tubulitis (t-score), and intimal arteritis (v-score) (Table 1).

Corticosteroids are the most common first-line treatment for episodes of acute cellular rejection. Intravenous methylprednisolone, 250–500 mg daily for 3 d followed by tapering steroid doses is the usual practice. Most initial episodes of Banff type 1 acute T-cell-mediated rejection respond to corticosteroids and show improvement in graft function back to baseline levels. Treatment of acute cellular rejection with antithymocyte globulin is more effective in restoring kidney function and preventing graft loss than treatment with corticosteroids, but is also associated with more adverse effects. The efficacy of thymoglobulins, especially those of rabbit origin, has been clearly established for the treatment of steroid-resistant acute cell-mediated rejection (Webster et al. 2006). Banff grade II–III patients show the greatest benefit with rejection reversal rates ranging from 88% to 97%. The outcome after rejection depends on the severity of the rejection and the response to therapy. Return to within 15% of the baseline serum creatinine is associated with the best long-term prognosis. If kidney function does not return to baseline or if there is a new decline in function after successful treatment of an acute rejection, a repeat biopsy should be considered to rule out additional rejection or other causes of graft dysfunction.

Acute Antibody-Mediated Rejection

Acute antibody-mediated rejection (AMR) tends to occur early, particularly in patients who were presensitized, and often presents solely with elevated serum creatinine. Kidney biopsies showing histological features of endothelial damage including neutrophil infiltration, necrosis, apoptosis, and thrombosis in patients with detectable donor-specific antibody (DSA) in serum have traditionally been termed vascular rejection. The presence of circulating DSA has been shown to play a key role in these findings. The 2003 update of the 1997 Banff criteria defined AMR on the basis of three criteria: the presence of anti-HLA DSA in serum, allograft histology (polymorphonuclear-rich glomerular and/or peritubular inflammation), and positive staining for C4d in peritubular capillaries. In

Table 1. Banff 2007 update of Banff 1997 diagnostic categories for renal allograft biopsies

Diagnostic category	Description	Criteria
1. Normal		
2. Antibody-mediated changes	C4d deposition without morphologic evidence of active rejection	$C4d^+$; circulating DSA present; no signs of acute or chronic TCMR or ABMR
	Acute antibody-mediated rejection	$C4d^+$; circulating DSA present; morphologic evidence of acute tissue injury such as acute tubular necrosis, capillaritis, and/or glomerulitis and/or thrombosis; severe arteritis (v3)
	Chronic active antibody-mediated rejection	$C4d^+$; circulating DSA present; morphologic evidence of chronic tissue injury such as glomerular double contours and/or peritubular capillary basement membrane multilayering; IF/TA; fibrous intimal thickening in arteries
3. Borderline change	"Suspicious" for acute T-cell-mediated rejection	No intimal arteritis (v0), but there are interstitial infiltrates and tubulitis insufficient to meet criteria for T-cell-mediated rejection (i0-i1, t1-t3 or i2-i3, t1)
4. T-cell-mediated rejection (TCMR)	Acute T-cell-mediated rejection	Type I: i2-3, t2-3 Type II: v1-v2 Type III: v3
	Chronic active T-cell-mediated rejection	Arterial intimal fibrosis with mononuclear cell infiltration in fibrosis, formation of neo-intima
5. Interstitial fibrosis and tubular atrophy	No evidence of any specific etiology	Severity graded I–III by the extent of tubulointerstitial fibrosis and atrophy
6. Other	Changes not considered to be due to rejection	For example, acute pyelonephritis, BK polyoma virus interstitial nephritis

Data adapted from Solez et al. (2008).

2011, the Banff meeting identified significant limitations of C4d as a diagnostic marker for AMR and acknowledged the existence of C4d-negative AMR as a phenotype (Sis et al. 2010; Mengel et al. 2012). The diagnostic criteria continue to evolve.

The reported incidence of AMR varies worldwide from 3.1% to 40%, depending on the diagnostic criteria, recipient sensitization, and the immunosuppressive regimen. Unlike cellular rejection, AMR responds poorly to corticosteroids and antithymocyte agents alone. The treatment of AMR often requires a combination of strategies that increase the clearance of circulating DSA and decrease antibody production (Vincenti et al. 2010). The most effective treatment so far appears to be a combination of plasmapheresis, IVIg, and rituximab, which is a monoclonal antibody to CD20 targeting B cells (Lefaucheur et al. 2009). Bortezomib, a proteasome inhibitor, directly targets plasma cells, although effects may be delayed because of the slow decay in levels of DSA. Eculizumab, a terminal complement inhibitor, has been used in refractory cases of AMR to arrest complement-mediated injury while other therapies directed at DSA take effect.

Although knowledge of the diagnosis and pathophysiology of AMR is advancing, evidence supporting existing treatments is poor (Roberts et al. 2012). Treatment decisions regarding dosing and efficacy are often made on the basis of data from desensitizing protocols. However, the observed effects may differ when treating AMR in the posttransplantation setting owing to ongoing antigenic stimulation. Further studies are required to determine the optimal treatment of AMR and to gain a better understanding of the risk–benefit ratio because the adverse effects and cost of treatments are not insignificant.

Thrombotic Microangiopathy

Posttransplant thrombotic microangiopathy (TMA) can present either de novo or as a recurrence in patients with atypical hemolytic uremic syndrome (HUS). Most patients with recurrent TMAwill eventually lose the graft. De novo posttransplant TMA is a rare but serious complication of transplantation, occurring in 0.8%–15% cases, and associated with graft loss in more than one-third of cases with systemic manifestations (Noris and Remuzzi 2010). Usually, de novo TMA occurs in the early posttransplant days, but it may also develop 2–6 yr after transplantation. The pathogenesis is not well understood, but it is speculated that the endothelial lesions caused by ischemia-reperfusion injury, viral infection, or rejection may be amplified by the endothelial injury caused by immunosuppressive agents such as calcineurin inhibitors (CNIs), which have direct prothrombotic and vasoconstrictive effects, or the TOR inhibitors, which are antiangiogenic. This form of TMA is commonly limited to the renal allograft and usually presents with renal dysfunction unaccompanied by thrombocytopenia or a hemolytic anemia. It is usually, but not always, reversible with discontinuation of the CNI or mTOR inhibitor. Other reported risk factors include transplantation from extended criteria donors, viral infections, antiphospholipid antibodies, complement regulation disorders, OKT3, and malignancy. TMA has also been associated with acute antibody-mediated rejection of the kidney transplant.

Treatment for de novo TMA has not been well defined, and even complete withdrawal of the CNI is associated with graft loss rates as high as 60%–100%. Addition of plasma exchange has been reported to salvage the graft in ~80% of cases. Switching the CNI to sirolimus may not lead to improvement because sirolimus itself has been associated with the development of posttransplant TMA. Cautious reintroduction of CNIs may be successful in those who recover graft good function. Replacement of CNIs and/or mTOR inhibitors with belatacept, a second-generation CTLA4-Ig that blocks T-cell costimulation, is an alternative strategy for the treatment of posttransplant TMA that has been reported to be successful without precipitating acute rejection or recurrence (Ashman et al. 2009). Eculizumab, a high-affinity humanized anti-C5 monoclonal antibody that prevents deployment of the terminal complement system and generation of the cytolytic membrane attack complex, has been described as rescue therapy in severe or resistant cases of systemic posttransplant TMA associated with complement gene abnormalities or antibody-mediated rejection.

THE FIRST YEAR

Recurrent Disease

Recurrent primary kidney disease is the third most frequently reported cause for graft failure 10 yr after kidney transplantation (Briganti et al. 2002; Canaud et al. 2012). The incidence of recurrent disease and resultant graft failure varies depending on the disease, duration of follow-up, whether or not the primary cause of ESRD was known, and the frequency of kidney biopsies. The most frequent recurrent diseases have been focal segmental glomerulosclerosis (FSGS), IgA nephropathy, diabetic nephropathy, membranoproliferative (MPGN) and membranous glomerulonephritis, hemolytic uremic syndrome/thrombotic thrombocytopenic purpura (HUS/TTP), and lupus nephritis. Recurrent disease may present as increased serum creatinine, new-onset or increased proteinuria, and/or hematuria. FSGS, HUS, and oxalosis may recur in the first few days to weeks posttransplantation, whereas the timing is variable but usually late in the others.

Idiopathic FSGS recurs in 20%–50% of cases, and in up to 80% if it has recurred in a prior kidney transplant. Risk factors for recurrence include age of onset of FSGS in the native kidneys between 6 and 15 yr, rapid course of the original disease, diffuse mesangial proliferation on histology, and non-African-American ethnicity. The strongest risk factor is recurrence in a previous transplant. Recurrent disease usually presents with nephrotic proteinuria. About 80% recur in the first 4 wk, and recurrence should be considered in all cases with delayed or slow graft function. The diagnosis is made by a kidney biopsy that shows foot process effacement on electron microscopy. Sclerosis may not be evident in early recurrence, and light microscopy may show normal glomerular architecture. Case reports and uncontrolled series have noted a substantial reduction in proteinuria after plasma exchange, possibly because of removal of circulating glomerular permeability factors. Predictors of response to plasma exchange include early recurrence and early initiation of treatment after recurrence. Unfortunately, proteinuria may recur after treatment and may require additional or even periodic, ongoing treatments. High-dose cyclosporine can induce remission of proteinuria and may be combined with plasmapheresis. A reduction in proteinuria with an ACE inhibitor and/or an ARB may also be attempted.

Recurrent IgA nephropathy is common after transplantation, with reported incidence varying from 13% to 53%. Recurrence is more common among recipients of living related-donor transplants as compared with deceased donor transplants and may be associated with reduced graft survival. Maintenance steroid use has been associated with a reduced risk of recurrence. There is no effective therapy for preventing recurrent IgA nephropathy, although fish oil, ACE-Is, and ARBs frequently reduce proteinuria and possibly preserve kidney function.

Membranous nephropathy (MN) recurs in ~40% of patients after kidney transplantation. Autoantibodies to the phospholipase A2 receptor (PLA2R) are found in >70% of those with idiopathic MN, and are associated with recurrence. Treatment with rituximab stabilized both the proteinuria and serum creatinine, and circulating anti-PLA2R became undetectable. The recurrence rate in idiopathic type I MPGN is 20%–30% and exceeds 80% in type 2 disease. The introduction of eculizumab, a terminal complement inhibitor, may provide the first effective therapy for this disease.

The risk of posttransplant recurrence of atypical HUS depends on the genetic abnormality involved and ranges from 15% to 20% in patients with gene mutations for membrane cofactor protein to 50%–100% in patients with gene mutations of circulating complement regulators such as factor H. Recurrence usually develops within 4 wk. Most patients develop microangiopathic anemia, thrombocytopenia, and kidney dysfunction, whereas others present with rapidly progressive graft dysfunction without showing the classic hematologic manifestations. Long-term graft survival is ~30% lower in those with recurrence. Treatment strategies have included plasmapheresis, IVIg, and rituximab. Eculizumab has recently emerged as an effective treatment for recurrent HUS.

Both ANCA-associated vasculitis and anti-GBM disease have low rates of recurrence if the disease is quiescent at the time of transplant. Most cases occur beyond the first posttransplant year with no significant impact on allograft function. Relapses of ANCA-associated vasculitis usually manifest as pauci-immune necrotizing glomerulonephritis, but graft function can also be affected by acute arteritis, ureteral stenosis, and obstructive uropathy due to granulomatous vasculitis. Pretransplantation disease course, disease subtype, ANCA type or titer, time of transplantation, or donor type does not predict recurrence. Kidney ANCA-associated vasculitis generally responds well to high-dose prednisolone and cyclophosphamide. Other treatment modalities that have been tried include MMF, plasmapheresis, IVIg, and rituximab. Clinical recurrence of anti-GBM disease is rare and consists of isolated case reports only. The incidence may be higher in those with circulating anti-GBM antibody at the time of transplantation. Treatment of clinically active anti-GBM disease may include pulse steroids, cyclophosphamide, and plasma exchange, particularly if there is potentially life-threatening pulmonary involvement.

Primary hyperoxaluria is caused by deficiency of hepatic peroxisomal alanine:glyoxylate aminotransferase, leading to increased synthesis and urinary excretion of oxalate, recurrent calcium oxalate urolithiasis, and nephrocalcinosis. In renal failure, insoluble oxalates accumulate throughout the body, especially in bone and arteries. Oxalate overproduction persists following isolated kidney transplantation, leading to recurrent calcium oxalate deposition in

>90% of transplanted kidneys and eventually graft loss. Total body oxalate burden is often very high in patients with advanced chronic kidney disease, and plasma and urine oxalate levels may remain high for months and years, respectively, even in patients undergoing simultaneous kidney and liver transplantation. Early posttransplant urinary dilution through aggressive fluid administration, and early and frequent dialysis in those with DGF, are essential to minimize the effects of hyperoxalemia and hyperoxaluria. In the long term, measures designed to increase oxalate excretion and reduce production include the maintenance of urine output >3 L/d, and the use of alkaline citrate, neutral phosphate, and magnesium oxide. Pharmacological doses of pyridoxine may also be helpful in reducing hyperoxaluria in some patients, especially those with a Gly170Arg mutation.

BK Virus

Polyomavirus BK (BKV) is ubiquitous among the general population, establishing a latent, asymptomatic infection in immunocompetent individuals. The cellular reservoir of latent BKV infection comprises numerous cell types, including urothelial cells. The adaptive immune response and, in particular, T cells play a crucial role in the clearance of most viral infections, and the development of both cellular and humoral immunological memory is important in the immune response against BKV.

In kidney transplant recipients, BKV reactivation is common and leads to nephropathy (BKVN) in ~5% of recipients within a year after transplantation. Rarely, infection has been noted to cause ureteral stenosis. Patients with BKVN generally present with an asymptomatic decrease in renal function. Although the development of BKVN has been linked in some studies to specific immunosuppressive agents such as tacrolimus, mycophenolate mofetil, polyclonal anti-T-cell antibodies, and corticosteroid pulses, it is likely that the cumulative intensity of the immunosuppressive regimen rather than one specific agent increases the risk for BKVN. Other factors such as cold ischemia time, cadaveric donor, and the number of HLA mismatches and rejection episodes have been inconsistently associated with BKVN (Vasudev et al. 2005).

The renal disease spectrum begins with viruria and sequentially proceeds through viremia and tubulointerstitial nephritis before ending in extensive irreversible kidney damage. Intervention at an early stage is therefore of paramount importance to prevent graft loss. Screening for active BKV replication involves the detection of viral DNA by quantitative PCR in urine and in blood and, in some cases, the detection of BKV-infected "decoy cells" or aggregates of BKV virions, the so-called *haufen* in urine. The diagnosis of BKVN can only be made by a kidney biopsy. Histopathological grades of severity have been defined as ranging from viral cytopathic changes of near-normal renal parenchyma and no or minimal tubular atrophy, interstitial fibrosis, or inflammation, to diffusely scarred renal tissue with extensive tubular atrophy, interstitial fibrosis, and inflammation. Because BKV affects the kidney in a random, multifocal manner, false-negative biopsy results may occur, especially in early stages of the disease.

Currently, reducing immunosuppression is the only established mode of therapy and aims to restore the antiviral immune response. There is no standard strategy, although reducing or eliminating MMF is often the first step (Hirsch et al. 2005). Upon in vitro infection, BKV activates the intracellular PI3K/Akt/mTOR pathway, and sirolimus reduced LTAg expression in a dose-dependent manner. Clinical efficacy of mTOR inhibitors in treating BKVN has not been convincingly shown.

Beyond tapering and/or altering immunosuppression, other antiviral agents have been proposed. Although polyomaviruses do not express the known target of cidofovir, viral DNA polymerase, they showed in vitro inhibition of BKV. Unfortunately, randomized controlled trials are lacking, and several confounders including the concomitant tapering of immunosuppression complicate the interpretation of the clinical effectiveness of cidofovir. Known side effects include nephrotoxicity and, rarely, anterior uveitis. Treatment of BKV infection with IVIg, the pyrimidine synthesis inhibitor leflunomide, and fluoroquinolones have been reported to be successful in small case series.

AFTER THE FIRST YEAR

Late Allograft Failure

Chronic allograft failure is second only to death with a functioning graft as a cause of late renal graft loss after transplantation. Despite advances in transplantation reducing early acute rejection rates to <15% and lifting 1-yr graft survival >90%, long-term graft attrition rates have remained unchanged at 4% loss per year (Meier-Kriesche et al. 2004). The most commonly reported pathological change is chronic interstitial fibrosis and tubular atrophy (IF/TA) accompanied by vascular changes and glomerulosclerosis. The term "chronic allograft nephropathy" or "CAN," initially coined as a general descriptor for fibrosis in the allograft representing the final pathway of nephron injury, was replaced in 2005 as per recommendations of the Banff Consensus Conference by more specific etiologically relevant diagnoses and the designation of nonspecific fibrosis as interstitial fibrosis and tubular atrophy (IF/TA), no cause identified (Solez et al. 2007). Moderate to severe IF/TA is present in a quarter of recipients at 1 yr after transplant, and in ~90% by 10 yr (Nankivell and Chapman 2006). It is believed to represent the cumulative effect of chronic immune- and nonimmune-mediated injury to the graft.

Chronic rejection due to alloimmunity in the allograft may be cell or antibody mediated and is now recognized in the majority of patients with late graft loss (El-Zoghby et al. 2009). Although early data indicated a strong association between acute and chronic rejection, recent studies using classic histological and gene expression analysis show that early T-cell-mediated rejection has no impact on term outcome (Reeve et al. 2013). The degree of HLA mismatches, the presence of preformed or de novo DSA, and nonadherence to immunosuppressive medications have been shown to increase the risk of late graft loss. Chronic transplant glomerulopathy and arterial changes of accelerated graft arteriosclerosis, especially with lymphocytes in the fibrosing intima, splintering of the internal elastic lamina, and formation of neomedia and neo-intima inside the vessel, are reliable features of chronic rejection (Fig. 3). The presence of markers of ongoing injury in allografts in the form of active lymphocytic tubulitis, arteritis, or positive peritubular capillary C4d staining, even in the context of significant fibrosis appears to be more deleterious to the graft than bland fibrosis alone (Fig. 4) (Mannon et al. 2010).

Nonimmunological factors associated with late allograft failure include increased donor

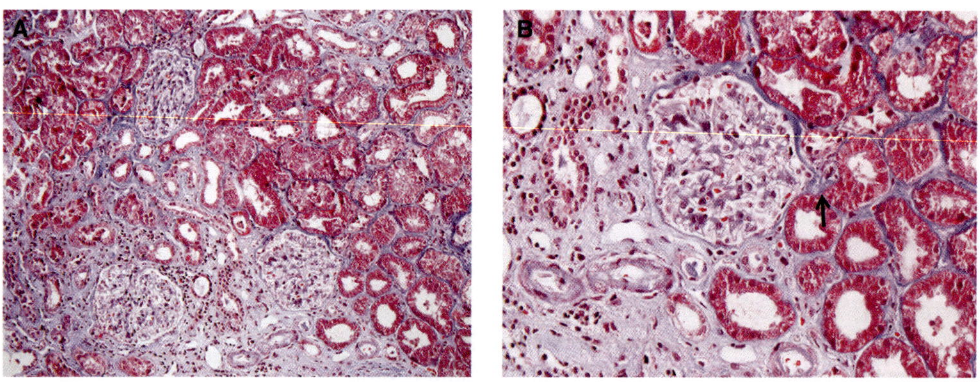

Figure 3. Pathological lesions on kidney allograft biopsies associated with chronic calcineurin inhibitor nephrotoxicity. (*A*) Areas of interstitial fibrosis and tubular atrophy alternating with preserved tubulointerstitium produce a picture of "striped fibrosis." (*B*) Peripheral arteriolar hyalinosis (arrow). Masson's trichrome stain. (Images courtesy of Dr. Zoltan Laszik.)

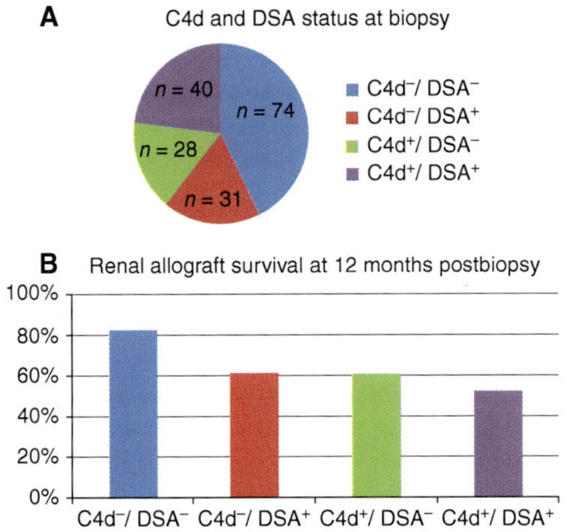

Figure 4. Impact of presence or absence of C4d and donor-specific antibody (DSA) on kidney allograft survival after for-cause biopsy. Both C4d positivity and the presence of detectable DSA increased the relative risk of graft failure with the best prognosis in the $C4d^-/DSA^-$ group and the worst graft survival in the $C4d^+/DSA^+$ group. (Figures created from data in Gourishankar et al. 2010.)

age, receipt of an ECD kidney, nephrotoxicity from CNIs, and systemic disorders such as diabetes, hypertension, and dyslipidemia. Histological features suggestive of CNI toxicity include obliterative arteriolopathy, ischemic collapse or scarring of the glomeruli, vacuolization of the tubules, and focal areas of tubular atrophy and interstitial fibrosis producing a picture of "striped" fibrosis (Fig. 5). The extent of IF has been correlated to the expression of transforming growth factor-β, which is up-regulated by CNIs. However, morphological lesions attributed to chronic CNI toxicity are nearly universal in long-functioning grafts, and it has been difficult to establish a definitive causal link, particularly in the modern era with the use of lower doses of CNIs as compared with when they were first introduced (Stegall et al. 2011). Therefore, the diagnosis of probable CNI toxicity is usually one of exclusion. Interestingly, the Deterioration of Kidney Allograft Function (DeKAF) study suggested that immune-mediated injury rather than CNI toxicity may be a more significant contributor to graft loss than previously thought; and in this study, patients with a histological diagnosis of CNI nephrotoxicity were found to be less likely to lose their allografts than those without such a diagnosis (Gourishankar et al. 2010).

Clinically, there is declining graft function, often accompanied by proteinuria and hypertension. A kidney biopsy is useful in determining whether specific features of immune-mediated injury are present and to rule out reversible causes of graft dysfunction such as BKVN, acute rejection, or recurrent kidney disease. When de-

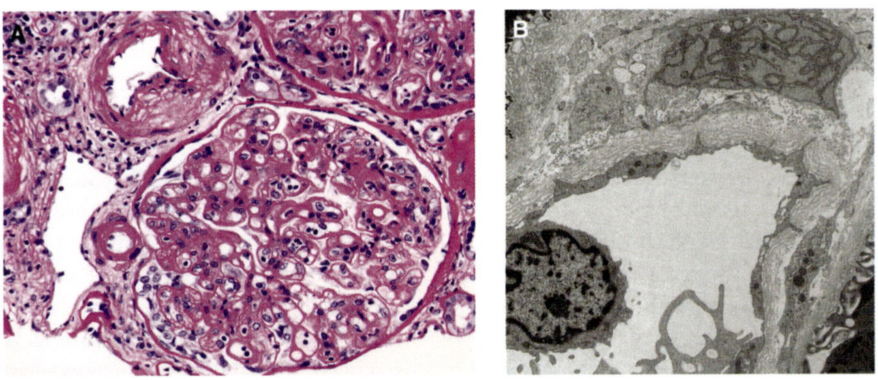

Figure 5. Pathologic lesions on kidney allograft biopsies characteristic of chronic antibody-mediated rejection. (*A*) Double contoured glomerular basement membranes accompanied by increased mesangial matrix and prominent glomerulitis. Periodic acid Schiff stain. (*B*) Peritubular capillary basement membrane multilayering on electron microscopy. (Images courtesy of Dr. Zoltan Laszik.)

teriorating function from CNI nephrotoxicity arises in low-immune-risk patients and ongoing immune injury is excluded, CNI withdrawal and maintenance with concentration-controlled mycophenolate mofetil and corticosteroids have been shown to improve GFR with a small risk of rejection and graft loss. In those with evidence of CNI toxicity and high immune risk, either CNI minimization or substitution with mTOR inhibitors can be considered, accompanied by appropriate monitoring. Augmentation of immunosuppression may be considered in patients whose biopsy shows chronic cell- or antibody-mediated rejection with the addition of ACE inhibitors or angiotensin-receptor blockers to reduce proteinuria. Plasmapheresis, IVIg, and rituximab have shown variable results, and benefits from bortezomib and eculizumab are unproven in this population.

Pregnancy after Kidney Transplantation

Kidney transplantation improves reproductive function in women with ESRD, often restoring fertility within a few months. Data from large voluntary registries and a recent meta-analysis (Deshpande et al. 2011) indicate overall live birth rates of 71%–79% among pregnant kidney transplant recipients. Despite this success, pregnancy poses risks to both fetus and graft, and thus, preconception counseling, family planning, and contraception are critical aspects of posttransplant care. A higher rate of obstetrical complications including miscarriage, gestational diabetes, preeclampsia, Cesarean section delivery, preterm labor, and low birth weight is seen in kidney transplant recipients. The transplant kidney is neither affected by, nor does it affect, a vaginal delivery. Younger maternal age is associated with favorable outcomes. Pregnant kidney transplant recipients should be preferably referred to an obstetrician with expertise in managing high-risk pregnancies.

Acute rejection rates of 3%–10% have been reported during pregnancy, although the impact of pregnancy on graft survival remains controversial, with several studies showing no statistically significant difference when compared with matched nonpregnant controls. Prepregnancy hypertension, elevated serum creatinine, and proteinuria have been associated with adverse graft and fetal outcomes. Extrapolating from studies of pregnant women with chronic kidney disease, an estimated GFR <40 mL/min per 1.73 m^2 and proteinuria >1 g/d are considered to be risk factors for accelerated decrease in GFR and potentially graft loss. KDIGO guidelines suggest waiting for at least 1 yr after transplantation before becoming pregnant and only attempting pregnancy when kidney function is stable with <1 g/d of proteinuria.

There are no prospective data on the safety of immunosuppressive agents during pregnancy. Mycophenolate is reported to cause severe structural malformations, and a characteristic phenotype includes cleft lip and palate, microtia, and absent external auditory canals. In September 2012, the FDA approved a single shared Risk Evaluation and Mitigation Strategy (REMS) for all mycophenolate-containing prescription drugs. The goals of the mycophenolate REMS are to prevent an unplanned pregnancy in patients using the drug, to minimize fetal exposure, and to inform patients about the serious risks associated with this class of medications. The REMS includes a medication guide, training for health-care professionals, and the establishment of a pregnancy registry for women who become pregnant and agree to participate in the registry.

Despite the FDA category D, azathioprine has been used safely over the years in pregnant transplant recipients. Thus, mycophenolate is generally changed to azathioprine, and a 6-wk window is recommended between discontinuing mycophenolate and attempting conception. Because of evidence of embryotoxicity in animal studies of sirolimus, it is recommended that mTOR inhibitors be discontinued or replaced before pregnancy is attempted. The incidence of fetal malformations has not been found to be higher in women taking cyclosporine. CNIs are usually continued during pregnancy with careful monitoring and dose adjustment as needed because drug levels can fall with the increase in plasma volume. Prednisone at doses low enough to prevent thymic aplasia (usually <15 mg/d) is considered safe in pregnant kidney transplant recipients.

REFERENCES

Annual Data Report of the Scientific Registry of Transplant Recipients. 2011. http://srtr.transplant.hrsa.gov/annual_reports/2011 (accessed May 1, 2013).

Ashman N, Chapagain A, Dobbie H, Raftery MJ, Sheaff MT, Yaqoob MM. 2009. Belatacept as maintenance immunosuppression for postrenal transplant de novo drug-induced thrombotic microangiopathy. *Am J Transplant* **9:** 424–427.

Briganti EM, Russ GR, McNeil JJ, Atkins RC, Chadban SJ. 2002. Risk of renal allograft loss from recurrent glomerulonephritis. *N Engl J Med* **347:** 103–109.

Canaud G, Audard V, Kofman T, Lang P, Legendre C, Grimbert P. 2012. Recurrence from primary and secondary glomerulopathy after renal transplant. *Transpl Int* **25:** 812–824.

Deshpande NA, James NT, Kucirka LM, Boyarsky BJ, Garonzik-Wang JM, Montgomery RA, Segev DL. 2011. Pregnancy outcomes in kidney transplant recipients: A systematic review and meta-analysis. *Am J Transplant* **11:** 2388–2404.

El-Zoghby ZM, Stegall MD, Lager DJ, Kremers WK, Amer H, Gloor JM, Cosio FG. 2009. Identifying specific causes of kidney allograft loss. *Am J Transplant* **9:** 527–535.

Gourishankar S, Leduc R, Connett J, Cecka JM, Cosio F, Fieberg A, Gaston R, Halloran P, Hunsicker L, Kasiske B, et al. 2010. Pathological and clinical characterization of the "troubled transplant": Data from the DeKAF study. *Am J Transplant* **10:** 324–330.

Hirsch HH, Brennan DC, Drachenberg CB, Ginevri F, Gordon J, Limaye AP, Mihatsch MJ, Nickeleit V, Ramos E, Randhawa P, et al. 2005. Polyomavirus-associated nephropathy in renal transplantation: Interdisciplinary analyses and recommendations. *Transplantation* **79:** 1277–1286.

Irish WD, Ilsley JN, Schnitzler MA, Feng S, Brennan DC. 2010. A risk prediction model for delayed graft function in the current era of deceased donor renal transplantation. *Am J Transplant* **10:** 2279–2286.

Kayler LK, Srinivas TR, Schold JD. 2011. Influence of CIT-induced DGF on kidney transplant outcomes. *Am J Transplant* **11:** 2657–2664.

Lefaucheur C, Nochy D, Andrade J, Verine J, Gautreau C, Charron D, Hill GS, Glotz D, Suberbielle-Boissel C. 2009. Comparison of combination plasmapheresis/IVIg/anti-CD20 versus high-dose IVIg in the treatment of antibody-mediated rejection. *Am J Transplant* **9:** 1099–1107.

Li L, Khatri P, Sigdel TK, Tran T, Ying L, Vitalone MJ, Chen A, Hsieh S, Dai H, Zhang M, et al. 2012. A peripheral blood diagnostic test for acute rejection in renal transplantation. *Am J Transplant* **12:** 2710–2718.

Mannon RB, Matas AJ, Grande J, Leduc R, Connett J, Kasiske B, Cecka JM, Gaston RS, Cosio F, Gourishankar S, et al. 2010. Inflammation in areas of tubular atrophy in kidney allograft biopsies: A potent predictor of allograft failure. *Am J Transplant* **10:** 2066–2073.

Meier-Kriesche HU, Schold JD, Srinivas TR, Kaplan B. 2004. Lack of improvement in renal allograft survival despite a marked decrease in acute rejection rates over the most recent era. *Am J Transplant* **4:** 378–383.

Mengel M, Sis B, Haas M, Colvin RB, Halloran PF, Racusen LC, Solez K, Cendales L, Demetris AJ, Drachenberg CB, et al. 2012. Banff 2011 Meeting Report: New concepts in antibody-mediated rejection. *Am J Transplant* **12:** 563–570.

Muthukumar T, Dadhania D, Ding R, Snopkowski C, Naqvi R, Lee JB, Hartono C, Li B, Sharma VK, Seshan SV, et al. 2005. Messenger RNA for FOXP3 in the urine of renal-allograft recipients. *N Engl J Med* **353:** 2342–2351.

Nankivell BJ, Chapman JR. 2006. Chronic allograft nephropathy: Current concepts and future directions. *Transplantation* **81:** 643–654.

Noris M, Remuzzi G. 2010. Thrombotic microangiopathy after kidney transplantation. *Am J Transplant* **10:** 1517–1523.

Pleass HC, Clark KR, Rigg KM, Reddy KS, Forsythe JL, Proud G, Taylor RM. 1995. Urologic complications after renal transplantation: A prospective randomized trial comparing different techniques of ureteric anastomosis and the use of prophylactic ureteric stents. *Transplant Proc* **27:** 1091–1092.

Reeve J, Sellarés J, Mengel M, Sis B, Skene A, Hidalgo L, de Freitas DG, Famulski KS, Halloran PF. 2013. Molecular diagnosis of T cell-mediated rejection in human kidney transplant biopsies. *Am J Transplant* **13:** 645–655.

Roberts DM, Jiang SH, Chadban SJ. 2012. The treatment of acute antibody-mediated rejection in kidney transplant recipients—A systematic review. *Transplantation* **94:** 775–783.

Siedlecki A, Irish W, Brennan DC. 2011. Delayed graft function in the kidney transplant. *Am J Transplant* **11:** 2279–2296.

Sis B, Mengel M, Haas M, Colvin RB, Halloran PF, Racusen LC, Solez K, Baldwin WM 3rd, Bracamonte ER, Broecker V, et al. 2010. Banff '09 Meeting Report: Antibody mediated graft deterioration and implementation of Banff working groups. *Am J Transplant* **10:** 464–471.

Solez K, Colvin RB, Racusen LC, Sis B, Halloran PF, Birk PE, Campbell PM, Cascalho M, Collins AB, Demetris AJ, et al. 2007. Banff '05 Meeting Report: Differential diagnosis of chronic allograft injury and elimination of chronic allograft nephropathy ("CAN"). *Am J Transplant* **7:** 518–526.

Solez K, Colvin RB, Racusen LC, Haas M, Sis B, Mengel M, Halloran PF, Baldwin W, Banfi G, Collins AB, et al. et al. 2008. Banff 07 classification of renal allograft pathology: Updates and future directions. *Am J Transplant* **8:** 753–760.

Stegall MD, Park WD, Larson TS, Gloor JM, Cornell LD, Sethi S, Dean PG, Prieto M, Amer H, Textor S, et al. 2011. The histology of solitary renal allografts at 1 and 5 years after transplantation. *Am J Transplant* **11:** 698–707.

Vasudev B, Hariharan S, Hussain SA, Zhu YR, Bresnahan BA, Cohen EP. 2005. BK virus nephritis: Risk factors, timing, and outcome in renal transplant recipients. *Kidney Int* **68:** 1834–1839.

Veale JL, Yew J, Gjertson DW, Smith CV, Singer JS, Rosenthal JT, Gritsch HA. 2007. Long-term comparative outcomes between two common ureteroneocystostomy techniques for renal transplantation. *J Urol* **177:** 632–636.

Vincenti F, Cohen SD, Appel G. 2010. Novel B cell therapeutic targets in transplantation and immune-mediated glomerular diseases. *Clin J Am Soc Nephrol* **5:** 142–151.

Webster AC, Pankhurst T, Rinaldi F, Chapman JR, Craig JC. 2006. Monoclonal and polyclonal antibody therapy for treating acute rejection in kidney transplant recipients: A systematic review of randomized trial data. *Transplantation* **81:** 953–965.

Facial and Hand Allotransplantation

Bohdan Pomahac[1], Ryan M. Gobble[1], and Stefan Schneeberger[2,3]

[1]Department of Surgery, Division of Plastic Surgery, Brigham and Women's Hospital, Harvard Medical School, Boston, Massachusetts 02115

[2]Department of Visceral, Transplant and Thoracic Surgery, Innsbruck Medical University, 6020 Innsbruck, Austria

[3]Department of Plastic and Reconstructive Surgery, Johns Hopkins University School of Medicine, Baltimore, Maryland 21205

Correspondence: bpomahac@partners.org

Vascularized composite allotransplantation (VCA) is a novel therapeutic option for treatment of patients suffering from limb loss or severe facial disfigurement. To date, 72 hand and 19 facial transplantations have been performed worldwide. VCA in hand and facial transplantation is a complex procedure requiring a multidisciplinary team approach and extensive surgical planning. Despite good functional outcome, courses after hand and facial transplantation have been complicated by skin rejection. Long-term immunosuppression remains a necessity in VCA for allograft survival. To widen the scope of these quality-of-life-improving procedures, minimization of immunosuppression to limit risks and side effects is needed.

Vascularized composite allotransplantation (VCA) is a novel therapeutic option for treatment of patients suffering from limb loss or severe disfigurement. According to the hand registry (www.handregistry.com) 41 hand transplantations have been performed worldwide. Of these, 21 were bilateral and 20 were unilateral transplantations. Overall graft survival and functional outcome are high, but diligent surgery together with carefully and individually adjusted immunosuppressive treatment, an intense rehabilitation protocol, and a high level of compliance and close patient follow-up is needed.

Since the first partial face transplant was performed in Amien, France by Drs. Dubernard and Devauchelle in 2005 (Devauchelle et al. 2006) a total of 19 face transplants (16 males, three females) have been reported in the literature. The largest number of transplants have been performed in France ($n = 9$) followed by the United States ($n = 6$), Spain ($n = 3$), and China ($n = 1$). Facial transplants were reported as partial transplants in 12 cases, near-total transplant in one case, and full transplants in six cases. Transplants incorporated osteomyocutaneous components in 12 cases and myocutaneous components in seven cases. The indication for transplantation included trauma in 15 cases (animal bite $n = 3$, burn $= 3$, ballistic $n = 9$), congenital deformities in three cases, and tumor ablation in one case. To date, there have been two postoperative deaths, one of which occurred in a patient who had concomitant bilateral hand transplants and succumbed to postoperative complications and a second

Cite this article as *Cold Spring Harb Perspect Med* doi: 10.1101/cshperspect.a015651

patient who was noncompliant with immunosuppressive therapy (Siemionow and Ozturk 2012). There have been no reported cases of graft loss, graft-versus-host disease, or hyperacute or chronic graft rejection.

Despite good functional outcome, courses after hand and face transplantation have been complicated by skin rejection (Schneeberger et al. 2009; Hautz et al. 2012a). Long-term immunosuppression remains a necessity in VCA for allograft survival. To widen the scope of these quality-of-life-improving procedures, minimization of immunosuppression to limit risks and side effects is needed.

PREPARATION AND SURGICAL PROCEDURE

In preparation for hand transplantation, the stumps of the recipients should be carefully assessed. Bone length, muscular, vascular, and nerve status can be analyzed using CT scan, angio-CT, angiography, magnetic resonance imaging, and ultrasound. This is to rule out a neuroma at a level far proximal to the amputation. Routine workup should include blood work, X rays, electrocardiography, echocardiography, and spirometry. For exclusion of comorbidities such as malignancies or infections patients undergo gastroscopy, colonoscopy, and dental and oropharyngeal examination (Brandacher et al. 2007).

Surgical strategies in hand transplantation have to be adapted to the level of amputation as well as the exact length and quality of structures in the recipient's stump. For each case a precise plan and in some cases development of novel reconstruction techniques are required (Schneeberger et al. 2011; Landin et al. 2012). At present, definite recommendations regarding the maximum duration of ischemia do not exist; however, ischemia time should be kept as short as possible and not exceed 10 hours. Although cold flush and preservation with histidine, tryptophan, ketoglutarat solution (HTK), or University of Wisconsin (UW) solution might limit myocyte damage, muscles are sensitive to ischemia and damage such as interstitial edema, microvascular constriction, and/or damage of myocyte membranes may result in muscle dysfunction after 2.5 h of (warm) ischemia (Nanobashvili et al. 2003).

All anatomic structures such as tendons, nerves, and vessels are dissected under tourniquet control in the donor and recipient. After release of the tourniquet, the forearm is perfused with cold HTK solution through the brachial artery. Next, all structures are transected and osteotomies of the radius and ulna are performed. After wound closure the donor is fitted with cosmetic prostheses.

Hand allotransplantation is then performed under tourniquet control. After bone preparation and osteosynthesis, revascularization is initiated by anastomosis of one main artery and two veins followed by tendon/muscle repair and definitive vessel repair. Next, nerve repair is performed before skin closure. Alternatively, revascularization can be performed later after the extensor tendons are repaired and rapid anastomosis of one to two dorsal veins and nerve repair. Both radial and ulnar arteries are repaired next, followed by reconstruction of the remaining veins. This is followed by flexor tendon repair and skin closure (Azari et al. 2012).

Facial transplantation is a complex endeavor that requires a multidisciplinary approach encompassing the most cutting-edge aspects of plastic surgery, transplantation immunology, and anesthesia. As no two faces are the same, no two operations are exactly the same. There are, however, guiding principles that should be followed. Each operation requires an anastomosis to create arterial inflow and venous outflow. Also nerve coaptation is required for both motor and sensory branches. Arterial and venous anastomosis as well as nerve coaptation may be either unilateral or bilateral depending on the amount of tissue to be replaced. Osseous portions from the donor may be included in the composite tissue allograft.

Arterial inflow has been established through the external carotid artery, and its branches including the facial artery and the external maxillary artery, whereas venous outflow has been achieved via the facial vein, thyrolingual facial trunk, external and internal jugular vein, and the retromandibular vein. Arterial and venous anastomoses are performed with the face flipped down onto the recipient's chest using a microsurgical technique. An end-to-end anastomosis is preferred whenever possible and is usually able to be performed given the similar caliber of donor and recipient vessel lumen as well as the ease in obtaining adequate length vessel length from the donor (Siemionow and Ozturk 2012). An end-to-side anastomosis between the right subclavian and the right common carotid arteries has been reported in a 42-yr-old male who underwent transplantation of 35% of the face including partial bilateral cheek, upper and lower lip, chin, tongue, and mandible (Cavadas et al. 2012).

Ultimately the choice of donor and recipient vessel inflow and outflow will depend on the tissues to be transplanted. It has been suggested that for reconstruction of tissues of the midface or upper face that the superficial temporal and facial vessels are preferred, whereas reconstruction of the lower face and mandible is best achieved by using the external carotid and its branches, and this would certainly seem to hold true for tissue allotransplantation (Takamatsu et al. 1996). When an osteomyocutaneous transplant is performed more proximal vessels are usually preferred (e.g., external carotid artery and facial artery); however, we have shown in a cadaver study that the facial artery itself is capable of maintaining perfusion to the entire midface, including the maxilla, anterior half of mandible from the masseter muscle insertion and zygoma (Pomahac et al. 2012b), as well as the total facial allograft including maxilla (Pomahac et al. 2012a).

Full transplants deserve particular attention with regard to inclusion of the superficial temporal vessels. Inclusion of these vessels requires inclusion of the parotid glands bilaterally, which results in unaesthetic fullness of the cheeks bilaterally unless a superficial parotidectomy is performed before transplantation, which significantly increases operative time. Including the parotid glands also limits facial nerve coaptation to the level of the trunk, as the branches of the donor facial nerve are inaccessible. We have developed a technique for full facial transplantation based on only the facial arteries, which allows for more target innervation of effector muscles without the unsightly bulges seen when the parotid gland is included (Pomahac et al. 2012b).

Most commonly, two arterial anastomoses and two or more venous anastomoses are performed; however, there is no definitive clinical or experimental evidence to support recommendations for which vessels should be used or how many anastomoses are required for each type of allotransplantation being performed. Given the high stakes of facial allotransplantation, one is better served to include more than one anastomosis when possible lest a single arterial or venous anastomosis become compromised risking loss of the entire graft. Also, given the nature of the recipient's vessels as most of these patients have undergone multiple prior operations and/or received prior radiation, it is wise to have a roadmap as to the nature and location of the recipient's vessels. As such, preoperative imaging including ultrasound and CT/MRI angiography should be obtained before operative intervention. Preoperative CT angiography is useful for identifying the best target vessels for vascular anastomoses thereby limiting operative time exploring unusable vessels. It can alert the surgical team for the potential need of vein grafts as a backup option if less than ideal vascular anatomy is identified preoperatively. The key aspects in choosing the ideal vessels are an optimal size match between recipient and donor vessels, an adequate amount of length between allograft pedicle and donor vessels, and accessibility of the pedicle (Soga et al. 2012). Magnetic resonance (MR) angiography may be particularly useful in both preoperative imaging with a large shrapnel component and/or postoperative follow-up once surgical plates have been introduced into the area. If temporal or parietal scalp need be included preservation of the transverse facial artery that connects the facial and superficial temporal arteries will provide adequate blood flow. If blood supply to these regions is not adequate dissection of the superficial temporal arteries in the superficial plane at the preauricular level may be performed for supercharging at the recipients corresponding location (Pomahac et al. 2012b).

CT imaging is useful not only in planning intraoperative vascular anastomoses but also in planning for osteotomies in cases in which an osteomyocutaneous transplant is performed. Brown et al. has described a cadaveric study for full face transplant including double jaw and tongue using CT-assisted techniques when performing osteotomies on both the donor and recipient (Brown et al. 2012). In this study both the donor and recipient cadavers were placed in intramaxillary fixation. Three-dimensional reconstructions were then superimposed on donor and recipient skulls using surgical modeling software. Leforte III and sagittal split osteotomies were planed based on these projections. Stereolithographic models were produced and cutting guides for Leforte III osteotomies were made for use during the mock transplantation. Planned osteotomies were then imported into intraoperative navigation software and the Leforte III osteotomies were performed using this intraoperative guidance. CT scans performed after the mock transplantations showed that accurate skeletal fixation and occlusion was noted to be normal. Dr. Rodriguez and colleagues were then able to translate the experienced gained performing these mock transplantations to complete the first total face, double jaw, and tongue transplant in a 37-yr-old male who sustained a central face high energy avulsive ballistic injury (Dorafshar et al. 2013). Use of CT imaging to plan osteotomies during osteomyocutaneous facial transplantation will likely be incorporated into future allotransplantations to improve the accuracy of skeletal fixation and decrease operative times. Furthermore, use of CT imaging may help us more accurately predict which donors are more likely to result in an optimal skeletal match for facial transplant recipients. However, limitations in donor availability will prevent passing up eligible donors despite a less than ideal skeletal match for the foreseeable future.

Sensory nerves used in facial transplantation include the trigeminal nerve and its branches (e.g., infraorbital nerve, supraorbital nerve, mental nerve, buccal sensory nerve, lingual nerve, inferior alveolar nerve, and cervicofacial nerve). Direct nerve coaptation of both motor and sensory nerves using fine monofilament suture under microscopic magnification is the preferred method; however, in cases in which trauma has destroyed the recipient sensory nerves motor coaptation alone has been performed (Guo et al. 2008; Siemionow et al. 2009, 2010; Lantieri et al. 2011). Also, in cases in which the mental nerves are unable to be repaired directly the donor mental nerves have been placed in close approximation of the recipient mental foramen (Lantieri et al. 2008). The facial nerve serves as the motor input for facial transplantation. Coaptation of facial motor nerve branches at more distal sites allows for more retention of native nerve length decreasing the chances of developing synkinesia (Pomahac et al. 2012b). Nerve grafts can be used when adequate length is not available (e.g., thoracodorsal nerve, greater auricular nerve).

Inset of the graft is typically performed in a central-to-lateral and inferior-to-superior sequence. Osseous fixation is achieved using plates as screws customized to the particular bone being fixed. Liberal use of Mitek (Burnsville, MN) anchors in different locations in the zygoma, supraorbital bar, and mandible are used for securing the tissues of the allograft to the underlying bony structure to provide some structure to the face and relieve tension on the vascular anastomoses (Pomahac et al. 2012b). It is essential to achieve watertight intraoral closure. The skin sutures are placed to promote the optimal facial contour and aesthetic outcome. Drains are placed as needed depending on the amount of tissue transplanted. At our institution we chose to also include a sentinel flap that can be used to monitor for rejection episodes. Sentinel flaps are placed at areas that require functional improvements. Rejection episodes were detected more accurately in the sentinel flap than in the facial allograft in two of our four face-transplant recipients (Pomahac et al. 2012a). The sentinel flap is less affected by confounding facial-skin conditions, including environmental exposure or from conditions such as dermatitis or rosacea, and is therefore felt to be more sensitive to rejection episodes (Saavedra et al. 2011).

OUTCOMES

Sensitivity

Nerve regeneration after hand transplantation progresses rapidly with nerve endings reaching the fingertips at 6–9 months depending on the level of amputation/transplantation (Petruzzo et al. 2010; Hautz et al. 2011). Protective sensation returned in all and discriminative sensation in the majority of patients. The transplanted hands were incorporated into the patient's body image in almost all patients. Protective sensation is routinely achieved.

The outcome after forearm transplantation differs significantly from the outcome after hand transplantation (Schneeberger et al. 2011). Particularly the time course of return of function is significantly prolonged when compared with hand transplantation. The functional outcome, therefore, needs to be read in the light of inherit differences between hand and forearm transplantation. Overall, however, despite the postoperative course being significantly more challenging when compared with hand transplantation, forearm transplantation has also resulted in good patient satisfaction (Schneeberger et al. 2007, 2011). In addition to restoration of body integrity, improved motor function and movement control and fewer everyday challenges when compared with myoelectric prostheses have been achieved. The long and intense rehabilitation protocol and the slow return of function require thorough and cautious patient selection.

Restoration of sensation in facial allotransplantation has been assessed using several different quantitative sensory tests, including two-point discrimination, Semmes-Weinstein monofilaments, pressure-specified sensory devices, and heat and cold tolerance (Siemionow and Ozturk 2012). Restoration of light touch and temperature sensation has been shown to occur between 3 and 8 months in the majority of cases (Dubernard et al. 2007; Guo et al. 2008; Lantieri et al. 2008; Siemionow et al. 2009; Barret et al. 2011; Pomahac et al. 2012a). Light touch and sensation to the oral mucosa returns around 3 months, whereas thermal sensation does not typically return until 6–8 months posttransplantation. The rate of return of sensation is similar to that reported with innervated free flaps (Kimata et al. 1999). Documentation of sensory recovery has been reported even in cases in which direct sensory nerve coaptation was not possible secondary to trauma (Lantieri et al. 2008). This may be owing to interconnections between the trigeminal and facial nerve branches that may make direct nerve coaptation not entirely necessary (Baumel 1974; Tohma et al. 2004). Although sensory recovery has been noted in cases in which direct nerve coaptation was not possible we still recommend its performance in all facial transplants when possible.

Scarring of the recipient bed has a clear negative impact on sensory nerve regeneration, which will affect the ultimate outcome (Hermanson et al. 1987). Inclusion of skin within a free flap improves sensory recovery, whereas the presence of muscle and bone may act as a barrier for neurontinization from the recipient bed (Siemionow et al. 2011). Patients with thinner transplants, less transplanted tissues, and younger patients may have better sensory recovery (Lahteenmaki et al. 1991); however, more clinical data with facial transplantation needs to be accumulated before this will be known.

Motor Function

In hand transplantation, motor function returns almost immediately after transplantation. The patient's own forearm muscles enable finger movement, which continues to increase in range and strength over time. In many cases, the return of intrinsic hand muscle function can be observed and further enhances motor function. The majority of patients are able to perform all major tasks of daily living including holding small objects, turning pages and door knobs, writing, and returning to work (Petruzzo et al. 2010).

As mentioned above, the functional outcome after transplantation at more proximal levels requires prolonged rehabilitation and is overall less consistent (Landin et al. 2012). The functional return in these cases may continue to improve during the first 5 years after transplantation. Although reinnervation of intrinsic hand muscles has been observed in some cases, it cannot be routinely expected and the level of function is inferior when compared with hand transplantation. Protective but not discriminative sensation can be observed in the majority of these patients. Return of hand function after forearm transplantation requires approximately 5 years to completion, as it is dependent on reinnervation and reactivation of graft forearm muscles. Overall, hand function in forearm transplantation remains inferior when compared with hand transplantation because reinnervation and reactivation of intrinsic hand muscles are unlikely.

In cases in which the level of amputation is proximal to the elbow area, the long distance for nerve regeneration and recovery is a major pace-limiting factor. Only a few such cases have been performed and arm function in these cases is still improving at this point. Elbow function and extrinsic finger flexion and extension have been satisfactory but the final and overall functional outcome in arm transplantation can only be estimated at this point (Brandacher et al. 2009). More patients and long-term follow-up is needed to ultimately determine the degree of function that can be regained in arm transplantation.

In all facial transplantations reported thus far, motor recovery has occurred later and often to a lesser extent than sensory recovery. This is unfortunate, as motor recovery will likely contribute more to a functionally useful and aesthetic face. As motor recovery takes longer to occur and motor recovery has not been documented in each case we do not know exactly the full extent of motor recovery to be expected following facial allotransplantation. However, reports from several different investigators do paint a picture of the extent of motor recovery possible. Dubernard et al. (2007) showed that return of motor activity occurred at 3 months after transplantation followed by the ability to move the lower lip at 4 months. Their patient was able to bring the lips into apposition at 6 months and to have a full smile at 18 months. Furthermore, the patient was able to move their nose and chin at 1-year posttransplantation (Dubernard et al. 2007). Lantieri et al. reported that their patient had no evidence of facial nerve reinnervation on electromyography at 3 months posttransplantation; however, at 1 year, there was documented evidence of reinnervation on electromyography (Lantieri et al. 2008). Siemionow et al. described an improvement in facial mimetics by 6 months posttransplantation as evidenced by symmetric smiling and upper lip occlusion; however, upper lip and lower eyelid movement were imperfect. By 1 year, the investigators noted almost complete motor recovery (Siemionow et al. 2009, 2010). Barrett et al. reported an ability to move the frontalis, zygomaticus, and upper orbicularis occuli muscles along with unrestricted masticatory movement; however, their patient was unable to close the eyes (Barret et al. 2011).

More recently we have reported on three patients who have undergone full facial transplantation (Pomahac et al. 2012a). All three patients recovered motor function of their face. Along with our first patient that received partial face transplant, three of four have an excellent symmetry (Figs. 1A,B, 2A,B, and 3A,B). One patient has an asymmetrical smile owing to previous injury of the recipient's facial nerve, perhaps owing to traction (Fig. 4A,B).

Lack of long-term follow-up in facial transplantation for both sensory and motor activity remains a problem. Other difficulties in monitoring for the return of motor activity remain an overall lack of reporting as well as no uniformity of measuring such progress. One investigator has suggested using electromyography and functional magnetic resonance imaging to monitor for facial nerve reanimation (Hui-Chou et al. 2010). Although these methods may become part of the standard for monitoring for motor recovery they are still untested in facial transplantation specifically, and, for now, physical examination will remain the gold standard.

Peripheral Nerve Regeneration and Cortical Integration

Sensory nerve recovery for free tissue reconstruction is greater in the face than for that of the trunk or lower extremities in both neuroti-

 Cite this article as *Cold Spring Harb Perspect Med* doi: 10.1101/cshperspect.a015651

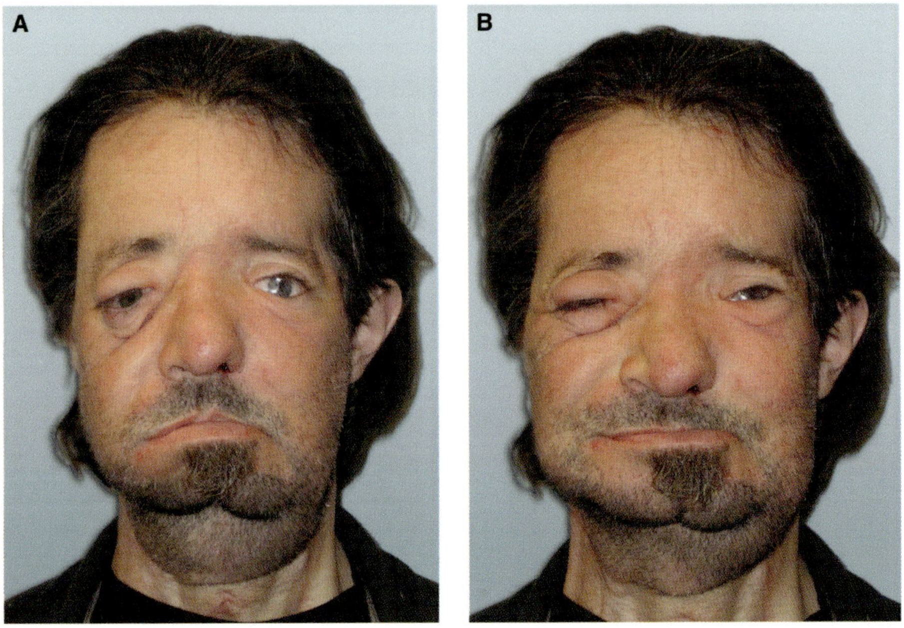

Figure 1. 26-yr-old male 12 months postfacial allotransplantation at rest (*A*), and smiling (*B*).

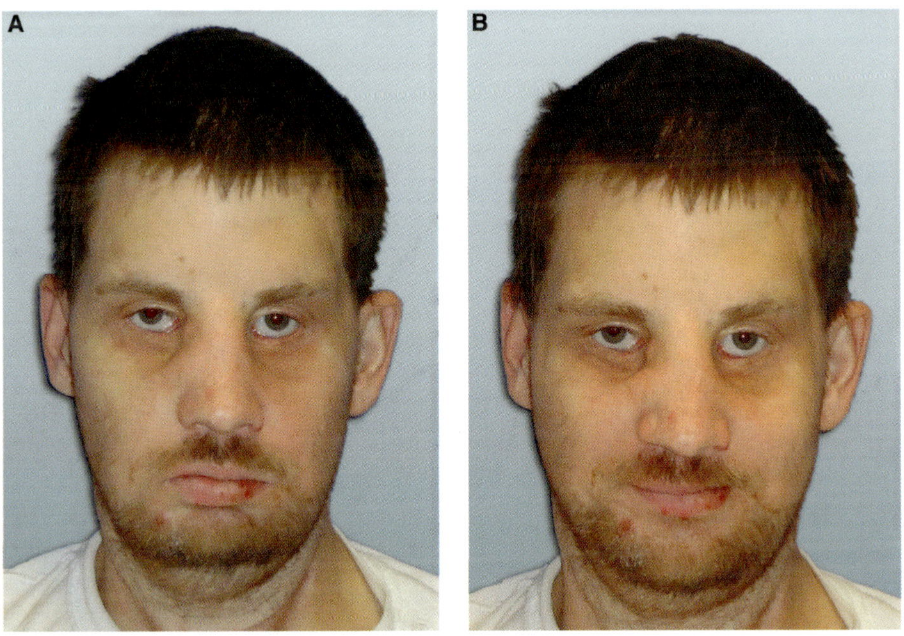

Figure 2. 31-yr-old male 12 months postfacial allotransplantation at rest (*A*), and smiling (*B*) .

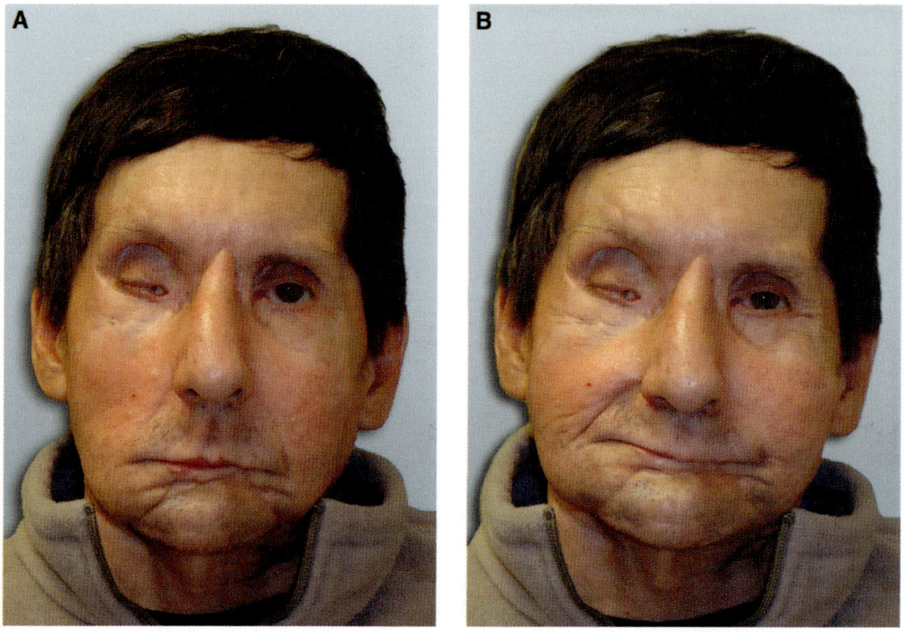

Figure 3. 62-yr-old male 36 months postfacial allotransplantation at rest (*A*), and smiling (*B*).

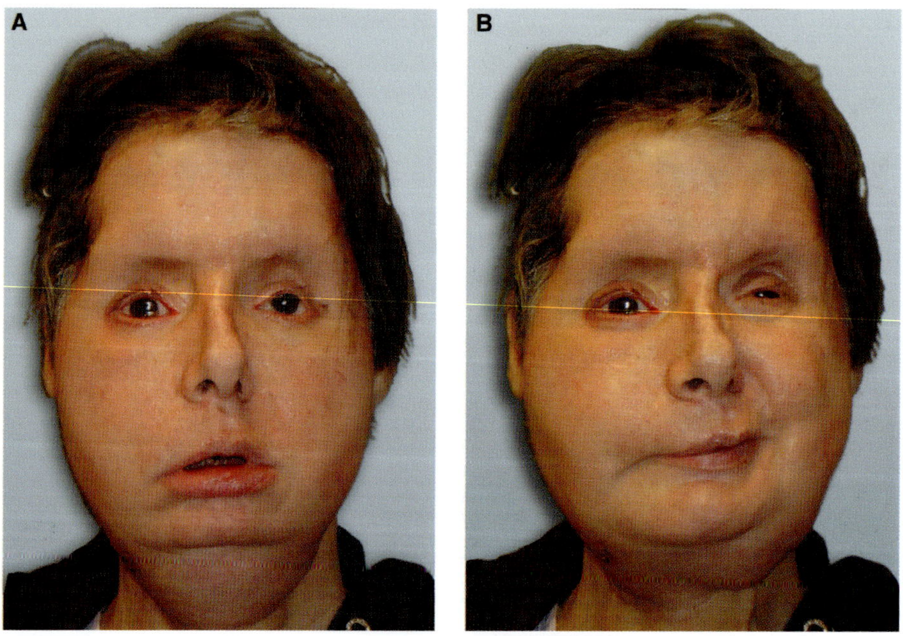

Figure 4. 58-yr-old female 12 months postfacial allotransplantation at rest (*A*), and smiling (*B*).

nized and non-neurotinized flaps, which shows the greater cortical representation the face receives in the human homunculus (Lahteenmaki et al. 1991; Kuriakose, et al. 2001; Kerawala et al. 2006; Bianchi et al. 2010).

Patient Satisfaction

Patient satisfaction is of the utmost importance in every operation, but no more so than in facial allotransplantation. These patients submit themselves to a long operation followed by an even longer hospital stay and ultimately lifelong immunosuppression. For a successful outcome patient buy in is essential. As previously shown restoration of sensation takes many months, whereas return of motor activity can take over a year. Despite all of these obstacles patient satisfaction remains very high. Several investigators have noted restoration of olfaction and an ability to eat and drink within a week postoperatively (Dubernard et al. 2007; Lantieri et al. 2008; Siemionow et al. 2009; Pomahac et al. 2012a). Patients who had been previously fed via gastrostomy tubes were able to have them removed and tracheostomy-dependent patients were able to be decanulated (Siemionow and Ozturk 2012). Furthermore, improvement in speech has been noted with transplantation of the hard palate and the use of an obturator (Siemionow et al. 2009; Pomahac et al. 2011).

Patient-specific questionnaire studies are yet to be performed, but most investigators report that their patients are happy with their outcomes. Results from a psychosocial analysis of quality-of-life measures using the MOS-SF-12 showed that patients at our institution who underwent facial allotransplantation had a significant improvement in their mental health in the 6 months posttransplantation (Chang and Pomahac 2012). Clearly for patients requiring facial transplantation the devastating nature of their injuries severely limits their ability to interact in society, and improvement both aesthetically and functionally is welcome. Ethical concerns remain, given the risk of death with the operation along with the risk of developing malignancy with the need for lifelong immunosuppression. Furthermore, access to care issues and an overall lack of donors make this life-altering procedure available only to a select group of patients.

IMMUNOLOGICAL ASPECTS

Immunosuppression

The immunology in VCA is characterized by the strong skin-directed immune response, the markedly different gravity of the immune response against the different tissues, and the ability to visually monitor the skin (Cendales et al. 2008; Hautz et al. 2012b). The failure of the first human hand transplantation reported in 1964 was caused by the strong immune response and the lack of modern immunosuppressive therapy (Gilbert 1964). Transplantation of the skin was considered impossible until hand and face transplantation was performed successfully after 1998. At this point, the side effects of long-term high-dose immunosuppression rather than rejection represents the remaining pace-limiting obstacle in this field (Hautz et al. 2010a; Brandacher et al. 2012). Two patients died following a combined hand and face transplantation and attempted bilateral upper and lower extremity transplantation (Lantieri et al. 2011; Daily Mail Reporter 2012). Rejection remains a challenge and threat, but the morbidity caused by the required drug treatment is limiting the larger-scale application of this nonlifesaving treatment. Concurrent with the advancements in solid organ transplantation, minimization or elimination of the need for long-term high-dose and/or multidrug immunosuppression is the primary goal of global research in this novel field. A reduction of maintenance immunosuppression has recently been achieved in a small clinical trial. Such findings need to be confirmed and advanced in a larger group of patients (Schneeberger et al. 2013).

Facial transplants, along with hand transplants, are unique among transplant types in that they allow for direct visual observation of rejection episodes. They are also unique in that they possess very large amounts of skin, which is one of the most immunoreactive and antigenic tissues in the human body (Lee et al. 1991).

Development of facial transplantation-specific protocols have adapted primarily from solid organ transplant protocols. Most centers use induction therapy with anti-interleukin 2 receptor antibodies (e.g., basiliximab, daclizumab) or antithymocyte globulins followed by a maintenance protocol that includes an antiproliferative agent (mycophenolatemofetil), a calcineurin inhibitor (tacrolimus), and a corticosteroid (prednisolone) (Siemionow and Ozturk 2012). Donor origin stem cells have also been used to attempt to induce donor-specific tolerance via the development of a mixed allogeneic chimerism (Devauchelle et al. 2006; Dubernard et al. 2007). The goal of maintenance therapy is to avoid graft rejection, while at the same time reducing immunosuppression to limit the adverse events, which include infections and malignancy (van Leeuwen et al. 2010).

Rejection Episodes

One patient in the United States has lost a hand allograft as a consequence of arterial intimal hyperplasia on day 275 and one patient in France requested amputation following progressive rejection of the skin after stopping immunosuppression at 29 months (Petruzzo et al. 2010). Furthermore, several unconfirmed cases of hand loss have occurred in China. Although rejection in hand and face transplantation remains incompletely understood at this point, it appears that both the skin and the vasculature are important targets for the immune response in VCA. In some elements, the molecular mechanisms and dynamics of skin rejection are similar to other inflammatory skin conditions. T cells are recruited to the skin via activation of adhesion molecules, cytokines, and chemokines (Hautz et al. 2010b, 2012b). In addition to T-cell recruitment, ectopic lymphoid structures within the allograft associated with chronic rejection in solid organ transplantation might contribute to the strong alloimmune response toward the skin. In preclinical trials, molecules involved in cell trafficking in the skin have been successfully targeted to prevent rejection in VCA. In human hand transplantation, episodes of skin rejection have been observed in the vast majority of cases. Skin rejection was reversible in all patients compliant with immunosuppression and graft loss was prevented (Hautz et al. 2010a).

All patients who have >1 year follow-up after facial transplantation have experienced at least one episode of acute rejection, which manifests as cutaneous changes including skin swelling and erythema along with the development of nodules and/or papules. Furthermore, seven patients have experienced two acute rejection episodes and two developed three rejection episodes (Siemionow and Ozturk 2012). Acute rejections have all been confirmed histologically via skin and/or mucosal biopsy. Acute rejection episodes were successfully treated using treatment algorithms from solid organ rejection including the use of high-dose parenteral steroids and the adjustment of oral immunosuppressive therapy. Unique to facial transplantation is the ability to use phototherapy and the topical application of steroid and tacrolimus creams.

To date, there are also no documented cases of chronic rejection, again likely owing to the short follow-up, or graft-versus-host disease. In kidney and liver transplantation chronic rejection may occur anywhere from months to years posttransplantation and is characterized by fibrosis of the graft parenchyma (Seetharam et al. 2010). In cardiac transplantation, chronic rejection occurs over the same time course; however, it typically manifests itself as a vasculopathy that ultimately leads to loss of the graft (Kalache et al. 2011). As follow-up times in facial and hand allotransplantation are limited it is unclear yet whether chronic rejection will manifest with more fibrotic changes to the skin and subcutaneous tissues or with a vasculopathy of the transplanted vessels. Furthermore, we do not know exactly over what time course chronic rejection will occur although we would expect it to be similar to that seen in solid organ transplantation.

COMPLICATIONS

A total of five patients have lost their grafts after hand transplantation. One patient lost his hand as a consequence of intimal hyperplasia in the arteries of the graft on day 275 (Louisville) and one patient as a result of progressive rejection of the skin after stopping immunosuppression at 29 months (Lyon) (Kanitakis et al. 2003; Kaufman et al. 2012). Two hands were lost subsequent to arterial thrombosis in Wroclaw (Jablecki 2011). Furthermore, one hand each was reamputated after transplantation in Atlanta and Boston (Pomahac et al. 2012a). Graft survival in hand transplantation 14 years after the first case is currently 82%.

In one case, a venous thrombosis was treated by microsurgical revision (Petruzzo et al. 2010). The thrombosis of an ulnar artery, which required thrombectomy, occurred in another case. There was also one case in which an arteriovenous shunt was observed requiring ligation. Skin necrosis was observed in four cases and required treatment with skin grafts (Petruzzo et al. 2010). A hematoma resulted in wound dehiscence on day 15 after transplantation in one case (Jablecki et al. 2009). In another case, a tenolysis for flexor tendon adhesions was required at 14 months (Schuind et al. 2007). Aseptic necrosis of a hip required treatment with bilateral arthroplasty in one patient (Petruzzo et al. 2010). In one case, *Staphylococcus aureus* osteomyelitis in the ulna was encountered and treated with antibiotics. CMV infection was reported in a large number of cases. Human papilloma virus infection, herpes simplex infection, fungal infections of the skin and mucosa, and a *Clostridium difficile* enteritis were reported in a smaller number of patients (Petruzzo et al. 2010). Frequently, transient hyperglycemia, hypertension, and impaired kidney function were observed but reversible with adequate medical treatment (Petruzzo et al. 2010). A basal cell carcinoma of the nose was treated by surgical excision (Landin et al. 2010).

Although psychiatric disorders like depression, fear, or denial of the transplanted hand, may appear especially early after transplantation, the overall adjustment to the new limb has been very good in most cases (Kumnig et al. 2012). In some patients the allografts were fully integrated in cerebral cortex and body image when sensitivity was established. Nevertheless, a suicide attempt was reported in one patient (Schuind et al. 2007), whereas the first recipient of a hand allograft in the recent era of hand transplantation requested reamputation when the hand showed signs of progressive rejection after cessation of immunosuppression (Schneeberger et al. 2011).

One significant problem in some cases after hand transplantation was compliance with medication and the demanding rehabilitation protocol. Although the existing experience with hand transplantation indicates that noncompliance is more common after unilateral than bilateral hand transplantation, this remains to be confirmed in a larger series of patients. A detailed psychological evaluation and risk assessment remains a prerequisite for patients enlisted for hand transplantation.

In face transplantation there have been no documented cases of loss of the allograft secondary to surgical complications such as arterial or venous thrombosis. One patient did require reoperation 3 days postoperatively secondary to a venous thrombosis; however, the anastomosis was revised with salvage of the allograft (Barret et al. 2011). Other documented postoperative surgical complications including ptosis of the eyelids, ectropion, Stenson's duct stenosis, and generalized wound healing problems have all been able to be managed conservatively.

The two most feared complications with immunosuppressive therapy are infections and malignancy. All facial transplantation patients received prophylaxis against opportunistic infections, which typically consists of coverage against herpes simplex (acyclovir), cytomegalovirus (ganciclovir), and *Pneumocystis jerovici* (trimethoprim-sulfamethoxazole). Despite these measures the majority of patients had at least one episode of an opportunistic bacterial, viral, or fungal infection including cytomegalovirus, Epstein–Barr virus, herpes simplex, herpes zoster, staphylococcal, *Pseudomonas*, *Enterobacter*, and Candida infections (Siemionow and Ozturk 2012). There was one death owing to *Pseudomonas* infection reported in a patient who underwent face as well as bilateral hand transplant (Lantieri et al. 2011). A second death was reported in the patient of Guo et al. (2008) after being noncompliant with immunosuppression therapy (Siemionow and Ozturk 2012). Malignancy has been already reported despite the relative infancy of facial transplantation. The first facial transplantation patient developed a cervical carcinoma in situ that was successfully treated with conization (Petruzzo et al. 2012).

OUTLOOK: BASIC AND CLINICAL SCIENCE

Hand and face allotransplantation are demanding and challenging operations and especially for face transplantation, mock transplantations via cadaveric dissections are recommended before performing the actual operation to ensure optimal outcomes. Several investigators have reported on the use of computerized tomography and cephalometric analysis to plan potential operations (Brown et al. 2012; Caterson et al. 2012). The major hurdle for wider range application of hand and face transplantation, however, remains the reduction or elimination of the need for long-term immunosuppressive treatment. The side effects of immunosuppression hamper the progress in this field as the variety and severity of side effects negatively impact on the risk–benefit balance of such non-lifesaving procedures.

REFERENCES

Azari KK, Imbriglia JE, Goitz RJ, Shores JT, Balk ML, Brandacher G, Schneeberger S, Gorantla V, Lee WP. 2012. Technical aspects of the recipient operation in hand transplantation. *J Reconstr Microsurg* **28:** 27–34.

Barret JP, Gavalda J, Bueno J, Nuvials X, Pont T, Masnou N, Colomina MJ, Serracanta J, Arno A, Huguet P, et al. 2011. Full face transplant: The first case report. *Ann Surg* **254:** 252–256.

Baumel JJ. 1974. Trigeminal-facial nerve communications. Their function in facial muscle innervation and reinnervation. *Arch Otolaryngol* **99:** 34–44.

Bianchi B, Copelli C, Ferrari S, Ferri A, Bailleul C, Sesenna E. 2010. Facial animation with free-muscle transfer innervated by the masseter motor nerve in unilateral facial paralysis. *J Oral Maxillofac Surg* **68:** 1524–1529.

Brandacher G, Schneeberger S, Margreiter R. 2007. Selection of candidates and waiting list, dealing with the media and the public, setting up a pilot study, clinical trial organisation, staff requisites for hand transplantation: The Innsbruck experience. In *Hand transplantation* (ed. Lanzetta M, Dubernard JM), pp. 149–157. Springer-Verlag Italia, Milan, Italy.

Brandacher G, Ninkovic M, Piza-Katzer H, Gabl M, Hussl H, Rieger M, Schocke M, Egger K, Loescher W, Zelger B, et al. 2009. The Innsbruck hand transplant program: Update at 8 years after the first transplant. *Transpl Proc* **41:** 491–494.

Brandacher G, Lee WP, Schneeberger S. 2012. Minimizing immunosuppression in hand transplantation. *Expert Rev Clin Immunol* **8:** 673–683; quiz 684.

Brown EN, Dorafshar AH, Bojovic B, Christy MR, Borsuk DE, Kelley TN, Shaffer CK, Rodriguez ED. 2012. Total face, double jaw, and tongue transplant simulation: A cadaveric study using computer-assisted techniques. *Plast Reconstr Surg* **130:** 815–823.

Caterson EJ, Diaz-Siso JR, Shetye P, Junker JP, Bueno EM, Soga S, Rybicki FJ, Pomahac B. 2012. Craniofacial principles in face transplantation. *J Craniofac Surg* **23:** 1234–1238.

Cavadas PC, Ibanez J, Thione A. 2012. Surgical aspects of a lower face, mandible, and tongue allotransplantation. *J Reconstr Microsurg* **28:** 43–47.

Cendales LC, Kanitakis J, Schneeberger S, Burns C, Ruiz P, Landin L, Remmelink M, Hewitt CW, Landgren T, Lyons B, et al. 2008. The Banff 2007 working classification of skin-containing composite tissue allograft pathology. *Am J Transpl* **8:** 1396–1400.

Chang G, Pomahac B. 2012. Psychosocial changes 6 months after face transplantation. *Psychosomatics* **54:** 367–371.

Daily Mail Reporter. 2012. Man, 27, who had world's first quadruple limb transplant dies days after operation. *Daily Mail Online*, February 28.

Devauchelle B, Badet L, Lengele B, Morelon E, Testelin S, Michallet M, D'Hauthuille C, Dubernard JM. 2006. First human face allograft: Early report. *Lancet* **368:** 203–209.

Dorafshar AH, Bojovic B, Christy MR, Borsuk DE, Iliff NT, Brown EN, Shaffer CK, Kelley TN, Kukuruga DL, Barth RN, et al. 2013. Total face, double jaw, and tongue transplantation: An evolutionary concept. *Plast Reconstr Surg* **131:** 241–251.

Dubernard JM, Lengele B, Morelon E, Testelin S, Badet L, Moure C, Beziat JL, Dakpe S, Kanitakis J, D'Hauthuille C, et al. 2007. Outcomes 18 months after the first human partial face transplantation. *N Engl J Med* **357:** 2451–2460.

Gilbert R. 1964. Transplant is successful with a cadaver forearm. *Med Trib Med News* **5:** 20–22.

Guo S, Han Y, Zhang X, Lu B, Yi C, Zhang H, Ma X, Wang D, Yang L, Fan X, et al. 2008. Human facial allotransplantation: A 2-year follow-up study. *Lancet* **372:** 631–638.

Hautz T, Brandacher G, Zelger B, Gorantla VS, Lee AW, Pratschke J, Schneeberger S. 2010a. Immunologic aspects and rejection in solid organ versus reconstructive transplantation. *Transpl Proc* **42:** 3347–3353.

Hautz T, Zelger B, Grahammer J, Krapf C, Amberger A, Brandacher G, Landin L, Muller H, Schon MP, Cavadas P, et al. 2010b. Molecular markers and targeted therapy of skin rejection in composite tissue allotransplantation. *Am J Transpl* **10:** 1200–1209.

Hautz T, Engelhardt TO, Weissenbacher A, Kumnig M, Zelger B, Rieger M, Rumpold G, Pierer G, Ninkovic M, Gabl M, et al. 2011. World experience after more than a decade of clinical hand transplantation: Update on the Innsbruck program. *Hand Clin* **27:** 423–431, viii.

Hautz T, Wolfram D, Grahammer J, Starzl R, Krapf C, Pratschke J, Lee WP, Brandacher G, Schneeberger S. 2012a. Mechanisms and mediators of inflammation: Potential models for skin rejection and targeted therapy in vascularized composite allotransplantation. *Clin Dev Immunol* **2012:** 757310.

Hautz T, Zelger B, Brandacher G, Mueller H, Grahammer J, Zelger B, Lee A, Cavadas P, Margreiter R, Pratschke J, et al. 2012b. Histopathologic characterization of mild rejection (grade I) in skin biopsies of human hand allografts. *Transpl Int* **25:** 56–63.

Hermanson A, Dalsgaard CJ, Arnander C, Lindblom U. 1987. Sensibility and cutaneous reinnervation in free flaps. *Plast Reconstr Surg* **79:** 422–427.

Hui-Chou HG, Nam AJ, Rodriguez ED. 2010. Clinical facial composite tissue allotransplantation: A review of the first four global experiences and future implications. *Plast Reconstr Surg* **125:** 538–546.

Jablecki J. 2011. World experience after more than a decade of clinical hand transplantation: Update on the Polish program. *Hand Clin* **27:** 433–42, viii.

Jablecki J, Kaczmarzyk L, Patrzalek D, Domanasiewicz A, Chelmonski A. 2009. A detailed comparison of the functional outcome after midforearm replantations versus midforearm transplantation. *Transpl Proc* **41:** 513–516.

Kalache S, Dinavahi R, Pinney S, Mehrotra A, Cunningham MW, Heeger PS. 2011. Anticardiac myosin immunity and chronic allograft vasculopathy in heart transplant recipients. *J Immunol* **187:** 1023–1030.

Kanitakis J, Jullien D, Petruzzo P, Hakim N, Claudy A, Revillard JP, Owen E, Dubernard JM. 2003. Clinicopathologic features of graft rejection of the first human hand allograft. *Transplantation* **76:** 688–693.

Kaufman CL, Ouseph R, Blair B, Kutz JE, Tsai TM, Scheker LR, Tien HY, Moreno R, Ozyurekoglu T, Banegas R, et al. 2012. Graft vasculopathy in clinical hand transplantation. *Am J Transpl* **12:** 1004–1016.

Kerawala CJ, Newlands C, Martin I. 2006. Spontaneous sensory recovery in non-innervated radial forearm flaps used for head and neck reconstruction. *Int J Oral Maxillofac Surg* **35:** 714–717.

Kimata Y, Uchiyama K, Ebihara S, Kishimoto S, Asai M, Saikawa M, Ohyama W, Haneda T, Hayashi R, Onitsuka T, et al. 1999. Comparison of innervated and noninnervated free flaps in oral reconstruction. *Plast Reconstr Surg* **104:** 1307–1313.

Kumnig M, Jowsey SG, Rumpold G, Weissenbacher A, Hautz T, Engelhardt TO, Brandacher G, Gabl M, Ninkovic M, Rieger M, et al. 2012. The psychological assessment of candidates for reconstructive hand transplantation. *Transpl Int* **25:** 573–585.

Kuriakose MA, Loree TR, Spies A, Meyers S, Hicks WL Jr. 2001. Sensate radial forearm free flaps in tongue reconstruction. *Arch Otolaryngol Head Neck Surg* **127:** 1463–1466.

Lahteenmaki T, Waris T, Asko-Seljavaara S, Astrand K, Sundell B, Jarvilehto T. 1991. The return of sensitivity to cold, warmth and pain from excessive heat in free microvascular flaps. *Scand J Plast Reconstr Surg Hand Surg* **25:** 143–150.

Landin L, Cavadas PC, Ibanez J, Roger I. 2010. Malignant skin tumor in a composite tissue (bilateral hand) allograft recipient. *Plast Reconstr Surg* **125:** 20e–21e.

Landin L, Bonastre J, Casado-Sanchez C, Diez J, Ninkovic M, Lanzetta M, del Bene M, Schneeberger S, Hautz T, Lovic A, et al. 2012. Outcomes with respect to disabilities of the upper limb after hand allograft transplantation: A systematic review. *Transpl Int* **25:** 424–432.

Lantieri L, Meningaud JP, Grimbert P, Bellivier F, Lefaucheur JP, Ortonne N, Benjoar MD, Lang P, Wolkenstein P. 2008. Repair of the lower and middle parts of the face by composite tissue allotransplantation in a patient with massive plexiform neurofibroma: A 1-year follow-up study. *Lancet* **372:** 639–645.

Lantieri L, Hivelin M, Audard V, Benjoar MD, Meningaud JP, Bellivier F, Ortonne N, Lefaucheur JP, Gilton A, Suberbielle C, et al. 2011. Feasibility, reproducibility, risks and benefits of face transplantation: A prospective study of outcomes. *Am J Transpl* **11:** 367–378.

Lee WP, Yaremchuk MJ, Pan YC, Randolph MA, Tan CM, Weiland AJ. 1991. Relative antigenicity of components of a vascularized limb allograft. *Plast Reconstr Surg* **87:** 401–411.

Nanobashvili J, Neumayer C, Fugl A, Punz A, Blumer R, Prager M, Mittlbock M, Gruber H, Polterauer P, Roth E, et al. 2003. Ischemia/reperfusion injury of skeletal muscle: Plasma taurine as a measure of tissue damage. *Surgery* **133:** 91–100.

Petruzzo P, Lanzetta M, Dubernard JM, Landin L, Cavadas P, Margreiter R, Schneeberger S, Breidenbach W, Kaufman C, Jablecki J, et al. 2010. The international registry on hand and composite tissue transplantation. *Transplantation* **90:** 1590–1594.

Petruzzo P, Testelin S, Kanitakis J, Badet L, Lengele B, Girbon JP, Parmentier H, Malcus C, Morelon E, Devauchelle B, et al. 2012. First human face transplantation: 5 years outcomes. *Transplantation* **93:** 236–240.

Pomahac B, Pribaz J, Eriksson E, Annino D, Caterson S, Sampson C, Chun Y, Orgill D, Nowinski D, Tullius SG. 2011. Restoration of facial form and function after severe disfigurement from burn injury by a composite facial allograft. *Am J Transpl* **11:** 386–393.

Pomahac B, Pribaz J, Eriksson E, Bueno EM, Diaz-Siso JR, Rybicki FJ, Annino DJ, Orgill D, Caterson EJ, Caterson SA, et al. 2012a. Three patients with full facial transplantation. *N Engl J Med* **366:** 715–722.

Pomahac B, Pribaz JJ, Bueno EM, Sisk GC, Diaz-Siso JR, Chandawarkar A, Westvik TS, Malin EW, Eriksson E. 2012b. Novel surgical technique for full face transplantation. *Plast Reconstr Surg* **130:** 549–555.

Saavedra AP, Bueno EM, Granter SR, Pomahac B. 2011. Transmission of donor-specific skin condition from donor to recipient of facial allograft. *Am J Transpl* **11:** 1340.

Schneeberger S, Ninkovic M, Gabl M, Ninkovic M, Hussl H, Rieger M, Loescher W, Zelger B, Brandacher G, Bonatti H, et al. 2007. First forearm transplantation: Outcome at 3 years. *Am J Transpl* **7:** 1753–1762.

Schneeberger S, Gorantla VS, Hautz T, Pulikkottil B, Margreiter R, Lee WP. 2009. Immunosuppression and rejection in human hand transplantation. *Transpl Proc* **41:** 472–475.

Schneeberger S, Landin L, Jableki J, Butler P, Hoehnke C, Brandacher G, Morelon E, ESOT CTA Working Group. 2011. Achievements and challenges in composite tissue allotransplantation. *Transpl Int* **24:** 760–769.

Schneeberger S, Gorantla VS, Brandacher G, Zeevi A, Demetris AJ, Lunz JG, Metes DM, Donnenberg AD, Shores JT, Dimartini AF, et al. 2013. Upper-extremity transplantation using a cell-based protocol to minimize immunosuppression. *Ann Surg* **257:** 345–351.

Schuind F, Abramowicz D, Schneeberger S. 2007. Hand transplantation: The state-of-the-art. *J Hand Surg Eur Vol* **32:** 2–17.

Seetharam A, Tiriveedhi V, Mohanakumar T. 2010. Alloimmunity and autoimmunity in chronic rejection. *Curr Opin Organ Transplant* **15:** 531–536.

Siemionow M, Ozturk C. 2012. Face transplantation: Outcomes, concerns, controversies, and future directions. *J Craniofac Surg* **23:** 254–259.

Siemionow M, Papay F, Alam D, Bernard S, Djohan R, Gordon C, Hendrickson M, Lohman R, Eghtesad B, Coffman K, et al. 2009. Near-total human face transplantation for a severely disfigured patient in the USA. *Lancet* **374:** 203–209.

Siemionow MZ, Papay F, Djohan R, Bernard S, Gordon CR, Alam D, Hendrickson M, Lohman R, Eghtesad B, Fung J. 2010. First U.S. near-total human face transplantation: A paradigm shift for massive complex injuries. *Plast Reconstr Surg* **125:** 111–122.

Siemionow M, Gharb BB, Rampazzo A. 2011. Pathways of sensory recovery after face transplantation. *Plast Reconstr Surg* **127:** 1875–1889.

Soga S, Pomahac B, Wake N, Schultz K, Prior RF, Kumamaru K, Steigner ML, Mitsouras D, Signorelli J, Bueno EM, et al. 2012. CT angiography for surgical planning in face transplantation candidates. *AJNR Am J Neuroradiol* doi: 10.3174/ajnr.A3268.

Takamatsu A, Harashina T, Inoue T. 1996. Selection of appropriate recipient vessels in difficult, microsurgical head and neck reconstruction. *J Reconstr Microsurg* **12:** 499–507; discussion 508–513.

Tohma A, Mine K, Tamatsu Y, Shimada K. 2004. Communication between the buccal nerve (V) and facial nerve (VII) in the human face. *Ann Anat* **186:** 173–178.

van Leeuwen MT, Webster AC, McCredie MR, Stewart JH, McDonald SP, Amin J, Kaldor JM, Chapman JR, Vajdic CM, Grulich AE. 2010. Effect of reduced immunosuppression after kidney transplant failure on risk of cancer: Population based retrospective cohort study. *BMJ* **340:** c570.

Opportunistic Infections—Coming to the Limits of Immunosuppression?

Jay A. Fishman

Transplant Infectious Disease and Compromised Host Program, Infectious Disease Division, MGH Transplantation Center, Harvard Medical School, Massachusetts General Hospital, Boston, Massachusetts 02114

Correspondence: jfishman@partners.org

Possible etiologies of infection in the solid organ recipient are diverse, ranging from common bacterial and viral pathogens to opportunistic pathogens that cause invasive disease only in immunocompromised hosts. The recognition of infectious syndromes in this population is limited by alterations in the clinical manifestations by immunosuppression. The risk of serious infections in the organ transplant patient is determined by the interaction between the patients' recent and distant epidemiological exposures and all factors that contribute to the patient's net state of immune suppression. This risk is altered by antimicrobial prophylaxis and changes in immunosuppressive therapies. In addition to the direct effects of infection, opportunistic infections and the microbiome may adversely shape the host immune responses with diminished graft and patient survivals. Antimicrobial therapies are more complex than in the normal host with a significant incidence of drug toxicity and a propensity for drug interactions with the immunosuppressive agents used to maintain graft function. Rapid and specific microbiologic diagnosis is essential. Newer microbiologic assays have improved the diagnosis and management of opportunistic infections. These tools coupled with assays that assess immune responses to infection and to graft antigens may allow optimization of management for graft recipients in the future.

If one were to design an experiment to study the effects of infection following organ transplantation, it is unlikely that one would use as subjects an outbred population with multiple, severe comorbidities, uncontrolled infectious exposures, and a wide variety of regimens for immunosuppression undergoing major surgery with implantation of organs from an equally diverse donor pool. Nonetheless, as regimens for immunosuppression have "stabilized," some general observations can be made based on this ongoing "experiment." Important clinical observations regarding infection in transplantation include:

1. The risk for infection is related to the nature and intensity of immunosuppression and the infectious exposures of the organ recipient and donor (Fishman and Rubin 1998; Fishman et al. 2007). An understanding of the relationship between the intensity of immunosuppression and infectious risk allows the development of effective prophylactic strategies.

Cite this article as *Cold Spring Harb Perspect Med* doi: 10.1101/cshperspect.a015669

2. Recognition of infection is more difficult in transplant recipients than in individuals with normal immune function given diminished signs and symptoms of infection and the array of noninfectious etiologies of fever (e.g., graft rejection, drug toxicity) (Fishman and Rubin 1998; Fishman 2007). Nonspecific signs and symptoms must be appreciated.
3. Infection progresses rapidly in the immunocompromised host. Thus, the prevention of infection is essential. Confronted with infection, early and effective therapy is required.
4. The spectrum of potential pathogens is broad; multiple pathogens may be present at the same time (e.g., virus and molds). Rapid and specific diagnoses are needed to guide antimicrobial therapies and to limit avoidable drug interactions and toxicities. Rapid and improved microbiologic assays have greatly improved care. Surgical debridement and invasive diagnostic procedures are often required.
5. There are no currently available assays that can determine an optimal immunosuppressive regimen while avoiding infection.

The nature of the clinical transplantation experiment has tended to obscure the impact of diverse infections on immune functions including tolerance induction and graft rejection. Recent studies in animal models have emphasized the importance of the microbiome, infectious exposures, and the innate immune system in determining aspects of immune function including the development of T-cell specificity and function, the risk for subsequent infections, and, to a degree, graft and patient survival (Fig. 1) (Selin et al. 1994; Yewdell and Bennink 1999; Adams et al. 2003a; Chong and Alegre 2012). This review focuses on more recent concepts in the impact of infection in solid organ transplantation.

RISK FOR INFECTION AND THE TIMELINE OF INFECTION

The risk for infection in an individual transplant recipient is determined by two factors:

1. The epidemiologic exposures of both the patient and the organ donor including those unrecognized by the patient or distant in time (Rubin et al. 1981; Fishman and Rubin 1998).
2. The patient's "net state of immunosuppression" including all factors contributing to the risk for infection (Table 1) (Fishman and Rubin 1998).

These factors are conceptual assessments similar to measuring the area under the curve of a formula relating epidemiology and immune function. At higher levels of immunosuppression, infection occurs at lower levels of infectious "exposure" or with organisms of lower levels of native virulence. With lower levels of immunosuppression, infection is less common, drug toxicities are less frequent, but graft rejection more prevalent. Specific immune deficits (genetic or acquired) and specific immunosuppressive regimens predispose to infection with specific classes of organisms (e.g., T-lymphocyte depletion activates latent cytomegalovirus [CMV] and allows viral replication to occur while corticosteroids predispose to *Pneumocystis* and other fungal infections) (Table 2) (Issa and Fishman 2009). Successful prophylaxis reduces the burden of organisms or protects against invasive infection (e.g., bacteremia) and allows more intensive immunosuppression. As a result, preventative strategies must be adapted to immunosuppression and to the individual's risk factors for infection.

EPIDEMIOLOGY

Epidemiologic exposures can be divided into four categories: donor-derived (often unknown exposures), recipient-derived infections and colonization, endemic community-derived infections, and health-care institution-derived pathogens. The importance of an otherwise unimportant organism on the skin of a transplant recipient is magnified in the face of surgery, devitalized tissues, hematoma or fluid collections, and immunosuppression.

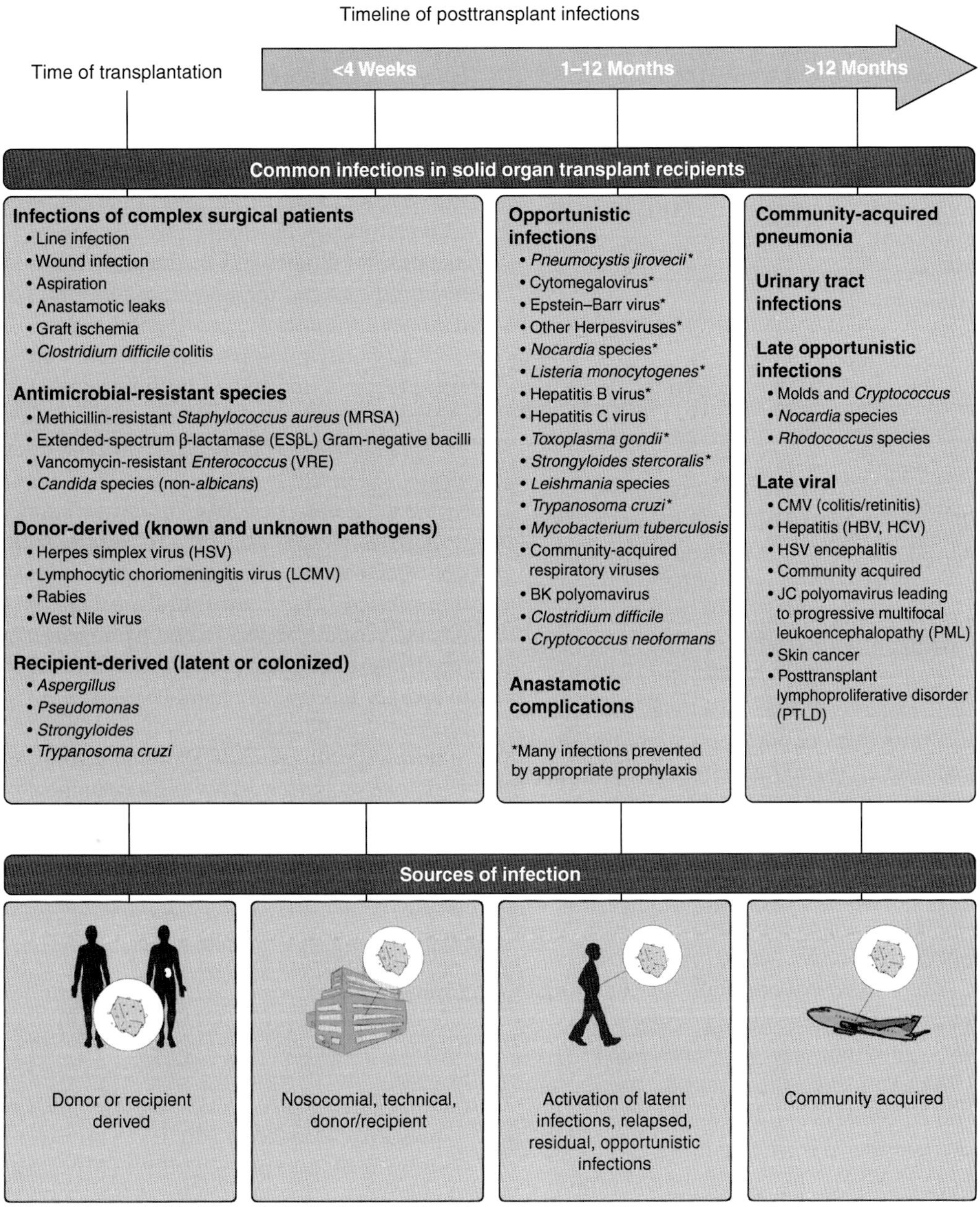

Figure 1. The timeline of infections following organ transplantation. The risk for infection following organ transplantation follows a standard pattern with routine immunosuppression and infectious exposures. The potential pathogens for which the risk is modified by prophylaxis, including vaccinations and antimicrobial agents, are indicated (*). Individual risk is modified by events such as surgery, treatment of graft rejection, or malignancy. Note that graft rejection and drug reactions may be among noninfectious causes of fever in transplant recipients. (From Fishman 2007; modified, with permission, from the author.)

Table 1. Factors contributing to the "net state of immunosuppression"

Immunosuppressive therapy: Type, temporal sequence, intensity
Prior therapies (chemotherapy or antimicrobials)
Mucocutaneous barrier integrity (intravenous access and urinary catheters, surgical drains)
Neutropenia, lymphopenia (often drug-induced)
Underlying immune deficiency (e.g., adrenal insufficiency, systemic lupus, complement deficiencies)
Metabolic conditions: Uremia, malnutrition, diabetes, alcoholism/cirrhosis
Viral infection (CMV, hepatitis B and C, RSV)

Donor-Derived Infections

Infection is transmitted efficiently with the transplantation of viable tissues; transmission is likely enhanced by immunosuppression (Fishman 2007). Data on allograft-associated disease transmissions are limited by difficulties in distinguishing allograft-derived infections from recipient-derived infections (e.g., because of latent infections or nosocomial infections) and by the failure to recognize or to report these events (Fishman 2007, 2008; Fishman et al. 2009; Grossi and Strong 2009; Grossi et al. 2009). In immunosuppressed hosts, the incidence of donor-derived infection may be underestimated owing to the absence of leukocytosis, pain (in a denervated graft), or erythema.

Some infections generally preclude organ donation (e.g., uncontrolled sepsis, HIV or HTLV infection, West Nile virus, Rabies virus, LCMV). Guidelines precluding the use of organs from HIV or HTLV-infected individuals are being reconsidered in some countries (Huang and Fishman 2011). The potential donor with microbiologically undiagnosed and untreated infection (e.g., sepsis, meningitis, or encephalitis) and/or for whom resolution has not been documented may represent a significant risk to the recipient (Satoi et al. 2001). Some common pathogens may limit the use of donor organs (to infected or immune recipients) and are included in donor (and recipient) screening panels (Table 3). These include serologic and other assays for hepatitis B and C viruses (HBV, HCV). The interpretation of serologic tests from donors must include consideration of the potential for false-positive assays after blood transfusion or false-negative assays after hemodilution of blood samples following infusion of colloids and crystalloids (Eastlund 2000). Similarly, testing for viral antibodies from newborns less than 1 mo of age is unreliable given exposure to maternal antibodies and the inconsistent antibody responses of the immature immune system.

Known, asymptomatic infections of donors are commonly transmitted including cytomegalovirus (CMV), Epstein–Barr virus (EBV), herpes simplex virus (HSV), and varicella zoster virus (VZV). Donor screening is then useful to develop posttransplant preventative or monitoring strategies for recipients. These same pathogens may cause significant disease if infection is active (i.e., viremia) at the time of procurement. Some donor-derived pathogens may require posttransplant therapy: *Treponema pallidum*, *M. tuberculosis*, nontuberculous mycobacteria, *Histoplasma capsulatum*, *Coccidioides immitis*, *Paracoccidioides* spp., *Blastomyces*. Donors from endemic regions merit screening for

Table 2. Immunosuppression and infection: Common associations

Antilymphocyte globulins (lytic depletion) and alloimmune response
T lymphocytes: Activation of latent (herpes)viruses, fever, cytokines
B lymphocytes: Encapsulated bacteria
Plasmapheresis: Encapsulated bacteria
Costimulatory blockade: Unknown so far; possible increased risk for EBV/PTLD
Corticosteroids: Bacteria, PCP, hepatitis B, possibly hepatitis C
Azathioprine: Neutropenia, possibly papillomavirus
Mycophenolate mofetil (MMF): Early bacterial infection, B cells, late CMV
Calcineurin inhibitors: Enhanced viral replication (absence of immunity), gingival infection, intracellular pathogens
mTOR inhibitors:
Poor wound healing
Excess infections in combination with other agents
Idiosyncratic interstitial pneumonitis

 Cite this article as *Cold Spring Harb Perspect Med* doi: 10.1101/cshperspect.a015669

Table 3. Common screening tests for organ donors

Human immunodeficiency virus antibody
Hepatitis B (HBV) serologies including HBV surface antigen, core antibody, surface antibody, and hepatitis δ antigen and/or antibody in HBsAg-positive donors
Hepatitis C antibody
Nontreponemal and treponemal testing (rapid plasma reagin [RPR] + TPHA or TPPA or FTA-Abs)
Human T-cell lymphotrophic virus (HTLV-I/II) antibody (less common currently given assay performance)
Toxoplasma antibody (notably in cardiac donors)
Cytomegalovirus antibody
Epstein–Barr virus (EBV) antibody panel (EBV viral capsid antigen, ± early antigen, and nuclear antigen antibody levels)
Herpes simplex virus antibody
Varicella-zoster virus antibody
Blood and urine cultures

Many procurement organizations supplement these tests with additional assays based on local epidemiology and/or use nucleic acid-based assays (NAT) (see also Grossi et al. 2009).

parasites (*Trypanosoma cruzi*, *Plasmodium* spp., *Strongyloides stercoralis*, *Schistosoma* spp., *Leishmania* spp.) as well as epidemic or endemic viruses (Chikungunya virus, West Nile virus, and human T-cell lymphotropic virus, HTLV-I/II). Donors also may be screened for infectious agents of importance in the transplantation of specific organs (e.g., *Toxoplasma gondii* in cardiac recipients).

Unintentional transmission of infection occurs in at least 1%–2% of recipients (Len et al. 2008; Ison et al. 2011). Clusters of infection in recipients of allografts from a single donor have included *Mycobacterium tuberculosis*, *Candida*, and *Aspergillus* (and other fungal) species, herpes simplex virus (HSV) and human herpes virus 8, lymphocytic choriomeningitis virus (LCMV), rabies virus, *Trypanosoma cruzi*, HIV, and hepatitis C virus. Detection of transmission requires clinical suspicion, access to advanced microbiologic testing including nucleic acid amplification technologies (NAT), recognition of epidemiologic exposures, and, where appropriate, assistance by public health authorities.

Significant controversy exists over the optimal screening panel and specific tests (e.g., nucleic acid tests or NAT) for donors (Table 3). This is a reflection of at least three areas of uncertainty. First, the degree to which it is possible to reduce infection. Second, the advantages and disadvantages of NAT assays compared with serologic (antibody-based) tests. Third, the actual risk of transmission posed by individuals with possibly increased risk of disease transmission, notably for HCV, HBV, or HIV, is unclear. HIV infectious risk is inferred from the donor's medical and social history, in addition to data from screening assays. The U.S. Public Health Service (PHS) published guidelines in 1994 describing the epidemiological risk factors identified for HIV transmission; those guidelines have also been applied widely for HCV and HBV prevention (USPH Service 1994). Given the low incidence of reported viral transmission associated with organ transplantation, these guidelines appear to have been effective, notably for HCV. Updated draft guidelines reflect the availability of, and controversy over, highly sensitive NAT and protein-based assays for HIV, HCV, and HBV (USPH Service 2011).

A limitation to serologic assays is the time required to develop positive assays compared with nucleic acid or protein detection assays. Transmission of infection may occur in this "window period" between donor infection and seroconversion (Schreiber et al. 1996; Pillonel et al. 1998; Kolk et al. 2002; Biswas et al. 2003; Fiebig et al. 2003; Kleinman et al. 2003; Busch et al. 2005; Kleinman and Busch 2006; Yao et al. 2007, 2008). HIV antibody assays can show seroconversion within approximately 22 d of infection (but can be as long as 6 mo), with NAT detection reducing the window period for detection from 5.6 to 10.2 d (i.e., 4–15 d in which infection is detected by NAT but not ELISA). HBV surface antigen (HBsAg) ELISA assays have a window period of 38.3 to 49.7 d, with NAT in the range of 20.4 to 25.7 d (Busch et al. 1995, 2005; Schreiber et al. 1996; Pillonel et al. 1998; Kolk et al. 2002; Biswas et al. 2003; Fiebig et al. 2003; Kleinman et al. 2003; Kleinman and Busch 2006; Yao et al. 2007, 2008). The use of HBV NAT testing may detect HBV infection in donors who are HBSAg negative. The HCV ELISA assays have window periods of 38 to

94 d, which is reduced to 6.1 to 8.7 d using NAT assays.

Decisions regarding the use of organs from donors with active or suspected infection take into account the urgency of transplantation for the recipient, microbiological data and treatment options for the donor or recipient, and the availability of alternatives. The key concept underlying donor screening is to provide data for clinicians and potential recipients that allow informed decisions balancing the risk of infection against the benefits of organ replacement.

Pretransplant Infections in Organ Recipients

Before donation, potential recipients (or donors) may become colonized by nosocomially acquired organisms carrying resistance to multiple antimicrobial agents (e.g., vancomycin-resistant *Enterococcus* [VRE], multidrug-resistant *Klebsiella pneumoniae*, azole-resistant *Candida* species). As a result, routine surgical prophylaxis may be inadequate to prevent transmission. The presence of, and microbiological susceptibility patterns for, such colonizing organisms should be discerned before transplantation when possible (notably in lung transplant candidates).

Any active infections in the recipient should be treated and, ideally, resolved before procurement or transplantation (Delmonico and Snydman 1998; Freeman et al. 1999; Lumbreras et al. 2001; Satoi et al. 2001; Singh 2002; D'Albuquerque et al. 2007; Fishman 2007; Nanni Costa et al. 2008). There are no data on which to base a recommendation for the optimal duration of therapy or the interval between resolution of infection and procurement. Clearance or control of infection should be documented and consideration given to surgical or posttransplant prophylaxis. The inflammation associated with bacterial peritonitis is a common source of diffuse intraoperative bleeding or of subsequent fibrosis in liver transplantation—increasing the surgical technical difficulty and the risk for infection. Spillage from infected renal or hepatic cysts may also contaminate a surgical field.

Community Exposures

A careful epidemiologic history may provide microbiologic clues to possible infectious presentations. Outbreaks of viral respiratory illness (e.g., influenza, respiratory syncytial virus) occur earlier in the season in immunocompromised individuals than in normal hosts. Travel, hobbies (gardening, hiking), or work (teachers) exposures may suggest exposures to contaminated food or water (*Listeria, Cryptosporidium*), soil (*Aspergillus* or *Nocardia*), birds (*Cryptococcus*), or geographically restricted mycoses (*Blastomyces dermatitidis, Coccidioides immitis, Paracoccidioides* sp., and *Histoplasma capsulatum*).

Nosocomial Exposures

Nosocomial infections are of increasing importance in transplant donors and recipients. As patients wait for organs and may have undergone multiple procedures and antimicrobial exposures, colonization may occur with hospital-acquired antimicrobial-resistant pathogens. Among the common pathogens are vancomycin-resistant enterococci (VRE), methicillin-resistant staphylococci (MRSA), fluconazole-resistant *Candida* species, *Aspergillus* species, and highly resistant Gram-negative bacteria (van Delden et al. 2009). Respiratory viral infections may be acquired from visitors and hospital staff.

THE NET STATE OF IMMUNOSUPPRESSION

The net state of immunosuppression is a conceptual measure of factors contributing to a patient's risk for infection (Table 1) (Fishman 2007). Among these are:

1. The specific immunosuppressive therapy, including dose, duration, and sequence of agents.
2. Surgical issues resulting from the transplant procedure, resulting in leaks (blood, lymph, urine) and fluid collections, devitalized tissue, poor wound healing, and surgical drainage catheters.
3. Prolonged ventilatory support.

 Cite this article as *Cold Spring Harb Perspect Med* doi: 10.1101/cshperspect.a015669

4. Broad-spectrum antibiotics.
5. Posttransplant renal, hepatic, pulmonary or cardiac dysfunction, or diabetes.
6. Prolonged use of urinary, vascular access, or dialysis catheters, surgical drains, or other breaks in skin or mucosal defenses.
7. Viral infection with resulting local or systemic immunosuppression and increased risk for superinfection. Of special importance in transplantation are the herpesviruses (CMV, EBV, HSV, VZV), HBV or HCV, or HIV.
8. Malnutrition.
9. Neutropenia, which may be related to medications such as mycophenolate mofetil, azathioprine, ganciclovir, valganciclovir (rarely to trimethoprim-sulfamethoxazole). Overestimates of renal function based on serum creatinine levels are often associated with drug toxicity.

Individual drugs are associated with increased risk for certain infections (Table 2), although the degree of immunosuppression associated with a specific regimen varies between individuals relative to the risks of both graft rejection and for infection. Assays that measure the level of "immunity" against specific pathogens (e.g., cellular immune function or antibody levels against specific organisms) are not yet predictive of infectious risk. Serologic assays determine past exposures and latent infections (herpesviruses) but are poorly predictive of the efficacy of the immune response to specific pathogens in immunosuppressed hosts. Measures of "global" immune function remain relatively crude (Husain et al. 2009). Few data exist on the integrity of immune reconstitution against infection after T- or B-lymphocyte depletion or treatment with other antibody preparations (e.g., antibodies to tumor necrosis factor, costimulatory blockade). Immune deficits associated with biologic agents are generally more severe and persist longer (months to years) than is generally appreciated. Similarly, the effects of organ dysfunction (e.g., cirrhosis, renal dysfunction) on systemic immune function resist precise quantification.

Among underappreciated "immune defects" are breaches in cutaneous barriers. As was noted, the combination of "sticky" organisms such as VRE, *Candida* spp., or *Staphylococcus aureus* from colonized skin often associated with biofilms on in-dwelling catheters or surgical drains leads to increased risk for bacteremia. These effects are amplified by neutropenia caused by drug effects (e.g., ganciclovir, mycophenolate mofetil, allopurinol, trimethoprim-sulfamethoxazole) and nutritional deficits. Surgical, invasive radiologic, or gastrointestinal procedures used to repair technical complications such as bile leaks or biliary or ureteric strictures, risk dissemination of organisms from those sites with distant seeding and abscess formation in joints or ischemic regions of grafts.

TIMELINE OF INFECTION

With more standardized immunosuppressive regimens, common infections are observed in a relatively consistent pattern based on the time elapsed since transplantation (Fig. 2). This is a reflection of changing risk factors over time: surgery/hospitalization, immune suppression, acute and chronic rejection, emergence of latent infections, and exposures in the community (Fishman 2007). A key concept is that predictable changes in the risk for infection occur with unique epidemiologic exposures (e.g., returning to work) and with alterations in the drug regimen (interactions, treatment of graft rejection, neutropenia). Infections appear later than usual when delayed by effective prophylaxis and as allograft function improves.

The timeline reflects three overlapping periods of risk for infection: (1) the perioperative period to approximately 30 d after transplantation, (2) the period 1–6 mo after transplantation (depending on the use of antilymphocyte "induction" therapy and the level of immunosuppression), and (3) the period beyond the 6–12 mo after transplantation. The timeline may be used in a variety of ways: (1) to establish a differential diagnosis for the transplant patient suspected of having an infection, (2) as a clue to the presence of an excessive environmental hazard for the individual, either within the hospital

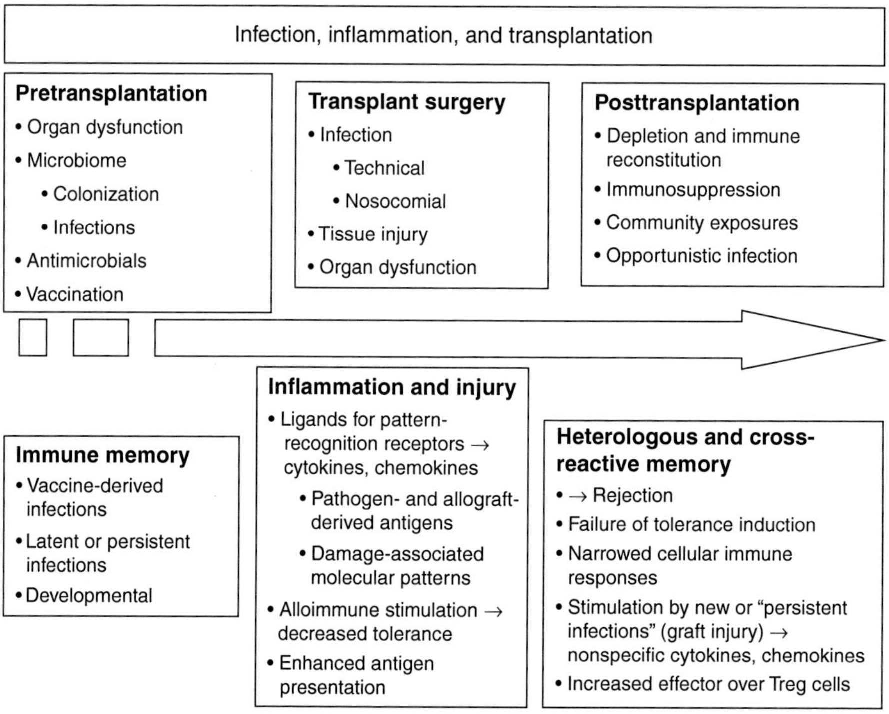

Figure 2. The impact of infectious exposures, inflammation, the microbiome, heterologous immunity, and innate immune stimulation on transplantation. The microbiome is the sum of colonizing organisms, acute and chronic infections, and reflects alterations in this flora by the immune system, vaccination, or antimicrobial agents. Antigens and activating molecular patterns from microbes or damaged tissues are released during all phases of transplantation and are present during graft rejection, immunosuppression, and/or immune reconstitution following lymphocyte depletion. These antigenic exposures and inflammatory mediators shape immune function via both the innate and adaptive immune systems and impact the outcome of transplantation (see also Chong and Alegre 2012).

or in the community, and (3) as a guide to the design of prophylactic antimicrobial strategies. Infections occurring outside the usual period or with unusual severity suggest the presence of an excessive epidemiologic hazard or of excessive immunosuppression (Fishman 2007). Routine preventative strategies from the Massachusetts General Hospital are outlined in Tables 4 and 5. It should be noted that such strategies serve only to delay the onset of infection in the face of epidemiologic pressure. The use of antibiotic prophylaxis, vaccines, and behavioral modifications (e.g., routine hand washing or advice against buying new kittens) may only result in a "shift to the right" of the timing of infection unless the intensity of immune suppression is reduced or immunity develops.

Phase 1: 1 Month Posttransplantation

During the first month after transplantation, the types of infection are generally related to complications of the transplant surgery (fluid collections, sepsis), infections present in the donor or recipient before transplantation, or nosocomial infections. Graft rejection may occur but is uncommon. Early removal of lines and drains, appropriate prophylaxis for surgery based on knowledge of the patient's colonization status, discontinuation of unneeded medications including antimicrobial agents, drainage of hematomas and other collections, and meticulous wound care are essential. Opportunistic infections are uncommon in this period despite intensive immunosuppression; sustained ad-

Table 4. Prophylaxis for *Pneumocystis jiroveci* pneumonia (PCP)

Background: Low dose trimethoprim-sulfamethoxazole prophylaxis (in adults: 1 single strength per day orally) is well tolerated and essentially eradicates *Pneumocystis* infection from this patient population. Lower doses (3 days per week) prevent PCP but may not prevent other infections such as urinary tract infection, including those attributable to susceptible *Nocardia* and *Listeria*, toxoplasmosis, and a variety of gastrointestinal and pulmonary infections.

Regimen: One single strength trimethoprim-sulfamethoxazole tablet (containing 80 mg trimethoprim, 400 mg sulfamethoxazole) po qhs for a minimum of 4–6 mo posttransplant. Patients infected with CMV, with chronic rejection, recurrent infections, and most lung, liver, and heart recipients may benefit from lifelong prophylaxis.

Alternative regimen: For patients proven not to tolerate trimethoprim-sulfamethoxazole, alternative regimens include (1) a combination of atovaquone 1500 mg po with meals once daily plus levofloxacin (or equivalent fluoroquinolone without antianaerobic spectrum) 250 mg once daily, (2) pentamidine (300 mg iv or inhaled q 3–4 weeks), and (3) dapsone (100 mg po qd to biw) ± pyrimethamine. Each of these agents has toxicities that must be considered including hemolysis in G6PD-deficient hosts with dapsone. None of these alternative programs offer the same broad protection of TMP-SMX.

ministration of immunosuppressive agents is generally required to allow organisms of low native virulence to establish invasive disease. When these occur, they reflect prior immune defects (pretransplant therapies) or environmental exposures.

In this period, the stage is also set for the emergence of a subgroup of patients—the "chronic ne'er do well"—individuals with early infectious complications and/or graft dysfunction and who may require sustained higher levels of immunosuppression. Such individuals show lifelong susceptibility to infection and merit prolonged (lifelong) prophylaxis (Tables 4 and 5).

Phase 2: 1 to 6 Months Posttransplant

The opportunistic infections emerge in the first 6 mo posttransplant coupled with technical issues remaining from the transplant admission. Use of anti-CMV therapy with broad antiherpes efficacy and trimethoprim–sulfamethoxazole (TMP-SMX) prophylaxis has altered the pattern of posttransplant infections. TMP-SMX eliminates *P. jiroveci* pneumonia (PCP) and, given daily, reduces the incidence of urinary tract infection and urosepsis, some respiratory and GI infections, *L. monocytogenes* meningitis, many *Nocardia* species infections, and *Toxoplasma gondii*. TMP-SMX also have activity against many strains of community-acquired MRSA. Other agents substituted for TMP-SMX in the sulfa-allergic or neutropenic patient do not provide the same spectrum of benefits. In practice, most "allergies" to TMP-SMX merit reconsideration unless well documented. Effective anti-CMV prophylaxis should prevent most CMV infections (and those caused by most herpesviruses) for the duration of therapy. Thus, the differential diagnosis of infectious syndromes in this period includes:

- Graft rejection.
- Persistent infection from the perisurgical period including relapsed *C. difficile* colitis, inadequately treated pneumonia, or infection related to graft-specific technical problems (e.g., pleural effusion, urine leak, cholangitis, lymphocele, infected hematoma, airway necrosis).
- Viral infections including CMV, herpes simplex virus (HSV), shingles (localized or disseminated zoster attributable to varicella zoster virus, VZV), human herpesvirus 6 or 7, BK polyomavirus, EBV, relapsed hepatitis (HBV, HCV), and the community-acquired respiratory viruses (adenovirus, influenza, parainfluenza, respiratory syncytial virus, metapneumovirus). Secondary postviral infections (bacterial or fungal) are common.
- Opportunistic infection as a result of *Pneumocystis jiroveci*, *Listeria monocytogenes*, *Toxoplasma gondii*, *Nocardia* species (in the absence

Table 5. Prophylaxis for herpes group viruses[a]

CMV universal antiviral prophylaxis[a]		
CMV serologic status ± antilymphocyte-globulin-induction therapy (ALG)	Prophylaxis	Monitoring (antigenemia)
D^+/R^- no ALG	Intravenous ganciclovir 5 mg/kg iv (loading dose) then po valganciclovir (900 mg/d corrected for renal function) or po ganciclovir (3 gm/d)[c] × 3 mo	Monthly for 6 mo after discontinuation of therapy[b]
D^+ or R^+ with ALG	Intravenous ganciclovir 5 mg/kg iv for first dose then either per renal function to discharge or switch to po valganciclovir (900 mg/d corrected for renal function) or po ganciclovir (3 gm/d)[c] × 6 mo	Monthly for 6 mo after discontinuation of therapy[b]
D^-/R^+ no ALG	Oral valganciclovir (900 mg/d corrected for renal function) × 3 mo	Symptoms only
D^-/R^-	Oral famciclovir 500 mg po qd × 3–4 mo (or valacyclovir 500 bid or acyclovir 400 tid). Use of CMV-negative or leukocyte-filtered blood	Symptoms, fever/neutropenia
Status unknown with ALG	Intravenous ganciclovir 5 mg/kg iv for first dose and QD (corrected for renal function) until sero-status determined	As above

The human herpes viruses are among the most important causes of infectious disease morbidity and mortality in the transplant recipient. Preventative regimens are determined by the clinical risk, the major determinants of which are the past experience of donor and recipient with the virus (as defined by the presence or absence of circulating antibody before transplant) and the nature of the immunosuppressive therapy. The doses of antiviral therapies are not, in general, reduced for neutropenia.

[a]Preemptive therapy: Preemptive therapy requires a carefully organized monitoring program and patient compliance. Either a molecular CMV viral load test or a pp65 antigenemia assay may be used for monitoring. Monitoring should be performed once weekly after transplantation for 12–24 wk. Infections indicated by positive assays are treated with either oral valganciclovir (900 mg 2 times a day) or intravenous ganciclovir (5 mg/kg 2 times a day). Full doses are used for loading after which dosing is corrected for renal function. Therapy is continued until viremia is undetectable. Mixed prophylaxis: Many centers prefer universal prophylaxis for highest risk recipients (D^+/R^- or R^+ with lymphocyte depletion) and preemptive therapy for other groups.

[b]ALG: Antilymphocyte antibodies include any of the lytic, lymphocyte-depleting antibody preparations.

[c]Valacyclovir (8 gm/d) has been used as an alternative agent in renal transplant recipients.

of TMP-SMX), *Aspergillus* species, and other agents. The specific opportunistic infections that occur reflect the specific immunosuppressive regimen used and the presence or absence of immunomodulating viral infection.

Phase 3: More than 6 to 12 Months Posttransplant

After the first 6–12 mo, transplant recipients with good graft function and reduced maintenance immunosuppression remain at some risk for reactivation of latent infections (e.g., VZV) or for primary CMV infection (socially acquired) but have community-acquired exposures as their main risk, which may predispose to more serious opportunistic superinfections. Respiratory viruses or infections related to underlying diseases (e.g., skin infections in diabetes) are common. The major challenges include EBV and posttransplant lymphoproliferative disorders, persistent BK polyomavirus infection, hepatitis C virus in liver recipients, and papillomavirus (warts, anogenital, and skin cancers). Advanced therapies for HIV and HCV have dramatically altered the impact of these infections. Viral infections are often the major factor in risk for infections or graft rejection.

The patients with less satisfactory allograft function or with renal dysfunction or diabetes as a result of calcineurin inhibitor toxicity tend to be relatively over-immunosuppressed. This group includes those identified early in the posttransplant course caused by delayed or poor initial graft function or complications and is at persistent risk for opportunistic infections such as *Pneumocystis*, *Zygomycetes*, or *Cryptococcus* (Fishman 2007). Lesions identified in this group such as of skin or lungs should be biopsied to exclude fungal infection or malignancy.

"INDIRECT EFFECTS," THE MICROBIOME, AND IMMUNE FUNCTION

Latent or acute infections are associated with adverse outcomes in the transplant recipient including an increased risk for opportunistic infection and graft rejection. The existence of these "indirect effects" has been implied based on clinical observations and the effects of effective prophylaxis. For CMV, in addition to an increased rate of opportunistic infections (fungal including aspergillosis, accelerated HCV infection), effects include coronary vaculopathy in heart recipients, bronchiolitis obliterans syndrome in lung recipients, and an increased rate of EBV-associated posttransplant lymphoproliferative disorders (PTLD) and increased mortality in the absence of prophylaxis. In liver transplant recipients with recurrent HCV hepatitis, there is an increased rate of infections, posttransplant diabetes, and mortality (Hodson et al. 2005; Kalil et al. 2005; Bloom and Lake 2006; Small et al. 2006). HCV infection may also enrich the hepatic regulatory T cell population, which may persist after liver transplantation (Spangenberg et al. 2005; Nellore and Fishman 2011).

These effects may be mediated by innate or adaptive immune mechanisms and should not be confused with "chronic allograft rejection" but are rather viral effects on cellular proliferation, functions of immune cells and inflammatory networks, and modulation of cellular gene activities in both the graft and host. These remain an important area for investigation.

Infection and the microbiome control many aspects of alloimmunity in transplantation (Chong and Alegre 2012). Thus, consistent with experience, "dirty" transplantation (e.g., lung or intestine) would be expected to have lower graft survival than cleaner (e.g., kidney, heart) procedures without exposure to organisms and the environment. This may also be a reflection of technical difficulty and differing immunologic challenges.

Heterologous immunity is the concept used to describe immune memory responses to previously encountered pathogens that alter subsequent immune responses to unrelated pathogens or to grafts. Some of these effects are attributed to the generation of immunological cross reactivity between viral epitopes and graft and other antigens. Virally induced alloreactive memory may create a barrier to transplantation tolerance or induce graft rejection or autoimmunity (Braciale et al. 1981; Adams et al. 2003a,b). The observed effects may be mediated via memory $CD8^+$ cells following viral infection or more broadly by inflammatory mediators such as TNF-α and IFN-γ. Such virally induced alloreactive memory may create a barrier to transplantation tolerance or induce graft rejection or autoimmunity. Such effects have been shown for both CMV and EBV. Viral reactivation may also compress the T-cell repertoire and diminish responses to new antigens with an increased risk for opportunistic infections (Yewdell and Bennink 1999; Brehm et al. 2002). All limbs of the host immune response may be altered by infection, including adaptive as well as NK cell, neutrophil, and macrophage responses, and may contribute to the risk for graft injury and opportunistic infections.

GENERAL APPROACHES TO INFECTION IN TRANSPLANT RECIPIENTS

The spectrum of infection in the immunocompromised host is broad. Given the potential toxicity of antimicrobial agents and the need for rapid interruption of infection, early and specific diagnosis is essential in this population. Advances in diagnostic modalities, including advanced imaging and molecular microbiologic techniques, greatly assist in this process. However, the need for invasive diagnostic approaches cannot be overemphasized. Whereas these approaches carry some risk, the failure to achieve a specific diagnosis often necessitates broad, empiric therapy without clear endpoints for therapy. Initial, often empiric, therapy will target a broad range of potential pathogens with rapid narrowing of the antimicrobial spectrum as data become available.

A central consideration for the patient with an "infectious syndrome" is whether to reduce the intensity of immunosuppression as a part of therapy, risking graft rejection or an immune reconstitution syndrome (IRIS). Given that infection and rejection are often linked, clear distinctions between these processes may be difficult. For example, recurrent HCV infection may increase the risk for liver graft rejection. Pneumonia, including aspiration, community-acquired respiratory viruses, CMV, and bacterial infections, will increase the rate of obliterative bronchiolitis in lung recipients (Bando et al. 1995; Kroshus et al. 1997; Avery 2006). CMV also contributes to cardiac and renal allograft rejection (Lowance et al. 1999; Potena and Valantine 2007). In practice, it is often possible to decrease the intensity of immune suppression in the face of significant infection; however, as the patient improves, rejection may occur. The selection of the specific agents to be reduced should depend on the organisms isolated—and often is not the calcineurin inhibitor. No prospective studies of the efficacy of reduced immunosuppression have been performed. Reversal of immune deficits (neutropenia, hypogammaglobulinemia) may be possible with adjunctive therapies (colony stimulating factors or antibody). Coinfection with virus (CMV) is a universal concern and merits therapy.

CONCLUSION

Transplant infectious disease has increasingly become a field characterized by preventative approaches and early therapies based on sensitive molecular diagnostic tests. The prevention of CMV, other herpesviruses, and *Pneumocystis* have been important advances in transplantation. Infection is often preventable with correction of technical problems (e.g., drainage of fluid collections before infection). Patients with viremia are over immunosuppressed relative to their immune systems. In the absence of assays useful for the individualization of global immunosuppression, sensitive molecular tests, and assays for pathogen-specific immune function may guide the modulation of the intensity of the individual's regimen. In the future, host risk factors will be amenable to investigation including genetic polymorphisms (e.g., toll-like receptors) and immunoregulatory elements to allow the individualization of transplant immunosuppression.

REFERENCES

Adams AB, Pearson TC, Larsen CP. 2003a. Heterologous immunity: An overlooked barrier to tolerance. *Immunol Rev* **196:** 147–160.

Adams AB, Williams MA, Jones TR, Shirasugi N, Durham MM, Kaech SM, Wherry EJ, Onami T, Lanier JG, Kokko KE, et al. 2003b. Heterologous immunity provides a potent barrier to transplantation tolerance. *J Clin Invest* **111:** 1887–1895.

Avery RK. 2006. Infections after lung transplantation. *Semin Respir Crit Care Med* **27:** 544–551.

Bando K, Paradis IL, Komatsu K, Konishi H, Matsushima M, Keena RJ, Hardesty RL, Armitage JM, Griffith BP. 1995. Analysis of time-dependent risks for infection, rejection, and death after pulmonary transplantation. *J Thorac Cardiovasc Surg* **109:** 49–57; discussion 57-49.

Biswas R, Tabor E, Hsia CC, Wright DJ, Laycock ME, Fiebig EW, Peddada L, Smith R, Schreiber GB, Epstein JS, et al. 2003. Comparative sensitivity of HBV NATs and HBsAg assays for detection of acute HBV infection. *Transfusion* **43:** 788–798.

Bloom RD, Lake JR. 2006. Emerging issues in hepatitis C virus-positive liver and kidney transplant recipients. *Am J Transplant* **6:** 2232–2237.

Braciale TJ, Andrew ME, Braciale VL. 1981. Simultaneous expression of H-2-restricted and alloreactive recognition by a cloned line of influenza virus-specific cytotoxic T lymphocytes. *J Exp Med* **153:** 1371–1376.

Brehm MA, Pinto AK, Daniels KA, Schneck JP, Welsh RM, Selin LK. 2002. T cell immunodominance and maintenance of memory regulated by unexpectedly cross-reactive pathogens. *Nat Immunol* **3:** 627–634.

Busch MP, Lee LL, Satten GA, Henrard DR, Farzadegan H, Nelson KE, Read S, Dodd RY, Petersen LR. 1995. Time course of detection of viral and serologic markers preceding human immunodeficiency virus type 1 seroconversion: Implications for screening of blood and tissue donors. *Transfusion* **35:** 91–97.

Busch MP, Glynn SA, Stramer SL, Strong DM, Caglioti S, Wright DJ, Pappalardo B, Kleinman SH. 2005. A new strategy for estimating risks of transfusion-transmitted viral infections based on rates of detection of recently infected donors. *Transfusion* **45:** 254–264.

Chong AS, Alegre ML. 2012. The impact of infection and tissue damage in solid-organ transplantation. *Nat Rev Immunol* **12:** 459–471.

D'Albuquerque LA, Gonzalez AM, Filho HL, Copstein JL, Larrea FI, Mansero JM, Peron G Jr, Ribeiro MA Jr, Oliveira e Silva A. 2007. Liver transplantation from deceased donors serologically positive for Chagas disease. *Am J Transplant* **7:** 680–684.

Delmonico FL, Snydman DR. 1998. Organ donor screening for infectious diseases: Review of practice and implications for transplantation. *Transplantation* **65:** 603–610.

Eastlund T. 2000. Hemodilution due to blood loss and transfusion and reliability of cadaver tissue donor infectious disease testing. *Cell Tissue Bank* **1:** 121–127.

Fiebig EW, Wright DJ, Rawal BD, Garrett PE, Schumacher RT, Peddada L, Heldebrant C, Smith R, Conrad A, Kleinman SH, et al. 2003. Dynamics of HIV viremia and antibody seroconversion in plasma donors: Implications for diagnosis and staging of primary HIV infection. *AIDS* **17:** 1871–1879.

Fishman JA. 2007. Infection in solid-organ transplant recipients. *N Engl J Med* **357:** 2601–2614.

Fishman JA. 2008. Rapporto dell'OMS sulla valutazione del programma nazionale trapianti in Italia. *Trapianti* **12:** 37–53.

Fishman JA, Rubin RH. 1998. Infection in organ-transplant recipients. *N Engl J Med* **338:** 1741–1751.

Fishman JA, Greenwald MA, Kuehnert MJ. 2007. Enhancing transplant safety: A new era in the microbiologic evaluation of organ donors? *Am J Transplant* **7:** 2652–2654.

Fishman JA, Strong DM, Kuehnert MJ. 2009. Organ and tissue safety workshop 2007: Advances and challenges. *Cell Tissue Bank* **10:** 271–280.

Freeman RB, Giatras I, Falagas ME, Supran S, O'Connor K, Bradley J, Snydman DR, Delmonico FL. 1999. Outcome of transplantation of organs procured from bacteremic donors. *Transplantation* **68:** 1107–1111.

Grossi P, Strong DM. 2009. The role of testing in determining suitability of donors and tissues. In *Tissue and cell donation: An essential guide* (ed. Warwick DFRM, Brubaker SA, Eastlund T). Blackwell, New York.

Grossi PA, Fishman JA, AST Infectious Disease Community of Practice. 2009. Donor-derived infections in solid organ transplant recipients. *Am J Transplant* **9:** S19–S26.

Hodson EM, Jones CA, Webster AC, Strippoli GF, Barclay PG, Kable K, Vimalachandra D, Craig JC. 2005. Antiviral medications to prevent cytomegalovirus disease and early death in recipients of solid-organ transplants: A systematic review of randomised controlled trials. *Lancet* **365:** 2105–2115.

Huang RC, Fishman JA. 2011. Screening of deceased organ donors: No easy answers. *Transplantation* **91:** 146–149.

Husain S, Raza K, Pilewski JM, Zaldonis D, Crespo M, Toyoda Y, Shutt K, Spichty K, Bentlejewski C, Pakstis DL, et al. 2009. Experience with immune monitoring in lung transplant recipients: Correlation of low immune function with infection. *Transplantation* **87:** 1852–1857.

Ison MG, Llata E, Conover CS, Friedewald JJ, Gerber SI, Grigoryan A, Heneine W, Millis JM, Simon DM, Teo CG, et al. 2011. Transmission of human immunodeficiency virus and hepatitis C virus from an organ donor to four transplant recipients. *Am J Transplant* **11:** 1218–1225.

Issa NC, Fishman JA. 2009. Infectious complications of antilymphocyte therapies in solid organ transplantation. *Clin Infect Dis* **48:** 772–786.

Kalil AC, Levitsky J, Lyden E, Stoner J, Freifeld AG. 2005. Meta-analysis: The efficacy of strategies to prevent organ disease by cytomegalovirus in solid organ transplant recipients. *Ann Intern Med* **143:** 870–880.

Kleinman SH, Busch MP. 2006. Assessing the impact of HBV NAT on window period reduction and residual risk. *J Clin Virol* **36:** S23–S29.

Kleinman SH, Kuhns MC, Todd DS, Glynn SA, McNamara A, DiMarco A, Busch MP. 2003. Frequency of HBV DNA detection in US blood donors testing positive for the presence of anti-HBc: Implications for transfusion transmission and donor screening. *Transfusion* **43:** 696–704.

Kolk DP, Dockter J, Linnen J, Ho-Sing-Loy M, Gillotte-Taylor K, McDonough SH, Mimms L, Giachetti C. 2002. Significant closure of the human immunodeficiency virus type 1 and hepatitis C virus preseroconversion detection windows with a transcription-mediated-amplification-driven assay. *J Clin Microbiol* **40:** 1761–1766.

Kroshus TJ, Kshettry VR, Savik K, John R, Hertz MI, Bolman RM 3rd. 1997. Risk factors for the development of bronchiolitis obliterans syndrome after lung transplantation. *J Thorac Cardiovasc Surg* **114:** 195–202.

Len O, Gavalda J, Blanes M, Montejo M, San Juan R, Moreno A, Carratala J, de la Torre-Cisneros J, Bou G, Cordero E, et al. 2008. Donor infection and transmission to the recipient of a solid allograft. *Am J Transplant* **8:** 2420–2425.

Lowance D, Neumayer HH, Legendre CM, Squifflet JP, Kovarik J, Brennan PJ, Norman D, Mendez R, Keating MR, Coggon GL, et al. 1999. Valacyclovir for the prevention of cytomegalovirus disease after renal transplantation. International Valacyclovir Cytomegalovirus Prophylaxis Transplantation Study Group. *N Engl J Med* **340:** 1462–1470.

Lumbreras C, Sanz F, Gonzalez A, Perez G, Ramos MJ, Aguado JM, Lizasoain M, Andres A, Moreno E, Gomez MA, et al. 2001. Clinical significance of donor-unrecognized bacteremia in the outcome of solid-organ transplant recipients. *Clin Infect Dis* **33:** 722–726.

Nanni Costa A, Grossi P, Gianelli Castiglione A, Grigioni WF. 2008. Quality and safety in the Italian donor evaluation process. *Transplantation* **85:** S52–S56.

Nellore A, Fishman JA. 2011. NK cells, innate immunity and hepatitis C infection after liver transplantation. *Clin Infect Dis* **52:** 369–377.

Pillonel J, Saura C, Courouce AM. 1998. Prevalence of HIV, HTLV, and hepatitis B and C viruses in blood donors in France, 1992–1996 (in French). *Transfus Clin Biol* **5:** 305–312.

Potena L, Valantine HA. 2007. Cytomegalovirus-associated allograft rejection in heart transplant patients. *Curr Opin Infects Dis* **20:** 425–431.

Rubin RH, Wolfson JS, Cosimi AB, Tolkoff-Rubin NE. 1981. Infection in the renal transplant recipient. *Am J Med* **70:** 405–411.

Satoi S, Bramhall SR, Solomon M, Hastings M, Mayer AD, de Goyet JV, Buckels JA, McMaster P, Mirza DF. 2001. The use of liver grafts from donors with bacterial meningitis. *Transplantation* **72:** 1108–1113.

Schreiber GB, Busch MP, Kleinman SH, Korelitz JJ. 1996. The risk of transfusion-transmitted viral infections. The Retrovirus Epidemiology Donor Study. *N Engl J Med* **334:** 1685–1690.

Selin LK, Nahill SR, Welsh RM. 1994. Cross-reactivities in memory cytotoxic T lymphocyte recognition of heterologous viruses. *J Exp Med* **179:** 1933–1943.

Singh N. 2002. Impact of donor bacteremia on outcome in organ transplant recipients. *Liver Transpl* **8:** 975–976.

Small LN, Lau J, Snydman DR. 2006. Preventing post-organ transplantation cytomegalovirus disease with ganciclovir: Meta-analysis comparing prophylactic and preemptive therapies. *Clin Infect Dis* **43:** 869–880.

Spangenberg HC, Viazov S, Kersting N, Neumann-Haefelin C, McKinney D, Roggendorf M, von Weizsacker F, Blum HE, Thimme R. 2005. Intrahepatic $CD8^+$ T-cell failure during chronic hepatitis C virus infection. *Hepatology* **42:** 828–837.

USPH Service. 1994. PHS guideline for preventing transmission of HIV through transplantation of human tissue and organs. In *Morbidity and mortality weekly report (MMWR)*, pp. 1–17.

USPH Service. 2011. (Draft) Public Health Service guideline for reducing transmission of human immunodeficiency virus (HIV), hepatitis B virus (HBV), and hepatitis C virus (HCV) through solid organ transplantation (ed. DoHaH Services), Docket No. CDC–2011–0011. CDC Federal Register.

van Delden C, Blumberg EA, AST Infectious Diseases Community of Practice. 2009. Multidrug resistant Gram-negative bacteria in solid organ transplant recipients. *Am J Transplant* **9:** S27–S34.

Yao F, Seed C, Farrugia A, Morgan D, Cordner S, Wood D, Zheng MH. 2007. The risk of HIV, HBV, HCV and HTLV infection among musculoskeletal tissue donors in Australia. *Am J Transplant* **7:** 2723–2726.

Yao F, Seed C, Farrugia A, Morgan D, Wood D, Zheng MH. 2008. Comparison of the risk of viral infection between the living and nonliving musculoskeletal tissue donors in Australia. *Transpl Int* **21:** 936–941.

Yewdell JW, Bennink JR. 1999. Immunodominance in major histocompatibility complex class I-restricted T lymphocyte responses. *Annu Rev Immunol* **17:** 51–88.

Cancer in the Transplant Recipient

Jeremy R. Chapman, Angela C. Webster, and Germaine Wong

Centre for Transplant and Renal Research, University of Sydney, Westmead Hospital, Westmead NSW 2145, Australia

Correspondence: Jeremy.chapman@sydney.edu.au

Malignancy has become one of the three major causes of death after transplantation in the past decade and is thus increasingly important in all organ transplant programs. Death from cardiovascular disease and infection are both decreasing in frequency from a combination of screening, prophylaxis, aggressive risk factor management, and interventional therapies. Cancer, on the other hand, is poorly and expensively screened for; risk factors are mostly elusive and/or hard to impact on except for the use of immunosuppression itself; and finally therapeutic approaches to the transplant recipient with cancer are often nihilistic. This article provides a review of each of the issues as they come to affect transplantation: cancer before wait-listing, cancer transmission from the donor, cancer after transplantation, outcomes of transplant recipients after a diagnosis of cancer, and the role of screening and therapy in reducing the impact of cancer in transplant recipients.

Malignancy has become one of the three major causes of death after transplantation in the past decade and is thus increasingly important in all organ transplant programs. Death from cardiovascular disease and infection are both decreasing in frequency from a combination of screening, prophylaxis, aggressive risk factor management, and interventional therapies. Cancer, on the other hand, is poorly and expensively screened for; risk factors are mostly elusive and/or hard to impact on except for the use of immunosuppression itself; and finally therapeutic approaches to the transplant recipient with cancer are often nihilistic. Here we review the issues as they come to affect transplantation: cancer before wait-listing, cancer transmission from the donor, cancer after transplantation, outcomes of transplant recipients after a diagnosis of cancer, and the role of screening and therapy in reducing the impact of cancer in transplant recipients.

THE POTENTIAL TRANSPLANT RECIPIENT WITH A PREEXISTING DIAGNOSIS OF CANCER

With the exception of a number of patients accepted for liver transplantation because they have a diagnosis of liver cancer (who will not be further considered in this work), the goal of most pretransplant assessment programs is to avoid transplantation of the patient who has had a cancer or who has an occult primary or secondary cancer. The two reasons for avoiding such patients are

(1) because transplantation does not improve and may reduce the patients' prognosis, and

Cite this article as *Cold Spring Harb Perspect Med* doi: 10.1101/cshperspect.a015677

(2) to avoid placing scarce donated organs into recipients with a limited prognosis.

Patients with chronic kidney, liver, or lung disease are at an increased risk of having had a primary cancer. The relative risk of a patient with chronic kidney disease developing cancer is increased with a glomerular filtration rate (GFR) of <50 mL/min, men having a 29% increase in cancer risk for every 10 mL reduction in GFR (Wong et al. 2009a). The incidence of cancer is also increased in patients who are commenced on dialysis therapy. A proportion of that excess risk of cancer is not because chronic kidney disease increases cancer risk but because cancer increases or directly causes end-stage kidney failure. Examples are multiple myeloma and renal cell cancers, both of which may lead to kidney failure or in the case of renal cell cancer, to bilateral nephrectomy. It is thus important to remove such end-stage renal failure-associated cancers and examine the risk of the remaining cancers. This has been undertaken in a number of national population-based studies in which the determination of kidney disease status is taken from the dialysis and transplant registries and the cancer data from the national or regional cancer databases, with the two databases then linked at an individual patient level. The standardized incidence ratio (SIR) is a ratio of the number of cancers seen in the study population compared with an age and sex match general population and is the best measure of increased risk because of the strong sex and age relationships for most cancers. An Australian analysis showed that the overall increased risk of a number of cancers in the Australian dialysis population (SIR 1.35), with specific cancers increased substantially such as Kaposi's sarcoma (SIR 19.6), lip cancer (SIR 1.87), and of course, the renal failure-associated cancers (Vajdic et al. 2006). Table 1 gives a list of the SIRs for different cancers in the predialysis, dialysis, and transplant populations in Australia. It is thus important to consider the risk profile of every patient being assessed for transplantation with particular attention being paid to the past medical history of cancer and any signs, symptoms, or tests that may suggest cancer, such as iron deficiency anemia, a breast lump, or past history of multiple skin cancers.

The main problem that transplant units must contend with is the patient with a definite history of cancer at some time before consideration for wait-listing for transplantation. Will the cancer return or metastasize and will immunosuppression increase that chance? These are the fundamental questions that must be considered for each individual patient. Each cancer is different and the stage of the cancer and type of treatment given, as well as the time from the diagnosis to the proposed transplantation need consideration. There are no certain answers to these questions, even when one knows the particular circumstances for an individual recipient, yet each set of transplant assessment clinical practice guidelines provides a clear answer for the most common cancers (Table 2) (Knoll et al. 2005; Batabyal et al. 2012). The word "guideline" must be respected carefully because these are merely guidelines to the normal behavior of a particular cancer type and the expected outcomes under immunosuppression. Cancers that are uninfluenced by immunosuppression can perhaps be considered differently to those that are heavily impacted, such as Kaposi's sarcoma or lymphomas, but there is no substitute for the potential recipient's oncologist defining the chance of recurrence or metastasis.

THE DONOR WITH PREEXISTING CANCER

There are sporadic reports of cancer being transmitted through organ donation in the 1970s, with the issue being championed by Penn (1993) and his colleagues through voluntary data collection and reporting. The rate at which different cancers were transmitted from donor to recipient gave rise to a donor selection strategy, which specifically included deceased organ donors with, for example, cerebral malignancy, but excluded all donors with a history of malignant cancer capable of metastasis. The experience of the last two decades has been driven, on the one hand, by selective risk taking, and on the other, by risk avoidance. The risk taking has

 Cite this article as *Cold Spring Harb Perspect Med* doi: 10.1101/cshperspect.a015677

Table 1. Risk of cancer in Australians predialysis, during dialysis, and after kidney transplant

	Up to 5 years before RRT		During dialysis		After transplantation	
Cancer site (ICD code[s])	SIR	95% CI	SIR	95% CI	SIR	95% CI
Lip (C00)	1.87	1.17–2.83	3.68	2.46–5.28	47.08	41.75–52.89
Tongue (C01–C02)	0.53	0.06–1.93	3.28	1.69–5.72	7.17	4.38–11.07
Mouth (C03–C06)	1.34	0.43–3.13	2.15	0.98–4.08	4.58	2.51–7.69
Salivary gland (C07–C08)	2.11	0.57–5.40	1.2	0.15–4.34	7.71	3.33–12.20
Esophagus (C15)	1.05	0.28–2.68	1.68	0.96–2.74	3.82	2.26–6.03
Stomach (C16)	0.81	0.35–1.60	1.52	1.01–2.19	1.84	1.07–2.94
Small intestine (C17)	1.25	0.15–4.53	3.06	1.12–6.67	1.73	0.21–6.25
Colon (C18)	1.33	1.06–1.65	1.18	0.93–1.47	2.36	1.87–2.92
Rectum (C19–C20)	1.33	0.98–1.77	1.02	0.72–1.40	0.63	0.33–1.07
Anus (C21)	0.33	0.07–0.96	0.23	0.03–0.82	2.76	1.51–4.64
Liver (C22)	2.87	0.78–7.34	2.25	1.23–3.77	3.19	1.53–5.87
Gallbladder (C23–C24)	0	-	1.55	0.67–3.05	4.34	2.16–7.76
Pancreas (C25)	2.16	0.87–4.45	1.17	0.69–1.85	1.21	0.56–2.30
Larynx (C32)	0.96	0.42–1.90	1.02	0.41–2.11	2.1	0.96–3.98
Trachea; bronchus and lung (C33–C34)	1.07	0.74–1.49	1.59	1.33–1.88	2.45	2.00–2.97
Melanoma (C43)	1.02	0.81–1.27	1.06	0.81–1.38	2.53	2.08–3.05
Mesothelioma (C45)	0.61	0.02–3.37	1.73	0.75–3.40	1.32	0.27–3.85
Kaposi's sarcoma (C46) (Nalesnik et al. 2011)	19.64	4.05–57.40	57.88	21.24–125.98	207.9	113.66–348.82
Connective and other soft tissue (C47–C49)	0.49	0.06–1.78	1.26	0.41–2.93	4.13	2.13–7.21
Breast (C50) (incl. males)	0.91	0.71–1.14	1.25	0.99–1.55	1.03	0.78–1.34
Vulva (C51)	1.57	0.19–5.67	1.59	0.19–5.73	24.54	14.55–38.79
Cervix uteri (C53)	1.6	0.80–2.86	2.58	1.38–4.42	2.49	1.33–4.27
Corpus uteri (C54)	1.53	0.92–2.40	1.07	0.53–1.91	1.74	0.92–2.97
Ovary (C56)	0.78	0.25–1.82	1	0.43–1.98	1.15	0.46–2.38
Penis (C60)	1.29	0.03–7.16	4.72	0.97–13.80	15.94	5.85–34.69
Prostate (C61)	1.16	0.98–1.36	0.66	0.52–0.83	0.95	0.68–1.29
Testis (C62)	2.1	0.77–4.57	0.71	0.02–3.94	1.25	0.34–3.20
Eye (C69)	2.1	0.68–4.91	1.22	0.15–4.39	7.57	3.46–14.36
Brain (C71)	0.19	0.00–1.07	1.1	0.59–2.05	0.57	0.16–1.46
Thyroid (C73)	2.57	1.44–4.24	9.23	6.53–12.67	6.9	4.69–9.79
Hodgkin disease (C81)	1.28	0.26–3.75	2.56	0.70–6.54	3.75	1.51–7.73
Non-Hodgkin lymphoma (C82–C85)	1.51	1.05–2.10	1.36	0.94–1.90	9.86	8.37–11.54
Leukemia (91–95)	0.89	0.51–1.44	1.14	0.74–1.77	2.46	1.65–3.67

Data adapted from Vadjic et al. 2006.

CI, confidence interval; RRT, renal replacement therapy; SIR, standardized incidence ratio.

been through active selection of donors with a known previous medical history of cancers that are likely to be cured; for example, a donor with a resected early-stage colon cancer several years before the proposed donation. The risk aversion, on the other hand, has been to separate the various cerebral malignancies into low and high risk and avoid donors with high-grade gliomas or with prior surgical interventions that increase the risk of extracerebral metastasis.

Deceased Donors with a History of Cancer

The major problem with an evidence-based strategy for avoiding the risk of cancer transmission from organ donation has been first that

Table 2. Guidelines for the minimum time interval between diagnosis and treatment of a cancer and the transplantation

Type	Stage	Guideline: AST	CARI	B&D	CST	EBPG	MMOH
Renal cell carcinoma	Small or discovered incidentally	○	○	○			○
	Symptomatic	◌	◉	◌		◌	
	Large or invasive	◉		◌			⊘
Bladder cancer	In situ or noninvasive papilloma	○	○	○	○	◌	○
	Invasive	◌	◌	◌	◌		◌
Breast cancer	Stage 0–2 (including early stage)	◌	⊘	⊘	◌	◉	⊘
	Stage 3–4 (advanced/invasive)	◉	●	⊘	●	◉	⊘
Colorectal cancer	Duke A or B1	◌	◌		⊘	◌	◌
	Duke C		◉			◌	◌
	Duke D		●			◌	◌
	Patients with a history of colorectal cancer	◉			◉		
Uterine cancer	Cancer of the uterine body	◌	◌			◌	◌
	Cervical cancer in situ	⊘	○	○	◌	◌	○
	Invasive cervical cancer		◉	⊘		◉	◉
Prostate cancer	Localized	○	○				
	Invasive	◌	◌	◌		◌	◌
Melanoma	In situ	◌	◌	◉	◌		◉
	Invasive	◉	◉	◉	◉	◌	◉
Nonmelanoma skin cancers	Basal cell carcinoma					◌	
	Squamous cell carcinoma						◌
Leukemia		◌			◌		
Lung cancer		◌			◌		
Lymphoma		◌	◌	⊘	◌		◌
Multiple myeloma			●		●		
Testicular cancer		◌	◌				◌
Thyroid cancer		◌	◌				◌
Wilms tumor		◌	◌	◌			◌

Symbol	Recommendation
○	0 years
◌	Minimum 2 years
⊘	2–5 years
◉	Minimum 5 years
●	Contraindicated
	No recommendation (or insufficient evidence)

Analysis adapted from Batabyal et al. (2012).

Early breast cancer was defined as Stages 0–2 and advanced breast cancer as Stages 3–4, according to the NHMRC clinical practice guidelines and International Union against Cancer's TMN classification system. B&D, EBPG, and MMOH guidelines did not define breast cancer stage. EBPG and MMOH guidelines did not define colorectal cancer stage.

AST, American Society of Transplantation (Kasiske et al. 2001); CARI, Caring for Australasians with Renal Impairment (CARI Guidelines 2011); B&D, Bunnapradist and Danovitch (2007); CST, Canadian Society of Nephrology (Knoll et al. 2005); EBPG, European Best Practice Guidelines (EBPG 2000); MMOH, Malaysia Ministry of Health (MMOH 2009).

Table 3. Cancers known to have been transmitted from donor to recipient on at least one occasion

Cancers
Breast cancer
Choriocarcinoma
Colon cancer
Glioblastoma multiforme
Liver cancer
Lung cancer
Lymphoma
Melanoma
Neuroendocrine
Ovarian carcinoma
Pancreas
Prostate
Renal cell carcinoma
Thyroid carcinoma

there will of course never be a randomized study, and second, the incidence of any cancer transmission is so low that sporadic case reports are the main source of information. The publication bias against single case reports means that this source of data is highly likely to underestimate the true incidence, and the low frequency and very variable stage of cancers mean that definitive risk calculations are impossible. A list of cancers that are reported to have been transmitted is shown in Table 3. The common features are cancers of the organ being transplanted—especially, renal cell carcinomas; and cancers with a high propensity for metastasis to the organs being donated.

There have been initiatives designed to ensure reporting of cancers that may have been transmitted by the donated organ in the U.S. and European environments. At a global level there is now a World Health Organization sponsored website (www.notifylibrary.org) to provide for a curated library of reported donor-related adverse events and incidents, including both transmitted infection and malignancy. These efforts have helped define the lexicon—for example, what evidence is required to impute that a cancer in a recipient has been transmitted from the donor, as opposed to developed in the transplanted organ? A renal cell cancer in a transplanted kidney may or may not have been present in the organ at donation. If there is DNA evidence that the tumor is of donor origin and other recipients of organs from the same donor have also suffered from the same renal cell cancer, then one may designate the event as a proven donor-transmitted cancer. If the time between transplantation and donation is short, for example, a matter of weeks, then it is highly probable that the cancer is donor transmitted. On the other hand, if the time delay is several years between the donation and the cancer diagnosis, it is more likely that it has developed in situ and thus donor derived.

There have been a number of approaches to guidance on the acceptance of risk of donor-derived disease including the U.S. Donor Transmitted Assessment Committee (DTAC) and the Council of Europe (Nalesnik et al. 2011). Table 4 shows the DTAC framework for consideration of risk of donor-transmitted disease that allows physicians and recipients to consider the specific issues related to a particular donor. The DTAC has assessed the stratification of different malignancies into the various risk bands and has pro-

Table 4. Framework for classification of the risk of transmission of donor disease, from the U.S. Donor Transmission Advisory Committee (DTAC)

Risk category	Recommendation
No significant risk	Standard
Minimal	Clinical judgment with informed consent
Low	Use in recipients at significant risk without transplant; informed consent required
Intermediate	Use of these donors is generally not recommended. On occasion, a lifesaving transplant may be acceptable in circumstances in which recipient expected survival without transplantation is short (e.g., a few days or less); informed consent required
High	Use of these donors is discouraged except in rare and extreme circumstances; informed consent required
Unknown risk	Use should be based on clinical judgment with informed consent

vided a useful resource for such decision making (Nalesnik et al. 2011).

Living Donor Assessment for Cancer Risk

The first reported case of cancer transmission from a living organ donor was a breast cancer transmitted from a donor wife to recipient husband, but lung, lymphoma, and renal cell cancer have also been recorded (Kauffman et al. 2002). The young age of most living organ donors provides a degree of protection against cancer transmission because of the strong age-associated risk of cancer; however, many of the common cancers such as breast, colon, prostate, cervix, and in Australia also melanoma, are sufficiently prevalent that one case might be expected to be transmitted unknowingly every 5000 living donations. Screening of living donors for these cancers should thus be part of routine assessment.

DE NOVO MALIGNANCY AFTER TRANSPLANTATION

Cancer is a major source of morbidity and mortality following solid organ transplantation. Overall risk of cancer is increased between two- and threefold compared with the general population of the same age and sex. Recipients of solid organ transplants typically experience cancer rates similar to nontransplanted people 20–30 years older, and risk is inversely related to age, with younger recipients experiencing a far greater relative increase in risk compared with older recipients (risk increased by 15–30 times for children, but twofold for those transplanted >65 years) (Webster et al. 2007). Posttransplant cancer risk is increased ~40% for those with a prior cancer, and similarly by white race. Overall, diabetics may experience fewer cancers, but this may be owing to the competing risk of an increased burden of cardiovascular disease (i.e., diabetics may succumb to heart disease before they develop cancer).

However, risk does vary by cancer site, and in applying evidence in clinical practice it is wise to be aware that when thinking about absolute risk, even a small relative risk increase for a common cancer may be more important to consider for some patients than a large relative risk of an uncommon cancer (Table 1). Reasons for the increased risk of most cancers after transplantation are likely owing to the interplay of several factors: the organ transplanted, prior and new exposure to viral infections, total load and duration of immunosuppression, perhaps the specific components of the immunosuppressive regimen, and the recipients accumulated baseline exposures known to increase cancer risk in the general population.

Variation in Cancer Risk According to Transplanted Organ

Table 5 summarizes the published standardized incidence ratios (SIR) for several different cancer sites, stratified by transplanted organ, taken from national population-based studies (Kyllönen et al. 2000; Adami et al. 2003; Vajdic et al. 2006; Villeneuve et al. 2007; Collett et al. 2010; Jiang et al. 2010; Engels et al. 2011; Na et al. 2013). SIR can be interpreted as relative risk, as they estimate risk for organ recipients relative to the cancer incidence experienced by the general population, after allowing for differences in age, sex, and year of diagnosis. For each cancer site, the magnitude of increased risk is largely similar across different countries, and where differences exist, this is largely attributable to era of study and methodology of calculation. However, although the overall risk for all-site cancer is of similar magnitude across organ recipients, the incidence of specific cancers does vary by transplanted organ. In general, kidney cancer risk is greatest for kidney recipients (increased approximately eightfold), and similarly, risk of lung cancer is highest for lung recipients (approximately fivefold risk), and liver cancer for liver recipients (although here, given that liver cancer is an indication for liver transplantation, interpretation is not straightforward). Risk of non-Hodgkins lymphoma, or posttransplant lymphoproliferative disease (PTLD), is elevated by >5 times for all organ recipients, but the magnitude of risk is by far the greatest for lung and heart recipients compared with those who receive other solid organs (Table 5) (Opelz and Döhler 2004).

Table 5. Risk of cancer by organ for some cancer sites, expressed as standardized incidence ratios (SIRs) relative to the general population of each country

Site and comparator national population	SIRs with 95% confidence intervals for different organ transplant recipients			
	Kidney	Liver	Heart	Lung
All site				
Australia	3.3 (3.1–3.7)	2.3 (1.9–2.7)	2.7 (2.4–3.0)	4.3 (3.5–5.2)
Canada	2.5 (2.3–2.7)		2.7 (2.3–3.2)	
Finland	3.3 (2.9–3.8)			
Sweden	3.9 (3.6–4.2)	4.9 (3.7–6.4)		
United Kingdom	2.4 (2.3–2.5)	2.2 (2.0–2.4)	2.5 (2.2–2.7)	3.6 (3.0–4.4)
United States	2.10 (2.06–2.14)			
Nonmelanoma skin cancer				
Finland	39.2 (29.3–51.4)			
Sweden	57.7 (51.0–65.1)	34.0 (17.0–60.6)		
United Kingdom	16.6 (15.9–17.3)	6.6 (5.8–7.5)	18.5 (16.9–20.3)	16.1 (13.1–19.6)
Lip				
Australia	47.1 (41.8–52.9)	14.0 (7.0–25.1)	27.5 (19.0–38.4)	41.9 (22.3–71.6)
Canada	31.3 (23.5–40.8)			
Finland	22.9 (12.6–38.6)			
Sweden	54.8 (39.0–74.9)	24.8 (0.6–138.6)		
United Kingdom	65.6 (49.9–84.6)	20.0 (5.4–51.2)	60.0 (31.0–104.8)	
United States				
Non-Hodgkins lymphoma				
Australia	9.9 (8.4–11.5)	5.6 (3.6–8.3)	7.0 (5.0–9.5)	16.8 (11.1–24.4)
Canada	8.8 (7.4–10.5)		22.7 (17.3–29.3)	
Sweden	3.8 (2.5–5.6)	37.3 (22.1–59.1)		
United Kingdom	28.3 (11.2–13.8)	13.3 (10.6, 16.6)	19.8 (16.1, 24.1)	30.0 (20.6, 42.1)
United States	7.54 (7.2–7.9)			
Colorectal				
Australia	2.4 (1.9–2.9)	2.4 (1.5–3.7)	1.0 (0.5–1.6)	2.6 (1.1–5.1)
Canada	1.4 (1.0–1.8)		0.6 (0.2–1.5)	
Finland	3.9 (2.1–6.7)			
Sweden—colon only	2.4 (1.5–3.5)			
United Kingdom	1.8 (1.2–2.1)	2.3 (1.7–3.0)	1.1 (0.7–1.7)	1.1 (0.3–2.9)
United States	1.24 (1.2–1.3)			
Lung				
Australia	2.5 (2.0–3.0)	0.5 (0.1–1.4)	2.2 (1.4–3.2)	3.8 (1.7–3.5)
Canada	2.1 (1.7–2.5)		2.0 (1.2–3.0)	
Sweden	1.7 (1.1–2.5)			
United Kingdom	1.4 (1.2–1.6)	1.6 (1.2–2.2)	2.1 (1.6–2.8)	5.9 (3.7–8.8)
United States	2.0 (1.9–2.1)			
Breast (female)				
Australia	1.0 (0.8–1.4)	1.3 (0.6–2.2)	0.5 (0.1–1.9)	0.8 (0.2–2.2)
Canada	1.3 (1.0–1.7)		1.1 (0.2–3.2)	
Sweden	1.0 (0.6–1.5)			
United Kingdom	1.0 (0.8–1.2)	0.8 (0.5–1.1)	0.8 (0.3–1.7)	0.3 (0.0–1.2)
United States	0.9 (0.8–0.9)			
Prostate				
Australia	1.0 (0.7–1.3)	0.6 (0.3–1.2)	1.1 (0.7–1.6)	0.7 (0.2–2.1)
Canada	0.9 (0.6–1.3)		1.3 (0.7–2.2)	
Sweden	1.1 (0.7–1.7)			
United Kingdom	1.1 (0.9–1.4)			
United States	0.9 (0.9–1.0)			

Australia (Vajdic et al. 2006; Na et al. 2013); Canada (Villeneuve et al. 2007; Jiang et al. 2010); Finland (Kyllönen et al. 2000); Sweden (Adami et al. 2003); United Kingdom (Collett et al. 2010); United States (Engels et al. 2011).

Variation in Risk by Cancer Site

The magnitude of cancer risk varies by cancer site. Registry data from around the world has established that the pattern of increase in risk for cancer at different sites is seen consistently (Kyllönen et al. 2000; Adami et al. 2003; Vajdic et al. 2006; Villeneuve et al. 2007; Collett et al. 2010; Jiang et al. 2010; Engels et al. 2011; Na et al. 2013). Cancers that are very greatly increased in the transplanted populations are nonmelanoma skin cancer and lip cancer (>10-fold increased risk), Kaposi's sarcoma (>50-fold increased risk), non-Hodgkin lymphoma (greater than eightfold increase), and cancers of the anogenital tract (greater than fourfold increase for vaginal, cervical, vulval, anal, and penile cancers). Other cancers have only a moderately increased relative risk in transplant recipients, but because they occur moderately frequently in the general population, the result of a small relative increase is clinically important. Examples are colorectal cancer (~40% increase), melanoma (approximately double risk), and lung cancer (>50% increase). Cancer at more than 20 other sites also show significant increased risk, including head and neck, thyroid, esophagus, stomach, leukemias, and plasma cell tumors. Notable cancers that do not show an increase in risk for organ transplant recipients are breast cancer and prostate cancer.

Skin cancers, specifically basal cell and squamous cell skin cancers, are common after transplantation, and incidence increases with time. Australian estimates in kidney recipients found 50% experienced at least one skin cancer by 10 years and 80% by 20 years after transplantation (Ramsay et al 2002). Accurate estimates of risk relative to the general population are challenging to generate, and most national cancer registries do not record skin cancers, thus understanding incidence in the general population is not straightforward. However, best estimates are that organ recipients risk of basal cell cancer is 10 fold, and squamous cell up to 100-fold; squamous cell cancers exceed basal cell cancers by 4:1, a reversal of the situation in the general population. This excess risk is modified by skin pigmentation, such that black recipients have much reduced risk compared with white recipients.

The Role of Viral Infection in Carcinogenesis after Transplantation

It is widely recognized that virus infection is implicated in a number of cancer sites in the general population, and also through study in patients with acquired immune dysfunction such as those with HIV and AIDS. Transplant patients are vulnerable to viral infection or reactivation of latent infection. Viruses implicated in carcinogenesis include Epstein-Barr virus (EBV), human herpes virus 8 (HHV-8), human papillomavirus (HPV), the Merkel cell polyomavirus, hepatitis B, and hepatitis C. The rapidity with which some malignancies develop after transplantation also is consistent with the concept that viral oncogenesis is involved because initiation of immunosuppression may promote unchecked viral replication. There is also evidence in kidney recipients that after transplant failure and reduction or cessation of immunosuppression, the risk of virus-related cancers decreases back to levels seen in pretransplant dialysis patients (Van Leeuwen et al. 2010).

EBV is implicated in the development of Hodgkins disease, non-Hodgkins lymphoma, and other manifestations of PTLD, as well as nasopharyngeal cancers and leiomyosarcomas. EBV is common in the general population, and in most parts of the world 90% of adults will show serological evidence of exposure. The classification of PTLD has changed over time as understanding has evolved. Current classification is the 2008 WHO system (Swerdlow et al. 2008). Lymphoma is among the most frequent cancers experienced by transplant recipients, and for the majority (but not all), EBV can be shown in the tumor tissue. Approximately 80% of all PTLD arises from B-cell proliferation induced by EBV. The remaining 20% of PTLD arises predominantly from T cells, about one-third of which have implicated EBV in causation. The majority of PTLD arise within two years of transplantation, with a further later peak of incidence beyond five years. Principal risk factors for PTLD include negative EBV serostatus at the time of transplantation (which means children are particularly vulnerable), and the degree of immunosuppression, particularly anti-T-cell agents, which impair EBV-specific T-cell-mediated immune surveillance. PTLD may arise in donor tissue, and it is not uncommon for PTLD to arise in proximity to the transplanted organ. Recipients of kidney transplants are more likely to get renal lymphoma than recipients of heart transplants; similarly heart transplant recipients more commonly get cardiac or thoracic PTLD than kidney recipients (Opelz and Döhler 2004).

HHV8 is associated with Kaposi's sarcoma, as well as primary effusion lymphoma, and has been implicated in multiple myeloma. HHV8 infection prevalence shows geographical variation, being more common in Mediterranean and African regions where 30%–50% of the general population shows serological evidence of infection, whereas prevalence is 20% in Southeast Asia and Northern Europe. In the immunocompetent, Kaposi's sarcoma is an indolent cutaneous disease that rarely disseminates; however, under immunosuppression disease is more aggressive, more frequently multicentric, and visceral involvement is more common. Visceral involvement is more likely in heart and lung recipients (up to 50% of cases) compared with kidney recipients. HHV8 is present in tumor tissue and there is usually serological evidence of infection. The main risk factor for posttransplant Kaposi's sarcoma is origin from an area of high seroprevalence, but there is also a male:female ratio of 3:1. Tumor cells may also be donor derived, and HHV8 infection may be transmitted from donor to recipient. Initial treatment of Kaposi's sarcoma is reduction of immunosuppression, which usually prompts regression.

HPV is associated with anogenital cancers including cervical, vaginal, vulval, penile, and anal cancers, head and neck cancer, and is implicated in squamous cell skin cancers. Cancer risk at all of these sites is increased in organ transplant recipients. HPV has multiple genotypes, and different subsets are associated with different cancers. The role of HPV is supported by HPV DNA being found in cancers and precancers at these sites. Anogenital cancers often occur together (for example, cervical and vulval, or penile and anal), and may manifest as extensive or multiple lesions. Transmission of HPV is by close personal contact, including sexual contact for anogenital infection, and is commonly asymptomatic.

Merkel cell carcinoma is increased in transplant recipients and people with HIV but is very rare in the general population. It is an aggressive, predominantly intradermal, neuroendocrine malignancy, known for local recurrence and lymph node metastases. Merkel cell polyomavirus can be detected in >80% of tumors.

Chronic hepatitis B infection is associated with development of hepatocellular carcinoma, risk is exacerbated by coinfection with hepatitis C, and increased further posttransplantation as immunosuppression promotes viral replication.

THE ROLE OF COMPONENTS OF THE IMMUNOSUPPRESSIVE REGIMEN

Overall cancer risk is moderated by duration and intensity of immunosuppression, rather than individual components of the drug regimen. This view is given weight by equivocal and contradictory findings in a range of studies of risk factors for cancer after transplantation, which have failed to consistently show association of overall increased cancer risk with any specific drug. Indirect evidence for this view is that some studies have shown that episodes of acute rejection within the first year posttransplant confer increased risk of subsequent malignancy. Those recipients experiencing acute rejection are treated by pulses of increased immunosuppression, thus increasing their immunosuppressive burden overall. It is possible that small differences of effect do exist among drugs, but that these are outweighed by the far greater effects of other known risk factors for cancer, such as age, sex, history of smoking, underlying disease leading to transplantation, and history of prior cancers (Webster et al. 2007; Gallagher et al. 2010). Given the large effect of these known risk factors and the need to balance long-term cancer risk against recipient and graft survival, any small differences among drugs are likely to be clinically unimportant.

In vitro studies suggest calcineurin inhibitors cyclosporine and tacrolimus promote carcinogenesis, potentially through production of cytokines that regulate tumor growth (such as transforming growth factor-β), metastasis, and angiogenesis. However, differences among recipients taking cyclosporine versus those that are not has not resulted in differences in cancer in long-term trials (Gallagher et al. 2010).

Azathioprine acts on DNA and RNA mechanisms, inhibiting repair of splicing and ultimately disrupting lymphocyte proliferation. When used as a single agent to treat autoimmune diseases, azathioprine is associated with an increased risk of lymphomas and an increased risk of a wide range of solid neoplasms, including squamous cell carcinomas, urinary bladder tumors, breast carcinomas, and brain tumors. However, in organ transplant recipients, comparing immunosuppression regimens with and without Azathipoprine has not resulted in differences in cancer incidence (Gallagher et al. 2010). Mycophenolate was originally developed as an anticancer drug, and acts through blockage of the purine biosynthesis, inhibiting inosine monophosphate dehydrogenase. In transplant registry studies mycophenolate-containing regimens have not shown differences in cancer rates compared with mycophenolate-free regimens, but do suggest lower rates of acute rejection.

Mammalian target of rapamycin inhibitors (MTORi) sirolimus and everolimus act by blocking T and B lymphocytes proliferation by preventing activation of the mTOR, which then halts the progression from the G1 to the S phase of the cell cycle. In addition, mTORIs also inhibit the translation of transcription factors resulting in reduced angiogenesis, preventing the multiprotein complexes (mTORC1 and mTORC2) pathway activation, thereby halting cell proliferation, particularly in the setting of cancer development (Guba et al. 2004). MTORi have antineoplastic properties, and show promise in reducing cancer recurrence while permitting ongoing immunosuppression (Law 2005). However, it is still not clear whether MTORi-containing regimens have any benefit in reducing de novo cancer risk.

For some specific cancers, lymphocyte-depleting antibodies such as antithymocyte globulin preparations may be implicated in causality. These agents appear to consistently increase risk of early EBV-driven PTLD (Opelz and Döhler 2004).

SURVIVAL AND OUTCOMES IN TRANSPLANT RECIPIENTS WITH CANCER

Not only does the risk of developing de novo cancer increase after kidney transplantation, the prognoses of recipients diagnosed with cancer is much worse than patients with transplants or cancer alone. Data from the Australian and New Zealand Dialysis and Transplant Registry have shown that transplanted women with breast cancer have an excess mortality of at least 40% compared with women with breast cancers in the general population. For men with colorectal cancers and kidney transplant, the overall 5-year survival is 27% compared with 75% for those in the general population with cancer but without transplants (Webster and Wong 2008).

In a Dutch kidney transplant population, the median patient survival after the diagnoses of cancer was only 2.7 years, compared with an average survival of recipients without cancer of 8.3 years ($p < 0.0001$) (van de Wetering et al. 2010). Cancers developed in transplant recipients were often more aggressive and developed at a much later stage than patients without transplants. Recent data from the Israel Penn Registry showed that the stage-specific survival for certain cancer types such as colon, lung, breast, prostate, and bladder cancers was significantly lower in patients with transplants compared with those in the general population (Miao et al. 2009).

On the contrary, using the United States Renal Data System (USRDS) cohort, a higher standardized cancer mortality ratio was observed only in younger recipients without a competing cardiovascular risk factor such as diabetes. Among older transplant recipients with diabetes, heart disease, and prior stroke, the overall standard mortality ratios for cancer are much lower than that in the age and gender matched general population. The interpretation is that the potential competing risk of death from cardiovascular disease dampens the effect of immunosuppression on cancer risk in older transplant recipients (Kiberd et al. 2009).

Reasons for the poorer cancer outcomes in kidney transplant recipients are unclear. Previous studies have shown that patients with chronic kidney disease are less likely to receive cancer screening because of the perceived reduced survival benefits compared with people without kidney disease, leading to the development of aggressive, more advanced stage disease at the time of cancer diagnoses. Transplant clinicians may also be reluctant to instigate intensive chemotherapeutic treatment and surgical intervention because of coexisting comorbidities such as cardiovascular disease. The fear of rejection and potential graft loss may also prevent the introduction of immunotherapy and withdrawal/reduction in immunosuppression.

INTERVENTIONS TO IMPROVE CANCER OUTCOMES AND SURVIVAL

In view of the higher cancer incidence and poorer prognoses, prevention and screening may play an important role in reducing the burden of cancer in kidney transplant recipients. Routine cancer screening is recommended for all transplant individuals. Recommendations for cancer screening in transplant recipients are mostly extrapolated from the general populations and are consistent with screening guidelines in the general population, with the exception of cervical, skin, colorectal, and renal cancers.

Screening for Cervical Cancer

The overall incidence of cervical cancers in women who received a kidney transplant is at least two to three times greater than the age and gender matched population. Despite the lack of evidence from screening trials in transplant recipients, current recommendations suggested more frequent cytologic screening (annual instead of biannual) because of the belief that precancerous lesions may progress more rapidly under the influence of immunosuppression. Apart from clinical effectiveness, recent modeled analyses reported that routine annual screening is cost-effective in women with kidney transplants, with an incremental cost-effectiveness ratio (ICER) in the order of $12,000 per life year saved and total gains of 0.05 life years per women screened. On the contrary, implementing HPV vaccination in HPV naïve women under the existing screening program is expensive, and may add only modest benefits. A program of HPV vaccination and screening would have to vaccinate a total of 1000 HPV naïve women before transplantation, to save one extra death from cervical cancers over a screening period of 50 years (Wong et al. 2009b). Although the bivalent and quadrivalent HPV vaccines appear promising in reducing the incidence of HPV-16 and -18 cervical dysplasia among the general population, there is a lack of trial data showing the immunogenicity and effectiveness of HPV vaccination in immunosuppressed individuals.

Screening for Breast Cancer

Biennial mammographic screening for breast cancer is standard practice in the general population. The American Transplant Society (AST) and the European Best Practice Guidelines (EBPG) recommend breast cancer screening in all female transplant recipients between 50 and 69 years of age (Kasiske et al. 2000, 2001; European Best Practice Guidelines for Renal Transplantation 2002). For women between 40 and 49 years of age, transplanted women could still undergo screening annually or biennially. However, evidence for or against screening in this group of transplanted women remains unclear. Transplanted women undergoing screening mammography should also be aware of the potential false-positive findings, particularly in women who developed large, dense, and multiple benign breast adenomas from long-term cyclosporine use. False-positive findings will lead to unnecessary and exhaustive diagnostic procedures such as fine-needle and core biopsies of the breasts.

Screening for Bowel Cancer

There is now consistent evidence showing an increased risk of colorectal cancer by at least two- to threefold among those with renal allografts. However, recommendations for screening bowel cancer are far from being standardized across the various different transplant practice guidelines groups. The AST recommends annual fecal occult blood testing (FOBT) and flexible sigmoidoscopy every 5 years in the United States (Kasiske et al. 2000, 2001). In Australia, biennial screening using the immunochemical FOBT is the recommended screening tool by the National Health and Medical Research Council (NHMRC). In Europe, the EBPG suggested annual screening for all transplant recipients using iFOBTs. All positive screening tests need to be followed by diagnostic colonoscopies (European Best Practice Guidelines for Renal Transplantation 2002).

Although there is to date, no trial-based evidence for cancer screening in this at-risk population, early cancer detection is effective. Previously modeled analyses have shown that colorectal cancer screening is probably effective in transplanted patients, but uncertainties exist; in particular, the costs, the test specificity, and the patient preferences for screening in patients with comorbidities (Wong et al 2008). A recent diagnostic test accuracy study conducted in a South Australian cohort of transplanted recipients suggested that the test sensitivity of immunochemical FOBT may be low at 36% compared with the expected test performance characteristics in the nontransplanted populations (Collins et al. 2012).

Screening for Skin Cancers

Skin cancer is the most common form of cancer in transplanted patients. Skin cancer prevention with sun-protective behaviors such as using sunscreen (SPF 15+), sun hats, avoidance of exposure to ultraviolet radiation during sun-peak hours, and covering up with pants and long-sleeve tops are effective measures to reduce the incidence of skin cancers and should be encouraged. Transplant follow-up combined with regular skin surveillance by experienced dermatologists should be advocated for all transplant units. Prophylactic treatment with retinoid acitretin or low-dose capecitabine in the secondary prevention of skin cancers appears to be efficacious in reducing the incidence rates of squamous cell and basal cell carcinoma in solid organ transplant recipients with manageable toxicity profile, but the use of these agents remains heterogenous.

Screening for Renal Cancers

Screening for urinary tract cancers using ultrasonography may be useful for high-risk patients such as those with a history of analgesic use or acquired cystic disease of the kidneys, but may not be cost-effective for all recipients of kidney transplants. A major concern associated with ultrasonographic screening is the test performance characteristics of the screening tool. The accuracy of ultrasonography is an important determinant of screening efficiency, but is uncertain in recipients of kidney transplants. Ultrasonography is not only operator dependent but performance varies with the size and morphology of the patient, the kidneys, and the tumor. The difficulties associated with ultrasonographic screening in people with ESKD include the effect of multicystic diseases and small scarred native kidneys on the overall test accuracy and poor reliability in differentiating small hyperechoic renal cancers from lesions such as adenomas and angiomyolipomas (Bunnapradist and Vincenti 2009; Wong et al. 2011).

MANAGEMENT OF CANCER IN KIDNEY TRANSPLANT RECIPIENTS

Surgical resection and radiotherapy remain the preferred treatment for most early-stage and locally invasive tumors among transplanted patients. Although there is a lack of trial-based evidence, judicious reduction in immunosuppression with regular monitoring of disease progression and graft function may be warranted, particularly among those with high-grade and advanced disease.

 Cite this article as *Cold Spring Harb Perspect Med* doi: 10.1101/cshperspect.a015677

Over the past decades, mammalian target of rapamycin inhibitors (mTORI) have been proposed to have a dual role in transplantation: immunosuppression and antioncogenic activities for transplanted patients. From the clinical perspective, retrospective analyses have shown reduced incidence of de novo cancer among those who received mTORIs as immunosuppression compared with recipients on calcineurin inhibitors (CNIs). Post hoc analyses of immunosuppression trials such as the CONVERT have shown a reduction in the incidence of skin cancer in the sirolimus cohort compared with those remaining on CNIs (Bunnapradist and Vincenti 2009; Alberú et al. 2011).

More recently, two randomized controlled trials have been conducted in kidney transplant recipients designed specifically to evaluate the effectiveness of mTORI as immunosuppressive agents in the context of malignancy. In the Australian study, a total of 96 kidney transplant recipients were randomized to receive maintenance CNIs or conversion to mTORIs. With a median follow-up period of 3.6 years, the overall incidence of new squamous cell skin cancer (SCC) among patients with a prior history of skin cancer was significantly lower than those on CNIs (0.88 versus 1.71 per-patient year, $p = 0.038$) (Campbell et al. 2012). A similar study, conducted in Lyons, France, randomized 120 high-risk transplanted patients with a prior history of SCCs reported a lower proportion of transplant recipients developed new SCCs in the mTORI group compared with those in the CNIs group. (47.6% vs. 70.5%, $p = 0.048$). However, poor tolerability of mTORI remains a major concern (Euvrard et al. 2012). More than 35% of transplant recipients discontinued mTORI treatment owing to significant side effects such as ankle swelling, acne, pneumonitis, and proteinuria, and infective complications have prevented the longer-term use of mTORI in these high-risk groups.

Therapeutic measures (both preventive and therapeutic) for melanomas are less well studied. Other novel treatment strategies for metastatic melanoma such as the antiangiogenic and immunomodulatory drugs, the proteasome inhibitors, and the specific targeted molecular therapies have been implicated in the general population (Simeone and Ascerto 2012). However, the efficacy and safety of using these newer agents are unclear and unproven in the transplant population.

mTORI have also been shown to be effective in achieving successful clinical and histological remission of Kaposi's sarcoma. Previous studies have shown that phase II and III trials have shown antitumor activity and survival advantage in patients with metastatic renal cancers treated with mTORI (Rao et al. 2004; Baldo et al. 2008; Amato et al. 2009; Wang et al. 2011). In the transplant population, the clinical effectiveness of mTORI for the prevention and treatment of renal cell carcinomas remained unclear.

REFERENCES

Adami J, Gabel H, Lindelof B, Ekstrom K, Rydh B, Glimelius B, Ekbom A, Adami HO, Granath F. 2003. Cancer risk following organ transplantation: A nationwide cohort study in Sweden. *Br J Cancer* **89:** 1221–1227.

Alberú J, Pascoe MD, Campistol JM, Schena FP, Rial Mdel C, Polinsky M, Neylan JF, Korth-Bradley J, Godberg-Alberts R, Maller ES, et al. 2011. Lower malignancy rates in renal allograft recipients converted to sirolimus-based, calcineurin inhibitor-free immunotherapy: 24-Month results from the CONVERT trial. *Transplantation* **92:** 303–310.

Amato RJ, Jac J, Giessinger S, Saxena S, Willis JP. 2009. A phase 2 study with a daily regimen of the oral mTOR inhibitor RAD001 (everolimus) in patients with metastatic clear cell renal cell cancer. *Cancer* **115:** 2438–2446.

Baldo P, Cecco S, Giacomin E, Lazzarini R, Ros B, Marastoni S. 2008. mTOR pathway and mTOR inhibitors as agents for cancer therapy. *Curr Cancer Drug Targets* **8:** 647–665.

Batabyal P, Chapman JR, Wong G, Craig JC, Tong A. 2012. Clinical practice guidelines on wait-listing for kidney transplantation: Consistent and equitable? *Transplantation* **94:** 703–713.

Bunnapradist S, Danovitch GM. 2007. Evaluation of adult kidney transplant candidates. *Am J Kidney Dis* **50:** 890–898.

Bunnapradist S, Vincenti F. 2009. Transplantation: To convert or not to convert: Lessons from the CONVERT trial. *Nat Rev Nephrol* **5:** 371–373.

Campbell SB, Walker R, Tai SS, Jiang Q, Russ GR. 2012. Randomized controlled trial of sirolimus for renal transplant recipients at high risk for nonmelanoma skin cancer. *Am J Transplant* **12:** 1146–1156.

CARI Guidelines. 2011. Recipient assessment for transplantation. Available from http://www.cari.org.au/trans_recipient_suitability_underdev.php (last accessed February 15, 2012).

Collett D, Mumford L, Banner NR, Neuberger J, Watson C. 2010. Comparison of the incidence of malignancy in recipients of different types of organ: A UK registry audit. *Am J Transplant* **10:** 1889–1896.

Collins MG, Teo E, Cole SR, Chan CY, McDonald SP, Russ GR, Young GP, Bampton PA, Coates PT. 2012. Screening for colorectal cancer and advanced colorectal neoplasia in kidney transplant recipients: Cross sectional prevalence and diagnostic accuracy study of faecal immunochemical testing for haemoglobin and colonoscopy. *BMJ* **345:** e4657.

Engels EA, Pfeiffer RM, Fraumeni JF Jr, Kasiske BL, Israni AK, Snyder JJ, Wolfe RA, Goodrich NP, Bayakly AR, Clarke CA, et al. 2011. Spectrum of cancer risk among U.S. solid organ transplant recipients: The Transplant Cancer Match Study. *JAMA* **306:** 1891–1901.

European Best Practice Guidelines (EBPG). 2000. Section I: Evaluation, selection and preparation of the potential renal transplant candidate. *Nephrol Dial Transplant* **15:** 3–38.

PG Expert Group on Renal Transplantation. 2002. European best practice guidelines for renal transplantation. Section IV: Long-term management of the transplant recipient. IV.6.3. Cancer risk after renal transplantation. Solid organ cancers: Prevention and treatment. *Nephrol Dial Transplant* **17:** 32, 34–36.

Euvrard S, Morelon E, Rostaing L, Goffin E, Brocard A, Tromme I, Broeders N, del Marmol V, Chatelet V, Dompmarin A, et al. 2012. Sirolimus and secondary skin-cancer prevention in kidney transplantation. *N Engl J Med* **367:** 329–339.

Gallagher MP, Kelly PJ, Jardine M, Perkovic V, Cass A, Craig JC, Eris J, Webster AC. 2010. Long-term cancer risk of immunosuppressive regimens after kidney transplantation. *J Am Soc Nephrol* **21:** 852–858.

Guba M, Graeb C, Jauch KW, Geissler EK. 2004. Pro- and anti-cancer effects of immunosuppressive agents used in organ transplantation. *Transplantation* **77:** 1777–1782.

Jiang Y, Villeneuve PJ, Wielgosz A, Schaubel DE, Fenton SSA, Mao Y. 2010. The incidence of cancer in a population-based cohort of Canadian heart transplant recipients. *Am J Transplant* **10:** 637–645.

Kasiske BL, Vazquez MA, Harmon WE, Brown RS, Danovitch GM, Gaston RS, Roth D, Scandling JD, Singer GG. 2000. Recommendations for the outpatient surveillance of renal transplant recipients. *J Am Soc Nephrol* **11:** S1–S86.

Kasiske BL, Cangro CB, Hariharan S, Hricik DE, Kerman RH, Roth D, Vazquez MA, Weir MR, American Society of Transplantation. 2001. The evaluation of renal transplantation candidates: Clinical practice guidelines. *Am J Transplant* **1:** 3–95.

Kauffman HM, McBride MA, Cherikh WS, Spain PC, Marks WH, Roza AM. 2002. Transplant tumor registry: Donor related malignancies. *Transplantation* **74:** 358–362.

Kiberd BA, Rose C, Gill JS. 2009. Cancer mortality in kidney transplantation. *Am J Transplant* **9:** 1868–1875.

Knoll G, Cockfield S, Blydt-Hansen T, Baran D, Kiberd B, Landsberg D, Rush D, Cole E, Kidney Transplant Working Group of the Canadian Society of Transplantation. 2005. Canadian Society of Transplantation: Consensus guidelines on eligibility for kidney transplantation. *Can Med Assoc J* **173:** S1–S25.

Kyllönen L, Salmela K, Pukkala E. 2000. Cancer incidence in a kidney-transplanted population. *Transpl Int* **13:** S394–S398.

Law BK. 2005. Rapamycin: An anti-cancer immunosuppressant? *Crit Rev Oncol Hematol* **56:** 47–60.

Miao Y, Everly JJ, Gross TG, Tevar AD, First MR, Alloway RR, Woodle ES. 2009. De novo cancers arising in organ transplant recipients are associated with adverse outcomes compared with the general population. *Transplantation* **87:** 1347–1359.

MMOH. 2009. *Renal replacement therapy—Clinical Practice Guidelines*, 3rd ed. Malaysian Ministry of Health, Kuala Lumpur.

Na R, Grulich AE, Meagher NS, McCaughan GW, Keogh AM, Vajdic CM. 2013. Comparison of de novo cancer incidence in Australian liver, heart and lung transplant recipients. *Am J Transplant* **13:** 174–183.

Nalesnik MA, Woodle ES, Dimaio JM, Vasudev B, Teperman LW, Covington S, Taranto S, Gockerman JP, Shapiro R, Sharma V, et al. 2011. Donor-transmitted malignancies in organ transplantation: Assessment of clinical risk. *Am J Transplant* **11:** 1140–1147.

Opelz G, Döhler B. 2004. Lymphomas after solid organ transplantation: A collaborative transplant study report. *Am J Transplant* **4:** 222–230.

Penn I. 1993. The effect of immunosuppression on pre-existing cancers. *Transplantation* **55:** 742–747.

Ramsay HM, Fryer AA, Hawley CM, Smith AG, Harden PN. 2002. Non-melanoma skin cancer risk in the Queensland renal transplant population. *Br J Dermatol* **147:** 950–956.

Rao RD, Buckner JC, Sarkaria JN. 2004. Mammalian target of rapamycin (mTOR) inhibitors as anti-cancer agents. *Curr Cancer Drug Targets* **4:** 621–635.

Simeone E, Ascierto PA. 2012. Immunomodulating antibodies in the treatment of metastatic melanoma: The experience with anti-CTLA-4, anti-CD137, and anti-PD1. *J Immunotoxicol* **9:** 241–247.

Swerdlow SH, Campo E, Harris NL, Jaffe ES, Pileri SA, Stein H, Thiele J, Vardiman JW. 2008. *WHO Classification of tumours of haematopoietic and lymphoid tissues*, 4th ed. International Agency for Research on Cancer, Lyon, France.

Vajdic CM, McDonald SP, McCredie MR, van Leeuwen MT, Stewart JH, Law M, Chapman JR, Webster AC, Kaldor JM, Grulich AE. 2006. Cancer incidence before and after kidney transplantation. *JAMA* **296:** 2823–2831.

van de Wetering J, Roodnat JI, Hemke AC, Hoitsma AJ, Weimar W. 2010. Patient survival after the diagnosis of cancer in renal transplant recipients: A nested case-control study. *Transplantation* **90:** 1542–1546.

Van Leeuwen MT, Webster AC, McCredie MRE, Stewart JH, McDonald SP, Amin J, Kaldor JM, Chapman JR, Vajdic CM, Grulich AE. 2010. Effect of reduced immunosuppression after kidney transplant failure on risk of cancer. Population based retrospective cohort study. *BMJ* **340:** c570.

Villeneuve PJ, Schaubel DE, Fenton SS, Shepherd FA, Jiang Y, Mao Y. 2007. Cancer incidence among Canadian kidney transplant recipients. *Am J Transplant* **7:** 941–948.

Wang Y, Wang XY, Subjeck JR, Shrikant PA, Kim HL. 2011. Temsirolimus, an mTOR inhibitor, enhances anti-tumour effects of heat shock protein cancer vaccines. *Br J Cancer* **104:** 643–652.

Webster AC, Wong G. 2008. *Cancer: ANZDATA registry 2008 report.* Australia and New Zealand Dialysis and Transplant Registry, Adelaide, South Australia.

Webster AC, Craig JC, Simpson JM, Jones MP, Chapman JR. 2007. Identifying high risk groups and quantifying absolute risk of cancer after kidney transplantation: A cohort study of 15,183 recipients. *Am J Transplant* **7:** 2140–2151.

Wong G, Howard K, Craig JC, Chapman JR. 2008. Cost-effectiveness of colorectal cancer screening in renal transplant recipients. *Transplantation* **85:** 532–541.

Wong G, Hayen A, Chapman JR, Webster AC, Wang JJ, Mitchell P, Craig JC. 2009a. Association of CKD and cancer risk in older people. *J Am Soc Nephrol* **20:** 1341–1350.

Wong G, Howard K, Webster A, Chapman JR, Craig JC. 2009b. The health and economic impact of cervical cancer screening and human papillomavirus vaccination in kidney transplant recipients. *Transplantation* **87:** 1078–1091.

Wong G, Howard K, Webster AC, Chapman JR, Craig JC. 2011. Screening for renal cancer in recipients of kidney transplants. *Nephrol Dial Transplant* **26:** 1729–1739.

Bioethics of Organ Transplantation

Arthur Caplan

Division of Medical Ethics, NYU Langone Medical Center, New York, New York 10016

Correspondence: arthur.caplan@nyumc.org

As the ability to transplant organs and tissues has grown, the demand for these procedures has increased as well, to the point at which it far exceeds the available supply creating the core ethical challenge for transplantation—rationing. The gap between supply and demand, although large, is worse than it appears to be. There are two key steps to gaining access to a transplant. First, one must gain access to a transplant center. Then, those waiting need to be selected for a transplant. Many potential recipients do not get admitted to a program. They are deemed too old, not of the right nationality, or not appropriate for transplant as a result of severe mental impairment, criminal history, drug abuse, or simply because they do not have access to a competent primary care physician who can refer them to a transplant program. There are also financial obstacles to access to transplant waiting lists in the United States and other nations. In many poor nations, those needing transplants simply die because there is no capacity or a very limited capacity to perform transplants. Although the demand for organs now exceeds the supply, resulting in rationing, the size of waiting lists would quickly expand were there to suddenly be an equally large expansion in the number of organs available for transplantation. Still, even with the reality of unavoidable rationing, saving more lives by increasing organ supply is a moral good. Current public policies for obtaining organs from cadavers are not adequate in that they do not produce the number of organs that public polls of persons in the United States indicate people are willing to donate.

PRESENT POLICIES REGARDING ORGAN PROCUREMENT ARE INADEQUATE

Why does the supply of organs lag behind demand? There are many reasons for believing that current systems of organ procurement are not as efficient as they could be in obtaining organs from cadaver sources. In the United States, the prevailing public policy is one of "encouraged voluntarism." This policy was established when states began adopting the Uniform Anatomical Gift Act in 1968 (Caplan 1984; DeVita and Caplan 2007). The United States as well as Canada, the United Kingdom, Denmark, Australia, and many other nations rely on altruistic donation reinforced by ongoing public education campaigns. In the case of Israel and Singapore, priority in access to transplants is an additional incentive to be a donor (Chandler 2005; Quigley et al. 2012). Individuals may use donor cards to make their wish to donate all or some organs known after their death. In some parts of the world and many states in the United States, computer registries facilitate organ donation.

Cite this article as *Cold Spring Harb Perspect Med* doi: 10.1101/cshperspect.a015685

Public surveys show support for organ donation in the United States at levels not reflected in actual donor rates (www.mayoclinic.org/news2013-rst/7428.html). Many organ donors are not identified as such at the time of death by health-care personnel. Although families in the United States do not have the legal standing to veto a signed request to donate by a deceased relative, in practice they are almost always accorded that right.

WHY HAS "ENCOURAGED VOLUNTARISM" FALLEN SHORT?

There are many reasons why the policy of encouraged voluntarism has not produced as many organs for transplant as might be expected. Many people still do not carry a donor card or other written directive specifying the disposition of their bodies when they die. Often people do not discuss their wish to donate with their family, partners, or friends. Rates of donation among the poor are low. Computer registries help but they are no substitute for discussions about donation before death. Many people are loath to contemplate their own death, much less make plans for the disposition of their bodies. This means that many people are not going to fill out a donor card just as they are not going to make out a will, buy a funeral plot, or complete an advance directive concerning the kind of medical care they would like to have if they become incompetent (Caplan 1984).

The fact that health care has in recent years become increasingly centralized in large and impersonal institutions undermines trust between patients and health-care providers. It is very difficult to trust in the good intentions of strangers. Some people are afraid that if they carry a donor card they may not receive aggressive medical care if they need it so that others can be given their organs. Still others believe that only the rich can get transplants and so see no reason to act as donors.

The failure to secure higher rates of compliance with respect to written directives and the mistrust of hospitals and health-care providers are, however, only part of the explanation for low rates of organ procurement. The reality in the United States, Canada, Germany, the United Kingdom, and many other nations is that when the subject of donation is raised it is done with the presumption that the family of the deceased must decide to opt into donation. It may be possible to obtain a higher rate of donation simply by changing the philosophy underlying requests to donate from an "opt in" approach to an "opt out" approach (Caplan 2013). Still other nations such as China are willing to execute "prisoners" to obtain a steady supply of organs—a grossly unethical approach to obtaining organs (Wang 2010; Caplan et al. 2012). There is some enthusiasm for using markets to increase the cadaver supply (Davis and Crowe 2009; Hughes 2009; Kerstein 2009; Taylor 2009; Lawlor 2011; Jaycox 2012; Slabbert 2012).

GETTING MORE ORGANS—MARKETS?

Two basic strategies have been proposed to provide incentives for people to sell their organs when they die. One strategy is simply to permit organ sale in the United States by changing the National Organ Transplant Act (NOTA), the 1986 federal law that bans organ sales. Then, individuals would be free to broker contracts with persons interested in selling at prices mutually agreed on by both parties (Lawlor 2011; Slabbert 2012). The other strategy is to create a regulated market in which the government would act as the purchaser of organs—setting a fixed price and enforcing conditions of sale (Matas 2006; Lawlor 2011). Both proposals have drawn heated ethical criticism.

Would markets really work in the United States or other economically advanced nations to increase organ supply? It is hard to imagine many people in wealthy countries eager to sell their organs upon their death who do not now donate them. In fact, even if compensation is relatively high, few will agree to sell. Polls show that the disincentive to cadaver donation has more to do with aesthetic, emotional, or religious concerns than a lack of payment. That has been the experience with markets in human eggs for research purposes and paid surrogacy in the United States with prices escalating through the roof and still relatively few sellers. If there is already a fear of premature pronouncement of death to obtain organs, enacting market incentives for cadaver donation is likely to exacerbate this fear (DeVita and Caplan 2007).

Perhaps living persons can be induced to sell nonvital organs. In fact, the United States now obtains more kidneys for transplant from those who donate a kidney to a friend or a family member than from cadaver sources. However, markets that incentivize living persons have their own ethical problems.

One problem is that only kidneys are likely to be available in a living seller market. Hearts will obviously not be. Lungs and liver involve such risky procurement surgery that they are not likely to attract sellers. Another moral challenge is that only the poor and very desperate will want to sell their body parts. If you need money, you might sell your kidney to try and feed your family or to pay back a debt. This may be a "rational" decision, but that does not make it a matter of free choice. Watching your child go hungry while you lack a job and a wealthy person waves a wad of bills in your face is not exactly a scenario that inspires confidence in the valid choices that the poor would make in a market for body parts. Talk of individual rights and autonomy is hollow if those with no options must "choose" to sell their organs to purchase life's necessities (Caplan and Prior 2009; Kerstein 2009; Taylor 2009; Lawlor 2011; Jaycox 2012). Choice requires options as well as information and some degree of freedom.

Major religious traditions oppose payment on the grounds that persons do not own but are, rather, the stewards of their bodies (Caplan and Prior 2009; Caplan 2013). Payment also introduces risks concerning the quality of organs obtained. And payment may give persons an inducement to kill or take organs from others to pay debts.

Another very important ethical challenge to proposals to permit markets is that selling organs, even in a tightly regulated market, violates the ethics of medicine. The core ethical norm of the medical profession is the principle "do no harm." The only morally defensible way to remove an organ from someone is if the donor chooses to undergo the harm of surgery solely to help another, and if there is sufficient medical benefit to the recipient. The creation of a market puts medicine and nursing in the position of removing body parts from people solely to abet those people's interest in securing compensation (Caplan 2004, 2013). In a market, even a regulated one, doctors and nurses would be using their skills to help people harm themselves for money. The resulting distrust and loss of professional standing may be too high a price to pay for a gamble that a market in living sellers may secure more organs (Tan et al. 2008).

DEFAULT TO DONATION

If public opinion surveys can be trusted, most Americans are willing to serve as a donor when they die. The problem is that "encouraged voluntarism" using donor cards does not seem to be able to fully tap this powerful sentiment. There is another way of modifying existing practice that would respect the dignity and value of individual choice while at the same time holding out the prospect of increasing the number of organs obtained from cadaver sources. It requires a shift in the ethical presumptions that prevail about organ donation.

If public policy were to be modified to a default that presumes that people do want to be donors, this might have a positive impact on donation rates. Some nations, such as France, Austria, Belgium, and Spain, have legislated versions of presumed consent with a consequent positive impact on their cadaver donor rates (Abadie and Gay 2006; Byk 2009; Neades 2009).

A policy of default to donation addresses many of the failings of the current approach to cadaver organ procurement. It allows opting out of cadaver donation while allowing the view of the majority toward donation to set policy.

There are two main worries about shifting to a default to donation policy in the United States, United Kingdom, Canada, and other nations. One is that forcing a choice will lead to many more people opting out than currently do. This has not been seen in nations with default to donation policies. The other is simple resistance to the shift toward donation for fear that objections will not be accurately tracked.

The increasing use of computer registries may help buffer this concern.

TOUGHENING "ENCOURAGED VOLUNTARISM"—CHARITY TO OBLIGATION?

For many years, the rhetoric in educating the public has emphasized charity. People are constantly being urged through public service advertisements in magazines, in newspapers, and on radio and television to "make the gift of life." What is interesting about "gift" to urge organ donation is that the moral force of such language is not particularly strong. Gifts are a form of charity. Most moral theories recognize a strong moral obligation or duty not to harm others. But few theories posit the existence of an obligation or duty to give gifts to other people. Gift-giving behavior is viewed as morally laudable, but it is not often seen as mandatory.

The question raised by using "gift" in this context is whether society really wants to view organ donation in this way. One answer to this question is to examine the ways in which positive obligations, or duties, to aid other people are usually generated. Some duties to help others arise as a result of particular roles or jobs in society. Firefighters, parents, and nurses all have special duties to render positive aid to others even when helping requires sacrifice and even risk. Another way in which duties to help others can arise is through the act of contracting or promising. I can voluntarily promise to help someone else if the need should arise, and can be blamed for my failure to do so under the appropriate circumstances (Caplan and Coehlo 1998).

Neither of these arguments for positive obligations is applicable to the situation with respect to cadaver organ donation. Unless I have promised to donate an organ to another and try to renege on my promise, there would seem to be no ground for saying that I ought to be an organ donor in any sense stronger than pure charity.

There is, however, another way in which duties to help others can go beyond exceptional cases of charity or gift giving. If it is possible for someone to do a great deal of good for another person without facing the prospect of a great deal of risk or even inconvenience, and if there is a strong likelihood of benefit for the recipient, then a duty exists which is stronger than the relatively weak obligations associated with supererogatory acts. If, for example, a strong swimmer can save the life of a drowning child merely by swimming 20 feet out into a calm lake, it would seem morally reprehensible and blameworthy for the swimmer not to do so. Indeed, it would seem odd to describe such a rescue as a charitable act, heroism, or even a gift from the swimmer to the drowning child (Caplan and Coehlo 1998). Is donation after death closer to an obligation of this sort? If it is true that many can be helped by an increase in the supply of organs, and if it is also true that the dead can suffer no harm by using their tissues and organs for transplantation, then is it correct to describe a decision to donate as a gift that is praiseworthy if offered but not blameworthy if withheld or, as an obligation that is both? Unless one holds religious views about the need to try and insure an intact body postdeath, it is hard to see how the ability to help others could overcome a reluctance to donate. Perhaps, shifting the rhetoric of donation to duty rather than gift would increase the supply of cadaver organs (Veatch 2002).

DEALING WITH SCARCITY—EQUITY IN TRANSPLANTATION

The trust that is essential for public support is a product of a key ethical value—equity. The values of altruism and autonomy—the foundations of organ procurement—rest on the presumption that organs, which are given freely, voluntarily and altruistically, will be distributed in a fair and impartial manner to those in need. Any policies, practices, or activities that suggest otherwise imperil the entire enterprise of voluntary and altruistic organ donation. There is sufficient evidence of inequity in the allocation of organs to raise doubts about the fairness of the existing system. If inequity is perceived then trust is imperiled and donation will be adversely impacted.

Cite this article as *Cold Spring Harb Perspect Med* doi: 10.1101/cshperspect.a015685

The ever-present news stories of families desperately seeking funds, begging for money to pay for transplants in many nations, leave an especially bitter taste in the mouths of a public that expects altruism. If organs are requested from the rich and poor alike but, given primarily to the rich, how can we expect the public to support transplantation with either organs?

An impressive number of papers have appeared in the past few years in professional journals maintaining that women, the elderly, the disabled, the retarded, and minorities are not represented in the ranks of those receiving transplants to the extent that they could and should be. Policies that give priority of access to patients on artificial heart assist devices (LVADs), or to those who require multiple organ transplants or retransplantation, leave plenty of room for doubt about what values are being used to allocate scarce organs. The tolerance of multiple listing (Sanaei Ardekani and Orlowski 2010)—going to more than one transplant center to be wait-listed—by those who can afford to do so and thus gain access to a larger share of the organ donor pool, a strategy that Steve Jobs and Lou Reed used to obtain livers, is blatantly unfair. Explicit policies at some centers that exclude some categories of patients from the prospective recipient pool because they have a history of drug abuse (including marijuana), crime, or simply because they have a mental disability give further reason for doubt about equity in the allocation of organs.

In addition, the fact that so many hospitals and medical centers have rushed into the transplant field raises legitimate doubts about the fairness of the system for distributing organs and tissues. The competition between transplant centers to secure organs is certainly not in the public interest especially when distribution is partly based on geography and there is little unmet need for access to a center in the United States in terms of program availability. The impression that the allocation of organs is based on bias, prejudice, favoritism, greed, geography, or ability to command publicity is enough to weaken public confidence that the system is equitable. Flat rates for cadaver organ donation may reflect skepticism about the fairness for rationing what is given.

IS IT FAIR TO GIVE EVERYONE IN NEED AN EQUAL CHANCE TO GET A TRANSPLANT?

To know whether the distribution of organs in the United States is fair and equitable, it is not sufficient to look at data on who did and did not get transplanted in any given year. This information is necessary but not sufficient for evaluating the fairness of the distribution. All that any pattern of distribution proves is that there may be reason for concern about inequity. The underrepresentation of minorities, the poor, the disabled, or the elderly in the ranks of transplant recipients might be unfortunate, but it might not be unfair.

To know whether the distribution of organs is fair, it is necessary to know the pattern of need for transplants. If every person is at equal risk of end-stage organ failure, or if those who actually suffer organ failure of various kinds represent a microcosm of the overall American population or any national population, then any deviation from this average in the distribution of organs gives reason for concern about fairness.

A final assessment of the equity of the distribution of organs will depend on an examination of the criteria and rules (or lack of them) being followed by transplant centers. It might be fair to give men more organs than women or whites fewer organs than blacks if the allocation were based on a set of criteria and rules that men and women and blacks and whites could all agree are fair, even if the rules work to the disadvantage of some persons or groups (Veatch 2002).

In asking what criteria, policies, and procedures are used to allocate organs, it is tempting to seek the answer by examining what happens to a particular organ that is donated or what decisions are made on any given day in the case of a particular transplant program. But looking only at such situations conveys a very misleading impression of how allocation decisions are actually made.

The question of who gets organs is a function of numerous factors: how many transplant programs there are, how many organ donors there are, what standards exist for evaluating the suitability of organs for transplant, how far organs and tissues can be shipped without damage, whether a person has health insurance or not, and a host of other variables. The fairness of organ allocation cannot be understood simply by watching what various surgeons or transplant centers actually do when they have the opportunity to transplant a donated organ. Transplant centers, and the surgeons who administer particular organ transplant programs within them, need to deal only with those patients who have actually made it through their doors. But many potential recipients never appear on any center's waiting list. The most important ethical decisions about allocation take place long before an organ actually arrives to be used at a particular transplant center.

JUSTICE AND RATIONING SCARCE ORGANS FOR TRANSPLANTATION

There is no consensus about what constitutes fair distribution with respect to scarce resources. There are theories of justice that would direct resources to those most likely to benefit from them, to those who are seen as most deserving, to those who are seen as likely to make the greatest social contribution in the future, to those willing to pay the highest price for them, to those who have the greatest responsibility for nurturing the lives of others, to those likely to enjoy the highest quality of life, or through a random system of allocation such as a lottery.

In health care, a strong case can be made for following a policy that maximizes the number of lives saved. Of course these lives must be of minimal quality since merely maintaining biological function in a body is not a standard of quality of life that most people deem worthy of medical resources. The criteria used to determine rationing schemes must be linked to the effectiveness of the resource, because the same standards may not be appropriate when one is trying to decide who gets access to a resource known to save lives (e.g., heart transplants), as opposed to granting access to resources whose effectiveness is poor or uncertain (e.g., pancreas transplants, intestinal transplants, and lung transplants for children).

If a form of transplantation works, then it would seem that efficacy in the saving of a life should drive allocation. After all, those who give organs say they do so to save lives. Those who transplant say they do so to save lives. And the reason organ transplantation has special status as a noncommodity in Western societies is that the goal of making the gift of an organ available is to save lives. If the goal is to save lives then giving organs to the sickest persons who need them is not necessarily the best way to achieve this goal. In the future more attention must be paid to outcomes as well as need and urgency in transplant allocation policies in the United States and other nations (Reese and Caplan 2011).

The current set of criteria in the United States reflects a complex set of moral considerations including sensitivity to the need for efficacy shown in weighting biological factors and moving pediatric patients to the head of many organ waiting lists, fairness shown in giving preference for prior donation as well as to waiting time, justice in permitting some of those with no access in their home nations to enter the system from foreign nations, and deservedness in allowing the better off greater access through multiple listing and to wealthy foreigners who constitute the majority of non-U.S. recipients. Although much is made of the weight given to waiting time, no single value governs the current system of rationing.

Proponents of change in rationing criteria to give more weight to efficacy make their case based on the waste in lives and life years lost created by too much emphasis on waiting time (Reese et al. 2010; Hippen et al. 2011; Reese and Caplan 2011). To improve survival with a transplant, a system of rules weighted toward "survival matching" could be used. This means allocating the highest quality kidneys to the candidates with the highest estimated posttransplant survival and allocating the remaining organs so that candidates who are within 10–15 years, older or younger, of the donor's age receive the highest priority (Reese and Caplan 2011).

It is true that the current ability to forecast graft and recipient survival is far from perfect

(Hippen et al. 2011). But, evidence for meaningful survival is not so deficient that it could not be invoked in a general way to help guide the more effective use of the scarce supply of organs. More to the point, criticism about the limits of current prognostication variables such as age or weight or comorbidity to predict who will do the best with a transplant does not meet the criticism that the system does not give sufficient weight to efficacy. It merely is an argument for more research toward obtaining evidence to achieve that end (Reese et al. 2010).

The case for moving toward greater efficacy in rationing makes ethical sense relative to distributing other scarce, valuable resources. In rationing situations saving the most lives and the greatest number of quality life years is a strategy that is followed in rescue scenarios or when facing public health emergencies (Reese et al. 2010). Giving the greatest weight to efficacy also seems consistent with what donors intend. Shifting toward efficacy defined as both increasing the odds of saving lives and saving the most life years, although not the exclusive value for rationing, would seem to make the most moral sense in confronting limited resources. Shifting public policy to expand those limited resources and minimize scarcity is a powerful requirement in any situation in which rationing exists, including transplantation.

REFERENCES

Abadie A, Gay S. 2006. The impact of presumed consent legislation on cadaveric organ donation: A cross-country study. *J Health Econ* **25:** 599–620.

Byk C. 2009. The European protocol on organ transplant: Key issues. *J Int Bioethique* **20:** 119–133.

Caplan AL. 1984. Organ procurement: It's not in the cards. *Hastings Center Report* October 14, pp. 6–9.

Caplan AL. 2004. Transplantation at any price? *Am J Transplant* **4:** 1933–1934.

Caplan AL. 2013. It is not morally acceptable to buy and sell organs for human transplantation. In *Contemporary debates in bioethics* (ed. Caplan AL, Arp R), pp. 59–67. Wiley Blackwell, Hoboken, NJ.

Caplan AL, Coehlo DH (ed.). 1998. *The ethics of organ transplants.* Prometheus, New York.

Caplan AL, Prior C. 2009. *Trafficking in organs, tissues and cells and trafficking in human beings for the purpose of the removal of organs.* Joint Council of Europe/United Nations Study, Strasbourg.

Caplan AL, Rockman HA, Turka LA. 2012. Editorial position on publishing articles on human organ transplantation. *J Clin Invest* **122:** 2.

Chandler JA. 2005. Priority systems in the allocation of organs for transplant: Should we reward those who have previously agreed to donate? *Health Law J* **13:** 99–138.

Davis FD, Crowe SJ. 2009. Organ markets and the ends of medicine. *J Med Philos* **34:** 586–605.

DeVita M, Caplan AL. 2007. Caring for organs or for patients? Ethical concerns about the Uniform Anatomical Gift Act. *Ann Intern Med* **147:** 876–879.

Hippen BE, Thistlethwaite JR, Ross LF. 2011. Risk, prognosis and unintended consequences in kidney allocation. *N Engl J Med* **364:** 1285–1287.

Hughes PM. 2009. Constraint, consent, and well-being in human kidney sales. *J Med Philos* **34:** 606–631.

Jaycox MP. 2012. Coercion, autonomy, and the preferential option for the poor in the ethics of organ transplantation. *Dev World Bioeth* **12:** 135–147.

Kerstein SJ. 2009. Autonomy, moral constraints, and markets in kidneys. *J Med Philos* **34:** 573–585.

Lawlor R. 2011. Organ sales needn't be exploitative (but it matters if they are). *Bioethics* **25:** 250–259.

Matas AJ. 2006. Why we should develop a regulated system of kidney sales: A call for action! *Clin J Am Soc Nephrol* **1:** 1129–1132.

Neades BL. 2009. Presumed consent to organ donation in three European countries. *Nurs Ethics* **16:** 267–282.

Quigley M, Wright L, Ravitsky V. 2012. Organ donation and priority points in Israel: An ethical analysis. *Transplantation* **93:** 970–973.

Reese PP, Caplan AL. 2011. Better off living—the ethics of the new UNOS proposal for allocating kidneys for transplantation. *Clin J Am Soc Nephrol* **6:** 2310–2312.

Reese PP, Caplan AL, Bloom RD, Abt PL, Karlawish JH. 2010. How should we use age to ration health care? Lessons from the case of kidney transplantation. *J Am Geriatr Soc* **58:** 1980–1986.

Sanaei Ardekani M, Orlowski JM. 2010. Multiple listing in kidney transplantation. *Am J Kidney Dis* **55:** 717–725.

Slabbert M. 2012. This is my kidney, I should be able to do with it what I want: Towards a legal framework for organ transplants in South Africa. *Med Law* **31:** 617–640.

Tan HP, Marcos A, Shapiro R, ed. 2008. *Living donor organ transplantation.* McGraw Hill, New York.

Taylor JS. 2009. Autonomy and organ sales, revisited. *J Med Philos* **34:** 632–648.

Veatch R. 2002. *Transplantation ethics.* Georgetown University Press, Washington, DC.

Wang M. 2010. Organ donation by capital prisoners in China: Reflections in Confucian ethics. *J Med Philos* **35:** 197–212.

Will Regenerative Medicine Replace Transplantation?

Giuseppe Orlando[1,2], Shay Soker[2], Robert J. Stratta[1], and Anthony Atala[1]

[1]Department of General Surgery, Section of Transplantation, Wake Forest School of Medicine, Winston Salem, North Carolina 27157

[2]Wake Forest Institute for Regenerative Medicine, Wake Forest School of Medicine, Winston Salem, North Carolina 27157

Correspondence: gorlando@wakehealth.edu

Recent groundbreaking advances in organ bioengineering and regeneration have provided evidence that regenerative medicine holds promise to dramatically improve the approach to organ transplantation. The two fields, however, share a common heritage. Alexis Carrel can be considered the father of both regenerative medicine and organ transplantation, and it is now clear that his legacy is equally applicable for the present and future generations of transplant and regenerative medicine investigators. In this review, we will briefly illustrate the interplay that should be established between these two complementary disciplines of health sciences. Although regenerative medicine has shown to the transplant field its potential, transplantation is destined to align with regenerative medicine and foster further progress probably more than either discipline alone. Organ bioengineering and regeneration technologies hold the promise to meet at the same time the two most urgent needs in organ transplantation, namely, the identification of a new, potentially inexhaustible source of organs and immunosuppression-free transplantation of tissues and organs.

Organ transplantation and regenerative medicine share the same heritage. Alexis Carrel is considered the father of transplant (and vascular) surgery, however, his seminal studies on cell culture and ex vivo organ preservation and growth anticipated organ bioengineering and regeneration concepts were not developed until decades later (Orlando et al. 2011b). Carrel's visionary investigations, the majority of which were performed in collaboration with engineer and celebrated aviator Charles Lindbergh, established the foundation of modern regenerative medicine. Notably, Carrel and Lindbergh's perfusion pump may be referred to as one of the greatest inventions in health sciences in the twentieth century. That machine, by allowing living organs to exist outside of the body during surgery, was a crucial step in the development of perfusion systems for cardiac and transplant surgery, as well as for any sort of device requiring extracorporeal circulation. Moreover, the perfusion pump may also be considered the precursor of the bioreactors commonly used nowadays in tissue and organ bioengineering, which are sealed mechanical chambers designed to provide appropriate environmental conditions

Cite this article as *Cold Spring Harb Perspect Med* doi: 10.1101/cshperspect.a015693

for cellular activity and welfare (Badylak et al. 2012). Carrel's pioneering developments on both fronts of organ transplantation and regenerative medicine occurred decades before either became clinical realities, and almost a century before the successful implantation in patients of bioengineered tissues or organs (Romagnoli et al. 1990; Pellegrini et al. 1997; Quarto et al. 2001; Shinoka et al. 2001; Warnke et al. 2004; Atala et al. 2006; Macchiarini et al. 2008; McAllister et al. 2009; Mertsching et al. 2009; Baiguera et al. 2010; Hibino et al. 2010; Rama et al. 2010; Junglebuth et al. 2011; Elliott et al. 2012; Olausson et al. 2012), or the early attempts to bioengineer more complex organs for transplant purposes (Ott et al. 2008, 2010; Ross et al. 2009; Petersen et al. 2010; Uygun et al. 2010; Wainwright et al. 2010; Baptista et al. 2011; Hammond et al. 2011; Soto-Gutierrez 2011; Barakat et al. 2012; Hata et al. 2012; Orlando et al. 2012a; Yagi et al. 2012).

THE URGENT NEED FOR NEW SOURCES OF TRANSPLANTABLE ORGANS

Organ transplantation may be referred to as one of the greatest achievements in the history of medicine. Short-, medium-, and long-term results are excellent and, when compared with patients who cannot receive a new organ, the impact on a patient's mortality, morbidity, and quality of life is tremendous. If we consider kidney transplantation as a paradigmatic example, it currently represents the gold standard for renal replacement therapy in patients affected by end-stage renal disease. In fact, when compared with maintenance peritoneal or hemodialysis, renal transplantation dramatically improves patient survival, quality of life, and is cost effective in terms of health care expenditures. According to a recent U.S. Renal Data System report, life expectancy from the time renal dialysis is initiated is approximately 8 years for patients between the ages of 40 and 44, and 4.5 years for those who are between 60 and 64 years of age (www.usrds.org). These figures are far surpassed by the increased survival rates following kidney transplantation, which are 85%, 70%, and 44% after 5, 10, and 20 years, respectively (information based on the Organ Procurement and Transplantation Network data as of July 27, 2012 [optn.transplant.hrsa.gov]) (Table 1). Similar data are reported by the European Renal Association—European Dialysis and Transplant Association (ERA-EDTA) Registry (www.era-edta-reg.org/files/annualreports/pdf/AnnRep2010.pdf), for which expected remaining lifetimes in years of dialysis versus transplant patients in the year ranges of 40–44 and 60–64 are 10.7 versus 27 years, and 5.7 versus 13.4, respectively. A consequence of the effectiveness of renal transplantation as renal replacement treatment is that, indications for kidney transplantation have expanded considerably in terms of recipient age, primary disease leading to end-stage renal failure, and degree of deterioration in clinical condition at the time of transplantation. In other words, we are now placing on the waiting list—and eventually transplanting—patients who are older and sicker than in earlier times, and are affected by diseases that years ago would have represented a contraindication for transplantation.

As a corollary, transplant waiting lists are rapidly increasing but, unfortunately, the numerous strategies that are currently being implemented to expand the donor organ pool are not keeping pace with the demand (Fig. 1). Some of these measures include expansion of living and deceased donor acceptance criteria such as use of organs from expanded criteria donors or donors after cardiac death, transplants across immunological barriers, paired donations, Good Samaritan or altruistic donations, etc., and have been very well discussed in other articles in this collection. Although the gap between increasing demand and supply of organs is dramatically widening, the cumulative probability to receive an organ in the critical time frame is dropping drastically and mortality on the waiting list is fearfully increasing (www.medscape.org/viewarticle/488926) (Table 2). At the end of 2011, 96,574 patients were waiting for a kidney in the United States, whereas only 16,813 kidney transplants were performed during the calendar year (optn.transplant.hrsa.gov). This corresponds to a cumulative probability to receive a renal graft at 1 year (from the time of registration on the waiting list) of only 9.65%. This figure increases at

Table 1. Actuarial patient survival rates after kidney transplant (KT)

	Year survival					
	1		5		10	
Organ	Sample size	Survival rate and 95% CI	Sample size	Survival rate and 95% CI	Sample size	Survival rate and 95% CI
(A) Patient survival rate after 1, 5, and 10 years for transplants performed between 1/1/1991 and 12/31/1999 by organ						
Heart	20,469	83.83 [83.31,84.34]	20,469	69.74 [69.10,70.38]	20,469	51.57 [50.85,52.29]
Heart-lung	498	65.79 [61.49,70.10]	498	42.14 [37.71,46.57]	498	26.73 [22.62,30.84]
Intestine	393	64.44 [59.39,69.49]	393	44.61 [39.33,49.89]	393	35.08 [29.90,40.27]
Kidney	99,844	95.29 [95.15,95.42]	99,844	84.90 [84.66,85.13]	99,844	69.90 [69.56,70.24]
Kidney-pancreas	6900	93.41 [92.81,94.02]	6900	83.68 [82.76,84.59]	6900	69.09 [67.86,70.33]
Liver	34,594	83.28 [82.87,83.69]	34,594	71.42 [70.91,71.92]	34,594	58.36 [57.78,58.94]
Lung	6712	74.36 [73.30,75.42]	6712	44.91 [43.70,46.13]	6712	24.28 [23.19,25.37]
Pancreas	1434	91.92 [90.39,93.44]	1434	79.63 [77.19,82.07]	1434	62.91 [59.45,66.37]

	Year survival			
	1		5	
Organ	Sample size	Survival rate and 95% CI	Sample size	Survival rate and 95% CI
(B) Patient survival rate after 1 and 5 years for transplants performed between 1/1/2000 and 12/31/2005 by organ				
Heart	12,753	86.78 [86.18,87.38]	12,753	73.74 [72.96,74.52]
Heart-lung	212	68.42 [61.80,75.04]	212	47.92 [41.00,54.84]
Intestine	748	76.60 [73.42,79.79]	748	56.75 [52.99,60.51]
Kidney	90,309	95.69 [95.56,95.83]	90,309	85.41 [85.17,85.66]
Kidney-pancreas	5367	94.73 [94.11,95.35]	5367	86.11 [85.13,87.09]
Liver	33,816	86.14 [85.76,86.52]	33,816	72.92 [72.42,73.42]
Lung	6723	81.35 [80.40,82.29]	6723	53.09 [51.87,54.31]
Pancreas	3116	94.64 [93.79,95.48]	3116	83.00 [81.49,84.51]

	Year survival	
	20	
Organ	Sample size	Survival rate and 95% CI
(C) Patient survival rate after 20 years for transplants performed between 10/01/1987 and 12/31/1990 by organ		
Heart	5843	18.45 [17.29,19.61]
Heart-lung	202	14.16 [8.35,19.98]
Kidney	28,987	43.69 [42.74,44.64]
Kidney-pancreas	993	38.41 [34.37,42.45]
Liver	6926	33.95 [32.51,35.40]
Pancreas	238	24.71 [15.03,34.39]

Data are stratified by year of transplantation (based on open access, official data from Organ Procurement and Transplantation Network data as of July 27, 2012).

CI, confidence interval.

Table 2. Cumulative probability to receive a renal graft at 1, 3, and 5 years from the time of registration on the waiting list, in two different time frames: 1/1/1995 to 12/31/2006 (A), and 1/1/2007 to 12/31/2012 (B).

		1		3		5	
Organ	Number of registrations	Percent (probability × 100) transplanted in 1 year	95% CI	Percent (probability × 100) transplanted in 1 year	95% CI	Percent (probability × 100) transplanted in 1 year	95% CI
(A) 1/1/1995–12/31/2006							
Heart	41,101	53.98%	[53.5%, 54.5%]	61.92%	[61.5%, 62.4%]	63.19%	[62.7%, 63.6%]
Heart-lung	1267	22.65%	[20.4%, 25.0%]	32.68%	[30.1%, 35.3%]	34.57%	[32.0%, 37.2%]
Intestine	2252	48.49%	[46.4%, 50.5%]	54.00%	[51.9%, 56.0%]	54.66%	[52.6%, 56.7%]
Kidney	287,194	14.21%	[14.1%, 14.3%]	28.88%	[28.7%, 29.0%]	36.00%	[35.8%, 36.2%]
Kidney-pancreas	19,832	35.49%	[34.8%, 36.2%]	52.80%	[52.1%, 53.5%]	54.97%	[54.3%, 55.7%]
Liver	117,489	39.93%	[39.7%, 40.2%]	48.67%	[48.4%, 49.0%]	50.49%	[50.2%, 50.8%]
Lung	22,800	32.63%	[32.0%, 33.2%]	50.28%	[49.6%, 50.9%]	53.07%	[52.4%, 53.7%]
Pancreas	9489	38.28%	[37.3%, 39.2%]	49.00%	[48.0%, 50.0%]	50.34%	[49.3%, 51.3%]
Pancreas islets	920	19.78%	[17.3%, 22.4%]	26.41%	[23.6%, 29.3%]	27.28%	[24.5%, 30.2%]
Total	502,344	25.80%	[25.7%, 25.9%]	38.62%	[38.5%, 38.8%]	43.47%	[43.3%, 43.6%]

Organ	Number of registrations	1		3	
		Percent (probability × 100) transplanted in 1 year	95% CI	Percent (probability × 100) transplanted in 1 year	95% CI
(B) 1/1/2007–12/31/2012					
Heart	13,534	59.23%	[58.4%, 60.0%]	67.34%	[66.5%, 68.1%]
Heart-lung	224	46.43%	[39.9%, 52.7%]	0	[[a], [a]]
Intestine	1049	59.49%	[56.5%, 62.4%]	65.51%	[62.5%, 68.3%]
Kidney	138,803	9.65%	[9.49%, 9.80%]	21.65%	[21.4%, 21.9%]
Kidney-pancreas	6335	34.98%	[33.8%, 36.2%]	50.32%	[49.1%, 51.6%]
Liver	45,517	45.17%	[44.7%, 45.6%]	51.84%	[51.4%, 52.3%]
Lung	8710	63.88%	[62.9%, 64.9%]	72.13%	[71.2%, 73.1%]
Pancreas	3288	34.91%	[33.3%, 36.5%]	44.02%	[42.3%, 45.7%]
Pancreas islets	311	31.83%	[26.8%, 37.0%]	44.29%	[38.6%, 49.9%]
Total	217,771	23.75%	[23.6%, 23.9%]	34.26%	[34.1%, 34.5%]

Noteworthy is that percentages are trending toward lower figures as years go by owing to the increasing gap between demand and offer (based on open access, official data from Organ Procurement and Transplantation Network data as of July 27, 2012).

[a]Value cannot be calculated owing to <10 at risk at time point.

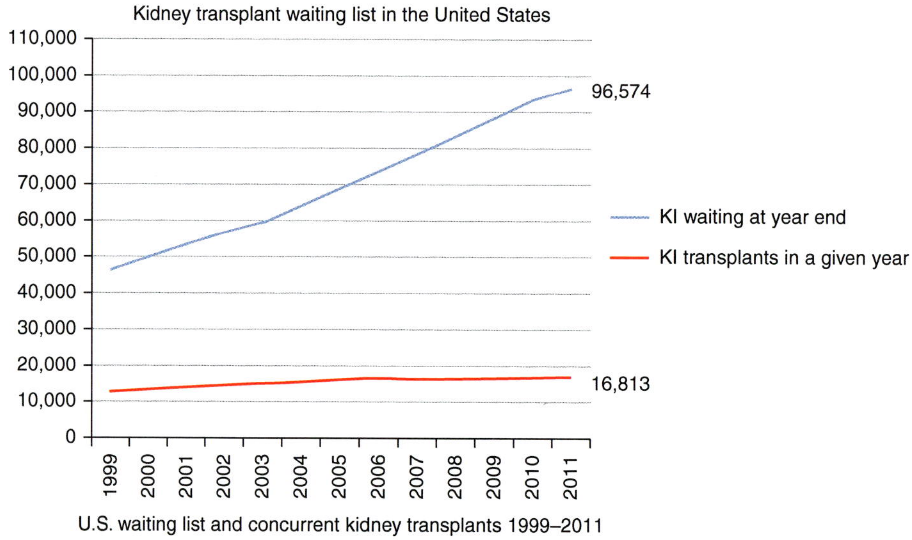

Figure 1. Representation of the comparison between the number of renal transplants performed yearly since 1999 and the number of patients on the waiting list for a kidney from a deceased donor. The gap between the two variables is dramatically increasing as a consequence of the constantly escalating number of patients registered on the waiting list and an unprecedented attenuation of the growth of kidney transplantation (based on official, open access data from Organ Procurement and Transplantation Network data as of August 24, 2012).

3 and 5 years to 21.65% and 36%, respectively (unos.org), a rate that is not sufficient to satisfy the demand for transplantable kidneys. For patients aged 60 or older who are placed on the waiting list, nearly half (46%) will die awaiting a potentially lifesaving kidney transplant. Mean annual mortality on the kidney waiting list is 6% per year, and increases to 10% in the presence of diabetes. Median waiting times range from 3 to 6 years depending on blood type, degree of sensitization, and geographic region. Even in this case, similar figures can be obtained from data analysis of the European population in which, in 2010, 91,798 patients were affected by end-stage renal failure but only 21,740 renal transplants could be performed (www.era-edta-reg.org/files/annualreports/pdf/AnnRep2010.pdf). It is critical to point out that the disparity between organ supply and demand is destined to increase because the number of uremic patients is reaching epidemic proportions worldwide as a consequence of widespread chronic noncommunicable diseases such as diabetes and hypertension, as well as infectious diseases and malnutrition in low socioeconomic countries (Meguid et al. 2005). At present, it is estimated that only 6% of the dialysis population will ever receive a kidney transplant. This data does not leave much to the imagination.

THE PURSUIT OF AN IMMUNOSUPPRESSION-FREE STATE

The ultimate goal of transplantation should be the achievement of an immunosuppression-free status, in which the recipient of an allograft is not receiving any immunosuppression while enjoying normal graft function (Orlando et al. 2009, 2010a; Orlando 2010). By definition, such an individual should have preserved the ability to mount an immune response to a third party and to infections, and should not present any signs of rejection at any clinical level (physical examination, laboratory or pathology testing, or genomic testing). Therefore, the induction of specific unresponsiveness to alloantigens would render lifelong immunosuppression unnecessary and would ultimately prevent all related complications and abate all costs, while dramatically improving the patient's quality of life. In fact, chronic immunosuppression that is required to prevent rejection places patients at risk for infectious and metabolic complications, as well as for malignancies and drug-specific toxicities.

If we remain in the context of kidney transplantation, infections are among the major causes of hospitalization in the first 2 years posttransplant and represent the main cause of hospitalization in the pediatric kidney transplant population (Dharnidharka et al. 2004). Moreover, cardiovascular events (which are related to long-term immunosuppression) represent the major cause of death (Dantal and Soulillou 2005). When compared with an age-matched healthy population, transplant recipients have an increased incidence of cancer, shown by the fact that after 20 years of immunosuppression, 40% of kidney transplant patients develop cancer and the risk is related to the amount of immunosuppressive drugs received (Dantal et al. 1998). In addition, about 50% of kidney grafts are lost within 10 years following transplant secondary to chronic allograft nephropathy, a process that depends on a combination of alloantigen-dependent and -independent factors, the latter of which includes the toxic effects of immunosuppression such as nephrotoxicity, hypertension, and hyperlipidemia.

Two more factors have fueled initiatives aiming to achieve immunosuppression-free transplantation. These include the high cost of immunosuppressive drugs and how they affect quality of life. Organ transplantation is one of the most expensive therapies currently available, and has a remarkable impact on hospital and health care systems expenditures. Countries in which the national health services are overburdened by the increasing medical needs of an aging population have to adopt strict cost optimization to limit access and expenses (Filipponi et al. 2003). Given the extremely high cost of lifelong antirejection therapy, a current challenge consists of devising management strategies that lower the expenses of transplantation without having an impact on clinical outcomes. In addition to patient and graft survival rates, it has been shown that improved quality of life is an important outcome measure after organ transplantation in addition to acute rejection and infection rates; these outcomes are important in determining the safety and efficacy of immunosuppressive regimens and function as a guide for other therapeutic choices (Karam et al. 2003). This occurs owing to the fact that patients are now informed about the results and outcomes achieved with transplantation from which they not only expect to be healed, but also achieve similar quality of life as nontransplant individuals. It is important to note that the cost of treatment along with the frequency and dosage of prescribed medications are among the factors that have the most impact on a patient's quality of life, meaning the more drugs and the higher the doses, the lower the perceived quality of life (Galbraith 2004). In addition, patient nonadherence—which is one of the most important determinants of treatment and major causes of organ failure—is significantly influenced by the number of medications as well. Therefore, the optimal immunosuppressive regimen should strive to use the fewest agents at the lowest effective doses to reduce toxicities and costs, but still prevent acute rejection and disease recurrence and maintain graft function while increasing patient satisfaction.

Currently, an immunosuppression-free status can be pursued by either using organs from genetically identical donors or by establishing a state of immunological tolerance. Unfortunately, transplantation among genetically identical individuals is exceptional, whereas true or complete tolerance can only be achieved in a small percentage of liver transplant recipients and remains an elusive goal after transplantation of other organs (Orlando et al. 2009, 2010a; Orlando 2010). After liver transplantation, an immunosuppression-free state can be attempted safely without exposing the patient to any harm, yet the candidate must meet very strict criteria and should have stable graft function and be at least 1 year following transplantation. Therefore, even in the case of liver transplantation, tolerance is never immediate, stable, or durable (Orlando et al. 2011a), and is attainable in only one out of four patients who undergo planned weaning of immunosuppression. When other transplantable organs are considered, the scenario is even worse. In fact, clinical tolerance has never been reported after intestinal, islet, or whole organ pancreas transplantation, whereas two exceptional cases of patients who developed operational tolerance have been described after lung and heart transplantation (Orlando et al. 2010a). In the case of renal transplantation, the achievement of a permanent and stable immunosuppression-free state is exceptional and all attempts to generate such a tolerant state reported to date have involved complex regimens with nonreproducible results. Overall, tolerogenic strategies are not yet available for daily clinical practice because they are not consistently effective, practical, or safe. Despite significant progress in the field of clinical transplant immunology, the understanding of the immune mechanisms underlying tolerance remains inadequate.

Ideally, and as anticipated above, tolerance should be immediate, stable, and durable. It should develop immediately after reperfusion without requiring any sort of induction or maintenance immunosuppression. It should be refractory to all insults that may alter the balance between self and nonself (stable). Finally, it should last as long as the graft itself (durable). Although none of the immunology-based presumed tolerogenic strategies have ever been able to achieve these goals, regenerative medicine technologies have permitted the bioengineering of relatively simple, hollow organs from a patient's own cells. These organs have been implanted in more than 160 patients suffering from different medical conditions without the need for antirejection treatment (Romagnoli et al. 1990; Pellegrini et al. 1997; Quarto et al. 2001; Shinoka et al. 2001; Warnke et al. 2004; Atala et al. 2006; Macchiarini et al. 2008; McAllister et al. 2009; Mertsching et al. 2009; Baiguera et al. 2010; Hibino et al. 2010; Rama et al. 2010; Junglebuth et al. 2011; Elliott et al. 2012; Olausson et al. 2012; Orlando et al. 2012b). Importantly, these figures far exceed the number of organ recipients who have been successfully weaned off all immunosuppression in the immediate postoperative period. Although the production of more complex modular organs of transplant interest—namely, kidney, liver, heart, lung, pancreas or islet, and intestine—have not been documented, it is incontrovertible that regenerative medicine is opening an exciting arena for investigations that may tremendously impact organ transplantation.

HOLLOW VERSUS COMPLEX MODULAR ORGANS

To date, progress in regenerative medicine has been defined by methods and technologies that have led to the production and implantation of hollow organs, which—from a bioengineering perspective—are less complex than the so-called complex modular organs. These latter are organs organized in functioning units referred to as modules that require reconnection to the vascular system of the recipient at the time of implantation to be viable. The heart, liver, lung, pancreas, and intestine are complex modular organs. On the other hand, hollow organs are either conduits (vessels, upper airways, urethras) or reservoirs (bladders), which theoretically are able to exert their function regardless of a direct reconnection to the recipient's bloodstream.

It is important to emphasize that traditional anatomy defines as hollow organs any "visceral organ that is a hollow tube or pouch (as the stomach or intestine) or includes a cavity (as the heart or urinary bladder)" (en.wikipedia.org/wiki/Organ_(28anatomy)). Nevertheless, from a regenerative medicine perspective, the definition is based on function rather than shape, and consequently intestine and heart are not considered hollow organs but rather complex modular organs. In fact, although the liver, kidney, and pancreas are intuitively considered to be complex modular organs, the heart and digestive tract may be referred to as hollow organs because of the fact that they clearly include cavities (in the case of the heart) or consist of a hollow tube (the intestine). Instead, in this specific context, myocardiocytes themselves and villi are considered the modules (namely, the functioning units) of heart and intestine, respectively. Vessels, bladders, upper airways, and urethras comply with the traditional definition of hollow organs.

Notably, all the bioengineered organs that, to date, have been reported to be implanted in humans were hollow organs, whereas the bioengineering of cardiac, hepatic, renal, pancreatic, and intestinal organoids is still far from the realm of possibility. The main technology that is currently being implemented for organ bioengineering consists of seeding cells on supporting scaffolding materials. We refer to this method as cell-scaffold technology, which is based on the idea—actually, on the dogma!—that cells require a supporting structure (which in vivo is represented by the extracellular matrix) to grow, expand, and exert their function. This notion is not new but has circulated for decades (Bissell et al. 1982). A paradigmatic example of the essential role of this interplay between cells and extracellular matrix is offered by clinical whole pancreas transplantation versus transplantation of pancreatic islets. Whole pancreas transplantation and islet transplantation are currently the only known therapies that reliably establish a long-term stable euglycemic state (Orlando et al. 2010b). Following these types of transplant, endogenous insulin secretion in response to normal feedback controls is restored and glycosylated hemoglobin levels normalize in the absence of the need for exogenous insulin therapy. However, if we compare the mid- and long-term outcome of these two different types of β-cell replacement therapies, whole pancreas transplantation overrides islet transplantation. In fact, despite a significant improvement in the overall results of islet transplantation recently reported by the Collaborative Islet Transplant Registry (CITR) (Barton et al. 2012), insulin independence rates, C-peptide and HbA1c serum levels, and long-term resolution of severe hypoglycemia recorded after islet transplantation remain inferior to whole pancreas transplantation, which remains the gold standard for diabetes treatment. These findings have been attributed to several factors, the most important of which are cold ischemia-derived stress and rejection. This notwithstanding, from a regenerative medicine (RM) perspective, the explanation is far more simple and lies in the fact that when islets—namely, clusters of cells—are removed from their microenvironment and therefore not supported by their natural supporting scaffold, they undergo tremendous stress that dramatically harms their welfare and viability.

Cellular material used in cell-scaffold technology may include adult cells, different types of progenitor cells per se, or progenitor cells that can be induced to differentiate into specific adult cell types. The scaffolds used may be either synthetic or natural, the latter being obtained via detergent-based decellularization of organs from animals or humans, a method that consistently allows for the complete clearance of the cellular compartment of almost any organ on any scale. Scaffolds obtained from innate organs seem to offer the best chances for success as they are naturally biocompatible and bioactive, while being relatively inexpensive and easy to access. They represent a biochemically, geometrically, and spatially ideal platform for bioengineering investigations, because they have both basic components (proteins and polysaccharides) and matrix-bound growth factors and cytokines preserved (Wang et al. 2011), they retain an intact and patent vasculature which—when implanted in vivo—sustains the physiologic blood pressure (Orlando et al. 2012a), and are able to drive differentiation of progenitor cells into an organ-specific phenotype (Bissell et al. 1982; Chung et al. 2005; Ross et al. 2009; Ng et al. 2011).

Although the above-mentioned hollow organs that have already been implanted in humans were produced mainly from synthetic scaffolds, natural scaffolds obtained from animals of various sizes and species have been used as a platform for cardiac, liver, pancreas, intestine, and renal bioengineering with some promising results and achievements. In the case of hollow organs, autologous cells were collected from the patients and expanded in vitro. In some cases, stem cells were isolated and then spontaneously differentiated into specific cell types before being seeded on supporting scaffolds (Junglebuth et al. 2011; Elliott et al. 2012). Afterward, the so-obtained cellular constructs were either allowed to mature in bioreactors or were implanted immediately after seeding without undergoing any maturation phase (Elliott et al. 2012). As such, cell-scaffold technology has made it possible to manufacture and implant relatively simple structures such as vessels, segments of upper airways, urethras, and neo-urinary conducts. However, current reports refer to very small series with short follow-up and insufficient discussion of the complications and limitations that may occur over time.

PLAYGROUNDS FOR FUTURE INVESTIGATIONS

Developmental and Evolutionary Biology

It seems apparent that the sequence of events represented by the seeding of cells on supporting scaffolding material → cell attachment to the extracellular matrix → growth → expansion followed by the so-called maturation phase that may take place either in bioreactors or in vivo, is not sufficient to allow effective and robust maturation of the new organoid, which is the sine qua non for the successful integration of the new construct with what will be its new environment following implantation. Theoretically what is expected to occur during the abovementioned sequence of steps is the dynamic recapitulation of all events occurring in utero under physiological conditions during organ ontogenesis. Unfortunately, most of these events remain unknown and the mechanisms underlying them, indefinite. Therefore, in-depth knowledge of the mechanisms through which organs develop is essential in organ bioengineering, as well as the acquisition of information on how cells, tissues, and organs respond to injury and implement repair and healing strategies.

It should be emphasized that an evolutionary approach to organ bioengineering and regeneration is needed as it is complementary to a genuinely developmental approach. We believe that investigators in the field should consider as a priority the understanding of why mammals, during phylogenesis, have lost the ability to regenerate themselves, a function that instead characterizes virtually all simple organisms of the animal and plant kingdoms. In fact, the notion that the more simply organized animals are capable of regenerating larger parts of their bodies when compared with animals with higher organization and complexity like mammals has been well known for almost a century, as it was formulated for the first time by the Russian embryologist P.P. Ivanov in 1937 (Fig. 2) (Ivanov 1937). If we take into account the fact that mammals in utero show formidable regenerative capacities that are lost after birth, we then infer that mammals—and so humans—do have innate regenerative abilities that nevertheless, for reasons that remain dramatically mysterious, are disabled or at least dramatically impaired ex utero. We therefore believe that the greatest challenge for RM researchers is to understand why and how this happens.

Bioreactors

Bioreactors represent a dynamic, controllable and reproducible culture environment designed to support the regeneration process and from which the bioengineered tissue/organ can be harvested for analysis or implantation. Bioreactors are devices that facilitate, monitor, and control biological, biochemical, and biophysical processes, and allow uniform scaffold cell distribution, ideal nutrient supply and waste removal, as well as hydrodynamic shear stress (Baiguera et al. 2010; Rauh et al. 2011). Shear stress is produced by forces inherent to the bioreactor-like rotation and other factors essential for tissue development, because it contributes to supporting metabolic activity and differentiation. Theoretically, bioreactors mimic in vivo conditions (temperature, oxygen concentration, pH, nutrient concentration, and biochemical and mechanical stimuli), which are the environment where the maturation phase follows cell seeding of the scaffold.

Unfortunately, present technology does not seem to produce adequate bioreactors. The recent report from the Great Ormond Street Hospital, London, UK, where a newly bioengineered segment of upper airways (Elliott et al. 2012) was implanted immediately after manufacturing (without undergoing any maturation phase) questions the quality and efficacy of currently available bioreactors. In fact, in all previous cases of implantation of bioengineered segments of the upper airways, the new construct was allowed to mature in a bioreactor before

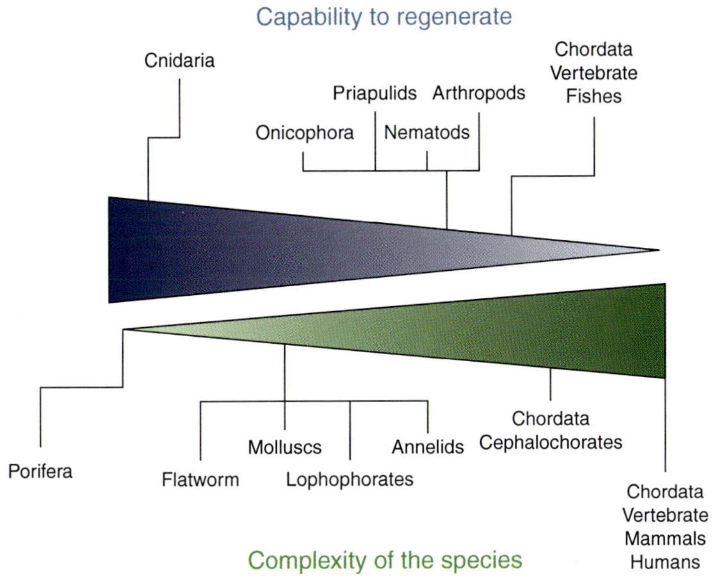

Figure 2. The ability to regenerate inversely proportional to the degree of complexity of a species. Phylum Chordata are on the extreme segment of the spectrum, yet some vertebrate species like zebrafish have maintained the capacity to fully regenerate organs after amputation and so represent a formidable experimental model for organ bioengineering and regeneration investigations. Humans are instead at the very end of the spectrum.

implantation. Seeding techniques, types and numbers of cells, reproduction of adequate biophysical forces mimicking those physiologically occurring in vivo, and composition and architecture of biomaterials are key factors for success in organ bioengineering and regeneration, and the lack or impairment of just one of these factors may lead per se to failure, regardless of whether all other factors are appropriate.

Vascularization

Importantly, all bioengineered constructs reported to date have been implanted without being reconnected to the systemic vascular system, thereby exposing the new organ to ischemia and ultimately to the risk of graft failure. All organs require a vascular supply, whereas it is well known that cells can survive only within an area approximately 1–3 mm away from a source of nutrients and oxygen (Folkman and Hochberg 1973). Therefore, the pivotal role of the reconnection of the vascular network to the systemic bloodstream to allow adequate oxygen and nutrients supply cannot be underestimated, unless the size of the body part in question is smaller than 0.3 cm. Interestingly, Macchiarini's group reported the occurrence of a ventral collapse of the most proximal 1 cm of the graft 8 months after the implantation of his first bioengineered trachea. Despite the investigators' speculation that this complication was owing to the pulsatile compression from the aortic arch superiorly and to the migration of the stem-cell-derived chondrocytes into the endoluminal surface of the graft, common sense suggests that it may simply have been caused by ischemia as the new organ did not have a vascular pedicle and was not reconnected to systemic circulation. For complex modular organs like kidneys, livers, and hearts, there is no question that vascular pedicles will have to be reconnected to the systemic circulation. For this reason, natural extracellular matrix scaffolds obtained from the decellularization of organs derived from animals and humans represent the ideal platform owing to their ability to maintain a well-preserved framework of their intrinsic vascular tree. In the case of renal extracellular matrix scaffolds obtained from kidneys discarded from human transplantation, or from pigs, the injection of regular contrast media within the renal artery depicts an intact vascular network that retains the hierarchical branching structures. Contrast media flows progressively from larger vessels to smaller capillaries to eventually drain out through the stump of the renal vein, without extravasation within the scaffold parenchyma (Orlando et al. 2012a).

RM AS THE NEW HOLY GRAIL FOR ORGAN TRANSPLANTATION

The Holy Grail is a sacred object found in literature and Christian traditions most often identified as a dish, plate, or cup allegedly used by Jesus at the Last Supper (en.wikipedia.org/wiki/Holy_Grail) (G Orlando, unpubl.). According to the legend, the Holy Grail was used by Joseph of Arimathea to collect Jesus' blood and sweat at the time of Jesus' crucifixion. After Christ's death, Joseph was imprisoned in a rock tomb and left to starve, nevertheless he survived because he could hide and bring with him the Holy Grail that gave him energy. Once out of prison, Joseph was sustained for several years by the power of the Grail, which provided him with fresh food and drink every morning (www.britannia.com/history/arthur/grail). In the Christian tradition, the Holy Grail is believed to possess miraculous powers and has therefore been the object of quest and intense research for centuries. However, no one has ever found where it has been carefully guarded, thus questioning whether the Holy Grail ever existed or whether it was a mere product of human fantasy.

In the literature and common language, the metaphor of the Holy Grail is often used to describe an exceptionally rare object or an ultimate, near-unattainable goal. In organ transplantation, the establishment of an immunosuppression-free state (otherwise said, tolerance) has represented the Holy Grail since the time when transplant investigators understood that the tool that allowed organ transplantation to become a formidable clinical treatment—namely, immunosuppression—should be considered also as our main foe as it is the primary determinant of patient mortality and morbidity from the second year posttransplant and beyond. Although tolerance has been the primary end point of innumerable investigations that have required massive investments since the early pioneering years, to date it remains an elusive goal. As we explained above, immediate, stable, robust, and durable tolerance cannot be obtained with available technology and the different strategies attempted to date are not yet reproducibly safe, effective, or easily applicable to large numbers of patients.

CONCLUSIVE REMARKS

Recent groundbreaking advances in organ bioengineering and regeneration have shown the immense potential that RM holds to change transplantation. It is now clear that the legacy that Alexis Carrel has left for the present and future generations of transplant and RM investigators is much richer than we thought and is not just about transplantation but also about RM. These two fields share a common ancestry and are destined to interdigitate in the future. Although RM has shown to the transplant field its tremendous potential, transplantation is destined to synergize with RM and foster further progress probably more than either field can achieve individually. The reason for this lies in the fact that organ bioengineering and regeneration technologies hold the promise to meet at the same time the two most urgent needs of organ transplantation, namely, the identification of a new, potentially inexhaustible source of organs and immunosuppression-free transplantation. The years ahead of us will be extraordinarily exciting and may rival the early pioneering years of transplantation.

REFERENCES

Atala A, Bauer SB, Soker S, Yoo JJ, Retik AB. 2006. Tissue-engineered autologous bladders for patients needing cystoplasty. *Lancet* **367:** 1241–1246.

Badylak SF, Weiss DJ, Caplan A, Macchiarini P. 2012. Engineered whole organs and complex tissues. *Lancet* **379:** 943–952.

Baiguera S, Birchall MA, Macchiarini P. 2010. Tissue-engineered tracheal transplantation. *Transplantation* **89:** 485–491.

Baptista PM, Siddiqui MM, Lozier G, Rodriguez SR, Atala A, Soker S. 2011. The use of whole organ decellularization for the generation of a vascularized liver organoid. *Hepatology* **53:** 604–617.

Barakat O, Abbasi S, Rodriguez G, Rios J, Wood RP, Ozaki C, Holley LS, Gauthier PK. 2012. Use of decellularized porcine liver for engineering humanized liver organ. *J Surg Res* **173:** e11–e25.

Barton FB, Rickels MR, Alejandro R, Hering BJ, Wease S, Naziruddin B, Oberholzer J, Odorico JS, Garfinkel MR, Levy M, et al. 2012. Improvement in outcomes of clinical islet transplantation: 1999–2010. *Diabetes Care* **35:** 1436–1445.

Bissell MJ, Hall HG, Parry G. 1982. How does the extracellular matrix direct gene expression? *J Theor Biol* **99:** 31–68.

Chung SY, Krivorov NP, Rausei V, Thomas L, Frantzen M, Landsittel D, Kang YM, Chon CH, Ng CS, Fuchs GJ. 2005. Bladder reconstitution with bone marrow derived stem cells seeded on small intestinal submucosa improves morphological and molecular composition. *J Urol* **174:** 353–359.

Dantal J, Soulillou JP. 2005. Immunosuppressive drugs and the risk of cancer after organ transplantation. *N Eng J Med* **352:** 1371–1373.

Dantal J, Hourmant M, Cantarovich D, Giral M, Blancho G, Dreno B, Soulillou JP. 1998. Effect of long-term immunosuppression in kidney-graft recipients on cancer incidence: Randomised comparison of two cyclosporin regimens. *Lancet* **351:** 623–628.

Dharnidharka VR, Stablein DM, Harmon WE. 2004. Post-transplant infections now exceed acute rejection as cause for hospitalization: A report of the NAPRTCS. *Am J Transpl* **4:** 384–389.

Elliott MJ, De Coppi P, Speggiorin S, Roebuck D, Butler CR, Samuel E, Crowley C, McLaren C, Fierens A, Vondrys D, et al. 2012. Stem-cell-based, tissue engineered tracheal replacement in a child: A 2-year follow-up study. *Lancet* **380:** 994–1000.

Filipponi F, Pisati R, Cavicchini G, Ulivieri MI, Ferrara R, Mosca F. 2003. Cost and outcome analysis and cost determinants of liver transplantation in a European national health service hospital. *Transplantation* **275:** 1731–1736.

Folkman J, Hochberg M. 1973. Self-regulation of growth in three dimensions. *J Exp Med* **138:** 745–753.

Galbraith CA, Hathaway D. 2004. Long-term effects of transplantation on quality of life. *Transplantation* **77:** S84–S87.

Hammond JS, Gilbert TW, Howard D, Zaitoun A, Michalopoulos G, Shakesheff KM, Beckingham IJ, Badylak SF. 2011. Scaffolds containing growth factors and extracellular matrix induce hepatocyte proliferation and cell migration in normal and regenerating rat liver. *J Hepatol* **54:** 279–287.

Hata T, Uemoto S, Fujimoto Y, Murakami T, Tateno C, Yoshizato K, Kobayashi E. 2012. Transplantation of engineered chimeric liver with autologous hepatocytes and xenobiotic scaffold. *Ann Surg* **257:** 542–547.

Hibino N, McGillicuddy E, Matsumura G, Ichihara Y, Naito Y, Breuer C, Shinoka T. 2010. Late-term results of tissue-engineered vascular grafts in humans. *J Thorac Cardiovasc Surg* **139:** 431–436.

Ivanov PP. 1937. *Obshchaia i sravnitel'naia embriologiia.* Moscow-Leningrad.

Jungebluth P, Alici E, Baiguera S, Le Blanc K, Blomberg P, Bozóky B, Crowley C, Einarsson O, Grinnemo KH, Gudbjartsson T, et al. 2011. Tracheobronchial transplantation with a stem-cell-seeded bioartificial nanocomposite: A proof-of-concept study. *Lancet* **378:** 1997–2004.

Karam VH, Gasquet I, Delvart V, Hiesse C, Dorent R, Danet C, Samuel D, Charpentier B, Gandjbakhch I, Bismuth H, et al. 2003. Quality of life in adult survivors beyond 10 years after liver, kidney, and heart transplantation. *Transplantation* **76:** 1699–1704.

Macchiarini P, Jungebluth P, Go T, Asnaghi MA, Rees LE, Cogan TA, Dodson A, Martorell J, Bellini S, Parnigotto PP, et al. 2008. Clinical transplantation of a tissue-engineered airway. *Lancet* **372:** 2023–2030.

McAllister TN, Maruszewski M, Garrido SA, Wystrychowski W, Dusserre N, Marini A, Zagalski K, Fiorillo A, Avila H, Manglano X, et al. 2009. Effectiveness of haemodialysis access with an autologous tissue-engineered vascular graft: A multicentre cohort study. *Lancet* **373:** 1440–1446.

Meguid El, Nahas A, Bello AK. 2005. Chronic kidney disease: The global challenge. *Lancet* **365:** 331–334.

Mertsching H, Schanz J, Steger V, Schandar M, Schenk M, Hansmann J, Dally I, Friedel G, Walles T. 2009. Generation and transplantation of an autologous vascularized bioartificial human tissue. *Transplantation* **88:** 203–210.

Ng SL, Narayanan K, Gao S, Wan AC. 2011. Lineage restricted progenitors for the repopulation of decellularized heart. *Biomaterials* **32:** 7571–7580.

Olausson M, Patil PB, Kuna VK, Chougule P, Hernandez N, Methe K, Kullberg-Lindh C, Borg H, Ejnell H, Sumitran-Holgersson S. 2012. Transplantation of an allogeneic vein bioengineered with autologous stem cells: A proof-of-concept study. *Lancet* **380:** 230–237.

Orlando G. 2010. Finding the right time for weaning off immunosuppression in solid organ transplant recipients. *Expert Rev Clin Immunol* **6:** 879–892.

Orlando G, Soker S, Wood K. 2009. Clinical operational tolerance after liver transplantation. *J Hepatol* **50:** 1247–1257.

Orlando G, Hematti P, Stratta RJ, Burke GW, Di Cocco P, Pisani F, Soker S, Wood KJ. 2010a. Clinical operational tolerance after renal transplantation: Current status and future challenges. *Ann Surg* **252:** 913–926.

Orlando G, Stratta RJ, Light J. 2010b. Pancreas transplantation for type 2 diabetes mellitus. *Curr Opin Organ Transplant* **16:** 110–115.

Orlando G, Wood KJ, Soker S, Stratta RJ. 2011a. How regenerative medicine may contribute to the achievement of an immunosuppression-free state. *Transplantation* **92:** e36–e38.

Orlando G, Wood KJ, Stratta RJ, Yoo J, Atala A, Soker S. 2011b. Regenerative medicine and organ transplantation: Past, present and future. *Transplantation* **91:** 1310–1317.

Orlando G, Farney A, Sullivan DC, Abouschwareb T, Iskandar S, Wood KJ, Atala A, Stratta RJ, Yoo JJ, Soker S. 2012a. Production and implantation of renal extracellular matrix scaffolds from porcine kidneys as a platform for renal bioengineering investigations. *Ann Surg* **256:** 363–370.

Orlando G, Wood KJ, De Coppi P, Baptista PM, Binder KW, Bitar KN, Breuer C, Burnett L, Christ G, Farney A, et al. 2012b. Regenerative medicine as applied to general surgery. *Ann Surg* **255:** 867–880.

Ott HC, Matthiesen TS, Goh SK, Black LD, Kren SM, Netoff TI, Taylor DA. 2008. Perfusion-decellularized matrix: Using nature's platform to engineer a bioartificial heart. *Nat Med* **14:** 213–221.

Ott HC, Clippinger B, Conrad C, Schuetz C, Pomerantseva I, Ikonomou L, Kotton D, Vacanti JP. 2010. Regeneration and orthotopic transplantation of a bioartificial lung. *Nat Med* **16:** 927–933.

Pellegrini G, Traverso CE, Franzi AT, Zingirian M, Cancedda R, De Luca M. 1997. Long-term restoration of damaged corneal surfaces with autologous cultivated corneal epithelium. *Lancet* **349:** 990–993.

Petersen TH, Calle EA, Zhao L, Lee EJ, Gui L, Raredon MB, Gavrilov K, Yi T, Zhuang ZW, Breuer C, et al. 2010. Tissue-engineered lungs for in vivo implantation. *Science* **29:** 538–541.

Quarto R, Mastrogiacomo M, Cancedda R, Kutepov SM, Mukhachev V, Lavroukov A, Kon E, Marcacci M. 2001. Repair of large bone defects with the use of autologous bone marrow stromal cells. *N Engl J Med* **344:** 385–386.

Rama P, Matuska S, Paganoni G, Spinelli A, De Luca M, Pellegrini G. 2010. Limbal stem-cell therapy and long-term corneal regeneration. *N Engl J Med* **363:** 147–155.

Rauh J, Milan F, Günther KP, Stiehler M. 2011. Bioreactor systems for bone tissue engineering. *Tissue Eng Part B Rev* **17:** 263–280.

Romagnoli G, De Luca M, Faranda F, Bandelloni R, Franzi AT, Cataliotti F, Cancedda R. 1990. Treatment of posterior hypospadias by the autologous graft of cultured urethral epithelium. *N Engl J Med* **323:** 527–530.

Ross EA, Williams MJ, Hamazaki T, Terada N, Clapp WL, Adin C, Ellison GW, Jorgensen M, Batich CD. 2009. Embryonic stem cells proliferate and differentiate when seeded into kidney scaffolds. *J Am Soc Nephrol* **20:** 2338–2347.

Shinoka T, Imai Y, Ikada Y. 2001. Transplantation of a tissue-engineered pulmonary artery. *N Engl J Med* **344:** 532–533.

Soto-Gutierrez A, Zhang L, Medberry C, Fukumitsu K, Faulk D, Jiang H, Reing J, Gramignoli R, Komori J, Ross M, et al. 2011. A whole-organ regenerative medicine approach for liver replacement. *Tissue Eng Part C Methods* **17:** 677–686.

Uygun BE, Soto-Gutierrez A, Yagi H, Izamis ML, Guzzardi MA, Shulman C, Milwid J, Kobayashi N, Tilles A, Berthiaume F, et al. 2010. Organ reengineering through development of a transplantable recellularized liver graft using decellularized liver matrix. *Nat Med* **16:** 814–820.

Wainwright JM, Czajka CA, Patel UB, Freytes DO, Tobita K, Gilbert TW, Badylak SF. 2010. Preparation of cardiac extracellular matrix from an intact porcine heart. *Tissue Eng Part C Methods* **16:** 525–532.

Wang Y, Cui CB, Yamauchi M, Miguez P, Roach M, Malavarca R, Costello MJ, Cardinale V, Wauthier E, Barbier C, et al. 2011. Lineage restriction of human hepatic stem cells to mature fates is made efficient by tissue-specific biomatrix scaffolds. *Hepatology* **53:** 293–305.

Warnke PH, Springer IN, Wiltfang J, Acil Y, Eufinger H, Wehmöller M, Russo PA, Bolte H, Sherry E, Behrens E, et al. 2004. Growth and transplantation of a custom vascularised bone graft in a man. *Lancet* **364:** 766–770.

Yagi H, Fukumitsu K, Fukuda K, Kitago M, Shinoda M, Obara H, Itano O, Kawachi S, Tanabe M, Coudriet GM, et al. 2012. Human-scale whole-organ bioengineering for liver transplantation: A regenerative medicine approach. *Cell Transplant* **22:** 231–242.

Index

I